# Green Energy and Technology

More information about this series at http://www.springer.com/series/8059

Zhong Lin Wang · Long Lin
Jun Chen · Simiao Niu · Yunlong Zi

# Triboelectric Nanogenerators

Springer

Zhong Lin Wang
School of Materials Science and Engineering
Georgia Institute of Technology
Atlanta, GA
USA

and

National Center for Nanoscience and Technology (NCNST)
Beijing Institute of Nanoenergy and Nanosystems, Chinese Academy of Sciences
Beijing
China

Long Lin
School of Materials Science and Engineering
Georgia Institute of Technology
Atlanta, GA
USA

Jun Chen
School of Materials Science and Engineering
Georgia Institute of Technology
Atlanta, GA
USA

Simiao Niu
School of Materials Science and Engineering
Georgia Institute of Technology
Atlanta, GA
USA

Yunlong Zi
School of Materials Science and Engineering
Georgia Institute of Technology
Atlanta, GA
USA

ISSN 1865-3529 ISSN 1865-3537 (electronic)
Green Energy and Technology
ISBN 978-3-319-40038-9 ISBN 978-3-319-40039-6 (eBook)
DOI 10.1007/978-3-319-40039-6

Library of Congress Control Number: 2016944334



Printed on acid-free paper

This Springer imprint is published by Springer Nature
The registered company is Springer International Publishing AG Switzerland

# Preface

The first organic materials-based triboelectric nanogenerator (TENG) was invented by my group in 2012. Using the electrostatic charges created on the surfaces of two dissimilar materials when they are brought into physical contact, the contact-induced triboelectric charges can generate a potential drop when the two surfaces are separated by a mechanical force, which can drive electrons to flow between the two electrodes built on the top and bottom surfaces of the two materials. The fundamental studies and technological applications of TENG are experiencing a rapid development and its applications cover a wide range of fields. This book provides a comprehensive review about the four modes of the TENGs, their theoretical modeling, and the applications of TENGs for harvesting energy from human motion, walking, vibration, mechanical triggering, rotating tire, wind, flowing water, and more. A TENG can also be used as a self-powered sensor for actively detecting the static and dynamic processes arising from mechanical agitation using the voltage and current output signals of the TENG, respectively, with potential applications as mechanical sensors and for touch pad and smart skin technologies. The potential of TENG for harvesting ocean wave energy is also discussed as a potential approach for the blue energy. The objective of writing this book is to systematically introduce the TENG, so that it can serve as a text book and a reference book for promoting the fundamental development and technological applications of TENG.

This book was written mainly based on the numerous papers we have authored since 2012, and many figures were adopted from our published work in the public domain. We like to thank my current and former group members and collaborators who have made outstanding contributions to the development of TENG (not in particular order): Fengru Fan, Guang Zhu, Sihong Wang, Ya Yang, Zong-Hong Lin, Long Lin, Jun Chen, Yusheng Zhou, Simiao Niu, Jin Yang, Weiqing Yang, Xiaonan Wen, Xia Cao, Jun Zhou, Sang-Woo Kim, Yong Qin, Xing Fan, Gang Cheng, Li Zheng, Yunlong Zi, Chi Zhang, Changbao Han, Wei Tang, Aifang Yu, Hulin Zhang, Yannan Xie, Peng Bai, Qingshen Jing, Shengming Li, Yuanjie Su, Zhong-Qun Tian, Min-Hsin Yeh, Fang Yi, Zhaoling Li, Hengyu Guo, Zhen Wen,

Caofeng Pan, Tao Jiang, Sangmin Lee, Ying Liu, Zhen Wen, Changsheng Wu, Jie Wang, Chenguo Hu, Yi Xi, Te-Chien Hou, Xiangyu Chen, Jr-Hau He, Jong-Min Bai, Yong Ding, Po-Kang Yang, Ken Pradel, Xiuhan Li, Xiaofeng Wang, Huifang Li, Yue Zhang, Ying Wu, Tao Zhou, Mengxiao Chen, Limin Zhang, and Weiming Du. We also like to thanks to our collaborators: Profs. Yue Zhang, Zhong-Qun Tian, Christian Falconi, Sang-Woo Kim, Jeong-Min Baik, Qing Zhang, Haixia Zhang, and Magnus Willander.

Lastly and most importantly, I thank my family members for their years of support and understanding. It was not possible to carry out such a research without their support.

Zhong Lin Wang
School of Materials Science and Engineering
Georgia Institute of Technology, USA
Beijing Institute of Nanoenergy and Nanosystems
National Center for Nanoscience
and Technology (NCNST)
Chinese Academy of Sciences, China
e-mail: zlwang@gatech.edu
Personal website:
http://www.nanoscience.gatech.edu/
SCI publication record:
http://www.researcherid.com/rid/E-2176-2011

# Contents

**Part II Applications as a Sustainable Power Source**

# Abbreviations

## High-Frequency Terminologies

| | |
|---|---|
| AC | Alternating current |
| DC | Direct current |
| FEP | Fluorinated ethylene propylene |
| ITO | Indium tin oxide |
| $J_{SC}$ | Short-circuit current density |
| Kapton | Polyimide |
| LED | Light-emitting diode |
| LIB | Lithium-ion battery |
| NG | Nanogenerator |
| OC | Open-circuit |
| PDMS | Polydimethylsiloxane |
| PE | Polyethylene |
| PET | Polyethylene terephthalate |
| PMMA | Poly(methyl methacrylate) |
| PNWs | Polymer nanowires |
| PTFE | Polytetrafluoroethylene |
| PVDF | Polyvinylidene fluoride |
| SC | Short-circuit |
| SEM | Scanning electron microscopy |
| TENG | Triboelectric nanogenerator |

## Chapter 1

| | |
|---|---|
| AFM | Atomic force microscopy |
| CMOS | Complementary metal–oxide–semiconductor field-effect transistor |
| MOSFET | Metal–oxide–semiconductor field-effect transistor |
| SKPM | Scanning Kevin probe microscopy |

**Chapter 2**

| | |
|---|---|
| MACRS | Minimum achievable charge reference state |
| NPs | Nanoparticles |
| PECVD | Plasma-enhanced chemical vapor deposition |

**Chapter 3**

| | |
|---|---|
| MACRS | Minimum achievable charge reference state |

**Chapter 4**

| | |
|---|---|
| F-TENG | Freestanding-triboelectric-layer-based triboelectric nanogenerator |
| MACRS | Minimum achievable charge reference state |
| SETENG | Single-electrode triboelectric nanogenerator |

**Chapter 5**

| | |
|---|---|
| CF-TENG | Contact-mode freestanding-triboelectric-layer-based triboelectric nanogenerator |
| FRD-TENG | Free-rotating disk triboelectric nanogenerator |
| F-TENG | Freestanding-triboelectric-layer-based triboelectric nanogenerator |
| GF-TENG | Grating-structured freestanding-triboelectric-layer-based triboelectric nanogenerator |
| ICP | Inductive coupling plasma |
| LE | Left-hand electrode |
| MACRS | Minimum achievable charge reference state |
| RE | Right-hand electrode |
| RIE | Reactive ion etching |
| R-TENG | Rolling triboelectric nanogenerator |
| SF-TENG | Sliding-mode freestanding-triboelectric-layer-based triboelectric nanogenerator |

**Chapter 6**

| | |
|---|---|
| FEM | Finite element method |
| FTENG | Freestanding-triboelectric-layer-based triboelectric nanogenerator |
| MACRS | Minimum achievable charge reference state |
| SETENG | Single-electrode triboelectric nanogenerator |
| SPICE | Simulation program with integrated circuit emphasis |

## Chapter 7

| | |
|---|---|
| 1S | 1-side side effects |
| 2S | 2-side side effects |
| Al | Aluminum |
| CEO | Cycles for energy output |
| CFT | Contact freestanding triboelectric layer |
| CMEO | Cycles for maximized energy output |
| CS | Contact-separation |
| Cu | Copper |
| FEM | Finite element method |
| FOM | Figure-of-merits |
| $FOM_{DM}$ | Dimensionless material figure-of-merits |
| $FOM_{M}$ | Material figure-of-merits |
| $FOM_{P}$ | Performance figure-of-merits |
| $FOM_{S}$ | Structural figure-of-merits |
| $FOM_{S,max}$ | Maximum structural figure-of-merits |
| FT | Freestanding triboelectric layer |
| LM-TENG | Liquid-metal-based TENG |
| LS | Lateral sliding |
| MACRS | Minimum achievable charge reference state |
| PENG | Piezoelectric nanogenerator |
| RH | Relative humidity |
| SE | Single-electrode |
| SEC | Single-electrode contact |
| SFT | Sliding freestanding triboelectric layer |
| wt | Weight |
| ZT | Figure-of-merits of thermoelectric materials |

## Chapter 8

| | |
|---|---|
| 3D | Three dimensional |
| Ag | Silver |
| Al | Aluminum |
| CCT | Coated cotton thread |
| CNT | Carbon nanotube |
| CS | Contact separation |
| Cu | Copper |
| EDX | Energy dispersive X-ray spectrum |
| EPD | Electric potential difference |
| FBG | Fiber-based triboelectric nanogenerator |
| FBHNG | Fiber-based hybrid nanogenerator |
| GD | Galvanostatically |

| | |
|---|---|
| GF-TENG | Grating-structured triboelectric nanogenerator |
| LCD | Liquid crystal display |
| Ni | Nickel |
| NR | Nanorods |
| PCCT | Carbon nanotube coated cotton thread |
| PENG | Piezoelectric nanogenerator |
| pNG | Paper-based nanogenerator |
| SFT | Sliding freestanding triboelectric layer |
| Ti | Titanium |
| ZnO | Zinc oxide |

**Chapter 9**

| | |
|---|---|
| 3D | Three dimensional |
| ICP | Inductively coupled plasma |
| OR | Open ratio |

**Chapter 10**

| | |
|---|---|
| Al | Aluminum |
| Cu | Copper |
| ICP | Inductive coupling plasma |
| R-TENG | Rotary-structured TENG |
| SFT | Sliding freestanding triboelectric layer |

**Chapter 11**

| | |
|---|---|
| AAO | Anodic aluminum oxide |
| EDLC | Electric double-layer capacitor |
| EMG | Electromagnetic generator |
| LCD | Liquid crystal display |
| TENG-NW | Triboelectric nanogenerator-network |

**Chapter 12**

| | |
|---|---|
| AAO | Anodic aluminum oxide |
| Ag | Silver |
| Al | Aluminum |
| $Al_2O_3$ | Aluminum oxide |
| Au | Gold |
| $BaTiO_3$ | Barium titanate |

| | |
|---|---|
| $Bi_2Te_3$ | Bismuth telluride |
| Cu | Copper |
| EC | Electrochemical cell |
| EMG | Electromagnetic generator |
| $H_2O$ | Water |
| Kapton | Polyimide |
| LCD | Liquid crystal display |
| MNDS | Micro/nano dual-scale |
| MO | Methyl orange |
| NaCl | Sodium chloride |
| PA | Polyamide |
| PENG | Piezoelectric nanogenerator |
| PFA | Perfluoroalkyoxy |
| PMN-PT | Lead magnesium niobate—lead titanate |
| PPENG | Pyroelectric-piezoelectric nanogenerator |
| PVDF | Polyvinylidene fluoride |
| PZT | Lead zirconate titanate |
| Si | Silicon |
| SiN | Silicon nitride |
| XRD | X-ray diffraction |
| ZnO | Zinc oxide |

**Chapter 13**

| | |
|---|---|
| AA | Ascorbic acid |
| ADC | Analog-to-digital converter |
| ALP | Alkaline phosphatase |
| $BaCl_2$ | Barium chloride |
| $CO_2$ | Carbon dioxide |
| CP | Cathodic protection |
| Cr | Chromium |
| Cu | Copper |
| $CuSO_4$ | Copper(II) sulfate |
| DMEM | Dulbecco's modified Eagle's medium |
| DNA | Deoxyribonucleic acid |
| ECD | Electrochromic device |
| ECG | Electrocardiograph |
| EDS | Energy dispersive X-ray spectrum |
| ELISA | Enzyme-linked immunosorbent assay |
| EMG | Electromagnetic generator |
| GC | Gas chromatography |
| $H_2O$ | Water |
| HCl | Hydrogen chloride |
| IC | Integrated circuit |

| | |
|---|---|
| ICCP | Impressed current cathodic protection |
| KOH | Potassium hydroxide |
| LCD | Liquid crystal display |
| MO | Methyl orange |
| MTT | 3-(4,5-dimethylthiazol-2-yl)-2,5-diphenyltetrazolium bromide |
| NaCl | Sodium chloride |
| $NaHSO_3$ | Sodium bisulfite |
| Ni | Nickel |
| $NO_2$ | Nitrogen dioxide |
| $NO_X$ | Nitrogen oxide |
| PB | Prussian blue |
| PCB | Print circuit board |
| PM | Particulate Matter |
| Pt | Platinum |
| RF-TENG | Rotational freestanding TENG |
| RKE | Remote keyless entry |
| rpm | Revolutions per minute |
| R-TENG | Rotational TENG |
| SACP | Sacrificial anode cathodic protection |
| SCPU | Self-charging power unit |
| $SO_2$ | Sulfur dioxide |
| SOC | State of charge |
| SPLC | Self-powered low-level laser cure |
| UV | Ultraviolet |
| XRD | X-ray diffraction |
| ZnHCF | Zinc hexacyanoferrate |

**Chapter 14**

| | |
|---|---|
| AgNWs | Silver nanowires |
| CNT | Carbon nanotube |
| DFT | Discreet Fourier Transformation |
| FAR | False Acceptance Rate |
| FRR | False Rejection Rate |
| IKB | Intelligent keyboard |
| KFE | Key functional element |
| resis-sensor | Resistive pressure sensor |
| ROC | Receiver Operating Characteristic |
| SNR | Signal-to-noise ratio |
| TEAS | Triboelectric active sensor |

## Chapter 15

| | |
|---|---|
| 3D | Three dimensional |
| AIx | Augmentation index |
| BMS | Bionic membrane sensor |
| EM | Euclidean metric |
| FAR | False acceptance rate |
| FDPW | First-derivatives-of-the-pulse-waves |
| FRR | False rejection rate |
| PWV | Pulse wave velocity |
| RI | Reflection index |
| STFT | Short-time Fourier transform |

## Chapter 16

| | |
|---|---|
| FEP | Fluorinated ethylene propylene |
| FRD | Free-rotating disk |
| ICP | Inductively coupled plasma |
| LED | Light-emitting diode |
| MCU | Microcontroller unit |
| PFA | Perfluoroalkoxy alkane |
| RMS | Root mean square |
| TES | Triboelectric sensor |

## Chapter 17

| | |
|---|---|
| 3-MPA | 3-mercaptopropionic acid |
| AAO | Anodic aluminum oxide |
| DNA | Deoxyribonucleic acid |
| ICP-MS | Inductively coupled plasma mass spectrometry |
| NPs | Nanoparticles |
| TNS | $TiO_2$ nanosheet |
| TNW | $TiO_2$ nanowire |
| UV | Ultraviolet |
| WD-TENG | Water-driven triboelectric nanogenerator |
| XRD | X-ray diffraction |
| β-CD | β-cyclodextrin |

# Symbols

## Chapter 1

| | |
|---|---|
| $z$ | Critical tunneling distance |
| $\sigma$ | The dielectric surface charge density |
| $\sigma_1$ | The induced charge density on the top metal |
| $\sigma_2$ | The induced charge density on the bottom metal |
| $\sigma_1/\varepsilon_0$ | Built-in electric field |
| $\Delta E_{vcc}$ | The change of vacuum energy level between the metal and the dielectric surface |
| $e$ | The elementary charge |
| $N_s(E)$ | The assumed surface density of states |
| $\Delta E_s$ | The range of filled surface states |

## Chapter 2

| | |
|---|---|
| $V_{OC}$ | Open-circuit voltage |
| $I_{SC}$ | Short-circuit current |
| $d_1$ | Thickness of dielectric 1 |
| $d_2$ | Thickness of dielectric 2 |
| $\varepsilon_0$ | Permittivity of vacuum |
| $\varepsilon_{r1}$ | Relative dielectric constant of dielectric 1 |
| $\varepsilon_{r2}$ | Relative dielectric constant of dielectric 2 |
| $x$ | The distance between two triboelectric surfaces |
| $\sigma$ | Surface triboelectric charge density |
| $V$ | The potential difference between two electrodes |
| $Q$ | The amount of charge transfer between two electrodes |
| $\eta_{CT}$ | Short-circuit charge transfer efficiency |
| $S$ | The area size of a triboelectric nanogenerator |
| $E_1$ | The electric field strength inside dielectric 1 |

| | |
|---|---|
| $E_2$ | The electric field strength inside dielectric 2 |
| $E_{air}$ | The electric field strength inside the air gap between two triboelectric surfaces |
| $d_0$ | Effective thickness constant |
| $Q_{SC}$ | The amount of charge transfer in short-circuit condition |
| $C$ | Capacitance |
| $C_1$ | Capacitance between Node 1 and Node 2 |
| $C_2$ | Capacitance between Node 2 and Node 3 |
| $v$ | Average velocity |
| $J_{SC}$ | Short-circuit current density |

## Chapter 3

| | |
|---|---|
| $V_{OC}$ | Open-circuit voltage |
| $I_{SC}$ | Short-circuit current |
| $d_1$ | Thickness of dielectric 1 |
| $d_2$ | Thickness of dielectric 2 |
| $\varepsilon_0$ | Permittivity of vacuum |
| $\varepsilon_{r1}$ | Relative dielectric constant of dielectric 1 |
| $\varepsilon_{r2}$ | Relative dielectric constant of dielectric 2 |
| $x$ | The distance between two triboelectric surfaces |
| $\sigma$ | Surface triboelectric charge density |
| $V$ | The potential difference between two electrodes |
| $Q$ | The amount of charge transfer between two electrodes |
| $\eta_{CT}$ | Short-circuit charge transfer efficiency |
| $\eta_{CT\text{-rectifieed}}$ | Rectified short-circuit charge transfer efficiency |
| $S$ | The area size of a triboelectric nanogenerator |
| $d_0$ | Effective thickness constant |
| $Q_{SC}$ | The amount of charge transfer in short-circuit condition |
| $w$ | Width of the triboelectric layer |
| $v$ | Average velocity |
| $n$ | Number of grating units |
| $L$ | Total length of the top plate |
| $l$ | Length of the one grating unit |
| $J_{SC}$ | Short-circuit current density |

## Chapter 4

| | |
|---|---|
| $V_{OC}$ | Open-circuit voltage |
| $I_{SC}$ | Short-circuit current |
| $J_{SC}$ | Short-circuit current density |
| $\varepsilon_0$ | Permittivity of vacuum |
| $x$ | The distance between two triboelectric surfaces |

| | |
|---|---|
| $\sigma$ | Surface triboelectric charge density |
| $V$ | The potential difference between two electrodes |
| $Q$ | The amount of charge transfer between two electrodes |
| $S$ | The area size of a triboelectric nanogenerator |
| $Q_{\mathrm{SC}}$ | The amount of charge transfer under short-circuit condition |
| $\eta_{\mathrm{CT}}$ | Short-circuit charge transfer efficiency |
| $C$ | Capacitance |
| $C_1$ | Capacitance that directly connects Node 1 and Node 2 |
| $C_2$ | Capacitance that directly connects Node 1 and Node 3 |
| $C_3$ | Capacitance that directly connects Node 2 and Node 3 |
| $C_{\mathrm{a}}$ | The total capacitance between Node 1 and Node 2 |
| $C_{\mathrm{b}}$ | The total capacitance between Node 1 and Node 3 |
| $C_0$ | The total capacitance between Node 2 and Node 3 |
| $v$ | Average velocity |
| $S$ | Area of the triboelectric surface |
| $l$, $w$ | Length and width of the triboelectric layer |
| $d_{\mathrm{m}}$ | Effective thickness of the metal electrodes |
| $g$ | Gap distance between the electrodes |

**Chapter 5**

| | |
|---|---|
| $V_{\mathrm{OC}}$ | Open-circuit voltage |
| $I_{\mathrm{SC}}$ | Short-circuit current |
| $J_{\mathrm{SC}}$ | Short-circuit current density |
| $l$, $L$ | Length of the triboelectric layer |
| $\varepsilon_0$ | Permittivity of vacuum |
| $x$ | The distance of the freestanding layer with reference to its initial position |
| $\sigma$ | Surface triboelectric charge density |
| $V$ | The potential difference between two electrodes |
| $Q$ | The amount of charge transfer between two electrodes |
| $S$ | The area size of a triboelectric nanogenerator |
| $Q_{\mathrm{SC}}$ | The amount of charge transfer under short-circuit condition |
| $\eta_{\mathrm{CT}}$ | Short-circuit charge transfer efficiency |
| $C$ | Capacitance |
| $C_i$ | Capacitance that directly connects two nodes |
| $v$ | Average velocity |
| $S$ | Area of the triboelectric surface |
| $w$ | Width of the triboelectric layer |
| $d_0$ | Effective thickness of the dielectric layer |
| $d_{\mathrm{i}}$ | Thickness of the ith dielectric layer |
| $d$, $g$ | Gap distance between the electrodes |
| $h$, $H$ | Freestanding height, distance between the triboelectric layers |
| $T$ | Period |

| | |
|---|---|
| $Q_{SC,max}$ | Maximum short-circuit charge transfers between two electrodes |
| $R$ | External load resistance |
| $\varepsilon_{ri}$ | Dielectric constant of the ith dielectric layer |
| $\Delta\sigma_{SC}$ | The amount of charge transfer density under short-circuit condition |
| $k$ | The spring constant of a single spring |
| $\eta$ | Energy conversion efficiency |
| $E_{output}$ | Electrical energy output |
| $E_{input}$ | Mechanical energy input |
| $F$ | Resistive force measured with a certain load resistance |
| $F_0$ | The resistive force measured with open-circuit condition |

## Chapter 6

| | |
|---|---|
| $\sigma$ | Triboelectric surface charge density |
| $V$ | Voltage between two electrodes |
| $Q$ | Charge transfers between two electrodes |
| $x$ | Relative displacement between the triboelectric layers |
| $x_{max}$ | Maximum relative displacement between the triboelectric layers |
| $V_{OC}(x)$ | Open-circuit voltage between two electrodes |
| $Q_{SC}(x)$ | Short-circuit charge transfers between two electrodes |
| $V_{OC,max}$ | Maximum open-circuit voltage between two electrodes |
| $Q_{SC,max}$ | Maximum short-circuit charge transfers between two electrodes |
| $T$ | Period |
| $Q_0$ | The initial charge on the electrode when the reference state is picked |
| $Q_1$ | The charge amount on metal 1 under short-circuit condition |
| $Q_2$ | The charge amount on metal 2 under short-circuit condition |
| $I$ | Output current |
| $I_{max}$ | The peak value of the output current |
| $Q_C$ | Initial charge on the load capacitance |
| $R$ | External load resistance |
| $R_{opt}$ | Optimum load resistance |
| $P_{max}$ | Peak instantaneous power |
| $C(x)$, $C_T(x)$ | Capacitance between two TENG electrodes |
| $C_L$ | Load capacitance |
| $v$ | Average velocity |
| $S$ | Area of the triboelectric surface |
| $g$ | Gap distance between the electrodes |
| $\varepsilon_0$ | The permittivity of vacuum |
| $l$, $w$ | Length and width of the triboelectric layer |

| | |
|---|---|
| $d_0$ | Effective thickness of the dielectric layer |
| $d_i$ | Thickness of the ith dielectric layer |
| $d_k$ | Thickness of the Kapton layer |
| $\varepsilon_{ri}$ | Dielectric constant of the ith dielectric layer |
| $E^C$ | Energy stored in the load capacitor |
| $Q^M{}_k$ | The total charges on the Node M at the beginning of the *k*th cycle |
| $C_{L,opt}$ | Optimum load capacitance |

## Chapter 7

| | |
|---|---|
| $\sigma$ | Triboelectric surface charge density |
| $V$ | Voltage between two electrodes |
| $Q$ | Charge transfers between two electrodes |
| $x$ | Relative displacement between the triboelectric layers |
| $x_{max}$ | Maximum relative displacement between the triboelectric layers |
| $V_{OC}(x)$ | Open-circuit voltage between two electrodes |
| $Q_{SC}(x)$ | Short-circuit charge transfers between two electrodes |
| $V'(x)$ | Absolute voltage at $Q = Q_{SC,max}$ at displacement $x$ |
| $V_{OC,max}$ | Maximum open-circuit voltage between two electrodes |
| $Q_{SC,max}$ | Maximum short-circuit charge transfers between two electrodes |
| $V'_{max}$ | Maximum achievable absolute voltage at $Q = Q_{SC,max}$ |
| $\overline{P}$ | Average output power |
| $T$ | Period |
| $E$ | Output energy per cycle |
| $I$ | Output current |
| $Q_C$ | Total cycling charge |
| $R$ | External load resistance |
| $C(x)$, $C_{Total}(x)$ | Capacitance between two electrodes |
| $E_m$ | The largest possible output energy per cycle |
| $\eta$ | Energy conversion efficiency |
| $\overline{v}$ | Average velocity |
| $\overline{F}$ | Average dissipative force |
| $A$ | Area of the triboelectric surface |
| $\varepsilon_0$ | The permittivity of vacuum |
| $l$, $w$ | Length and width of the triboelectric layer |
| $d_0$ | Effective thickness of the dielectric layer |
| $d_i$ | Thickness of the ith dielectric layer |
| $\varepsilon_i$ | Dielectric constant of the ith dielectric layer |
| $d$, $g$ | Gap distance |
| $\sigma_N$ | Normalized triboelectric surface charge density |

## Chapter 8

| | |
|---|---|
| $V_{OC}$ | Open-circuit voltage |
| $I_{SC}$ | Short-circuit current |
| $J_{SC}$ | Short-circuit current density |
| $N_{Total}$ | Total number of unit cells |
| $n$ | The number of unit cells along the edge length |
| $\theta$ | Separation angle |
| $\theta_{max}$ | Maximum separation angle |
| $V_A$, $V_B$ | Electric potential of A and B electrodes |
| $\sigma$ | Triboelectric surface charge density |
| $\Delta\sigma$, $\sigma_2$ | Induced surface charge density |
| $\Delta\sigma_{SC}$ | Short-circuit surface charge density transfer |
| $\Delta\sigma_{SC\text{-}Rec}$ | Accumulated short-circuit surface charge density transfer |
| $\varepsilon_0$ | The permittivity of vacuum |
| $\varepsilon_r$ | Dielectric constant of the dielectric layer (PTFE) |
| $l$ | Edge length of one unit cell |
| $d_1$ | Thickness of the first dielectric layer |
| $d$ | Distance between electrodes |
| $R$ | External load resistance |
| $\Delta Q$ | Charge transfers between two electrodes of TENG |

## Chapter 9

| | |
|---|---|
| $V_{OC}$ | Open-circuit voltage |
| $I_{SC}$ | Short-circuit current |
| $f_0$ | Natural frequency |
| $m_0$ | Mass |
| $k$ | Stiffness coefficient |
| R | External resistance |
| $P_d$ | Instantaneous peak power density |
| $n$ | The number of pinned fingers |
| $d_0$ | The thickness of PTFE |
| $\varepsilon_r$ | The relative permittivity of PTFE |
| $s$ | The effective contact area |
| $v$ | The relative velocity |
| $S$ | Cross-sectional area |
| $L'$ | Effective length of the neck |
| $V$ | Cavity volume |
| $c$ | Speed of sound |

## Chapter 10

| | |
|---|---|
| $\Delta x_0$ | Original tensile deformation (of the spring) |
| $\Delta x_{max}$ | Maximum vertical separation distance |
| $k$ | Spring constant |
| $F(X_0)$, $F(Y_0)$ | The force on the tilted rectangle blades along the tangential or axial direction |
| $n$, $a$, $b$ | The number, length, and width of the rectangle fan blades |
| $\rho_{air}$ | The density of the surrounding air |
| $\theta$ | The tilted angle of each blade |
| $\Delta x$ | The extra tensile deformation caused by the vertical separation |
| $v_1$, $v_2$, $v_3$ | The boundary wind speeds |
| $v_{wind}$ | The wind speed |
| $\Delta Q_{SC}$, $\Delta Q$ | Short-circuit (total) charge transfers |
| $V_{OC}$ | Open-circuit voltage |
| $C$ | Capacitance between two electrodes of TENG |
| $I_{SC}$ | Short-circuit current |

## Chapter 11

| | |
|---|---|
| $V_{OC}$ | Open-circuit voltage |
| $I_{SC}$ | Short-circuit current |
| $J_{sc}$ | Short-circuit current density |
| PTFE | Polytetrafluoroethylene |
| $Q_{SC}$ | Short-circuit transferred charges |
| $\Delta t$ | Peak width |
| $E_0$ | Average power |
| $E_c$ | Generated average power in one collision |
| $E_{cv}$ | Generated power per second per unit volume |
| $\beta$ | Volume ratio |
| $f$ | Collision frequency |

## Chapter 12

| | |
|---|---|
| $E$ | The induced electrodynamic potential |
| $E_m$ | The maximum induced electrodynamic potential |
| $B$ | The magnetic flux density |
| $l$ | The length of the conductor stick; the width of the friction surface perpendicular to the sliding direction |
| $v$ | The velocity of the conductor stick cutting the magnetic induction lines; the sliding velocity of the top metal |
| $\sigma$ | The triboelectric charge density of the friction surface |

| | |
|---|---|
| $n$ | The turns of coils; the segments number of the disk |
| $S$ | The area of a single turn of the coil; the area of the friction disk |
| $\omega$ | The rotational angular velocity of the coil/top metal |
| $\Delta\Phi/\Delta t$ | The change rate of magnetic flux in each coil |
| $\Delta Q/\Delta t$ | The transfer rate of electric charge in each segment |
| $E_{\mathrm{electricity}}$ | Generated electrical energy |
| $I$ | The output current of the nanogenerator |
| $R$ | The loading resistance |
| $V_{\mathrm{OC}}$, $V_{\mathrm{OC(TENG)}}$ | Open-circuit voltage of TENG |
| $\varepsilon_0$ | The permittivity of vacuum |
| $d_{\mathrm{S}}$, $d$ | Distance between electrodes; thickness of PVDF or PDMS |
| $V_{\mathrm{OC(Pyro)}}$ | Pyroelectric open-circuit voltage |
| $p$ | Pyroelectric coefficient |
| $D$ | The thickness of the pyroelectric device |
| $\Delta T$ | The change in temperature |
| $\varepsilon_{\mathrm{r}}$ | Dielectric constant of the dielectric layer (PTFE) |
| $A'$ | The total area of the PVDF in PPENG |
| $C$ | The capacitance of the PVDF in PPENG |
| $A$ | The frictional area |
| $\Delta Q$ | Transferred charge |
| $\Delta V$ | The voltage variation |
| $\Delta\sigma$, $\sigma_1$ | Induced surface charge density variation; induced surface charge density |
| $\Delta P$ | The polarization density variation |
| $\mathrm{d}T/\mathrm{d}t$ | Changing rate of the temperature |
| $\sigma$, $\sigma_0$ | Triboelectric surface charge density |
| $J$ | Current density |
| $J_{\mathrm{SC}}$ | Short-circuit current density |
| $\phi_{\mathrm{f}}$ | The volume ratio of $BaTiO_3$ nanoparticles in the composite material |

## Chapter 13

| | |
|---|---|
| $V_{\mathrm{OC}}$ | Open-circuit voltage |
| $I_{\mathrm{SC}}$, $I_{\mathrm{TENG}}$ | Short-circuit current |
| $Q_{\mathrm{SC}}$ | Short-circuit transferred charges |
| $\eta_{\mathrm{board}}$ | Board efficiency |
| $\eta_{\mathrm{total}}$ | Total efficiency |
| $V_{\mathrm{temp}}$ | The voltage across the temporary capacitor |
| $V_{\mathrm{store}}$ | The voltage across the final storage capacitor |
| $V_{\mathrm{battery}}$ | The voltage across the battery |

| | |
|---|---|
| $C_{temp}$ | The temporary capacitor |
| $C_{store}$ | The storage capacitor |
| $C$ | Weight loss |
| $t$ | Periodic test time |
| $\Delta T$ | Transmittance |
| $C_{1,2}$ | Capacitors |
| $C_T$ | Total capacitor |
| $R_{1,2}$ | Resistors |
| $D_{1,2}$ | Diodes |
| $T_{h,l}$ | Higher/lower time |
| $q$ | Pulse duty cycle |
| $I_P$ | Current output of a single stimulation pulse |
| $t_P$ | Pulse width |

## Chapter 14

| | |
|---|---|
| $V_{OC}$ | Open-circuit voltage |
| $\Delta\sigma$ | Transferred charge density |
| $J_{SC}$ | Short-circuit current density |
| $d$ | Vertical gap distance between the two triboelectric layers |
| $d_0$ | Maximum gap distance between the two triboelectric layers without pressure |
| $\varepsilon_0$ | Permittivity in vacuum |
| $\sigma$ | Triboelectric charge density |
| $Q$ | The amount of induced charges on one electrode with pressure |
| $Q_0$ | The amount of induced charges on one electrode without pressure |
| $S$ | Effective area of electrodes |
| $d_{PDMS}$ | Thickness of the PDMS membrane |
| $\varepsilon_{r,PDMS}$ | Relative dielectric constant of the PDMS membrane |
| $p$ | Applied pressure |
| $k$ | Elastic property of the materials in the TEAS |
| $p_0$ | Low-end detection limit |
| $V_{pi}$ | The maximum peak value of the output voltage from channel $i$ |
| $V_{th}$ | The threshold voltage |
| $n$ | The total number of channels |
| $i$ | Integral from 1 to n |
| $s_{4,k}$ and $d_{j,k}$ | Wavelet coefficients |
| $j = 1, 2, 3, 4\ldots k$ | The number of translations of the wavelet for any given scale |
| $\varphi_{4,k}(t)$ | The father wavelets |
| $\psi_{j,k}(t)$ | The mother wavelets |

| | |
|---|---|
| $f(\mathrm{t})$ | Either voltage or current signal of the typing patterns |
| $S_4$ | Smooth |
| $D_4$, $D_3$, $D_2$, and $D_1$ | A set of voltage or current components that provide representations of the original signals at different resolution levels |
| S | Sensitivity of the triboelectric active sensor |
| I | The current when applied pressure on the sensor |
| $\mathrm{I_{off}}$ | The current of sensor with no pressure |
| V | The voltage when applied pressure on the sensor |
| $\mathrm{V_{off}}$ | The voltage of sensor with no pressure |

## Chapter 15

| | |
|---|---|
| $V_{\mathrm{OC}}$ | Open-circuit voltage |
| $I_{\mathrm{SC}}$ | Short-circuit current |
| $P_{\mathrm{S}}$ | Systolic peak |
| $P_{\mathrm{i}}$ | Point of inflection |
| $P_{\mathrm{D}}$ | Dicrotic wave |
| $\Delta t$ | Time delay |

## Chapter 16

| | |
|---|---|
| $V_{\mathrm{OC}}$ | Open-circuit voltage |
| $I_{\mathrm{SC}}$ | Short-circuit current |

## Chapter 17

| | |
|---|---|
| $((I_0 - I)/I_0$ | Short-circuit current ratio |
| $((V_{oc} - V_{oc}^{o})/V_{oc}^{o})$ | Open-circuit voltage ratio |
| $((J_{sc} - J_{sc}^{o})/J_{sc}^{o})$ | Short-circuit current density ratio |
| $V_{\mathrm{OC}}$ | Open-circuit voltage |

# Abstract

Triboelectric nanogenerator (TENG) was first invented by Zhong Lin Wang's group in 2012 for converting small-scale mechanical energy into electricity by a conjunction of triboelectrification and electrostatic induction. TENG is a paradigm shift technology and has unpresented performances, with output area power density up to 500 W/m$^2$ and an instantaneous conversion efficiency of $\sim$70 %. TENG uses neither magnets nor coils; it is light in mass, low in density, low in cost, and can be fabricated using most of the organic materials. Most importantly, in contrast to classical electromagnetic generator, TENG works the best at low frequency (<5–10 Hz); thus, it is the unique choice for harvesting low-frequency energy from body motion and ocean wave (the blue energy). TENG can also be used as a self-powered sensor for actively detecting the static and dynamic processes arising from mechanical agitation using the voltage and current output signals, respectively, with applications as mechanical sensors, physiological detection, motion sensing, touch pad, and electronic skin technologies. This book provides a comprehensive review about the four modes of the TENGs, their theoretical modeling, and the applications of TENGs for harvesting energy from human motion, walking, vibration, mechanical triggering, rotating tire, wind, flowing water and more, for applications in portable/wearable electronics, biomedical devices, sensor networks, Internet of things, environmental sensing, and infrastructure monitoring and security.

# Chapter 1
# Triboelectrification

The massive development of the world electronic technology follows a general trend of miniaturization, portability and functionality. The development of computers is a typical example of miniaturization, from the vacuum tube based huge-size machine, to solid state metal–oxide–semiconductor field-effect transistor (MOSFET) based main frame computer and later laptop computer. The tremendous increase of popularity of handhold cell phones is a typical example of portable/mobile electronics. The next few decades will be about building functionality on existing electronics, which inevitably involves developing a range of sensors including but not limited to navigation, motion, chemical, biological and gas sensors. The near future development is about electronics that are much smaller than the size of a cell phone, so that each person on average can have at least dozens to hundreds of such small electronics. Such small size electronics operates at ultralow power consumption, making it possible to be powered by the energy harvested from our living environment [1]. It will become impractical if sensor networks have to be powered entirely by batteries because of the huge number of devices, large scope of distribution, and difficulty to track and recycle to minimize environmental impact and possibly health hazardous. Therefore, power sources are desperately needed for independent and continuous operations of such small electronics, which could be used widely for ultrasensitive chemical and bimolecular sensors, nanorobotics, micro-electromechanical systems, remote and mobile environmental sensors, homeland security and even portable/wearable personal electronics.

New technologies that can harvest energy from the environment as sustainable self-sufficient micro/nano-power sources are newly emerging field of nano energy, which is about the applications of nanomaterials and nanotechnology for harvesting energy for powering micro/nano-systems. In the last decade, we have been developing nanogenerators (NGs) for building self-powered systems and as active sensors [2]. We have mainly utilized two physics effects for harvesting small-magnitude mechanical energies: piezoelectric effect and triboelectric effect. A basic introduction about piezoelectric NG has been given in a recent book [2] and



Z.L. Wang et al., *Triboelectric Nanogenerators*,
Green Energy and Technology, DOI 10.1007/978-3-319-40039-6_1

a few review articles [3, 4]. The objective of this book is to give a systematic and comprehensive introduction about triboelectric nanogenerator (TENG) from theory to experiments, from fundamental operation modes to technological application, from single devices to system integration, both as a new energy technology and as self-powered active sensors.

## 1.1 Nano Energy and Mega Energy

In general, energy usually means the power required to run a factory, a city or a country. This is generally referred to as mega energy, which is measured in the scale of Giga Watt or Mega Watt. General characteristics of a technology for macro-scale energy are the total power output, stability, conversion efficiency and cost. In many cases, cost is the most important measure, such as for solar cell (Fig. 1.1).

On the other hands, with the tremendous increase in the number of portable electronics, developing energy storage related technologies is vitally important because most of which are run by batteries. Although the power consumption of each is rather small, the total number of devices is extremely huge. Over three billion people around the world have cell phones. With the implementation of sensor networks around the globe, a gigantic number of sensors will be distributed worldwide; powering of such a horrendous network consisting of trillions of sensors would be impossible using batteries, because one has to find the location, replace batteries and

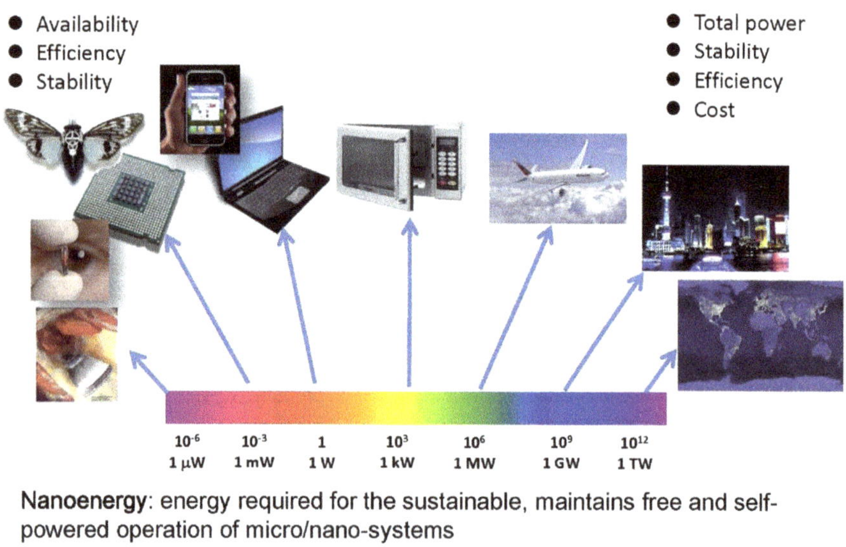


**Fig. 1.1** Magnitude of power and its corresponding applications. Macro-scale energy is for powering a city and even a country, the nano-scale energy is to power tiny small electronics, both of such applications are measured by different characteristics. Reproduced with permission from Wiley [5]

inspect the proper operation of batteries from time to time. Energy harvesting from the environment in which the senor is employed is a possible solution. This is the field of nano energy, which is the power for sustainable, maintains free and self-powered operation of micro/nano-systems [5]. The general characteristics for nano energy power sources are the availability, efficiency and stability (Fig. 1.1). In a case that a device is used under the light, the use of solar energy would be a natural choice. In a case of a device is used near an engine possibly in dark, harvesting mechanical vibration energy would be the best choice. As for biological application, harvesting deformation energy from muscle stretching would be a good approach. Although we may have a super high efficient solar cell, the condition under which the device will work may have little light, the high efficient solar cell is not the choice for this device. Therefore, the type of energy to be harvested depends on the working environment of the device. This is what we mean by the availability of the energy source for a particular application. The stability of the energy source is also important because it guarantee the long-term operation of the device. Take solar cell as an example, it has strong dependence on the day or night, weather or even season. This is the reason that we have been developing technologies for converting mechanical energy into electricity for self-powered sensors.

## 1.2 Triboelectric Effect

The triboelectric effect is a contact induced electrification in which a material becomes electrically charged after it is contacted with a different material through friction. Triboelectric effect is a general cause of every day's electrostatics. The signed of the charges to be carried by a material depends on its relative polarity in comparison to the material to which it will contact.

Triboelectric effect is probably the only a few effects that have been known for thousands of years. Although this is one of the most frequently experienced effect that each and every one of us inevitably uses every day, the mechanism behind triboelectrification is still being studied possibly with debate [6, 7]. It is generally believed that after two different materials coming into contact, a chemical bond is formed between some parts of the two surfaces, called adhesion, and charges move from one material to the other to equalize their electrochemical potential. The transferred charges can be electrons or may be ions/molecules. When separated, some of the bonded atoms have a tendency to keep extra electrons, and some a tendency to give them away, possibly producing triboelectric charges on surfaces.

Materials that usually have strong triboelectrification effect are likely less conductive or insulators, thus, they usually capture the transferred charges and retain them for an extended period of time, building up the electrostatic charges, which are usually attributed to as negative effect in our daily life and technology developments. We can use the following examples to illustrate the damages that can be caused by triboelectrification. Aircraft flying will develop static charges from air friction on the airframe, which will interfere with radio frequency communication. Electrostatic charges are an important concern for safety, due to the fact that it can cause explosion and ignite flammable vapors. Carts/cars that may carry volatile liquids, flammable gasses, or explosive chemicals have to be discharged properly to avoid fire. Some electronic devices, most notably complementary metal-oxide semiconductor (CMOS) integrated circuits and MOSFET transistors, can be accidentally destroyed by high-voltage static discharge that may be carried by gloves. Therefore, tribo-electrification is mostly taken as a negative effect in our daily life, industrial manufacturing and transportation. Therefore, by surprise, although triboelectrification is known for thousands of years, it has not been used for many positive applications. It is until recently that triboelectric effect has been widely used for converting mechanical energy into electricity and as self-powered active mechanical sensors.

## 1.3 Quantification of Triboelectrification

Although the triboelectrification effect is known for thousands of years, a fundamental understanding about it is rather limited. Research has been conducted to characterize the triboelectrification process using various methods such as rolling sphere tool-collecting induced charges from rolling spheres on top of dielectric disk [8, 9], and using atomic force microscopy (AFM) to measure surface electrostatic force or potential on surfaces contacted by micro-patterned materials [10–12]. However, these methods either lack an accurate control of the electrification process and/or cannot directly reveal the triboelectric interface, thus hardly achieving a quantitative understanding about the in situ triboelectric process.

We have demonstrated an in situ method to quantitatively characterize the triboelectrification at nano-scale via a combination of contact-mode AFM and scanning Kevin probe microscopy (SKPM) [13]. Benefited from the capability of controlled charge transferring and in situ measurement, AFM can be used to investigate the triboelectric charge transfer at surfaces. As a model system, a $SiO_2$ thin film was rubbed for multiple cycles at the same area with constant contact force. The corresponding SKPM images after each friction cycle are shown in Fig. 1.2a and the extracted potential profiles are presented in Fig. 1.2b. Within eight cycles of friction, the magnitude of the potential increased from 0.1 to 0.7 V at a slowing rate. As shown in Fig. 1.2c, there is a clear trend for the surface charge accumulation and saturation process. By quantitatively fitting the experimental data, surface charge density before triboelectric process is $\sigma_0 = (-12 \pm 3)$ μC/m$^2$, and saturation charge density after infinite numbers of cycles of friction is $\sigma_\infty = (-150 \pm 8)$ μC/m$^2$. The electric field at the vicinity of the surface is $\sim 1.7 \times 10^7$ V/m, which can easily generate a high voltage. The contact induced charge transfer is the experimental base of our TENG.

The charges to be delivered to the surface of material can be manipulated by applying a bias at the tip. Here we present the effect of an extrinsically applied electric field on contact electrification between metals and dielectric films [14]. Figure 1.3a schematically displays the charge transfer between a metal and a

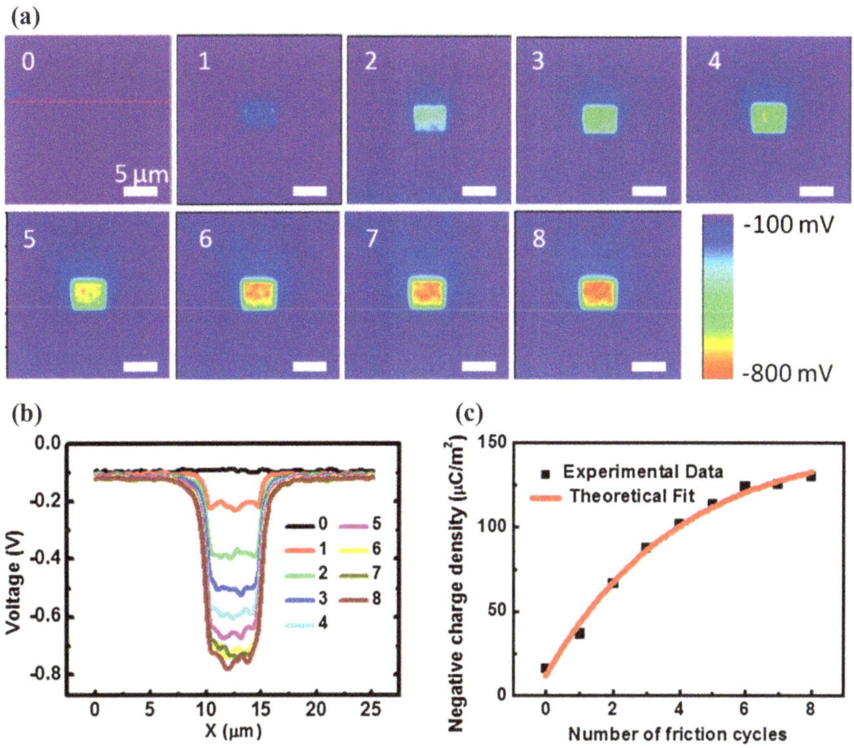


**Fig. 1.2** Triboelectric charge accumulation on the $SiO_2$ surface with the increase of the number of repeated rubbing at the same area. **a** Series of surface potential images taken in the same area from intact status to the one after 8th rubbing cycles, and **b** their corresponding potential profiles. **c** Derived surface charge density as a function of the number of friction cycles, and our fit based on charge accumulation theory. Reproduced with permission from American Chemical Society [13]

dielectric layer in contact electrification. When the top electrode contacts the dielectric layer, according to the electron transfer mechanism [15, 16], electrons will transfer from one to the other due to the difference in the effective work functions of two materials. Subsequently, the transferred contact electric charges can also induce opposite charges in the two metal electrodes due to electrostatic induction. Although the equivalent band gap in dielectric materials are large (usually larger than 8 V), it usually has surface states within the bandgap that can accommodate electrons [14]. Here we utilized the energy band diagrams to illustrate the case of a triboelectrically negative dielectric material (as compared to the metal). Accordingly, the highest filled surface energy states of the dielectric material is below the Fermi level of the metal, as illustrated in Fig. 1.3b. When the two materials are in contact, electrons in the metal will flow from the top metal electrode onto the dielectric surface to fill up the surface states as high as the metal's Fermi level (Fig. 1.3c). When two materials are separated from each other, an electric field is built up due to the transferred charge on the dielectric surface and image charges on the metal side, as depicted in Fig. 1.3d. The strength of the

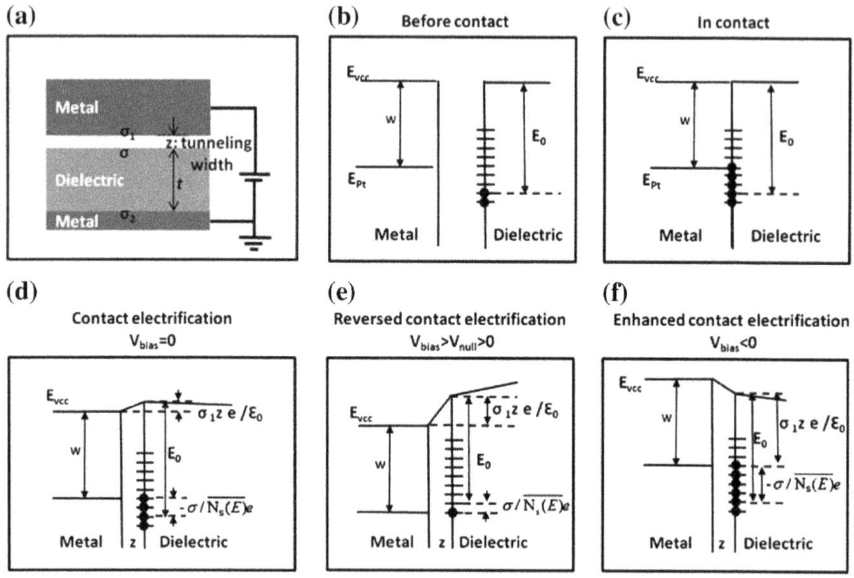


**Fig. 1.3** **a** Illustration of the contact electrification process with a tunneling width z, the charge transferred to the dielectric surface with a density of $\sigma$, induced charge density $\sigma_1$ and $\sigma_2$. **b–f** Energy band diagrams for the metal and dielectric materials in the situations of pre-contact (**b**), in contact with no bias (**c**), in separation equilibrium with no (**d**), positive (**e**) and negative (**f**) bias. Reproduced with permission from American Chemical Society [14]

electric field is proportional to the induced charge density $\sigma_1$ in the metal side. Due to this locally built-up electric potential, a portion of the charges on the dielectric surface can flow back to the metal. The potential difference is determined by the separation distance and electric field strength. At the same time, the separation process also creates an energy barrier between two surfaces that hinders the back-flow of electrons. For simplicity, we assign a critical tunneling distance $z$, and assume that below $z$ the electrons can flow freely between two surfaces to maintain a constant Fermi level and above $z$ the barrier is large enough to prevent any tunneling. For a parallel-plates model, at the distance $z$, the dielectric surface charge $\sigma$ induces $\sigma_1$ on the top metal and $\sigma_2$ on the bottom metal, which should satisfy

$$\sigma + \sigma_1 + \sigma_2 = 0 \tag{1.1}$$

Due to built-in electric field $\sigma_1/\mathcal{E}_0$, the vacuum energy level between the metal and the dielectric surface changes by $\Delta E_{vcc}$:

$$\Delta E_{\mathrm{vcc}} = \sigma_1 z e / \varepsilon_0 \tag{1.2}$$

where $e$ is the elementary charge.

When the system is in equilibrium, the dielectric surface state will be filled up as high as the Fermi energy level in the metal. If we assume that the surface density of states is $N_s(E)$ and the range of filled surface states is $\Delta E_s$, we have

$$\sigma = -e \int_{E_0}^{E_0 + \Delta Es} N_s(E)dE \tag{1.3}$$

Here, we use the averaged surface density of states as defined below,

$$\overline{N_s(E)} = \frac{\int_{E_0}^{E_0 + \Delta Es} N_s(E)dE}{\Delta E_s} \tag{1.4}$$

Therefore, the range of filled surface states $\Delta E_s$ can be described as

$$\Delta E_s = -\sigma / \overline{N_s(E)} e \tag{1.5}$$

Combining Eqs. (1.2) and (1.5), we have,

$$E_0 - W = \Delta E_{vcc} + \Delta E_s = \sigma_1 z e / \varepsilon_0 - \sigma / \overline{N_s(E)} e \tag{1.6}$$

An external potential applied between the top and bottom (electrode underneath the dielectric film) metals will change the relative energy band height, which modulates the charge transfer accordingly. A positive bias applied to the top metal can lower its Fermi energy level, thus reducing the number of electrons transferred from the metal to the dielectric. At certain bias, the Fermi energy level will be as low as the highest filled surface states in the dielectric material. As a result, no charge will transfer between the metal and the dielectric surface, suggesting the contact electrification will be nullified. When the bias is more positive than the nullified bias, electrons will flow in a reversed way from the dielectric to the metal, leaving dielectric surface positively charged, as illustrated in Fig. 1.3e. On the contrary, a negative bias to the metal can raise its Fermi energy level, driving more electrons flow to the dielectric surface to fill up higher surface energy states, resulting in the dielectric surface to be more negatively charged (Fig. 1.3f). Under the parallel-plate assumption, the bias $V$ between two metal electrodes can be written using Poisson equation as:

$$V = \frac{\sigma_1}{\varepsilon_0} z - \frac{\sigma_2}{\varepsilon \varepsilon_0} t \tag{1.7}$$

Combining Eqs. (1.1), (1.6) and (1.7), we can derive the surface charge density on the dielectric surface $\sigma$ as [14]:

$$\sigma = \frac{V + \frac{(W - E_0)}{e}(1 + t/\varepsilon z)}{t/\varepsilon\varepsilon_0 + \frac{1}{N_s(\mathrm{E}) e^2}(1 + t/\varepsilon z)} \tag{1.8}$$

This equation provides a guideline on how the external electric field quantitatively modulates the contact electrification.

From Eq. (1.8), the transferred charge density $\sigma$ should change with the externally applied bias $V$. To quantify the electric field influence, fresh areas of a Parylene film (2 μm in thickness) were rubbed with different bias from −10 to 10 V and then measured in the SKPM mode. Figure 1.4a displays the surface potential distribution in the areas that were rubbed with biases from −2 to 5 V: a negative bias can enhance the negative charge density; a positive bias of 2 V almost nullified

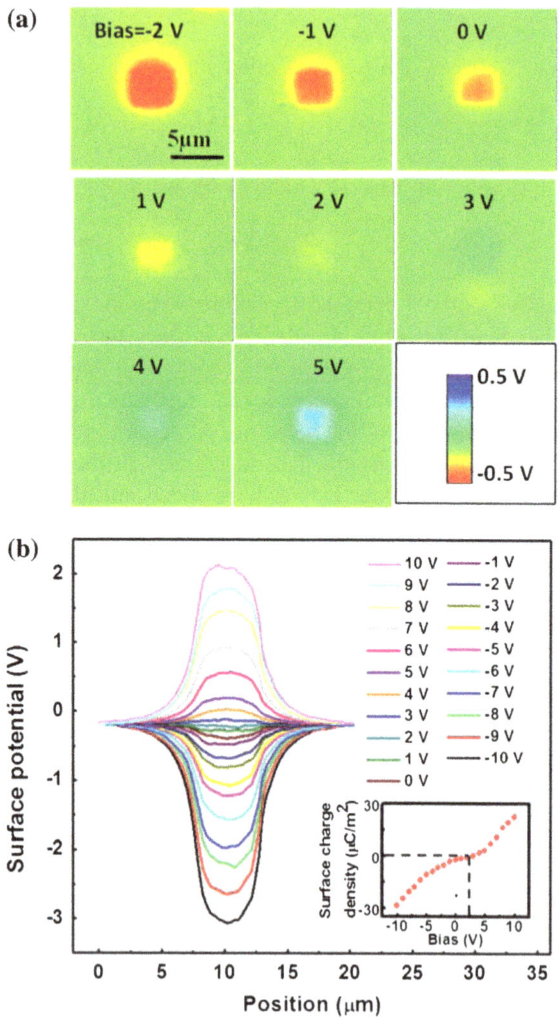


**Fig. 1.4** **a** Surface potential distributions of the Parylene film including the areas that were rubbed by Pt coated AFM tip at different bias from −2 to 5 V. **b** Cross section profiles of the surface potential of the Parylene film rubbed with bias from −10 to 10 V. The *inset* is the calculated surface charge density as a function of bias. With a bias of about 2.5 V, the surface charge density is zero, indicating the contact electrification is completely canceled out by the applied bias. Reproduced with permission from American Chemical Society [14]

the charges; a bias of higher than 2 V brought positive charges to the rubbed surface. A cross section profile of the surface potential distribution of each experiment was plotted in Fig. 1.4b. The surface potential of the rubbed area is adjusted monotonically by the applied bias, where the positive bias brings the surface potential to be more positive, and the negative bias exhibits the opposite effect. The corresponding surface charge density calculated from the surface potential profile is plotted as a function of the applied bias during the rubbing process in the inset of Fig. 1.4b. It can be seen that the surface charge density increases with increasing the applied external bias with a nonlinear characteristic, which is associated with the different energy state densities in the middle of band gap and closer to the conduction/valence band.

## 1.4 Materials for Triboelectrification

Almost any materials we know have triboelectrification effect, from metal, to polymer, to silk and to wood, almost everything. All of these materials can be candidates for fabricating TENGs, so that the materials choices for TENG are huge. However, the ability of a material for gaining/losing electron depends on its polarity. John Carl Wilcke published the first triboelectric series in a 1757 on static charges [16, 17]. Table 1.1 gives such a series for some conventional materials. A material towards the bottom of the series, when touched to a material near the top of the series, will attain a more negative charge. The further away two materials are from each other on the series, the greater the charge transferred. Recently, using the contact between a solid and liquid metal, such as Hg, a methodology has been established to quantitatively measure the surface charge density as a result of triboelectrification (see Sect. 7.4).

Beside the choice of the materials in the triboelectric series, the morphologies of the surfaces can be modified by physical techniques with the creation of pyramids-, square- or hemisphere-based micro- or nano-patterns, which are effective for enhancing the contact area and possibly the triboelectrification. The surfaces of the materials can be functionalized chemically using various molecules, nanotubes, nanowires or nanoparticles, in order to enhance the triboelectrification effect. Surface functionalization can largely change the surface potential. The introduction of nanostructures on the surfaces can change the local contact characteristics. The contact materials can be made of composites, such embedding nanoparticles in polymer matrix. This not only changes the surface electrification, but also the permittivity of the materials so that they can be effective for electrostatic induction.

**Table 1.1** Triboelectric series for some commonly materials following a tendency of easy losing electrons (positive) to gaining electrons (negative) [7]

| | | | |
|---|---|---|---|
| | Polyformaldehyde 1.3-1.4 | (continued) | |
| | Etylcellulose | Polyester (Dacron) | |
| | Polyamide 11 | Polyisobutylene | |
| Positive | Polyamide 6-6 | Polyuretane flexible sponge | |
| | Melanime formol | Polyethylene Terephthalate | |
| | Wool, knitted | Polyvinyl butyral | |
| | Silk, woven | Polychlorobutadiene | |
| | Aluminum | Natural rubber | |
| | paper | Polyacrilonitrile | |
| | Cotton, woven | Acrylonitrile-vinyl chloride | |
| | Steel | Polybisphenol carbonate | |
| | Wood | Polychloroether | |
| | Hard rubber | Polyvinylidine chloride (Saran) | |
| | Nickel, copper | Polystyrene | |
| | Sulfur | Polyethylene | |
| | Brass, silver | Polypropylene | |
| | Acetate, Rayon | Polyimide (Kapton) | |
| | Polymethyl methacrylate (Lucite) | Polyvinyl Chloride (PVC) | Negative |
| | Polyvinyl alcohol | Polydimethylsiloxane (PDMS) | |
| | (continued) | Polytetrafluoroethylene (Teflon) | |

(continued)

**Table 1.1** (continued)

| | | | |
|---|---|---|---|
| | Aniline-formol resin | Polyvinyl alcohol | |
| | Polyformaldehyde 1.3-1.4 | Polyester (Dacron) (PET) | |
| | Etylcellulose | Polyisobutylene | |
| Positive | Polyamide 11 | Polyuretane flexible sponge | |
| | Polyamide 6-6 | Polyethylene terephthalate | |
| | Melanime formol | Polyvinyl butyral | |
| | Wool, knitted | Formo-phenolique, hardened | |
| | Silk, woven | Polychlorobutadiene | |
| | Polyethylene glycol succinate | Butadiene-acrylonitrile copolymer | |
| | Cellulose | Nature rubber | |
| | Cellulose acetate | Polyacrilonitrile | |
| | Polyethylene glycol adipate | Acrylonitrile-vinyl chloride | |
| | Polydiallyl phthalate | Polybisphenol carbonate | |
| | Cellulose (regenerated) sponge | Polychloroether | |
| | Cotton, woven | Polyvinylidine chloride (Saran) | |
| | Polyurethane elastomer | Poly(2,6-dimethyl polyphenyleneoxide) | |
| | Styrene-acrylonitrile copolymer | Polystyrene | |
| | Styrene-butadiene copolymer | Polyethylene | |
| | Wood | Polypropylene | |
| | Hard rubber | Polydiphenyl propane carbonate | Negative |
| | Acetate, Rayon | Polyimide (Kapton) | |
| | Polymethyl methacrylate (Lucite) | Polyethylene terephtalate | |
| | Polyvinyl alcohol | Polyvinyl Chloride (PVC) | |
| | (continued) | Polytrifluorochloroethylene | |
| | | Polytetrafluoroethylene (Teflon) | |

## 1.5 Van de Graaff Generator

Traditional triboelectric generator is a mechanical device that produces static electricity, or electricity at high voltage by contact charging. The most popular ones are the Wimshurst machine and Van de Graaff generator, which were invented in ~1880 and 1929, respectively. Both machines use the accumulated static charges generated by triboelectrification. A simple Van de Graaff generator consists of a belt of rubber (or a similar flexible dielectric material) running over two rollers of differing material, one of which is surrounded by a hollow metal sphere, as shown in Fig. 1.5 [17]. Two electrodes, (2) and (7), in the form of comb-shaped rows of sharp metal points, are positioned near the bottom of the lower roller and inside the sphere, over the upper roller. Comb (2) is connected to the sphere, and comb (7) to ground. The rubber of the belt will become negatively charged while the acrylic

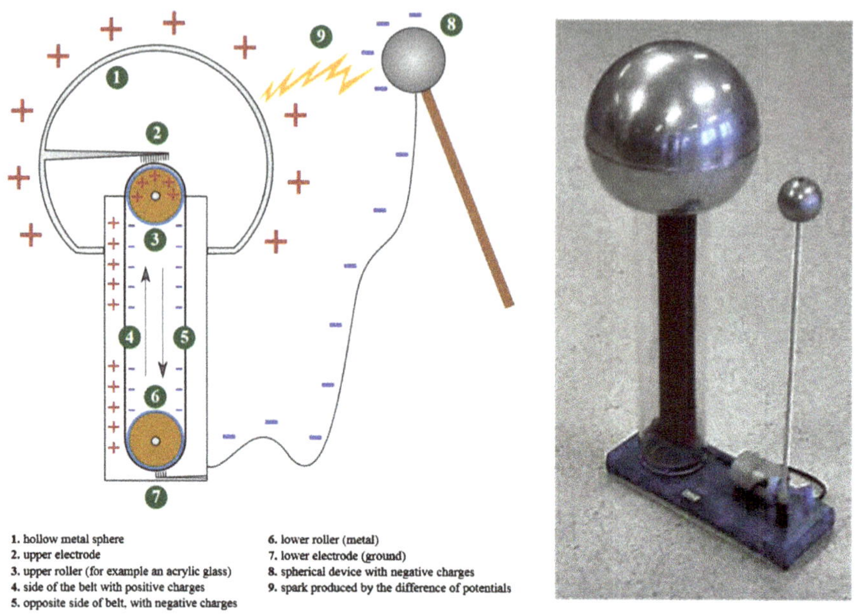


**Fig. 1.5** The Van de Graaff generator and its working mechanism. Reproduced from Ref. [17]

glass of the upper roller will become positively charged. The belt carries away negative charge on its inner surface while the upper roller accumulates positive charge. Next, the strong electric field surrounding the positive upper roller (3) induces a very high electric field near the points of the nearby comb (2). At the points, the field becomes high enough to ionize air molecules, and the electrons are attracted to the outside of the belt while positive ions go to the comb. At the comb (2) they are neutralized by electrons that were on the comb, thus leaving the comb and the attached outer shell (1) with fewer net electrons. The excess positive charge is accumulated on the outer surface of the outer shell (1), leaving no field inside the shell. Electrostatic induction by this method continues, building up very large amounts of charge on the shell.

In the example, the lower roller (6) is metal, which picks negative charge off the inner surface of the belt. The lower comb (7) develops a high electric field at its points that also becomes large enough to ionize air molecules. In this case the electrons are attracted to the comb and positive air ions neutralize negative charge on the outer surface of the belt, or become attached to the belt. As the belt continues to move, a constant 'charging current' travels via the belt, and the sphere continues to accumulate positive charge until the rate that charge is being lost (through leakage and corona discharges) equals the charging current. The metal sphere (8) is connected to ground, as is the lower comb (7); electrons are drawn up from ground due to the attraction by the positive sphere, and when the electric field is large enough (see below) the air breaks down in the form of an electrical discharge spark (9). Since the material in the belt and rollers can be selected, the accumulated charge on the hollow metal sphere can either be made positive (electron deficient) or negative (excess electrons).

Van de Graaff generator is a high voltage source, and there is no current unless there is a discharging. For a polished spherical electrode of 30 cm in diameter, it could be expected to develop a maximum voltage of about 450 kV.

## 1.6 Triboelectric Nanogenerators

Today's electronics is mostly associated with human activities for the purpose of health, safety and communication. The most abundant energy associated with a human is mechanical energy as a result of body motion. Recently, using the electrostatic charges created on the surfaces of two dissimilar materials when they are brought into physical contact, the contact induced triboelectric charges can generate a potential drop when the two surfaces are separated by mechanical force, which can drive electrons to flow between the two electrodes built on the top and bottom surfaces of the two materials. This is the first triboelectric nanogenerator (TENG) invented by Wang's group in 2012, aiming at harvesting small scale mechanical energy [18–20]. TENG has four basic modes as elaborated in details below.

### *1.6.1 Vertical Contact-Separation Mode*

We use the simplest design of TENG as an example (Fig. 1.6a) [21, 22]. Two dissimilar dielectric films face with each other, and there are electrode being deposited on the top and the bottom surfaces of the stacked structure. A physical contact between the two dielectric films creates oppositely charged surfaces. Once the two surfaces are separated by a small gap under the lifting of an external force, a potential drop is created. If the two electrodes are electrically connected by a load, free electrons in one electrode would flow to the other electrode to build an opposite potential in order to balance the electrostatic field. Once the gap is closed, the triboelectric charge created potential disappears, the electrons flow back.

### *1.6.2 Lateral Sliding Mode*

The structure to start with is the same as that for the vertical contact-separation mode. When two dielectric films are in contact, a relative sliding in parallel to the surface also creates triboelectric charges on the two surfaces (Fig. 1.6b) [23, 24]. A lateral polarization is thus introduced along the sliding direction, which drives the electrons on the top and bottom electrodes to flow in order to fully balance the field

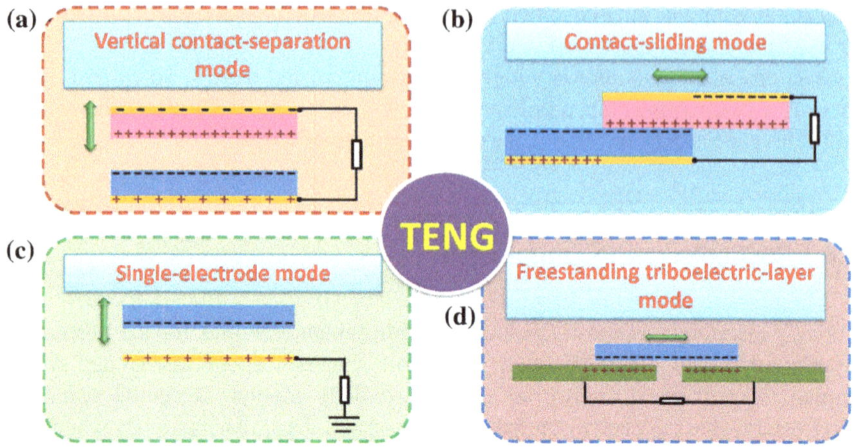


**Fig. 1.6** The four fundamental modes of triboelectric nanogenerators: **a** vertical contact-separation mode; **b** in-plane contact-sliding mode; **c** single-electrode mode; and **d** freestanding triboelectric-layer mode

created by the triboelectric charges. A periodic sliding apart and closing generates an AC output. This is the sliding mode TENG. The sliding can be a planar motion, a cylindrical rotation, or disc rotation. Related theoretical studies have been carried out for understanding the basic mode and grating structured TENG.

### 1.6.3 *Single-Electrode Mode*

The two modes introduced in Sects. 1.6.1 and 1.6.2 have two electrodes interconnected by a load. Such TENGs can freely move so that it can work for mobile cases. In some cases, the object that is part of the TENG cannot be electrically connected to the load because it is a mobile object, such as a human walking on a floor. In order to harvest energy from such a case, we introduced a single electrode TENG, in which the electrode on the bottom part of the TENG is grounded (Fig. 1.6c). If the size of the TENG is finite, an approaching or departing of the top object from the bottom one would change the local electrical field distribution, so that there are electron exchanges between the bottom electrode and the ground to maintain the potential change of the electrode. This energy harvesting strategy can be in both contact-separation mode [25] and contact-sliding mode [26, 27].

### 1.6.4 *Freestanding Triboelectric-Layer Mode*

In nature, a moving object is naturally charged due to its contact with air or other object, such as our shoes walking on floors that are usually charged. The charges remain on surface for hours and the contact or friction is unnecessary within this period of time because the charge density reaches a maximum. If we make a pair of symmetric electrode underneath a dielectric layer and the size of the electrodes and the gap distance between the two are of the same order as the size of the moving object, the object's approaching to and/or departing from the electrodes create an asymmetric charge distribution in the media, which causes the electrons to flow between the two electrodes to balance the local potential distribution (Fig. 1.6d) [28]. The oscillation of the electrons between the pair electrodes produces power. The moving object does not have to be directly touch the top dielectric layer of the electrodes, so that, in rotation mode, a free rotation is possible without direct mechanical contact, so that the wearing of the surfaces can be drastically reduced. This is a good approach for extend the durability of the TENGs.

Based on the four modes illustrated above, we have fabricated various TENGs depending on specific applications. Figure 1.7 shows a collection of photographs of

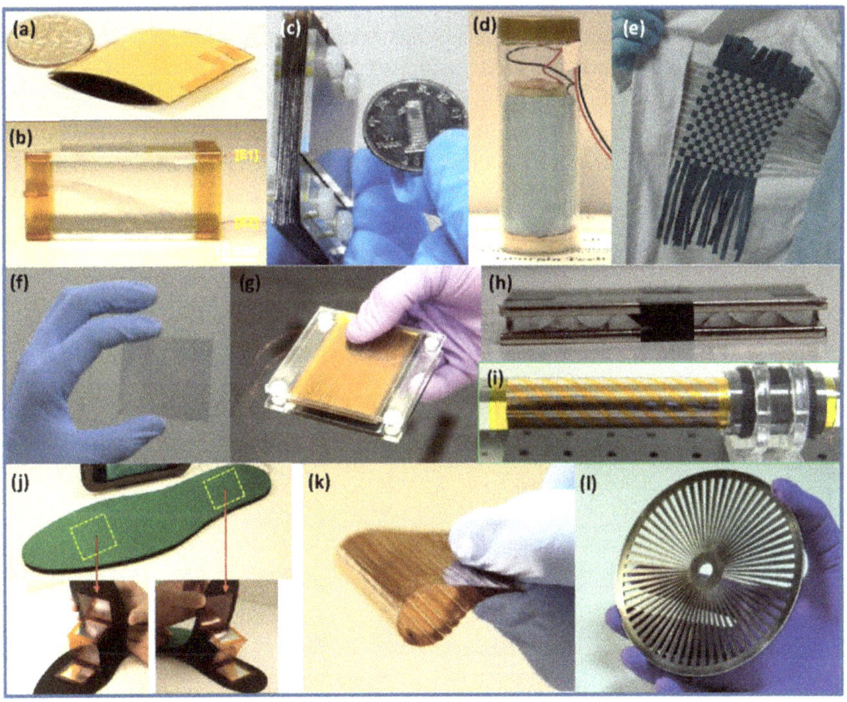


**Fig. 1.7** Typical photographs of some triboelectric nanogenerators fabricated for harvesting **a** finger tapping energy; **b** air-flow/wind energy; **c** relative in-plane sliding energy; **d** enclosed cage for harvesting oscillating/disturbing energy in water or mechanical vibration; **e** fabric for harvesting body motion energy; **f** transparent TENG for harvesting energy in touch pad; **g** foot/hand pressing energy; **h** water impact energy; **i** cylindrical rotation energy; **j** shoe insole for walking energy; **k** flexible grating structure for harvesting sliding energy; and **l** disc shape rotation energy. Reproduced with permission from Royal Society of Chemistry [32]

TENGs we have fabricated for harvesting various types of energy. These structures are the fundamental units for providing micro-scale power for small electronics, the assembly and integration of them can be the basis of harvesting mega-scale energy.

## 1.7 Perspectives

The discovery of triboelectric nanogenerator (TENG) is a major milestone in the field of converting mechanical energy into electricity for building self-powered systems. It offers a completely new paradigm for effectively harvesting mechanical energy using organic and inorganic materials. As of today, the area power density produced by a TENG has reached as high as 500 W $m^{-2}$, volume power density reaches 15 MW $m^{-3}$, and an instantaneous conversion efficiency of ~70 % has been demonstrated (Fig. 1.8) [29]. For low frequency agitation and if the energy generated by all the residual vibrations are acquired, a total energy conversion efficiency of up to 85 % has been shown experimentally [30]. A standard for measuring the performance of a TENG has been established. The TENG can be applied to harvest all kinds of mechanical energies that are available but wasted in our daily life, such as human motion, walking, vibration, mechanical triggering, rotating tire, wind, flowing water and more [19, 20]. This is the first major area of applications of TENG as micro-/nano-scale power source (Fig. 1.9).

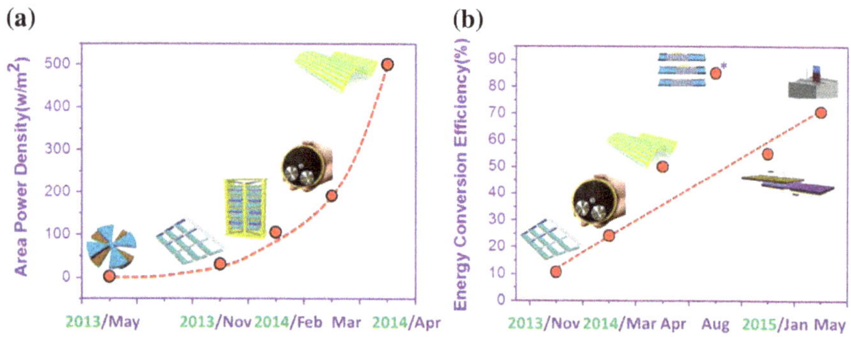


**Fig. 1.8** Summary on the progress in the performance of TENG made in the **a** area output power density and **b** energy conversion efficiency within the last 12 months. In May 2015, an instantaneous energy conversion efficiency of ~70 % has been demonstrated using liquid metal as a contact material. For low frequency agitation and if the energy generated by all the residual vibrations are acquired, a total energy conversion efficiency of up to 85 % has been shown experimentally in 2014. * Total energy conversion efficiency; the rest of data are instantaneous energy conversion efficiency. Reproduced with permission from Royal Society of Chemistry [20]

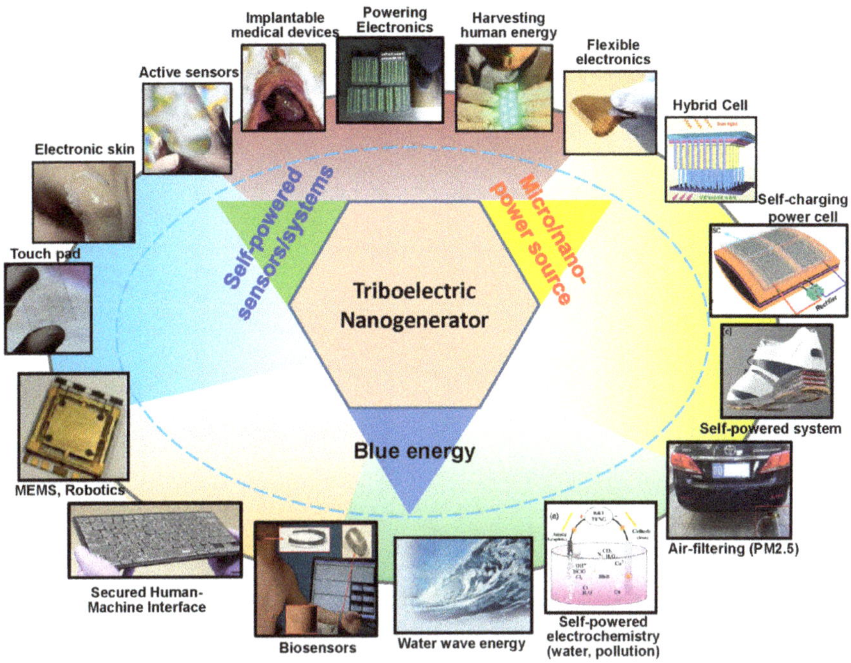


**Fig. 1.9** Summary on the three major field of applications of TENG in micro/nano-power source, self-powered sensors/systems (or active sensors) and mega-scale blue energy. The selected examples were from the research of Wang's group in the field of mobile/wearable electronics, internet of things, environmental science, medical sciences human-machine interfaces, infrastructure monirtoring and security

Alternatively, a TENG can also be used as a self-powered sensor for actively detecting the static and dynamic processes arising from mechanical agitation using the voltage and current output signals of the TENG, respectively, with potential applications as mechanical sensors and for touch pad and smart skin technologies [31] (Fig. 1.9). Such sensor does not need an external power source, but creates electric signal itself in responding to the external triggering or stimulation.

Furthermore, by integrating TENGs into network structure, it has the potential to harvest water wave energy in ocean, which could be a new paradigm for mega-scale blue energy [32–34] (Fig. 1.9). TENG has the incompatible efficiency for harvesting low-frequency mechanical energy in comparison to traditional electromagnetic generator of equivalent volume, so that it has unique and killer applications for the exploration of ocean wave energy—*blue energy*, which has the potential of making a huge contribution to the sustainable energy for the world. Figure 1.10 shows a proposed technology development road map for nanogenerators and its commercial applications. Furthermore, TENG can serve as a

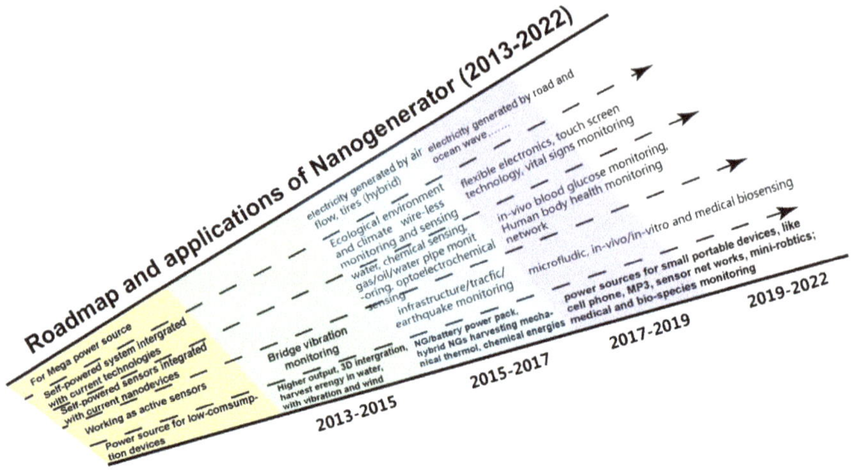


**Fig. 1.10** Proposed technology roadmap for nanogenerators (2012) by Wang. Reproduced with permission from Royal Society of Chemistry [32]

self-powered sensor for sensing mechanical triggering, pressure, muscle stretching and more, which may become commercially available sooner than energy harvesters.

# References


1. Z.L. Wang, J. Song, Piezoelectric nanogenerators based on zinc oxide nanowire arrays. Science **312**(5771), 242–246 (2006)
2. Z.L. Wang, Nanogenerators for self-powered devices and systems. Georgia Institute of Technology, SMARTech digital repository (2011). http://hdl.handle.net/1853/39262
3. Z.L. Wang, ZnO nanowire and nanobelt platform for nanotechnology. Mater. Sci. and Eng. Rep. **64**(3), 33–71 (2009)
4. Z.L. Wang, R. Yang, J. Zhou, Y. Qin, C. Xu, Y. Hu, S. Xu, Lateral nanowire/nanobelt based nanogenerators, piezotronics and piezo-phototronics. Mater. Sci. and Eng. Rep. **70**(3), 320–329 (2010)
5. Z.L. Wang, W. Wu, Nanotechnology-enabled energy harvesting for self-powered micro-/nanosystems. Angew. Chem. Int. Ed. **51**(47), 11700–11721 (2012)
6. J. Henniker, Triboelectricity in polymers. Nature **196**(474) (1962)
7. D. Davies, Charge generation on dielectric surfaces. J. Phys. D Appl. Phys. **2**(11), 1533 (1969)
8. R. Elsdon, F. Mitchell, Contact electrification of polymers. J. Phys. D Appl. Phys. **9**(10), 1445 (1976)
9. L.S. McCarty, G.M. Whitesides, Electrostatic charging due to separation of ions at interfaces: contact electrification of ionic electrets. Angew. Chem. Int. Ed. **47**(12), 2188–2207 (2008)
10. J.J. Cole, C.R. Barry, R.J. Knuesel, X. Wang, H.O. Jacobs, Nanocontact electrification: patterned surface charges affecting adhesion, transfer, and printing. Langmuir **27**(11), 7321–7329 (2011)
11. H.T. Baytekin, B. Baytekin, S. Soh, B.A. Grzybowski, Is water necessary for contact electrification? Angew. Chem. Int. Ed. **50**(30), 6766–6770 (2011)

12. H. Baytekin, A. Patashinski, M. Branicki, B. Baytekin, S. Soh, B.A. Grzybowski, The mosaic of surface charge in contact electrification. Science **333**(6040), 308–312 (2011)
13. Y.S. Zhou, Y. Liu, G. Zhu, Z.-H. Lin, C. Pan, Q. Jing, Z.L. Wang, In situ quantitative study of nanoscale triboelectrification and patterning. Nano Lett. **13**(6), 2771–2776 (2013)
14. Y.S. Zhou, S. Wang, Y. Yang, G. Zhu, S. Niu, Z.-H. Lin, Y. Liu, Z.L. Wang, Manipulating nanoscale contact electrification by an applied electric field. Nano Lett. **14**(3), 1567–1572 (2014)
15. J. Lowell, A. Rose-Innes, Contact electrification. Adv. Phys. **29**(6), 947–1023 (1980)
16. http://owlsmag.wordpress.com/2010/01/20/a-natural-history-devin-corbin/
17. https://en.wikipedia.org/wiki/Van_de_Graaff_generator
18. F.-R. Fan, Z.-Q. Tian, Z.L. Wang, Flexible triboelectric generator. Nano Energy **1**(2), 328–334 (2012)
19. Z.L. Wang, Triboelectric nanogenerators as new energy technology for self-powered systems and as active mechanical and chemical sensors. ACS Nano **7**(11), 9533–9557 (2013)
20. Z.L. Wang, J. Chen, L. Lin, Progress in triboelectric nanogenerators as a new energy technology and self-powered sensors. Energy Environ. Sci. **8**(8), 2250–2282 (2015)
21. G. Zhu, C. Pan, W. Guo, C.-Y. Chen, Y. Zhou, R. Yu, Z.L. Wang, Triboelectric-generator-driven pulse electrodeposition for micropatterning. Nano Lett. **12**(9), 4960–4965 (2012)
22. S. Wang, L. Lin, Z.L. Wang, Nanoscale triboelectric-effect-enabled energy conversion for sustainably powering portable electronics. Nano Lett. **12**(12), 6339–6346 (2012)
23. S. Wang, L. Lin, Y. Xie, Q. Jing, S. Niu, Z.L. Wang, Sliding-triboelectric nanogenerators based on in-plane charge-separation mechanism. Nano Lett. **13**(5), 2226–2233 (2013)
24. G. Zhu, J. Chen, Y. Liu, P. Bai, Y.S. Zhou, Q. Jing, C. Pan, Z.L. Wang, Linear-grating triboelectric generator based on sliding electrification. Nano Lett. **13**(5), 2282–2289 (2013)
25. Y. Yang, Y.S. Zhou, H. Zhang, Y. Liu, S. Lee, Z.L. Wang, A single-electrode based triboelectric nanogenerator as self-powered tracking system. Adv. Mater. **25**(45), 6594–6601 (2013)
26. Y. Yang, H. Zhang, J. Chen, Q. Jing, Y.S. Zhou, X. Wen, Z.L. Wang, Single-electrode-based sliding triboelectric nanogenerator for self-powered displacement vector sensor system. ACS Nano **7**(8), 7342–7351 (2013)
27. S. Niu, Y. Liu, S. Wang, L. Lin, Y.S. Zhou, Y. Hu, Z.L. Wang, Theoretical investigation and structural optimization of single-electrode triboelectric nanogenerators. Adv. Funct. Mater. **24** (22), 3332–3340 (2014)
28. S. Wang, Y. Xie, S. Niu, L. Lin, Z.L. Wang, Freestanding triboelectric-layer-based nanogenerators for harvesting energy from a moving object or human motion in contact and non-contact modes. Adv. Mater. **26**(18), 2818–2824 (2014)
29. W. Tang, T. Jiang, F.R. Fan, A.F. Yu, C. Zhang, X. Cao, Z.L. Wang, Liquid—metal electrode for high-performance triboelectric nanogenerator at an instantaneous energy conversion efficiency of 70.6 %. Adv. Funct. Mater. (2015)
30. Y. Xie, S. Wang, S. Niu, L. Lin, Q. Jing, J. Yang, Z. Wu, Z.L. Wang, Grating-structured freestanding triboelectric-layer nanogenerator for harvesting mechanical energy at 85 % total conversion efficiency. Adv. Mater. **26**(38), 6599–6607 (2014)
31. S. Wang, L. Lin, Z.L. Wang, Triboelectric nanogenerators as self-powered active sensors. Nano Energy **11**, 436–462 (2015)
32. Z.L. Wang, Triboelectric nanogenerators as new energy technology and self-powered sensors —Principles, problems and perspectives. Faraday Discuss. **176**, 447–458 (2014)
33. J. Chen, J. Yang, Z. Li, X. Fan, Y. Zi, Q. Jing, H. Guo, Z. Wen, K.C. Pradel, S. Niu, Networks of triboelectric nanogenerators for harvesting water wave energy: a potential approach toward blue energy. ACS Nano **9**(3), 3324–3331 (2015)
34. X. Wang, S. Niu, Y. Yin, F. Yi, Z. You, Z.L. Wang, Triboelectric nanogenerator based on fully enclosed rolling spherical structure for harvesting low-frequency water wave energy. Adv. Energy Mater. **5**(24), (2015)

# Part I
# Fundamental Operation Modes

# Chapter 2
# Triboelectric Nanogenerator: Vertical Contact-Separation Mode

Triboelectric nanogenerators (TENGs) can operate in four fundamental modes: vertical contact-separation [1–3], lateral sliding [4], single-electrode [5], and free-standing triboelectric-layer [6]. The working mechanism, the device fabrication and performance characterization of the TENGs in these modes will be introduced respectively in this chapter and Chaps. 3–5. In the basic mode of contact-separation, two dissimilar dielectric films face with each other, and there are electrodes deposited on the top and the bottom surfaces of the stacked structure. A physical contact between the two dielectric films creates oppositely charged surfaces. Once the two surfaces are separated by a mechanical force to introduce a small gap, during which the work done in overcoming the electrostatic attraction as well as the elasticity of the materials is the input energy, a potential drop is created. If the two electrodes are electrically connected by a load, free electrons in one electrode would flow to the other electrode to balance the potential drop created by triboelectric charges. Once the gap is closed, the triboelectric-charge-created potential disappears and the electrons flow back. Herein, the gap separating two triboelectric surfaces could be created by a variety of approaches, and several original device structures have been developed, including the spacer structure [1], the arch-shaped structure [2], the spring-assisted structure [3], and many more advanced designs [7, 8], all of which will be presented in this chapter.

## 2.1 Basic Principle

The operating principle of TENGs with vertical contact-separation mode can be described by the coupling of contact electrification and electrostatic induction [1]. Respectively, Fig. 2.1a, b depict electric output of open-circuit voltage ($V_{OC}$) and short-circuit current ($I_{SC}$). At original state, no charge is generated or inducted, with no electric potential difference between the two electrodes (Fig. 2.1a-I). With an



Z.L. Wang et al., *Triboelectric Nanogenerators*,
Green Energy and Technology, DOI 10.1007/978-3-319-40039-6_2

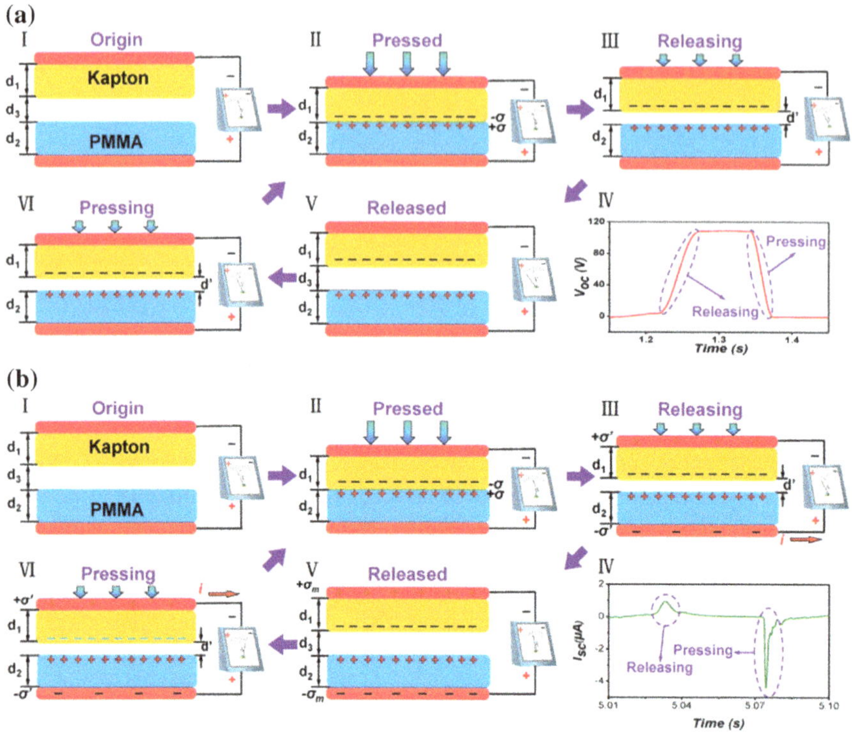


**Fig. 2.1** Operating principle of the generator. **a** Open-circuit condition. **b** Short-circuit condition. Reproduced with permission from American Chemical Society [1]

externally introduced displacement, the two polymers are brought into contact with each other. Surface charge transfer then takes place at the contact area due to triboelectric effect [9–11]. According to the triboelectric series [12] that is a list of materials based their tendency to gain or lose charges, electrons are injected from polymethyl methacrylate (PMMA) into Kapton, resulting in net negative charges at the Kapton surface and net positive charges at the PMMA surface, respectively. It is worth noting that the insulating property of the polymers allows a long-time retention of triboelectric charges for hours or even days [13]. Since they are only confined on the surface, charges with opposite signs coincide at almost the same plane, generating practical no electric potential difference between the two electrodes (Fig. 2.1a-II).

As the generator starts to be released, the Kapton film intends to revert back to its original position due to its own resilience. Once the two polymers separate, a potential difference is then established between the two electrodes under the open circuit condition since the opposite triboelectric charges are separated (Fig. 2.1a-III). As the generator is being released, $V_{OC}$ keeps increasing until reaching the maximum value when the Kapton film fully reverts to the original position (Fig. 2.1a-IV, V). Such a signal will remain constant provided that the input impedance of the electrometer is infinitely large. If pressing is immediately followed, the potential difference starts diminishing as the two polymer layers get closer to each other. As a result, $V_{OC}$ drops from the maximum value to zero when a full contact is made again between the two polymers (Fig. 2.1a-V, VI).

If the two electrodes are shorted, any established potential difference as the two polymers separate drives electrons from the top electrode to the bottom electrode (Fig. 2.1b-III), resulting in an instantaneous positive current during the releasing process (Fig. 2.1b-IV). Once the generator is pressed again, reduction of the interlayer distance would make the top electrode possess a higher electric potential than the bottom electrode. As a consequence, electrons are driven from the bottom electrode back to the top electrode, reducing the amount of induced charges (Fig. 2.1b-VI). This process corresponds to an instantaneous negative current (Fig. 2.1b-V). When the two polymers are in contact again, all induced charges are neutralized (Fig. 2.1b-II).

## 2.2 Fundamental Theory

The basic structure for contact-separation TENG is shown in Fig. 2.2 [14]. From electrostatic induction point of view, it belongs to the attached-electrode category, so it is also named as attached-electrode contact-mode TENG [15]. Based on the materials for the triboelectric pairs, the contact-separation TENG is mainly divided into two categories: dielectric-to-dielectric and conductor-to-dielectric types. The model built for a dielectric-to-dielectric contact-mode TENG is shown in Fig. 2.2a.

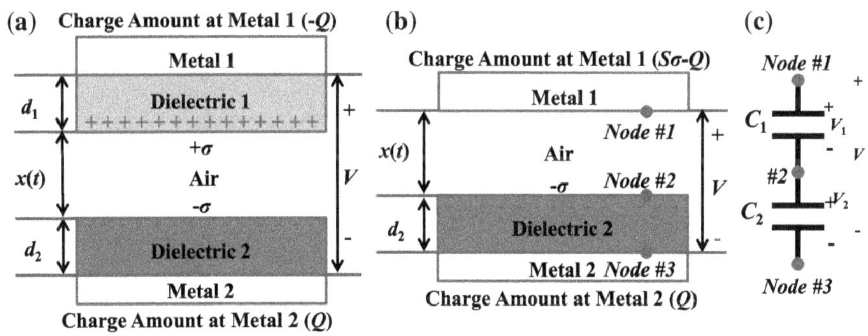


**Fig. 2.2** Theoretical models for **a** dielectric-to-dielectric attached-electrode parallel-plate contact-mode TENG and **b** conductor-to-dielectric attached-electrode parallel-plate contact-mode TENG. **c** Equivalent circuit diagram for Conductor-to-dielectric attached-electrode parallel-plate contact-mode TENG. Reproduced with permission from Royal Society of Chemistry [14]

The two dielectric plates, with thicknesses of $d_1$ and $d_2$ and the relative dielectric constants $\varepsilon_{r1}$ and $\varepsilon_{r2}$, respectively, are stacked face to face as two triboelectric layers. At the outer surface of these two dielectrics, two metal layers are deposited as two electrodes. The distance ($x$) between the two triboelectric layers can be varied under the agitation of mechanical force. After being forced into contact with each other, the inner surfaces of the two triboelectric layers will have opposite static charges (tribo-charges) with equal density of $\sigma$, as a result of contact electrification. For insulators as discussed, it is reasonable to assume that the tribo-charges are uniformly distributed at the two surfaces with negligible decay. When the two triboelectric layers start to separate from each other, with increased $x$, a potential difference ($V$) between the two electrodes will be induced. The amount of transferred charges between the two electrodes, as driven by the induced potential, is defined as $Q$. We choose the minimum achievable charge reference state (MACRS, details will be discussed in Chap. 6), so the instantaneous amount of charges on the electrode is $Q$ and $-Q$, respectively.

Similarly, the model for conductor-to-dielectric type is built, as shown in Fig. 2.2b. In this structure, we no longer have the Dielectric 1 and now the metal 1 plays two roles: both as the top triboelectric layer and as the top electrode. Therefore, the total charges in the metal 1 now have two parts: one is the triboelectric charges ($S\cdot\sigma$), the other is the transferred charges between the two electrodes ($-Q$). Thus, the total charges in the metal 1 are ($S\sigma - Q$).

The most important theoretical equation for representing the real-time power generation of a TENG is a relationship among three parameters: the voltage ($V$) between the two electrodes, the amount of transferred charge ($Q$) in between, and the separation distance ($x$) between the two triboelectric charged layers, which can be named as the $V$-$Q$-$x$ relationship. With the above model, the $V$-$Q$-$x$ relationship of such contact-mode TENG can be derived based on electrodynamics. Since the area size ($S$) of the metals is several orders of magnitude larger than their separation distance ($d_1 + d_2 + x$) in the experimental case, it is reasonable to assume that the two electrodes are infinitely large. Under this assumption, the charges on the metal electrodes will uniformly distribute on the inner surfaces of the two metals. Inside the dielectrics and the air gap, the electric field only has the component in the direction perpendicular to the surface, with the positive value pointing to metal 2. From Gauss theorem, the electric field strength at each region is given by [14]:

Inside Dielectric 1:

$$E_1 = -\frac{Q}{S\varepsilon_0\varepsilon_{r1}} \tag{2.1}$$

Inside the air gap:

$$E_{air} = \frac{-\frac{Q}{S} + \sigma(t)}{\varepsilon_0} \tag{2.2}$$

Inside Dielectric 2:

$$E_2 = -\frac{Q}{S\varepsilon_0\varepsilon_{r2}} \tag{2.3}$$

The voltage between the two electrodes can be given by:

$$V = E_1 d_1 + E_2 d_2 + E_{air} x \tag{2.4}$$

We can get obtain the *V-Q-x* relationship for dielectric-to-dielectric TENG in contact-mode, which is given by:

$$V = -\frac{Q}{S\varepsilon_0}\left(\frac{d_1}{\varepsilon_{r1}} + \frac{d_2}{\varepsilon_{r2}} + x(t)\right) + \frac{\sigma x(t)}{\varepsilon_0} \tag{2.5}$$

Similarly, for the conductor-to-dielectric category, the *V-Q-x* relationship is given by:

$$V = E_2 d_2 + E_{air} x = -\frac{Q}{S\varepsilon_0}\left(\frac{d_2}{\varepsilon_{r2}} + x(t)\right) + \frac{\sigma x(t)}{\varepsilon_0} \tag{2.6}$$

Equation 2.6 is exactly the same equation as Eq. 2.5 if $d_1 = 0$. We define the effective thickness constant $d_0$ as the summation of all the thickness of the dielectric $d_i$ between the two metal electrodes divided by its relative effective thickness $\varepsilon_{ri}$, as shown below:

$$d_0 = \sum\nolimits_{i=1}^{n} \frac{d_{\mathrm{i}}}{\varepsilon_{\mathrm{ri}}} \tag{2.7}$$

With such definition of effective dielectric thickness, the *V-Q-x* relationship for both of these two types can be unified as the following equation.

$$V = -\frac{Q}{S\varepsilon_0}(d_0 + x(t)) + \frac{\sigma x(t)}{\varepsilon_0} \tag{2.8}$$

The intrinsic output characteristics ($V_{OC}$, $Q_{SC}$, and $C$) of such a contact-mode TENG can be derived based on electrodynamics, as shown below.

$$V_{OC} = \frac{\sigma x(t)}{\varepsilon_0} \tag{2.9}$$

$$Q_{SC} = \frac{S\sigma x(t)}{d_0 + x(t)} \tag{2.10}$$

$$C = \frac{\varepsilon_0 S}{d_0 + x(t)} \tag{2.11}$$

Besides utilizing the Gauss Theorem, we can derive its *V-Q-x* relationship from the detailed electrostatic induction mechanism, through which we can show the fundamental working principle of such triboelectric generators [15].

We first need to define the nodes in this electrostatic system. A node in an electrostatic system corresponds to a surface/volume with the same electrical potential. Since the area size is assumed to be infinitely large and all the dielectric layers are uniform, the whole tribo-charged dielectric 2 surface is with the same electrical potential and can be thought as one node (Node 2). Since the electrical potential of the whole metal electrode is also the same, the electrode 1 and electrode 2 are other two nodes in this electrostatic system (Node 1 and Node 3, respectively), as shown in Fig. 2.2b. Since every two nodes are connected by electric field lines, an equivalent capacitance is formed between each of them. However, since dielectric 2 surface is assumed to be infinitely large, the electrical line connection of Node 1 and Node 3 is fully blocked by the Node 2. Therefore only two capacitances exist in this electrostatic system, as shown in Fig. 2.2c.

Now we consider this system under short-circuit conditions, when the Node 1 and Node 3 has the same electrical potential. Under this condition, the total amount of charges on Node 1 and Node 3 is $\sigma S$ while that on Node 2 is $-\sigma S$. From charge conservation and Kirchhoff's Law, the amount of charges on Node 3 is easily given by:

$$Q_{\text{node3}} = \frac{\sigma S}{1 + \frac{C_1(x)}{C_2(x)}} \tag{2.12}$$

So when $x$ is changing from 0, the short circuit transferred charges from Node 3 to Node 1 ($Q_{SC}$) can be given by:

$$Q_{SC} = \frac{\sigma S}{1 + \frac{C_1(x)}{C_2(x)}} - \frac{\sigma S}{1 + \frac{C_1(x=0)}{C_2(x=0)}} \tag{2.13}$$

Utilizing the parallel-plate capacitance mode, Eq. (2.9) can be easily derived to be the same with Eq. 2.6. Through the above derivation, it is clearly observed that the core working mechanism of TENGs is that the change of the capacitance ratio $C_1/C_2$ due to the change of the position of tribo-charged surface will induce electron transfer between the electrodes under SC condition.

When the $x$ is sufficiently large, the charge transfer efficiency $\eta_{CT}$ (defined as the ratio between the final transferred charges and the total triboelectric charges) can be given by:

$$\eta_{CT} = \frac{Q_{SC,\,\mathrm{final}}}{\sigma S} = \frac{1}{1 + \frac{C_1(x = x_{max})}{C_2(x = x_{max})}} - \frac{1}{1 + \frac{C_1(x=0)}{C_2(x=0)}} \tag{2.14}$$

For the above attached-electrode contact-mode TENGs, when $x = 0$, $C_1$ ($\varepsilon_0 S/x$) is close to infinity while $C_2$ is a finite number $\varepsilon_0 S/d_0$, which ensures the amount of charge on Node 3 at this time is 0. When $x$ is sufficiently large (larger than 10 $d_0$), $C_1$ ($\varepsilon_0 S/x$) is close to 0 while $C_2$ is still $\varepsilon_0 S/d_0$, which ensures the amount of charge on Node 3 at this time is $\sigma S$. Therefore, the maximum charge transfer efficiency for an attached-electrode contact-mode TENG can be reach 100 %.

From the above derivation, the intrinsic output characteristics of attached-electrode contact-mode TENGs have the following properties. Their open-circuit voltage is linearly proportional to the separation distance $x$ while their inherent capacitance is inversely proportional to the separation distance $x$. For short-circuit transferred charges, it has a saturation trend. $Q_{SC}$ quickly reaches 90 % of the final saturation value $\sigma S$ when $x$ only changes from 0 to $10d_0$. Since the dielectric thickness is usually very small, the minimum separation distance must be effectively reduced to 0 to reach high charge transfer efficiency. If the minimum separation distance is comparable to $10d_0$, $C_1(x)$ ($\varepsilon_0 S/x$) will always be much smaller than $C_2(x)$ ($\varepsilon_0 S/d_0$). Under this condition, the capacitance ratio $C_1(x)/C_2(x)$ will always be close to 0 and changes little when $x$ further increases, which will lead to no charge transfer between the electrodes. Besides, when $x$ is already comparable to the size of the device, the device can be no longer seen as infinitely large and edge effect will dominate the performance. Under this condition, the open-circuit voltage will deviate from the linear relationship.

## 2.3 Basic Device Structures

### 2.3.1 *Spacer Structure*

The first design of contact-separation mode TENG is schematically shown in Fig. 2.3 [1]. Between two polymer layers, a cavity is formed, which is sustained by a spacer and designed for the charge generation and separation processes. The fabrication process started with a square glass sheet, which was deposited with a thin layer of aluminum as the bottom electrode using electron beam evaporator. Then a thin layer of PMMA was spin-coated, followed by adding a spacer layer at

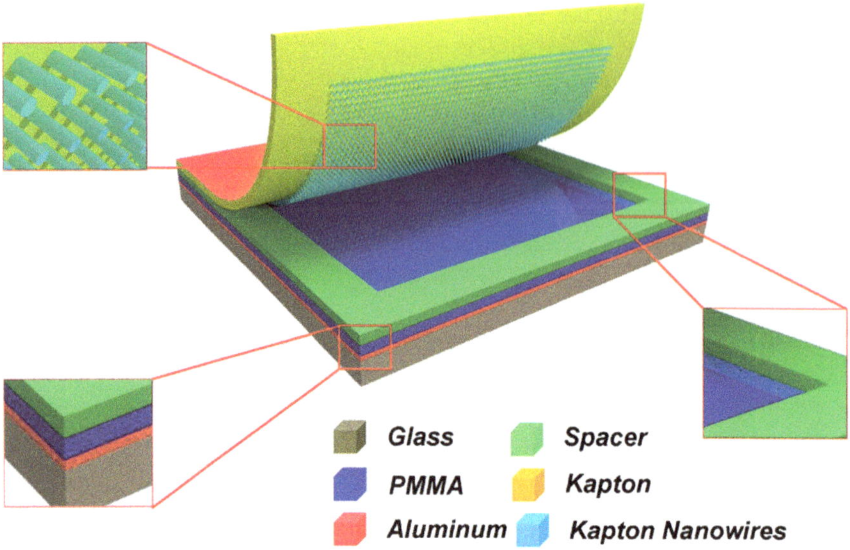


**Fig. 2.3** Sketch of the generator's structure and materials selection. The polymer nanowires (PNWs) are depicted with a different color from the substrate for clear visualization even though they are the same material. Reproduced with permission from American Chemical Society [1]

the edges, leaving a square cavity at the center. One side of a Kapton film was deposited with a layer of aluminum as the top electrode, while the other side was dry-etched to create vertically aligned polymer nanowires (PNWs) [16]. Then it was anchored on the spacer with the top electrode facing up. The spacer was made of insulating polymer with double-sided adhesive, keeping the Kapton film at a fixed distance away from the PMMA layer underneath.

As triggered by a vibration source with controlled frequency and amplitude, the generator produced an open-circuit voltage (Fig. 2.4a) and a short-circuit current (Fig. 2.4b) as predicted in the above analytical model. Electric output with opposite sign was obtained by switching the polarity for electric measurement. The peak value of the $V_{OC}$ and $I_{SC}$ were up to 110 V and 6 μA, respectively. Substituting the experimentally determined $V_{OC}$ into Eq. (2.9), we obtained a theoretical triboelectric charge density of 97.39 μC m$^{-2}$, and the maximum short-circuit transfered charge density of 73.72 μC m$^{-2}$. Therefore, electrons are pumped back and forth between the two electrodes as a result of contact electrification and electrostatic induction. For one cycle of contacting-separating the integration of current over time for releasing has the same value as that for pressing, indicating that equal amount of electrons flow in the opposite direction. It is observed that the current peak corresponding to releasing has a smaller magnitude but lasts longer than that for pressing (insets in Fig. 2.4b). Such an observation can be explained by the fact that pressing is caused by the external vibration source while it is the resilience of the Kapton film that leads to releasing. Therefore, it is very likely that releasing corresponds to a slower process and thus a smaller but wider current signal.

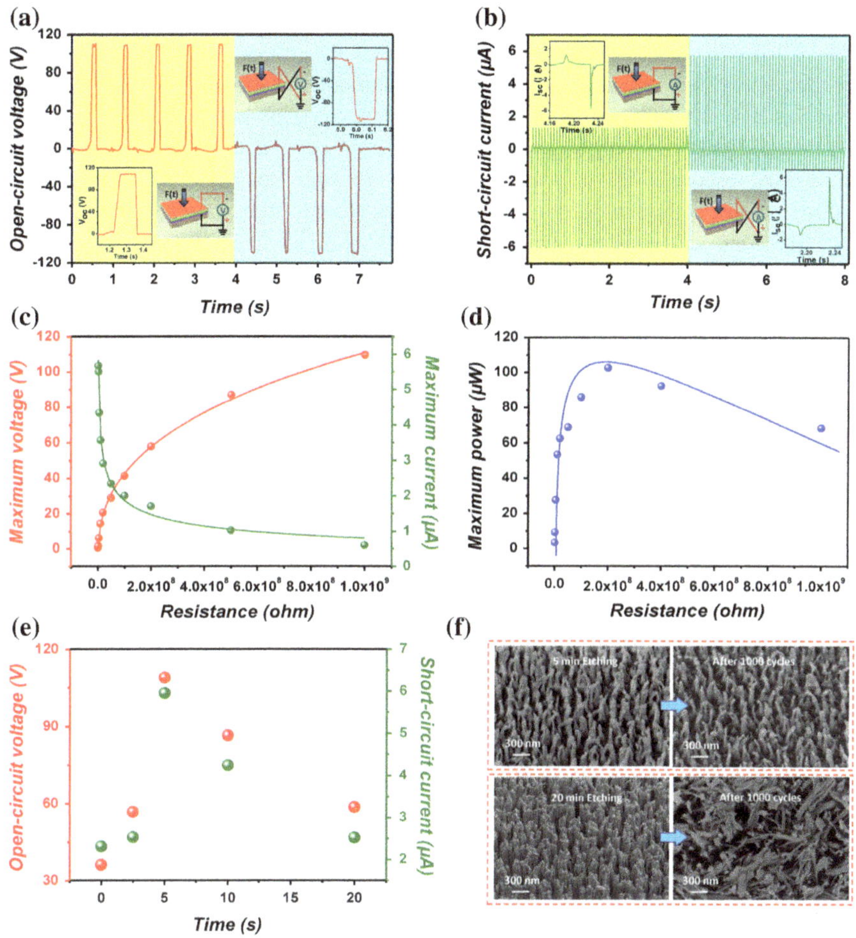


**Fig. 2.4** Investigation on the performance of the generator. **a** Open-circuit voltage with forward connection on the *left half* (*bottom-left insets*) and reverse connection on the *right half* (*upper-right insets*). **b** Short-circuit current with forward connection on the *left half* (*upper-left insets*) and reverse connection on the *right half* (*bottom-right insets*). **c** Electric voltage (*left axis*) and current (*right axis*) as a function of the load resistance. The dots are experimentally measured values, while the *lines* are fitted results. **d** Instantaneous electric power as a function of the load resistance. The *dots* are experimentally measured values, while the line is a fitted result. **e** The effect of time for etching polymer nanowires on open-circuit voltage (*left axis*) and short-circuit current (*right axis*). **f** SEM images of the PNWs before (*left pairs*) and after (*right pairs*) the generator operates for 1000 cycles to explain why PNWs with proper length (*top pairs*) contribute to much more enhanced electric output than do excessively long PNWs (*bottom pairs*). Reproduced with permission from American Chemical Society [1]

External load matching for the generator has also been investigated. With an increase in the load resistance, the maximum current decreases due to Ohmic loss, while the maximum voltage across the load has an opposite trend (Fig. 2.4c).

Accordingly, the electric power exhibits an instantaneous peak value of 110 μW (Fig. 2.4d), in correspondence to a power density of 31.2 $mW/cm^3$. The measurement results reveal that the generator is particularly efficient provided that the load has a resistance on the order of mega ohms. This demonstrated generator features simple fabrication/implementation, low cost, and strong performance. Previously reported electrostatic generators usually required pre-charging by external power source in order to introduce electrostatic force [17]. On the contrary, the TENG is self-charged, with triboelectric charges instantly generated upon contact. Besides, the fabrication processing is highly straightforward with neither material synthesis nor micromachining. The selection of materials is cost-oriented. Extremely low cost for fabrication and materials is thus anticipated, which is likely to be further reduced for mass production.

Several factors may be attributed to this largely enhanced performance. The first one is appropriate selection of materials. Substantial polarity difference exists between Kapton and PMMA according to the triboelectric series, leading to very effective charge transfer upon contact. In addition, the spacer structure plays an important role. The spacer's height not only relates to $V_{OC}$ but also affects $I_{SC}$. Without the spacer, only very little electric output could be detected. Furthermore, surface morphology modification is another key factor. Here, selective dry etching was applied on the Kapton surface to create vertically aligned polymer NWs, with longer etching corresponding to larger aspect ratio of the NWs. Enhanced surface roughness as a result of the NWs is expected to introduce additional friction as the two polymers are brought into contact. Such frictional movement is likely to generate more triboelectric charges; and thus enhanced electric output is achieved. However, experimental observation shows that the electric output substantially drops with excessive etching time (Fig. 2.4e). For NWs with proper length (hundreds of nanometers), the elastic property ensures retained orientation and morphology of the NWs even after numerous contacts. For overly long NWs, strain generated at the root is likely to exceed the elastic limit of the polymer material, resulting in permanent deformation (Fig. 2.4e). As a result, they are no longer able to introduce effective friction, which explains the less enhanced electric output.

## 2.3.2 *Arch-Shaped Structure*

The arch-shaped TENG is based on the contact electrification between patterned polydimethylsiloxane (PDMS) as the top plate and patterned Al foil as the bottom plate (Fig. 2.5a) [2]. According to the triboelectric series [12], the purposely-chosen PDMS and Al are almost at the two ends with very large difference in ability to attract and retain electrons. The unique arch-shaped structure of the TENG from the naturally-bent top plate, which helps to carry out the action of effective charge separation and contact using the elasticity of the film, is achieved by the following innovative fabrication process (Fig. 2.5b, c): The top part starts from a piece of flat Kapton film (Fig. 2.5b–i). A layer of 500 nm $SiO_2$ film is deposited using

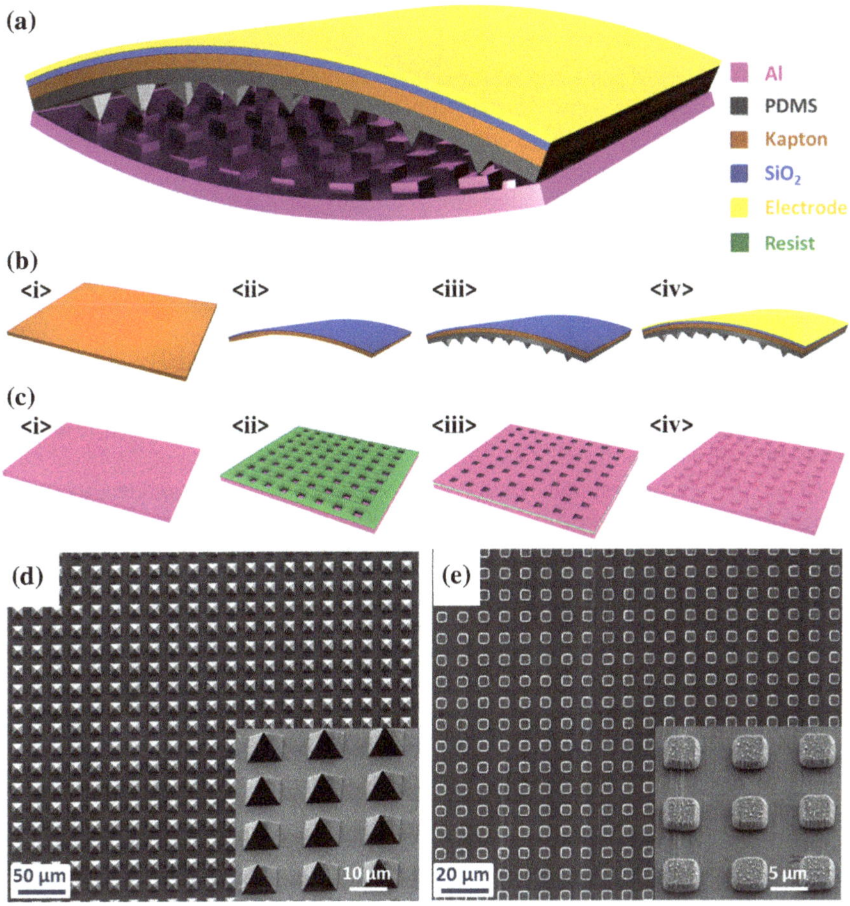


**Fig. 2.5** Structure and fabrication process of the *arch-shaped* triboelectric nanogenerator (TENG). **a** Schematic diagram showing the structural design of the arch-shaped TENG; the *inset* is the photograph of a typical arch-shaped TENG device. **b**, **c** Fabrication flowchart for **b** the *top plate* and **c** *bottom plate* of the TENG. **d**, **e** *Top view* SEM images of **d** the PDMS surface with pyramid patterns and **e** Al surface with cubic patterns; the *insets* are high-magnification images at a tilted angle. Reproduced with permission from American Chemical Society [2]

plasma-enhanced chemical vapor deposition (PECVD) at 250 °C (Fig. 2.5b-ii). Upon cooling down to room temperature, the Kapton will shrink to a much larger extent than the $SiO_2$ film because of the large difference in thermal expansion coefficients, so that thermal stress across the interface will make the plate bent naturally towards the $SiO_2$ side. Then, the pre-fabricated PDMS film with pyramid patterns is glued to the inner surface through a thin PDMS bonding layer (Fig. 2.5b-iii). Finally, the electrode is deposited on top (Fig. 2.5b-iv). As for the bottom plate, an aluminum foil (Fig. 2.5c i) is patterned with a typical photolithography process: defining the photoresist to the array of square windows

(Fig. 2.5c-ii); depositing a layer of aluminum on top (Fig. 2.5c-iii), and finally lift-off, leaving the patterned Al cubes on the foil (Fig. 2.5c-iv). At last, the two as-fabricated plates of the same size are attached face to face and sealed at the two ends. The soft Al plate will be forced to bend outward under the contraction from the other plate, so that a gap will form naturally in between. The patterned surfaces of PDMS film (Fig. 2.5d) and Al foil (Fig. 2.5e) are fabricated to enhance the triboelectric charging and are characterized using scanning electron microscopy (SEM). Both the arrays are uniform and regular across a very large area. The operation of the arch-shaped TENG is realized by applying a cycled compressive force onto the whole area of the device, so that the bending plates will be periodically pressed to flatten and contact closely with each other. Once released, the two plates will separate apart due to the stored elastic energy and revert instantaneously back to their original arch shape due to resilience. Herein, the deposition of $SiO_2$ layer by PECVD plays a key role in realizing a separation between the two layers through internal thermal residual stress.

Under the above described periodic deformation scenario, the electric output measurement was performed on an arch-shaped TENG device in a size of 3 cm × 2.8 cm, with the triggering frequency of 6 Hz and controlled amplitude. Since the accumulation of the triboelectric charges increases and reaches equilibrium in a certain period of time after multiple cycles, the output will gradually go up in the first stage upon deformation. Then, the open-circuit voltage ($V_{OC}$) will stabilize at 230 V (Fig. 2.6a), measured by an electrometer with infinite input resistance. From the inset of Fig. 2.6a, when the bottom Al is connected to the positive probe of the electrometer, upon the release of the pressing force a positive voltage is generated because of the immediate charge separation. Since in an open-circuit condition the electrons cannot flow to screen the induced potential difference between the two electrodes, the voltage will hold at a plateau until the subsequent pressing deformation in the second half cycle. As shown in Fig. 2.6b, the peak value of the short-circuit current ($I_{SC}$) reaches 94 μA, corresponding to the half cycle of pressing that is in higher straining rate than releasing. The integration of the each current peak gives the total charges transferred in a half cycle of deformation. Moreover, when the connection polarity to the electrometer is switched, both the voltage and current signal will be completely reversed. The gap from the arch-shaped structure is a key factor for the enhanced output, because without the $SiO_2$-film-introduced bending, there will be much smaller and unstable electrical output.

Besides the triboelectric charge density and the separation distance of the two plates, another factor that could make an influence on the output of the TENG is the deformation frequency, which affects the straining rate. Since the mechanical energy from the environment is always irregular and varies in frequencies, it is necessary to study the dependence of TENG's output on the frequency. Thus, we tested the TENG device in a series of 9 different frequencies, from 2 to 10 Hz, with the amplitude of the triggering motor remaining constant. As shown in Fig. 2.6c, $V_{OC}$ almost remains the same at different frequencies. The probable reason is that at the open-circuit condition, it doesn't involve the dynamic process of charge

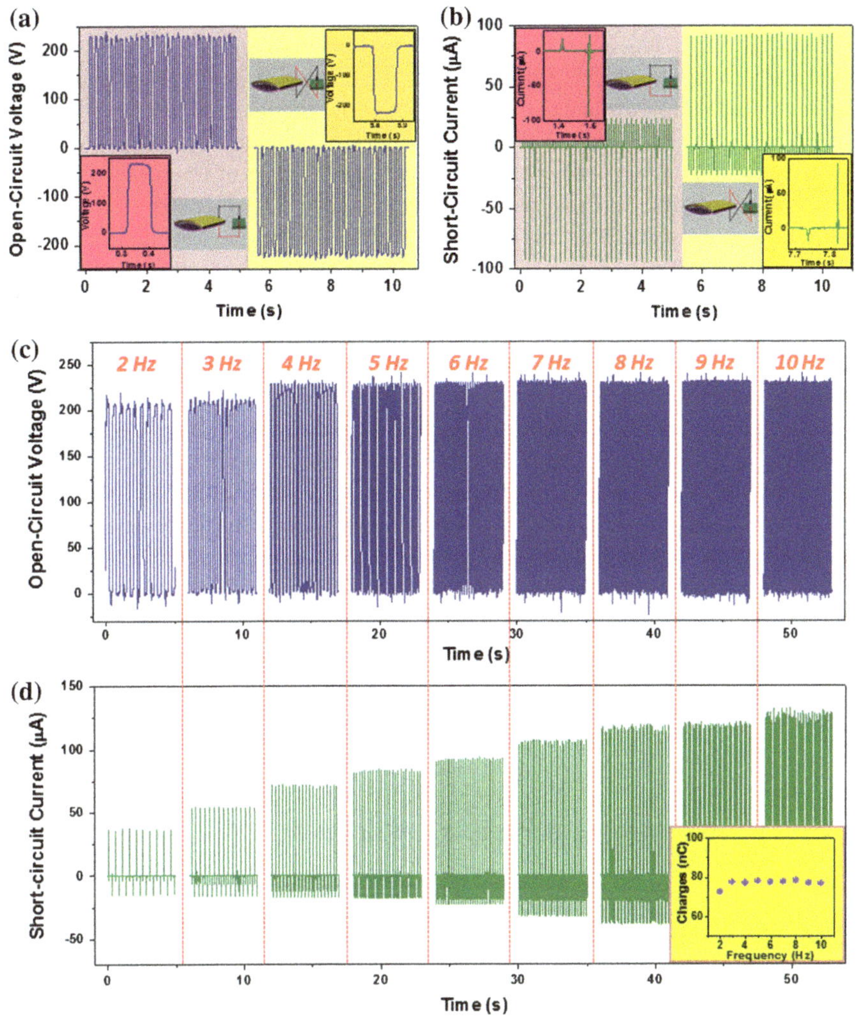


**Fig. 2.6** Characterizing the performance of the *arch-shaped* TENG. **a** Open-circuit voltage and **b** short-circuit current of the TENG under the deformation frequency of 6 Hz. These are studies under both the forward connection (Al connected to the positive probe) shown at the *left-hand side* and reverse connection shown at the *right-hand side*. The *insets* are the magnified output curve in one cycle and the sketch of the corresponding connection polarity. **c**, **d** Influence of the deformation frequency on **c** the open-circuit voltage and **d** the short-circuit current. The *inset* in **d** is the integration of a single current peak from each frequency, which gives the total charges transferred in a half cycle, showing that the total contact charges generated are almost independent of the deformation frequency. Reproduced with permission from American Chemical Society [2]

transfer. The voltage is only determined by the triboelectric charge density and the plate separation at any given time. But for the short-circuit current, it presents a very clear increasing trend with the increase of frequency, from 35 μA at 2 Hz to

130 μA at 10 Hz (Fig. 2.6d), because the deformation rate increases with deformation frequency, which leads to a higher flow rate of charges, i.e. higher current, but the total amount of charges transferred is constant at given triboelectric condition and separation distance. This is confirmed by the integration of every single current peak from each of the 9 different frequencies (inset of Fig. 2.6d). Therefore, the instantaneous power output increases with the increase of frequency, so that it will be more eligible to drive electronic devices with larger power consumption. From the above results, when the frequency reaches 10 Hz, the instantaneous power output reaches up to 3.56 and 128 mW/cm$^3$. The instantaneous power on the load will reach the maximum value of 9 mW at a load resistance of ~3 MΩ. Based on the above results, the energy conversion efficiency for this structure has been estimated. Since the thickness of the substrate used here has little influence on the electricity generation process but does consume a lot of mechanical energy, the output efficiency can be dramatically high if the thickness of the plate is reduced according to the feasible engineering fabrication while still maintain the mechanical integrity of the TENG. Based on experimentally achievable design and fabrication, the efficiency of this structure with the same gap distance could be as high as 10–39 %. With such high instantaneous output current and power, the arch-shaped TENG is enabled to directly power up portable electronics including light-emitting diodes, and it is further integrated with a lithium-ion battery to drive commercial electronic components like a cell phone or wireless sensor networks. These unique and first-time demonstrations confirm the remarkable performance of this energy conversion approach.

### 2.3.3 *Spring-Assisted Separation Structure*

Here in this work, a new type of TENG is developed that has not only much simplified structure but also substantially higher power output enabled by nanoparticle-based surface modification [3]. The TENG has a layered structure with two substrates, as schemed in Fig. 2.7a. PMMA was selected as the material for substrates due to its decent strength, light weight, easy processing, and low cost. On the lower side, a layer of contact electrode is prepared. The contact electrode plays dual roles of electrode and contact surface. It consists of a gold thin film and gold nanoparticles coated on the surface. Alternatively, nanoparticles of non-precious metals can also be used as replacements. Figure 2.7c shows uniform distribution of the nanoparticles on the surface. They modify the surface both physically and chemically, which will be discussed in detail later. On the other side, a thin film of gold is laminated between the substrate and a layer of PDMS. This electrode is called back electrode for later reference. The two substrates are connected by four springs installed at the corners, leaving a narrow spacing between the contact electrode and the PDMS. The as-fabricated NG is exhibited in Fig. 2.7b. As sketched in Fig. 2.7d, the fabrication flow is straightforward without sophisticated equipment and process. In this structure, the elastic springs are responsible for

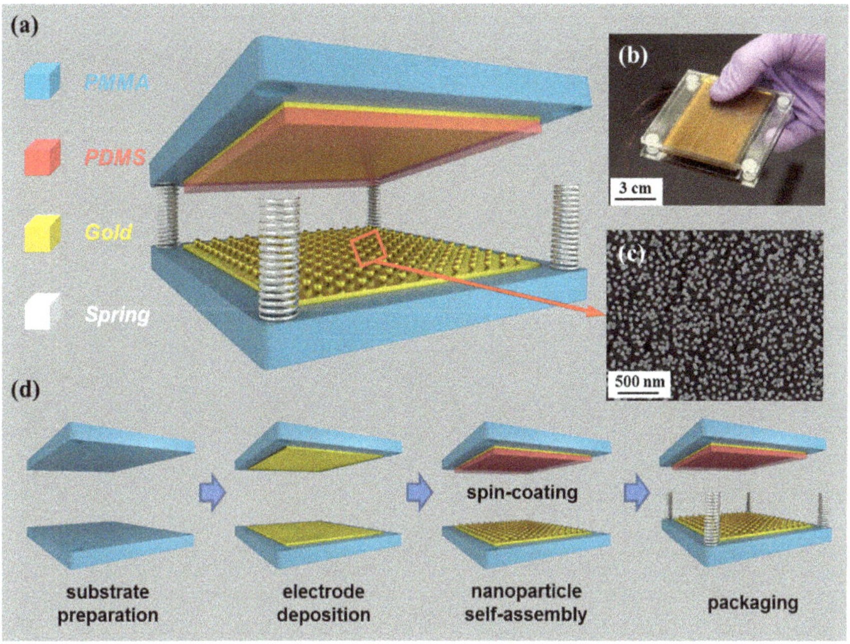


**Fig. 2.7** **a** Schematic and **b** photograph of a fabricated triboelectric NG. **c** SEM image of gold nanoparticles coated on gold surface. **d** Process flow for fabricating the NG. Reproduced with permission from American Chemical Society [3]

moving apart the two contacting surfaces when the impacting force is withdrawn from the TENG.

To trigger the TENG, a mechanical shaker was used to apply impulse impact. Here, the interaction force generated between the gold and the PDMS is defined as contacting force, which was monitored by a force sensor. The $V_{OC}$ and $I_{SC}$ were measured to characterize the TENG's electric performance. With a contacting force of 10 N, the measured $V_{OC}$ and the $I_{SC}$ are presented in Fig. 2.8a, b, respectively. The $V_{OC}$ switched between zero and a plateau value, respectively corresponding to the contact position and the original position. The $I_{SC}$ exhibits alternating current (AC) behavior, with equal amount of electrons flowing in opposite directions within one cycle. It is observed that the current signal for separation has a smaller magnitude but longer duration than that for contact (inset of Fig. 2.8b). It can be explained by faster contact resulting from external impact compared to slower separation caused by restoring force of the springs. The polarity of the measured electric signals can be reversed upon switching the connection polarity between the TENG and the measurement instrument. Furthermore, the AC output could be transferred to pulse output in the same direction simply by a full-wave rectifying bridge (Fig. 2.8c).

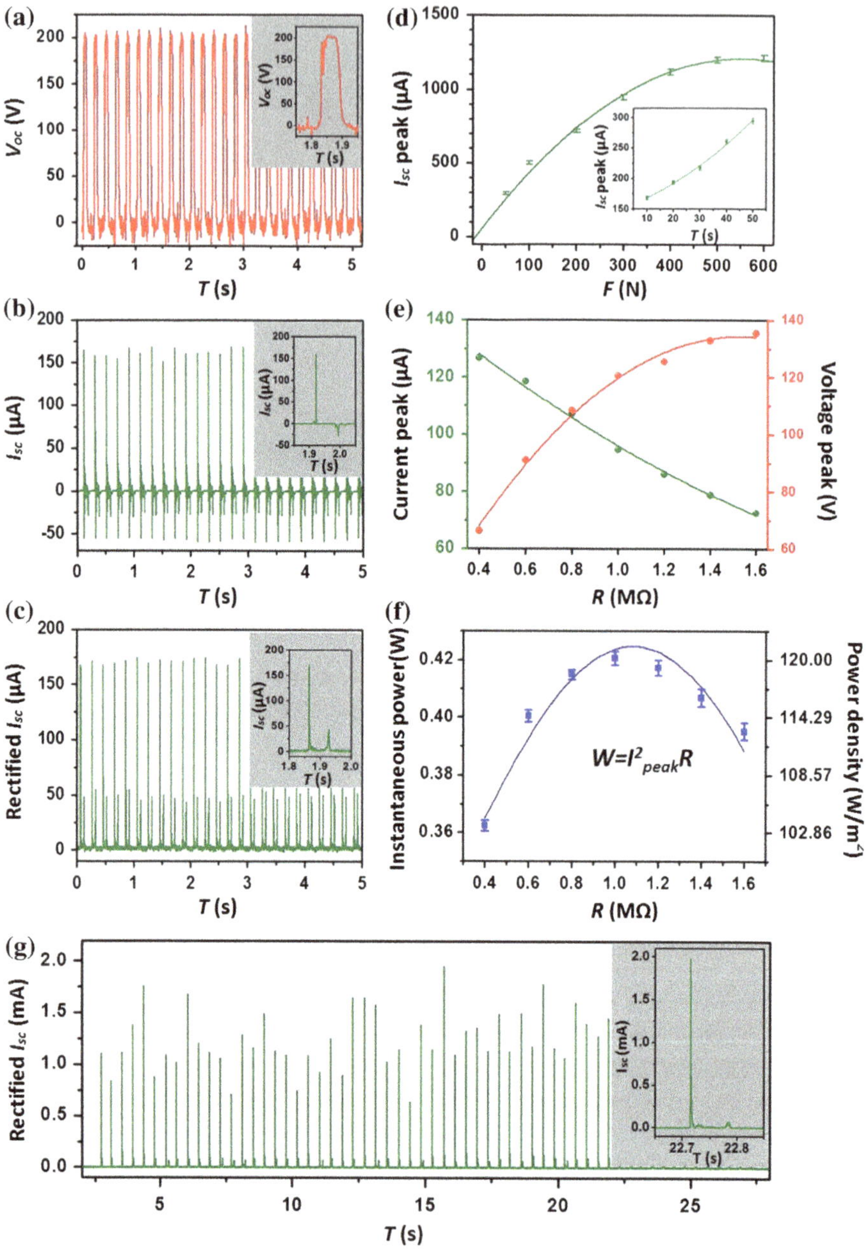
(a)
$V_{oc}$ (V)
T (s)
(d)
$I_{sc}$ peak (μA)
F (N)
(b)
$I_{sc}$ (μA)
T (s)
(e)
Current peak (μA)
Voltage peak (V)
R (MΩ)
(c)
Rectified $I_{sc}$ (μA)
T (s)
(f)
Instantaneous power(W)
Power density (W/m²)
$W=I^2_{peak}R$
R (MΩ)
(g)
Rectified $I_{sc}$ (mA)
T (s)

◀ **Fig. 2.8** **a** Open-circuit voltage ($V_{OC}$) at contacting force of 10 N. *Inset* Enlarged view of one cycle. Separation causes rising of the $V_{OC}$ to a plateau value and contact make it fall back to zero. **b** Short-circuit current ($I_{SC}$) at contacting force of 10 N. *Inset* Enlarged view of one cycle. Contact and separation correspond to a positive current pulse and a negative current pulse, respectively. **c** Rectified short-circuit current ($I_{SC}$) by a full-wave bridge at contacting force of 10 N. *Inset* Enlarged view of one cycle, showing two current pulses in the same direction. **d** Dependence of the $I_{SC}$ on the contacting force. The larger contacting force corresponds to higher current until saturation occurs. *Inset* $I_{SC}$ at contacting force less than 50 N. The *data points* represent the peak value of current pulses while the line is the fitted result. **e** Dependence of the current output on external load resistance at contacting force of 10 N. Increased load resistance results in lower current output but higher voltage output. The *points* represent peak value of electric signals while the *line* is the fitted result. **f** Dependence of the power output on external load resistance at contacting force of 500 N, indicating maximum power output when R = 1 MΩ. The curve is a fitted line. **g** Short-circuit current ($I_{SC}$) when triggered by human footfall that can generate contacting force between 500 and 600 N. *Inset* enlarged view of the highest current pulse. Reproduced with permission from American Chemical Society [3]

The TENG's electric output is strongly related to the contacting force, yielding higher output with a larger force. At a force as small as 10 N, the TENG can still produce $I_{SC}$ ranging from 160 to 175 μA (Fig. 2.8d). When the force increases to 500 N, the electric output reaches a saturated value, producing a peak $I_{SC}$ of 1.2 mA. This result is due to increased contact area with larger force. The two contacting surfaces are neither absolutely flat nor smooth. Surface roughness is caused by inherent curvature of the substrates, nanoparticle modification, and fabrication defects such as particle contamination from the air. At small contacting force, the surface roughness prevents fully intimate contact between the contact electrode and the PDMS, leaving some areas untouched. With larger force, due to elastic property, the PDMS can deform and fill more vacant space, thus leading to larger contact area. As a result, the electric output increases until all the vacant spacing is completely filled by the PDMS, reaching a saturated limit.

Resistors were connected as external loads to further investigate the effective electric power of the TENG. As demonstrated in Fig. 2.8e, the instantaneous current drops with increasing load resistance due to Ohmic loss, while the voltage builds up. Consequently, the instantaneous power output reached the maximum at a load resistance of 1 MΩ. At a contacting force of 500 N, a power output of 0.42 W was achieved (Fig. 2.8f), corresponding to a power density of 109 W/m$^2$ for one layer of NG. This value is substantially larger than previously demonstrated ones. If the mechanical shaker was replaced by human footfalls which can generate a contacting force between 500 to 600 N, the maximum $I_{SC}$ could reach up to 2 mA, as demonstrated in Fig. 2.8g. It corresponds to an instantaneous current of 1.1 mA at a load of 1 MΩ, instantaneous output power of 1.2 W, and power density of 313 W/m$^2$. The power density can be further multiplied by stacking layers of the NGs in vertical direction. Integrating the largest $I_{SC}$ signal (inset of Fig. 2.8g) with time gives the maximum inductive charges of 2.3 μC that flow between the two electrodes, giving a maximum triboelectric charge density of 594.2 μC/m$^2$. The apparent deviation shown in Fig. 2.8g is probably caused by impact from footfall which lacks control on position, force, and direction. However, it is noted that the mean value is comparable to that obtained with the mechanical shaker (1.2 mA in Fig. 2.8d). As another important figure of merit, energy conversion efficiency of the TENG was calculated to be 9.8 % at the device level, and 14.9 % at the energy conversion level.

### 2.3.4 Multiple Layer Integration

Previously demonstrated TENGs only had a single layer, making it challenging to substantially scale up the area density of the output power. Therefore, enhancing the output power inevitably involved largely increased area of a TENG. Here in this work [7], an innovative design of a TENG with multiple layers of units fabricated on a single flexible substrate is developed. The structure provides a novel and simple method to stack multiple TENGs without either expanding the area or considerably complicating the fabrication process. As a consequence, it is possible to achieve multi-folds enhancement of the output power while keeping the TENG's area constant. The structure of a flexible multilayered TENG from different views of angle is schemed in Fig. 2.9a. A Kapton thin film with a thickness of 125 μm was selected as the substrate due to its availability, flexibility, light weight and proper stiffness. The Kapton film was shaped to a zigzag structure by making deformation at evenly-spaced intervals (Fig. 2.9b). The zigzag-shaped design enables that both sides of the Kapton film can serve as substrates of the TENG units. As sketched in Fig. 2.9a, the TENG consists five layers of units, with three of them shown in the front view and the other two in the back view. For each layer, it incorporates an aluminum-coated polytetrafluoroethylene (PTFE) thin film and an aluminum foil with nano-pores-based surface modification as the contact electrode. Figure 2.9c shows the distribution of the nano-pores on the surface of aluminum foil, which play an important role in enhancing the output power. A thin film of aluminum was deposited on the back of PTFE as a back electrode. The whole TENG was packaged by fixing the deformed edges of the substrate together, as illustrated by a photograph of the as-fabricated TENG in Fig. 2.9d. Demonstrated in Fig. 2.9e, the fabricated TENG shows great flexibility, making it applicable to harvest energy not just from press but also from deformation.

For each of the layer fabricated on the zigzag-shaped substrate, it can be electrically connected with other layers in parallel through external wiring to enhance the output current. To characterize the output power of the multilayered TENG, a compressive contact force of around 400 N generated from a human palm was applied. Rectified $I_{SC}$ of individual layers were measured separately. Figure 2.10a displays the output current for each of these layers from the top to the bottom (Fig. 2.10$a_1$–$a_5$), respectively. It is noticed that the most top layer delivers the highest output current. It is suggested that this observation is attributed to the damping effect of the TENG's structure. The compressive force may be damped when it is transmitted from the top layer to the bottom layer. When all of the layers

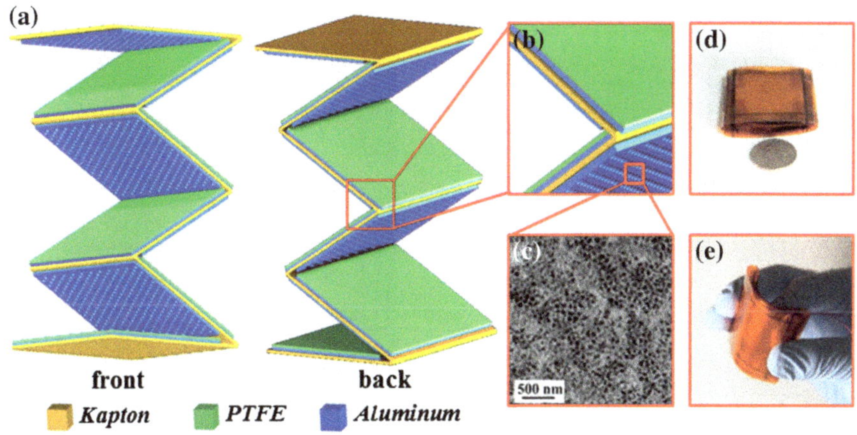


**Fig. 2.9** Structure and photographs of a flexible multilayered TENG with five layers of units. **a** Schematic and **b** an enlarged view of the zigzag-shaped structure of the TENG. **c** SEM image of nanopores on aluminum foils. **d** Photograph of a fabricated flexible multilayered TENG and **e** photograph of a bent flexible multilayered TENG by human fingers. Reproduced with permission from American Chemical Society [7]

are put into parallel connection, an $I_{SC}$ of 0.66 mA and a current density ($J_{SC}$) of 45.19 μA/cm$^2$ were achieved as shown in Fig. 2.10a$_6$. Likewise, we measured the $V_{OC}$ of each layer (Fig. 2.10b$_1$–b$_5$) and the TENG as a whole (Fig. 2.10b$_6$). The voltage of the entire TENG is not the sum of but comparable to the voltages from individual layers. The polarity of the measured electric signals above can be reversed by switching the connection polarity between the TENG and the measurement instrument. Based on the measurement results, it is effective to scale up the electric output by incorporating more layers on a longer substrate while keeping the area of the TENG constant. Furthermore, The maximum output power of 4.2 mW and a corresponding power density of 2.9 W/m$^2$ can be achieved at a load resistance of 1 MΩ. The contact force is a critical factor that affects the TENG's electric output.

### 2.3.5 *Microcavity-Nanoparticle Assembled Structure*

In this work, a facile and scalable synthesis of mesoporous films impregnated with gold nanoparticles (NPs) was employed as effective dielectrics for enhancing the TENG's performance based on vertical contact-separation mode [8]. A schematic diagram representing the fabrication of the TENG and the resulting free-standing mesoporous film is shown in Fig. 2.11a. The scanning electron microscopy (SEM) images in Fig. 2.11b show the sponge-like mesoporous structures of the PDMS film after removal of distilled water and gold NPs inside the pores. PDMS is

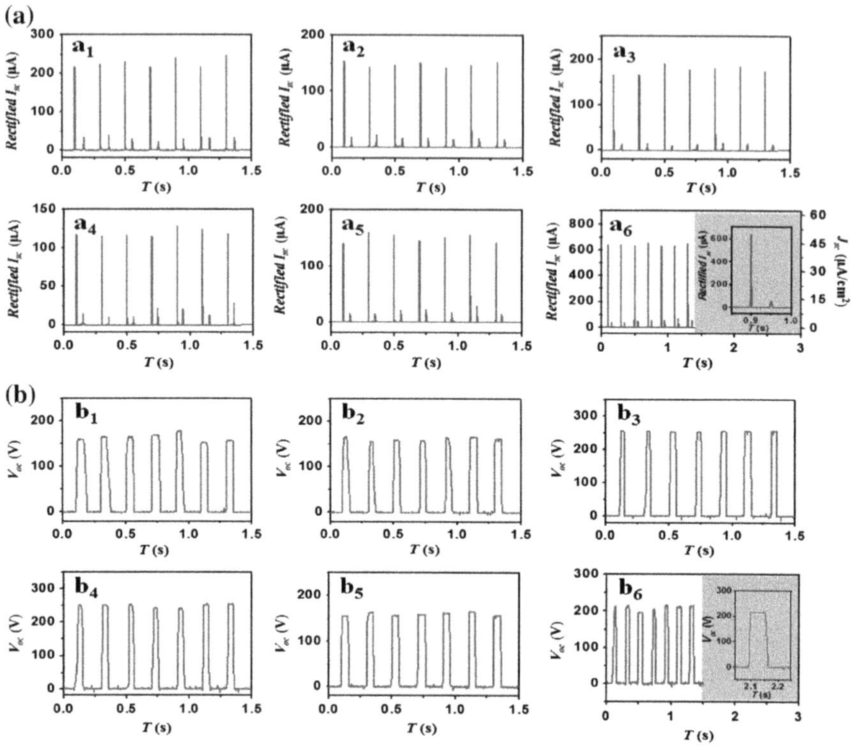


**Fig. 2.10** Output current and output voltage of the TENG with five layers of units. **a** Rectified short-circuit current ($I_{SC}$) under a contact force of 400 N applied by a human palm. *Inset of panel a6* enlarged view of one cycle, showing two current pulses in the same direction but contact corresponds to a larger current pulse. **b** Open-circuit voltage ($V_{OC}$) under contact force applied by a human palm. *Inset of panel b6* enlarged view of one cycle. Contact causes rising of the $V_{OC}$ to a plateau value and separation makes it fall back to zero. Reproduced with permission from American Chemical Society [7]

inherently hydrophobic, thus, the water in the PDMS solution attempts to form a spherical shape in the PDMS/water mixture. As pore initiation is random, the pores with a broad size distribution are randomly arranged. However, this approach is an easy fabrication method of scalable TENG based on sponge-like mesoporous films for developing practical self-powered electronic systems. The porosity increases to approximately 59 % as the amount of water increases to 50 %, as shown in Fig. 2.11c. To calculate the porosity of the mesoporous film, the SEM images in the inset of Fig. 2.11c were first converted into binary images, and the MATLAB image processing toolbox was used to sum total the volume of pores in the film. The porosity was calculated by dividing the volume of pores formed in the mesoporous film by the total volume of film. Using this method, a large-area mesoporous PDMS thin film (30 cm × 30 cm) was fabricated (Fig. 2.11d), demonstrating the capability of producing large-scale films. The photograph in the

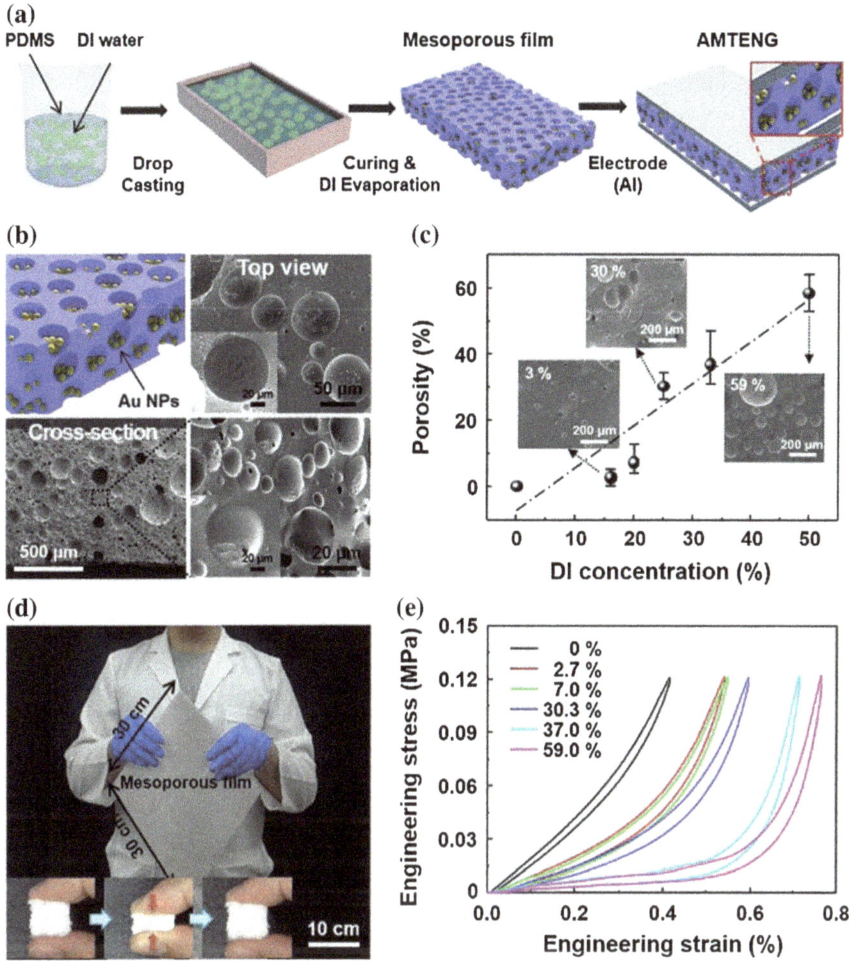


**Fig. 2.11** **a** Schematic diagrams of the fabrication process for the AMTENG. **b** *Top* and *cross-sectional* SEM images of a mesoporous PDMS film with Au NPs (0.28 wt%). The *insets* show the morphology of Au NPs and cross-sectional SEM images of the mesoporous film. **c** Porosity changes as a function of the DI water concentration ranging from 0 to 50 %. The *insets* show SEM images of mesoporous films with various porosities. **d** Photographs of a fabricated large-area mesoporous PDMS thin film (30 cm × 30 cm). The *insets* show manual compression with a volume reduction of over 50 %. **e** The loading and unloading stress–strain curves as a function of the porosity ranging from 0 to 59 %. Reproduced with permission from Royal Society of Chemistry [8]

inset of Fig. 2.11d, representing a manual compression test of the mesoporous film with a volume reduction of over 50 %, shows that the as-fabricated thin film almost perfectly recovers its original shape after the release of the compressive force, remarkably compliant and springy. To measure the compressibility of the mesoporous films with porosities from 0 to 59 % and evaluate resistance to deformation by external forces, compressibility measurements were conducted on the films, as shown in Fig. 2.11e. It shows the loading and unloading stress-strain curves for mesoporous films with porosities from 0 to 59 %. It is obvious that the loading and unloading curves for all samples are almost identical. This indicates that plastic collapse in the mesoporous films does not occur during loading and unloading. As the porosity increases, it is clearly seen that the elastic modulus decreases, thereby, the mesoporous film becomes more flexible. The TENG based on the mesoporous film was fabricated by attaching metal electrodes (aluminum foil in this case) to both sides of the film. Most of the TENGs have an air gap between a polymer film and a metal electrode, which not only helps in successive contact and separation, but also affects the electrical potential and charge density. The gap distance, which varies considerably from 1 to 3 mm, significantly affects the electrical output performance of the device. Without an air gap, a very poor electrical output performance was seen in TENGs. However, the presence of the gap may require a complicated fabrication effort and may limit the large-scale fabrication of practical devices. More importantly, an air gap left between the two layers makes such devices sensitive to moisture, and their packaging becomes a major issue. This strongly limits their applications in liquids.

In this design, the presence of such an air gap is no longer necessary, as the periodic contact and separation between gold nanoparticles and PDMS can be obtained through the porous structure. Under open-circuit conditions, the charge generation of the TENG under cycled compressive force can be understood from the coupling of the triboelectric effect and electrostatic induction. Before the contact of Au NPs and the PDMS inside pores, there is no charge transfer, and thus no potential generated. By the compressive force, friction between two materials occurs and electrons are transferred from gold NPs to PDMS film, resulting in positive and negative charges on both surface. As the force is withdrawn, the contacting surfaces are separated. Based on the above mechanism, the output voltage is usually shown as a direct current (DC) signal. However, the TENG shows a sharp alternating current (AC) signal, which may imply that charge dissipation of surface charges in pores occurs when measured using a low resistance measurement system. Under short-circuit conditions, the negative charges on PDMS inside pores induce positive charges on the top electrode, while the positive charges on gold NPs induce negative charges on the bottom electrode, and thus electrons in the top electrode move to the bottom electrode through an external circuit to neutralize the positive triboelectric charges in the top electrode, reaching an electrostatic equilibrium due to the electrostatic induction and the conservation of charges, generating a negative current signal. When the top electrode fully reverts to the initial position, there is no electron flow through the circuit. Once the TENG is pressed again, the gold NPs will contact with PDMS again, causing the charges to disappear. As a consequence, electrons are driven from the bottom electrode back to the top electrode, reducing the amount of induced charges. This process corresponds to a positive current signal. Gold NPs and the PDMS inside the pores are in contact again and all induced charges are neutralized as shown in Fig. 2.12a.

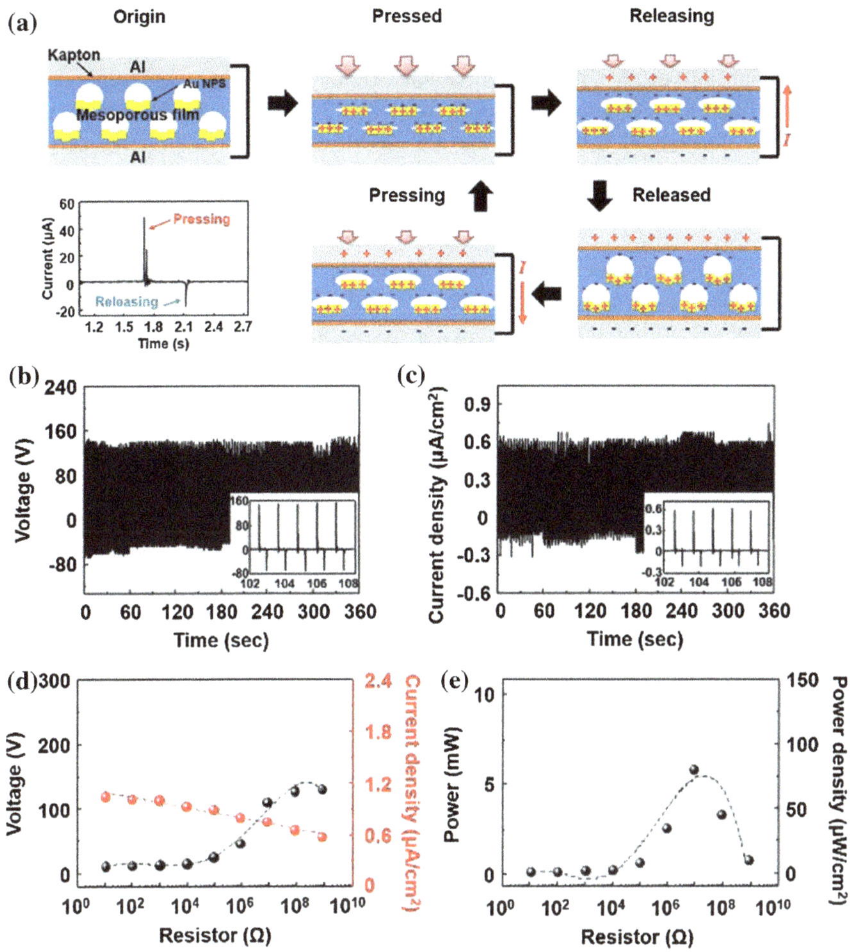


**Fig. 2.12** **a** The charge generation mechanism of the AMTENG under an external force and short-circuit conditions. **b** The output voltage and **c** output current density of the AMTENG under an external force during 360 s. The detailed output voltage and current density of the device are also show in the *inset*. **d** The output voltage and current density, and **e** the output power of the AMTENG without the gap with the resistance of external loads from 10 to 109 Ω. Reproduced with permission from Royal Society of Chemistry [8]

The electrical output performance also increases almost linearly with the active area. Additionally, the electrical output performance does not appear to change significantly in 360 s, indicating that the TENG is a reliable design for large-scale manufacturing (Fig. 2.12b, c). To investigate the output power of the TENG, resistors were used as external loads from 10 Ω to 1 GΩ, the instantaneous power of the external resistance for the TENG reaches a peak value of 6 mW (78 mW $cm^{-2}$) at a resistance of 10 MΩ, as shown in Fig. 2.12d, e. The pores in the mesoporous film play a very important role in determining the electrical output performance of the TENG. In detail, the output voltage and current density produced by the TENG at a fixed gold content (0.28 wt%) show that as the porosity increases to 59 %, the electrical output performance is significantly enhanced. Also, as the applied force increases from 0 to 70 mN, the mesoporous film shows a more rapid increase in capacitance, compared with that of the flat PDMS film. It is also clearly seen that the rate of the capacitance increases with the porosity. Furthermore, the dramatic increase in the output signals of the nanogenerator can be also attributed to the increase of the surface area-to-volume ratio in the mesoporous PDMS films. The ratio is increased up to several hundred times, as the porosity increases to 59 %. Therefore, the contact area is increased by increasing the porosity, increasing the surface charges during the contact of the films, and enhancing the total output power of the TENG.

## 2.4 Summary

In this chapter, the basic working principle, theoretical analysis, and several representative prototype structures of the TENG in vertical contact-separation working mode are introduced. This working mode has facile and scalable fabrications, simple design strategy, high instantaneous output power, and easy to scale up with multiple layer integrations. The TENGs at this working mode are usually driven by an external mechanical impact, and a critical factor affecting its output voltage is the amplitude of the separation distance between the two triboelectric layers (or contact surfaces), while the output current is dictated by the speed at which the two surfaces being contacted or separated. Hence, most of advanced designs are focused on the way to realize spontaneous separation by introducing a spacer, or taking advantage of the elastic resilience of the materials.

With this working mode, an open-circuit voltage of up to 1200 V, a short-circuit current of up to 2 mA, an instantaneous power density of 313 $W/m^2$, and an energy conversion efficiency of up to 39 % has been delivered. These output performances represent the state-of-the-art progress of the TENGs. Other than the prototypes mentioned above, a lot of unique structures have been developed for various purposes, such as powering portable electronics and self-powered active sensors, which will be covered in the following chapters.

## References

1. G. Zhu, C.F. Pan, W.X. Guo, C.Y. Chen, Y.S. Zhou, R.M. Yu, Z.L. Wang, Triboelectric-generator-driven pulse electrodeposition for micropatterning. Nano Lett. **12**(9), 4960–4965 (2012)
2. S.H. Wang, L. Lin, Z.L. Wang, Nanoscale triboelectric-effect-enabled energy conversion for sustainably powering portable electronics. Nano Lett. **12**(12), 6339–6346 (2012)

3. G. Zhu, Z.H. Lin, Q.S. Jing, P. Bai, C.F. Pan, Y. Yang, Y.S. Zhou, Z.L. Wang, Toward large-scale energy harvesting by a nanoparticle-enhanced triboelectric nanogenerator. Nano Lett. **13**(2), 847–853 (2013)
4. S.H. Wang, L. Lin, Y.N. Xie, Q.S. Jing, S.M. Niu, Z.L. Wang, Sliding-triboelectric nanogenerators based on in-plane charge-separation mechanism. Nano Lett. **13**(5), 2226–2233 (2013)
5. Y. Yang, Y.S. Zhou, H.L. Zhang, Y. Liu, S.M. Lee, Z.L. Wang, A single-electrode based triboelectric nanogenerator as self-powered tracking system. Adv. Mater. **25**(45), 6594–6601 (2013)
6. S.H. Wang, Y.N. Xie, S.M. Niu, L. Lin, Z.L. Wang, Freestanding triboelectric-layer-based nanogenerators for harvesting energy from a moving object or human motion in contact and non-contact modes. Adv. Mater. **26**(18), 2818–2824 (2014)
7. P. Bai, G. Zhu, Z.H. Lin, Q.S. Jing, J. Chen, G. Zhang, J. Ma, Z.L. Wang, Integrated multilayered triboelectric nanogenerator for harvesting biomechanical energy from human motions. ACS Nano **7**(4), 3713–3719 (2013)
8. J. Chun, J.W. Kim, W.S. Jung, C.Y. Kang, S.W. Kim, Z.L. Wang, J.M. Baik, Mesoporous pores impregnated with Au nanoparticles as effective dielectrics for enhancing triboelectric nanogenerator performance in harsh environments. Energy Environ. Sci. **8**(10), 3006–3012 (2015)
9. J. Henniker, Triboelectricity in Polymers. Nature **196**(4853), 474 (1962)
10. L.S. McCarty, G.M. Whitesides, Electrostatic charging due to separation of ions at interfaces: Contact electrification of ionic electrets. Angew. Chem. Int. Ed. **47**(12), 2188–2207 (2008)
11. D.J. Lacks, R.M. Sankaran, Contact electrification of insulating materials. J. Phys. D Appl. Phys. **44**(45), 453001 (2011)
12. A.F. Diaz, R.M. Felix-Navarro, A semi-quantitative tribo-electric series for polymeric materials: the influence of chemical structure and properties. J. Electrostat. **62**(4), 277–290 (2004)
13. Y. Tada, Experimental characteristics of electret generator, using polymer film electrets. Jpn. J. Appl. Phys. **31**(3), 846–851 (1992)
14. S.M. Niu, S.H. Wang, L. Lin, Y. Liu, Y.S. Zhou, Y.F. Hu, Z.L. Wang, Theoretical study of contact-mode triboelectric nanogenerators as an effective power source. Energy Environ. Sci. **6**(12), 3576–3583 (2013)
15. S.M. Niu, Z.L. Wang, Theoretical systems of triboelectric nanogenerators. Nano Energy **14**, 161–192 (2015)
16. H. Fang, W.Z. Wu, J.H. Song, Z.L. Wang, Controlled growth of aligned polymer nanowires. J. Phys. Chem. C **113**(38), 16571–16574 (2009)
17. A. Crossland, P. Wyllie, L. Ran, Mechanical to electrical energy conversion in a hybrid liquid-solid dielectric electrostatic generator. J. Appl. Phys. **106**(4), 044108 (2009)

# Chapter 3
# Triboelectric Nanogenerator: Lateral Sliding Mode

## 3.1 Basic Principle

The sliding-induced electricity generation mechanism is schematically depicted in Fig. 3.1 [1]. Here, a pair of polymer material, such as Nylon and PTFE, is taken as an example to illustrate the general working principle that can be applied to any other materials. In the original position (Fig. 3.1a), the two polymeric surfaces fully overlap and intimately contact with each other. Because of the large difference in the ability to attract electrons, the triboelectrification will leave the Nylon surface with net positive charges and the PTFE with net negative charges with equal density. Since the tribo-charges on the insulators will only distribute in the surface layer and will not be leaked out for an extended period of time [2], the separation between the positively-charged surface and negatively-charged surface is negligible at this overlapping position, and thus there will be little electric potential drop across the two electrodes. Once the top plate with the positively-charged surface starts to slide outward (Fig. 3.1b), the in-plane charge separation is initiated due to the decrease in contact surface area. The separated charges will generate an electric field pointing from the right to the left almost parallel to the plates, inducing a higher potential at the top electrode. This potential difference will drive a current flow from the top electrode to the bottom electrode in order to generate an electric potential drop that cancels the tribo-charge-induced potential. Under ideal conditions, the vertical distance between the electrode layer and the tribo-charged polymeric surface is negligible compared to the lateral charge separation distance, so the amount of the transferred charges on the electrodes approximately equals to the amount of the separated charges at any sliding displacement. Thus, the current flow will continue with the continuation of the on-going sliding process that keeps increasing the separated charges, until the top plate fully slides out of the bottom plate and the tribo-charged surfaces are entirely separated (Fig. 3.1c). The measured current should be determined by the rate at which the two plates are being slid apart.



Z.L. Wang et al., *Triboelectric Nanogenerators*,
Green Energy and Technology, DOI 10.1007/978-3-319-40039-6_3

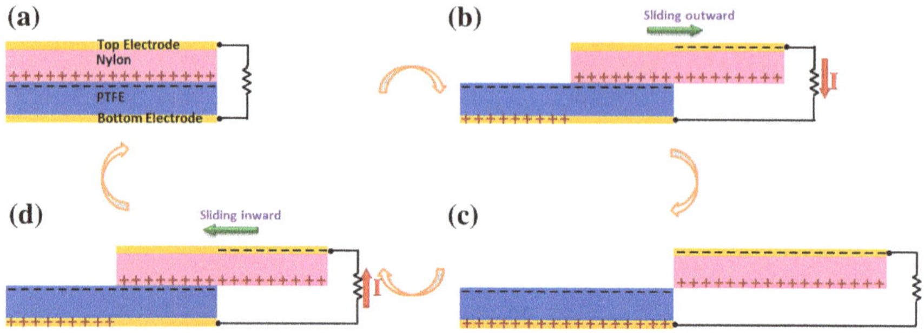


**Fig. 3.1** Working mechanism of the TENG based on in-plane charge separation. The sketches that illustrate the electricity generation process in a full cycle of the sliding motion under quasi-short-circuit condition. Reproduced with permission from American Chemical Society [1]

Subsequently, when the top plate is reverted to slide backwards (Fig. 3.1d), the separated charges begins to get in contact again but no annihilation due to the insulator nature of the polymer materials. The redundant transferred charges on the electrodes will flow back through the external load with the increase of the contact area, in order to keep the electrostatic equilibrium. This will contribute to a current flow from the bottom electrode to the top electrode, along with the second half cycle of sliding. Once the two plates reach the overlapping position, the charged surfaces get into fully contact again. There will be no transferred charges left on the electrode, and the device returns to the state in Fig. 3.1a. In this entire cycle, the processes of sliding outwards and inwards are symmetric, so a pair of symmetric alternating current peaks should be expected.

## 3.2 Fundamental Theory

### *3.2.1 Sliding-Mode TENG with Only One Unit*

The basic structure for sliding-mode TENG is shown in Fig. 3.2 [3]. The structure is very similar to the contact-separation mode discussed in Chap. 2, and the only difference is now the separation direction of the tribo-charges parallel to the triboelectric surfaces. Therefore, its electrostatic induction process is also similar to the contact-separation mode, so we call both of them attached-electrode mode. Therefore, this mode can also be named as attached-electrode sliding-mode [4]. We first analyze its working principle under ideal conditions, that is the length and width is much larger than the dielectric thickness so the edge effect can be neglected. Considering the materials to be used as the triboelectric layers, the attached-electrode sliding-mode TENG also has two types: dielectric-to-dielectric and conductor-to-dielectric. We first take the dielectric-to-dielectric category as an example. The model for this type of TENG was built with dimensions close to those

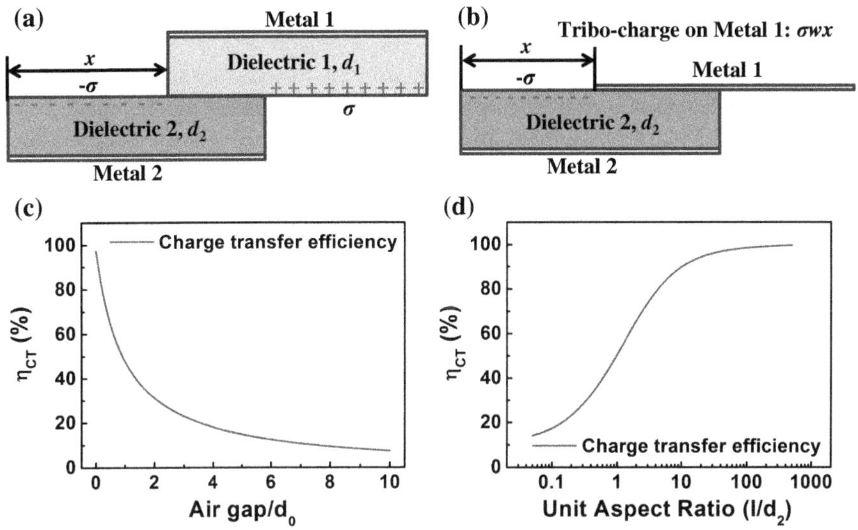


**Fig. 3.2** Theoretical models for **a** dielectric-to-dielectric attached-electrode parallel-plate sliding-mode TENG and **b** conductor-to-dielectric attached-electrode parallel-plate sliding-mode TENG. **c–d** Influence of **c** air gap size and **d** dielectric aspect ratio on the charge transfer efficiency of conductor-to-dielectric sliding-mode TENG when a full separation has been reached. The dielectric constant is 2 in this calculation. Reproduced with permission from Wiley [3]

of the real device, as shown in Fig. 3.2a. Considering that the width w of the whole structure is much larger than the thickness, a 2D model is utilized to simplify the calculation. The geometric size of the two dielectrics is l in longitudinal direction and $d_i$ (i = 1, 2) in thickness direction. Two metal electrodes are bonded with the two dielectrics layers. The bottom part is fixed while the top can slide through the longitudinal direction. The lateral separation distance is defined as x. When the two dielectrics are separated, the lower surface of Dielectric 1 and the upper surface of Dielectric 2 have charges with different signs at the non-overlapped regions, which are because of the triboelectric effect. Same as attached-electrode contact-mode TENGs, it is reasonable to assume that those triboelectric charges are uniformly distributed on these two surfaces, which is an excellent approximation for insulators. Also, the decay of tribo-charges with time can be neglected on these insulating polymers. The surface density of the tribo-charges is defined as σ at surface of Dielectric 1 and –σ at surface of Dielectric 2. The overall density of tribo-charges on the overlapped surface can be regarded as 0 because the distance between the center of the positive charges and the negative charges are in atomic level. Under MACRS and OC conditions, the total charge at each metal electrode was assigned as 0. If the transferred charges amount is defined by Q, the total charge amount on metal 1 and metal 2 is −Q and Q, respectively. For conductor-to-dielectric attached-electrode sliding-mode TENGs (Fig. 3.2b), the only geometrical structure difference is the absence of Dielectric 1. In this case, Metal 1 is not only the top triboelectric layer, but also the top electrode itself. When Metal 1 and Dielectric 2 are separated, due to the triboelectric effect, the upper surface at the separated region of Dielectric 2 has the uniform tribo-charges (surface density is $-\sigma$) and the same amount of tribocharges with opposite sign will exist on the surface of Metal 1. Therefore, at the OC condition (MACRS), the total amount of charges on the bottom electrode is 0. For the ease of physical understanding, the tribo-charges on the dielectric surface in the overlapped region has been deemed to cancel the opposite tribo-charge portions on the top metal layer, with the density of $\sigma$ and amount of $\sigma w(l - x)$. Thus, in our model, the total amount of charges on the top electrode is $\sigma wx$.

The analytical equation for the general case of attached-electrode sliding-mode TENGs cannot be derived and rigorous theoretical analysis can only base on the numerical method. But in the practical applications, since l is always much larger than $d_1$ and $d_2$ and x is always smaller than 0.9 l (because of the difficulty of perfect alignment of the two dielectric surfaces), an approximate analytical V-Q-x relationship can be derived when the edge effect can be neglected under the above conditions.

To derive the V-Q-x relationship, the analytical estimation of C(x) and $V_{OC}(x)$ must first be obtained. We take the dielectric-to-dielectric case as an example. The capacitance C is first deducted. Since the thickness of the dielectrics is much smaller than their length, the capacitor between the overlapped region is the dominate part of the total capacitance as long as the two dielectrics are not close to thoroughly separated. Therefore, utilizing the parallel-plate capacitor model, the total capacitance C can be estimated by the following equation:

$$C = \frac{\varepsilon_0 w(l - x)}{d_0} \tag{3.1}$$

where $d_0$ is the effective dielectric thickness.

$V_{OC}$ can be estimated by understanding the charge distribution. Since the length of the dielectrics is much larger than the thickness, we can assume in each region, the metal electrodes work as infinite plates. Under this approximation, the charge distribution is uniform in each region, and the electric field is uniform along the Y-direction inside the dielectrics. (This conclusion can also be verified through FEM calculation results [3]). Therefore, in an ideal case, the absolute value of surface charge density is $\sigma$ for the non-overlapped region. For the overlapped region, the charge density is still uniform. Thus, the charge density at the overlapped region can be calculated through the 0 total charges at each electrode under OC conditions. The ideal charge distribution under OC condition (utilizing MACRS) can be approximated by the following equations:

$$\text{For the non-overlapped region of the bottom electrode: } \rho = \sigma \tag{3.2a}$$

$$\text{For the overlapped region of the bottom electrode: } \rho = -\frac{\sigma x}{l - x} \tag{3.2b}$$

$$\text{For the non-overlapped region of the top electrode: } \rho = -\sigma \tag{3.2c}$$

$$\text{For the overlapped region of the top electrode: } \rho = \frac{\sigma x}{l - x} \tag{3.2d}$$

Utilizing the charge distribution shown above and Gauss Theorem, $V_{OC}$ can be easily estimated as:

$$V_{OC} = \frac{\sigma x d_0}{\varepsilon_0 (l - x)} \tag{3.3}$$

So the V-Q-x relationship for attached-electrode sliding-mode TENGs when the edge effect can be neglected can be shown as:

$$\mathrm{V} = -\frac{1}{\mathrm{C}}\mathrm{Q} + V_{OC} = -\frac{d_0}{\mathrm{w}\varepsilon_0 (l - x)}\mathrm{Q} + \frac{\sigma d_0 x}{\varepsilon_0 (l - x)} \tag{3.4}$$

For conductor-to-dielectric TENGs when the edge effect can be neglected, similar analysis shows the same results.

The electrostatic induction process of attached-electrode sliding-mode TENGs can also be thoroughly understood through the node discussion similar to the attached-electrode contact-mode TENGs. For simplicity, we take the conductor-to-dielectric attached-electrode sliding-mode TENGs as an example mainly because it only has one tribo-charged surface and we only analyze the capacitance distribution at the two ends ($x = 0$ and $x = x_{max} > l$). When $x = 0$, the capacitance between the tribo-charged dielectric surface and the top electrode is infinity while the capacitance between the tribo-charged dielectric surface and the bottom electrode is a finite large number, so the ratio between these two capacitances is infinity. When a full separation has already been reached, the capacitance between the tribo-charged dielectric surface and the top electrode is nearly 0 while the capacitance between the tribo-charged dielectric surface and the bottom electrode maintains still same, so the ratio between these two capacitances is now 0. This change of capacitance ratio due to the change of the relative position of tribo-charged surface will induce electron transfer between the electrodes under SC condition. From the above discussion, the detailed working mechanism of attached-electrode sliding-mode TENGs is almost the same as that of attached-electrode contact-mode TENGs and the maximum charge transfer efficiency for this kind of TENG can also reach 100 %.

From the above derivation, the intrinsic output characteristics of attached-electrode sliding-mode TENGs have the following properties. Their short-circuit transferred charge is linearly proportional to the separation distance x while their inherent capacitance is linearly proportional to the length of the overlapped region (l-x). For open-circuit voltage, it increases dramatically when x is close to l due to the quick decrease of the capacitance.

One important consideration for attached-electrode sliding-mode TENGs is that their output is very sensitive to the gap between the surfaces of the two triboelectric layers. For example, if the thickness of such gap is the same as $d_0$, the capacitance between the tribo-charged dielectric surface and the top electrode when $x = 0$ is the same with the other capacitance. And the capacitance ratio can only change from 1 to 0 when x changes from 0 to $x_{max}$. Therefore, the charge transfer efficiency will be degraded to only 50 %, as shown in Fig. 3.2c. Since $d_0$ is always a small number, the gap tolerance of attached-electrode sliding-mode TENGs is poor, which can be improved by utilizing sliding-mode freestanding TENGs described in the following section.

Besides, another important consideration for attached-electrode sliding-mode TENGs is the edge effect [3]. When the thickness of the dielectrics is comparable to their length, strong edge effect will dominate the output and the charge transfer efficiency will be strongly degraded. Figure 3.2d shows the FEM simulation results for a conductor-to-dielectric attached-electrode sliding-mode TENG whose dielectric constant is 2. When the aspect ratio of length and dielectric thickness reduces from 500 to 0.05, the charge transfer efficiency reduces from nearly 100 % to about 14 %. Usually the edge effect can be ignored in practical attached-electrode sliding-mode TENGs with only one unit. However, this effect will become dominant in attached-electrode grating-structured TENGs when the grating is fine enough, which we will discuss in detail in the following section.

## 3.2.2 *Grating TENGs*

Grating structure is an advanced structural design based on the above discussed sliding-mode TENGs [5–7]. Utilizing grating structure, the in-plane charge separation cycles can by multiplied. However, the output characteristics of grating structure are very complicated mainly because the edge effect can no longer be neglected. The length of one grating unit is usually comparable to the dielectric thickness when the grating is fine. In this section, in-depth theoretical models of two types of grating TENGs, one with equal length plates and the other with unequal length plates, are discussed in detail [8]. Through theoretical and computational methods, their output characteristics are obtained and analyzed. With this basic understanding of the output performance, the effect of finer pitches is outlined for both grating structures. In addition, structural optimization strategies are provided to maximize the power output.

### 3.2.2.1 Grating Structure with Equal-Length Plates

#### Influence of Electrode Structure

The grating structure with two equal-length plates is first considered. In a grating TENG, the patterning of the triboelectric layer is essential in order to enable multiple charge separation cycles. On the other hand, patterning is not necessary for

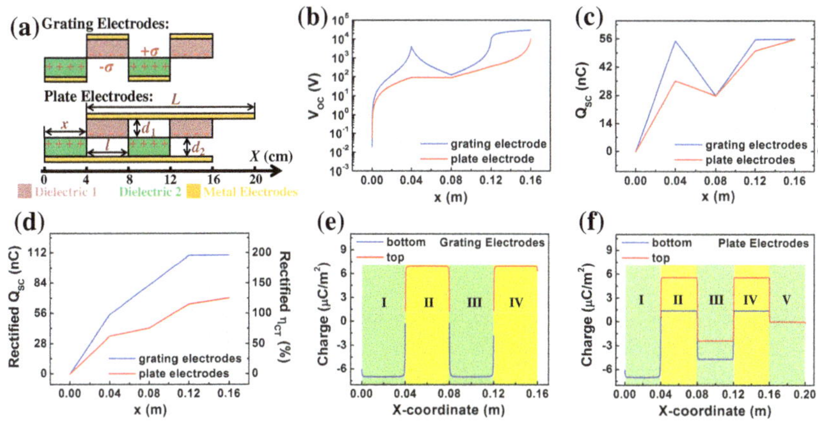


**Fig. 3.3** Comparison of the electrode structure. **a** Structure of the FEM model for grating TENGs with grating electrodes and plate electrodes. **b–d** Calculated **b** open circuit voltage, **c** transferred charges at short circuit condition, and **d** rectified transferred charges at short circuit condition for grating TENGs with both grating electrodes and plate electrodes. **e–f** Charge distribution at the metal electrodes for **e** grating electrodes and **f** plate electrodes at the case x = l. Reproduced with permission from Royal Society of Chemistry [8]

the attached electrodes. As shown in Fig. 3.3a, there are two choices available: a continuous plate electrode and a grating electrode, in which the electrodes are fabricated with the same grating as the dielectric layer. In this case, the electrical connection is realized through external circuits or electrode bars. These two electrode configurations will lead to different output performance.

To compare the output characteristics of the two grating structures with different electrode configurations, their finite element method (FEM) models were built and calculated using COMSOL. Since the width of grating TENGs is usually much larger than their thickness, 2D models were utilized to simplify the calculation. Two grating dielectrics with 50 % duty cycle are placed as tribo-pairs (a pair of materials which will undergo contact electrification). Their half pitch is defined as l and the number of grating units in the top plate is defined as n. The total length of the top plate is L, such that L = 2 nl. As an example, the case with two grating units (n = 2) was studied to illustrate the difference. (Detailed calculation parameters are listed in Table 3.1.) Due to contact electrification, different signs of static charges (called tribo-charges) are distributed at the lower surface of Dielectric 1 and the upper surface of Dielectric 2. We assume that the tribo-charges are uniformly distributed on these surfaces with a density of σ. In order to simulate the relative-sliding motion in operation, the bottom dielectric is fixed and the top dielectric slides in the lateral direction, with the lateral displacement defined as x.

The profiles of $V_{OC}$ and $Q_{SC}$ generated by these two TENGs (Under MACRS) with different electrode structures are completely different, as shown in Fig. 3.3b–d. $V_{OC}$, $Q_{SC}$, and the charge transfer efficiency $\eta_{CT}$ of both structures show an oscillating trend, but those of the plate electrodes are always smaller than those of

**Table 3.1** Parameters utilized in FEM calculations for comparing grating and plate electrodes

| Structure component | Parameter utilized |
|---|---|
| Dielectric 1 | $\varepsilon_{r1} = 4$, $d_1 = 220$ μm |
| Dielectric 2 | $\varepsilon_{r2} = 2$, $d_2 = 220$ μm |
| Width of dielectrics (w) | 0.1 m |
| Total length of the top plate (L) | 0.16 m |
| Tribo-charge surface density (σ) | 7 $\mu Cm^{-2}$ |
| Velocity (v) | 1 $ms^{-1}$ |

the grating electrodes. If the AC output is rectified to DC signal for storage, the TENG with grating electrodes would contribute a higher amount of accumulated charge. $\eta_{CT\text{-rectified}}$ after a full displacement cycle from the grating electrode structure can reach 200 % while that from the plate electrode only reaches 126 %. It should be noted that the reason that $\eta_{CT\text{-rectified}}$ is larger than 100 % for both structures comes from the multiple cycles of charge separation, which is the key advantage of the grating structure. As for the open circuit voltage, $V_{OC}$ after a full displacement cycle from the grating electrode can reach around 31 kV while that from the plate electrode only reaches around 10 kV. This set of comparisons clearly shows that the grating electrode is more effective in electricity generation than the plate electrode.

This difference in output characteristics results from the difference in charge distribution at the metal electrodes in these two electrode structures, as shown in Fig. 3.3e, f for the example case of x = l under SC conditions. $Q_{SC}$ equals half of the difference between the total amount of charge on top electrodes and on the bottom electrodes, which can be regarded as the sum of the contributions in all sub-regions (as marked in Fig. 3.3e, f). As an example, we can analyse the charge distribution in Region II and its contribution to $Q_{SC}$ to elucidate the difference between these two electrode structures.

For the grating electrode, only the top electrode exists in this region. Therefore, the induced charge density on the top electrode is σ and its contribution to $Q_{SC}$ is σwl/2. For the plate electrode, both the top and bottom electrode exist in this region. A charge density of $\sigma_T$ will be induced on the top electrode, and $\sigma_B$ on the bottom. Thus its contribution to $Q_{SC}$ is $(\sigma_T - \sigma_B)$wl/2. Since the electric field inside the metal electrodes is 0, the following relationship exists:

$$\sigma_T + \sigma_B = \sigma \tag{3.5}$$

In addition, because the voltage between the top and bottom electrodes is 0 under SC conditions, we will have:

$$\frac{\sigma_T}{\varepsilon_0}\frac{d_1}{\varepsilon_{r1}} = \frac{\sigma_B}{\varepsilon_0}d_2 \tag{3.6}$$

This equation indicates that $\sigma_T$ and $\sigma_B$ will have the same sign. From Eq. (3.5), $(\sigma_T - \sigma_B)$ will be obviously smaller than $\sigma$. Thus, the contribution of the plate electrode in region II to $Q_{SC}$ is smaller than that of the grating electrode structure. Likewise, in regions III and IV, their contribution to $Q_{SC}$ of the plate electrode is also smaller than that of the grating electrode. As a result, the total $Q_{SC}$ generated by the plate electrode will be smaller. Therefore, through comparison of plate and grating electrode structures, it can be concluded that grating electrode structure is favourable for high output power. Thus, in the following discussion, we will mainly focus on TENGs with grating electrodes.

### Influence of Number of Grating Units

The number of grating units (n) has a direct impact on the total output characteristics. When n starts to increase from 1, the approximate ideal charge distribution is still satisfied because l is still much larger than the thickness of dielectrics. However, when n is increased to fairly large values and l is comparable with $d_1$ or $d_2$, the non-ideal edge effect is significant and cannot be neglected any longer.

The influence of n in relatively small regions which satisfies the ideal conditions is first discussed. In this case, the instantaneous $Q_{SC}$ and the short circuit current ($I_{SC}$) have the following relationships with n:

$$Q_{SC} = \frac{\sigma w L}{2}\left[\frac{k}{n} + \frac{2(n-k)}{L}(x-2kl)\right], 2kl \le x \le (2k+1)l \quad (k \in N) \tag{3.7a}$$

$$Q_{SC} = \frac{\sigma w L}{2}\left[1 - \frac{2(n-k-1)}{L}(x-2kl-l)\right], (2k+1)l \le x \le (2k+2)l \ (k \in N) \tag{3.7b}$$

$$I_{SC} = \frac{dQ_{SC}}{dx}\frac{dx}{dt} = \sigma w(n-k)\frac{dx}{dt}, 2kl \le x \le (2k+1)l \quad (k \in N) \tag{3.8a}$$

$$I_{SC} = \frac{dQ_{SC}}{dx}\frac{dx}{dt} = -\sigma w(n-k-1)\frac{dx}{dt}, (2k+1)l \le x \le (2k+2)l \quad (k \in N) \tag{3.8b}$$

Therefore, the following equations give the accumulated charges ($Q_{SC\text{-rectified}}$) and charge transfer efficiency ($\eta_{CT\text{-rectified}}$) after rectification under SC conditions when a full displacement is finished:

$$Q_{SC,rectified}(x = L) = \frac{\sigma w L}{2} n \tag{3.9}$$

$$\eta_{CT,rectified} = \frac{Q_{SC,rectified}(x = L)}{Q_{tribo-total}} = n \tag{3.10}$$

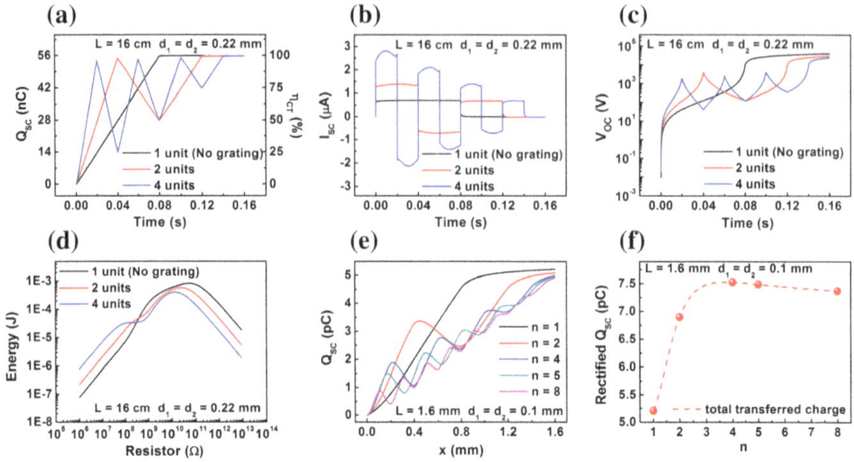

**Fig. 3.4** Influence of number of grating for equal-length grating TENGs. **a–d** Influence of n in the ideal condition. **a** The relationship between transferred charges at short circuit and time at different n. **b** The relationship between current at short circuit and time at different n. **c** The relationship between open circuit voltage and time at different n. **d** The relationship between generated energy and load resistance at different n. **e–f** Influence of n in the non-ideal condition. **e** Transferred charges at short circuit profile with moving distance at different n. **f** Total transferred charges relationship with n. Reproduced with permission from Royal Society of Chemistry [8]

From the above derivation, the peak value of $I_{SC}$ and $Q_{SC\text{-rectified}}$ in the grating structure can be enhanced by n times through the subdivision process under ideal conditions, which is the most significant advantage of fabricating finer pitch structures.

To verify these results, a numerical calculation for a grating TENG under ideal conditions (L/d = 727.3, detailed calculation parameters are the same as shown in Table 3.1) is performed and the results are shown in Fig. 3.4a–d. The numerically calculated results are consistent with the theoretical analysis above. When n increases, the number of charge transfers increases as a response, which significantly elevates the amount of accumulated charges after rectification. In addition, the slope of $Q_{SC} - t$ curve also increases with n, leading to an enhancement of $I_{SC}$, as shown in Fig. 3.4b. But as for voltage, since finer pitch results in a significant increase in capacitance when the top dielectrics and bottom dielectrics are fully separated, the peak value of $V_{OC}$ at larger n significantly drops while the peak value of $Q_{SC}$ remains almost the same, which is determined by the general $V_{OC}$, $Q_{SC}$ and C equation.

Such a change in $V_{OC}$ is shown in Fig. 3.4c. However, the total energy generated is more complicated, as it is determined by both the current and the voltage. Figure 3.4d shows the energy generated by 3 TENG structures with different n in one full back-and-forth cycle under different load resistances. The following conclusions can be reached by comparing the curves shown in Fig. 3.4d. First, a finer pitch can generate more energy in the low resistance range, where the energy is

**Table 3.2** Parameters utilized in the FEM calculation for illustrating non-ideal effect for the equal-length grating TENGs

| Structure component | Parameter utilized |
|---|---|
| Dielectric 1 | $\varepsilon_{r1} = 4$, $d_1 = 100\ \mu m$ |
| Dielectric 2 | $\varepsilon_{r2} = 2$, $d_2 = 100\ \mu m$ |
| Width of dielectrics (w) | 1 mm |
| Total length of the top plate (L) | 1.6 mm |
| Tribo-charge surface density (σ) | $7\ \mu Cm^{-2}$ |
| Velocity (v) | $1\ ms^{-1}$ |

mainly dominated by $I_{SC}$. However, a wider pitch can generate more energy in the high resistance range, where the generated energy is mainly determined by $V_{OC}$. The optimum resistance (the resistance at which the total energy is maximized) decreases with n, due to the increase of the inherent TENG capacitance. Considering the peak value of the harvested energy (in Fig. 3.4d), a finer pitch doesn't significantly improve the output, as $V_{OC}$ decreases more significantly than $I_{SC}$ increases for the equal-length grating TENGs.

When n continues to increase to very large values, the aspect ratio of each unit will be further lowered and the non-ideal edge effect cannot be neglected. To study the influence of this non-ideal effect, a numerical calculation for grating TENGs with a small L/d ratio of 16 was conducted, as shown in Fig. 3.4e, f. (Detailed calculation parameters are shown in Table 3.2.) At this aspect ratio, although the times of charge transfer increases with increasing values of n, the peak value of $Q_{SC}$ in each period is far smaller than that of $Q_{tribo}$. Especially when n is larger than 5, the transfer of charges is unidirectional after the first few cycles, so $Q_{SC\text{-rectified}}$ doesn't have a multiplication effect in these cycles. As a result, $Q_{SC\text{-rectified}}$ starts to drop with an increasing value of n. Because of the reverse influence of n in the two different regions, there is an optimum value of n that yields the maximum $Q_{SC\text{-rectified}}$. As shown in Fig. 3.4f, for a total geometric aspect ratio, the optimum value of n is about 4, in which the geometric aspect ratio of each unit is 2. This calculation shows that it is critically important to have this optimum value of n to generate the optimum charge when this kind of TENG is being used to charge a battery or capacitor.

#### 3.2.2.2 Grating Structure with Unequal-Length Plates

In the above section, we mainly discussed the design of grating TENGs with two plates of equal length. In such a structure, the top units will gradually slide out from the bottom grating due to lateral displacement. This feature is favourable for a high open-circuit voltage (because of the small side capacitance), but results in non-periodic signals. To avoid this situation, the length of the bottom plate that is stationary during operation can be increased to fully cover the sliding region of the top plate, which forms another structure of grating TENGs with plates of unequal length, as shown in Fig. 3.5a. As a typical example, the length of the bottom part is

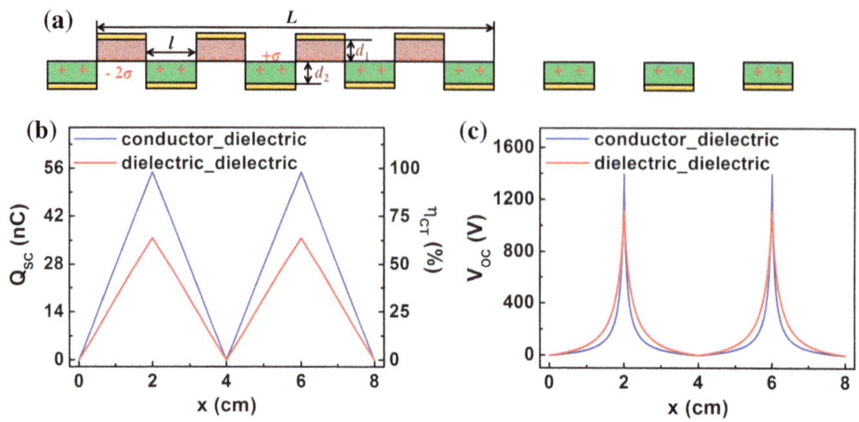


**Fig. 3.5** Influence of dielectric thickness on the performance of the unequal-length grating TENGs. **a** FEM model for the unequal-length grating TENGs. **b–c** Comparison of conductor-to-dielectric and dielectric-to-dielectric unequal-length grating TENGs on **b** short circuit transferred charges and **c** open circuit voltage. Reproduced with permission from Royal Society of Chemistry [8]

set to 2L, which is twice the length of top part. Thus, the sliding displacement is maintained as L. In this structure, because of the charge conservation in contact electrification, the tribo-charge density of Dielectric 1 ($-2\sigma$) is twice as high as that of Dielectric 2 ($\sigma$). Because Dielectric 1 is always sliding within the range of Dielectric 2, $Q_{SC}$, $V_{OC}$ and $C$ will be periodic. In addition, the $Q_{SC}$, $V_{OC}$ and $C$ curves all have mirror symmetry in each single period, which can be mathematically shown by the following set of equations.

$$V_{OC}(x + 2\text{kl}) = V_{OC}(x) \tag{3.11a}$$

$$V_{OC}(2\text{kl} - x) = V_{OC}(x) \tag{3.11b}$$

$$C(x + 2\text{kl}) = C(x) \tag{3.11c}$$

$$C(2\text{kl} - x) = C(x) \tag{3.11d}$$

$$Q_{SC}(x + 2\text{kl}) = Q_{SC}(x) \tag{3.11e}$$

$$Q_{SC}(2\text{kl} - x) = Q_{SC}(x) \tag{3.11f}$$

Because of the periodicity and mirror symmetry, we only need to simulate the TENG for its first half period. As for the electrode structure, the grating electrode still provides a better performance than the plate electrode in this case, which can be obtained from a similar discussion and derivation as above. Therefore, in the following discussion, we mainly focus on the unequal-length grating TENGs with the grating electrode.

### Influence of Dielectric Thickness

Unlike equal-length grating TENGs, the choice of dielectric thickness can significantly influence $Q_{SC}$. Under ideal conditions (l/d is sufficiently large), $Q_{SC}$ and $\eta_{CT}$ of this grating TENG can be given by:

$$Q_{SC} = \left(2 - \frac{1}{1 + \frac{d_1 \varepsilon_{r2}}{d_2 \varepsilon_{r1}}}\right) n\sigma w(x - 2kl), 2kl \le x \le (2k+1)l \quad (k \in N) \tag{3.12a}$$

$$Q_{SC} = \left(2 - \frac{1}{1 + \frac{d_1 \varepsilon_{r2}}{d_2 \varepsilon_{r1}}}\right) n\sigma w(2kl + 2l - x), (2k+1)l \le x \le (2k+2)l \quad (k \in N) \tag{3.12b}$$

$$\eta_{CT} = \left(1 - \frac{1}{2}\frac{1}{1 + \frac{d_1 \varepsilon_{r2}}{d_2 \varepsilon_{r1}}}\right) \frac{x - 2kl}{l}, 2kl \le x \le (2k+1)l \quad (k \in N) \tag{3.13a}$$

$$\eta_{CT} = \left(1 - \frac{1}{2}\frac{1}{1 + \frac{d_1 \varepsilon_{r2}}{d_2 \varepsilon_{r1}}}\right) \frac{2kl + 2l - x}{l}, (2k+1)l \le x \le (2k+2)l \quad (k \in N) \tag{3.13b}$$

where n is defined as the number of units in top dielectrics. From the above equations, it is observed that $Q_{SC}$ and $\eta_{CT}$ decrease dramatically if the ratio of $d_2/d_1$ increases, which is because of the non-zero value of $Q_0$ from the not totally overlapped surface when x = 0. To minimize this effect, as shown in Eq. (3.12a, b), the ratio $d_2/d_1$ needs to be reduced. An effective design is to eliminate the dielectric layer on the bottom plate and use a conductive material as both electrode and tribo-layer, in which $d_2$ can be regarded as 0. In order to validate this, the performance of this conductor-to-dielectric structure ($d_2 = 0$) where n = 4 is compared with the dielectric-to-dielectric TENG structure ($d_2 = d_1$) as shown in Fig. 3.5b, c. The detailed parameters for this calculation are listed in Table 3.3. The peak values of $Q_{SC}$, $\eta_{CT}$ and $V_{OC}$ from the conductor-to-dielectric TENG are all higher than the dielectric-to-dielectric TENG. Thus, the conductor-to-dielectric design is favourable for efficient charge transfer, from the point of view of device structure. Therefore, we will mainly focus on this design in the following discussion.

**Table 3.3** Parameters for FEM calculation for unequal-length grating TENGs

| Structure component | Dielectric-to-dielectric unequal-length grating TENGs | Conductor-to-dielectric unequal-length grating TENGs |
|---|---|---|
| Dielectric 1 | $\varepsilon_{r1} = 4$, $d_1 = 220\ \mu m$ | $\varepsilon_{r1} = 4$, $d_1 = 220\ \mu m$ |
| Dielectric 2 | $\varepsilon_{r2} = 2$, $d_2 = 220\ \mu m$ | Metal, $d_2 = 0$ |
| Width of dielectrics (w) | 0.1 m | 0.1 m |
| Total length of the top plate (L) | 0.16 m | 0.16 m |
| Tribo-charge surface density at the bottom dielectric surface (σ) | 3.5 $\mu Cm^{-2}$ | 3.5 $\mu Cm^{-2}$ |
| Velocity (v) | 1 $ms^{-1}$ | 1 $ms^{-1}$ |

### Influence of Number of Grating Units

With the above basic understanding of the output characteristics of the unequal-length grating TENG, we can discuss the influence of the most important design parameter: the number of grating units. The output performance of a TENG with different n is calculated and plotted in Fig. 3.6. When n increases in the small region, since the aspect ratio of each grating unit ($l/d_1$) is still large enough, the

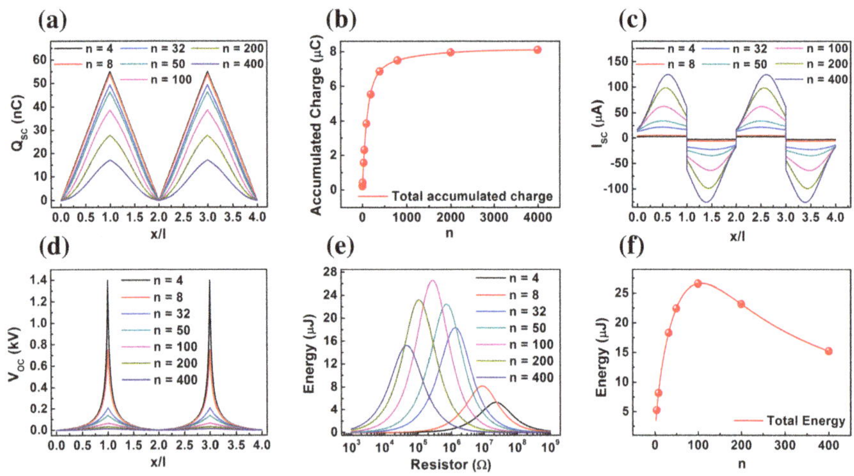


**Fig. 3.6** Influence of number of grating for unequal-length grating TENGs. **a** The relationship between transferred charges at short circuit in one period at different n. **b** Extracted total transferred charges when x = L. **c** The current profile at short circuit in one period at different n. **d** The voltage profile at open circuit in one period at different n. **e** The relationship between total generated energy and load resistance at different n. **f** Influence of n on the total generated energy. Reproduced with permission from Royal Society of Chemistry [8]

transferred charge amount in one half-cycle stays almost the same as $Q_{tribo}$ (Fig. 3.6a). Therefore, since the total accumulated charges ($Q_{SC\text{-rectified}}(x = L)$) equals to 2n times the amount of transferred charge in one half-cycle, the total accumulated charges and average current magnitude increase quasi-linearly with n in its small value region, as shown in Fig. 3.6b. However, when n continues to increase, the non-ideal edge effect becomes much more significant due to the decrease in the aspect ratio in each unit, resulting in a decreased number of transferred charges in each half-cycle (Fig. 3.6a). Although the number of charge transfer cycles still increases, the elevation slope of the total accumulated charges decreases dramatically. When n is sufficiently large, the slope of the total accumulated charges is close to 0 and the increase slope of the average current quickly decreases (Fig. 3.6c). Additionally, finer pitches also contribute to a significant increase in the side capacitance between the top and bottom electrodes. Therefore, from the fundamental TENG relationship, $V_{OC}$ will significantly drop with increased n, as shown in Fig. 3.6d. Unlike the case in equal-length grating TENG, the $V_{OC}$ and $Q_{SC}$ curves in each half-cycle are always monotonic, even when n is sufficiently large. The peak value of the $V_{OC}$ and $Q_{SC}$ curves are always in the middle position of each period. With these characteristics, the unequal-length grating TENGs can have applications as a self-powered sensor for actively measuring the displacement and velocity with high accuracy and resolution.

With the above basic output characteristics, the load performance of these TENGs was numerically calculated, from which the total harvested energy was plotted corresponding to different load resistances, as shown in Fig. 3.6e, f. Since finer pitches yield a larger inherent TENG capacitance, the optimum resistance shifts significantly to lower values. In addition, unlike the equal-length grating TENGs, an optimum value of n exists (about 108 through interpolation) for the unequal-length grating TENGs, which yields the maximum total harvested energy. With this optimum n, the aspect ratio is 3.37 for each grating unit. This optimum n originates from the complicated behavior of $V_{OC}$ and $I_{SC}$. When n first increases, the average value of $I_{SC}$ increases more significantly than $V_{OC}$ decreases, so the harvested energy will increase. However, when n continues increasing, the growth rate of $I_{SC}$ decreases dramatically due to the non-ideal edge effect. Therefore, the total harvested energy starts to decrease when the decreasing $V_{OC}$ becomes the dominant factor. This critical aspect ratio for the individual dielectric unit that yields the largest total energy mainly depends on the motion process and the relative dielectric constant of Dielectric 1. We can take the unequal-length grating TENG at constant velocity as an example. Under this condition, it can be mathematically proved that this critical aspect ratio is only dependent on the relative dielectric constant of Dielectric 1 (independent on $d_1$, L, and v).

## 3.3 Basic Device Structures

### 3.3.1 Plain-Sliding Structure

The sliding-triboelectric nanogenerator is structurally composed of two plates (71 mm × 50 mm) with glass slides as the supporting substrates to ensure the surface flatness (Fig. 3.7a) [1]. Polyamide 6,6 (Nylon) and polytetrafluoroethylene (PTFE) films, the two polymers at the opposite ends of the triboelectric series [9], are purposely chosen as the triboelectric layers adhered on surfaces of the glass slides, respectively, for effective electrification during sliding. On each polymer film, a layer of metal electrode was deposited on the side next to the glass slide (named as top electrode for the one on Nylon and bottom electrode for the one on PTFE). The two plates are kept in parallel to each other and the polymeric surfaces are in intimate contact. As driven by the mechanical motion/vibration along the long-edge of the plate, the two plates slide against each other, with the contact area changing periodically. In order to enhance the triboelectric charge density on the surface [10, 11], the PTFE film was dry-etched using inductive coupled plasma (ICP) to create aligned nanowire-structures [12] (Fig. 3.7b), which will increase the surface roughness and the effective surface area. As shown in the scanning electron microscopy (SEM) images (Fig. 3.7c), the PTFE surface is uniformly covered with nanowire-structures, with an average length of ~1.5 μm. As illustrated in

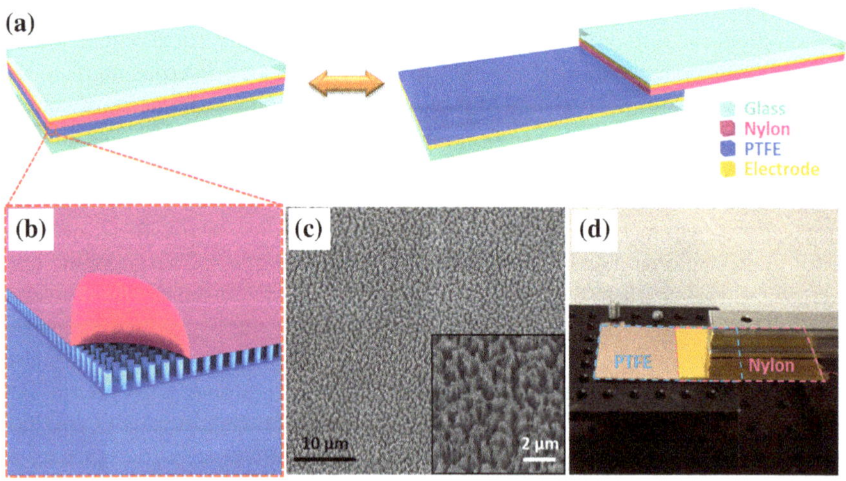


**Fig. 3.7** Device structure of a sliding-triboelectric nanogenerator (TENG). **a** The schematic diagram showing the structural design of the TENG in two sliding states: the overlapping position (*on the left*) and the separation position (*on the right*). **b** The magnified schematic of the surface between the two polymeric films, showing the fabricated nanowire array on PTFE surface. **c** SEM image of the PTFE surface with etched nanowire structure at the tilted view of 30°, the *inset* is the SEM image in higher magnification. **d** The photograph of a typical sliding-driven TENG on the measurement stage. Reproduced with permission from American Chemical Society [1]

Fig. 3.7d, during the real-time measurement, the PTFE-covered plate was bonded to a stationary stage, while the Nylon side was fastened to a parallel flat-end rail guide, which was connected to a linear motor for inducing an in-plane sliding motion.

The electrical output of the sliding-TENG was measured, with one plate guided by the linear motor in the direction parallel to the long-edge of the plates. The sliding displacement was 71 mm, which was the same with the length of the plate and thus covered the entire effective region for generating electricity (as mentioned above). The sliding movement was in a symmetric acceleration-deceleration mode (the instantaneous sliding velocity was plotted as the pink curve of Fig. 3.8d), with the acceleration rate of $\pm 20$ m/s$^2$. The open-circuit voltage ($V_{OC}$) was measured by an electrometer with very large input resistance. The electrode at the back of Nylon film was connected to the positive probe. When the plates in the TENG slid from the contact position to the separated position, the $V_{OC}$ jumped from 0 to $\sim$1300 V (Fig. 3.8a), which reflects the induced potential difference between the two electrodes by the in-plane charge separation. At the separation position, the $V_{OC}$ decayed a little bit, due to the slow charge leakage through the electrometer. When the TENG slid back to the contact position, the $V_{OC}$ jumped back to 0. The density of the transferred charge ($\Delta\sigma$) was also measured at the short-circuit condition. As shown in Fig. 3.8b, the charges with a density of $\sim$105 μC/m$^2$ transferred back and forth between the two electrodes, while the plates slid in and out. Since the plates got fully separated in each cycle, the transferred charge density should approximately equal to the triboelectric charge density on the polymeric surfaces. The transfer of the charges between the electrodes in the outer circuits produced an alternating-current output, with the peak short-circuit current density ($J_{SC}$) of 4.1 mA/m$^2$ at a maximum sliding velocity of 1.2 m/s (Fig. 3.8c). The area under each current density peak is the amount of charge density transferred in each sliding motion. The enlarged $J_{SC}$ curve in one single cycle is displayed in Fig. 3.8d, together with the corresponding curve of the sliding velocity. We can find that the current output has a symmetric profile and synchronizes very well with the velocity of the sliding motion. With such a power output converted from the sliding motion, hundreds of commercial light-emitting diodes (LEDs) can be instantaneously driven by a single TENG device (Fig. 3.8e).

In practice, the TENG's effective output power to the loads depends on the match with the resistance of the load. As shown in Fig. 3.8f, when the resistance is below $\sim$1 MΩ, the output current density and the voltage have little changes from the short-circuit condition: the current density remains at a value of $J_{SC}$ while the voltage stays close to 0. This is because the TENG has infinitely-large inner resistance. When the resistance goes beyond 1 MΩ, the current density will drop with the increase of the resistance; while the voltage on the load rises up. Consequently, the instantaneous power density on the load (Fig. 3.8g) remains close to 0 with the resistance below 1 MΩ, and reaches the maximum value of

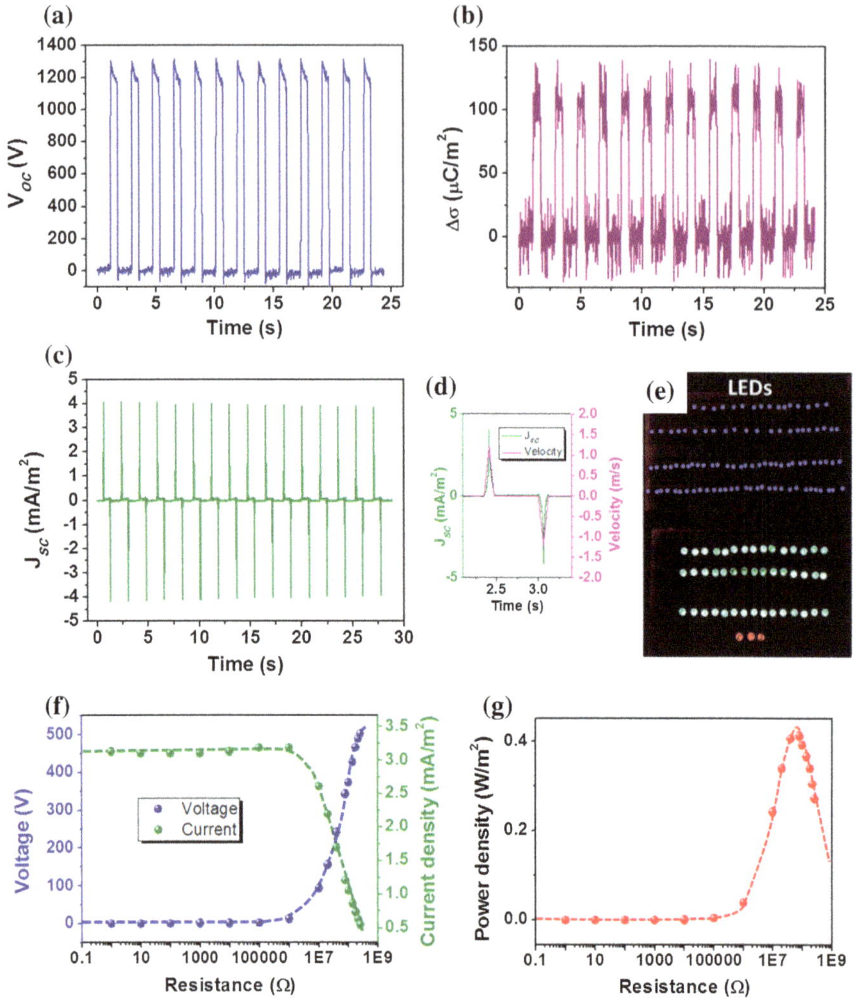


**Fig. 3.8** Performance of the TENG driven by the cycled sliding motion. **a** The open-circuit voltage ($V_{OC}$), **b** the density of transferred charges (Δσ), and **c** the short-circuit current density ($J_{SC}$) under the in-plane sliding with the displacement of 71 mm and the acceleration of 20 m/s$^2$. **d** The enlarged profile of $J_{SC}$ within one cycle (*green curve*), with the corresponding *curve* of the instantaneous sliding velocity (*pink*). **e** The snapshot of ~160 commercial LEDs in serial-connection directly driven by a TENG under sliding. **f–g** The dependence of **f** the output voltage (*blue*), current density (*green*) and **g** the power density on the resistance of the external load. Reproduced with permission from American Chemical Society [1]

0.42 W/m$^2$ at a resistance of ~50 MΩ. This is the maximum output power density on the device.

This basic structure can be integrated into three-dimensional multiple layered structure to improve its output performance, especially the output current [13].

### 3.3.2 Linear-Grating Structure

Introducing linear grating on the sliding surfaces enables the new principle to become an extremely efficient means for energy harvesting [5]; and it is far superior to the previously demonstrated ones in total output charge, current frequency and efficiency. Linear grating with uniform period is fabricated on both sliding surfaces. The rows of grating units have the same size as intervals in between, with all rows being electrically connected at both ends by two buses. The grating patterns on both sliding surfaces are identical so that they can match well with each other when aligned. Although the grating design reduces the total contact area by half thus seemingly sacrifices half of the triboelectric charges, it increases the percentage of the mismatched area to 100 % for a displacement of only a grating unit length rather than the entire length of the TEG so that it dramatically increase the transport efficiency of the inductive charges. Inductive free electrons can be pumped back and forth between electrodes for multiple times due to the grating structure, providing multi-folds of output charge compared to a non-grating TENG. Every row of the grating units can be considered as a reduced-sized TENG; and it is in parallel connection with all other rows through buses. In contrast to a non-grating TENG that needs to be fully displaced in order to complete pumping of the inductive charges for one time, the grating TENG only requires a displacement of a unit length to completely transport the inductive charges, largely improving the energy conversion efficiency. With further displacement of another length of the unit, back flow of the inductive charges can be realized. Therefore, for a one-way sliding process across the whole length of the TEG, the inductive charges can be pumped for (2N – 1) times in total, where N is the number of grating units. If we take into account that the contacting area decreases as the two surfaces slide apart, the following equation represents the total inductive charges Q that the grating TEG can provide for a single sliding across the entire length of the TEG:

$$\begin{aligned} Q &= Nq' + \left|-q' \times (N-1)\right| + q' \times (N-1) + \cdots + \left|-q'\right| + q' \\ &= Nq' + 2q' \times \sum\nolimits_{i=1}^{N-1} i\,d \\ &= (2q'N)N/2\,d \end{aligned} \tag{3.13}$$

where $q'$ is the inductive charges generated from a single grating unit for a displacement of the unit length.

The total maximum inductive charge generated by a non-grating TENG of the same length is $2q'N$. Therefore, Eq. (3.13) indicates that the total inductive charges linearly increase with the grating density. Experimental data from grating TENGs with more units further confirm the validity of our theoretical analysis. For a single sliding process, the $I_{SC}$ data produced by grating TENGs having 6, 8 and 10 units are presented in Fig. 3.9a–c, corresponding to a unit length of 0.5, 0.4 and 0.3 cm, respectively, with a constant full length of the TEG (6.4 cm). The total counts of AC peaks are in accordance with Eq. (3.13). The enhancement of charge output by

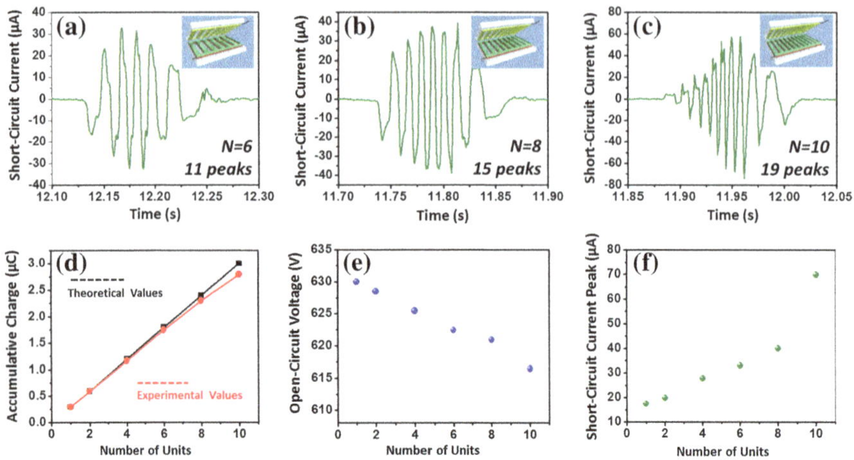


**Fig. 3.9** Electrical measurement results of grating TEGs with multiple grating units within a fixed total length. **a** Short-circuit current of a grating TEG with six grating units for a single sliding process across the full length of the TEG. *Inset* schematic of the TEG structure. **b** Short-circuit current of a grating TEG with eight grating units for a single sliding process across the full length of the TEG. *Inset* schematic of the TEG structure. **c** Short-circuit current of a grating TEG with 10 grating units for a single sliding process across the full length of the TEG. *Inset* schematic of the TEG structure. **d** Accumulative induced charges generated for a single sliding process across the full length of a TEG as a function of grating units. **e** Open-circuit voltage as a function of grating units. **f** Maximum values of short-circuit current as a function of grating units. Reproduced with permission from American Chemical Society [5]

the grating structure is plotted in Fig. 3.9d. With 10 grating units, a total of 2.8 μC of inductive charges can be generated within 6.4 ms at a sliding velocity of 10 m/s, which is equivalent to a continuous DC source of 0.44 mA. Ideally, according to Eq. (3.13), the total inductive charges are proportional to the density of the grating units. If the grating size is further scaled down to 30 μm, an ideal enhancement of another 100 times is expected, making the TEG comparable to various types of solar cells in output current and in current density. An estimation for the $N = 10$ gives an output conversion efficiency of ~8–31 %.

Though the experimental data closely match the theoretical values for $N \leq 4$, slight deviation starts to appear if N further increases (Fig. 3.9d). The deviation may be attributed to two probable reasons. The first is non-ideal orientation mismatch. The two sliding surfaces are likely to have an angle of mismatch that is introduced by experimental operation. As the length of the grating units shrinks, the angle of mismatch will exert more substantial effect on the effective contact area between the two surfaces, leading to increasingly reduced output charge. This effect can be largely eliminated by more precise alignment during experiment operation. Secondly, we suggest that assumptions made in the analytical theory may become imprecise for very fine grating units due to the edge effect of each unit. The assumption of infinite charged plane will no longer hold once the grating edges can dominantly affect the electric field distribution, leading to reduced quantity of inductive charges. Further systematic investigation by both experiment and simulation is required to quantitatively understand this non-ideal effect. As diagramed in Fig. 3.9e, the $V_{OC}$ is only weakly related to the number of grating units, though slight reduction can be observed. The reduction can also be attributed to the increasingly stronger effect from the mismatched angle. Besides, as revealed in Eq. (3.13), finer grating units will theoretically give a smaller $V_{OC}$, though this effect may be dominant only when the grating unit has a length approaching the PTFE's thickness. The advantages of the grating structure include not only the enhanced total transported charges but also improved output current. With constant sliding velocity, finer grating shortens the time to transport inductive charges between the electrodes for one time, resulting in higher output current (Fig. 3.9f). However, the peak of $I_{SC}$ does not increase linearly with the number of units, which may result from non-uniform velocity during sliding process.

To improve the output performance of the linear-grating TENG, the size of the grating features has been further narrowed down to millimeter or even micrometer scale [7]. In this work, we report an ultrahigh-power micro-grating triboelectric nanogenerator (MG-TENG) based on thin-film materials for harnessing triboelectrification between two sliding surfaces. Enabled by two sets of complementary micro-sized electrode gratings on thin-film polymers and by surface modification by nanoparticles, the MG-TENG offers an unprecedentedly high level of output power, which completely solves the major concern for triboelectrification in electricity generation. Operating at an in-plane sliding velocity of 10 m $s^{-1}$, a MG-TENG having a contact area of 20 $cm^2$ could generate sinusoidal-like AC current at an amplitude of 9.8 mA and at a frequency of 5 kHz. Under the matched load, an average effective power of 3 W was achieved, corresponding to power volume density of 15 W $cm^{-3}$. Having an efficiency of nearly 50 %, it successfully powered multiple types of light bulbs, demonstrating the capability of the MG-TENG as a power supply for regular electronics. Due to the shape-adaptive design based on thin-film materials, the MG-TENG could be even applied onto curved surfaces, providing a unique and straightforward solution in harnessing relative sliding motions, in which other existing technologies cannot be implemented. As an approach that is cost-effective, simple-implementing, and scalable, MG-TENG is suited to harvest a variety of mechanical energy not only for self-powered electronics but also for possible electricity generation in large scale.

A MG-TENG is composed of polytetrafluoroethylene (PTFE) thin films with a pair of metal gratings on opposite sides. As schemed in Fig. 3.10a, the grating is a collection of metal strips separated by equal-sized intervals. All of the strips are electrically connected by a bus at one end. The paired gratings are identical but complementary with relative displacement of half pitch (zoom-in view in Fig. 3.10a). Figure 3.10b exhibits a photograph of the PTFE thin film with double-sided metal gratings. On the top surface, a layer of PTFE nanoparticles is applied as surface modification (Fig. 3.10c). Two of the thin films with different length are prepared, which have a total area of 60 $cm^2$, a total volume of 0.2 $cm^3$, and a total weight of 0.6 g. They are then respectively applied onto the surfaces of

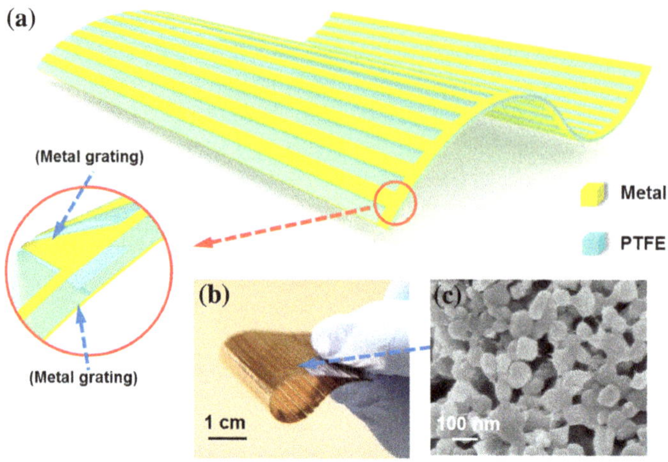


**Fig. 3.10** Device design of the MG-TENG. **a** Schematic illustrations of the thin film that composes of the MG-TENG. The zoom-in illustration (*bottom left*) reveals that the film consists of a PTFE layer and a pair of metal gratings that have complementary patterns. **b** Photograph of a flexible PTFE film with a pair of complementary metal gratings on both sides. The paired gratings are complementary, making central part of the film opaque. **c** SEM image of PTFE nanoparticles applied onto the surfaces on which friction takes place. They play a key role in enhancing energy conversion efficiency and mechanical robustness of the MG-TENG. Reproduced with permission from Wiley [7]

two objects that have relative sliding, i.e., a slider and a guide. The motion direction is perpendicular to the metal strips. Since the two metal gratings in the middle keep in contact, they form a common electrode called base electrode.

The MG-TENG operates in a unique principle that relies on the coupling between triboelectric effect and electrostatic induction. As schemed in Fig. 3.11, components labeled in yellow and green belong to the slider and the guide, respectively. The entire MG-TENG is equivalent to two sets of units in parallel connection. The first set consists of the top electrode, the top film, and part of the base electrode on the guide. The second set is composed of the bottom electrode, the bottom film, and part of the base electrode on the slider. As the slider moves away from the aligned position in Fig. 3.11a–I, free electrons are driven from both the top electrode and the bottom electrode to the base electrode by uncompensated triboelectric charges on the PTFE films. The two streams of electrons converge at the base electrode and add up because they are synchronized. The flow of electrons lasts until the base electrode is completely misaligned with respect to the bottom electrode and the top electrode (Fig. 3.11a–III).

Further motion in the same direction starts to bring the base electrode back towards alignment (Fig. 3.11b) because the grating is a collection of identical repetitions. As a result, accumulated free electrons on the base electrode redistribute, generating two separate streams of electrons towards the top electrode and the bottom electrode until the aligned position is again achieved. Consequently, a

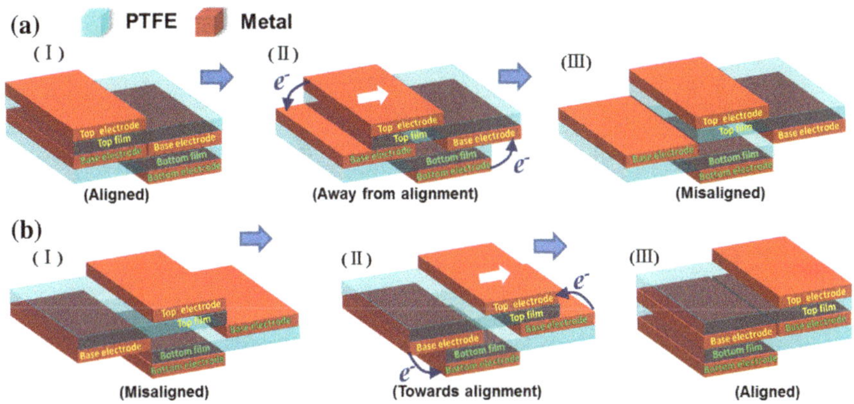


**Fig. 3.11** Overall process of electricity generation of the MG-TENG when both the *top* electrode and the *bottom* electrode are connected to the base electrode. **a** First half cycle of electricity generation process. Components marked in *yellow* and *green labels* belong to the slider and the guide, respectively. The sliding from the aligned state *I* to the misaligned state *II* corresponds to a flow of electrons towards the base electrode from the *top* and *bottom* electrodes. **b** Second half cycle of electricity generation process. Transition from the misaligned state *I* to the aligned state *II* is accompanied by the divergence of electron flow from the base electrode to the other two electrodes. Since the gratings are collections of repeated periodic patterns, continuous motion of the slider on the guide generates alternating current output. Reproduced with permission from Wiley [7]

cycle of electricity generation can be achieved by motion distance of a grating pitch. Therefore, it is the micro-grating design that enables alternating charge transport for numerous times in a small time frame, realizing the breakthrough in output current. The output charge from the MG-TENG is quantified by the following equation.

$$Q = (L/l)\sigma_{\text{induced}}A \tag{3.14}$$

where $Q$ is the output charge defined as the overall amount of induced charges that can transport between electrodes regardless of the current direction, $L$ is the sliding distance of the slider, $l$ is the grating width that equals half pitch, $\sigma_{\text{induced}}$ is the maximum density of induced charges on electrodes, and $A$ is the contact area.

To quantitatively characterize the output power of the MG-TENG, a linear motion was connected to the slider to provide mechanical force, while the guide keeps stationary. Driven by the linear motor that controls the sliding velocity, the slider makes reciprocating linear motion at a direction perpendicular to the metal strips. At a sliding velocity of 2 m $s^{-1}$, short-circuit current ($I_{SC}$) has continuous AC output at an average amplitude of 2 mA and constant frequency of 1 kHz (Fig. 3.12a). For open-circuit voltage ($V_{OC}$), it oscillates between 0 and a maximum value of 500 V at the same frequency as $I_{SC}$ (Fig. 3.12b). With a bridge rectifier, the output charge without external load reaches 13.2 μC in 10 ms (Fig. 3.12c), corresponding to an effective current ($I_{\text{effective}} = \Delta Q/\Delta t$) of 1.32 mA in short-circuit

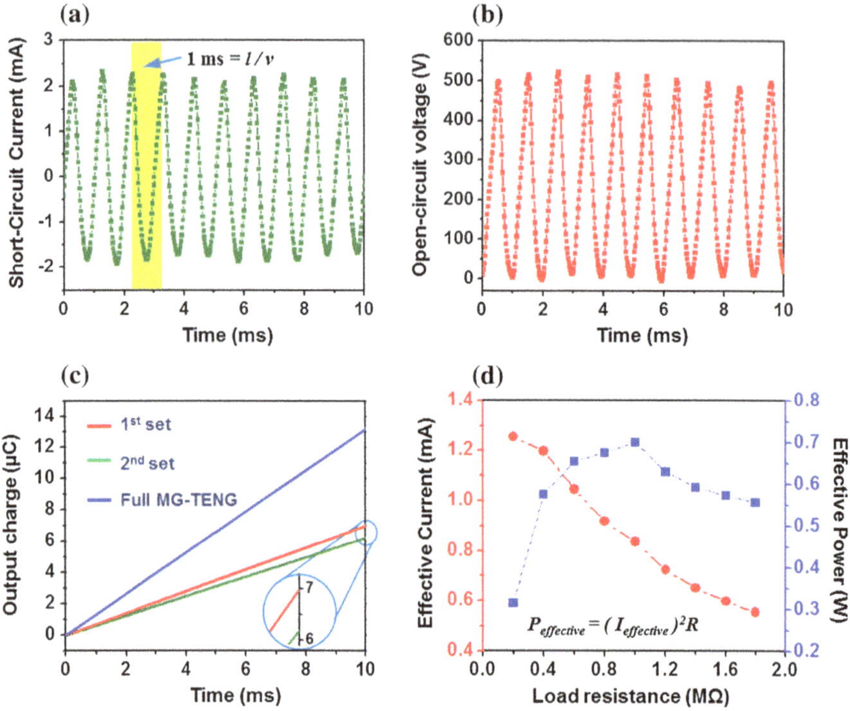


**Fig. 3.12** Results of electric measurements for a MG-TENG having a total area of 60 $cm^2$ and an effective contact area of 20 $cm^2$. **a** Short-circuit current ($I_{sc}$) at a sliding velocity of 2 m $s^{-1}$. **b** Open-circuit voltage ($V_{oc}$) at a sliding velocity of 2 m $s^{-1}$. **c** Output charge at a sliding velocity of 2 m $s^{-1}$. The overall output charge (*blue curve*) is composed of the output charge from two separate sets that generate electricity independently (*red curve and green curve*). **d** Load matching test at a sliding velocity of 2 m $s^{-1}$. Maximum effective power is obtained at the matched load of 1 MΩ. Reproduced with permission from Wiley [7]

condition. As indicated in the operating principle, two sets of units in the MG-TENG operate independently without interference. They produce synchronized currents that can add up. The inset in Fig. 3.12c clearly exhibits separate output charge of the two sets. Each contributes approximately half of the overall output charge respectively.

It is noticed that the time span of a current cycle is determined by the ratio between the grating width and the sliding velocity (Fig. 3.12a). Once an external load is applied, it brings about resistance and thus reduces the amount of electrons that can transport between electrodes within such a fixed time frame. This effect is revealed by the reduced $I_{effective}$ with increasing load resistance, as shown in Fig. 3.12d. At the matched load of 1 MΩ, effective power ($P_{effective} = I^2_{effective}R$, where R is the load resistance) reaches the optimum value of 0.76 W at a sliding velocity of 2 m $s^{-1}$.

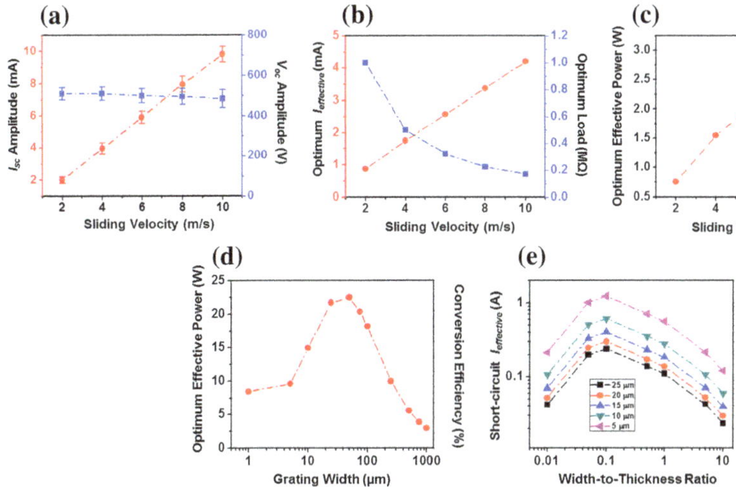


**Fig. 3.13** The effect of sliding velocity on the electric output of the MG-TENG. **a** Amplitude of $I_{sc}$ and $V_{oc}$ with varying sliding velocity. **b** The optimum effective current and corresponding load with varying sliding velocity. **c** The optimum effective power of the MG-TENG with varying sliding velocity. **d** The optimum effective power with scale-down of the grating width at a sliding velocity of 10 m $s^{-1}$. The PTFE thickness remains at 25 μm. The x-axis is log-scale. **e** Effective current in short-circuit condition as a function of the ratio between the grating width and the PTFE thickness. At a fixed ratio, thinner PTFE corresponds to higher short-circuit $I_{effecitve}$. Both axes are log-scale. Reproduced with permission from Wiley [7]

Sliding velocity is a major determining factor in electric output of the MG-TENG. A nearly linear relationship between the amplitude of $I_{SC}$ and the sliding velocity can be obtained while the amplitude of $V_{OC}$, independent of the sliding velocity, remains at a stable value (Fig. 3.13a). The sliding velocity also influences the optimum $I_{effective}$ (effective current at the matched load) as well as the corresponding matched load. Based on a series of load matching tests, Fig. 3.13b exhibits a linear-like relationship between the optimum $I_{effective}$ and the sliding velocity. However, the corresponding matched load is approximately reversely proportional to the sliding velocity (Fig. 3.13b). Therefore, the resultant optimum $P_{effective}$ is roughly linearly related to the sliding velocity (Fig. 3.13c). The minor deviation from linear behavior is likely attributed to finite inner resistance of the MG-TENG from the micro-sized metal gratings. As shown in Fig. 3.13c, the optimum $P_{effective}$ of 3 W is achieved at a sliding velocity of 10 m $s^{-1}$, corresponding to a power density of 50 mW $cm^{-2}$ (based on overall area) and 15 W $cm^{-3}$.

Another major factor that influences the electric output of the MG-TENG is the dimension of the grating design, especially grating width and thickness of the dielectric film. Further scale-down of the grating width results in higher frequency of the output current. However, as the grating width approaches the thickness of the PTFE film, the amount of electrons that can transport in a single current cycle considerably reduces because of weakened in-plane polarization. Based on the measured optimum $P_{effective}$ at grating width of 1 mm and numerical calculation results using finite element analysis, the optimum $P_{effective}$ at different values of grating width can be derived (Fig. 3.13d). It is to be pointed out that the grating width is expected to linearly affect the matched load because it is in reverse proportion with the time span of a current cycle. As a consequence, if the grating width shrinks down to 50 μm, the optimum $P_{effective}$ is expected to reach a maximum value of 22.5 W at a sliding velocity of 10 m $s^{-1}$. Furthermore, scale-down of the PTFE thickness can also tremendously enhance the output charge. When the PTFE thickness shrinks down to 5 μm, $I_{effective}$ in short-circuit condition is expected to reach a maximum value of 1.2 A (Fig. 3.13e). However, the thickness is also determined by the fabrication process with a number of practical issues considered, such as manipulability and mechanical robustness of the thin film. For the currently adopted fabrication process that is easily scalable in size and does not demand sophisticated tools for patterning, the commercial cast PTFE film that is 25 μm in thickness is the smallest possible choice.

### 3.3.3 Rotation-Disk Structure

Other than sliding electrification, rotary electrification can also be employed for power generation using a segmentally-patterned disk structure, and this type of TENG is developed for harvesting energy from rotation motions [14]. Such a design aims at lowering the output voltage but enhances the output current, so that the total output power will not be sacrificed. The basic structure of the disk TENG is composed of two disk-shaped components with four sectors each, as schematically illustrated in Fig. 3.14a. In the fabrication of the TENG device, two polymethyl methacrylate (PMMA) sheets were firstly processed by laser cutting to form the desired four-sector-structured cyclostyle that served as the templates for the effective contacting parts of the TENG. Then, a 50-μm-thick Kapton film with Au electrode deposited on the back side was manually-patterned into this exact shape and then securely attached onto one of the templates, while a piece of Al foil tailored into the same shape was attached on the other template. The Al part and Kapton part was brought to a face-to-face intimate contact and the Al part was driven to spin on the surface of the Kapton around their common axis. To promote the triboelectrification and increase the effective contact area between the two layers, Kapton nanorod (NR) array was created on the Kapton surface by a top-down approach [12] through inductive coupled plasma (ICP) reactive ion etching, as displayed in the inset of Fig. 3.14a. Scanning electron microscopy (SEM) images of the Kapton NRs are shown in Fig. 3.14b. After a two-minute ICP etching, the NRs were uniformly distributed on the surface of Kapton, with an average diameter and length of 150 and 600 nm, respectively. Figure 3.14c is a photograph of the two parts of the disk-shaped TENG with a diameter of 4 in. The total effective area of the TENG device is 40.54 $cm^2$. The working principle of the

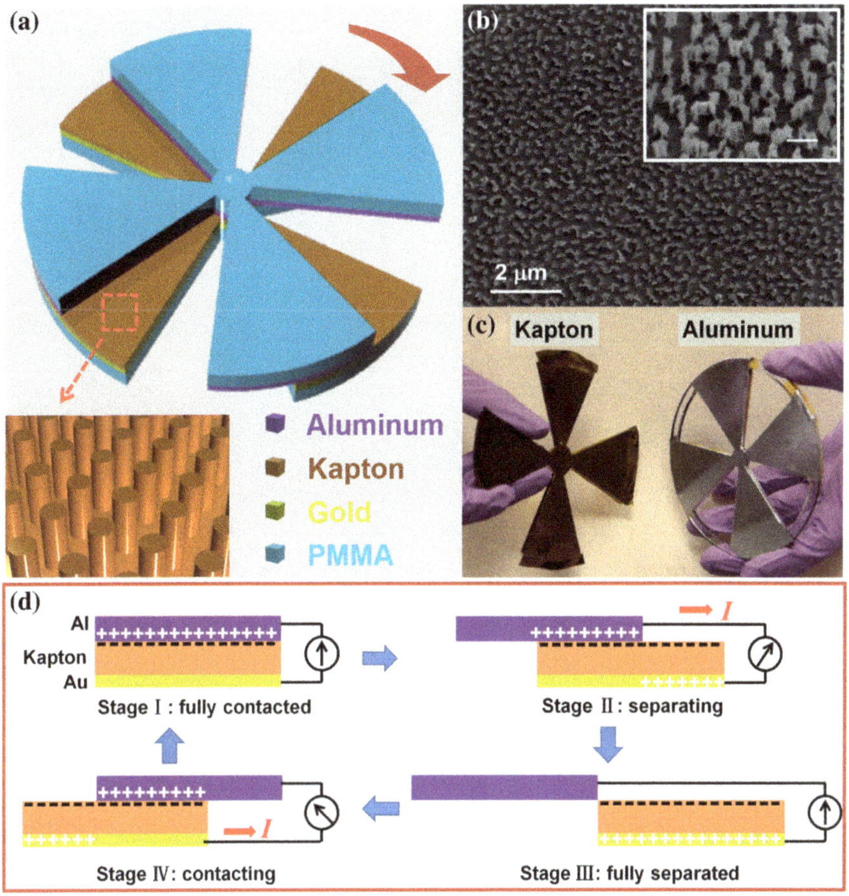


**Fig. 3.14** Basic structure and proposed working principle of the disk TENG. **a** A schematic illustration showing the structure design of the disk TENG. The *inset* (*bottom left*) is an enlarged figure showing the Kapton nanorod array created on the surface area. **b** A top-view SEM image of the Kapton nanorods showing its uniformity in a large range. The *inset* is a high magnification SEM image of the Kapton nanorods in 30°-tilted view. The scale bar is 500 nm. **c** A photograph showing the two parts of a real disk TENG. **d** Schematic illustrations showing the proposed working principle of the disk TENG with the electrons flow diagram in four consecutive stages within a full cycle of electricity generation. Please note that only one pair of sectors (the cross-section area entangled in part **a**) was shown with surface charges for clarity of illustration, and the surface charges on the interface area between Al foil and Kapton film are hidden and are not drawn for easy presentation. Reproduced with permission from American Chemical Society [14]

disk TENG is based on rotary electrification and the relative-rotation-induced cyclic in-plane charge separation between Al and Kapton.

The electricity generation process of the disk TENG relies on the magnitude and the rate of the triboelectric charge separation. In this regard, the configuration of the

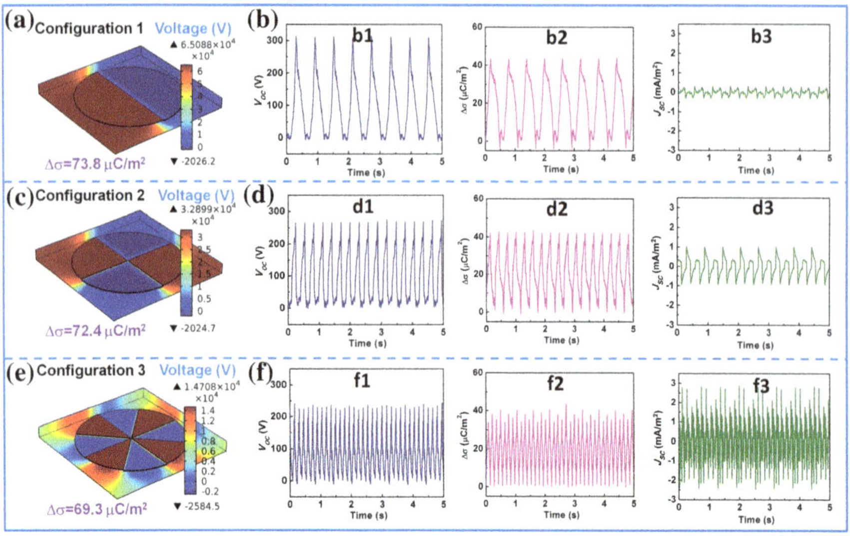


**Fig. 3.15** Configuration dependence of the output performance of the disk TENG. **a** The calculated open-circuit voltage and transferred charge density for the disk TENG with Configuration 1, in which the whole disk was divided into one sector (*half of the full round shape*). **b** The measured (*b1*) open-circuit voltage, (*b2*) transferred charge density, and (*b3*) short-circuit current density of Configuration 1 at 100 rpm. **c** The calculated open-circuit voltage and transferred charge density for the disk TENG with Configuration 2, in which the whole disk was divided into two sectors. **d** The measured (*d1*) open-circuit voltage, (*d2*) transferred charge density, and (*d3*) short-circuit current density of Configuration 2 at 100 rpm. **e** The calculated open-circuit voltage and transferred charge density for the disk TENG with Configuration 3, in which the whole disk was divided into four sectors (equivalent to the structure discussed in Figs. 3.1 and 3.2). **f** The measured (*f1*) open-circuit voltage, (*f2*) transferred charge density, and (*f3*) short-circuit current density of Configuration 3 at 100 rpm. Reproduced with permission from American Chemical Society [14]

device, which is the number of sectors (N) that the entire disk is divided into, can be expected to play a significant role in the output performance of the TENG. For this purpose, we studied three types of TENG devices with the disks divided into 1 sector (Configuration 1, semi-circle), 2 sectors (Configuration 2), and 4 sectors (Configuration 3, equivalent to the structure discussed in Fig. 3.14), respectively. Both theoretical calculations and measurements (with a rotation speed of 100 rpm) were carried out to demonstrate the effect of configurations on the output performance, as given in Fig. 3.15. From the numerical calculation results in Fig. 3.15a, c, e, it could be found that as N increases from 1 to 4, both the calculated potential difference and transferred charge density drop to some extent. This trend is also observed in the measurement results of the $V_{OC}$ and the $\Delta\sigma$ (Fig. 3.15b, d, f), in which the $V_{OC}$ falls from 315 to 230 V, and the $\Delta\sigma$ decreases slightly from 43.6 to 40.8 $\mu C/m^2$. The decreases of the $V_{OC}$ and the $\Delta\sigma$ may result from the lower magnitude of polarization due to the shrunk separation distance for finer segments. On the contrary, the $J_{SC}$ shows a phenomenal ascending trend from Configuration 1 to Configuration 3, mainly owing to the obvious increasing of the charge transferring rate due to the smaller rotation angle from fully contact to fully separation. Above all, it could also be found that the frequencies of all the three types of output signals double with the doubled segmentations. This group of comparisons shows a clear trend that with finer segmentation patterns, the amount of electricity generated by the TENG within a single rotation cycle will be largely increased owing to the multiple times (N) of charge transferred across the external load, which can greatly boost up the power conversion efficiency. Based on this finding, through patterning the disk with much finer segmentations using the techniques of photolithography and etching in the future work, the current output could be greatly improved. On the other hand, when there is no segmentation but just two full-round-shaped disks, the output will be much lower: 8 V, 1.2 $\mu C/m^2$, and 1.5 $\mu A/m^2$ as the values of $V_{OC}$, $\Delta\sigma$, and $J_{SC}$, respectively, at a rotating speed of 100 rpm, which are about two orders of magnitude lower than the results in Fig. 3.15. Ideally, if the contacting surfaces were perfectly flat and there were no wobbling during the rotation, there should be no electric output for this case. The observed electric output is likely due to the roughness of the two disk surfaces, which causes a fluctuation in contact area in the vertical direction.

Besides, an intimate contact in the vertical direction is also highly desired in the current proposed mechanism. To elucidate this issue, another set of experiments was carried out with a series of vertical distances between the pair of concentric disks of Configuration 3, at the rotating speed of 100 rpm. It could be found that the output performance will rapidly decay to zero as the vertical displacement between the two plates increases from 0 to 2 mm. These results provide us with the two important necessities for the high output of the disk TENG, and offer a strong validation for the proposed in-plane charge-separation-based working mechanism of the TENG in this work. Based on theoretical analysis, the $J_{SC}$ will increase linearly with the rotating speed n, while the $V_{OC}$ and the $\Delta\sigma$ should not be influenced by the velocity, as long as the configuration of the device is fixed and the initial surface charge density keeps constant. To verify these relationships, a group of electrical measurements with variable rotating speeds were conducted on the TENG with Configuration 3, which is presented in Fig. 3.16. Figure 3.16a, c show the measured open-circuit voltage and transferred charge density with different rotating speeds. Both the $V_{OC}$ and the $\Delta\sigma$ almost keep constant with the rotating speed increased from 50 to 500 rpm. Though a slight decay at high speed occurs, which might result from the relatively low sampling rate of the voltage/charge meter, it is negligible with the $V_{OC}$ of around 230 V and $\Delta\sigma$ of about 40 $\mu C/m^2$, as summarized in Fig. 3.16b, d. Besides, the averaged slopes of the $V_{OC}$ curve were also analyzed and shown in Fig. 3.16b, with an increasing trend with ascending rotating speed because of the rising of the charge separation rate. Different from the $V_{OC}$ and the $\Delta\sigma$, the $J_{SC}$ is strongly enhanced with higher rotating speed (Fig. 3.16e), and the good linear fitting in Fig. 3.16f is coherently consistent with

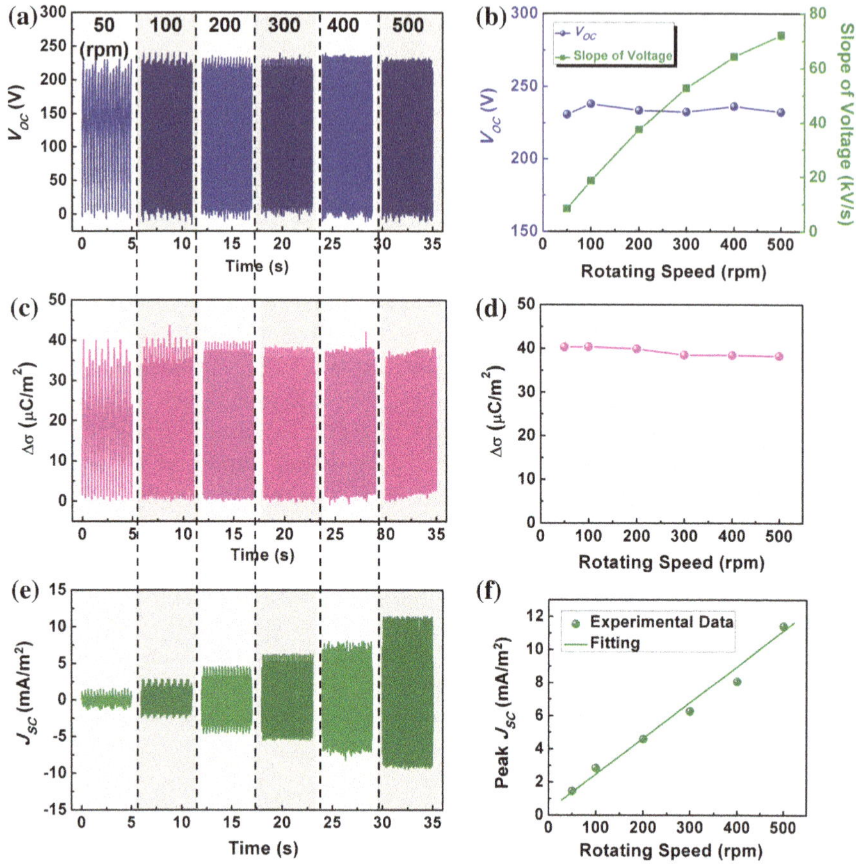


**Fig. 3.16** Influence of the rotating speed on the output performance of the disk TENG. **a** The measured open-circuit voltage with different rotating speeds from 50 to 500 rpm. **b** The summarized relationship between the open-circuit voltage/slope of the voltage and the rotating speed. **c** The measured transferred charge density with different rotating speeds from 50 to 500 rpm. **d** The summarized relationship between the transferred charge density and the rotating speed. **e** The measured short-circuit current density with different rotating speeds from 50 to 500 rpm. **f** The summarized relationship between the short-circuit current density and the rotating speed. Reproduced with permission from American Chemical Society [14]

the behavior predicted by theoretical analysis. This linear relationship between the $J_{SC}$ and the rotating speed implies that the magnitude of the output current of the disk TENG could be utilized to actively measure the angular speed of a rotating disk structure, which has potential applications in automobile braking systems or transmitting instruments.

### 3.3.4 Rotation-Cylinder Structure

Here in this work, a coaxial cylindrical structured rotating-TENG is developed to harvest mechanical energy from rotation in analogous to an electromagnetic induction based generator [15]. A schematic diagram of a rotating-TENG with 6 strip units is shown in Fig. 3.17a. The TENG has a core-shell structure that is composed of a column connected to a rotational motor and a hollow tube fixed on a holder. Here, acrylic was selected as the structural material due to its light weight,

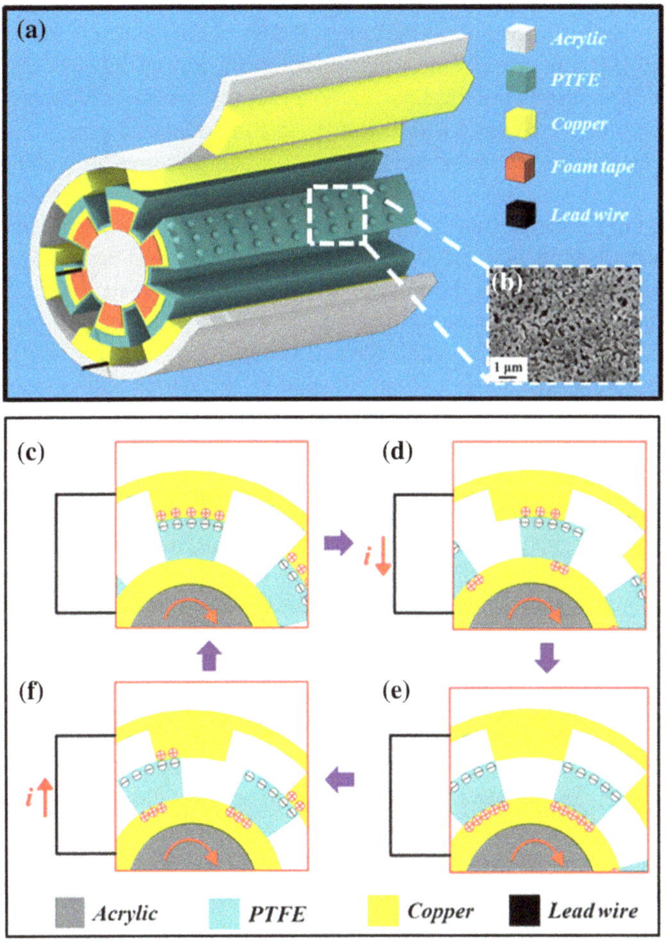


**Fig. 3.17** **a** Schematic of the rotating TENG with 6 strip units. **b** SEM image of PTFE nanoparticles on the surface of PTFE film. Working principle of the rotating TENG. Distributions of charges at **c** fully aligned position, **d** two surfaces are sliding apart, **e** fully mismatched position, and **f** two surfaces are sliding back together. Reproduced with permission from American Chemical Society [15]

good machinability and proper strength. On the inner surface of the fixed tube, metal strips made of cooper are evenly distributed with each having a central angle of 30°, creating equal-sized intervals in between. These metal strips, connected by a bus at one end, play dual roles as a sliding surface and as a common electrode. On the surface of the rotatable column, a collection of separated strips of foam tape were adhered as a buffer layer with a central angle of 30° for each unit. The buffer layer is a critical design that ensures robustness and tolerance on off-axis rotation, which will be discussed in details later. On top of the foam tape, a layer of copper and a layer of polytetrafluoroethylene (PTFE) film are conformably attached in sequence. The copper layer serves as back electrode of the TENG. The outer surface of the PTFE film is modified by spreading a layer of PTFE nanoparticles for enhancing energy conversion efficiency. PTFE nanoparticles with an average diameter of 100 nm spread uniformly on the surface of PTFE film as shown in Fig. 3.17b. Demonstrated in Fig. 3.17a, the PTFE film on the column and the metal strips on the tube were configured to match for a full contact during rotation. These two components and the motor are set to be coaxial to minimize wobbling during rotation.

The working principle of a rotating-TENG can be described by the coupling of contact electrification and electrostatic induction. The design of the cylindrical rotating TENG is based on the relative sliding motion of grated surfaces. Here, a pair of sliding units is selected to illustrate the process of electricity generation, as schemed in Fig. 3.17c–f. The foam tape and nanoparticles on the PTFE film are not presented for simplification and clear illustration. At the aligned position, the two parts of the sliding pair completely match because of the same central angle. Upon contact between the PTFE film and the metal strips, charge transfer takes place. Due to the difference on triboelectric polarities, electrons are injected from metal into the surface of PTFE. Since the two sliding surfaces are completely aligned, triboelectric charges with opposite polarities are fully balanced out, making no electron flow in the external circuit (Fig. 3.17c). Once a relative sliding occurs as a result of rotation, triboelectric charges on the mismatched areas cannot be compensated. The negative ones on the PTFE will drive free electrons on the back electrode to the sliding electrode through the external circuit due to electrostatic induction, neutralizing positive triboelectric charges on the sliding electrode and leaving positive induced charges behind on the back electrode (Fig. 3.17d). The flow of the induced electrons lasts until the mismatch between the two sliding surfaces reaches the maximum when the positive triboelectric charges are fully screened by induced electrons (Fig. 3.17e). As the relative rotation continues, the PTFE film will come into contact with an adjacent metal strip (Fig. 3.17f). Thus, the induced electrons will flow back in an opposite direction until the fully aligned position is restored (Fig. 3.17c). Therefore, in a cycle of electricity generation process, AC electric output is generated. Since all of the sliding units are electrically connected in parallel, output current from different units are synchronized to constructively add up.

To characterize the electric output, $I_{SC}$ and $V_{OC}$ of a rotating-TENG with 6 strip units were measured at a rotation rate of 1000 r/min, corresponding to an equivalent

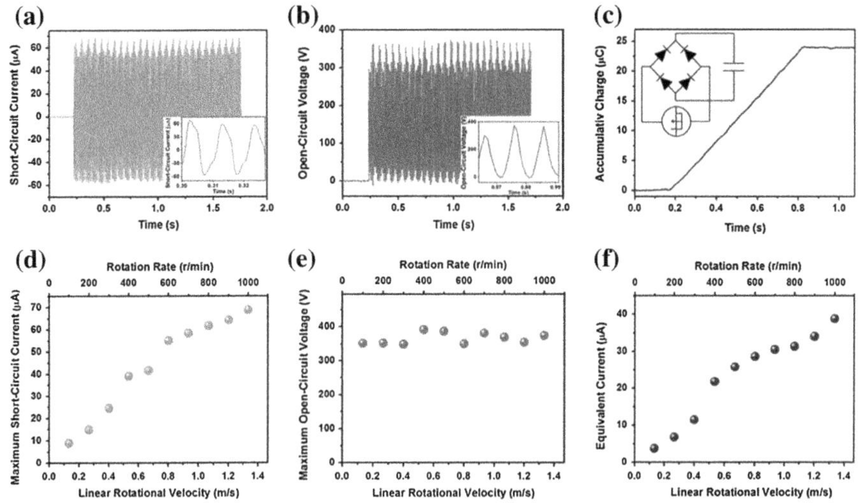


**Fig. 3.18** Electricity output of a rotating TENG with 6 strip units. **a** $I_{sc}$ of a rotating TENG at a linear rotational velocity of 1.33 m/s (a rotation rate of 1000 r/min). *Inset* enlarged view of the current peaks. **b** $V_{oc}$ of a rotating TENG at a linear rotational velocity of 1.33 m/s (a rotation rate of 1000 r/min). *Inset* enlarged view of the voltage peaks. **c** Accumulative inductive charges generated by a rotating TENG. *Inset* electric circuit diagram. **d** Maximum $I_{sc}$ as a function of linear rotational velocity/rotation rate. **e** Maximum $V_{oc}$ as a function of linear rotational velocity/rotation rate. **f** Equivalent direct current as a function of linear rotational velocity/rotation rate. Reproduced with permission from American Chemical Society [15]

linear rotational velocity of 1.33 m/s. As shown in Fig. 3.18a, $I_{SC}$ of 60 μA was achieved. An enlarged view in the inset exhibits continuous AC output current with a frequency of 100 Hz. The frequency is consistent with the one derived from the aforementioned working principle. The $V_{OC}$ reached a maximum value of 373 V (Fig. 3.18b), and the enlarged view in the inset of Fig. 3.18b shows the oscillating behavior between zero and the peak value at a frequency that is the same as the output current. With a diode bridge (as inserted in Fig. 3.18c), output current of different directions can be added up constructively, leading to accumulative induced charges. Illustrated in Fig. 3.18c, the accumulative charges can reach 24.5 μC in 0.63 s, which corresponds to an effective direct current of 38.9 μA. To investigate the influence of rotation rate on the output of the rotating-TENG, $I_{SC}$ and $V_{OC}$ at different rotation rate were measured. At a rotation rate of 100 r/min (a linear rotational velocity of 0.13 m/s), the amplitude of the AC current generated by a rotating-TENG with 6 strip units is 9.0 μA. When the rotation rate increases to 1000 r/min (a linear rotational velocity of 1.33 m/s), the amplitude reaches to 68.8 μA. An approximately liner relationship between the linear rotational velocity and the current amplitude can be derived from the results show in Fig. 3.18d. The liner relationship between the velocity and output current can be explained by the increasing frequency. Because the linear rotational velocity has no influence on the transferred charge density under constant displacement, a higher linear rotational velocity results in a faster electrodes transfer rate.

$$I = \frac{dq}{dt} = \frac{dq}{dl} \cdot v \tag{3.15}$$

where q is the quantity of the triboelectric charges on one sliding surface; t is the time needed for charge transfer, which is the time needed for achieving the mismatched displacement; l is the mismatched displacement between the two sliding surfaces; v is the linear rotational velocity. Thus, an increased $I_{SC}$ is expected. However, the $V_{OC}$ stays stable at different rotation rate because the voltage is only a function of percentage of mismatch as shown in Fig. 3.18e. For output charge, a higher current frequency produces a larger amount of accumulative charges within the same period of time, resulting in a larger equivalent direct current. As shown in Fig. 3.18f, the equivalent direct current increases from 3.7 to 38.9 μA when the linear rotational velocity increases from 0.13 to 1.33 m/s.

### 3.3.5 *Case-Encapsulated Structure*

Here a compact case-encapsulated triboelectric nanogenerator (TENG) is reported that targets reciprocating motions driven by either direct or inertia forces [16]. The TENG relies on electrode pairs that have fine-grating patterns on cylindrical sliding surfaces to generate alternating current. The structural configuration of the TENG is depicted in Fig. 3.19 by using two concentric cylinders that can slide one against the other. The basic principle of the TENG relies on sliding triboelectrification at the interface. The case wrap applied tightly around the outer cylinder allows the two cylindrical structures to move in a coaxial 1-D motion relative to each other. Polyimide film (Kapton) was selected as an electrification material for generating triboelectric charges and also, as the supporting substrate. The Kapton film was chosen as it possesses good electrification and tensile strength properties, leading to a high-performance in electricity generation and durability. On both sides of the Kapton film, copper stripes (width of 2 mm) were deposited with a linear pitch of 4 mm. The electrode patterns on different sides exhibited a linear shift of half pitch so that they are complementary to each other (Fig. 3.19b, c). Two Kapton films having the same electrode patterns formed a pair of contact surfaces (Fig. 3.19b). All copper stripes on the two contact surfaces were connected by bus electrodes to form a common inner electrode (IE), while those on back surfaces of the Kapton films constituted an outer electrode (OE) (Fig. 3.19c). Spherical polytetrafluoroethylene (PTFE) nanoparticles were spread between the contact surfaces (Fig. 3.19f) as lubricant to reduce friction and to further improve energy conversion efficiency. The two prepared Kapton films with electrodes were applied on the outer surface of a PMMA tube and the inner surface of a foam tube, respectively (Fig. 3.19d). The fully assembled TENG is shown in Fig. 3.19e.

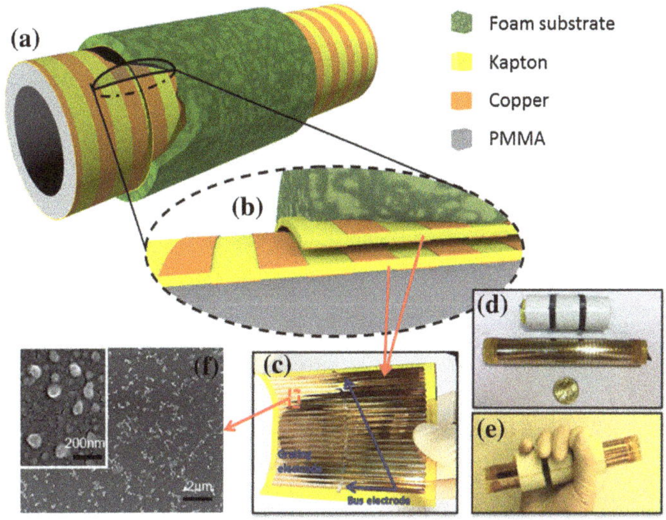


**Fig. 3.19** Case-encapsulated triboelectric nanogenerator (cTENG). **a** 3D model of cTENG and **b** cross-section view. **c** Kapton film deposited with grating copper electrode on both sides. A cTENG is shown (**d**) before and (**e**) after assembly, with 1 U.S. dollar coin as reference. **f** SEM image with different scales showing PTFE particles spread on the surface of Kapton. Reproduced with permission from American Chemical Society [16]

To characterize the electric output of the TENG, $I_{SC}$ and $V_{OC}$ were measured at different sliding velocities. Reciprocating motions were introduced by a velocity-controllable and programmable linear motor, which operated at a stroke of 35 mm. Velocities were programed from 0.3 to 1.0 m/s, with an increment of 0.1 m/s. Accelerations were set at relatively high values to ensure sliding at a uniform velocity within most of the stroke. As shown in Fig. 3.20a, the amplitude of the $I_{SC}$ rises with increasing sliding velocities. Statistical values plotted in Fig. 3.20b exhibit a linear relationship between the current amplitude and the sliding velocities. The maximum value of 668 μA is achieved at a sliding velocity of 1.0 m/s. Such linearity is anticipated because the larger velocity shortens the time (time = l/v, where l refers to the sliding distance and v the velocity) of a single charge-transport process given that the amount of the transferrable induced charges between electrodes is constant. At the same time, the frequency of the output current is also linearly enhanced. On the other hand, the $V_{OC}$ does not have an apparent dependence on the sliding velocity, as shown in Fig. 3.20c, d. In Fig. 3.20c, the $V_{OC}$ densely switched between zero and maximum values, which corresponds to fully aligned and fully displaced positions, staggered. The maximum values only show a minor fluctuation of 110–130 V with a deviation of less than ±8.3 %. This is because the $V_{OC}$ is a function of displacement/position instead of the sliding velocity, which is revealed in the simulation results.

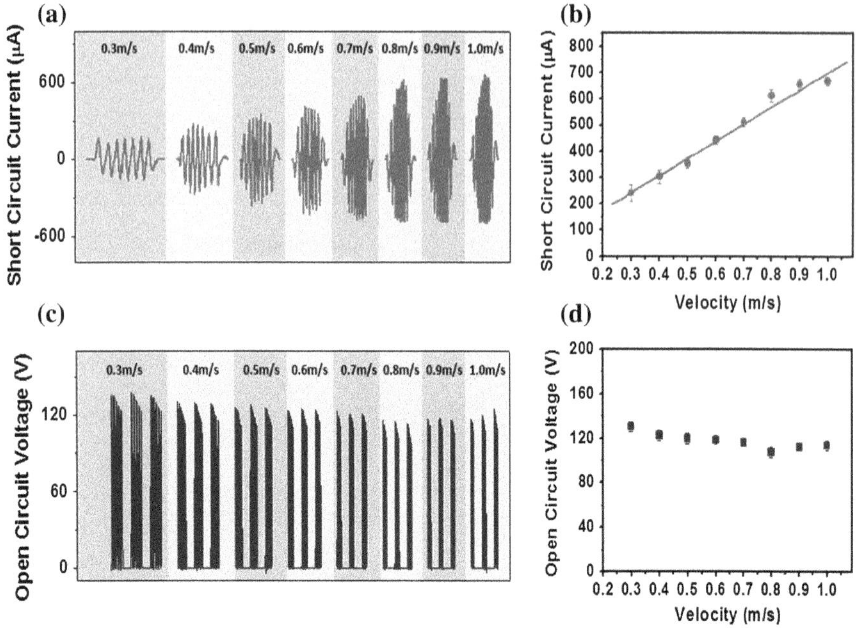


**Fig. 3.20** Current and voltage output characters. **a** Short circuit current over different sliding velocities. **b** Current peak values show linear relationship over sliding velocity. **c** Open circuit voltage over different sliding velocities. **d** Voltage peak values do not change much over sliding velocity. Reproduced with permission from American Chemical Society [16]

### *3.3.6 Liquid-Metal Structure*

Most of the TENGs reported today are normally based on solid materials, so that the effectiveness of contact, especially to the nanometer level, can be largely affected by the roughness of the two surfaces and the match between the two. According to the literature, the present TENG's charge density is about 100 μC $m^{-2}$. Meanwhile, the solid–solid friction will result in heat generation and dissipation. Those two largely limit the efficiency of the TENG.

The current mostly used low-cost electrodes are aluminum and copper. As a metal material, liquid metal is widely investigated for its outstanding physical capabilities, such as high conductivity and favorable flexibility, which initiates promising applications in chip cooling, printed electronics, and energy science (lithium battery, thermoelectric cell). Since triboelectrification is a surface charging effect and the liquid–solid contact will potentially introduce larger contact area, higher contact intimacy and lower friction coefficient, with respect to the solid–solid contact, the liquid metal would be an ideal contact material for TENG's contact electrode.

In this work, a new type of liquid-metal-based triboelectric nanogenerator (LM-TENG) is developed [17]. Operating at a separating velocity of 0.25 m $s^{-1}$,

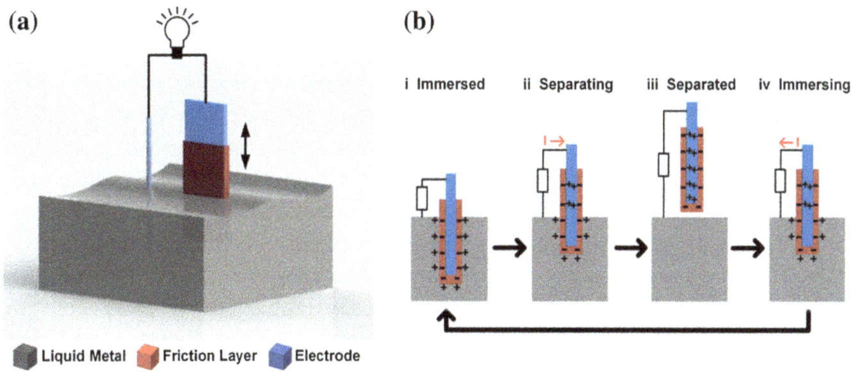


**Fig. 3.21** Working principle of the liquid-metal-based triboelectric nanogenerator. **a** Schematic illustration showing the device configuration of the LM-TENG. **b** Step-by-step illustration showing the working principle of the LM-TENG. Reproduced with permission from Wiley [17]

the LM-TENG having a contact area of 15 $cm^2$ could generate a voltage of 679 V and a current of 9 μA. More importantly, its output charge density reaches 430 μC $m^{-2}$, which is over 4 times higher than that of the solid–solid-contact TENG, and its power density reaches 6.7 W $m^{-2}$ and 133 kW $m^{-3}$. Owing to the shape-adaptive advantage, with a 20-layer configuration, the LM-TENG achieves a high current output above 130 μA. In addition, the liquid-metal-based TENG's energy conversion efficiency is demonstrated to be 70.6 %. Furthermore, the LM-TENG is applied for vibration energy harvesting. At amplitude of 1.2 mm and a frequency of 10 Hz, the LM-TENG outputs a voltage of 145 V, a current of 5.8 μA, and charges a 100 μF capacitor to 1 V in 65 s. This work presents a new approach for enhancing the TENG's output efficiency and further broadening its applications.

A LM-TENG is composed of two parts, as schemed in Fig. 3.21a. One part is a slice, composed of friction material and its induction electrode, and the other is the liquid metal. The friction material could be any insulating material with different electron-attractive ability compared to the liquid metal, such as polyimide (Kapton), polytetrafluoroethylene (PTFE), polyethylene terephthalate (PET), and so on. The induction electrode's one end is fully covered by the friction material, avoiding any shortcut between the induction electrode and the liquid metal. As for the liquid metal, many choices are valid, including mercury, gallium, etc.

Figure 3.21b depicts the working principle of the LM-TENG step by step, which can be explained as a result of triboelectrification and electrostatic induction. When the slice is partly immersed into the liquid metal, the friction material makes contact with the liquid metal. According to the literature [18–20], due to their different abilities in attracting electrons, the electrons will be injected from the liquid metal to the surface of the friction layer, so that there will be net negative charges on the surface of the friction layer and net positive charges on the interface of the liquid metal (Fig. 3.21b–i). As the slice moves out of the liquid, the tribo-charges in the interface regions are separated, which will induce a higher potential in the liquid metal than the induction electrode; thus the electrons in the induction electrode will be driven to flow to the liquid metal through an external load (form a current flow in the reverse direction, Fig. 3.21b–ii). In this process, electrons keep flowing until the slice is fully taken out, which is represented by Fig. 3.21b–iii. At this moment, both the induced potential difference and the amount of transferred charges between the induction electrode and the liquid metal reach the maximum values. As the slice moves back into the liquid metal, the potential difference will drop with the increase of the slice's wetting area. As a result, the electrons will flow back in the opposite direction from the liquid metal to the induction electrode (Fig. 3.21b–iv). Thus, the entire process will result in an alternating-current (AC) pulse output.

The electric characteristics of the LM-TENG are measured at the slice's moving velocity of 0.25 m $s^{-1}$. The Kapton (50 μm thick), copper and mercury serve as the friction material, induction electrode and liquid metal, respectively. The measured results of the open-circuit voltage ($V_{OC}$), short-circuit current ($I_{SC}$), and amount of transferred charges (Q) are displayed in Fig. 3.22a–c. It can be observed that the LM-TENG can deliver a high output performance as $V_{OC}$ = 679 V, $I_{SC}$ = 9 μA, and Q = 0.645 μC. Taking the contacting area of 15 $cm^2$ into account, the output

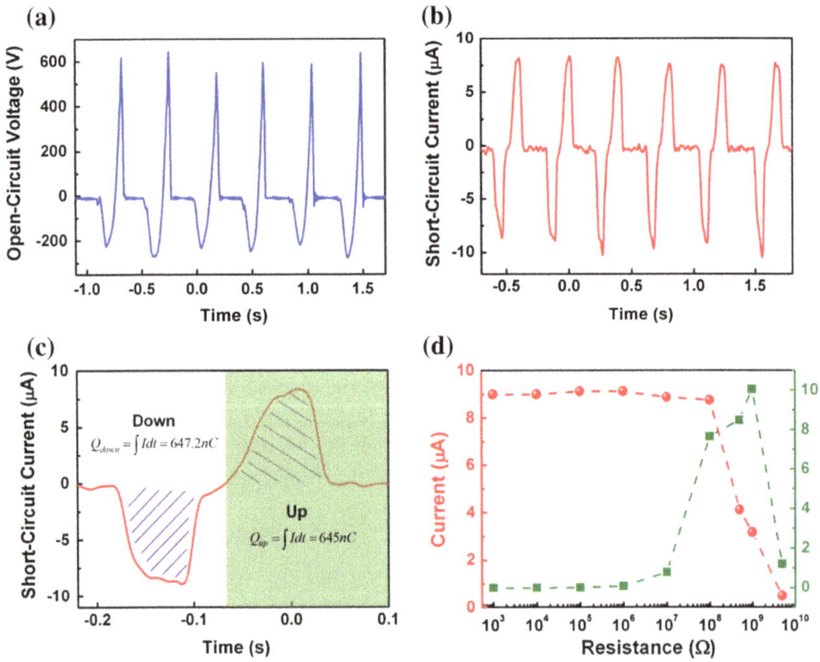


**Fig. 3.22** Electric characteristics of the mercury-based TENG with 50 μm thick Kapton as the friction material. **a–c** Measurement results of the open-circuit voltage, short-circuit current, and transferred charges of the TENG, respectively. **d** Output current and power under variable load resistances. Reproduced with permission from Wiley [17]

charge density reaches 430 μC $m^{-2}$, which is much higher than the previous solid–solid contact-induced charge density that is around 100 μC $m^{-2}$. The output current is also measured with different resistances connected in the load. It remains stable when the resistance is smaller than 100 MΩ, and then decreases as the resistance increases. An optimum output power of 10 mW (6.7 W $m^{-2}$ and 133 kW $m^{-3}$) is delivered at a resistance of 1 GΩ (Fig. 3.22d).

To further demonstrate the concept of the liquid-metal-based TENG, the influence of liquid metal, friction materials, electrode materials, as well as surface roughness has been systematically studied. The scalability of the LM-TENG's output in two dimensions is proved by changing the contacting area, as discussed above. Multi-layer integration is thereby employed in the vertical direction to achieve multi-fold enhancement, which is previously reported to be very effective. In this work, the liquid metal is continuously deformable, so that the experiment is performed expediently by stacking the slices together. As the layer number increases from 1 to 20, the output current is significantly enhanced from 9 to 131 μA. Due to the uneven liquid metal level over all the slices, the achieved enhancement is not as high as 20 times bigger. As for the output voltage, it increases considerably, and then saturates around 1400 V. It is owing to that the voltage meter in our experiment is similar to a load resistance. Therefore, with the increase of the current, the voltage value also increases to some degree.

In this work, the liquid-metal-based triboelectric nanogenerator was developed to deliver high energy conversion efficiency. Owing to the outstanding properties of the liquid metal contact, such as enhanced effective contact, shape-adaptability and low friction coefficient with solid, the LM-TENG exhibits an output charge density of 430 μC $m^{-2}$, which is 4–5 times higher than that in the case if the electrode is solid film. With a 20-layer configuration, the LM-TENG achieves a high current output above 130 μA. In addition, the liquid-metal-based TENG's energy conversion efficiency is demonstrated to be 70.6 % at the best matched load (to be discussed in Sect. 3.4.2). Moreover, the easy-fluctuating property of the liquid metal makes it inherently suitable for vibration energy harvesting. Given its high output performance and other significant advantages in volume, scalability, the LM-TENG is a very promising approach in harvesting mechanical motions for self-powered electronics.

## 3.4 Energy Conversion Efficiency

### *3.4.1 Solid–Solid*

Typical examples of high-efficiency sliding-mode TENG based on solid–solid electrification are the linear-grating TENGs, as we discussed in Sect. 3.3.2. The energy conversion efficiency for the plain sliding TENG is estimated to be ~23 % [1]. Then, an estimation for the grating TENG with 10 grating numbers is calculated as ~8–31 % [5]. This energy conversion efficiency is further optimized by the micro-grated TENG (MG-TENG) with much finer grating features, as discussed in [7].

The efficiency of the MG-TENG is defined as the ratio between input power from mechanical motion and electric power that is delivered to the load. To quantify the mechanical energy input, the slider was connected to the linear motion through an additional force sensor, which could measure the lateral force applied to the slider in the direction of sliding. Such a force is equivalent to the shear force between the two surfaces during sliding. By doing so, we were able to experimentally obtain the total input mechanical energy to the MG-TENG. With the electric energy that was experimentally measured, the efficiency reached nearly 50 % when a matched load was connected. Such a high efficiency is attributed to not only large output power but also low-level losses resulting from nanoparticle-enabled surface modification. The nanoparticles shown in Fig. 3.10c do not affect electric output, but significantly reduces the effective dynamic friction coefficient due to the following possible reasons. First, the sphere-shaped nanoparticles may partially convert sliding friction to rolling friction. Second, they play a role of interface layer by introducing a nano-sized gap between PTFE and metal. Such spacing is expected to considerably lower the electrostatic attraction between triboelectric charges. The substantially lowered friction then benefits durability of the MG-TENG, enabling long-term durability against wear.

### 3.4.2 Solid–Liquid

The efficiency of the LM-TENG is defined as the ratio between the input mechanical energy and the generated electric energy that is delivered to the load [17]. When the slice is pushed down to the bottom of the liquid metal, the buoyancy and surface tension will force it to rise up. Under the action of the buoyancy, surface tension and its own weight, the slice is accelerated to some speed and meanwhile delivers out the electric energy. Therefore, the efficiency is determined by the following equation:

$$\eta = \frac{E_{\text{ele}}}{W_{\text{total}}} = \frac{\int I^2 R dt}{W_{\text{B}} + W_{\text{S}} - W_{\text{G}}} \tag{3.16}$$

where, $E_{\text{ele}}$, $W_{\text{total}}$ stand for the electric energy and the total work done by the ambient, $W_{\text{B}}$, $W_{\text{S}}$, $W_{\text{G}}$ represent the work done by the buoyancy, surface tension and gravity, respectively. $E_{\text{ele}}$ is measured under various load resistances, and shows an optimum energy output of 0.149 mJ at a load of 1 GΩ (Fig. 3.23a, b). Whereas, the $W_{\text{total}}$ is calculated to be 0.211 mJ. Consequently, the efficiency reaches 70.6 % at the best matched load (Fig. 3.23c).

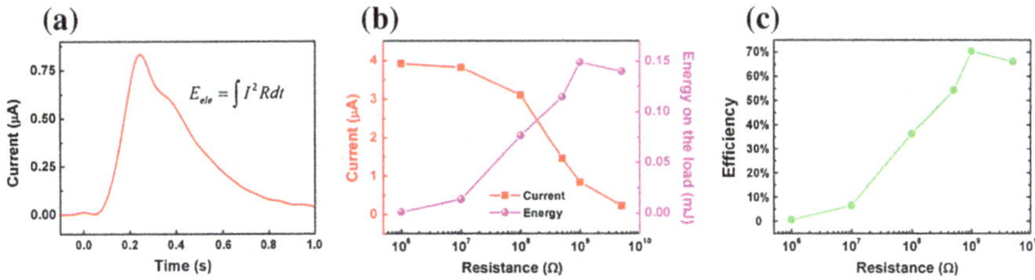


**Fig. 3.23** LM-TENG's energy efficiency measurement. **a** Output current of the LM-TENG under free rising process at a load resistance of 1 GΩ. **b** The output current and energy as a function of the load resistance. **c** The energy efficiency of the LM-TENG at various load resistances. Reproduced with permission from Wiley [17]

Such a high efficiency is attributed to the extremely low friction coefficient and large effective contacting surface of the liquid–solid contact. In addition, it is worth noting that, the slice is accelerated to some speed and thus possesses some kinetic energy. This is not dissipated during the energy generation process and should be excluded from the total input work. In our calculation, such a kinetic energy is ignored, which means that we underestimate the efficiency. The energy loss could be owing to that, during the surface tension's working process, the liquid requires inner motion and thus consumes some energy.

## 3.5 Summary

In this chapter, we present a fundamental working mode of TENGs, which is the lateral sliding mode. This working mode has superior advantage over the vertical contact-separation mode in that no air gap is required to bring apart the two triboelectric surfaces, which is beneficial for its subsequent packaging, and it can be operated in planar sliding or rotation modes. The plain-sliding structure delivers a large open-circuit voltage of 1200 V. Furthermore, the basic device structure could be modified to extend its functionality and improve the output performance. For example, both triboelectric surfaces could be patterned with grating structure to multiply number of energy conversion cycles and elevate the instantaneous current. Segmentally-patterned disk structure and rotary cylinder structure have been designed to fulfill rotational mechanical energy harvesting, and case-encapsulated structure is also developed for harvesting energy from a reciprocating motion in a tube-shaped channel. Finally, a liquid-metal based structure has been investigated that deliver a record-high energy conversion efficiency, $\sim 70\ \%$.

## References


1. S.H. Wang, L. Lin, Y.N. Xie, Q.S. Jing, S.M. Niu, Z.L. Wang, Sliding-triboelectric nanogenerators based on in-plane charge-separation mechanism. Nano Lett. **13**(5), 2226–2233 (2013)
2. F. Saurenbach, D. Wollmann, B. Terris, A. Diaz, Force microscopy of ion-containing polymer surfaces: morphology and charge structure. Langmuir **8**(4), 1199–1203 (1992)
3. S.M. Niu, Y. Liu, S.H. Wang, L. Lin, Y.S. Zhou, Y.F. Hu, Z.L. Wang, Theory of Sliding-Mode Triboelectric Nanogenerators. Adv. Mater. **25**(43), 6184–6193 (2013)
4. S.M. Niu, Z.L. Wang, Theoretical systems of triboelectric nanogenerators. Nano Energy **14**, 161–192 (2015)
5. G. Zhu, J. Chen, Y. Liu, P. Bai, Y.S. Zhou, Q.S. Jing, C.F. Pan, Z.L. Wang, Linear-grating triboelectric generator based on sliding electrification. Nano Lett. **13**(5), 2282–2289 (2013)
6. Y.S. Zhou, G. Zhu, S.M. Niu, Y. Liu, P. Bai, Q.S. Jing, Z.L. Wang, Nanometer resolution self-powered static and dynamic motion sensor based on micro-grated triboelectrification. Adv. Mater. **26**(11), 1719–1724 (2014)
7. G. Zhu, Y.S. Zhou, P. Bai, X.S. Meng, Q.S. Jing, J. Chen, Z.L. Wang, A Shape-adaptive thin-film-based approach for 50 % high-efficiency energy generation through micro-grating sliding electrification. Adv. Mater. **26**(23), 3788–3796 (2014)
8. S.M. Niu, S.H. Wang, Y. Liu, Y.S. Zhou, L. Lin, Y.F. Hu, K.C. Pradel, Z.L. Wang, A theoretical study of grating structured triboelectric nanogenerators. Energy Environ. Sci. **7**(7), 2339–2349 (2014)
9. A.F. Diaz, R.M. Felix-Navarro, A semi-quantitative tribo-electric series for polymeric materials: the influence of chemical structure and properties. J. Electrostat. **62**(4), 277–290 (2004)
10. F.R. Fan, L. Lin, G. Zhu, W.Z. Wu, R. Zhang, Z.L. Wang, Transparent triboelectric nanogenerators and self-powered pressure sensors based on micropatterned plastic films. Nano Lett. **12**(6), 3109–3114 (2012)
11. G. Zhu, C.F. Pan, W.X. Guo, C.Y. Chen, Y.S. Zhou, R.M. Yu, Z.L. Wang, Triboelectric-generator-driven pulse electrodeposition for micropatterning. Nano Lett. **12**(9), 4960–4965 (2012)
12. H. Fang, W.Z. Wu, J.H. Song, Z.L. Wang, Controlled growth of aligned polymer nanowires. J. Phys. Chem. C **113**(38), 16571–16574 (2009)
13. W.M. Du, X. Han, L. Lin, M.X. Chen, X.Y. Li, C.F. Pan, Z.L. Wang, A three dimensional multi-layered sliding triboelectric nanogenerator. Adv. Energy Mater. **4**(11), (2014)
14. L. Lin, S.H. Wang, Y.N. Xie, Q.S. Jing, S.M. Niu, Y.F. Hu, Z.L. Wang, Segmentally structured disk triboelectric nanogenerator for harvesting rotational mechanical energy. Nano Lett. **13**(6), 2916–2923 (2013)
15. P. Bai, G. Zhu, Y. Liu, J. Chen, Q.S. Jing, W.Q. Yang, J.S. Ma, G. Zhang, Z.L. Wang, Cylindrical rotating triboelectric nanogenerator. ACS Nano **7**(7), 6361–6366 (2013)
16. Q.S. Jing, G. Zhu, P. Bai, Y.N. Xie, J. Chen, R.P.S. Han, Z.L. Wang, Case-encapsulated triboelectric nanogenerator for harvesting energy from reciprocating sliding motion. ACS Nano **8**(4), 3836–3842 (2014)
17. W. Tang, T. Jiang, F.R. Fan, A.F. Yu, C. Zhang, X. Cao, Z.L. Wang, Liquid-metal electrode for high-performance triboelectric nanogenerator at an instantaneous energy conversion efficiency of 70.6 %. Adv. Funct. Mater. **25**(24), 3718–3725 (2015)
18. F.-R. Fan, Z.-Q. Tian, Z.L. Wang, Flexible triboelectric generator. Nano Energy **1**(2), 328–334 (2012)
19. Z.L. Wang, Triboelectric nanogenerators as new energy technology for self-powered systems and as active mechanical and chemical sensors. ACS Nano **7**(11), 9533–9557 (2013)
20. Z.L. Wang, J. Chen, L. Lin, Progress in triboelectric nanogenerators as a new energy technology and self-powered sensors. Energ Environ. Sci. **8**(8), 2250–2282 (2015)

# Chapter 4
# Triboelectric Nanogenerator: Single-Electrode Mode

## 4.1 Basic Principle

The operating principle of SETENGs can be described by the coupling of contact electrification and electrostatic induction, as shown Fig. 4.1 (we use a conductor-to-dielectric contact-mode SETENG as an example) [1, 2]. At original state, the aluminum and fluorinated ethylene propylene (FEP) layers are brought into contact with each other. Surface charge transfer then takes place at the contact area due to triboelectric effect. According to the triboelectric series that is a list of materials based their tendency to gain or lose charges, electrons are injected from aluminum into FEP, resulting in net negative charges at the FEP surface and net positive charges at the aluminum surface, respectively. It is worth noting that the insulating property of the polymers allows a long-time retention of triboelectric charges for hours or even days. Since they are only confined on the surface, charges with opposite signs coincide at almost the same plane, generating practical no electric potential difference between the two electrodes (Fig. 4.1a).

Once the FEP layer is separated from the aluminum layer, a potential difference is then established between the two electrodes (or between the primary electrode and the ground if ground is considered as the second electrode) under the open circuit condition since the opposite triboelectric charges are separated. As the FEP layer is being separated, $V_{OC}$ keeps increasing until reaching the maximum value when the FEP layer goes to its maximum separation position. If the two electrodes are shorted, any established potential difference as the two triboelectric layers separate drives electrons from the reference electrode (copper layer) to the primary electrode (aluminum layer, Fig. 4.1b), resulting in an instantaneous current. When the FEP layer is far away enough, about a half of the negative charge is transferred from reference electrode to primary electrode, making them have almost the same amount of positive charges (Fig. 4.1c). Once the FEP layer is driven to contact the aluminum electrode again, reduction of the interlayer distance would make the primary electrode possess a lower electric potential than the reference electrode.



Z.L. Wang et al., *Triboelectric Nanogenerators*,
Green Energy and Technology, DOI 10.1007/978-3-319-40039-6_4

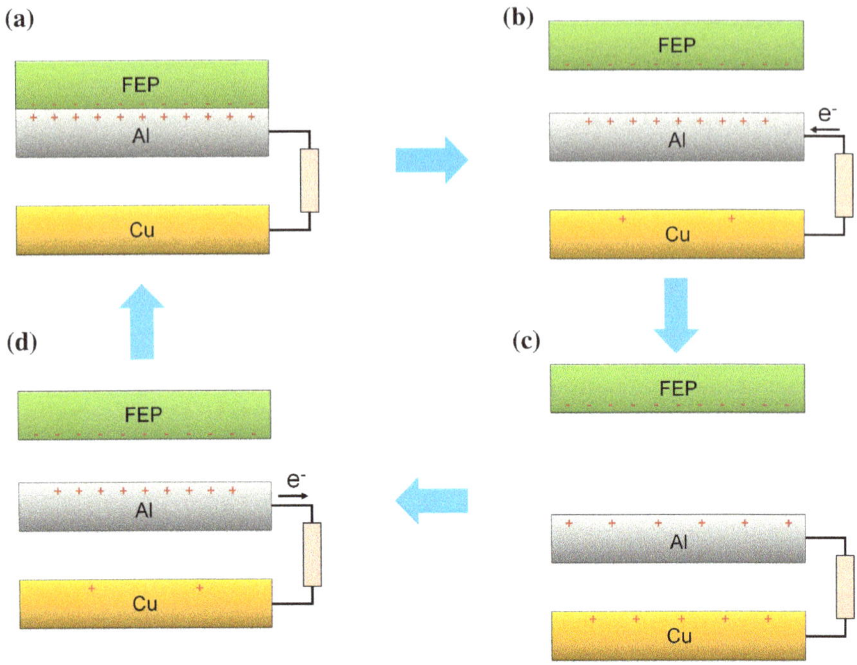


**Fig. 4.1** Working mechanism of the conductor-to-dielectric single-electrode TENG based on vertical contact separation. The sketches illustrate the electricity generation process in a full cycle of the contact motion under quasi short-circuit condition. The requirement here is that the size of the device is small to minimize the electrostatic shield effect from the aluminum layer, otherwise the generated current/voltage is small

As a consequence, electrons are driven from the primary electrode back to the reference electrode. (Figure 4.1d) This process corresponds to an instantaneous current as well. When the two triboelectric layers are in contact again, all positive charges flow back to the aluminum layer (Fig. 4.1a).

## 4.2 Fundamental Theory

### *4.2.1 Basic Working Principle and Electrostatic Shield Effect*

Similar to attached-electrode triboelectric generators discussed in previous two chapters, SETENGs also have contact-mode and sliding-mode, which shows almost the same characteristics. Therefore, we mainly focus on the contact-mode SETENGs in the following discussions.

We first take the conductor-to-dielectric contact-mode SETENGs as an example. Its finite element model was built in two dimensions to simplify the numerical calculation without losing its inherent physics properties, as shown in Fig. 4.2a. One dielectric plate and one metal electrode (primary electrode) are stacked face to face to form the triboelectric pair, with the same length of $l$ and width of $w$. The thickness is $d_1$ for Dielectric 1 and $d_m$ for the primary electrode. As one typical configuration, the reference electrode of the same geometric size locates underneath the primary electrode, with a gap of $g$. The two electrodes are fixed and Dielectric 1 can move along the vertical direction under external mechanical force, with the distance between the two triboelectric layers defined as $x$. As a result of contact-electrification with the primary electrode, Dielectric 1 carries net negative charges with a density of $-\sigma$ on its bottom surface. Considering the excellent insulation properties of the polymer and the nature of contact electrification, it can be assumed that the charges are uniformly distributed on the surface at macro scale with negligible decay. At the same time, there will be equal amount ($\sigma wl$) of positive charges injected to the primary electrode through contact electrification. With $Q$ defined as the transferred charges from the primary electrode to the reference electrode, the total amount of charges at the primary electrode and reference electrode under minimum achievable charge reference state (MACRS, discussed in detail in Chap. 6) are $\sigma wl - Q$ and $Q$, respectively [2].

The FEM calculation results of $C$, $V_{OC}$, and $Q_{SC}$ for the contact-mode SETENG are plotted in Fig. 4.2b–d, in comparison with the corresponding results of the attached-electrode contact-mode TENG with the same size (detailed calculation parameter is listed in Table 4.1). Their characteristics are completely different. First, due to the immobility of the SETENG's electrodes, its $C$ is nearly a constant ($C_0$) with the increase of $x$. However, for the attached-electrode structure, its $C$ decreases quickly when the distance between the two electrodes is increasing. Secondly, in the SETENG, its $V_{OC}$ quickly gets saturated at a value much lower than that of the attached-electrode structure. This is because when Dielectric 1 is far away from the electrodes, the influence of Dielectric 1 on the electric field distribution around the electrodes is little. Therefore, further increase in $x$ will contribute little to the increase of $V_{OC}$. Finally, in the SETENG, $Q_{SC}$ slowly reached its saturation value, which is only half of the value for the attached-electrode structured TENG. When Dielectric 1 is far away from the electrodes, the tribo-charges generated are equally distributed between the two electrodes to maintain the two electrodes at the same electric potential. The maximized $\eta_{CT}$ can only reach 50 %, while $\eta_{CT}$ for the attached-electrode TENGs can reach nearly 100 % even when $x$ is relatively small.

The degradation of output performance of SETENGs mainly comes from the electrostatic shield effect of the primary electrode, which can be analyzed from their equivalent circuit model [2]. The nodes are still first analyzed in the electrostatic system. For the above system in OC condition, the electric potential across the whole bottom surface of Dielectric 1 is nearly constant. Thus, the whole surface can be assumed as a node (Node 1). Similarly, the primary electrode and the reference electrode also serve as two nodes, which can be named as Node 2 and Node 3, respectively. Since every two nodes are connected by electric field lines, an

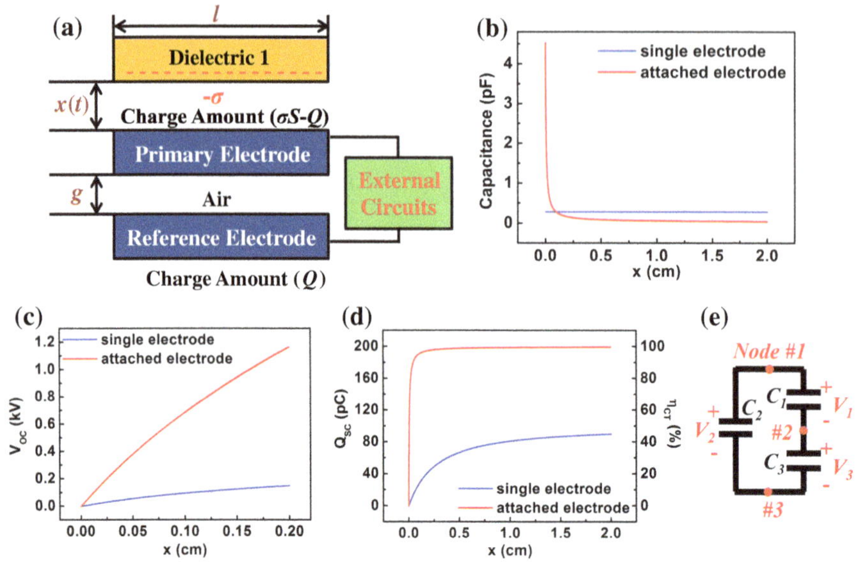


**Fig. 4.2** Intrinsic output characteristics of conductor-to-dielectric contact-mode SETENGs. **a** Structure of the FEM model. **b–d** Calculated **b** capacitance between the two electrodes, **c** open circuit voltage, and **d** transferred charges at short circuit condition of SETENGs at different $x$, together with comparison with results of attached-electrode TENGs. **e** Equivalent circuit model containing three capacitances for the SETENGs under open-circuit (OC) condition. Reproduced from reference [2]. Copyright from 2014 Wiley

**Table 4.1** Utilized parameters in the calculation of SETENG's output characteristics

| | |
|---|---|
| Dielectric 1 | $\varepsilon_{r1} = 2$, $d_1 = 100$ μm |
| Thickness of electrodes $d_m$ | 1 μm |
| Width of the structrue $w$ | 5 mm |
| Length of dielectrics $l$ | 5 mm |
| Gap distance between electrodes $g$ | 1 mm |
| Tribo-charge surface density $\sigma$ | 8 μC $m^{-2}$ |
| Maximum separation distance $x_{max}$ | 0.01 m |
| Average velocity $v$ | 1 $ms^{-1}$ |

equivalent capacitance is formed between each of them, as shown in Fig. 4.2e. Their capacitances $C_1$, $C_2$, and $C_3$ in the equivalent circuit only represent the capacitive effects from the direct electric line connection between every two nodes without crossing any charged objects (without electrostatic shield of the third obeject), so they are not actual capacitances that reflects all the electric line connection (both direct and non-direct) between the two nodes and can be directly measured.

As a quantitative relationship between the two categories of capacitance, the actual capacitance is the combination of these three capacitances. For example, the actual capacitance ($C_b$) between Node 1 and Node 3 contains two parts: $C_2$ and the serial connection of $C_1$ and $C_3$ that shows the capacitance effect of non-direct electric lines connection through Node 2, as given by:

$$C_b = C_2 + \frac{C_1 C_3}{C_1 + C_3} \tag{4.1}$$

Similarly, the actual capacitances between Node 1 and Node 2 ($C_a$) and between Node 2 and Node 3 ($C_0$) can also be given as below:

$$C_a = C_1 + \frac{C_2 C_3}{C_2 + C_3} \tag{4.2}$$

$$C_0 = C_3 + \frac{C_1 C_2}{C_1 + C_2} \tag{4.3}$$

Under OC conditions (MACRS), the total charges on Node 1, 2, and 3 are $-\sigma wl$, $\sigma wl$, and 0, respectively. Thus, from basic characteristics of capacitances and charge conservation on each node, $V_{OC}$, $Q_{SC}$ and $\eta_{CT}$ can be given by ($C_1(x = 0)$ is close to infinity in the following equation) [2]:

$$Q_{SC} = \frac{\sigma wl}{1 + \frac{C_1(x)}{C_2(x)}} - \frac{\sigma wl}{1 + \frac{C_1(x=0)}{C_2(x=0)}} = \frac{\sigma wl}{1 + \frac{C_1(x)}{C_2(x)}} \tag{4.4}$$

$$\eta_{CT} = \frac{Q_{SC}}{\sigma wl} = \frac{1}{1 + \frac{C_1(x)}{C_2(x)}} - \frac{1}{1 + \frac{C_1(x=0)}{C_2(x=0)}} = \frac{1}{1 + \frac{C_1(x)}{C_2(x)}} \tag{4.5}$$

$$V_{OC} = \frac{\sigma wl C_2}{C_1 C_2 + C_2 C_3 + C_3 C_1} \tag{4.6}$$

Equations 4.4 and 4.5 are the same as the above equation in the attached-electrode case, which shows that the core electrostatic induction principle of SETENGs is the same with attached-electrode TENGs. However, in SETENGs, the trend of the capacitance is different from the attached-electrode TENG case. When $x = 0$, $C_a$ goes to infinity while $C_b$ and $C_0$ become equal with each other. Therefore, from Eqs. (4.1) to (4.3), $C_1$ goes to infinity while $C_2$ and $C_3$ are equal to $C_0/2$, which is a finite number. Therefore, the ratio $C_1/C_2$ is infinity. When $x$ approaches to infinity, both $C_a$ and $C_b$ approach to 0 at the same rate. Therefore, both $C_1$ and $C_2$ approach to 0 with the ratio of $C_1/C_2$ approaching to 1. Thus, under this condition, the transferred charges is only half amount of the tribo-charges and the maximum charge transfer efficiency is only 50 %.

From Eqs. (4.4) to (4.6), we observe that the capacitance $C_2$ is critically important for both $V_{OC}$ and $Q_{SC}$. $C_2$ represents the portion of the electric lines directly connecting Node 1 and Node 3, without being electrostatically shielded by Node 2. If all the electric lines connections between the reference electrode and

Dielectric 1 needs to come through the primary electrode (electrostatic shielded by the primary electrode), $C_2$ will be equal to 0, which will lead to no output from the SETENG. Therefore, SETENG with infinite area size will lead to zero output due to the fully electrostatic shield condition.

Given a lot of similarities between the SETENG and attached-electrode TENGs, they still have a few differences both structurally and electrically. Among all of the differences, the influences of the electrode gap distance and their area size on the output performance are most important ones and need to be systemically discussed.

### 4.2.2 *Effect of Electrode Gap Distance*

Unlike attached-electrode TENGs, SETENGs have a fixed gap between their two electrodes, which are independent of the motion of Dielectric 1. This leads to an approximately fixed inherent capacitance $C_0$, which is critical to the performance of the whole device. To understand the influence of the gap ($g$), the fundamental properties of the SETENG, i.e. $V_{OC}$ and $Q_{SC}$, are calculated for SETENGs with different gap distances ($g$), as shown in Fig. 4.3a, b. In this calculation, all of the other parameters used are the same as shown in Table 4.1 except the gap distance, which is specified in each figure. At the same moving distance ($x$), $V_{OC}$ rises up monotonically with the increase of $g$ while $Q_{SC}$ decreases. When $g$ is large enough, the raise speed $V_{OC}$ is quite slow and $V_{OC}$ approaches to its saturation value. Similarly, when $g$ approaches to 0, $Q_{SC}$ also approaches to its saturation value.

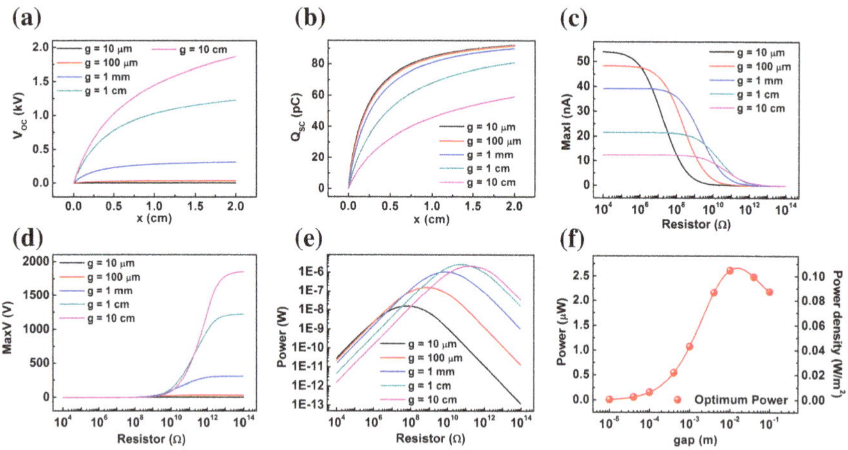


**Fig. 4.3** The influence of gap distance ($g$) on the output characteristics of the SETENG. **a** The relationship between open circuit voltage and separation distance at different gap distances. **b** The relationship between transferred charges at short circuit and separation distance at different gap distance. **c–e** Maximum output **c** current profile, **d** voltage profile, and **e** power profile with load resistance at different gap distances. **f** Extracted optimum power profile with gap distance. Reproduced from reference [2]. Copyright from 2014 Wiley

This influence of $g$ can be interpreted utilizing the equivalent circuit built above. When $g$ increases, $C_b$ and $C_0$ decreases significantly while $C_a$ is barely affected. Therefore, $C_2$ and $C_3$ decrease significantly while $C_1$ maintains almost the same value. In addition, the decrease of $C_3$ is in a higher rate than that of $C_2$. Therefore, from Eqs. (4.4) to (4.6), $V_{OC}$ monotonically increases with the increase of $g$ while $Q_{SC}$ monotonically decreases to 0.

Besides the basic trend of $V_{OC}$ and $Q_{SC}$, their saturation behavior can be theoretically understood as well. First, the saturation behavior of $V_{OC}$ when $g$ is sufficiently large is discussed. When $g$ is sufficiently large, $C_b$ and $C_0$ approach to 0 at the same rate while $C_a$ is almost unaffected. Therefore, $C_2$ and $C_3$ approach to 0 with its ratio approaching to 1 while $C_1$ approaches to $C_a$. Thus, $V_{OC}$ saturates to the value of $\sigma wl/(2C_a)$ when $g$ is large enough.

The saturation behavior of $Q_{SC}$ can also be interpreted. When $g$ decreases to 0, $C_o$ is getting close to infinity and $C_a$ equals to $C_b$. Thus, both $C_1$ and $C_2$ are approaching to $C_a/2$. As a result, $Q_{SC}$ approaches to its saturation value $\sigma wl/2$.

Besides the fundamental property, the influence of the gap on the load characteristics is investigated as well. The output current, voltage, and transit power peak were calculated under different load resistances for SETENGs with different $g$, as shown in Fig. 4.3c–e. A smaller gap provides a larger inherent capacitance, leading to the decrease of the optimum resistance. The relationship between the maximum transit power and the gap distance is plotted in Fig. 4.3f. An optimum gap about 1.54 cm is observed to generate the maximized transit power.

### 4.2.3 Effect of Area Size (Length)

Besides the gap between the electrodes, another important design parameter is the area size (represented by the length $l$ in this 2D model) of the SETENG. To systematically study its influence, $V_{OC}$, $Q_{SC}$, and $\eta_{CT}$ of the SETENG with different length $l$ is calculated through the COMSOL software, as shown in Fig. 4.4a–c. All of the parameters utilized in this calculation are the same as listed in Table 4.1, except that the value of $x_{max}$ utilized in the calculation is 2 cm. From the calculation results, for any given $x$, there exists a specific $l$ to generate the largest $V_{OC}$. In addition, when $x$ increases, this $l$ for the largest $V_{OC}$ also increases. $Q_{SC}$ should have been proportional to the length $l$, but this is true only when $l$ is sufficiently small. When $l$ is large enough, the $Q_{SC} - l$ curve shifts downwards dramatically, which shows that $\eta_{CT}$ decreases significantly when $l$ is large.

The complicate behavior of the influence of $l$ to outputs of SETENGs can be understood from the equivalent circuit built above. The capacitances $C_a$, $C_b$, and $C_0$ are first calculated, which are all 2D parallel electrode capacitances. For an non-ideal 2D parallel electrode capacitance considering the edge effect, its capacitance value can be shown as the following equation [3],

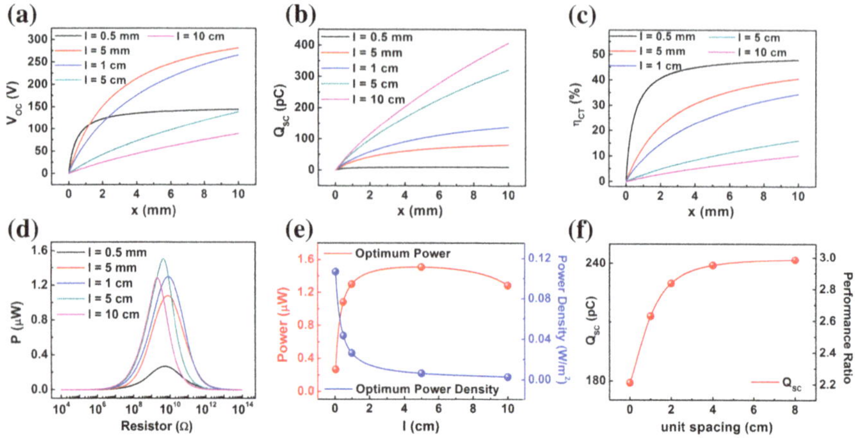

Fig. 4.4 The influence of area size/length on the output characteristics of the SETENG. **a** open circuit voltage profiles with separation distances for SETENGs with different lengths. **b** short circuit transferred charges profile with separation distances for SETENGs with different lengths. **c** short circuit charge transfer efficiency profile with separation distances for SETENGs with different lengths. **d** Maximum output power profile with load resistance for SETENGs with different lengths. **e** Extracted optimum power and power density profile with length of SETENG. **f** The influence of unit spacing on the short circuit transferred charges in a three-SETENG system Reproduced from reference [2]. Copyright from 2014 Wiley

$$C = \frac{\varepsilon_0 wl}{d}\left\{1 + \frac{d}{\pi l}\left\{1 + \ln\left[1 + \frac{2\pi l}{d} + \ln\left(1 + \frac{2\pi l}{d}\right)\right]\right\}\right\} = \frac{\varepsilon_0 wl}{d}\left[1 + \alpha\left(\frac{d}{l}\right)\right] \quad (4.7)$$

where $l$ is its length, $w$ is its width and $d$ is the separation distance between the two parallel electrodes. Inside Eq. (4.7), the edge effect is reflected by the function $\alpha(y)$:

$$\alpha(y) = \frac{y}{\pi}\left\{1 + \ln\left[1 + \frac{2\pi}{y} + \ln\left(1 + \frac{2\pi}{y}\right)\right]\right\} \quad (4.8)$$

With $C_a$, $C_b$, and $C_0$ calculated through the above equation, $C_1$, $C_2$, and $C_3$ can be analytically calculated utilizing Eqs. (4.1)–(4.3). Thus, from the calculated $C_1$, $C_2$, and $C_3$, $V_{OC}$, $Q_{SC}$, and $\eta_{CT}$ can be analytically calculated through Eqs. (4.4)–(4.6). Their behaviors at the two $l$ limits are shown below.

When $l$ approaches to 0 ($x/l$ and $g/l$ is very large):

$$V_{OC} = \frac{\sigma}{2\varepsilon_0}\pi l \quad (4.9a)$$

$$Q_{SC} = \frac{\sigma wl}{2} \quad (4.9b)$$

$$\eta_{CT} = \frac{1}{2} \tag{4.9c}$$

When $l$ approaches to infinity ($x/l$ and $g/l$ is close to 0):

$$V_{OC} = \frac{\sigma g x \ln(l)}{\pi \varepsilon_0 l} \tag{4.10a}$$

$$Q_{SC} = \frac{\sigma w x}{\pi} \ln(l) \tag{4.10b}$$

$$\eta_{CT} = \frac{x}{\pi l} \ln(l) \tag{4.10c}$$

From Eqs. 4.9a and 4.10a, the behavior of $V_{OC}$ can be easily understood. When $l$ is small, $V_{OC}$ is proportional to $l$ and increases with the increase of $l$. But when $l$ is large enough, $V_{OC}$ satisfies Eq. 4.10a and starts to drop. Therefore, there is an optimum $l$ at which $V_{OC}$ reaches its maximum. When $x$ or $g$ increases, a larger range of $l$ can satisfy the condition for Eq. (4.9a, b, c), so that the $l$ providing the maximum $V_{OC}$ rises. For $Q_{SC}$, when $l$ is small (condition of Eq. (4.9a, b, c)), $Q_{SC}$ is proportional to $l$ and increases dramatically with the increase of $l$. When $l$ is sufficiently large (condition of Eq. (4.10a, b, c)), $Q_{SC}$ is proportional to ln($l$), resulting in a much lower slope of $Q_{SC} - l$ curve. From the $Q_{SC}$ behavior, $\eta_{CT}$ is close to 50 % when $l$ is small and decays to 0 when $l$ is sufficiently large. This analysis clearly explains the FEM calculation results shown in Fig. 4.4.

Besides these fundamental characteristics, the resitive load characteristics of the contact-mode SETENG are studied as well. Because $V_{OC}$ and $\eta_{CT}$ of SETENGs with infinite $l$ are 0, an optimum $l$ of 4.7 cm is observed to generate the maximum total transit power, as shown in Fig. 4.4d, e. In addition, the optimum resistance increases first then starts to decrease when $l$ increases. This is because when $l$ first increases, $V_{OC}$ increases at a faster speed than $I_{SC}$, Region II of the single electrode shifts to a higher resistance range and the optimum resistance increases [2]. When $l$ further increases, $V_{OC}$ starts to decrease and $I_{SC}$ continues increasing. Thus, the optimum resistance starts to decrease [2]. Moreover, the peak power density is monotonically lowered when $l$ increases, which is because larger area $l$ lowers $\eta_{CT}$. Thus, unlike the attached-electrode TENGs, SETENGs are not optimized for direct scaled-up processing.

### 4.2.4 Effect of Spacing Between Units for Scale up

Finally, the effect of connecting several SETENGs in parallel is discussed, which is another way to scale-up the output from SETENGs besides increasing the area size. For a three-SETENG system, a simulation was performed at OC condition through FEM calculation, in order to study the influence of the spacing between each

SETENG. The total $Q_{SC}$ generated by the three SETENGs with different spacing is shown in Fig. 4.4f. When the unit spacing is small, electric field from each SETENG will interfere, resulting in a decrease of $\eta_{CT}$ for each SETENG, which is similar to the effect of increasing $l$ on one SETENG. Only when their spacing is large enough, the mutual influence will be minimized, so that there will be little decrease on $\eta_{CT}$. From Fig. 4.4f, to reach the total charge transfer amount more than 270 % of that for each SETENG, a minimum unit spacing of 1.24 cm is observed, about 2.5 times the length of an individual SETENG. Therefore, a large enough spacing is critical for the scale-up of the output through parallel connection.

## 4.3 Basic Device Structures

### *4.3.1 Contact-Separation Structure*

The first design of the contact-separation structure based SETENG is shown in Fig. 4.5. A schematic diagram of the fabricated TENG was presented in Fig. 4.5a [1]. Due to the good strength and light weight, acrylic used as substrate material was covered by an Al foil, where the Al plays dual roles of a triboelectric surface and an electrode. A triboelectric polymer film, polyamide (PA) or polytetrafluoroethylene (PTFE) was attached on a Kapton film, which was manually bent to an arch shape and the two sides were fixed on a planar acrylic substrate using a Kapton tape. This device architecture helps to facilitate effective separation/contact between PA (or PTFE) and Al. The optical images of the fabricated TENG were shown in Fig. 4.5b, where only one electrode was fabricated as the output terminal (the other was ground). The titled optical image clearly shows an arch-shaped structure of TENG. Figure 4.5c shows the AFM images of the PA and PTFE films with the measured area of 20 μm × 20 μm. As compared with the PTFE film, the rougher surface of PA film can increase the effective surface area of device and give a better output performance, where the nanowire-structures have the average height of about 100 nm. The working mechanism of the TENG has been thoroughly discussed in Sect. 4.1–4.2. We found that the peak width of the measured output voltage is much larger than that of the output current. It is related to the different RC relaxation time $\tau = RC(t)$. Much larger loading resistance for the voltage measurement results in the larger peak width than that in the current measurement.

The output voltage and current of the single electrode-based TENG should satisfy linear superposition criterion in the basic circuit connections. Two TENGs were connected in parallel to examine the linear superposition of output voltage and current, as shown in Fig. 4.6a. The results show that both the total output voltage and current can be enhanced and approximately equal to the sum of the individual output voltage and current (Fig. 4.6b). The increase of the total output voltage in parallel is determined by the increase of the current to flow through the external load. These results not only rule out the possible artifacts, but also indicate that the

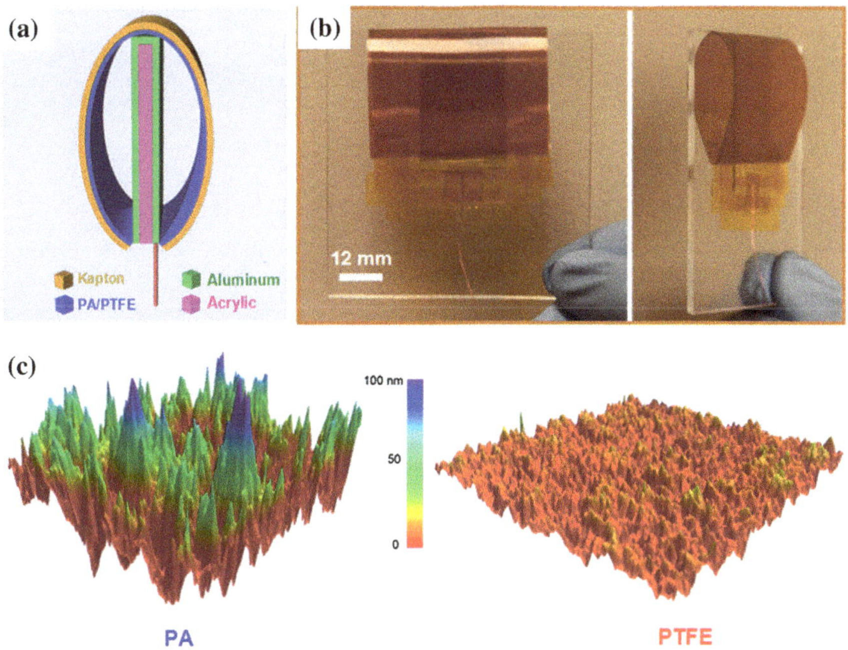


**Fig. 4.5** **a** Schematic diagram of a contact-separation single-electrode mode TENG. **b** A photograph of the as-fabricated device. **c** AFM images of the PA and PTFE films with the measured area of 20 μm × 20 μm. Reproduced with permission from Wiley [1]

output performance of the device can be greatly improved by connecting multiple TENGs in parallel. Moreover, to demonstrate the capability of the single electrode-based TENG as a direct power source, tens of green LEDs were utilized as operating load. To clearly demonstrate the AC output without rectification of the output, LEDs were divided into two groups, which were connected in parallel with the reversed polarity. When the TENG was compressed by the finger, the produced voltage/current peak was capable of simultaneously lighting up one group of the LEDs, as shown in Fig. 4.6c. When the compressive force was released, the produced energy can light up the other group of the LEDs (Fig. 4.6d). The results are consistent with the AC output of the TENG.

In addition, a single-friction-surface triboelectric nanogenerator (STENG) was also reported based on a contact separation structure single electrode mode [4]. This device incorporates only one single micro-structured PDMS or flat PET friction surface. Figure 4.7a shows a schematic diagram of the STENG. The device is fabricated on a 125 μm thick PET substrate. A PDMS film patterned with an array of micro-pyramids serves as the friction surface. A transparent ITO induction electrode is coated on the back side of the PET substrate. A 100 μm thick copper foil was employed as the reference electrode, and was grounded (as an alternative, a

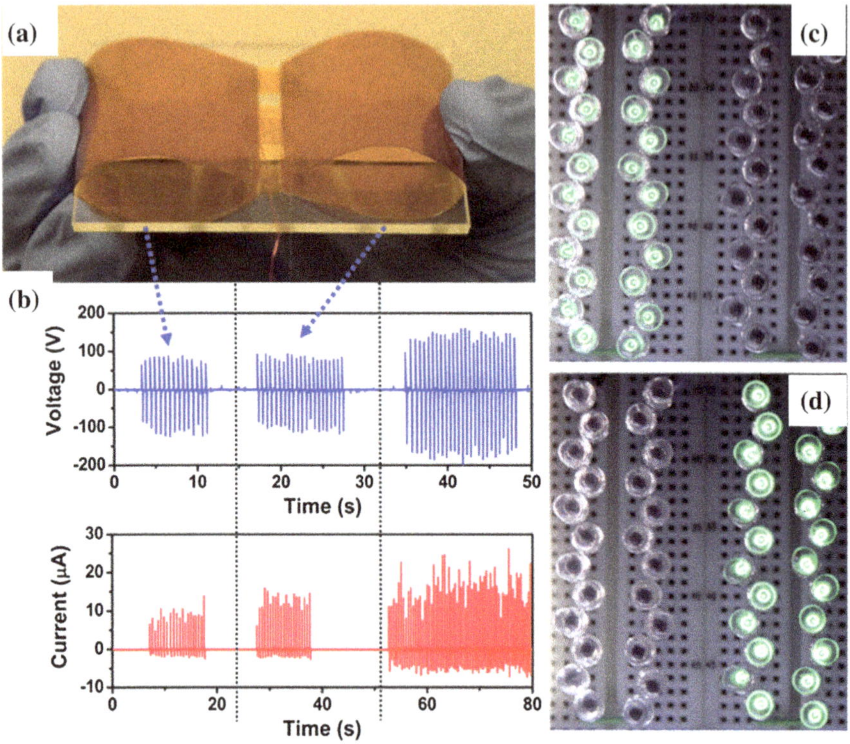


**Fig. 4.6** **a** Optical image of the fabricated two TENGs, which are connected in parallel. **b** The corresponding output voltage and current of the TENGs. **c**, **d** Optical images of the lighted *green LEDs*, where the alternant flash of *green LEDs* shows the AC output voltage/current of the TENG. Reproduced with permission from Wiley [1]

large-size reference electrode could be used as the equivalent ground). Thus, the location of reference electrode is irrelevant to the operation of the device. As required for different applications, the reference electrode can be placed beside (Fig. 4.7a-i) or beneath (Fig. 4.7a-ii) the induction electrode. For comparison, we also developed a simplified TENG that employs a PET substrate as the single friction surface. When a finger or other object is brought into contact with the PDMS or PET surface of the TENG, the surface of this object will serve as an active friction surface and compose a friction pair with the fixed friction surface. The effective friction area of the TENG is about 1.5 cm × 2.5 cm, which is similar in size to the contact area of a human finger. Figure 4.7b shows the SEM image of the micro-patterned PDMS film. The side length of the micro-pyramids is 10 μm, and the period is 20 μm. The photographs of the fabricated TENG in Fig. 4.7c illustrate the flexibility and transparency of the device.

Figure 4.8a shows the open-circuit voltage, short-circuit current and the amount of charge transferred in a half cycle of the TENG with the micro-patterned PDMS

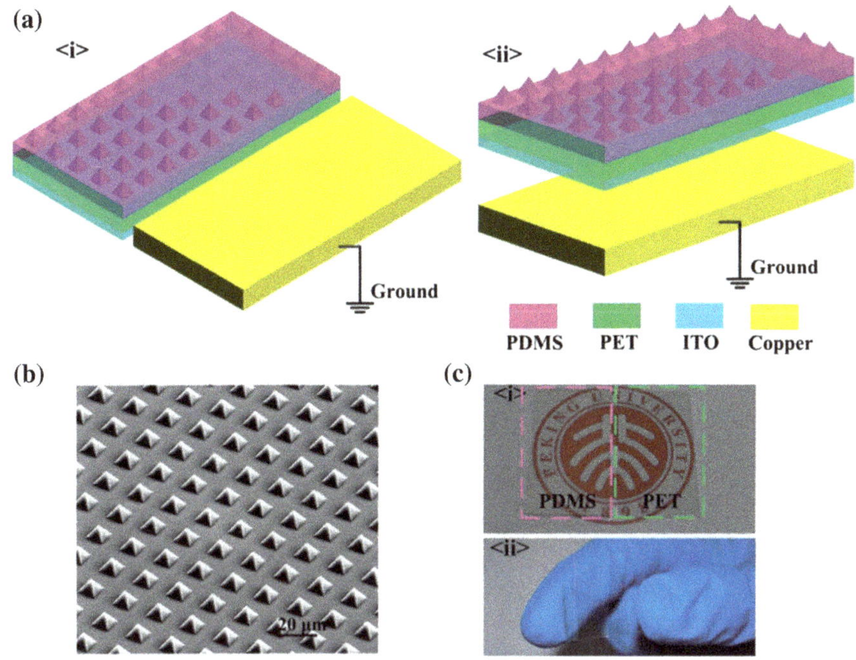


**Fig. 4.7** **a** Schematic of the single-friction-surface triboelectric generator (TENG) using a micro-structured PDMS friction surface with the grounded reference electrode placed (*i*) beside or (*ii*) beneath the induction electrode. **b** SEM image of the micro-patterned PDMS film. **c** Photographs of the TENGs with a PDMS surface and PET surface showing their high transparency and flexibility. Reproduced with permission from the Royal Society of Chemistry [4]

surface when tapped with a bare finger (Fig. 4.8a) and tapped with a finger covered in a PE glove (Fig. 4.8b). It is clear that considerable contact electrification occurred under both two conditions tested. The PDMS surface shows a tendency to attract electrons, while the PET surface shows a tendency to donate electrons. With transferred charge of about 4–5 nC in a half cycle, the TENG with the micro-patterned PDMS surface shows similar performance in charging regardless of whether it is tapped by a bare finger or by the PE glove. Though PDMS and PE exhibit similar abilities in attracting electrons, the micro-structure on the PDMS surface significantly enhances the contact electrification. When tapped with a bare finger, the TENG with the micro-patterned PDMS surface achieved an open-circuit voltage of over 130 V with a short-circuit current density of about 1 $\mu A\ cm^{-2}$. This TENG can be fabricated on transparent and flexible materials using a very simple process and is low cost. The single-friction-surface device achieves a significantly higher output power density than a previously reported inseparable-structure transparent TENG. Since this TENG is flexible and totally transparent, it could be a promising power source for flexible electronic devices and portable electronic products with displays such as smartphones and tablets.

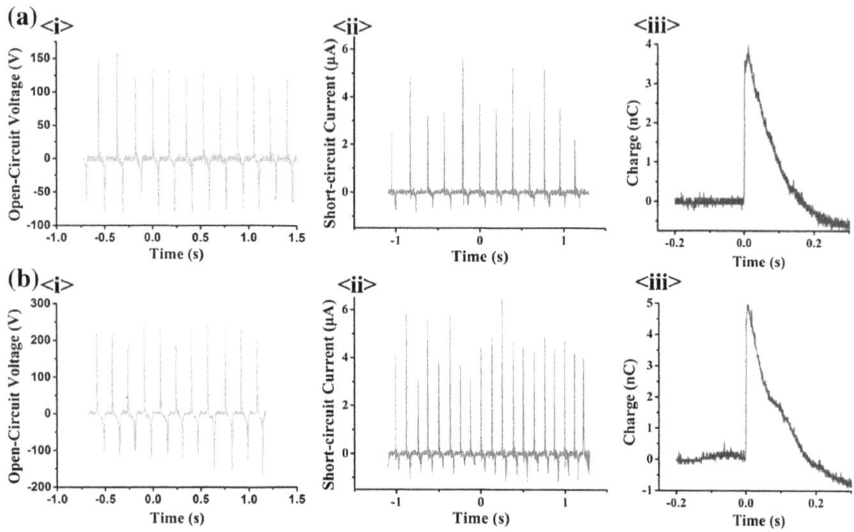


**Fig. 4.8** Characterization of the TENG. The open-circuit voltage, short-circuit current and charge transferred in a half cycle of the TENG with the micro-patterned PDMS surface **a** when tapped with a bare finger and **b** when tapped with a finger covered in a PE glove and the same measurements for the TENG with the flat PET surface. Reproduced with permission from the Royal Society of Chemistry [4]

### 4.3.2 Lateral Sliding Structure

Figure 4.9a illustrates a schematic diagram of a lateral sliding structure based single electrode mode TENG, which consists of a triboelectric PTFE patch and an Al plate, where the Al plays dual roles of a triboelectric surface and an electrode [5]. The PTFE patch and the Al plate are kept in parallel to each other, where the inner surfaces are in intimate contact. The two plates slide against each other with the contact area changing periodically by the mechanical motion along the short-edge of the plates. Usually, the output performance of the TENG can be enhanced by increasing the surface roughness and the effective surface area of the triboelectric materials to induce a larger triboelectric charge density. To increase the triboelectric charge density on the surface of the PTFE, it can be dry-etched using inductively coupled plasma (ICP) to create the nanoparticle structures. Figure 4.9b shows a scanning electron microscopy (SEM) image of the etched PTFE surface, which is uniformly covered with the nanoparticle-structures with an average diameter of about 200 nm. The etched surface of PTFE was further characterized by the atomic force microscope (AFM), showing the same nanoparticle structures on the surface, as illustrated in Fig. 4.9c.

Figure 4.10a shows the output performance of the single-electrode based TENG, where the $V_{OC}$ can reach 1100 V with a peak $J_{SC}$ of 6 mA/m$^2$. To confirm that the obtained output signals in Fig. 4.10b were generated by the TENG, the produced

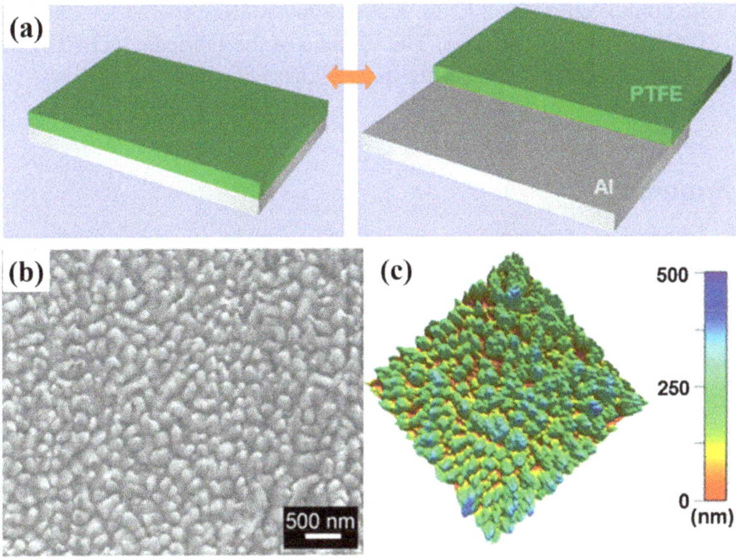


**Fig. 4.9** **a** Schematic diagram of the lateral sliding single-electrode mode TENG. **b** SEM image of the PTFE surface with etched nanoparticle structure. **c** AFM image of the etched PTFE surface. Reproduced with permission from American Chemical Society [5]

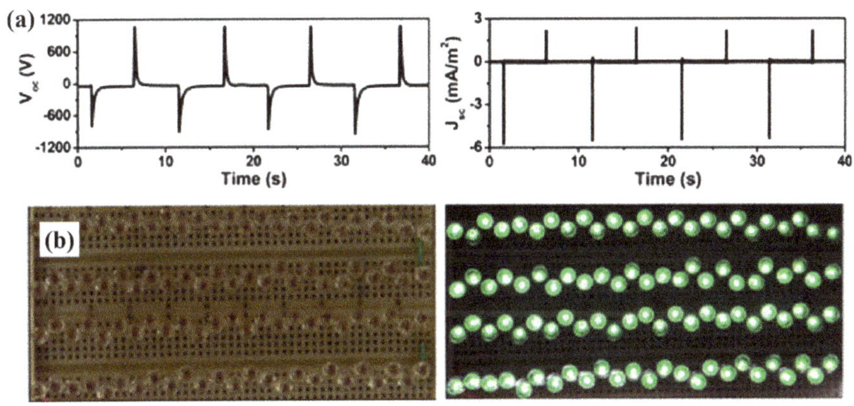


**Fig. 4.10** Electric output performance of the single-electrode based sliding triboelectric nanogenerator. **a** Open-circuit voltage ($V_{OC}$) and the short-circuit current density ($J_{SC}$) of the TENG. **b** Snapshot of 100 commercial green LEDs directly driven by the TENG under a sliding motion. Reproduced with permission from American Chemical Society [5]

energy was used to directly drive one hundred of green LEDs under the fast sliding motion of the PTFE on the Al plate, as shown in Fig. 4.10c. In order to clearly demonstrate the AC output, these LEDs were divided into two groups in series, which were connected to the TENG. Under a low sliding frequency, the two LED groups were alternately lighted up. The lifetime of the fabricated TENG can be up to several tens of years by increasing the thickness of PTFE film.

## 4.4 Summary

In this chapter, a systematic introduction of the basic working principle, theoretical analysis, and several representative prototypes of the single-electrode mode TENG was presented. This mode is unique in a way that it does not need to attach a wire to a moving object in order to harvest energy from mechanical motions. Although the induced electrons transfer across the electrode is not the most effective one due to the electrostatic screening effect, one of the triboelectric layers can move freely without any restriction, allowing it for a range of applications. The single-electrode mode has been applied to harvest energy from air flow [6], rotating tire [7], rain drop [8], turning the book pages [9]. It was also applied as self-powered displacement vector sensor [5], visualized touch sensor [4], active tactile sensor [10, 11], self-powered trajectory, velocity sensor [12, 13], angle measurement sensor [14], acceleration sensors [15], biosensor [16], water/ethanol sensor [17], pressure sensor [18], sensor for healthcare monitoring [19], body motion sensors [20, 21], self-powered identification systems [22], and self-powered distress signal emitter [23]. For some of these applications, we will discuss them in Part 3 of this book.

## References


1. Y. Yang, Y.S. Zhou, H. Zhang, Y. Liu, S. Lee, Z.L. Wang, A single-electrode based triboelectric nanogenerator as self-powered tracking system. Adv. Mater. **25**(45), 6594–6601 (2013)
2. S.M. Niu, Y. Liu, S.H. Wang, L. Lin, Y.S. Zhou, Y.F. Hu, Z.L. Wang, Theoretical investigation and structural optimization of single-electrode triboelectric nanogenerators. Adv. Funct. Mater. **24**(22), 3332–3340 (2014)
3. Y. Li, Y.H. Li, Q.X. Li, Y.Y. Zi, Computation of electrostatic forces with edge effects for non-parallel comb-actuators. J. Tsinghua Univ. (Sci. & Tech.) **43**(8), 1024–1026 (2003)
4. B. Meng, W. Tang, Z.-H. Too, X. Zhang, M. Han, W. Liu, H. Zhang, A transparent single-friction-surface triboelectric generator and self-powered touch sensor. Energy Environ. Sci. **6**, 3235–3240 (2013)
5. Y. Yang, H. Zhang, J. Chen, Q. Jing, Y.S. Zhou, X. Wen, Z.L. Wang, Single-electrode-based sliding triboelectric nanogenerator for self-powered displacement vector sensor system. ACS Nano **7**(8), 7342–7351 (2013)
6. Y. Yang, G. Zhu, H. Zhang, J. Chen, X. Zhong, Z.-H. Lin, Y. Su, P. Bai, X. Wen, Z.L. Wang, Triboelectric nanogenerator for harvesting wind energy and as self-powered wind vector sensor system. ACS Nano **7**(10), 9461–9468 (2013)
7. H. Zhang, Y. Yang, X. Zhong, Y. Su, Y. Zhou, C. Hu, Z.L. Wang, Single-electrode-based rotating triboelectric nanogenerator for harvesting energy from tires. ACS Nano **8**(1), 680–689 (2014)

8. Q. Liang, X. Yan, Y. Gu, K. Zhang, M. Liang, S. Lu, X. Zheng, Y. Zhang, Highly transparent triboelectric nanogenerator for harvesting water-related energy reinforced by antireflection coating. Sci. Rep. **5**, 9080 (2015)
9. Q. Zhong, J. Zhong, B. Hu, Q. Hu, J. Zhou, Z.L. Wang, A paper-based nanogenerator as a power source and active sensor. Energy Environ. Sci. **6**, 1779–1784 (2013)
10. Y. Yang, H. Zhang, Z.-H. Lin, Y.S. Zhou, Q. Jing, Y. Su, J. Yang, J. Chen, C. Hu, Z.L. Wang, Human skin based triboelectric nanogenerators for harvesting biomechanical energy and as self-powered active tactile sensor system. ACS Nano **7**(10), 9213–9222 (2013)
11. G. Zhu, W. Yang, T. Zhang, Q. Jing, J. Chen, Y.S. Zhou, P. Bai, Z.L. Wang, Self-powered, ultrasensitive, flexible tactile sensors based on contact electrification. Nano Lett. **14**(6), 3208–3213 (2014)
12. F. Yi, L. Lin, S. Niu, J. Yang, W. Wu, S. Wang, Q. Liao, Y. Zhang, Z.L. Wang, Self-powered trajectory, velocity, and acceleration tracking of a moving object/body using a triboelectric sensor. Adv. Funct. Mater. **24**(47), 7488–7494 (2014)
13. Y. Su, G. Zhu, W. Yang, J. Yang, J. Chen, Q. Jing, Z. Wu, Y. Jiang, Z.L. Wang, Triboelectric sensor for self-powered tracking of object motion inside tubing. ACS Nano **8**(4), 3843–3850 (2014)
14. Y. Wu, Q. Jing, J. Chen, P. Bai, J. Bai, G. Zhu, Y. Su, Z.L. Wang, A self-powered angle measurement sensor based on triboelectric nanogenerator. Adv. Funct. Mater. **25**(14), 2166–2174 (2015)
15. H. Zhang, Y. Yang, Y. Su, J. Chen, K. Adams, S. Lee, C. Hu, Z.L. Wang, Triboelectric nanogenerator for harvesting vibration energy in full space and as self-powered acceleration sensor. Adv. Funct. Mater. **24**(10), 1401–1407 (2014)
16. H. Zhang, Y. Yang, T.-C. Hou, Y. Su, C. Hu, Z.L. Wang, Triboelectric nanogenerator built inside clothes for self-powered glucose biosensors. Nano Energy **2**(5), 1019–1024 (2013)
17. H. Zhang, Y. Yang, Y. Su, J. Chen, C. Hu, Z. Wu, Y. Liu, C.P. Wong, Y. Bando, Z.L. Wang, Triboelectric nanogenerator as self-powered active sensors for detecting liquid/gaseous water/ethanol. Nano Energy **2**(5), 693–701 (2013)
18. P.-K. Yang, Z.-H. Lin, K.C. Pradel, L. Lin, X. Li, X. Wen, J.-H. He, Z.L. Wang, Paper-based origami triboelectric nanogenerators and self-powered pressure sensors. ACS Nano **9**(1), 901–907 (2015)
19. P. Bai, G. Zhu, Q. Jing, J. Yang, J. Chen, Y. Su, J. Ma, G. Zhang, Z.L. Wang, Membrane based self-powered triboelectric sensors for pressure change detection and its uses in security surveillance and healthcare monitoring. Adv. Funct. Mater. **24**(37), 5807–5813 (2014)
20. F. Yi, L. Lin, S. Niu, P.-K. Yang, Z. Wang, J. Chen, Y. Zhou, Y. Zi, J. Wang, Q. Liao, Y. Zhang, Z.L. Wang, Stretchable rubber based triboelectric nanogenerator and its application as self-powered body motion sensors. Adv. Funct. Mater. **25**(24), 3688–3696 (2015)
21. W. Yang, J. Chen, X. Wen, Q. Jing, J. Yang, Y. Su, G. Zhu, W. Wu, Z.L. Wang, Triboelectrification based motion sensor for human-machine interfacing. ACS Appl. Mater. Interfaces **6**(10), 7479–7484 (2014)
22. P. Bai, G. Zhu, Q. Jing, Y. Wu, J. Yang, J. Chen, J. Ma, G. Zhang, Z.L. Wang, Transparent and flexible barcode based on sliding electrification for self-powered identification systems. Nano Energy **12**, 278–286 (2015)
23. Y. Su, X. Wen, G. Zhu, J. Yang, J. Chen, P. Bai, Z. Wu, Y. Jiang, Z.L. Wang, Hybrid triboelectric nanogenerator for harvesting water wave energy and as a self-powered distress signal emitter. Nano Energy **9**, 186–195 (2014)

# Chapter 5
# Triboelectric Nanogenerator: Freestanding Triboelectric-Layer Mode

## 5.1 Sliding Freestanding Triboelectric-Layer TENG

### *5.1.1 Basic Principle*

The first type of the F-TENG is based on the triboelectric effect between a freestanding dielectric layer and two metal films that serve not only as the counter triboelectric material, but also as two electrodes (Fig. 5.1a) [1–6]. In order to maximize the electrification between the two layers, we purposely chose materials with a large difference in triboelectric polarity—Al as the conducting films and fluorinated ethylene propylene (FEP) as the freestanding triboelectric layer [7]. The FEP film was tailored into a 7 cm × 5 cm rectangle, and then adhered on the supporting substrate (made of acrylic) of the same size. In order to further improve the triboelectric charge (tribo-charge) density, the downside surface of the FEP film was treated by inductive coupling plasma (ICP) to create nanorod structures [8]. On the other side, two rectangular Al electrodes of this exact size were deposited on another acrylic substrate perpendicular to the sliding direction. The length (*L*) of the electrode along the sliding direction is 5 cm, and the in-plane distance (*d*) between the two electrodes is purposely controlled as one of the most important parameters in the device structure (Fig. 5.1b<i>). Driven by a tangential force, the freestanding FEP layer can slide back and forth between the two electrodes (Fig. 5.1), through which alternating-current is provided to the load connected between the two electrodes.

When the FEP slides against the Al electrodes with surfaces in contact, the triboelectric effect will render FEP surface with negative charges, and Al electrodes with positive charges. There are two sceneries. In the first case, if the dielectric FEP and the electrodes are uncharged at the first place, all of the static charges are then to be generated by the triboelectrification after their physical contact. The negative charges on the FEP surface should have an equal amount with the positive charges in the electrodes. When the FEP is at the position fully overlapped with the first



Z.L. Wang et al., *Triboelectric Nanogenerators*,
Green Energy and Technology, DOI 10.1007/978-3-319-40039-6_5

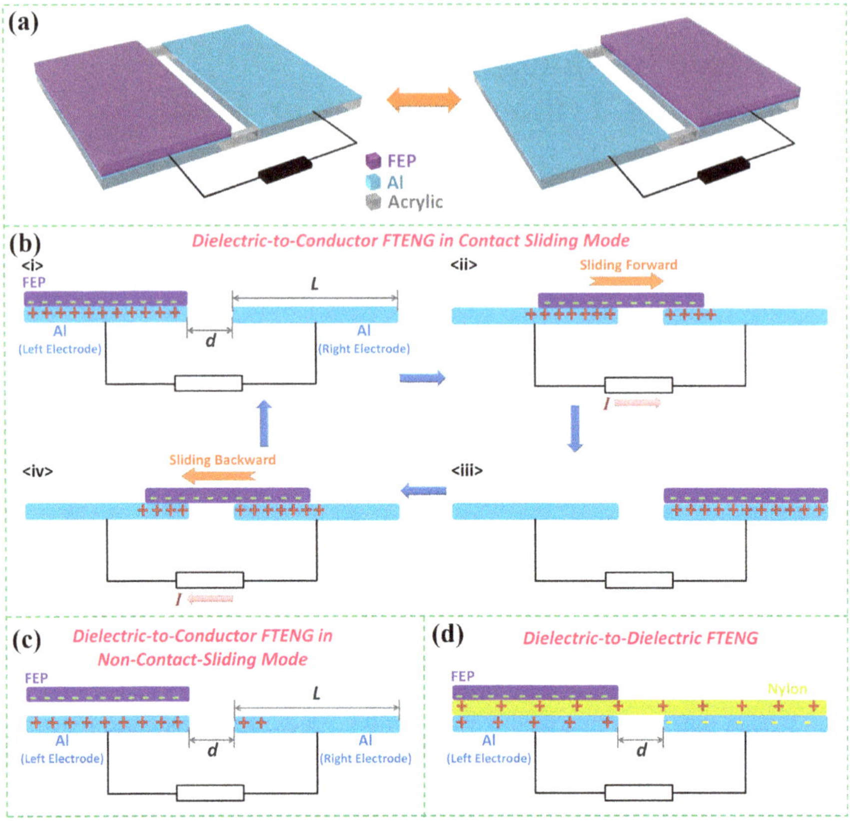


**Fig. 5.1** Device structure, basic operations and working principles of the freestanding-triboelectric-layer based nanogenerator (FTENG). **a** Typical device structure of a conductor-to-dielectric FTENG. **b** Schematic working principle of a conductor-to-dielectric FTENG in contact sliding mode. **c** Schematic diagram of a conductor-to-dielectric FTENG in non-contact sliding mode. **d** Schematic diagram of a dielectric-to-dielectric FTENG

electrode (e.g., the left electrode as shown in Fig. 5.1b<i>), all of the positive charges in the loop will be attracted to the upper surface of the left-hand electrode (LE). Then, when the FEP layer slides towards the right-hand electrode (RE) (Fig. 5.1b<ii>), the positive charges in the loop will flow from the LE to the RE via the load to screen the local field of the non-mobile negative charges on the dielectric, which is the first half cycle of electricity generation. When the FEP reaches the overlapping position of the RE (Fig. 5.1b<iii>), all of the positive charges will then be driven to the RE. Subsequently, a backward sliding of the FEP layer from the RE to the LE should drive the flow of the positive charges in the same direction, forming a reverse current in the load (Fig. 5.1b<iv>). This is the second half of electricity generation process.

In the second case, if the dielectric FEP is already charged due to prior-triboelectrification before approaching the electrodes, which may have or may not have electrostatic charges, electricity can be generated by swing the FEP between the two electrodes even without direct contact (Fig. 5.1c). This is feasible because triboelectric charges can be preserved on insulator surfaces for hours and even days. As long as the vertical separation is much smaller than the relative displacement of the two electrodes (which could be on the order of several centimeters or even larger), the sliding of the non-mobile charges on the FEP film still can induce a significant portion of charges to flow. Thus, unlike existing TENGs, this new mode of F-TENG should be able to operate in non-contact sliding mode.

Besides, the other basic design of F-TENGs is based on the triboelectrification between two different dielectric films, e.g. FEP as the freestanding layer and polyamide (Nylon) as the film fully covering the two stationary electrodes (Fig. 5.1d). When the freestanding FEP film is driven to slide on top of the Nylon film, negative charges will be injected from the Nylon surface to the FEP surface. As for the positive static charges on the Nylon surface, since they are stationary all the time, their induced potentials on the two electrodes remain as constant, which cannot provide any driving force to the charge flow in the external load. Thus, all of the driving force of the current still comes from the sliding of the negatively-charged FEP film, which is the same as the demonstrated dielectric-conductor F-TENG structure. Therefore, this dielectric-dielectric structure should generate electricity in a similar manner.

The electrical outputs were measured from 5 devices with different electrode distances (0.1, 0.3, 1, 3, and 5 cm). As shown in Fig. 5.2a, the 5 sets of $\Delta\sigma_{SC}$ all stay at the level around 65 μC/m$^2$, which is very similar to the theoretical expectation that $d$ doesn't have an influence on the charge transfer. There is only a minor decreasing trend when $d$ gets larger, probably because of a larger tribo-charge density ($\sigma_{Tribo}$) on FEP surface resulted from a more intensive friction when two electrodes get closer. However, the measured $V_{OC}$ has a very different trend from the simulation results (Fig. 5.2b). Rather than showing a significant increase along with the elevation of $d$, the magnitudes of $V_{OC}$ from these 5 devices are all around the level of 15 kV, only presenting a bit enhancement from the increase of $d$. Actually, the measured results are much smaller than the simulated values of $V_{OC}$ by almost an order of magnitude. The possible cause of this difference is that the inner resistance of the electrometer is not infinitely large, so that real $V_{OC}$ cannot be accurately recorded due to the internal leakage under such a high voltage. Thus, the measured results could not reflect the influence of $d$. As for the short-circuit current densities ($J_{SC}$) provided by these 5 F-TENGs, since the total amounts of transferred charge densities are about the same for one sliding motion, $J_{SC}$ should be mostly determined by the time span of one sliding. Therefore, the increase of $d$, which will lead to a longer sliding time, should be unfavorable for the magnitude of $J_{SC}$. This agrees with the measurement results of $J_{SC}$ (Fig. 5.2c): the device with the smallest $d$ of 0.1 cm generates the highest $J_{SC}$ of $\sim$3.5 mA/m$^2$. Thus, from this influence of $d$, a smaller electrode distance is preferred in structural design for a high output power.

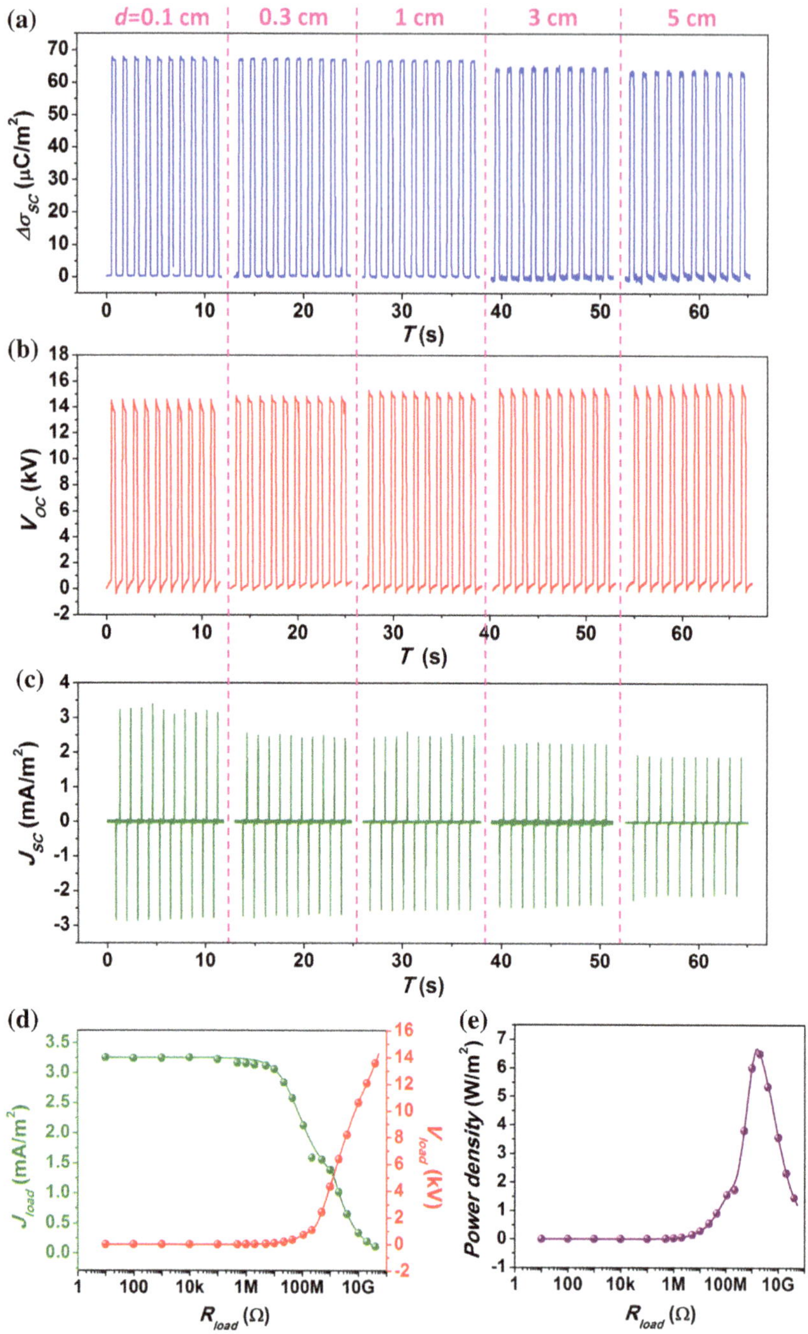
(a)
d=0.1 cm
0.3 cm
1 cm
3 cm
5 cm
Δσ_SC (μC/m²)
T (s)
(b)
V_OC (kV)
T (s)
(c)
J_SC (mA/m²)
T (s)
(d)
J_load (mA/m²)
V_load (kV)
R_load (Ω)
(e)
Power density (W/m²)
R_load (Ω)

◀ **Fig. 5.2** Measurement results of FTENGs' electrical outputs. **a** The transferred charge densities ($\Delta\sigma_{SC}$) of 5 FTENGs with different electrode distances (0.1, 0.3, 1, 3, 5 cm). **b** The open-circuit voltage ($V_{OC}$) of the above 5 FTENGs. **c** The short-circuit current density ($J_{SC}$) of the above 5 FTENGs. **d**, **e** The dependence of the electrical outputs on the load resistance, obtained from the FTENG with the electrode distance of 0.1 cm: **d** the voltage and the current density; **e** the power density

For a comprehensively characterization of the demonstrated F-TENG as a power source, we measured the actual voltages and current densities of an F-TENG with $d = 0.02L$ under a series of different load resistance. As shown in Fig. 5.2d, at the range below 10 MΩ, both the voltage and the current density have little variation from the short-circuit condition. Then, a further increase of the resistance beyond 10 MΩ will lead to the increase of the voltage and the decrease of the current density. At the resistance of ~1 GΩ, the F-TENG provides the maximum power density to a load, which is ~6.7 W/m$^2$ (Fig. 5.2e).

In order to design an efficient TENG, it is desirable to maintain an intimate contact between the freestanding triboelectric layer and the electrode films during sliding. However, in real applications of such sliding-based TENGs, it is sometimes difficult to keep a good in-plane contact during sliding, due to possible irregularities of input mechanical motions. Thus, a good tolerance of a vertical separation between the two layers in electricity generation is critical for expanding TENGs' applications. As we have discussed in Fig. 5.1c, the F-TENG demonstrated here should be able to operate in non-contact sliding mode. We theoretically studied the influence of the vertical gap distance ($H$) between the two facing surfaces. As shown in Fig. 5.3a, when $H$ is increased from 0 to 10 mm, the calculated transferred charge densities ($\Delta\sigma_{SC}$) in one sliding motion still preserve a considerable portion of the maximum value ($\Delta\sigma_o$): 61 % for the structure with $d = 0.1$ cm, and 77 % for $d = 5$ cm. This set of calculation results gives us a good prediction that the presence of a vertical gap during sliding will not have a severe influence on the effectiveness of mechanical energy harvesting. In order to verify this prediction experimentally, the transferred charge densities were measured from those two devices, under a series of different $H$ between 0 and 10 mm. From the data in Fig. 5.3b, although the measured outputs show a faster decay than the simulation results, the $\Delta\sigma_{SC}$ from the case of $H = 10$ mm still retains more than 20 % of the maximum value (24 % for $d = 0.1$ cm, and 29 % for $d = 5$ cm). This higher decay rate of the measured output possibly comes from the reverse electrostatic-induction effect of the upper side of the acrylic substrate (with a thickness of 6.3 mm), which could carry some positive static charges due to its triboelectric polarity. This tolerance of non-contact sliding is much better than traditional sliding-mode TENG in which one electrode is attached with the sliding triboelectric layer. As shown in Fig. 5.3c, when $H$ increases to merely 2 mm, the $\Delta\sigma_{SC}$ from the traditional sliding TENG dramatically decays to almost 0. Therefore, a good tolerance of vertical separation is a unique feature of F-TENGs.

With this capability of generating electricity by non-contact sliding, if the charges on the dielectric surface can preserve for a sufficiently long time, the

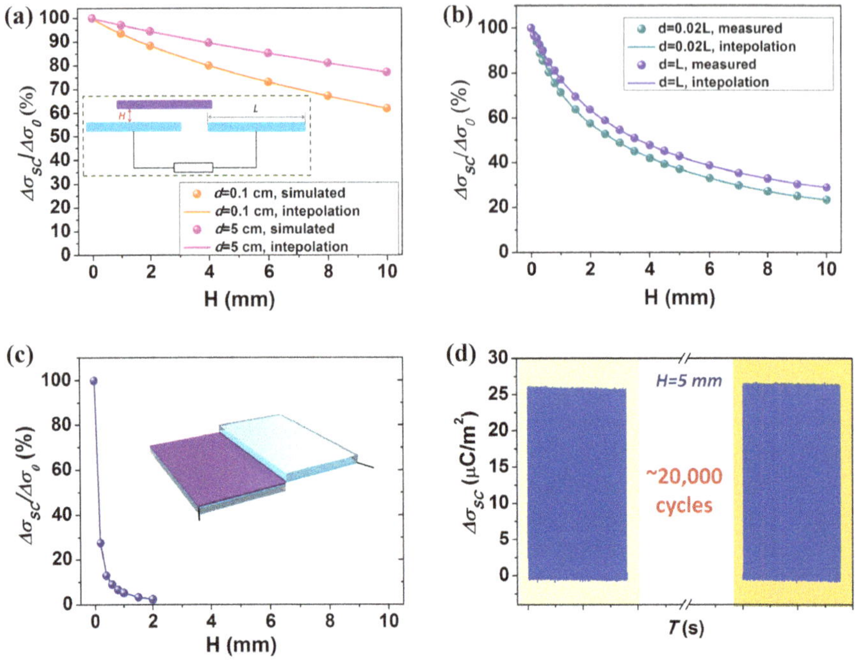


**Fig. 5.3** FTENGs' tolerance of vertical separation between the sliding triboelectric layer and the electrode surface. **a** The FEM simulated influence of the vertical separation ($H$) on the $\Delta\sigma_{SC}$ of two FTENG structures with $d$ of 0.1 and 5 cm respectively. The dots are simulated $\Delta\sigma_{SC}$ at certain $H$, and the lines are the interpolation results. The *inset* is a 2-dimensional diagram showing the FTENG with a vertical gap. **b** The experimentally measured influence of the vertical separation ($H$) on the $\Delta\sigma_{SC}$ of the above two FTENGs. The dots are measurement results, and the lines are the interpolation results. **c** The influence of vertical separation on the traditional sliding TENG, in which one electrode is attached with the sliding triboelectric layer. The *inset* shows the structure of the tested sliding TENG of the same size and materials. In all the above three figures (**a–c**), $\Delta\sigma_0$ is the maximum transferred charge density of each FTENG in contact-sliding mode. **d** The output stability of the FTENG working in non-contact sliding mode, over ~20,000 continuous cycles

F-TENG will be able to continuously operate in this non-contact condition, under which there is no friction so that the energy conversion efficiency could achieve theoretically expected 100 % although the actual output power is lowered. In the F-TENG structure demonstrated in this paper, the material used for the freestanding triboelectric layer-FEP-is a type of electret (the category of materials that can quasi-permanently retain the static charges on them) [9–11]. Thus, once the tribo-charges are generated on FEP by the contact sliding in the first few cycles, they can stay on the FEP surface to drive the flow of electricity without any further contact or friction between the two surfaces needed. In order to test the stability of the F-TENG in the non-contact mode, we continuously ran one F-TENG for ~20,000 cycles with a vertical distance of 5 mm. As shown in Fig. 5.3d, the generated short-circuit charge density of ~25 $\mu Cm^{-2}$ didn't have an obvious decay after these ~20,000 cycles. The capability of working in non-contact mode provides a number of advantages for mechanical energy harvesting: extremely-high energy conversion efficiency (theoretically 100 %) due to a low mechanical energy input that doesn't need to overcome the sliding friction; the minimized surface wearing from sliding friction; expanded types of applicable mechanical energies.

### *5.1.2 Fundamental Theory*

Based on the freestanding material, SF-TENGs have two subcategories, dielectric SF-TENGs whose freestanding materials only contain dielectrics and metal SF-TENGs whose freestanding materials contains metal [12]. The basic model of these two SF-TENGs is shown in Fig. 5.4a and b. In the SF-TENGs, Metal 1 and 2 are placed in the same plane with a gap $g$ to form two electrodes. A freestanding layer with the same size of Metal 1 stands on the top of the metal electrodes and the

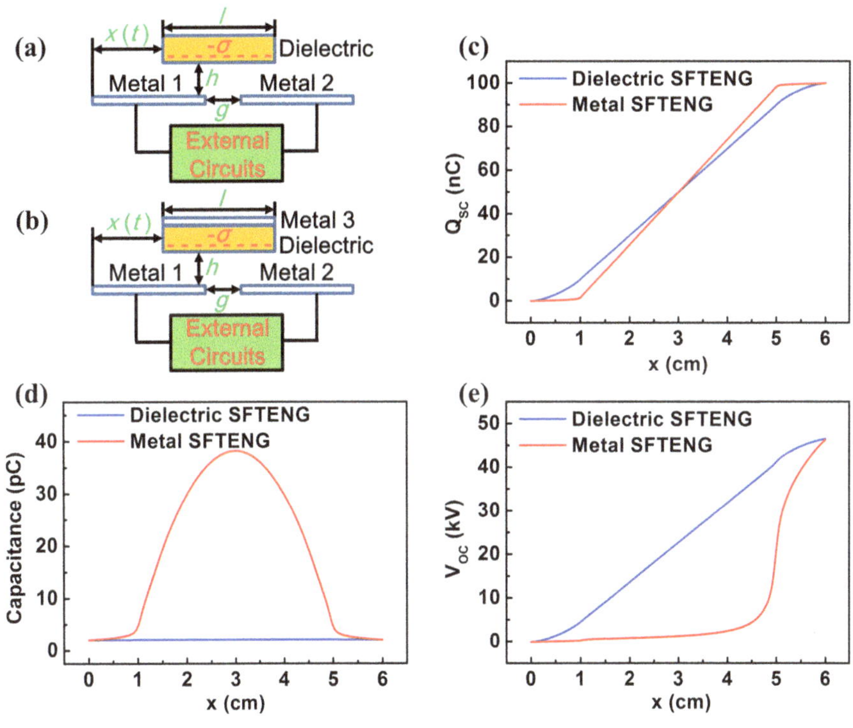


**Fig. 5.4** Models and basic output characteristics of SFTENGs. **a** FEM model of a typical dielectric SFTENG. **b** FEM model of a typical metal SFTENG. **c**–**e** Influence of freestanding layer material on the basic output characteristics: **c** short-circuit transferred charges under MACRS, **d** total capacitance between the two electrodes, and **e** open-circuit voltage under MACRS

distance between the bottom surface of freestanding layer and Metal 1 is defined as the freestanding height $h$. The width of this structure is defined as $w$. For dielectric SF-TENGs shown in Fig. 5.4a, the freestanding layer only contains a layer of dielectrics. And for metal SF-TENGs, a thin layer of metal is deposited at the backside of the dielectric freestanding layer. These two structures look very similar, but unlike CF-TENGs, the freestanding material will strongly influence the output characteristics of the SF-TENGs.

We can still use the node concept to analyze their working principle. We first analyze dielectric SF-TENGs. In contrast to CF-TENGs (discussed in Sect. 5.2.2), since the electrical potential of the dielectric bottom surface is not a constant, this surface could not be seen as a single node. The electric potential superposition principle is utilized to solve this problem. First, we assume that only a small region of $\mathrm{d}k$ in the bottom dielectric surface (the distance of this region to the left edge of the bottom dielectric surface is $k$) contains the tribo-charges with a density of $-\sigma$, and correspondingly the total charges on metal 1 and 2 are $\sigma w \mathrm{d}k$. Utilizing the similar derivation method we utilized above, the total charges on metal 1 and 2 ($\mathrm{d}Q_1$ and $\mathrm{d}Q_2$) under SC conditions because of these tribo-charges can be given by the following equations, where $C_i(k)$ states for the capacitance between this small surface $\sigma w \mathrm{d}k$ and metal i:

$$\mathrm{d}Q_1 = \frac{\sigma w \mathrm{d}k}{1 + \frac{C_2(k)}{C_1(k)}} \tag{5.1a}$$

$$\mathrm{d}Q_2 = \frac{\sigma w \mathrm{d}k}{1 + \frac{C_1(k)}{C_2(k)}} \tag{5.1b}$$

Because of the superposition principle of the electrostatic field, the total charges on metal 1 and metal 2 considering the whole charges on the dielectric surface is the integration of each small tribo-charged region, which can be shown as:

$$Q_1 = \sigma w \int_0^l \frac{\mathrm{d}k}{1 + \frac{C_2(k)}{C_1(k)}} \tag{5.2a}$$

$$Q_2 = \sigma w \int_0^l \frac{\mathrm{d}k}{1 + \frac{C_1(k)}{C_2(k)}} \tag{5.2b}$$

Therefore, $Q_{\mathrm{SC,final}}$ can be shown as:

$$Q_{\mathrm{SC,final}} = \int_0^l \frac{\sigma w \mathrm{d}k}{1 + \left(\frac{C_2(k)}{C_1(k)}\right)_{x=g+l}} - \int_0^l \frac{\sigma w \mathrm{d}k}{1 + \left(\frac{C_2(k)}{C_1(k)}\right)_{x=0}} \tag{5.3}$$

When $x = 0$, the tribo-charged dielectric surface is much closer to metal 1 than to metal 2, so the ratio $C_2(k)/C_1(k)$ is close to 0 for all the $k$ values. Therefore, $Q_1$ will be approximately $\sigma wl$ and $Q_2$ will be approximately 0 from Eq. 5.2. On the contrast, the ratio $C_2(k)/C_1(k)$ is close to infinity for all the $k$ values when $x = g + l$. Therefore, $Q_1$ will be approximately 0 and $Q_2$ will be approximately $\sigma wl$. Thus, $Q_{SC,final}$ can reach $\sigma wl$ and the charge-transfer efficiency ($\eta_{CT}$) can reach 100 %. Same as CF-TENGs, the change of the ratio of these two capacitances with the change of $x$ serves as the driving force for electrons to transfer between the two electrodes.

For metal SF-TENGs, the basic electrostatic induction principle is the same as the dielectric SF-TENGs. Therefore, as shown in Fig. 5.4c, the numerically calculated trend of $Q_{SC}$ profile (under MACRS) is quite similar to that of dielectric SF-TENGs. (Detailed numerical calculation parameter is listed in Table 5.1). However, unlike CF-TENGs, their characteristics of capacitance and $V_{OC}$ are completely different from each other, as shown in the numerical results presented in Fig. 5.4d and e. The capacitance profile of dielectric SF-TENGs is nearly constant when the position of the freestanding layer changes. However, for metal SF-TENGs, the capacitance when the freestanding layer is in the middle is several orders larger than that when the freestanding layer is at the two ends. This capacitance difference leads to the difference of their $V_{OC}$ profile. Since the capacitance of metal-SF-TENGs in the middle region is much larger than that of dielectric-SF-TENGs, so from Eq. 2.3, the open-circuit voltage of metal-SF-TENGs in this region will be much smaller than that of dielectric-SF-TENGs. And when $x$ is close to $g + l$, since the capacitance of the metal-SF-TENG quickly dropped to the value close to the dielectric-SF-TENG, the open-circuit voltage of the metal-SF-TENG also quickly increase to a value close to the dielectric-SF-TENG. Therefore, although the final open-circuit voltage values of the two SF-TENGs are almost the same, the metal-SF-TENG curve is much more non-linear than the dielectric-SF-TENG curve and the maximum slope of the metal-SF-TENG curve is much larger than that of the dielectric-SF-TENG curve.

**Table 5.1** Parameters utilized in the calculation of output characteristics for both dielectric and metal SF-TENGs

| Structure component | Parameter utilized |
|---|---|
| Dielectric | $d = 500$ μm, $\varepsilon_r = 2$ |
| Width of the structure $w$ | 0.1 m |
| Length of electrode $l$ | 0.05 m |
| Thickness of metal electrode and freestanding layer | $d_m = 10$ μm |
| Tribo-charge surface density $\sigma$ | 20 μCm$^{-2}$ |
| Electrode gap $g$ | 1 cm |
| Freestanding height $h$ | 0 mm |
| Motion average velocity $v$ | 1 ms$^{-1}$ |

These differences in capacitance and $V_{OC}$ profiles are mainly from the charge redistribution inside the metal freestanding layer. In dielectric-SF-TENGs, the charges are not free to move. Since the relative dielectric constant of the commonly utilized dielectrics is not much larger than the relative dielectric constant of the air (always in the range of 2–4), the change of the capacitances due to the change of the freestanding-layer position is little. However, the charge on the surface of metal freestanding layer can redistribute to keep the electrical potential of the metal freestanding layer constant, which will lead to huge disturbance of the electrostatic field distribution. Therefore, the capacitance of metal-SF-TENGs has two parts. The first part is the direct parasitic capacitance between these two electrodes, which can be analogous to the capacitance of dielectric SF-TENGs. The second part is the serial connection of the capacitance between metal 1 and metal 3 ($C_{f1}$) and metal 3 and metal 2 ($C_{f2}$), which is the unique capacitance part of metal SF-TENGs. When both the dielectric thickness and the freestanding height $h$ are not too large, the contribution of this serial connection is very significant to the total capacitance. At the two ends, either $C_{f1}$ or $C_{f2}$ will close to 0 because the metal freestanding layer only have overlapping region with one electrode, so the total capacitance of metal SF-TENGs is close to the capacitance of dielectric SF-TENGs. When $x$ equals to $(g + l)/2$, $C_{f1}$ and $C_{f2}$ will be the same and their serial connection gets its maximum, leading to the largest total capacitance under this condition. At the same time, strong charge redistribution has been observed in the inner surface of the metal freestanding layer.

The most significant advantage of the SF-TENG structure is its excellent tolerance to the freestanding height $h$ [12]. This loosens the requirement of the sliding object and reduces the energy loss due to the friction. The influence of the freestanding height on the output performance of dielectric SF-TENGs is first investigated. Besides the parameter specified in the figure, the value of all the other parameters are the same as listed in Table 5.1. From Fig. 5.5a, $Q_{SC,final}$ decreases when $h$ increases, which is mainly because of the reduce of the difference between $(C_2(k)/C_1(k))_{x=0}$ and $(C_2(k)/C_1(k))_{x=g+l}$. Specifically, when $h = 0$, $C_2(k)/C_1(k)$ equals to 0 at the position of $x = 0$ and equals to infinity at the position of $x = g + l$, so $Q_{SC,final}$ can reach $\sigma wl$. When $h$ is large enough, the average distance between the tribo-charge surface and two electrodes will be approximately the same. Thus both $(C_2(k)/C_1(k))_{x=0}$ and $(C_2(k)/C_1(k))_{x=g+l}$ are approximately 1, resulting in a zero $Q_{SC,final}$. But the decreasing slope with height is much flatter than the attached-electrode STENG case. From the numerical calculation results, the half-life height that $Q_{SC,final}$ reduces to its half-value for attached-electrode sliding-mode TENGs is 250 μm, while that for SF-TENG structure is about 2 cm. As for the capacitance profile shown in Fig. 5.5b, the increase of $h$ only slightly decreases the total capacitance since the capacitance in the dielectric SF-TENGs is mainly from the direct capacitance between metal 1 and metal 2. For the $V_{OC}$ profile of the dielectric SF-TENG shown in Fig. 5.5c, it almost has the same trend with $Q_{SC}$, for the capacitance almost has no dependence with $h$. As for the total energy-$R$ curve, this degradation of both $Q_{SC}$ and $V_{OC}$ will lead to the decrease of the maximum harvested energy when $h$ increases. However, the optimum resistance

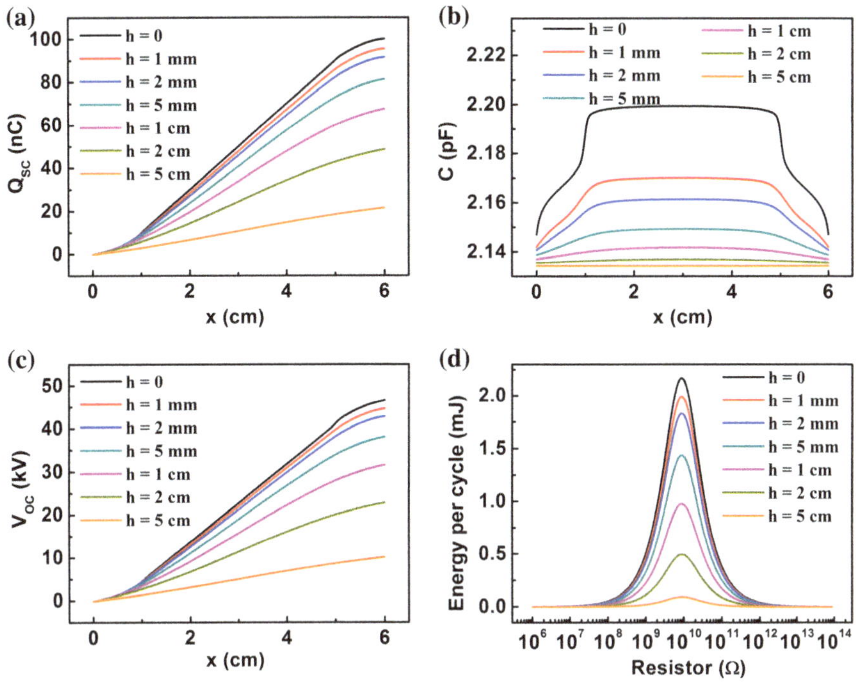


**Fig. 5.5** **a–c** Influence of the freestanding height $h$ on **a** $Q_{SC}$ curves under MACRS, **b** total capacitance curves, and **c** $V_{OC}$ curves under MACRS of the dielectric SFTENG. **d** The harvested energy by the dielectric SFTENG in one cycle under different load resistance and different $h$

stays almost the same, which is because of the nearly same inherent capacitance for different $h$.

For metal SF-TENGs, their height characteristics have some similarity with dielectric SF-TENGs, but it will still have several differences. Since the basic electrostatic induction mechanism in the metal SF-TENG system is very similar to that of dielectric SF-TENG system, the change of $Q_{SC}$ profiles with height is similar to that of dielectric SF-TENG system (Fig. 5.6a). However, there are still many differences. When $h$ increases, the influence of the metal layer on the electrostatic system reduces and both $C_{f1}$ and $C_{f2}$ significantly decrease. Thus, as shown in Fig. 5.6b, the total capacitance at the middle position is largely affected when $h$ first increases from 0. When $h$ is large enough, the total capacitance starts to be dominated by the direct parasitic capacitance between the two electrodes and is nearly independent of $x$. At this time, the metal SF-TENGs are indistinguishable with the dielectric SF-TENGs. As for the open-circuit voltage shown in Fig. 5.6c, the metal SF-TENGs also show a very complicated profile, which is because both $Q_{SC}$ and $C$ have strong dependence on the freestanding height. $V_{OC\text{-final}}$ will decrease because of a reduced $Q_{SC\text{-final}}$ and an almost non-changed $C$ in that region. However, at the middle region, $V_{OC}$ will increase first when $h$ increases then

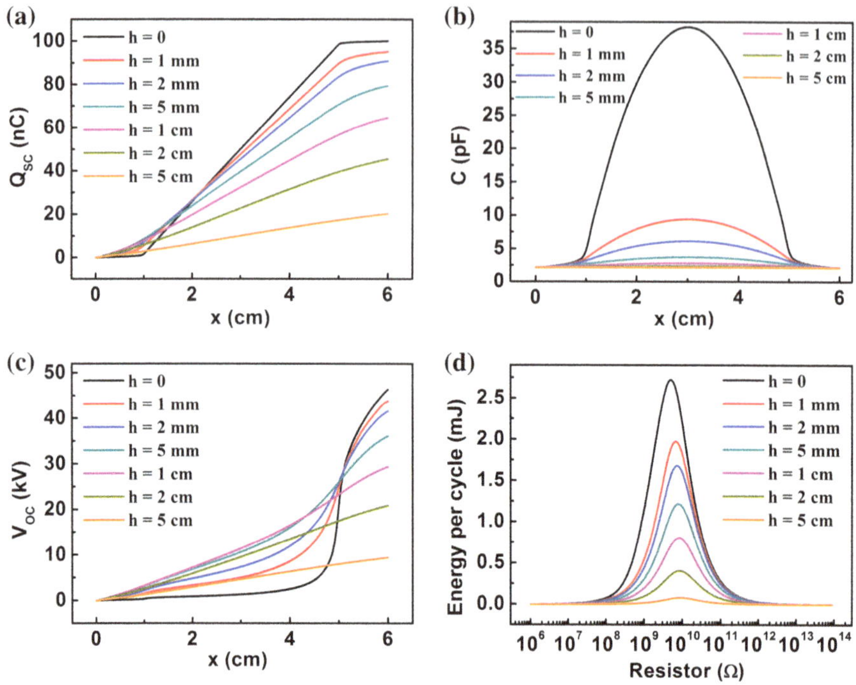


**Fig. 5.6** **a–c** Influence of the freestanding height $h$ under $g$ = 1 cm condition on **a** $Q_{SC}$ curves under MACRS, **b** total capacitance curves, and **c** $V_{OC}$ curves of the metal SFTENG under MACRS. **d** The harvested energy by the metal SFTENG in one cycle under different load resistance and different $h$

decreases when $h$ is already large enough. This is because the decreasing speed of the serial connection of $C_{f1}$ and $C_{f2}$ is faster than the decreasing speed of $Q_{SC}$ in this region. When $h$ is small and $x$ is in the middle range, this serial connection dominates the total capacitance and $V_{OC}$ has this unusual increase. As for the total harvested energy, it also has the similar decay trend as the dielectric SF-TENGs (Fig. 5.6d). However, since the average total capacitance drops with the increase of $h$, the optimum resistance shifts to the right when $h$ first increases and finally stays on the value which is almost the same as the dielectric SF-TENGs.

Besides the freestanding height $h$, another important design parameter is the gap distance between the electrodes ($g$). The capacitance between the electrodes is strongly dependent on the gap distance. Therefore, the gap distance will have strong effect on the total output characteristics.

The effect of $g$ on the total output performance for dielectric SF-TENGs under $h = 0$ conditions is studied first and plotted in Fig. 5.7. Besides the parameters specified in the figure, all the other parameters is the same as listed in Table 5.1. Since $h$ equals to 0, the ratio $C_2(k)/C_1(k)$ is still 0 when $x = 0$ and infinity when $x = g + l$. Therefore, the charge transfer efficiency $\eta_{CT}$ can still reach 100 % for all

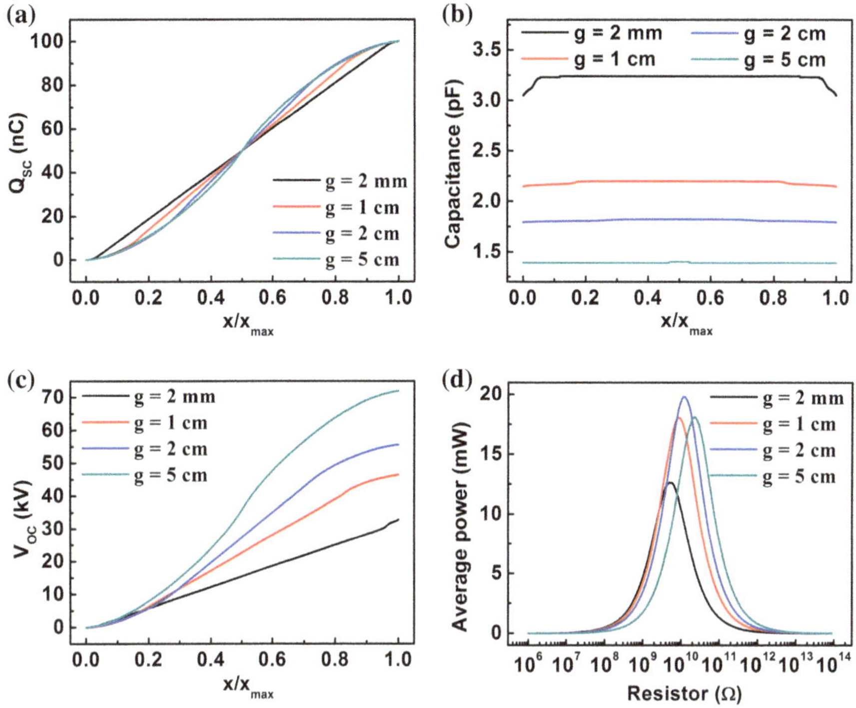


**Fig. 5.7** **a–c** Influence of the electrode gap $g$ under $h = 0$ condition on **a** $Q_{SC}$ curves under MACRS, **b** total capacitance curves, and **c** $V_{OC}$ curves under MACRS of the dielectric SFTENG. **d** The average power by the dielectric SFTENG in one cycle under different load resistance and different $g$

the $g$ values and $Q_{SC\text{-final}}$ will not change with $g$ (shown in Fig. 5.7a). However, the capacitance between the two electrodes decreases when $g$ increases, as shown in Fig. 5.7b. Thus, through fundamental $V_{OC}$, $Q_{SC}$, and $C$ relationship, $V_{OC\text{-final}}$ when a full separation is reached will increase with $g$ (shown in Fig. 5.7c). Therefore, considering the total energy harvested in one cycle, the dielectric SF-TENG with a higher $g$ will generate a larger total energy with a higher optimum resistance. However, for dielectric SF-TENGs with a higher $g$, their cycle ($T$) is also longer when the average velocity stays the same. Therefore, the average power (defined as Eq. 5.4a) will have a different trend with the total harvested energy, as shown in Fig. 5.7d.

$$P_{avg} = \frac{R\int_0^T I^2 dt}{T} \tag{5.4a}$$

$$T = \frac{2(l+g)}{v} \tag{5.4b}$$

When $g$ first increases from 0, the cycle length increases only a little because $l$ still dominates the total cycle length. Therefore, the increase of $V_{OC\text{-final}}$ dominates the total average power and average power increases with $g$. However, when $g$ is already very large, the decreasing slope of the total capacitance is flattened. Therefore, the increase of $V_{OC\text{-final}}$ cannot catch up with the increasing speed of $T$ and the average power begins to drop. Thus, an optimum gap is observed to generate the largest average power.

For metal SF-TENGs under $h = 0$ conditions, similar results have also been observed. The charge transfer efficiency $\eta_{CT}$ can still reach 100 % for all the $g$ values and $Q_{SC\text{-final}}$ will not change with $g$. The total capacitance of metal SF-TENGs also decrease with the increase of $g$, but this decrease contributes from two reasons. The first reason is the same as dielectric SF-TENGs, which will lead to the decrease of the direct parasitic capacitance between these two electrodes. Besides, when $g$ increases, the region when the freestanding layer are overlapped with both of the two electrodes decreases, leading a decrease of the serial connection of $C_{f1}$ and $C_{f2}$. When $g$ is larger than $l$, the region that the freestanding layer is overlapped with both electrodes no longer exists and the serial connection of $C_{f1}$ and $C_{f2}$ can be neglected. The influence of $g$ on $V_{OC}$ of metal SF-TENGs is very similar to the case in dielectric SF-TENGs. $V_{OC\text{-final}}$ will increase with the increase of $g$. Also, similar to the dielectric SF-TENGs, the optimum resistance still increases when $g$ increases and there is an optimum gap to reach the highest average power.

Besides the $h = 0$ case we discussed above, the influence of the gap distance will be much more complicated if there is a non-zero freestanding height. When $h$ is non-zero, the capacitance ratio $C_2(k)/C_1(k)$ is no longer 0 when $x = 0$ and no longer infinity when $x = g + l$. And if the gap decreases, the difference of the capacitance ratio $C_2(k)/C_1(k)$ between $x = 0$ position and $x = g + l$ position will reduce as a response, resulting in a decrease of $Q_{SC\text{-final}}$. The 3d plot of $Q_{SC\text{-final}}$ under different $h$ and $g$ for dielectric SF-TENG is shown in Fig. 5.8a. Consistent with our theoretical analysis, when $g$ increases, the slope of $Q_{SC\text{-final}}$-$h$ curve decreases and the dielectric SF-TENG has more tolerance to the height $h$. In addition, $Q_{SC\text{-final}}$ will increase with the increase of $g$ when $h$ is a non-zero constant and the slope of $Q_{SC\text{-final}}$-$g$ curves increase when $h$ goes up. As for the average power shown in Fig. 5.8b, there is still an optimum gap value when $h$ increases. This optimum gap at which the highest average power is reached increases with $h$, as a result of better tolerance of height for larger gap devices. For metal SF-TENGs, similar results have also been observed.

## 5.2 Contact Freestanding Triboelectric-Layer TENG

Besides the SF-TENGs, another important category of F-TENGs is contact-mode freestanding-TENGs (CF-TENG) based on vertical charge separation, in which the vibration of the freestanding triboelectric layer between the two electrodes

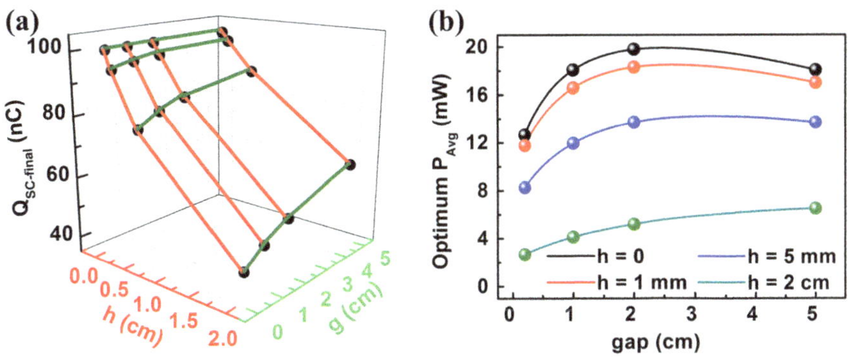


**Fig. 5.8** **a** Dependence of the short circuit transferred charge when a full separation is reached on the freestanding height and the electrode gap. **b** Influence of the freestanding height on the optimum gap at which the maximum average power is reached

periodically changes the induced potential difference between the two electrodes and thus generate electricity in external load [2]. In such a structure with the electrodes' lateral dimension much larger than their vertical separation distance, the electricity generation has a linear relationship with the moving distance of the freestanding layer. Based on this unique linearity both theoretically and experimentally revealed in this paper, the CF-TENG is developed as the first triboelectrification-enabled self-powered vibration sensor that can quantitatively measure both the vibration amplitude and frequency. Besides, the CF-TENG is also demonstrated as an effective energy harvester for vibration with the capability of non-contact operation.

### 5.2.1 *Basic Principle*

The structural design of the vibration-enabled CF-TENG is schematically shown in Fig. 5.9a. Its skeleton is constructed by laser-cut acrylic sheets. Two Al-deposited acrylic plates are supported in a face-to-face configuration with a distance of 2 cm, which serve as the two stationary electrodes of the CF-TENG. They both have a lateral dimension of 10 cm. In between, another acrylic sheet (with the thickness of ~1.6 mm) is employed as the vibration resonator in the CF-TENG through having its four corners connected by 8 springs (all with the spring constant of 33.3 N/m) to the two ends of the acrylic skeleton. Two 50-μm fluorinated ethylene propylene (FEP) films that are purposely chosen as the freestanding triboelectric layers due to its distinctively opposite triboelectric polarity with Al are laminated onto the two sides of this acrylic sheet. They have the same size with the electrode. Triggered by the external vibration source that the CF-TENG is attached to, the two triboelectric layers are brought by the resonating acrylic sheet to alternatively approach the two

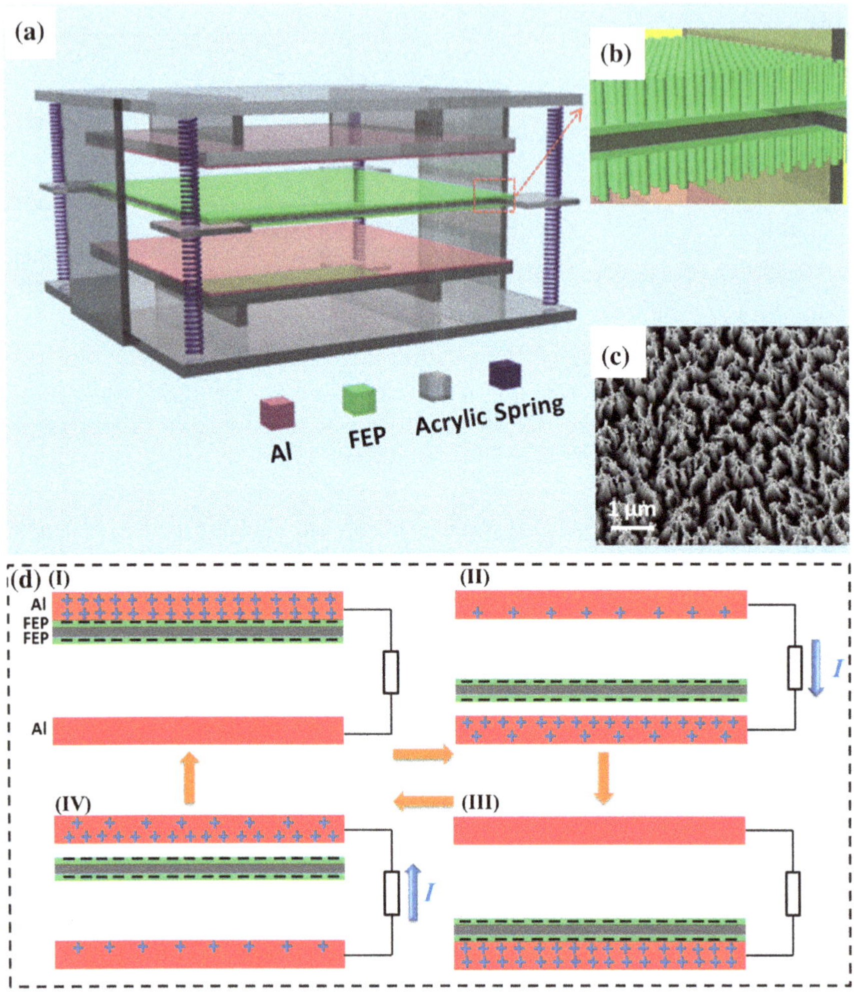


**Fig. 5.9** Device structure and working principle of the contact-mode freestanding triboelectric layer based nanogenerator (CF-TENG). **a** Schematic diagram showing the typical device structure of a CF-TENG. **b** Enlarged view of the device structure showing the nanowire-structure on the surface of the FEP films attached on the resonator plate. **c** SEM image of the nanowire structures on the FEP films. **d** Schematic diagram showing the basic working principle of the CF-TENG

electrodes in a vertical-to-plane manner. When its vibration amplitude is large enough, the contact between the FEP layer and the Al surface will generate triboelectric charges. Since the FEP is an electret material, the triboelectric charges can preserve on its surface for a very long time to induce the electricity generation, even if there is no physical contact between the FEP and the Al in some of the following operation cycles. In order to further enhance the triboelectric charge density through nanostructures, the FEP surface is etched by the inductive coupling plasma (ICP) reactive ion etching to create the nanowire-structures (Fig. 5.9b), which helps to increase the surface roughness as well as the effective surface area. As shown in the scanning electron microscopy (SEM) image (Fig. 5.9c), these vertically-aligned nanowires have an average diameter of ~100 nm and a length of ~1 μm.

As a result of the triboelectrification, the total amount of negative triboelectric charges on these two pieces of FEP films attached to the two surfaces of the resonator should ideally be the same with the positive charges in the electrode loop composed by the two Al plates. When the negatively-charged resonator vibrates between the two electrodes, the positive charges in the electrode loop will be induced to transfer back and forth through the external load, as shown in Fig. 5.9d. This electricity generation process can be described in details as follows. When the resonator gets into contact with the top Al plate (as shown in Fig. 5.9d<i>), all of the positive charges will be attracted to this top electrode (assuming that the distance between the two electrodes is much larger than the thickness of the resonator plate). Subsequently, when this negatively-charged resonator vibrates downwards from the top Al to the bottom Al, the top electrode will be induced to have a higher potential compared to the bottom electrode, which will drive the positive charges in the electrode loop to transfer through the external load in the same direction with the resonator's movement. This generates a transient current flow (Fig. 5.9d<II>). Once the resonator reaches the bottom Al, all of the positive charges will be then present at this bottom electrode, as shown in Fig. 5.9d<III>. Subsequently, the reversed movement of the resonator will lead to a reversed transfer of the positive charges as the second current peak through the external load (Fig. 5.9d<IV>). Until the resonator reaches the status shown in Fig. 5.9d<I>, a full cycle is complete.

The CF-TENG can work as an energy harvesting device for the environmental vibration, when the generated electricity is utilized to power other electronic devices/systems. Under the condition that the vibration source has the amplitude of 1.5 cm and the frequency of 15 Hz (i.e. the resonant frequency of the CF-TENG), a continuous AC output is generated. The $V_{OC}$ has a peak-to-peak value of 165 V (Fig. 5.10a), while the -short-circuit current density ($J_{SC}$) reaches 0.35 mA/m$^2$ amplitude (Fig. 5.10b). Once the CF-TENG is connected to an external load, the current density obtained by the load decreases as the load resistance increases, while the voltage shows a rising trend (Fig. 5.10c). The power density received by the load reaches the maximum value of 17 mW/m$^2$ at the load resistance of 20 MΩ (Fig. 5.10d). This generated electricity can directly drive 60 commercial light-emitting diodes (LEDs) simultaneously, as shown in Fig. 5.10e. Owing the linearity in the electricity generation behavior, the physical contact between the triboelectrically-charged resonator and the electrodes is not mandatory for the effective mechanical energy conversion. Similarly with the sliding-based freestanding-TENG, the CF-TENG's capability of non-contact operation also brings about the advantages of improved energy conversion efficiency and better

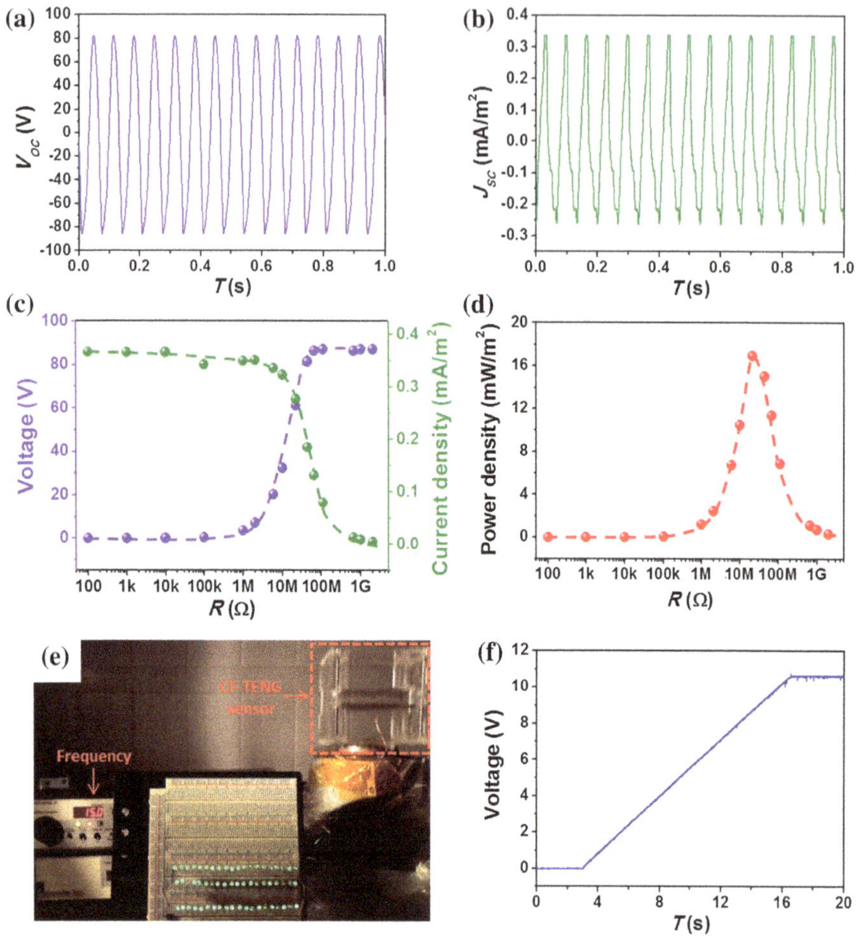


**Fig. 5.10** Performance of the CF-TENG when it serves as an energy harvester for the vibration with the amplitude of 1.5 cm and the frequency of 15 Hz. **a** Output $V_{OC}$, and **b** output short-circuit current density ($J_{SC}$) of the CF-TENG. **c** Voltage and current density obtained by the load with different resistances. **d** Power density obtained by the load with different resistances. **e** 60 LEDs instantaneously driven by the CF-TENG. **f** Storage of the electricity generated by the CF-TENG in a 14.7-μF capacitor

stability. Besides directly driving electronics, the electricity generated by the CF-TENG can also be stored in energy storage units (such as batteries or capacitors) in order to get a regulated power. As shown in Fig. 5.10f, the CF-TENG under the above described vibration condition can charge a 1.47-μF capacitor from empty to 10.5 V in only 13.5 s.

### 5.2.2 Fundamental Theory

Two typical structures of CF-TENGs are shown in Fig. 5.11 [12]. Figure 5.11a shows a typical structure for dielectric-freestanding-layer CF-TENGs. A dielectric plate (thickness: $d_1$, relative dielectric constant: $\varepsilon_{r1}$) and two metal plates are stacked face to face, forming two triboelectric pairs. The two metal plates also serve as two electrodes. The total air gap thickness between these two metal plates is defined as $g$. After the dielectric plate being forced to contact with the two metal plates, both the top and the bottom surfaces of Dielectric 1 will have static triboelectric charges due to contact electrification. For simplicity, we assume the triboelectric charge density of both surfaces is the same ($-\sigma$). At the same time, the two metal plates will have the same amount of positive charges in total because of charge conservation.

The above system can be easily solved through Gauss's Law. But to fully show its inherent physics and working behavior, we still try to utilize the concept of nodes to analyze this electrostatic system. In practical applications, the size of the metal electrode is always much larger than the air gap. Therefore, the area size of the freestanding contact-mode TENG ($S$) is seen as infinitely large and the edge effect can be ignored. As a result, the electric potential on metal 1, metal 2, the top and bottom surface of dielectric 1 are all constant, so these 4 surfaces can be seen as 4 nodes in this electrostatic system. The electrical line connection between every two nodes forms an equivalent capacitance between them. Since the area size ($S$) is assumed as infinitely large, the electrical line connection between every two non-adjacent nodes (for example, Node 1 and 3) is fully blocked by the intermediate node (for example, Node 2). Therefore, only three capacitances exist in the equivalent circuit model of this electrostatic system, as shown in Fig. 5.11b. The total capacitance between the two electrodes (Node 1 and 4) is the serial connection of $C_1$, $C_2$, and $C_3$, which can be easily given by the following equation.

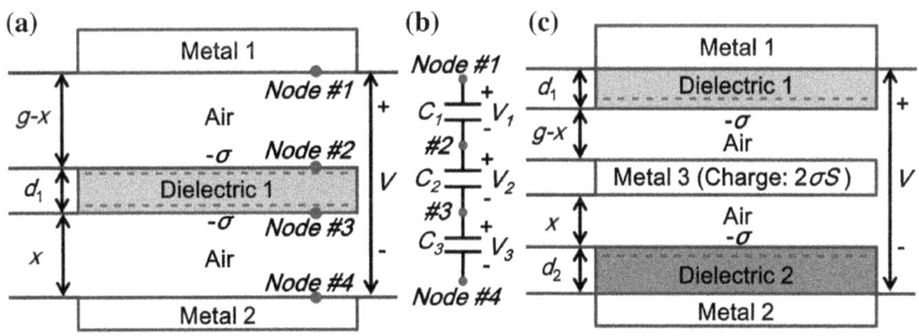


Fig. 5.11 Theoretical models of CFTENGs **a** model of a typical dielectric freestanding layer CFTENG. **b** equivalent circuit model of the dielectric CFTENG electrostatic system. **c** model of a typical metal freestanding layer CFTENG

$$C = \frac{1}{\frac{1}{C_1} + \frac{1}{C_2} + \frac{1}{C_3}} = \frac{\varepsilon_0 S}{d_0 + g} \tag{5.5}$$

Inside Eq. (5.5), the definition of the effective dielectric thickness $d_0$ is the same with the previous definition.

The charge distribution under short circuit (SC) condition when Node 1 and Node 4 have the same electrical potential is then analyzed. The charges on each node need to be deducted to solve this electrostatic system. For both Node 2 and 3, the amount of charges on the node is $-\sigma S$. In addition, because of charge conservation, we only know that the total charges on Node 1 and 4 are $2\sigma S$, but the detailed charge distribution is still unknown. Under SC condition, we set the total charges on metal 1 (Node 1) is $Q_1$ and the total charges on metal 2 (Node 4) is $Q_2$ ($Q_2 = 2\sigma S - Q_1$). From the basic electrodynamics theory and charge conservation, the total charges on metal 1 and 2 can be easily solved as:

$$Q_1 = \sigma S \frac{\frac{1}{C_2} + \frac{2}{C_3}}{\frac{1}{C_1} + \frac{1}{C_2} + \frac{1}{C_3}} \tag{5.6a}$$

$$Q_2 = \sigma S \frac{\frac{2}{C_1} + \frac{1}{C_2}}{\frac{1}{C_1} + \frac{1}{C_2} + \frac{1}{C_3}} \tag{5.6b}$$

In practical applications, the effective dielectric thickness of the dielectrics $d_1/\varepsilon_{r1}$ is always negligible compared to the air gap and $C_2$ can be seen as infinitely large. Thus, Eq. (5.6a and b) can be further simplified to:

$$Q_1 \approx \sigma S \frac{\frac{2}{C_3}}{\frac{1}{C_1} + \frac{1}{C_3}} = \frac{2\sigma S}{1 + \frac{C_3}{C_1}} \tag{5.7a}$$

$$Q_2 \approx \frac{2\sigma S}{1 + \frac{C_1}{C_3}} \tag{5.7b}$$

From Eqs. (5.7a) and (5.7b), the basic working mechanism of CF-TENGs can be easily observed. When $x = 0$, $C_3$ is infinity, as a result, $Q_1$ is close to 0 and $Q_2$ is approximately $2\sigma S$. Under this condition, all the positive tribo-charges are attracted to the bottom electrode by the negative charges on the surface of dielectric 1. While when $x = g$, $C_1$ is infinity and all the positive charges on the electrodes are attracted to Metal 1. Therefore, if Dielectric 1 is under a vibration inside the air gap, charges will alternatively flow between Metal 1 and Metal 2 due to the change of $C_1/C_3$, forming an AC short circuit current. Same as all previous TENGs, the change of this capacitance ratio induced by the change of Dielectric 1 position can drive electrons to flow under SC conditions between the two electrodes.

Under MACRS, the short circuit transferred charges ($Q_{SC}$) and open-circuit voltage can be easily obtained as:

$$Q_{SC} = Q_1(x) - Q_1(x = 0) = \frac{2\sigma Sx}{d_0 + g} \tag{5.8}$$

$$V_{OC} = \frac{2\sigma x}{\varepsilon_0} \tag{5.9}$$

Therefore, the governing equation for this CF-TENG, which is its *V*-*Q*-*x* relationship, can be shown as [12]:

$$V = -\frac{1}{C}Q + V_{OC} = -\frac{d_0 + g}{\varepsilon_0 S}Q + \frac{2\sigma x}{\varepsilon_0} \tag{5.10}$$

Besides the dielectric-freestanding-layer CF-TENGs, another important category is metal-freestanding-layer CF-TENGs, as shown in Fig. 5.11c. The working principle of metal-freestanding-layer CF-TENGs is very similar to the dielectric-freestanding-layer case. Similar to the previous discussed attached-electrode contact-mode TENGs, metal can be simply seen as dielectrics with 0 thickness to calculate $d_0$ and the derived equations above are all applicable to metal CF-TENGs.

With the above information, the comparison of the intrinsic output characteristics of CF-TENGs and attached-electrode contact-mode TENGs can be provided. Their $V_{OC}$ characteristics are similar to each other, which are both linearly proportional to $x$. However, their inherent capacitance and short-circuit transferred charges characteristics are completely different. First, the inherent capacitance of CF-TENGs is constant and independent of $x$ while that of contact-mode attached-electrode TENGs decreases when $x$ increases. In addition, $Q_{SC}$ of CF-TENGs is also linearly proportional to $x$ while that of attached-electrode TENGs has a saturation trend.

The most advantage of this CF-TENG structure is its superior linearity [12]. The linear dependence of both $V_{OC}$ and $Q_{SC}$ on $x$ makes it an ideal medium to transform mechanical vibration signal to electrical signal. Besides, the constant inherent capacitance makes it a linear and time-invariant device. Thus, the electrical signal output of this device is compatible with all the currently available signal processing techniques, making it easy to accomplish post-processing of the obtained electrical signal. Therefore, this structure could have wide applications in self-powered vibration detection.

Besides the self-powered sensor application, this superior linearity is also beneficial in vibration energy harvesting applications as well. To fully uncover its advantages, we design a device structure that can harvest vibration energy under both freestanding-mode and traditional attached-electrode mode, as shown in Fig. 5.12a. When metal 1 and metal 2 are directly utilized as two electrodes and metal 3 and 4 are suspended with no output terminal, this device is working in the freestanding-mode (shown in the right solid line). If metal 1 and metal 4 are

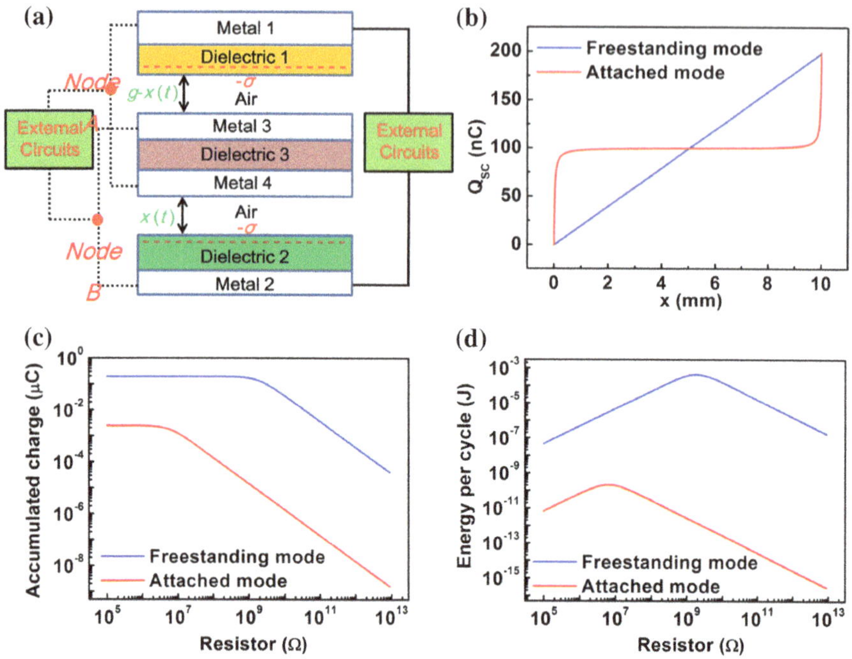


**Fig. 5.12** **a** Structure of the carefully designed TENG which can work in both freestanding mode and traditional attached-electrode mode. **b** Compare of short-circuit transferred charges characteristics under these two modes. **c–d** When the vibration amplitude is only a quarter of the air gap, the comparison of these two modes considering **c** the amount of the transferred charges and **d** harvested vibration energy in one cycle under different load resistance

connected to form Node A, metal 2 and metal 3 are connected to form Node B and then Node A and Node B are utilized as the output terminals to connect with external load circuits, the device is working in traditional attached-electrode mode, which is equivalent to two parallel-connected attached-electrode TENGs. The reason to connect metal 1 and 4 rather than metal 1 and 2 to form the output node is to ensure the polarity of these two TENGs is consistent.

As a comparison, $Q_{SC}$ under both two operation-modes are numerically calculated utilizing the same parameters and plotted in Fig. 5.12b. Although for both of these two-modes, their $Q_{SC\text{-final}}$ (short-circuit transferred charge amount when a full separation is reached) can reach $2\sigma S$, the shape of $Q_{sc}$ is completely different. For the freestanding mode, $Q_{SC}$ changes linearly with $x$, ensuring a relative high slope in the whole $x$ range. However, for the traditional attached-electrode mode, $Q_{SC}$ only has a huge slope at the two ends. When $x$ is between 0.05 and 0.95 g, $Q_{SC}$ is nearly a constant. Therefore, when the traditional attached-electrode mode is utilized to harvest vibration energy with vibration amplitude which is not large enough to ensure the freestanding layer to fully contact with the two electrodes (which is always the case in practical application), the total short-circuit transferred charges will be strongly limited. As an example, this device is utilized to harvest a harmonic vibration whose vibration amplitude is $g/4$ and center position is $x = g/2$. The steady state electrical outputs under these two modes are numerically calculated through the TENG simulator utilizing the periodic boundary condition. As shown in Fig. 5.12c, the amount of transferred-charges of the freestanding mode is more than 70 times higher than the attached-electrode mode in all the resistance ranges. In addition, the optimum energy harvested from the freestanding-mode is several orders larger than the attached-electrode mode. From the above comparison, we can conclude that the linearity of the freestanding mode is highly beneficial for harvesting the mechanical energy from the vibration that cannot ensure good contact of the moving object and the two electrodes, which will also save energy loss from the inelastic collision and increase the energy conversion efficiency.

After we finish the discussion of all TENG fundamental modes, we have known that the fundamental working mechanism of TENGs is a conjugation of contact electrification and electrostatic induction. Through contact electrification, electrostatic charges can be generated and kept on the dielectric surface. There are two equivalent capacitances that will form between the tribo-charged dielectric surface and the two metal electrodes, respectively. The change of the ratio of these two capacitances due to the change of the position of the dielectric will induce electrons to transfer between the metal electrodes under SC condition. The difference of the capacitances' changing behavior gives rise to the different fundamental modes of TENGs. As a summary, the capacitance changing trend of all the TENG fundamental modes is summarized in Table 5.2.

**Table 5.2** Comparison of different TENG fundamental modes

| | Attached-electrode TENGs (including vertical-separation and lateral-sliding TENGs) | Single-electrode TENGs | Freestanding TENGs |
|---|---|---|---|
| Core working mechanism | $\eta_{CT} = \frac{1}{1+\frac{C_1(x=x_{max})}{C_2(x=x_{max})}} - \frac{1}{1+\frac{C_1(x=0)}{C_2(x=0)}}$ | | |
| $C_1\ (x = 0)$ | Infinity | Infinity | Infinity |
| $C_2\ (x = 0)$ | Finite non-zero value | Finite non-zero value | Finite non-zero value |
| $C_1\ (x = x_{max})$ | Infinitesimal | Infinitesimal | Finite non-zero value |
| $C_2\ (x = x_{max})$ | Finite non-zero value | Infinitesimal | Infinity |
| $C_1\ (x = 0)/C_2\ (x = 0)$ | Infinity | Infinity | Infinity |
| $C_1\ (x = x_{max})/C_2\ (x = x_{max})$ | 0 | 1 | 0 |
| $\eta_{CT}$ | 100 % | 50 % | 100 % |

## 5.3 Advanced Device Structures

### 5.3.1 *Linear-Grating Structure*

As we have discussed above, the freestanding-mode TENG device requires a relative large sliding displacement (the whole length of one electrode) to realize an effective electricity generation, which is unfavorable for its achievable power output and energy scavenging in practical use. Since mechanical vibration in our environment is always irregular and varying in amplitude and frequency, we herein report a linear grating-structured freestanding triboelectric-layer nanogenerator (GF-TENG) with largely-elevated output, high efficiency, and broad range of applications [3]. This new design is composed of a grating-segmented triboelectric layer and two pairs of interdigitated electrodes, all in the same periodicity. Both of the theoretical analysis and experimental study indicated significant enhancement in the amount of collectable charges, current density, and output frequency through introducing finer grating segments. As driven by a linear motor at an acceleration of 30 m/s$^2$, the GF-TENG with 16 grating segments delivered an open circuit voltage ($V_{OC}$) of ~135 V and a short-circuit current density ($J_{SC}$) of ~9 mA/m$^2$, corresponding to a maximum power density of ~1.2 W/m$^2$. Moreover, we also demonstrated that the device working in non-contact mode (that is free of sliding friction) exhibited supreme stability for a long period and achieved extremely-high total conversion efficiency of 85 % at low operation frequency. With these unique features, the GF-TENG was utilized to generate electricity from a wide range of ambient mechanical motion, such as sliding of human hand, acceleration or deceleration of vehicles, and people walking. All these demonstrations reveal a great potential of GF-TENG as a high-efficiency device for practically-applicable mechanical energy harvesting.

The structure of the GF-TENG is depicted in Fig. 5.13a. The device mainly consists of two groups of components: a freestanding triboelectric layer with grating segments and two interdigitated metal electrodes. To enable an intensive electrification during contact, fluorinated ethylene propylene (FEP) and Al were chosen as the materials for the freestanding triboelectric layer and metal electrodes, respectively, according to the triboelectric series. In the fabrication of the freestanding part, an acrylic sheet was laser-cut into grating structures with $N$ segments ($N$ = 8 for Fig. 5.13a) as the supporting substrate for the triboelectric layer. Then, the FEP film was adhered to the substrate and cut into the same configuration. To enhance the contact electrification, nanorod structures (Fig. 5.13b) was introduced onto the surface of FEP film through inductive coupling plasma reactive ion etching. As for the metal-electrode part, another acrylic was patterned to $N$ units of interdigitated structures with the same periodicity as the freestanding layer. After the deposition of aluminum film, two groups of grated-electrodes with the complimentary shape were obtained. The spacing between two adjacent electrode fingers was 1 mm in all the devices. For the GF-TENGs with different $N$, we maintained the total effective lengths ($l_0$) of the freestanding layer and each electrode group with the same value

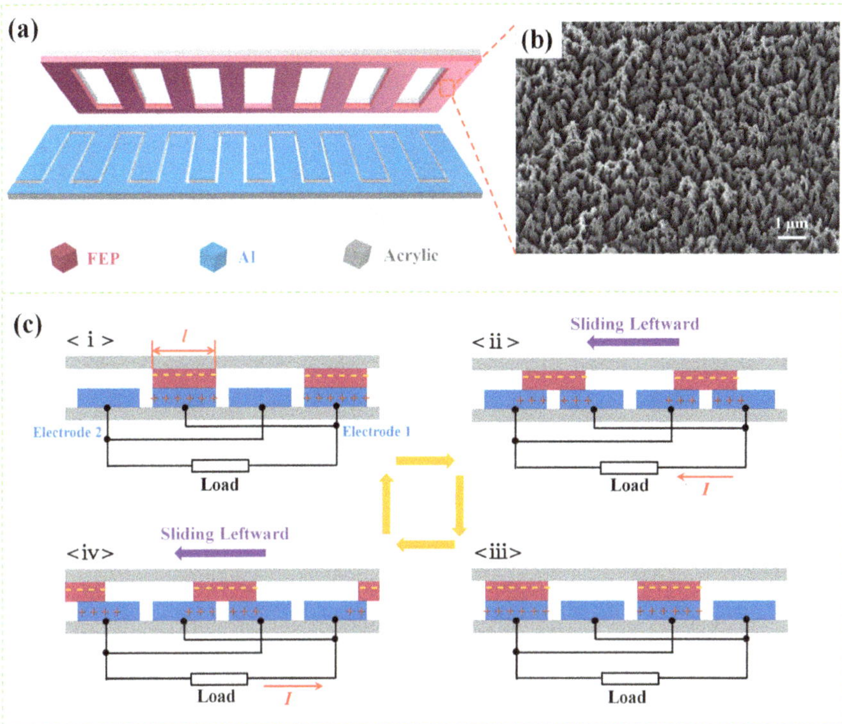


**Fig. 5.13** Device structure and operation mechanism of grating-structured freestanding triboelectric-layer nanogenerator (GF-TENG). **a** Typical structure design of GF-TENG. **b** SEM image of nanorod structure on the FEP surface. **c** Working mechanism of the GF-TENG

of 6.4 cm, which makes sure these devices having the same total effective area (38.4 $cm^2$). Thus, the larger number of segments only brings about the smaller length ($l$) of each grating unit, with a relationship of $l = l_0/N$.

The operation principle of the GF-TENG can be explained as the coupling of contact electrification and in-plane-sliding-induced charge transfer, as shown in Fig. 5.13c. Initially, the grating rows of the FEP layer are brought into contact with the fingers of the electrode group (EG) 1 at a fully overlapping position (Fig. 5.13c<i>). Since FEP is more triboelectrically negative than Al, contact electrification between two surfaces occurs, which results in negative charges on the FEP surface and positive charges on the EG 1 with the equal amount. In this state, there is no charge flowing between the two electrodes due to the electrostatic equilibrium. When the FEP layer starts to slide leftward, this grated layer gradually moves from the overlapping position of EG 1 to that of EG 2 (Fig. 5.13c<ii>). In this case, an electric potential drop is generated and drives positive charges flow from the EG 1 to the EG 2 along with the sliding motion, which produces a transient current in the external load. Once the FEP layer reaches the fully overlapping position with EG 2 (Fig. 5.13c<iii>), all of the positive charges transfer to this electrode, with another electrostatic equilibrium being achieved. This is one unit step of sliding. When the FEP continues to slide leftward to the EG 1 by another step (Fig. 5.13c<iv>), the positive charges will be electrostatically attracted to flow back to EG 1, bringing about a reverse current in the circuit. Therefore, an electricity generation cycle is completed. If the FEP is driven to slide rightward, a reverse process will take place.

To experimentally study the influence of the grating segmentations on the output performance, five devices with different grating segments ($N$ = 1, 2, 4, 8, 16) but the same effective area were fabricated, and their electrical characteristics were measured. The $V_{OC}$, $\Delta\sigma_{SC}$, and $\Delta\sigma_{SC\text{-Rec}}$ were measured by Keithley 6514 system, and the short-circuit current density ($J_{SC}$) was measured by SR570 low noise current amplifier (Stanford Research System). Figure 5.14a shows the transferred charge density of the GF-TENGs with insets showing the enlargement profiles of one sliding cycle of each structure. A full sliding cycle is composed of two parts with the left part corresponding to a one-way sliding and the right one representing the backward process. It can be observed that tribo-charges are transferred $(2N - 1)$ times for the $N$ segments device (e.g., 31 times for 16 segments) in a one-way process, which is accordant with the simulation work, although the maximum $\Delta\sigma_{SC}$ from the first step declines slightly with the increase of $N$. The trends may be ascribed to the uneven surface induced by the procedure of laser cutting of grating structure, which affects the effective contact area. By connecting the GF-TENGs with a bridge rectifier, all of the transferred charges can be accumulatively collected, e.g. in an energy storage device. As shown in Fig. 5.14b, the accumulated charges have a largely-enhanced amount by increasing segment numbers because of multiplied charge transfer cycles. When 16 segments are introduced, the $\Delta\sigma_{SC\text{-Rec}}$ goes up to 640 μC/m$^2$ which is over 10 times larger than that of 1 segment structure, demonstrating a huge enhancement in performance for the multiple segmentation structured device. As for the $V_{OC}$, the profiles display similar oscillation behaviors as the $\Delta\sigma_{SC}$ with a much more severe decaying tendency by the increase of $N$, as displayed in Fig. 5.14c. This reflects the prediction in the theoretical analysis. The characteristics of the $J_{SC}$ were shown in Fig. 5.14d, which were measured at a sliding acceleration of 10 m/s$^2$. For 1 segment device, only one low and wide current peak was generated by one unidirectional sliding step; when it comes to 16 segments structure, 31 sharp and narrow peaks are produced with a peak value of ~4.0 mA/m$^2$ from a similar one-way sliding step. The mean frequencies of the two GF-TENGs are about 7 and 130 Hz, respectively. The enhanced performance can be explained by the following equation:

$$J_{SC} = \frac{\mathrm{d}\Delta\sigma_{SC}}{\mathrm{d}t} \tag{5.11}$$

where $t$ is the time. Since both of the length and the time for each unit step will be shorter when the segmentation number increases, tribo-charges can be transferred faster and hence the higher amplitude and frequency are achieved for the current.

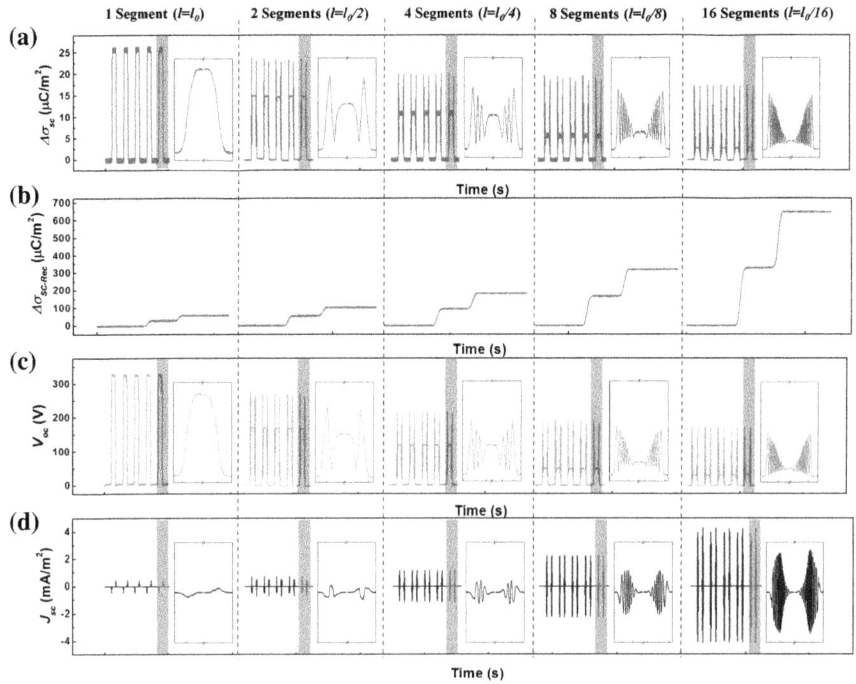


**Fig. 5.14** Electrical measurement results of GF-TENGs with different segment structures ($N = 1$, 2, 4, 8, 16). **a** The measured $\Delta\sigma_{SC}$. **b** The measured $\Delta\sigma_{SC\text{-}Rec}$. **c** The measured $V_{OC}$. **d** The measured short-circuit current density ($J_{SC}$). The *insets* show the enlargement of the corresponding highlighted parts

Therefore, introducing more grating segments in the structure will be favorable for the accumulation of charges, current density and frequency.

With the unique advantages such as high output performance, and ability to work at non-contact operations, the GF-TENGs can be employed to harvest numerous kinds of mechanical energy existing in our environment and act as direct power sources for electronics. As shown in Fig. 5.15a, after attaching a free-standing FEP film to a human hand, the hand can slide freely across the electrodes without any constrain. This movement can effectively generate electricity to light up 60 commercial LEDs (Nichia NSPG500DS) instantaneously even under a slow speed and a small displacement. The GF-TENG can also be potentially utilized to harvest the vibration energies from moving vehicles. For the demonstration, a GF-TENG is installed to a remote control car. When the car was in the acceleration or deceleration, the GF-TENG can generate electricity for powering electronic components on the vehicle (Fig. 5.15b). The GF-TENG could also be used as a self-powered sensor for detecting the acceleration of moving vehicles. If the generator is fully sealed, it can be further applied to harvest wave/tidal energy from the sea.

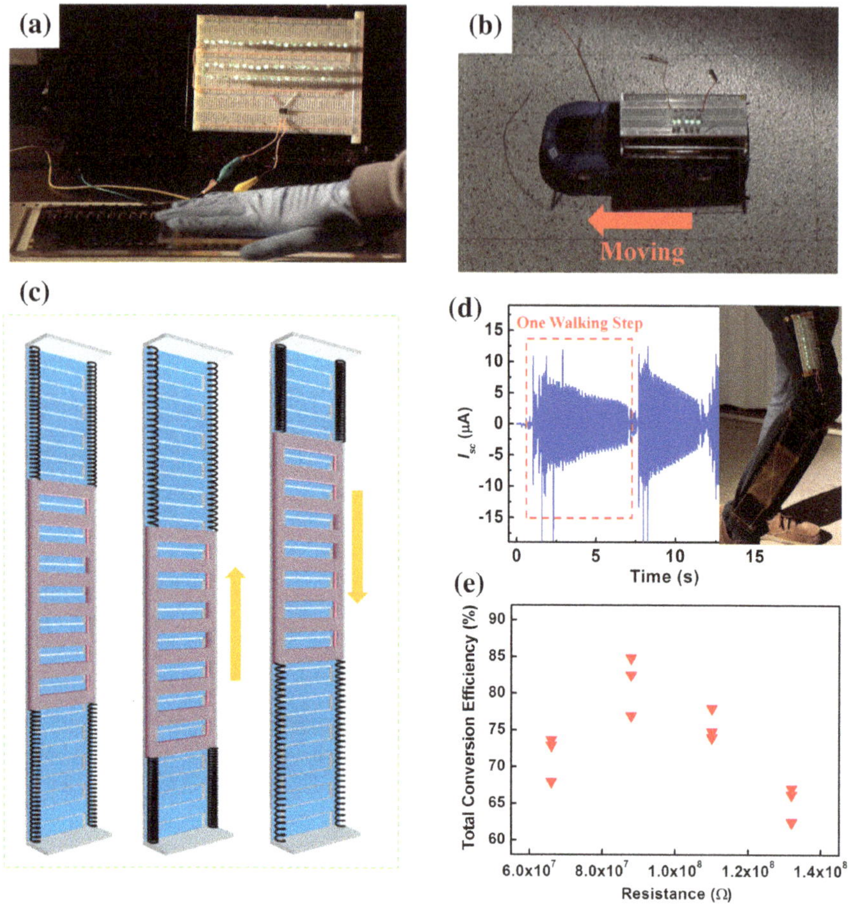


**Fig. 5.15** Applications of GF-TENG for harvesting a wide range of mechanical energy. **a** Harvesting energy from sliding of a human hand. **b** Harvesting energy from acceleration or deceleration of a remote control car. **c** Device structure for non-contact GF-TENG. **d** Harvesting energy from people walking by non-contact GF-TENG and the real-time measurement of short-circuit current ($I_{SC}$). **e** Total conversion efficiency of non-contact GF-TENG for harvesting slight vibration under different load resistances

Furthermore, the GF-TENG working in non-contact mode can harvest mechanical energy with high conversion efficiency. As shown in Fig. 5.15c, four identical extension springs (McMaster-Carr 9654K511) are used to position the FEP within the parallel plane to the surface of the electrodes with a distance of 1.6 mm. When the external mechanical motion applies onto the device along the spring direction, the FEP will oscillate around the equilibrium position for a long term without extra mechanical input, converting the stored mechanical energy in the spring into electricity. As shown in Fig. 5.15d, this design can scavenge the mechanical energy from people's walking motion when it is bonded to human legs.

It is noticed that each walking step will produce lasting and decaying electrical outputs in a certain period, which corresponds to the damped oscillation of the FEP around the equilibrium position.

This spring-based device can also be utilized to estimate the conversion efficiency of the GF-TENG in non-contact mode. Here, total conversion efficiency is introduced and defined as the ratio of the total generated electrical power (by the instantaneous triggering and the residual mechanical vibration after the triggering) to the input mechanical power. Ideally, if there is not any friction in the environment, the theoretical total conversion efficiency of non-contact mode is expected to be 100 %. However, in the real situation, air friction, a non-neglectable factor, will bring about an energy loss for the device working against it. To reduce the impact of air resistance, we can make the FEP oscillate at a low speed (slight vibration). Therefore, high total conversion efficiency would be achieved in this situation. During the measurement, the FEP is initially pulled to a position with a small displacement ($x$ = 1 mm) relative to the equilibrium point, through which the mechanical energy is stored in the springs. When the FEP layer is released at this point, it will oscillate around the equilibrium position with gradually damping amplitude for ~4 min. In this process, the stored mechanical energy is gradually converted firstly to kinetic energy of the FEP and then to the electrical energy. Through measuring the electrical output on the external load, the total amount of electrical energy can be calculated. The electrical energy obtained by the load reaches the maximum values at a certain resistance. Thus, we measured the current on a series of loads with different resistances in order to get the highest-achievable electrical energy and hence the efficiency. Therefore, the total generated electrical power can be calculated as:

$$E_{\text{electrical}} = \int I^2 \cdot R \cdot dt \tag{5.12}$$

where $I$ is the instantaneous current which can be recorded by electrometer, $R$ is the load resistance. As for the input mechanical energy, it will be determined by the difference between potential energy at initial position ($x$ = 1 mm) and final position (equilibrium point), which can be calculated as (supporting information):

$$E_{\text{mechanical}} = 2 \cdot k \cdot x^2 \tag{5.13}$$

where $k$ is the spring constant of a single spring ($k$ = 15.76 N/m), $x$ is its displacement ($x$ = 1 mm). Figure 5.15e shows the estimated total conversion efficiency under different resistances from a number of repeated measurements. It can be observed that the highest efficiency reaches 85 % at an external load of 88 MΩ. It clearly demonstrates the capability of harvesting mechanical energy at a very high efficiency by the freestanding-layer-mode triboelectric nanogenerator.

In summary, we have demonstrated a new design of triboelectric nanogenerator, which consists of a freestanding triboelectric layer with grating segments and two stationary electrodes with interdigitated patterns. With grating units sweep across electrode fingers, multiple alternating currents are generated between the two electrodes due to electrostatic induction in both contact and non-contact mode. According to the theoretical and experimental investigation, the GF-TENG with finer grating units exhibits improved performance in the respect of the amount of collectable charges, current density, and output frequency. When the device is operated in non-contact mode, it shows an excellent stability and high total conversion efficiency up to 85 % at low operation frequency. On basis of these advantages, the GF-TENG is capable of harvesting energy from a variety of ambient sources, such as sliding of human hand, acceleration of vehicles, and people walking. This work represents a significant progress of TENG and exhibits huge potential of the GF-TENG as a high-efficiency energy harvester for practical applications.

### 5.3.2 *Rotation-Disk Structure I*

A novel type of disk TENG with both groups of patterned electrodes attached onto a stationary disk is designed, together with a freestanding triboelectric layer on a rotational disk [4]. With such a structure, there is no necessity for electrode deposition or electrical connection for the rotational part, which dramatically improves the operating facility of the energy harvester. Moreover, owing to the unique feature of this new electricity-generation mechanism, the non-contact free-rotating disk triboelectric nanogenerator (FRD-TENG) can be operated without friction after initial contact electrification, with little lost on performance but superior durability enhancement, because the surface triboelectric charges preserve on insulator surfaces for hours. Under the non-contact condition, an open-circuit voltage ($V_{OC}$) of 220 V and a short-circuit current density ($J_{SC}$) up to 13 mA/m$^2$ was produced, with an effective output power density of 1.2 W/m$^2$, which is enough for instantaneously and continuously driving 100 serially-connected light-emitting diodes (LEDs). The high output power remains stable upon continuous working for 500,000 cycles and is capable of fully charging a lithium ion battery (LIB) within 1.5 h, with an equivalent DC charging current of 7.5 μA. Finally, the FRD-TENG can be employed as a self-powered mechanical sensor for simultaneously detecting the vertical separation distance between the two disks and the rotation speed, through reading the magnitude and the frequency of the measured $J_{SC}$. This work presents an outstanding progress towards the practical applications of nanogenerators in harvesting rotational mechanical energy, and boosts the development of self-powered sensor networks, especially for application in the braking systems of automobile industry.

As schematically illustrated in Fig. 5.16a, the basic structure of the FRD-TENG consists of two parts: the freestanding rotational part of tribo-charged layer, and the stationary part of metal electrodes. The rotational part was fabricated from a piece of fluorinated ethylene propylene (FEP) thin film (~50 μm in thickness), which was tailored into a four-segment structure and attached onto an acrylic supporting

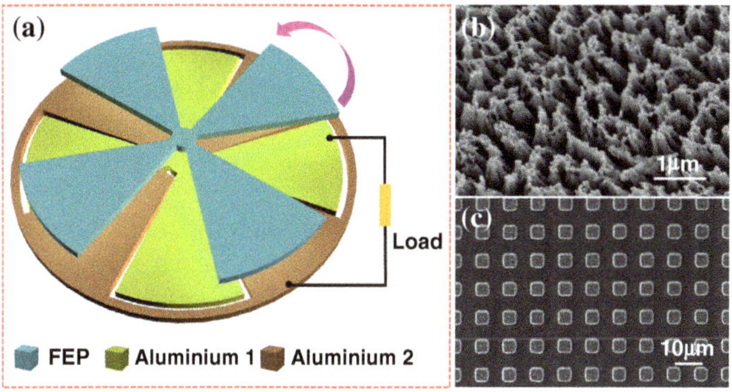


**Fig. 5.16** Device structure of the noncontact free-rotating disk triboelectric nanogenerator. **a** A schematic of the basic structure of the FRD-TENG composed of the freestanding FEP layer and the stationary aluminum layer. The *bottom inset* is the figure legend. **b** A 30°-tilted-view SEM image of the nanorod structure created on the surface of the FEP thin film. **c** A top view SEM image of the cubic micropatterned structures on the aluminum foil

substrate with same shape. In this new structure, there is no need to deposit electrode on the back surface of the FEP membrane, indicating that versatile energy harvesting is achievable with an arbitrary moving object without processing (e.g. rotating tires). The stationary part is composed of two separated aluminum foils ($\sim 50$ μm in thickness and 4 inches in diameter) with complementary four-segment shapes attached on a round-shape acrylic disk, which act as the two stationary electrodes. To increase the density of triboelectric charges through enhanced electrification process, micro-patterned structures were introduced on the inner surface of both triboelectric layers, respectively. On one side, the inductively coupled plasma (ICP) process was utilized to create nanorods on the surface of the FEP film by reactive ion etching (RIE); on the other side, we employed the photolithography approach to selectively deposit aluminum patterns of micro-scale cubic structure onto the flat aluminum foil. The typical scanning electron microscopy (SEM) images of both structures are displayed in Fig. 5.16b, c. The two triboelectric layers were first brought into contact to produce triboelectric charges through the transfer of electrons from the surface of aluminum to FEP. The surface charges on the dielectric FEP would remain almost unchanged for days, and then the device could be continuously working even when the two layers were not in contact. This is the initial step to have the two surfaces being electrostatically charged with opposite signs of charges, respectively. The length of time that the electrostatic charges are preserved on the surfaces depends on the nature of the materials and condition in environment such as humidity. The surface triboelectric charges could even be regenerated by bringing the two surfaces back to contact after a long term of non-contact operation. In the experiment, the rotational part was driven by a rotation motor with tunable rotation speed ($r$), and the stationary part was fixed on a three-dimensional stage to control the vertical distance ($d$) between the two parallel triboelectric surfaces.

The typical output characteristics of the FRD-TENG operated at 500 rpm are presented in Fig. 5.17b, and were compared with the corresponding output performances at contact-mode (Fig. 5.17a). Figure 5.17a1-3 shows the open-circuit voltage ($V_{OC}$), the transferred charge density ($Q_{SC}$), and the short-circuit current density ($J_{SC}$) of the TENG under contact mode ($d = 0$), respectively. Under this condition it can be found that $V_{OC} = 250$ V, $Q_{SC} = 21$ μC/m$^2$, and $J_{SC} = 2.0$ mA/m$^2$. The corresponding measurement results of the non-contact mode ($d = 0.5$ mm) are shown in Fig. 5.17(b1–b3). The output performances for the non-contact mode are only slightly lower than those for the contact mode, with $V_{OC} = 220$ V, $Q_{SC} = 20$ μC/m$^2$, and $J_{SC} = 1.8$ mA/m$^2$. These results elucidate the FRD-TENG's capability of effectively generating electricity without further friction, indicating that a tight contact during the energy conversion process is no-longer necessary. Specifically, it can be found that the output profiles of both the $V_{OC}$ and the $Q_{SC}$ are triangular waveform, while the $J_{SC}$ yields square waveform (Insets of Fig. 5.17b), which reflects the time-differentiation relationship between

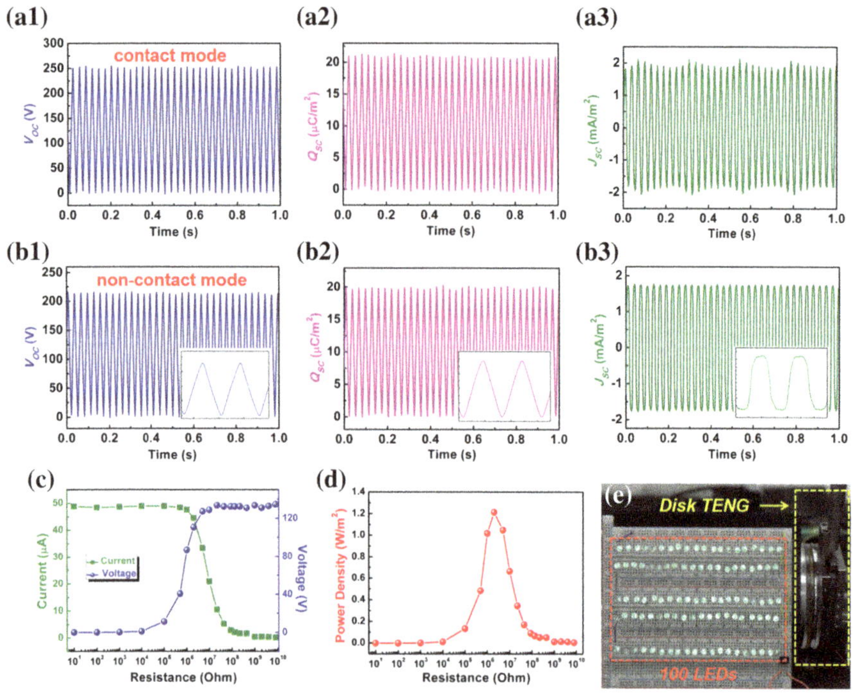


**Fig. 5.17** Electrical output characteristics of the FRD-TENG. **a1–a3** The measured $V_{OC}$, $Q_{SC}$, and $J_{SC}$ of the FRD-TENG at contact mode, respectively. **b1–b3** The measured $V_{OC}$, $Q_{SC}$, and $J_{SC}$ of the FRD-TENG at noncontact mode, respectively. The *insets* are the enlarged view showing the detailed shape of each output profile. **c** The measured output voltage and current across an external load with variable resistances. **d** The calculated effective power output of the FRD-TENG with variable resistances. **e** A snapshot from the video showing that the FRD-TENG is able to power up 100 serially connected LED instantaneously and continuously

the transferred charges and the current. To gain a comprehensive view of the effective output power on the external load, we conducted the measurement of the output voltage and current with variable load resistances, as shown in Fig. 5.17c. Generally, the output voltage increases with the loading resistance while the output current exhibits an opposite trend. Both curves have a quasi-linear region between 0.1 and 100 ΩMΩ, where their variations with the loading resistance become significant. The output power with different resistances could be calculated by simply multiplying the as-measured data of the output current and voltage, as summarized in Fig. 5.17d. Based on this result, it can be found that the maximum output power density of the FRD-TENG is 1.22 W/m$^2$, at a load resistance of ~2 ΩMΩ. The high output power of the FRD-TENG can be utilized for instantaneously and continuously driving 100 LEDs in serial connection, as shown in Fig. 5.17e, implying its applications as a self-lighting disk for security monitoring, wind detection, or amusement purposes.

## 5.3.3 Rotation-Disk Structure II

Besides the previous disk design, another rotation-disk triboelectric nanogenerator is designed for producing energy from rotary surfaces with unprecedented performance [5]. Enabled by a design of two radial-arrayed fine electrodes that are complementary on the same plane, the planar-structured TENG generates periodically changing triboelectric potential that induces alternating currents between electrodes. Operating at a rotation rate of 3000 rmin$^{-1}$, a TENG having a diameter of 10 cm can produce an open-circuit voltage ($V_{OC}$) of ~850 V and a short-circuit current ($I_{SC}$) of ~3 mA at a frequency of 3 kHz. Under the matched load of 0.8 M Ω, an average output power of 1.5 W (an area power density of 19 mWcm$^{-2}$) can be delivered to an external load at an efficiency of 24 %, representing a gigantic leap in terms of output power by orders of magnitude compared to previous reports. The TENG is demonstrated as an effective measure in harvesting a variety of ambient mechanical motions, such as light wind, water flow, and body movement. More importantly, a complete power-supplying system is built through integrating a power management circuit with the TENG, which can provide a continuous direct current (DC) source at a constant voltage for sustainably driving as well as charging various commercial electronics and thus immediately demonstrate the feasibility of the TENG as a practical technology. Given other compelling features of the TENG including small volume, light weight, low cost, and proven scalability, it is not only suited to harvest mechanical energy for self-powered electronics, but also it can be potentially applied to large-scale energy generation.

A TENG has a multi-layered structure, which consists of mainly two parts, i.e., a rotator and a stator, as sketched in Fig. 5.18a. The rotator is a collection of radially-arrayed sectors separated by equal-degreed intervals in between. With each sector unit having a central angle of 3°, the rotator has a total of 60 units. For the stator, it is divided into three components. A layer of fluorinated ethylene propylene

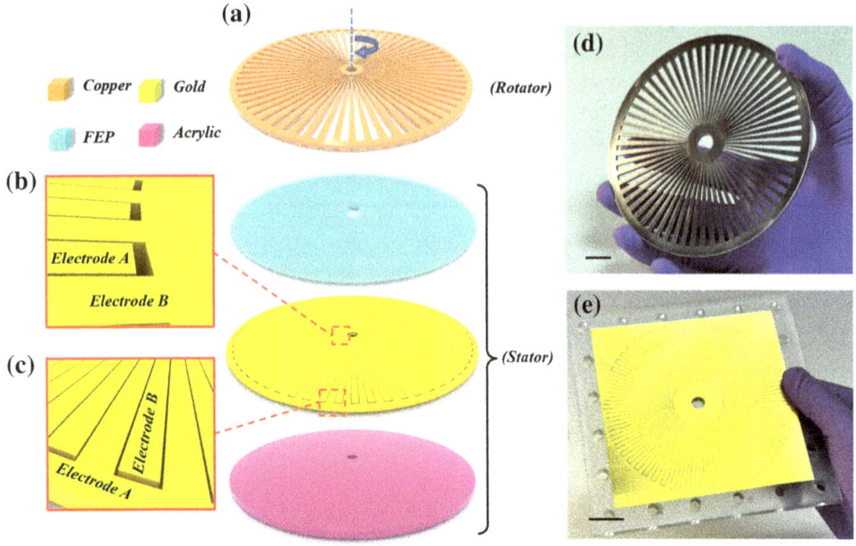


**Fig. 5.18** Structural design of the triboelectric generator. **a** Schematic illustrations of the triboelectric generator, which has two parts, that is, a rotator and a stator. The zoomed-in illustrations at the inner end (**b**) and the outer end (**c**) reveal that the two electrodes have complementary patterns, which are separated by fine trenches in between. It is noted that these drawings do not scale. **d** Photograph of a rotator (scale bar, 1 cm). **e** Photograph of a stator, in which the through-holes along edges are for mounting purpose (scale bar, 2 cm)

(FEP) as an electrification material, a layer of electrodes, and an underlying substrate are laminated along the vertical direction. The electrode layer is composed of two complementary-patterned electrode networks that are disconnected by fine trenches in between (Fig. 5.18b, c). Having the same pattern as that of the rotator, each network is formed by a radial array of sectors that are mutually connected at one end. The electrode layer is fully imbedded and stationary. This rational design not only leads to structural simplicity but also accounts for excellent robustness, making the TENG practically reliable and durable. As exhibited in Fig. 5.18d, e, both the rotator and the stator have two-dimensional planar structures, respectively, resulting in a small volume of the TENG. Detailed fabrication process is discussed in Methods.

To control the rotation rate for quantitative measurement, a programmable rotary motor was connected to the rotator that was in co-axial alignment with the stator. At a rotating rate of 500 rmin$^{-1}$, short-circuit current ($I_{SC}$) has a continuous AC output at an average amplitude of 0.5 mA (Fig. 5.19a). For open-circuit voltage ($V_{OC}$), it oscillates at the same frequency as that of $I_{SC}$ with a peak-to-peak value of 870 V (Fig. 5.19b), which corresponds well to the theoretical value though the slight deviation that is likely attributed to the fact that the actual contact area is less than the apparent device area because of surface roughness. In short-circuit condition, the amount of electrons in a single electron-transport process reaches 0.32 μC

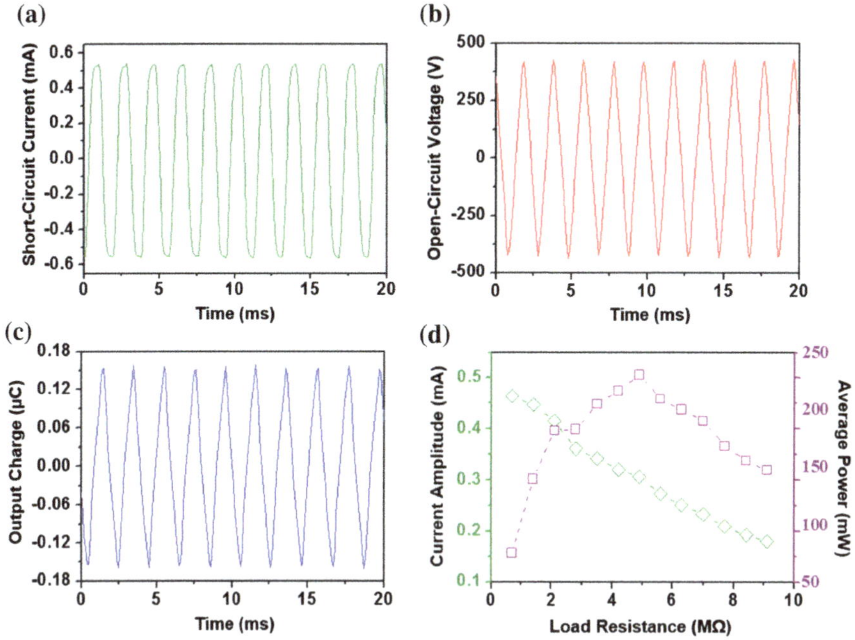


**Fig. 5.19** Results of electric measurements. **a** Short-circuit current (Isc) at a rotation rate of 500 rmin$^{-1}$. **b** Open-circuit voltage ($V_{oc}$) at a rotation rate of 500 rmin$^{-1}$. **c** Output charge at a rotation rate of 500 rmin$^{-1}$. **d** Load matching test at a rotation rate of 500 rmin$^{-1}$. Maximum average output power is obtained at the matched load of 4.9 MΩ

(Fig. 5.19c), which corresponds to an effective DC current ($I_{DC} = \Delta Q/\Delta t$) of 0.32 mA. It is noticed that the duration of a current peak is determined by the ratio between the central angle of a sector and the rotation rate (Fig. 5.19a). Once an external load is applied, the amplitude of the output current drops as the load resistance increases, as shown in Fig. 5.19d. The average output power is equivalent to the Joule heating of the load resistor, which can be calculated as $I^2_{\text{effective}} \cdot R$, where $I_{\text{effective}}$ is the effective current defined as the root mean square value of the current amplitude, and $R$ is the load resistance. At the matched load of 4.9 MΩ, the average output power reaches 0.23 W at a rotation rate of 500 rmin$^{-1}$ (Fig. 5.19d)·

Rotation rate is a major factor that determines electric output of the TENG. A linear relationship can be derived from Fig. 5.20a between the amplitude of $I_{SC}$ and the rotation rate since higher rate linearly shortens the duration of a current peak and thus boosts the current amplitude. In comparison, the amplitude of $V_{OC}$ remains stable regardless of the rotation rate (Fig. 5.20a) because it is only dependent on the position of the rotator. It is found that the matched load is also a variable value, exhibiting a reversely proportional relationship with the rotation rate, as diagramed in Fig. 5.20b. Consequently, linearly rising output power can be obtained at higher rotation rate (Fig. 5.20b). Given the linear behavior of the

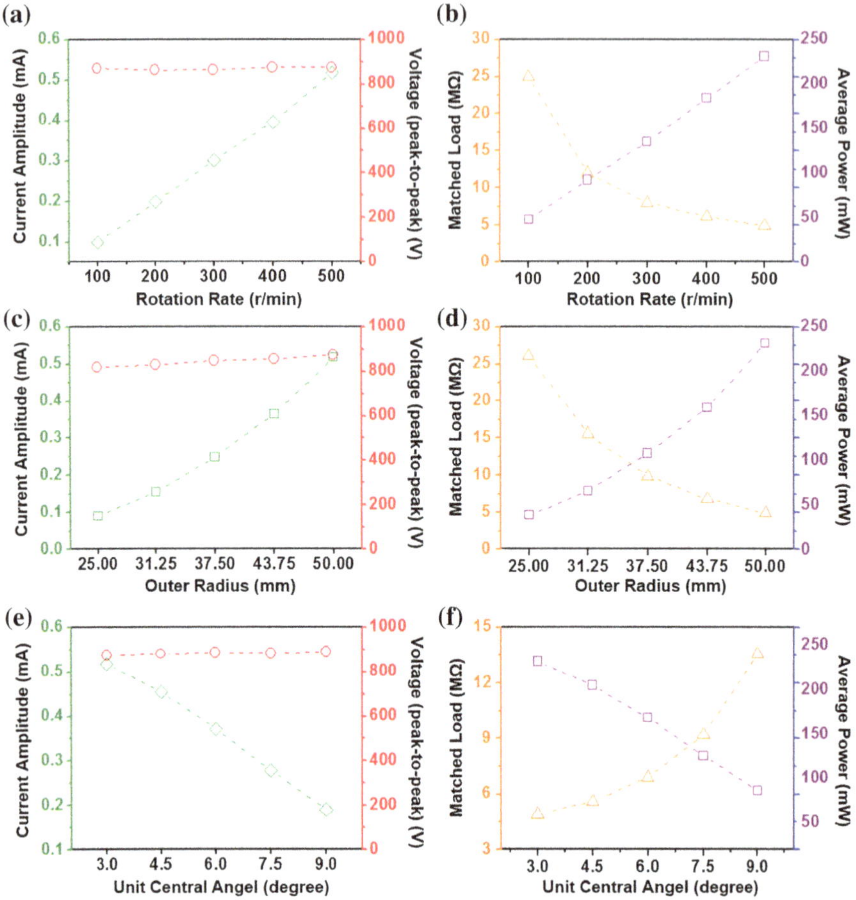


**Fig. 5.20** Factors that influence the electric output of the triboelectric generator. **a** Amplitude of $I_{sc}$ and peak-to-peak value of $V_{oc}$ with increasing rotation rate. **b** Matched load resistance and average output power with increasing rotation rate. **c** Amplitude of $I_{sc}$ and peak-to-peak value of $V_{oc}$ with increasing outer radius of the triboelectric generator (500 $rmin^{-1}$). **d** Matched load resistance and average output power with increasing outer radius of the triboelectric generator (500 $rmin^{-1}$). **e** Amplitude of $I_{sc}$ and peak-to-peak value of $V_{oc}$ with increasing central angle of a unit sector (500 $rmin^{-1}$). **f** Matched load resistance and average output power with increasing central angle of a unit sector (500 $rmin^{-1}$)

TENG, it delivers an optimum average output power of ~1.5 W at the matched load of ~0.8 MΩ when operating at a rotation rate of 3000 $rmin^{-1}$, which corresponds to an average output power density of 19 $mWcm^{-2}$. For the first time, the output power from triboelectrification-based nanogenerators boosts to a level where it is sufficiently powerful to drive daily used electronics, immediately resolving the most critical concern for the concept of power generation via triboelectrification.

Design parameters, especially the size and the unit central angle, can largely influence the output power of the TENG. Figure 5.20c shows an approximately quadratic dependence of the $I_{SC}$ and independence of the $V_{OC}$ on the radius of the TENG, which are consistent with the results in Eqs. (5.4a) and (5.5), respectively. With the matched load decreasing with the radius, the average output power exhibits a roughly quadratic relationship with the radius (Fig. 5.20d). In other words, the output power linearly scales with the area of the TENG since the triboelectrification is a surface charging effect that is area-dependent. Compared to the device size, the unit central angle reversely affects the output power. As revealed in Fig. 4e, the $I_{SC}$ linearly drops as the central angle increases, while the $V_{OC}$ still remains stable. Again, the measured results fit well with the theoretical model. Consequently, the average output power decreases in almost a linear way if devices with larger central angle are used. Therefore, fine feature size of the unit sector plays a key role in achieving high output power.

The efficiency of the TENG is defined as the ratio of the input mechanical power from the motor to the electric power that is delivered to the load. When driving the TENG at a rotation rate of 3000 $\text{rmin}^{-1}$, the motor exhibits a load factor of approximately 20 %, corresponding to an actual torque of 0.02 N m delivered to the TENG. Then we can derive the power input from the motor to be 6.28 W by using the above values. Given the electric output power of 1.5 W at the same rotation rate, the efficiency is calculated to be ~24 %. The reliability of the TENG, especially the resistance against mechanical wear, is important in evaluating its performance. Here, adhesive wear that occurs when two nominally flat solid bodies are in sliding contact applies to the TENG. Therefore, the adhesion of the deposited metal on the rotator largely determines the wear resistance. Special treatment was taken in fabricating the TENG, including adding an adhesion layer and surface plasma treatment before metal deposition. As a result, the TENG shows excellent stability and durability. After continuously producing more than 10 million cycles of alternating current, the output current does not even exhibit any measurable decay or degradation, which firmly proves the reliability of the TENG as a feasible approach for practical applications.

## 5.4 Rolling Friction Operation Mode

Up to now, four different working modes of the TENGs have been developed to accommodate applications in various types of mechanical motions, including the contact mode, the sliding mode, the single-electrode mode, and the freestanding mode [13, 14]. However, as an energy harvesting approach, the energy conversion efficiency and device durability are still essential issues, which may be limited by the relatively large frictional resistive force between triboelectric surfaces during the operation of the TENG, especially for the working modes based on sliding electrification. To improve the efficiency of the TENG and the robustness of the surface for anti-wearing purpose, most of efforts have been focused on enhancing the output power through sophisticated structural designs or surface modification [5, 15].

Here in this chapter, we introduced a novel working mode of rolling triboelectric nanogenerators (R-TENG) that can deliver ultrahigh energy conversion efficiency without scarifying its robustness and stability [6]. For the first time, rolling electrification between cylinder-shaped steel rods and planar thin films of fluorinated ethylene propylene (FEP) was employed for converting the kinetic energy of rolling rods into electric power. The rolling triboelectric nanogenerator (R-TENG) has a sandwiched structure composed of two FEP thin films and steel rods rolling between them. The rolling motion of the steel rods between the FEP thin films introduces triboelectric charges on both surfaces and led to the change of potential difference between each pair of electrodes on the back of the FEP layer, which drives the electrons flow in the external load. As power generators, each pair of output terminals can work independently that delivers an open-circuit voltage of 425 V, an instantaneous charge transfer of 0.145 μC, and a short-circuit current density of 5 $mA/m^2$. The two sets of output terminals can also be integrated to present an overall power density of up to 1.6 $W/m^2$. The impact of variable structural factors were investigated to achieve optimization of the output performance, such as the velocity of the rolling motion, the gap distance between adjacent electrodes, and the diameter of the steel rods. Owing to the low friction coefficient of the rolling movement, a high energy conversion efficiency of up to 55 % has been demonstrated, with much smaller wearing of triboelectric surfaces as compared with sliding friction. Based on the basic concept of rolling electrification, many other prototypes of R-TENGs have been successfully derived, like the rotating disk structure, the grating structure, and other structures based on rolling balls. This work demonstrates a new working mode of high-efficient and robust TENGs towards large scale energy harvesting.

The device structure of the R-TENG is schematically illustrated in Fig. 5.21a, which is composed of a group of rolling steel rods sandwiched by two layers of FEP thin films. Each FEP thin film was deposited with two separated copper electrodes on its back side, respectively. Polymer nanowire structures were created on both inner surfaces of the FEP thin film through inductively coupled plasma (ICP) etching to enhance the tribo-charge density upon rolling electrification. During the operation of the R-TENG, the bottom layer FEP film was securely attached onto a three-dimensional stage, and the top layer FEP film was driven by a linear motor to oscillate back and forth, which also initiated the rolling motion of the steel rods. As the top layer FEP film moved from the left-hand end to the right-hand end of the bottom layer FEP film, the steel rods would also move from the left-hand part to the right-hand part of the bottom layer FEP film, until the top layer FEP film stopped and started to move back. Figure 5.21b–e depicts the working principle of the R-TENG based on a step-by-step analysis of the charge transfer process. As the steel rods gets contacted with the entire area of both FEP films, due to their different abilities in attracting electrons, there will be net negative charges on the inner surface of the FEP films, and net positive charges on the surface of the steel rods. By assuming even distribution of the negative charges on

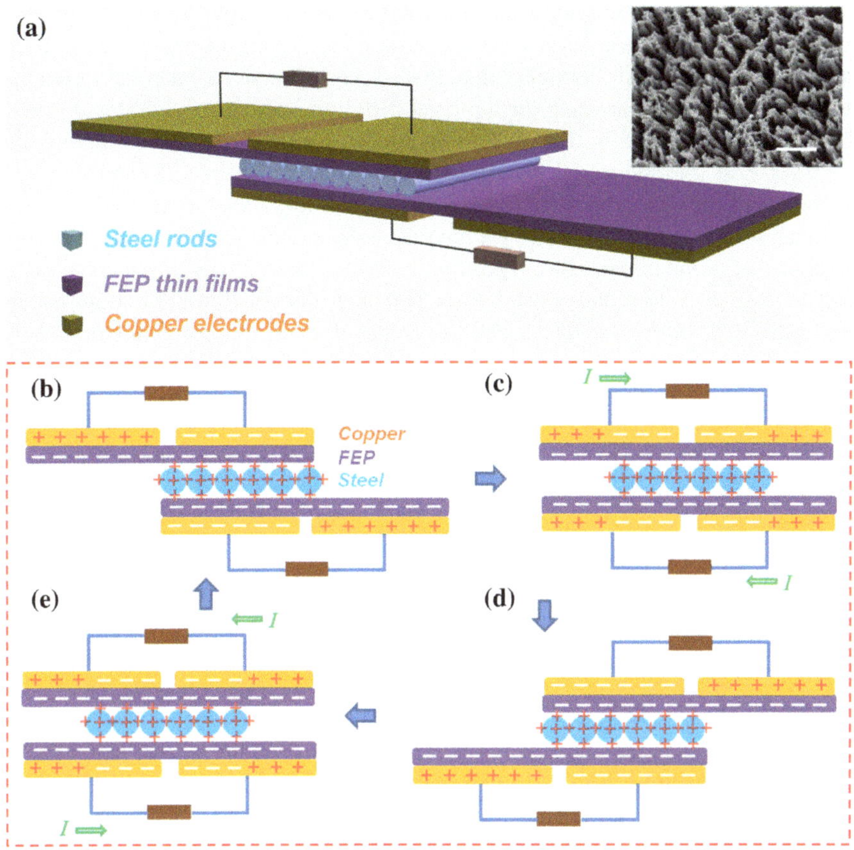


**Fig. 5.21** Structure and working principle of the rolling triboelectric nanogenerator. **a** Schematic illustration showing the device structure of the RTENG. *Inset* (*top right*) the scanning electron microscopy (SEM) image of the nanowire structure created on the inner surface of the FEP thin film. The scale bar is 1 µm. **b–e** Step-by-step illustration showing the working principle of the RTENG

the FEP surface and the symmetric layout of the electrodes, the negative charges will not lead to electric output, and the potential difference between each pair of electrodes will only rely on the relative location of the positively-charged steel rods. Initially (Fig. 5.21b), the positively-charged steel rods are located beneath the right-hand electrode (RE) of the top layer FEP film, and over the left-hand electrode (LE) of the bottom layer FEP film. Hence, for the top layer, the RE poses a higher electrical potential than the LE, leading to an instantaneous electrons flow from the LE to the RE to reach an equilibrium state. The opposite potential polarity (the potential of the LE is higher than that of the RE) exists in the bottom layer. As the top layer FEP film starts to slide towards the right-hand side along with the positively-charged steel rods, the potential difference between the RE and the LE on the top layer will be gradually reduced, until it reaches the equal potential condition in Fig. 5.21c. As the top layer FEP film continues moving forward, the electrical potential on the LE will be higher than that of the RE, and the potential difference will reach the maximum when the top layer FEP film stops at the right-hand side of the bottom layer FEP film (Fig. 5.21d). In the next half cycle of rolling motion, the top layer FEP film moves back from the right-hand side to the left-hand side, and the potential difference between the LE and the RE first drops to zero (Fig. 5.21e) and then becomes negative until reaching the original position (Fig. 5.21b). In this process, due to the moving of the positively-charged rolling rods in reference to the bottom layer FEP film, the revolution of potential difference in the bottom pair of electrodes will follow a reverse trend as compared to the top layer. In terms of the short-circuit condition, the change of potential difference in the first half cycle of motion will lead to electrons flow from the RE to the LE in the top layer, and the electrons will flow back from the LE to the RE in the second half cycle. Similar to the revolution of potential difference, the short-circuit electrons flow follows an opposite direction in the bottom layer as compared to the top layer.

In order to achieve a comprehensive understanding about the working performance of the R-TENG, the dependences of the output performance on various structural parameters were investigated, including the sliding velocity, the electrode separation distance, and the diameter of the rolling rods, as shown in Fig. 5.22. For convenience purpose, all of the measurements in this section were carried out on the bottom output terminals. First, with other parameters kept constant, the output performances with variable sliding velocities were measured (Fig. 5.22a, b), and it can be found that while the $V_{OC}$ remains unchanged with the sliding velocity, the $J_{SC}$ is raised with increasing velocities from 0.1 to 0.5 m/s. The different trends can be attributed to the fact that the $V_{OC}$ is a static signal that only depends on the amount of tribo-charges and the magnitude of the charge separation; on the other hand, the $J_{SC}$ is a dynamic signal that is also highly proportional to the rate of the charge transfer process. Hence, the maximum output power also increases with the sliding velocity. Second, with the horizontal separation distance between two electrodes on each FEP thin film increasing from 1.6 to 25.4 mm, the $V_{OC}$ shows a slight enhancement, but the $J_{SC}$ decays drastically (Fig. 5.22c, d). Theoretically, the $V_{OC}$ should have increased with the elevation of gap distance, since a larger distance between the two electrodes leads to a smaller capacitance, which is inversely proportional to the $V_{OC}$ with constant amount of transferred charges; however, this increasing trend can hardly be recorded due to the non-ideal inner resistance of the voltage meter. The increase of the separation distance also results in longer time for the charge transfer process and hence lower $J_{SC}$, which is consistent with the measurement results. Based on these results, it is easy to understand that total power output declines with larger gap distance. Third, as displayed in Fig. 5.22e, f, both the $V_{OC}$ and $J_{SC}$ decrease with the increase of the diameter of the rolling rods. Assuming constant charge density on the steel rods with variable diameters, the surface charges on each single rod can be considered as equivalent charge centers at axial line of the cylinder structure. Hence, as the diameter of the rolling rods grows larger, the distance between the charge center and the electrodes keeps increasing,

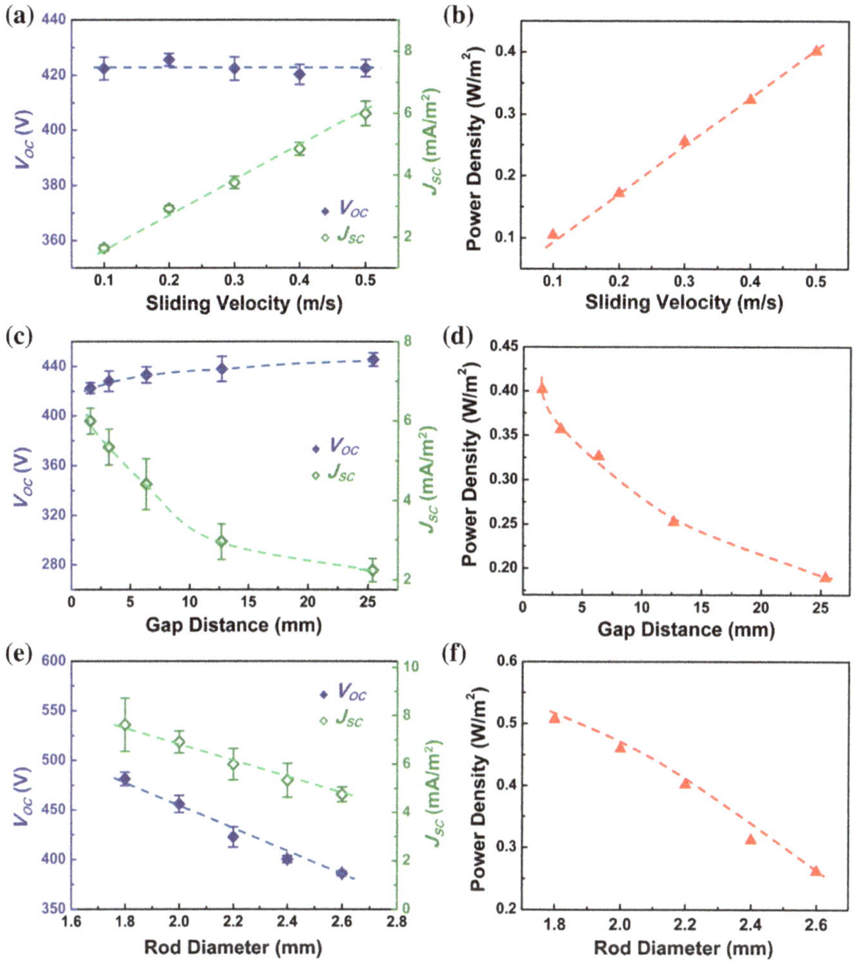


**Fig. 5.22** Dependences of the output performance of the RTENG on variable structural parameters. **a** Impact of the sliding velocity on the open-circuit voltage and the short-circuit current density of the RTENG. **b** Impact of the sliding velocity on the optimum output power density of the RTENG. **c** Impact of the gap distance between two electrodes on the same FEP thin film on the open-circuit voltage and the short-circuit current density of the RTENG. **d** Impact of the gap distance on the optimum output power density of the RTENG. **e** Impact of the rod diameter on the open-circuit voltage and the short-circuit current density of the RTENG. **f** Impact of the rod diameter on the optimum output power density of the RTENG

which renders lower potential difference than that with the smaller rolling rods. All in all, combining the results demonstrated by Fig. 5.22c, f, it can be concluded that lower rod size and gap distance are favorable for the overall output performance of the R-TENG.

## 5.5 Energy Conversion Efficiency

Rolling rods structure was utilized to minimize the frictional force in the power generator without scarifying the output power, and thus it may enhance the energy conversion efficiency. In this regard, the energy conversion efficiency was evaluated by measurement of the resistive force in both open-circuit condition and variable load resistances. It can be found that the resistive force became larger when load resistances were connected (Fig. 5.23a), for the reason that the current flow or energy consumption in the load will yield higher electrostatic resistance to the rolling motion. Hence, by assuming constant frictional energy loss from heat dissipation, the energy conversion efficiency could be calculated by the amount of additional work done with external load divided by the total work through the following equation:

$$\eta = \frac{E_{output}}{E_{input}} = \frac{\int Fdt - \int F_0 dt}{\int Fdt} \tag{5.14}$$

Here, $\eta$ is the energy conversion efficiency, $E_{\text{output}}$ is the electrical energy output, $E_{\text{input}}$ is the mechanical energy input, $F$ is the resistive force measured with a certain load resistance, and $F_0$ is the resistive force measured with open-circuit condition. The calculation results are displayed in Fig. 5.23b, indicating that the instantaneous efficiency can be up to 55 % when the external resistance of the load is compatible with the internal impedance of the R-TENG at current motion status. The high energy conversion efficiency can be attributed to the low frictional coefficient between the rolling rods and the planar FEP surface, which is considered a unique advantage of this design. From the measurement of normalized friction coefficient (Fig. 5.23c), it can be found that the frictional coefficient of the rolling rod structure is substantially lower than that of the planar structures. This high-efficient R-TENG could be employed for harvesting energy from gentle finger movement, which was then utilized for driving portable electronics, like light-emitting diodes (Fig. 5.23d). Furthermore, the durability of the R-TENG was verified by monitoring the surface morphology of the polymer nanowire structures over a long period of operations, as shown in Fig. 5.23e. Two samples of the nanostructured FEP thin films were set up to undergo the rolling friction and sliding friction for 1000 cycles, respectively. The scanning electron microscopy (SEM) images of both samples were taken before and after the rolling or sliding frictions. It is clearly indicated that the sample applied with rolling friction had little degradation in the nanowire structures, while the most of the nanowires on the other sample enduring sliding friction were destroyed by the frictional force. With this comparison, the high durability of the R-TENG was successfully demonstrated, which confirmed another unique advantage of the R-TENG as compared to the traditional sliding TENGs.

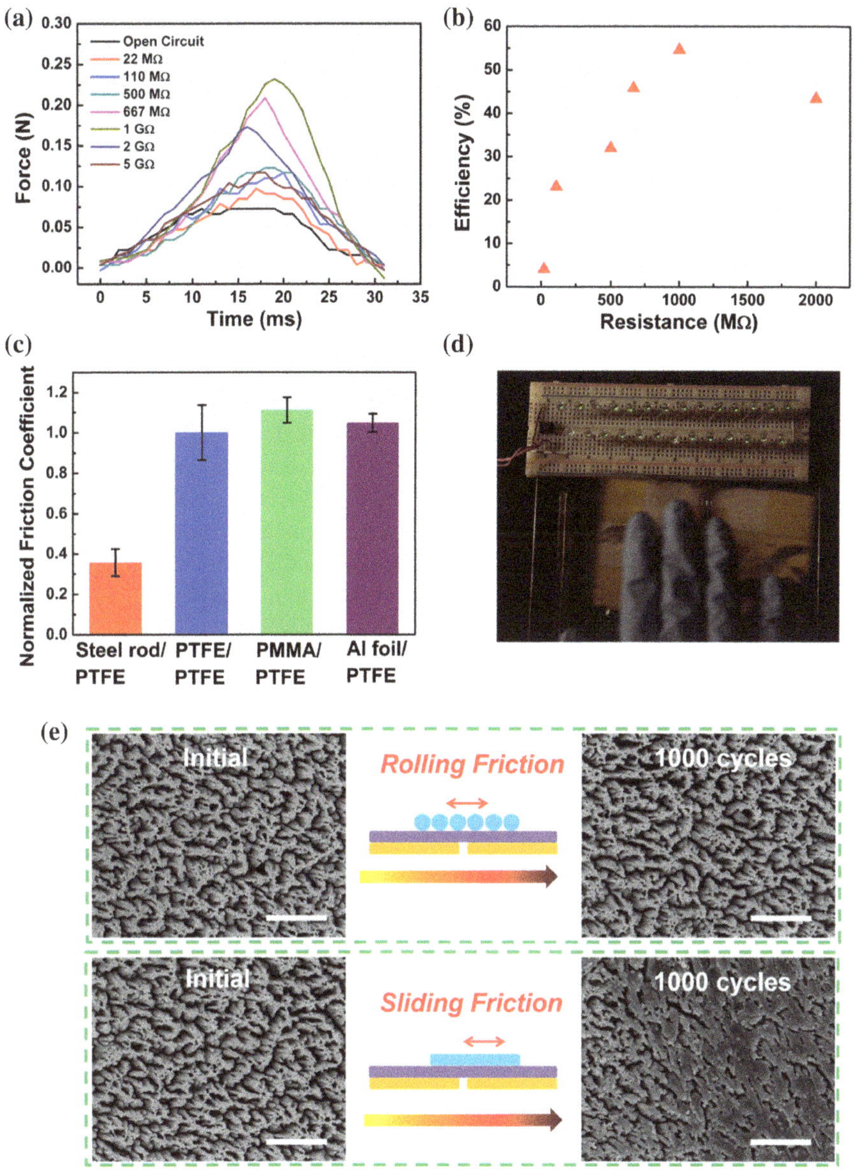


**Fig. 5.23** Characterization of the energy conversion efficiency of the RTENG. **a** Measurement of the resistive force during the energy conversion process of the RTENG with variable loading resistances. **b** Calculated energy conversion efficiency with variable load resistances. **c** Normalized friction coefficient between different types of materials customized for triboelectric nanogenerators. The friction coefficient between PTFE films were set as the reference value. **d** Capture of live demonstration indicating the instantaneous powering of portable electronics by the RTENG. **e** Comparison of the impact of surface morphology between rolling friction steel rods (*top*) and sliding friction by steel foils (*bottom*). The *left* two SEM images show the surface morphologies of the two samples before the friction and the *right* two SEM images show the surface morphologies after the friction, respectively. The scale bars are 2 μm for all images

Other than this structure, by using the non-contact operation of the linear-grating structure, the highest total energy conversion efficiency of up to 85 % was demonstrated, as we discussed in Fig. 5.15. This energy conversion efficiency is different from the instantaneous energy conversion efficiency, as it accounts all of the electricity generated by the residual vibrations of the device. In addition, with the rotation disk structure, a high instantaneous efficiency of up to 24 % was conveyed.

## 5.6 Summary

The freestanding triboelectric-layer mode usually consists of a freestanding triboelectric layer and a pair of stationary electrodes. The triboelectric layer is driven to move in between the two electrodes to induce periodical change of potential difference between them. The potential difference then drives the cyclic electrons flow between the two electrodes to complete the power generation process. Similar with the single-electrode mode, this working mode does not require electrode deposition on the moving part, thus improving the convenience of fabrications and operations. Furthermore, this working mode could deliver a much higher efficiency of charge transfer than that of the single-electrode mode, for the screening effect in single-electrode mode is diminished. Hence high output performance could be delivered.

Several unique structures have been developed for this working mode, including the plain sliding structure, the contact-separation structure, the linear grating structure, the rotation disk structure. The last two prototypes were designed to improve the output power by incorporating multiple energy conversion cycles in a single cycle of mechanical motion. Moreover, non-contact energy harvesting was enabled for this specific working modes, with almost the same level of output power as that of the contact operations, but with largely-reduced frictional loss. Also, rolling electrification instead of sliding electrification could realize the same function to improve the robustness and energy conversion efficiency of the triboelectric nanogenerators.

In summary, the freestanding triboelectric-layer working mode of TENG could deliver yet the highest output current (3 mA), output power (1.5 W) and total energy conversion efficiency (85 %). The converted electrical energy was sufficient to drive home appliance like light bulbs, small electronics, and commercialized cell phones. This design is ideal for harvesting energy from a moving object without connecting a wire or even without direct physical contact, such as walking on floor and a moving automobile or train.

# References


1. S.H. Wang, Y.N. Xie, S.M. Niu, L. Lin, Z.L. Wang, Freestanding triboelectric-layer-based nanogenerators for harvesting energy from a moving object or human motion in contact and non-contact modes. Adv. Mater. **26**(18), 2818–2824 (2014)
2. S. Wang, S. Niu, J. Yang, L. Lin, Z.L. Wang, Quantitative measurements of vibration amplitude using a contact-mode freestanding triboelectric nanogenerator. ACS Nano **8**(12), 12004–12013 (2014)
3. Y.N. Xie, S.H. Wang, S.M. Niu, L. Lin, Q.S. Jing, J. Yang, Z.Y. Wu, Z.L. Wang, Grating-structured freestanding triboelectric-layer nanogenerator for harvesting mechanical energy at 85 % total conversion efficiency. Adv. Mater. **26**(38), 6599–6607 (2014)
4. L. Lin, S.H. Wang, S.M. Niu, C. Liu, Y.N. Xie, Z.L. Wang, Noncontact free-rotating disk triboelectric nanogenerator as a sustainable energy harvester and self-powered mechanical sensor. ACS Appl. Mater. Interfaces **6**(4), 3031–3038 (2014)
5. G. Zhu, J. Chen, T.J. Zhang, Q.S. Jing, Z.L. Wang, Radial-arrayed rotary electrification for high performance triboelectric generator. Nat. Commun. **5**, 3426 (2014)
6. L. Lin, Y.N. Xie, S.M. Niu, S.H. Wang, P.K. Yang, Z.L. Wang, Robust triboelectric nanogenerator based on rolling electrification and electrostatic induction at an instantaneous energy conversion efficiency of 55 %. ACS Nano **9**(1), 922–930 (2015)
7. A.F. Diaz, R.M. Felix-Navarro, A semi-quantitative tribo-electric series for polymeric materials: the influence of chemical structure and properties. J. Electrostat. **62**(4), 277–290 (2004)
8. H. Fang, W.Z. Wu, J.H. Song, Z.L. Wang, Controlled growth of aligned polymer nanowires. J. Phys. Chem. C **113**(38), 16571–16574 (2009)
9. M. Eguchi, XX. On the permanent electret. Philos. Mag. **49**(289), 178–192 (1925)
10. G. Sessler, J. Hillenbrand, Electromechanical response of cellular electret films. Appl. Phys. Lett. **75**(21), 3405–3407 (1999)
11. W.J. Kong, L. Cheng, X.D. He, Z.H. Xu, X.Y. Ma, Y.D. He, L.J. Lu, X.P. Zhang, Y.Q. Deng, Electret-based microfluidic power generator for harvesting vibrational energy by using ionic liquids. Microfluid. Nanofluid. **18**(5–6), 1299–1307 (2015)
12. S.M. Niu, Y. Liu, X.Y. Chen, S.H. Wang, Y.S. Zhou, L. Lin, Y.N. Xie, Z.L. Wang, Theory of freestanding triboelectric-layer-based nanogenerators. Nano Energy **12**, 760–774 (2015)
13. Z.L. Wang, Triboelectric nanogenerators as new energy technology for self-powered systems and as active mechanical and chemical sensors. ACS Nano **7**(11), 9533–9557 (2013)
14. Z.L. Wang, J. Chen, L. Lin, Progress in triboelectric nanogenerators as a new energy technology and self-powered sensors. Energy Environ. Sci. **8**(8), 2250–2282 (2015)
15. G. Zhu, Y.S. Zhou, P. Bai, X.S. Meng, Q.S. Jing, J. Chen, Z.L. Wang, A shape-adaptive thin-film-based approach for 50 % high-efficiency energy generation through micro-grating sliding electrification. Adv. Mater. **26**(23), 3788–3796 (2014)

# Chapter 6
# Theoretical Modeling of Triboelectric Nanogenerators

Practical applications still requires huge improvement of the output performance of TENGs, so rational design and careful optimization of both materials and structures is completely necessary especially when the current performance of TENG is already very high. Moreover, similar to the development of complementary metal-oxide-semiconductor field-effect-transistor based integrated circuits and systems, the fully-integrated energy harvesting systems that contain TENGs, power management circuits, signal processing circuits, energy storage elements, and/or load circuits are essential. Theoretical simulation plays a key role in understanding the working mechanism and analyzing the output performance of the entire system. Finally, from methodology point of view, simulation is always a necessary step in the whole device design process, because performing control experiments is usually time-consuming and not cost-effective. Thus a thorough theoretical understanding of TENGs is completely urgent in the whole research field. This knowledge can help choosing the appropriate TENG structure and materials, avoid designs which will greatly harm the output performance, and choose suitable system-level topologies for integrated energy harvesting systems.

The objective of this chapter is to give a summary about the fundamental theory research of triboelectric nanogenerators [1]. First, we will discuss the governing equation, equivalent circuit model, and simulation method. Then, we will discuss the resistive load and capacitive load characteristics in detail. The theory presented in this chapter can provide a guideline for understanding the TENG and its practical implication in self-powered systems.

 
Z.L. Wang et al., *Triboelectric Nanogenerators*,
Green Energy and Technology, DOI 10.1007/978-3-319-40039-6_6

## 6.1 Inherent Capacitive Behavior and Governing Equations: *V*-*Q*-*x* Relationship

The fundamental working principle of TENGs is a conjugation of contact electrification and electrostatic induction. Contact electrification provides static polarized charges and electrostatic induction is the main mechanism that converts mechanical energy to electricity. Since the most fundamental device based on electrostatics is a capacitor, fundamentally TENG will have inherent capacitive behavior [2].

An arbitrary TENG is analyzed to unveil its inherent capacitive behavior. For any triboelectric generators, there are pair of materials which are face to each other (called tribo-pairs). The distance ($x$) between these two triboelectric layers can vary under the agitation of mechanical force. After brought into contact with each other, the contact surfaces of the two triboelectric layers will have opposite static charges (tribo-charges), as a result of contact electrification. Besides the tribo-pair layers, there are two electrodes that are carefully insulated inside the TENG system, which ensures the charges can only transfer between the two electrodes through external circuits. If we define the transferred charges from one electrode to another is $Q$, one electrode will have the transferred charge $-Q$ and the other electrode will have the transferred charge $+Q$.

The electrical potential difference between the two electrodes of any TENGs mainly contributes to two parts. The first part is from the polarized triboelectric charges and their contribution to the voltage is $V_{OC}(x)$, which is a function of separation distance $x$. Besides, the already transferred charges $Q$ will also contribute to an electric potential difference. If we assume no triboelectric charges exist in this structure, this structure is completely a typical capacitor, so the contribution of these already transferred charges is $-Q/C(x)$, where $C$ is the capacitance between the two electrodes. Therefore, due to the electrical potential superposition principle, the total voltage difference between the two electrodes can be given by [2]:

$$V = -\frac{1}{C(x)}Q + V_{OC}(x) \tag{6.1}$$

Equation (6.1) (named as *V*-*Q*-*x* relationship) is the governing equation of any TENGs, clearly explaining its inherent capacitive behavior [2]. The separation of the polarized tribo-charges will generate an electrical potential difference between the two electrodes. If the external circuit exists between the two electrodes, this electrical potential will drive electrons to flow from one electrode to another. These already-transferred electrons can further screen the electrical potential between the two electrodes [3]. Under short-circuit (SC) conditions, these transferred charges ($Q_{SC}$) fully screen the electrical potential generated from polarized triboelectric charges. Therefore, the following equation can be easily derived for TENGs under short-circuit conditions [4].

$$0 = -\frac{1}{C(x)} Q_{SC}(x) + V_{OC}(x) \tag{6.2}$$

Thus, the fundamental relationship among $Q_{sc}$, $C$, and $V_{oc}$ can be given by [4]:

$$Q_{SC}(x) = C(x) V_{OC}(x) \tag{6.3}$$

## 6.2 First-Order Lumped-Parameter Equivalent Circuit Model

From the governing equation (*V*-*Q*-*x* relationship) of triboelectric nanogenerators, we can easily derive their lumped parameter equivalent circuit model [5]. From Eq. (6.1) shown above, there are two terms at the right side, which can be represented by two circuit elements in the equivalent circuit model. The first one is a capacitance term, which is originated from the inherent capacitance between the two electrodes and can be represented by a capacitor (*C*). The other is an open-circuit voltage term, which is originated from the separation of the polarized tribo-charges and can be represented by an ideal voltage source ($V_{OC}$). With combination of these two terms, the whole lumped parameter equivalent circuit model can be represented by a serial connection of an ideal voltage source and a capacitor, as shown in Fig. 6.1 [5].

Besides, it is valuable to note that the inherent impedance of TENG mainly comes from its inherent capacitance. For a rationally designed TENG, the internal resistance is close to infinity mainly for the excellent insulation between the two electrodes. So in the first-order equivalent circuit model, this resistance was not included.

Finally, we want to briefly discuss the difference between TENGs and traditional electromagnetic induction generators. First, their fundamental operating principles are different: TENGs rely on the conjugation effect of contact electrification and electrostatic induction while electromagnetic induction generators rely on the

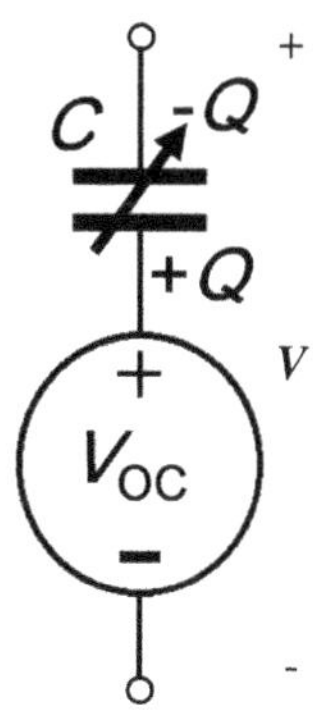


**Fig. 6.1** First-order lumped-parameter equivalent circuit model of any triboelectric nanogenerators. Reproduced with permission from Elsevier [5]

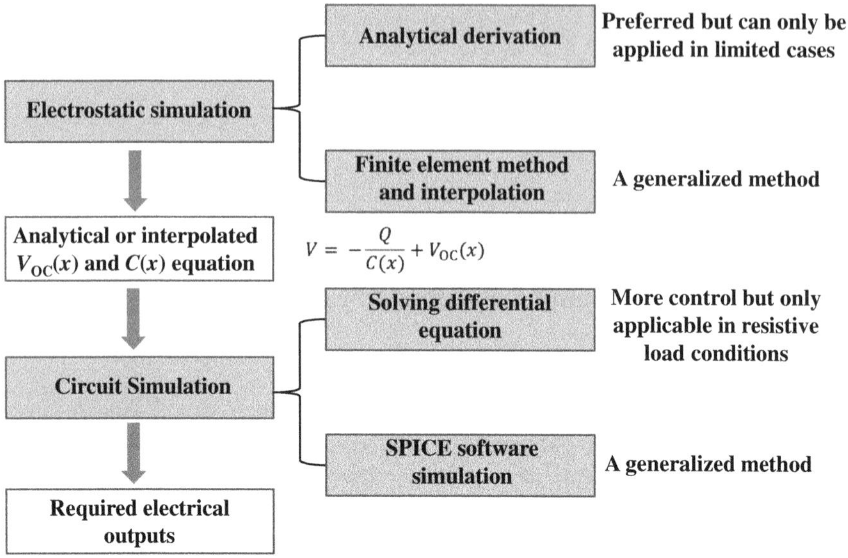


**Fig. 6.2** Flow chart of typical TENG simulation calculation

electromagnetic induction. Second, TENGs are inherently capacitive and the internal resistance for rational designed TENGs is close to infinity. Their internal high impedance is mainly from their small inherent capacitance. However, electromagnetic induction generators are inherently resistive and their internal resistance is much lower than the impedance of TENGs. Third, the equivalent circuit model for TENGs is a serial connection of an ideal voltage source with a capacitor while that for electromagnetic induction generators is a serial connection of an ideal voltage source with a resistor. In summary, TENGs are high-impedance charge sources and electromagnetic generators are low-resistance voltage sources.

From the equivalent circuit model, we can summarize the fundamental simulation method for TENGs. The simulation of TENG-based energy harvesting systems mainly contains two parts, including electrostatic simulation and load circuit simulation, as summarized in Fig. 6.2 [5].

The main objective of the electrostatic simulation is to obtain the $V_{OC}(x)$ and $C(x)$ functions in the $V$-$Q$-$x$ relationship. Two methods are currently developed to specify the $V_{OC}(x)$ and $C(x)$ relationships. Analytical derivation is a preferred method but it only works for certain geometry features such as parallel-plate attached-electrode contact-mode and sliding-mode TENGs with appropriate approximations. A more general method which works for all TENGs is numerical calculation based on finite element method (FEM). The finite element calculation can be easily performed utilizing common FEM simulation software such as COMSOL and ANSYS. In this numerical method, the value of $V_{OC}$ and $C$ is first obtained at certain values of $x$. Then, from continuous fraction interpolation [6], a numerical $V_{OC}(x)$ and $C(x)$ relationship can be generated in the entire $x$ region.

Once the numerical $V_{OC}(x)$ and $C(x)$ relationship is obtained, the output characteristics can be numerically calculated through load circuit simulation. When the load is very simple, the governing differential equation of whole system can be easily derived through Kirchhoff's law. For example, when the load is purely resistive, the governing differential equation containing the TENG and the resistor can be given by [5]:

$$R\frac{\mathrm{d}Q}{\mathrm{d}t} = V = -\frac{1}{C}Q + V_{OC} \tag{6.4}$$

Such a differential equation can be solved by numerical integration and for some specific TENG systems and motion conditions, even analytical results can also be obtained. However, this kind of differential equation can only be obtained for limited cases with very simple load circuits. When the load circuit is complicated, solving the differential equation set is nearly an impossible task.

A general method that works for all TENG systems is to utilize SPICE (simulation program with integrated circuit emphasis) software [5]. From the above TENG equivalent circuit model, the TENG can be embedded into the SPICE software as a basic element consisting of a voltage source in serial connection with a capacitor. Then any other circuit elements, including resistors, capacitors, inductors, diodes, and transistors can be put into the software to form the entire TENG system. After specifying the motion process and the initial condition, the powerful SPICE software can be easily utilized to simulate the real-time output of any TENG systems.

## 6.3 Charge Reference State

The reason to define a charge reference state for TENG is simple. For any TENGs under short-circuit conditions, the charge distribution on the two electrodes is unique. However, the cases are completely different under open-circuit conditions. We only know the total charges amount on the two electrodes as a combination and the charge amount on each electrode is a constant because of no charge transfer between the two electrodes. However, the detailed charge distribution on one electrode can be arbitrary and therefore the voltage between the two electrodes is arbitrary. Therefore, to facilitate the theoretical calculation, we need to assign a charge reference state that one electrode has specific amount of charge ($Q_0$) and under this condition, we can calculate the basic characteristics of TENGs [7].

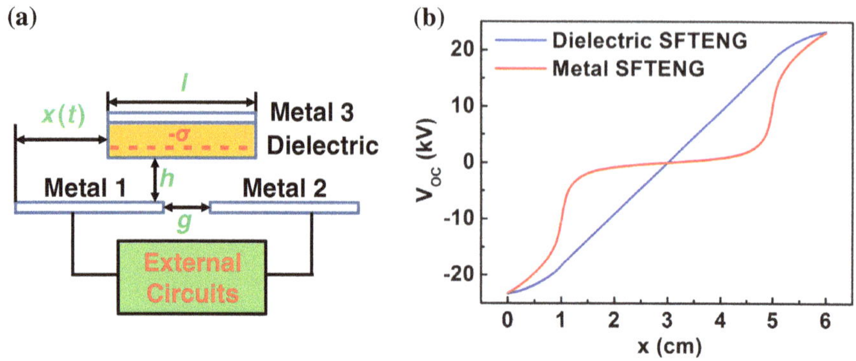


**Fig. 6.3** **a** Structure of a dielectric freestanding-layer sliding-mode TENG. **b** Open-circuit profile of Dielectric freestanding sliding-mode TENG and metal freestanding sliding-mode TENG under symmetric reference state. Reproduced with permission from Elsevier [7]

### *6.3.1 Influence of Charge Reference State on the Intrinsic Characteristics of TENGs*

For any TENGs, their intrinsic characteristics are their $V_{OC}$-$x$, $Q_{SC}$-$x$ and $C$-$x$ curves. Of course, the $C$-$x$ curve is completely independent of the charge reference state. In fact, the charge reference state pick-up will not affect the magnitude and the shape of the $Q_{SC}$-$x$ curve, but only makes them shift upward or downward. But the chosen of the charge reference state will strongly affect both the magnitude and the shape of the $V_{OC}$-$x$ curve if the inherent TENG capacitance is not a constant and varies with $x$ [7].

For example, we can analyze the metal sliding-mode freestanding-layer-based TENG electrostatic system shown in Chap. 5. (Figure 6.3a). In the system under OC condition, we only know that the total charge amount on metal 1 and metal 2 as a combination is $\sigma wl$ and the total charge amount on metal 3 is 0. Under SC condition, the charge distribution on the two electrodes is unique, so the charge amount on metal 1 and metal 2 under short circuit conditions ($Q_1$ and $Q_2$, respectively) is a specific number if the position of the freestanding layer is given.

But the voltage between metal 1 and 2 relies on the charge reference state. If we assign a charge reference state A that the amount of charge on metal 2 is $Q_{0A}$ (at the same time, the amount of charge on metal 1 is $\sigma wl - Q_{0A}$), we can calculate the open-circuit voltage between metal 1 and 2 ($V_{OCA}$) through FEM calculation. Besides, the short-circuit transferred charges under charge reference state A ($Q_{SCA}$) can also be calculated as $Q_2 - Q_{0A}$.

If we choose a new reference state B that the amount of charge on metal 2 is $Q_{0B}$, we can analyze its influence on the calculated $V_{OC}$ and $Q_{SC}$. Under this condition, the voltage is contributed from two parts of the charges. One part is the same as the case of charge reference state A and the voltage generated by these charges is also $V_{OCA}$. The other part is the extra $Q_{0A} - Q_{0B}$ charges on metal 1 and the extra $Q_{0B} - Q_{0A}$ charges on metal 2 and the voltage generated by these charges is $-(Q_{0B} - Q_{0A})/C$, where $C$ is the total capacitance between metal 1 and metal 2. Therefore, because of the electric potential superposition principle, the open-circuit voltage between metal 1 and 2 ($V_{OCB}$) under charge reference state B can now be given by:

$$V_{OCB} = V_{OCA} - \frac{Q_{0B} - Q_{0A}}{C} \tag{6.5}$$

Besides, the short-circuit transferred charges under charge reference state B ($Q_{SCB}$) can be given by:

$$Q_{SCB} = Q_2 - Q_{0B} = Q_2 - Q_{0A} - (Q_{0B} - Q_{0A}) = Q_{SCA} - (Q_{0B} - Q_{0A}) \tag{6.6}$$

Therefore, from the above two equations, we can fully understand the influence of charge reference state on the $Q_{SC}$-$x$ and $V_{OC}$-$x$ curve. From Eq. (6.6), we know that the $Q_{SC}$-$x$ curve under charge reference state B is just a vertical translation of the curve under charge reference state A. However, the $V_{OC}$-$x$ curve change is much more complicated because that the capacitance may also be a function of $x$. Therefore, both the magnitude and the shape of the $V_{OC}$-$x$ curve may change if a different charge reference state is picked up. But if the capacitance is a constant, the $V_{OC}$-$x$ curve under charge reference state B will also be a vertical translation of the curve under charge reference state A.

### 6.3.2 *Influence of Charge Reference State on the Output Characteristics of TENGs*

The chosen of the charge reference state will not affect the calculated voltage and current output when a resistance load is connected with the TENG as long as right boundary condition for solving the differential equation is picked up as well.

For example, we can still analyze the above case. To calculate the output characteristics, we need to specify the motion condition which is $x = x(t)$ and the boundary condition which is the amount of charges on metal 1 when $t = 0$ ($Q_{\text{initial}}$).

If we choose the charge reference state A, the differential equation is:

$$R\frac{dQ_A}{dt} = -\frac{1}{C}Q_A + V_{OCA} \tag{6.7}$$

And the corresponding boundary condition that the amount of charges on metal 2 when $t = 0$ is $Q_{\text{initial}}$ is mathematically shown as the following equation under charge reference state A.

$$Q_A(t=0) = Q_{\text{initial}} - Q_{0A} \tag{6.8}$$

Under charge reference state B, the differential equation and corresponding boundary condition can be given by:

$$R\frac{\mathrm{d}Q_B}{\mathrm{d}t} = -\frac{1}{C}Q_B + V_{\text{OCB}} \tag{6.9}$$

$$Q_B(t=0) = Q_{\text{initial}} - Q_{0B} = Q_A(t=0) - (Q_{0B} - Q_{0A}) \tag{6.10}$$

And utilizing Eqs. (6.5), (6.9) can be simplified to:

$$R\frac{\mathrm{d}(Q_B + (Q_{0B} - Q_{0A}))}{\mathrm{d}t} = R\frac{\mathrm{d}Q_B}{\mathrm{d}t} = -\frac{1}{C}Q_B + V_{\text{OCA}} - \frac{Q_{0B} - Q_{0A}}{C} \tag{6.11}$$

Thus,

$$R\frac{\mathrm{d}(Q_B + (Q_{0B} - Q_{0A}))}{\mathrm{d}t} = -\frac{1}{C}(Q_B + (Q_{0B} - Q_{0A})) + V_{\text{OCA}} \tag{6.12}$$

Compare Eqs. (6.12) and (6.7) with Eqs. (6.10) and (6.8), we can easily obtain that:

$$Q_B = Q_A - Q_{0B} + Q_{0A} \tag{6.13}$$

Since the current and voltage output on the resistor is the differential of $Q_A$ and $Q_B$ by $t$, the calculated current and voltage output will be independent of the choice of charge reference state. Similarly, if the periodic boundary condition is utilized to solve the equation, the calculated current and voltage output will also be independent of the choice of charge reference state.

### 6.3.3 *Typical Charge Reference States*

The most commonly utilized charge reference state is called minimum achievable charge reference state (MACRS), which utilizes the amount of charge on metal 2 under SC condition when $x = 0$ as $Q_0$ (which is also the minimum achievable amount of charges on metal 2 under SC conditons) [7]. The main advantage to utilize the minimum achievable charge reference state is to set both $Q_{SC}$ ($x = 0$) and $V_{OC}$ ($x = 0$) to 0 and this time $Q_{SC}$ stands for the total transferred charges amount from one electrode to another under short-circuit conditions.

FEM results from other charge reference states can be easily converted to the MACRS. From the above derivation, we can easily get the below conversion equation.

$$V_{OC,MACRS}(x) = V_{OC,other}(x) - \frac{V_{OC,other}(0)C(0)}{C(x)} \tag{6.14}$$

Although the minimum achievable charge reference state is the most commonly utilized charge reference state in TENGs, utilizing this charge reference state in freestanding TENGs will cause the calculated $V_{OC}$ nonsymmetrical. However, an inherent symmetry can be easily observed in the structure for both contact-mode freestanding TENGs and sliding-mode freestanding TENGs. The reason that the calculated $V_{OC}$ are nonsymmetrical under MACRS mainly because the amount of charges assigned on metal 1 and metal 2 are different. To maintain the symmetry, we can assign both metal 1 and metal 2 have the same amount of charges ($\sigma wl/2$). Then the calculated $V_{OC}$ will be fully symmetric. Figure 6.3b shows the FEM calculation result of $V_{OC}$ under symmetric charge reference state for both dielectric SFTENG and metal SFTENG. If we compare the current results with the results shown in Chap. 5 that are under MACRS, the curve shape of the dielectric SFTENG is similar mainly because its capacitance is nearly a constant. The curve shape of the metal SFTENG changes a lot and now has odd symmetry.

## 6.4 Resistive Load Characteristics

### 6.4.1 *Resistive Load Characteristics and "Three-Working-Region" Behavior*

Resistive load is the simplest load condition for TENGs. When TENGs are connected with a resistive load $R$, the equivalent circuit of the whole system is shown in Fig. 6.4a. Utilizing Kirchhoff's law, the governing equation of the whole system can be easily given by [2, 3]:

$$R\frac{dQ}{dt} = -\frac{1}{C}Q + V_{OC} \tag{6.15}$$

The above equation is a first-order differential equation and can be easily solved by specifying the initial boundary condition $Q(t = 0)$. There are two types of boundary conditions that are most commonly utilized. First, we can consider a very common case that the device has stopped at the equilibrium position ($x = 0$) for a long time and electrostatic equilibrium has been reached before the motion process. If $V_{OC}$ are utilized its value under MACRS, the boundary condition is $Q\ (t = 0) = 0$. Second, if the TENG is under a periodic motion with a specific period, output from any initial boundary condition will gradually converge to a periodic output wave after the first few periods, which is the steady-state. To solve the steady-state solution, periodic boundary condition must be utilized, which is $Q\ (t = 0) = Q\ (t = T)$, where $T$ is the period of the motion process. For simplicity,

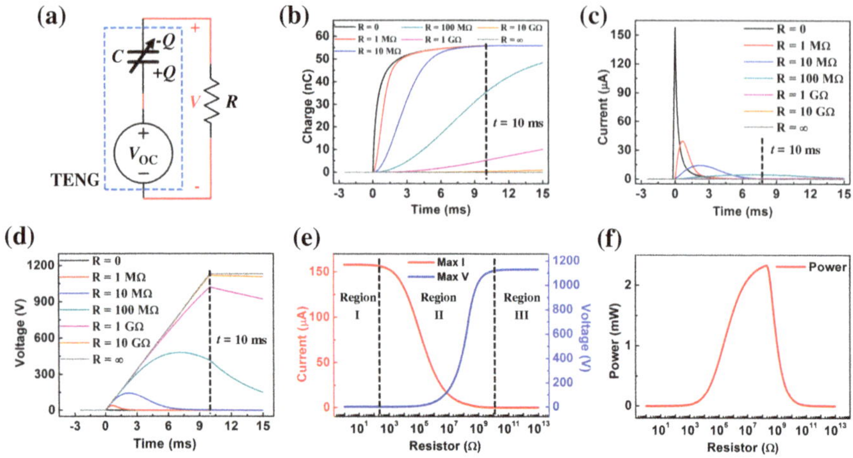


**Fig. 6.4** Theoretical calculated TENG resistive-load output characteristics under uniform velocity separation. The *top* plate reaches the maximum separation distance and stops moving at $t$ = 10 ms. **a** Equivalent circuit model of the whole TENG system. **b** Real-time transferred charge-time relationship at different load resistances. **c** Real-time current-time relationship at different load resistances. **d** Real-time voltage-time relationship at different load resistances. **e** The influence of the load resistance on the magnitude of the output current and voltage. Three working regions are marked. **f** The influence of the load resistance on the instantaneous power output. Reproduced with permission from Royal Society of Chemistry [3]

in this section, we mainly utilized the boundary condition $Q\ (t = 0) = 0$, which will not affect the main conclusion. Under this boundary condition, Eq. (6.15) can be solved as:

$$Q(t) = \frac{1}{R}\exp\left[-\frac{1}{R}\int_0^t \frac{1}{C(x(t))}\mathrm{d}t\right]\int_0^t V_{OC}(x(t))\exp\left[\frac{1}{R}\int_0^t \frac{1}{C(x(t))}\mathrm{d}t\right]\mathrm{d}t \quad (6.16a)$$

$$I(t) = \frac{V_{OC}}{R} - \frac{1}{R^2C}\exp\left[-\frac{1}{R}\int_0^t \frac{1}{C(x(t))}\mathrm{d}t\right]\int_0^t V_{OC}(x(t))\exp\left[\frac{1}{R}\int_0^t \frac{1}{C(x(t))}\mathrm{d}t\right]\mathrm{d}t \quad (6.16b)$$

$$V(t) = V_{OC} - \frac{1}{RC}\exp\left[-\frac{1}{R}\int_0^t \frac{1}{C(x(t))}\mathrm{d}t\right]\int_0^t V_{OC}(x(t))\exp\left[\frac{1}{R}\int_0^t \frac{1}{C(x(t))}\mathrm{d}t\right]\mathrm{d}t \quad (6.16c)$$

The above equations are a bit complicated. As a typical example, we consider an attached-electrode contact-mode TENG under a constant velocity motion process [3]. For attached-electrode contact-mode TENG, the above equations can be simplified as:

$$Q(t) = \sigma S - \sigma S \exp\left[-\frac{1}{RS\varepsilon_0}\left(d_0 t + \int_0^t x(t)\mathrm{d}t\right)\right]$$
$$-\frac{\sigma d_0}{R\varepsilon_0}\exp\left[-\frac{1}{RS\varepsilon_0}\left(d_0 t + \int_0^t x(t)\mathrm{d}t\right)\right] \times \int_0^t \exp\left[\frac{1}{RS\varepsilon_0}\left(d_0 z + \int_0^t x(z)\mathrm{d}z\right)\right]\mathrm{d}z \tag{6.17a}$$

$$\begin{aligned} I(t) = & -\frac{\sigma d_0}{R\varepsilon_0} + \frac{\sigma(d_0 + x(t))}{R\varepsilon_0}\exp\left[-\frac{1}{RS\varepsilon_0}\left(d_0 t + \int_0^t x(t)\mathrm{d}t\right)\right] \\ & + \frac{\sigma d_0}{R\varepsilon_0} \times \frac{d_0 + x(t)}{RS\varepsilon_0}\exp\left[-\frac{1}{RS\varepsilon_0}\left(d_0 t + \int_0^t x(t)\mathrm{d}t\right)\right] \\ & \times \int_0^t \exp\left[\frac{1}{RS\varepsilon_0}\left(d_0 z + \int_0^t x(z)\mathrm{d}z\right)\right]\mathrm{d}z \end{aligned} \tag{6.17b}$$

$$\begin{aligned} V(t) = & -\frac{\sigma d_0}{\varepsilon_0} + \frac{\sigma(d_0 + x(t))}{\varepsilon_0}\exp\left[-\frac{1}{RS\varepsilon_0}\left(d_0 t + \int_0^t x(t)\mathrm{d}t\right)\right] \\ & + \frac{\sigma d_0}{\varepsilon_0} \times \frac{d_0 + x(t)}{RS\varepsilon_0}\exp\left[-\frac{1}{RS\varepsilon_0}\left(d_0 t + \int_0^t x(t)\mathrm{d}t\right)\right] \\ & \times \int_0^t \exp\left[\frac{1}{RS\varepsilon_0}\left(d_0 z + \int_0^t x(z)\mathrm{d}z\right)\right]\mathrm{d}z \end{aligned} \tag{6.17c}$$

In a special case of uniform motion at a constant velocity $v$ (shown as Eq. (6.18a, 6.18b)), the above equations can be further simplified to Eq. (6.19a–6.19c), as shown below.

$$x = vt\left(t < \frac{x_{max}}{v}\right) \tag{6.18a}$$

$$x = x_{\max}\left(t \geq \frac{x_{\max}}{v}\right) \tag{6.18b}$$

$$\begin{aligned} Q(t) = \sigma S\Big[ & 1 - \exp(-At - Bt^2) + \sqrt{2}F\exp(-At - Bt^2) \\ & \times \mathrm{Dawson}\left(\frac{F}{\sqrt{2}}\right) - \sqrt{2}F \times \mathrm{Dawson}\left(\frac{F}{\sqrt{2}} + \sqrt{B}t\right)\Big] \end{aligned} \tag{6.19a}$$

$$\begin{aligned} I(t) = \sigma S\Big[ & \exp(-At - Bt^2)(A + 2Bt) - \sqrt{2}F \times \exp(-At - Bt^2)(A + 2Bt) \\ & \times \mathrm{Dawson}\left(\frac{F}{\sqrt{2}}\right) - A + 2A\left(\frac{F}{\sqrt{2}} + \sqrt{B}t\right) \times \mathrm{Dawson}\left(\frac{F}{\sqrt{2}} + \sqrt{B}t\right)\Big] \end{aligned} \tag{6.19b}$$

$$V(t) = \sigma SR\Big[\exp(-At - Bt^2)(A + 2Bt) - \sqrt{2}F \times \exp(-At - Bt^2)(A + 2Bt) \times \text{Dawson}\left(\frac{F}{\sqrt{2}}\right) - A + 2A\left(\frac{F}{\sqrt{2}} + \sqrt{B}t\right) \times \text{Dawson}\left(\frac{F}{\sqrt{2}} + \sqrt{B}t\right)\Big] \tag{6.19c}$$

The constants $A$, $B$, and $F$ are defined by:

$$A = \frac{d_0}{RS\varepsilon_0} \tag{6.20a}$$

$$B = \frac{v}{2RS\varepsilon_0} \tag{6.20b}$$

$$F = \frac{A}{\sqrt{2B}} = \frac{d_0}{\sqrt{RS\varepsilon_0 v}} \tag{6.20c}$$

The Dawson's integral (Dawson($x$)) in the above expressions is shown as:

$$\text{Dawson}(x) = \exp(-x^2)\int_0^x \exp(y^2)dy \tag{6.21}$$

The numerical calculated results are shown in Fig. 6.4b–f. (Parameters utilized are listed in Table 6.1) The detailed profiles of the charge, current, and voltage relationships at different loads are shown in Fig. 6.4b–d. Under short-circuit conditions, the charge transfers at the fastest speed. For a relatively small $R$, $Q$ can still get its saturation value when the top electrode stops moving ($t$ = 10 ms). However, when $R$ is more than 100 MΩ, at $t$ = 10 ms, the charge cannot get saturated due to the limit of the charge transfer rate by the resistor, resulting in the unstopped charge transfer from metal 1 to metal 2 after $t$ = 10 ms. Therefore, the current is a peak-shape when $R$ is small while the current continues increasing during the plate movement when $R$ is large. The voltage has the same profile with the current, but a different trend in magnitude. The peak values of the voltage, current and power on different loads are displayed in Fig. 6.4e, f. It can be clearly observed that the operation of the TENG can be divided into three working regions. First, In Region I

**Table 6.1** Parameters utilized in the resistive-load characteristic calculation

| Structure component | Parameter utilized |
|---|---|
| Dielectric 1 | Metal, $d_1$ = 0 μm |
| Dielectric 2 | $d_2$ = 125 μm, $\varepsilon_{r2}$ = 3.4 |
| Area size of dielectrics ($S$) | 58.0644 cm$^2$ (9 in.$^2$) |
| Tribo-charge surface density ($\sigma$) | 10 μCm$^{-2}$ |
| Maximum separation distance ($x_{max}$) | 0.001 m |
| Average velocity ($v$) | 0.1 ms$^{-1}$ |

where the resistance is low (0.1–1000 Ω), the peak current has little drop from the short-circuit condition, which is due to the similar charge transfer process in comparison with that of the short-circuit condition. As a result, the maximum voltage is approximately proportional to the external resistance. On the other hand, when the resistor is larger than 1 GΩ (Region III), the output characteristics are close to the open-circuit condition, in which the maximum voltage saturates at $V_{OC}$. The medium range of the resistance is Region II, where the maximum current drops dramatically but the maximum voltage increases at a reverse trend. Around 100 MΩ inside this region, the TENG reaches its maximum instantaneous output power.

This unique output characteristic of TENGs can be interpreted both physically and mathematically [3]. In physics, we can understand this unique characteristic from two different views. As an intuitive sense, we can explain this three-working-region behavior from the charge transfer rate point of view. The working principle of the TENG is a conjunction of the triboelectric effect and electrostatic induction. When the tribo-charges are separated, an induced voltage will be generated between the two electrodes, which will drive the electrons to flow from electrode 2 to electrode 1 to screen the original electric field from the tribo-charges. The charge accumulation rate at SC condition is the maximum charge transfer rate, which is determined by the mechanical motion. However, as there is an external load with certain resistance, the resistor will limit the real charge transfer rate, making it lower than that at the SC condition. When the load resistance is very small, this limitation is not obvious so that the rate can still catch up with the maximum rate determined by the moving mode, enabling the $Q$ to stay close to $Q_{SC}$ in Region I. When the load resistance continues to increase and gets into Region II, it begins to significantly limit the charge flow rate, making the charge accumulation curve deviate downward from $Q_{SC}$. As the load resistance is large enough, the transfer of electrons from metal 2 to metal 1 is rather slow, which results in a very small current. The induced voltage between the two electrodes will hold for a long time due to the little screening effect, with the magnitude approaching $V_{OC}$. This is the Region III of the TENG.

Besides the explanation from charge transfer rate, this unique three-working-region behavior can also be understood from the impedance match between the inherent TENG capacitance and the load resistance, although now there is no strict definition of the impedance of the inherent TENG capacitance because the TENG capacitance is time-variant and $V_{OC}$ is also not a pure harmonic function. When the load resistance is small, its impedance is much smaller than the impedance of the TENG capacitance. Thus, the total impedance is dominated by the TENG inherent capacitance and the output current remains almost unchanged when the resistance increases. When the load resistance is large enough, its impedance is much larger than the impedance of the TENG capacitance. Thus, almost all of the $V_{OC}$ is applied on the load resistance and the TENG is working under a quasi-OC condition. When the load resistance is in the middle region (matches the TENG impedance), the maximum transit power can be reached.

Mathematically, this can be interpreted by considering the two limitation conditions. When $R$ approaches 0, Eq. (6.16a) can be easily simplified to be $CV_{OC}$, which is just the same equation with $Q_{SC}$. The TENG is now in the quasi-SC condition (Region I). When $R$ approaches infinity, Eq. (6.16c) can be easily simplified to be $V_{OC}$, which is just the open-circuit voltage. The TENG is now in the quasi-OC condition, which is the Region III. When $R$ is neither too large nor too small, the behavior of the TENG is in the transitional region between short-circuit and open-circuit conditions (Region II), within which the maximum power is reached.

### 6.4.2 *Optimum Resistance*

In this three-working-region behavior of the TENG, the optimum resistance for the maximum instantaneous power is a key parameter to maximize the effective efficiency of the TENG. Therefore, the position of the optimum resistance needs to be systematically studied to find its relationship with the TENG's structural parameters and operational conditions, so that the TENG can be rationally controlled for different application purposes.

From the impedance match process we discussed above, the influence of structural parameters and motion condition on the optimum resistance can be easily addressed. For a TENG, its average impedance is approximately $1/(\omega C_{avg})$, where $\omega$ is the angular frequency of the signal source $V_{OC}$ and $C_{avg}$ is the average TENG inherent capacitance. Any structural parameters that can increase $C_{avg}$ will lower the TENG impedance and lower the optimum resistance. Increasing the motion velocity is equivalent to the increase of $\omega$, which will also lower the optimum resistance.

As a typical example, we can analyze the optimum resistance of an attached-electrode contact-mode TENG. [3] The optimum resistance can be mathematically investigated under an arbitrary motion for an attached-electrode TENG model built above. We first decompose an arbitrary smooth moving mode with an average velocity $v$ into a series of polynomial moving-mode, which is shown in the equations below.

$$x = \sum_{n=1}^{\infty} \alpha_n \frac{v^n}{x_{max}^{n-1}} t^n \quad \left(\mathrm{t} < \frac{x_{max}}{\mathrm{v}}\right) \tag{6.22a}$$

$$x = x_{max} \left(\mathrm{t} \geq \frac{x_{max}}{\mathrm{v}}\right) \tag{6.22b}$$

$\alpha_n$ is the coefficient of $n$th polynomial moving-mode, and satisfies the following equation.

$$\sum_{n=1}^{\infty} \alpha_n = 1 \tag{6.22c}$$

If we substitute this equation into the general equation of the current (Eq. 6.17b), the current under arbitrary moving mode can be given as:

$$\begin{aligned}
I(t) = & -\frac{\sigma d_0}{R\varepsilon_0} + \frac{\sigma d_0}{R\varepsilon_0}\left(1+\sum_{n=1}^{\infty}\frac{\alpha_n v^n}{d_0 x_{\max}^{n-1}}t^n\right)\exp\left[-\frac{1}{RS\varepsilon_0}\left(d_0 t+\sum_{n=1}^{\infty}\frac{\alpha_n v^n}{(n+1)x_{\max}^{n-1}}t^{n+1}\right)\right] \\
& + \frac{\sigma d_0}{R\varepsilon_0}\times\frac{d_0}{RS\varepsilon_0}\left(1+\sum_{n=1}^{\infty}\frac{\alpha_n v^n}{d_0 x_{\max}^{n-1}}t^n\right) \\
& \exp\left[-\frac{1}{RS\varepsilon_0}\left(d_0 t+\sum_{n=1}^{\infty}\frac{\alpha_n v^n}{(n+1)x_{\max}^{n-1}}t^{n+1}\right)\right] \\
& \times\int_0^t \exp\left[\frac{1}{RS\varepsilon_0}\left(d_0 z+\sum_{n=1}^{\infty}\frac{\alpha_n v^n}{(n+1)x_{\max}^{n-1}}z^{n+1}\right)\right]dz\left(t\le\frac{x_{\max}}{\mathrm{v}}\right)
\end{aligned} \tag{6.23}$$

Since the current starts to decay exponentially when $t$ is larger than $x_{max}/v$ (the external motion stops), the peak value of the current can only exist inside the closed interval $[0, x_{\max}/v]$. In order to find the time ($t_{\max}$) of the current peak, we take the differentiation of the $I(t)$ and the time $t_0$ satisfies the following equation.

$$\left(\frac{\mathrm{d}I}{\mathrm{d}t}\right)_{t=t_0} = 0 \tag{6.24}$$

When $R$ is small ($R < 100\ \mathrm{M\Omega}$ in Fig. 6.4), the peak value appears between $t = 0$ and $t = x_{\max}/v$ and $t_{\max}$ is equal to $t_O$. When $R$ is large enough ($R > 1\ \mathrm{G\Omega}$ in Fig. 6.4), $t_O$ is larger than $x_{max}/v$, so the current will increase monotonically between $t = 0$ and $t = x_{max}/v$. Then, $t_{\max}$ is equal to $x_{\max}/v$. Thus, in the entire region of $R$, $t_{\max}$ is the minimum value of $t_O$ and $x_{\max}/v$, which can be given by:

$$t_{\max} = \min\left(t_0, \frac{x_{\max}}{v}\right) \tag{6.25}$$

With the thickness factor $y$ defined as Eq. (6.26), $t_O$ is easy to be proven as a function ($f$) of $F$ and $y$, as shown in Eq. (6.27): [3]

$$y = \frac{x_{\max}}{d_0} \tag{6.26}$$

$$t_0 = \frac{x_{\max}}{v}f(F, y) \tag{6.27}$$

Therefore, we have:

$$t_{\max} = \min\left(t_0, \frac{x_{\max}}{v}\right) = \frac{x_{\max}}{v}\min(f(F, y), 1) = \frac{x_{\max}}{v}M(F, y) \tag{6.28}$$

where $M$ is an another function. Substituting Eq. (6.28) into Eq. (6.23), the peak value of the current $I_{\max}$ can be given by: [3]

$$I_{\max} = \frac{\sigma d_0}{R\varepsilon_0} G(F, y) \tag{6.29}$$

where $G$ is a new function. Then, the peak instantaneous power is:

$$P_{\max} = I_{\max}^2 R = \frac{(\sigma d_0)^2}{\varepsilon_0^2} \times \frac{1}{R} G^2(F, y) = \frac{\sigma^2 S v}{\varepsilon_0} F^2 G^2(F, y) \tag{6.30}$$

The optimum load resistance $R_{\text{opt}}$ satisfies the following equation:

$$\frac{\partial P_{\max}}{\partial \text{R}} = 0 \tag{6.31}$$

Therefore, combining Eqs. (6.30) and (6.31), the optimum resistance satisfies the following equation:

$$\text{G}\left(F_{\text{opt}}, y\right) + F_{\text{opt}} \left(\frac{\partial \text{G}}{\partial \text{F}}\right)_{F=F_{\text{opt}}} = 0 \tag{6.32}$$

Equation (6.32) shows that the optimum value of $F$ is only a function of $y$. Therefore,

$$F_{\text{opt}} = H(y) = \frac{d_0}{\sqrt{R_{opt} S \varepsilon_0 v}} \tag{6.33}$$

where the function $H$ only relies on the moving mode.

Thus, the expression of $R_{\text{opt}}$ can be given by:

$$R_{\text{opt}} = \frac{d_0^2}{F_{\text{opt}}^2 S v \varepsilon_0} = \frac{d_0^2}{[H(y)]^2 S v \varepsilon_0} = \frac{d_0^2}{\left[H\left(\frac{x_{\max}}{d_0}\right)\right]^2 S v \varepsilon_0} \tag{6.34}$$

This equation can be easily understood from the fundamental impedance match process. An increase in the area size of the TENG will increase the TENG capacitance than will lower its impedance. The optimum resistance reduces as a response. Similarly, increasing average velocity is equivalent to increase the high frequency component of $V_{\text{OC}}$, which will reduce the TENG impedance as well. As a result, the optimum resistance will decrease. If the effective dielectric thickness increases, the inherent TENG capacitance will reduce and its impedance will increase, so the optimum resistance increases. In addition, increasing the gap will reduce the average capacitance and increase the average impedance, so the optimum resistance also increases. From the similar analysis, since the TENG

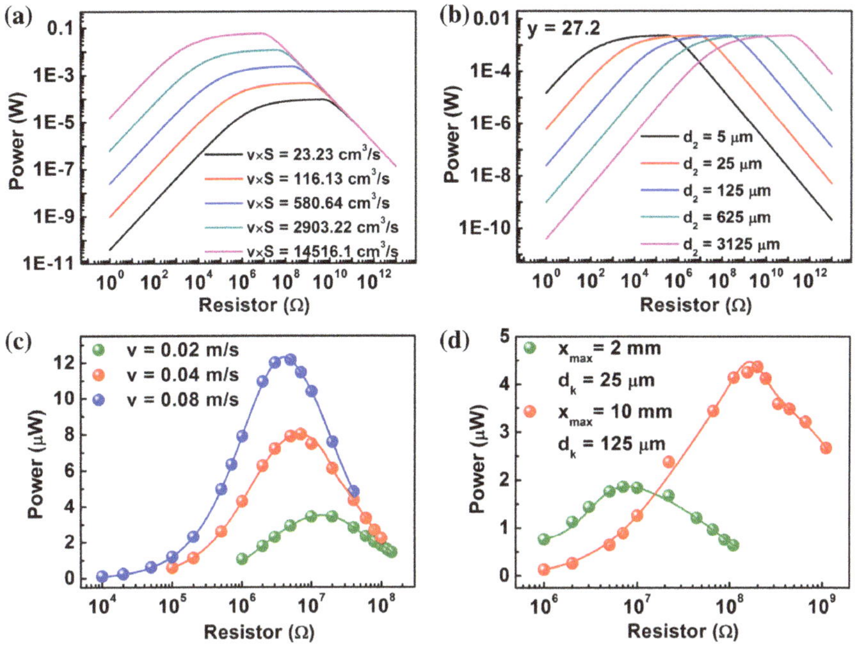


**Fig. 6.5** **a–b** The relationship of optimum resistance with different TENG parameters under uniform velocity motion. **a** Maximum output power profile with load resistance and at different velocities or area sizes. **b** Maximum output power profile with load resistance under different effective dielectric thickness and gap distances with maintaining their ratio as a constant of 27.2. **c–d** Experimental comparison of the optimum resistances from TENG devices with different parameters. The dots are measured values and the lines are obtained through interpolation of the experimental data. **c** Maximum output power profile with load resistance at different average velocities. **d** Maximum output power profile with load resistance at different Kapton film thickness and gap distances with their ratio (*y*) remaining as constant. Reproduced with permission from Royal Society of Chemistry [3]

impedance is independent of the tribo-charge density, the optimum resistance will not be affected if the tribo-charge density changes.

These conclusions of optimum resistance can also be validated through numerical calculation and experimental studies [3]. As shown in Fig. 6.5a, $R_{\mathrm{opt}}$ is inversely proportional to either the area size ($S$) or the average velocity ($v$). On the other hand, the influences of effective dielectric thickness and air gap distance on $R_{\mathrm{opt}}$ are displayed in Fig. 6.5b. If we keep the ratio ($y$) of the above two thicknesses ($d_0$ and $x_{max}$) as constant of 27.2, $R_{\mathrm{opt}}$ is proportional to $d_0^2$. Two set of experiments were also performed to validate the theoretical results. In the first set of experiments, the same device was measured under different average velocities, as shown in Fig. 6.5c. When the average velocities are 0.02, 0.04, and 0.08 m/s, the derived optimum load resistance (through interpolation of the experiment data) is 14.2, 6.66, and 4.08 MΩ, respectively, which shows the same trend with the theoretical estimation. In the other set of measurements, the influence of the gap and the dielectric thickness is experimentally studied. The thickness of the Kapton film ($d_k$) in device 1 is 25 μm while that of device 2 is 125 μm, and the corresponding gaps ($x_{max}$) are 2 and 10 mm, respectively, which maintains the same $y$. Through the measurement on a series of resistors with all the other parameters kept the same, the obtained optimum resistance for device 1 is 7.69 MΩ while that of device 2 is 162.3 MΩ. The experimental optimum resistance ratio is 21.1, which is quite close to 25, as the theoretical estimation value.

## 6.5 Capacitive Load and Charging Characteristics

In the above section the resistive load characteristics of TENG are discussed. In this section, we will discuss another important load condition: capacitive load [8]. We will start from the simplest unidirectional motion charging characteristics. Then a much more complicated multiple-cycle charging process was analyzed to show the unique TENG charging characteristics and an optimum load capacitance was observed for maximized energy storage. To obtain the optimized design strategy for such system, the analytical solution of the optimum load capacitance was deducted and its dependence on the charging cycle numbers and TENG structural parameters was shown.

### *6.5.1 TENG Charging Characteristics Under Unidirectional Mechanical Motion*

We start our discussion with the simplest case, which is utilizing a single-electrode contact-mode TENG to charge a load capacitor ($C_L$) under a unidirectional mechanical motion. In this process, no external rectifier is needed because the charging current of the TENG is also unidirectional. Besides, the inherent capacitance ($C_T$) of a single-electrode TENG is nearly constant, which will also simplify our discussion. The model of the single-electrode contact-mode TENG is the same as we described above in Chap. 4. Utilizing the above equivalent circuit model of TENG, the equivalent circuit model of this whole system is shown as Fig. 6.6a. In the discussion, $Q$ was defined as the transferred charges from the primary electrode to the reference electrode and $Q^C$ was defined as the charges on the top plate of the load capacitor.

To solve the above equivalent circuit, the initial condition must be specified. As a typical example, we consider the case that at $t = 0$, the top dielectric has stopped at $x = 0$ position for a long time and just starts moving. Besides, no initial charges are stored on the load capacitor, so initial charges on both $C_T$ and $C_L$ are 0 ($Q(t = 0) = Q^C(t = 0) = 0$). Thus, the following equations can be obtained from

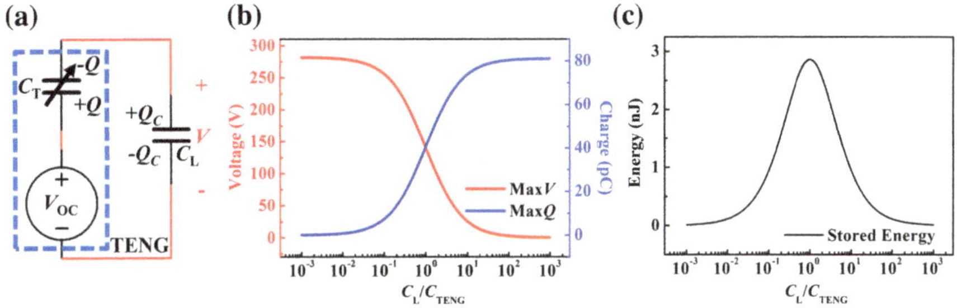


**Fig. 6.6** **a** Equivalent circuit model of TENG-capacitor system of the unidirectional charging calculation. **b** The influence of the load capacitance on the final voltage and charge stored in the load capacitor. **c** Final stored energy profile with load capacitance. Reproduced with permission from IEEE [8]

Kirchhoff's Law and node charge conservation, where $V$ stands for the voltage across $C_L$ and $V_{OC}$ stands for the open circuit voltage of the TENG.

$$V = -\frac{1}{C_T}Q + V_{OC} = \frac{1}{C_L}Q^C \tag{6.35}$$

$$Q^C - Q = Q^C(t = 0) - Q(t = 0) = 0 \tag{6.36}$$

In practical applications, the motion of the dielectric always has a maximum separation distance, which is defined as $x_{max}$. Thus, the final voltage and charge on the capacitor when a full separation is reached ($x$ reaches $x_{max}$) can be obtained by solving the above two equations. ($V_{OC,max}$ and $Q_{SC,max}$ stands for the open circuit (OC) voltage and short circuit (SC) transferred charge of the TENG when $x = x_{max}$)

$$V(x = x_{max}) = \frac{C_T V_{OC,max}}{C_L + C_T} = \frac{Q_{SC,max}}{C_L + C_T} \tag{6.37}$$

$$Q_C(x = x_{max}) = \frac{C_L Q_{SC,max}}{C_L + C_T} \tag{6.38}$$

Thus, the total energy stored in the capacitor ($E^C$) can be given by:

$$E_C = \frac{1}{2}C_L[V(x = x_{max})]^2 = \frac{C_L[Q_{SC}(x = x_{max})]^2}{2(C_L + C_T)^2} \tag{6.39}$$

To obtain an intuitive sense of such equations, numerical calculation for the single-electrode TENG under constant velocity motion condition is performed and output profiles as a function of time with different load capacitances are shown in Fig. 6.6. (The detailed calculation parameters are listed in Table 6.2). When $C_L$ is small, its impedance is much larger than the impedance of $C_T$. Thus, almost all of the $V_{OC}$ is applied on $C_L$ and the TENG is working under a quasi-OC condition.

**Table 6.2** Parameters utilized in the calculation of the unidirectional charging behaviors for a single-electrode TENG

| Structure component | Parameters utilized |
|---|---|
| Relative dielectric constant ($\varepsilon_r$) | 2 |
| Thickness of the dielectric ($d_r$) | 100 μm |
| Width of the structure ($w$) | 5 mm |
| Length of dielectrics ($l$) | 5 mm |
| Gap distance between the electrodes ($g$) | 1 mm |
| Tribo-charge surface density ($\sigma$) | 8 $\mu Cm^{-2}$ |
| Maximum separation distance ($x_{max}$) | 0.02 m |
| Average velocity ($v$) | 1 $ms^{-1}$ |

Therefore, the voltage charging curves for small $C_L$ are all close to the $V_{OC}$ curve. However, the stored charge is still close to 0 because of the small $C_L$, leading to a limited final stored energy. On the other hand, when $C_L$ is very large, its impedance is much smaller than the impedance of $C_T$, so the voltage applied on $C_L$ is approximately 0. At this moment, the TENG is working under a quasi-SC condition and the stored-charge curves for large $C_L$ are all close to the $Q_{SC}$ curve. However, the total stored energy is still limited because of the low voltage across $C_L$. From the above analysis, only at the transition region the stored energy can reach its maximum. Mathematically, it can be easily derived from 6.39 that the optimum load capacitance ($C_{L,opt}$) at which the maximum stored energy is reached is equal to $C_T$, as shown in Fig. 6.6c. Physically, this means that the impedance match between the TENG and the load is reached.

### 6.5.2 *TENG Charging Performance Under Periodic Mechanical Motion*

Above we analyzed the charging behavior of a single-electrode TENG to charge a load capacitor under a unidirectional motion. However, in practical applications, the mechanical motion is almost never unidirectional, resulting in an AC charging current. Therefore, a full-bridge diode rectifier is necessary to prevent the charge leaking back from $C_L$ to the TENG. Besides, the inherent capacitance of most TENGs is time-variant [1]. Therefore, the real case is much more complicated than the above discussion. As a typical example, the case we analyzed in the following section is a contact-mode attached-electrode TENG under a periodic harmonic motion process (shown in Eq. (6.40)) [3].

$$x = x_{max}\left[\frac{1}{2} - \frac{1}{2}\cos\left(\frac{\pi v}{x_{max}}t\right)\right] \tag{6.40}$$

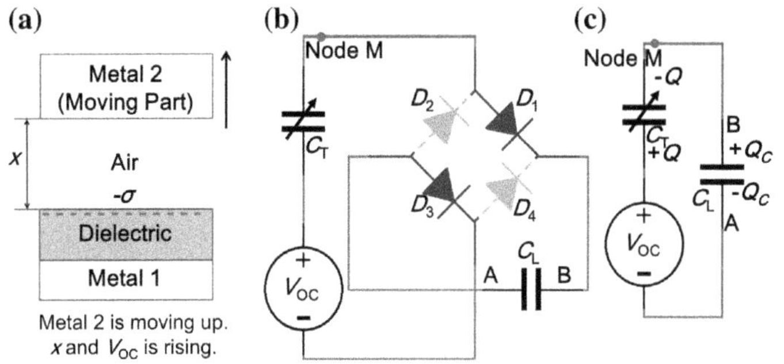


**Fig. 6.7** Equivalent circuit diagram for TENG charging performance calculation under periodic mechanical motion. **a** Structure of the contact-mode attached-electrode TENG used in the calculation. **b** Circuit diagram showing the conduction condition of full-bridge rectifier in the first half cycle. **c** Simplified circuit diagram under ideal diode approximation in the first half cycle. Reproduced with permission from IEEE [8]

**Table 6.3** Parameters utilized in the calculation of the multiple-cycle charging behaviors for a contact-mode attached-electrode TENG

| Structure component | Parameters utilized |
|---|---|
| Relative dielectric constant ($\varepsilon_r$) | 2.1 |
| Thickness of the dielectric ($d_r$) | 50 μm |
| Area of the dielectric ($S$) | 0.005 $m^2$ |
| Tribo-charge surface density ($\sigma$) | 40 $\mu Cm^{-2}$ |
| Maximum separation distance ($x_{max}$) | 0.002 m |
| Average velocity ($v$) | 0.1 $ms^{-1}$ |

The structure of this TENG is shown in Fig. 6.7a and the working principle has been discussed in detail in Chap. 2. Utilizing the parameters listed in Table 6.3, we can obtain the numerical results of this non-linear time-variant system shown in Fig. 6.7b from the TENG-simulator [5]. The simulation results are plotted in Fig. 6. 8. As shown in Fig. 6.8a, a saturation charging curve is observed for all the load capacitors, which is very similar to a typical RC charging curve in shape. At $t = 0$, the load capacitor is charging at a maximum speed. Then the charging speed gradually slows down and finally the same saturation voltage about 104.8 V is reached for all the different load capacitors. It takes less time to charge a smaller $C_L$ to reach its saturation voltage. However, a different trend is observed for the stored charges. At beginning, the curves for all load capacitors converge to one linear curve, with the charging rate of $2Q_{SC,max}$ per cycle. However, when time increases, the curves from small $C_L$ shift downwards first and finally get saturated. At this moment, very few charges can be pumped into the capacitor. The voltage and the stored charge of the load capacitor after 1000 charging cycles are shown in Fig. 6. 8c, which have the similar trend as the previous shown unidirectional charging case. When $C_L$ is small enough, the voltage on the load capacitor is approximately its saturation voltage while the stored charges are close to 0 and proportional to $C_L$. When $C_L$ is large enough, the stored charges are approximately $2kQ_{SC,max}$ while the

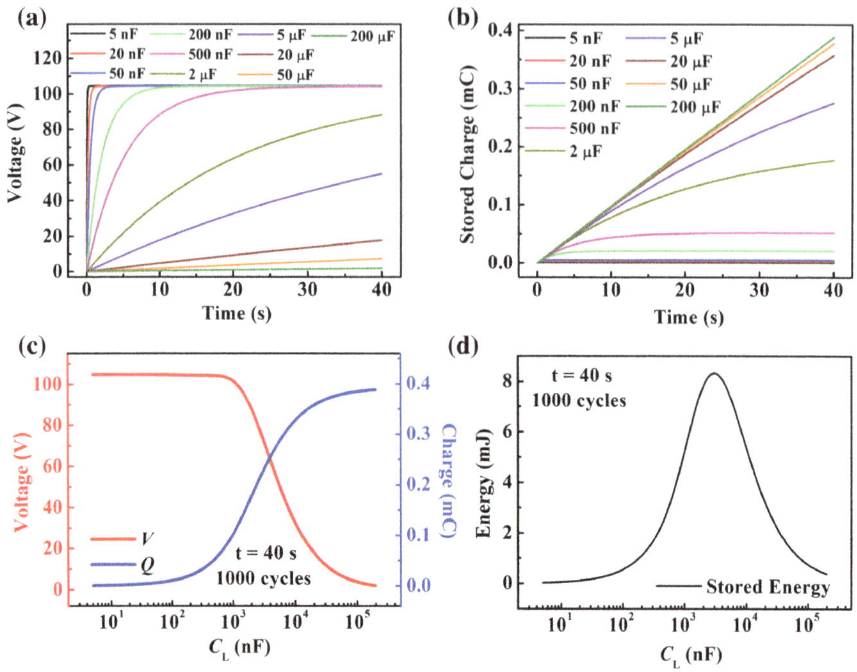


**Fig. 6.8** TENG charging characteristics under periodic mechanical motion. **a** Voltage-time relationship at different load capacitances. **b** Stored charge-time relationship at different load capacitances. **c** The influence of the load capacitance on the final voltage and charge stored in the load capacitor. **d** Final stored energy profile with load capacitance. Reproduced with permission from IEEE [8]

voltage is close to 0 and inversely proportional to $C_L$. Similar to the previous unidirectional charging, an optimum capacitance for maximum stored energy is also observed in the transition region of $C_L$.

In addition to the demonstrated numerical calculation, an analytical solution is preferred to better understand the underlying physics despite the fact that only approximate solution can be derived. To address this non-linear time-variant system, linearizing the diodes of the bridge rectifier is a critical step. As shown in Fig. 6.7a–b, in the first half of each cycle, the top plate is moving upward and $V_{OC}$ is increasing. Neglecting the reverse leakage current of the diode, we can assume only diode 1 and 3 are conducting in this step and diode 2 and 4 are fully open-circuit. Besides, the conducting voltage of these two diodes can be neglected, for it is much smaller than the open-circuit voltage of the TENG. With the above assumptions, the complex circuit shown in Fig. 6.7b can be simplified to the linear circuit shown in Fig. 6.7c, which is similar to the unidirectional circuit shown in Fig. 6.6a. Similarly, in the second half of each cycle when $V_{OC}$ is decreasing, the circuit can also be simplified, but the direction of $C_L$ is reversed because of the function of the full-bridge rectifier.

The initial condition we utilized is still the same with the analysis of unidirectional charging, that is the initial charges on both $C_T$ and $C_L$ are 0. So at the beginning of the first cycle, the total charges stored on Node M (shown in Fig. 6.7b, c) are 0, which is shown in mathematics as $Q^M{}_1 = 0$. ($Q^M{}_k$ stands for the total charges on the Node M at the beginning of the $k$th cycle).

Now we consider the $k$th charging cycle. During its first half, it is a unidirectional charging step. Similar to the previous derivation, at the end of this half cycle ($x = x_{max}$), the charges stored on $C_T$ ($Q$) and $C_L$ ($Q^C$) can be given by:

$$Q_{k,1\text{end}} = \frac{C_L Q_{SC,\max} - C_{\min} Q^M{}_k}{C_L + C_{\min}} \tag{6.41}$$

$$Q^C{}_{k,1end} = C_L \frac{Q_{SC,\max} + Q^M{}_k}{C_L + C_{\min}} \tag{6.42}$$

where the subscript in $Q_{k,1\text{end}}$ stands for its value at the end of the 1st half in the $k$th charging cycle and $C_{min}$ stands for the value of $C_T$ ($x = x_{max}$).

From the end of the first half cycle to the beginning of the second half cycle, the polarity of $C_L$ is reversed due to the function of the full bridge rectifier. Thus, the charges stored on Node M at the beginning of the second half of the $k$th cycle $\left(Q^M{}_{k,\text{mid}}\right)$ can be given by:

$$Q^M{}_{k,\text{mid}} = -Q_{k,2\text{begin}} + Q^C{}_{k,2\text{begin}} = -Q_{k,1\text{end}} - Q^C{}_{k,1\text{end}} \tag{6.43}$$

The second half cycle is also a unidirectional charging step ($x$ is from $x_{max}$ back to 0). Similar to the previous derivation, at the end of this half cycle ($x = 0$), the charges stored on $C_T$ ($Q$) and $C_L$ ($Q^C$) can be obtained as ($C_{max}$ stands for the value of $C_T(x = 0)$)

$$Q_{k,2\text{end}} = -\frac{C_{\max} Q^M{}_{k,\text{mid}}}{C_L + C_{\max}} \tag{6.44}$$

$$Q^C{}_{k,2\text{end}} = \frac{C_L Q^M{}_{k,\text{mid}}}{C_L + C_{\max}} \tag{6.45}$$

At the end of this second half-cycle, the direction of $C_L$ is reversed again. Thus, the charges stored on Node M at the beginning of the ($k$ + 1)th cycle ($Q^M{}_{k+1}$) can be given by:

$$Q^M{}_{k+1} = -Q_{k+1,1\text{begin}} + Q^C{}_{k+1,1\text{begin}} = -Q_{k,2\text{end}} - Q^C{}_{k,2\text{end}} \tag{6.46}$$

Through the above equations, a recursion relationship for $Q^M{}_k$ can be derived below.

$$Q^M{}_{k+1} = \frac{C_L - C_{\max}}{C_L + C_{\max}} \frac{C_L - C_{\min}}{C_L + C_{\min}} Q^M{}_k + 2\frac{C_L - C_{\max}}{C_L + C_{\max}} \frac{C_L}{C_L + C_{\min}} Q_{SC,\max} \tag{6.47}$$

With the boundary condition of $Q^{\mathrm{M}}{}_1 = 0$, the above recursion relationship can be easily solved as:

$$Q^M{}_k = \frac{C_L - C_{\max}}{C_{\min} + C_{\max}} Q_{SC,\max} - \frac{C_L - C_{\max}}{C_{\min} + C_{\max}} Q_{SC,\max} \left[\frac{(C_L - C_{\max})(C_L - C_{\min})}{(C_L + C_{\max})(C_L + C_{\min})}\right]^{k-1} \tag{6.48}$$

Therefore, the voltage on $C_{\mathrm{L}}$ at the end of $k$th charging cycle $\left(|V^{\mathrm{C}}_{k,2\mathrm{end}}|\right)$ can be easily derived as:

$$\left|V^C{}_{k,2\mathrm{end}}\right| = \frac{Q_{SC,\max}}{C_{\min} + C_{\max}} \left\{1 - \left[1 - \frac{2(C_{\min} + C_{\max})C_L}{(C_L + C_{\max})(C_L + C_{\min})}\right]^k\right\} \tag{6.49}$$

It is useful to note that the above derivation is also applicable to other categories of TENGs because we didn't utilize any TENG category information in our derivation. In practical applications, $C_{\mathrm{L}}$ is usually much larger than both $C_{\max}$ and $C_{\min}$. Under this condition, Eq. (6.49) can be further simplified to (where $f$ stands for the frequency of the periodic motion):

$$\begin{aligned} \left|V^C{}_{k,2\mathrm{end}}\right| &= \frac{Q_{\mathrm{SC},\max}}{C_{\min} + C_{\max}} \left\{1 - \exp\left[-\frac{2(C_{\min} + C_{\max})k}{C_L}\right]\right\} \\ &= \frac{Q_{\mathrm{SC},\max}}{C_{\min} + C_{\max}} \left\{1 - \exp\left[-\frac{2(C_{\min} + C_{\max})ft}{C_L}\right]\right\} \end{aligned} \tag{6.50}$$

Equation (6.50) clearly shows that a TENG together with a bridge rectifier under periodic external motion is completely comparable to a DC voltage source in series with a resistor regarding the capacitor charging characteristics. They all follow the same exponential saturation trend. The value of the DC source is the saturation voltage ($V_{\mathrm{sat}}$) shown in Fig. 6.8a and its value can be easily calculated as:

$$V_{\mathrm{sat}} = \lim_{k\to\infty} \left|V^C{}_{k,2\mathrm{end}}\right| = \frac{Q_{SC,\max}}{C_{\min} + C_{\max}} \tag{6.51}$$

Same as a first order resistor-capacitor charging circuit, $V_{\mathrm{sat}}$ is only a function of TENG parameters but is independent of $C_{\mathrm{L}}$, which is consistent with the above numerical results shown in Fig. 6.8a. Utilizing the parameters shown in Table 6.3, we can calculate that the saturation voltage for the previous TENG system is 105.06 V, which is very close to the simulation results (104.8 V).

Besides the saturation voltage, the charging speed is another important charging parameter. Analogous to a first-order RC charging circuit, time constant ($\tau$) is utilized to characterize the charging speed, which can be easily given by:

$$\tau = \frac{C_L}{2(C_{\min} + C_{\max})f} \tag{6.52}$$

The time constant is directly proportional to $C_L$. Therefore, the value of the effective resistor in the first-order RC charging circuit can be easily extracted as $1/[2f(C_{\max} + C_{\min})]$, which shows that the TENG with a larger inherent capacitance can charge $C_L$ with a higher speed.

Moreover, when $C_L$ is large enough, linear charging behavior is observed and the TENG is working on quasi-SC condition. When $C_L$ is close to infinity, the final stored charges at the end of the $k$th charging cycle can be given by:

$$\lim_{C_L \to \infty} \left| V^C{}_{k,2\text{end}} \right| = \lim_{C_L \to \infty} \frac{2kQ_{SC,\max}}{C_L} = 0 \tag{6.53}$$

$$\lim_{C_L \to \infty} \left| Q^C{}_{k,2\text{end}} \right| = C_L \lim_{C_L \to \infty} \left| V^C{}_{k,2\text{end}} \right| = 2kQ_{SC,\max} \tag{6.54}$$

Since the maximum charge generated from the TENG is $Q_{SC,\max}$ for each half cycle, the above results show that all the charges generated from the TENG are stored in $C_L$. At the same time, the voltage difference across the TENG is 0, which is completely consistent with the numerical results shown in Fig. 6.8b.

Finally and most importantly, there also exists an optimum load capacitance ($C_{L,\text{opt}}$) at which the stored energy reaches its maximum value. This optimum load capacitance is critically important for optimized design for such energy harvesting system and needs to be analytically derived to show its dependence on the TENG structural parameters and cycle numbers.

From Eq. (6.50), the energy stored in the load capacitor at the end of the $k$th charging cycle $\left(E^C{}_{k,\text{end}}\right)$ is given by:

$$E^C_{k,\text{end}} = \frac{C_L V^C{}_{k,2\text{end}}{}^2}{2} = \frac{C_L Q_{SC,\max}{}^2}{2(C_{\min} + C_{\max})^2} \left\{ 1 - \exp\left[ -\frac{2(C_{\min} + C_{\max})k}{C_L} \right] \right\}^2 \tag{6.55}$$

At the optimum capacitance, $E^C{}_{k,\text{end}}$ reaches its maximum value. Therefore,

$$\left( \frac{dE^C{}_{k,\text{end}}}{dC_L} \right)_{C_L = C_{L,opt}} = 0 \tag{6.56}$$

Thus, an analytical solution for $C_{\mathrm{L,opt}}$ can be obtained as:

$$C_{\mathrm{L,opt}} = 1.592k(C_{\mathrm{min}} + C_{\mathrm{max}}) = 1.592ft(C_{\mathrm{min}} + C_{\mathrm{max}}) \tag{6.57}$$

Equation (6.57) is an excellent approximation of $C_{\mathrm{L,opt}}$ only when $k$ is larger than 10 because the approximation that $C_{\mathrm{L}}$ is much larger than both $C_{\mathrm{max}}$ and $C_{\mathrm{min}}$ is utilized during the derivation of Eq. (6.50). However, the charging cycle number in almost all cases is more than 10, so Eq. (6.57) is always a good estimation in practical applications.

The optimum load capacitance is linearly related to the charging time, which makes this conclusion not difficult to operate in real application. However, if we put Eq. (6.57) back to Eq. (6.50), we can find that under this optimum charging condition, all the load capacitor is charging to the same voltage 0.72 $V_{\mathrm{sat}}$, which is only dependent on the intrinsic TENG properties. This information can help design the optimum charging cycles to maximize the charging efficiency.

To verify the above theoretical anticipations, numerical calculations are carried out with contact-mode attached-electrode TENGs with different parameters and cycle numbers, as shown in Fig. 6.9. Without special notation in the figure, the values utilized for the parameters are exactly the same as those in Table 6.3. First, the influence of the charging cycle number on $C_{\mathrm{L,opt}}$ is investigated. As shown in Fig. 6.9a, b, when the number of charging cycles increases, $C_{\mathrm{L,opt}}$ and the maximum energy stored both increases and an accurately linear trend between $C_{\mathrm{L,opt}}$ and the charging cycle numbers is observed. In addition, the influence of several

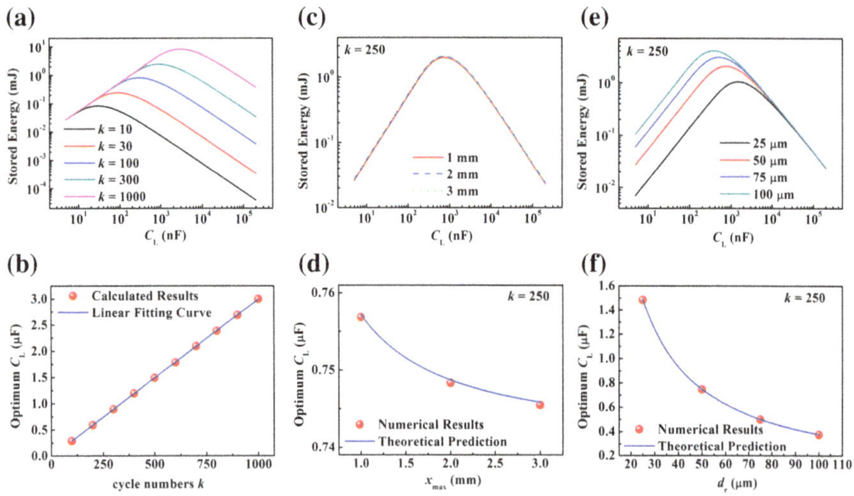


**Fig. 6.9** Optimum load capacitance characteristics. **a** The relationship between stored energy and load capacitance at different cycle numbers. **b** Extracted optimum capacitance with cycle numbers. **c** The relationship between stored energy and load capacitance at different $x_{\mathrm{max}}$. **d** Extracted optimum capacitance with $x_{\mathrm{max}}$. **e** The relationship between stored energy and load capacitance at different $\mathrm{d_r}$. **f** Extracted optimum capacitance with $d_r$. Reproduced with permission from IEEE [8]

structural parameters on $C_{L,opt}$ is investigated as well. For practical contact-mode TENGs with acceptable charge generation efficiency ($x_{max}$ should be much larger than $d_0$) shown in Fig. 6.7a [3], Eq. (6.57) can be further simplified utilizing the parallel plate capacitance model.

$$C_{L,opt} = 1.592k\left(\frac{\varepsilon_0 S}{d_0 + x_{max}} + \frac{\varepsilon_0 S}{d_0}\right) \approx \frac{1.592k\varepsilon_0 S}{d_0} \tag{6.58}$$

First, we investigate the influence of $x_{max}$. Intuitively, increasing this maximum moving distance will increase the open-circuit voltage of the TENG [3] and may have an obvious influence on $V_{sat}$ and $C_{L,opt}$. However, from Eq. (6.50), $V_{sat}$ and $C_{L,opt}$ are mainly affected by the sum of $C_{min}$ and $C_{max}$, in which $C_{min}$ is negligible compared to $C_{max}$ for a contact-mode attached-electrode TENG [3]. Thus, since $x_{max}$ can only affect $C_{min}$, the whole charging behavior and $C_{L,opt}$ is almost independent of $x_{max}$. As shown in Fig. 6.9c, the total stored energy as a function of $C_L$ for different $x_{max}$ almost coincide with each other. When $x_{max}$ increases from 1 to 3 mm, $C_{L,opt}$ only decreases from 756.9 to 745.5 nF, which is completely consistent with the theoretical estimation from 6.58. In contrast with $x_{max}$, since the dielectric thickness ($d_r$) will have significant impact on $C_{max}$, the total charging behavior and $C_{L,opt}$ is obviously dependent on $d_r$. As shown in Fig. 6.9e and f, when $d_r$ increases from 25 to 100 μm, the maximum stored energy increases from 1.04 to 4.05 mJ and $C_{L,opt}$ decreases from 1486 to 379 nF, which is completely consistent with the theoretical inverse ratio relationship.

To further validate the theoretical equation presented above, experiments were carried out to compare with the theoretical expectations. A contact-mode attached-electrode TENG with the structure shown in Fig. 6.7a was utilized in this experiment. The triboelectric pair material in the TENG was Al and FEP. Al was utilized as one electrode and at the backside of the FEP, copper was deposited as another electrode. The whole structure was supported by two Acrylic sheets. The bottom plate was bonded onto the measurement desk and the top plate was driven by a motor with a 2 Hz harmonic motion. Through bridge rectifier made from four low-leakage and high-breakdown diodes, different low-leakage capacitors were charged by this TENG and a voltage-meter (Keithley Model 6514) was utilized to measure the real-time voltage across the load capacitor. The measured charging curves for different capacitors are shown in Fig. 6.10a, b, which have the same shape and trend as the theoretical prediction shown in Fig. 6.8. The relationship between the stored energy and the load capacitance is also plotted and an optimum capacitance for the maximized stored energy is also observed as the theoretical prediction. If the interpolated optimum capacitance was extracted and plotted with the cycle number, a very good linear relationship is observed, which is consistent with our theoretical prediction. All of the above experimental results further validate the conclusions from our theoretical derivation, clearly showing its potential to guide future experimental designs.

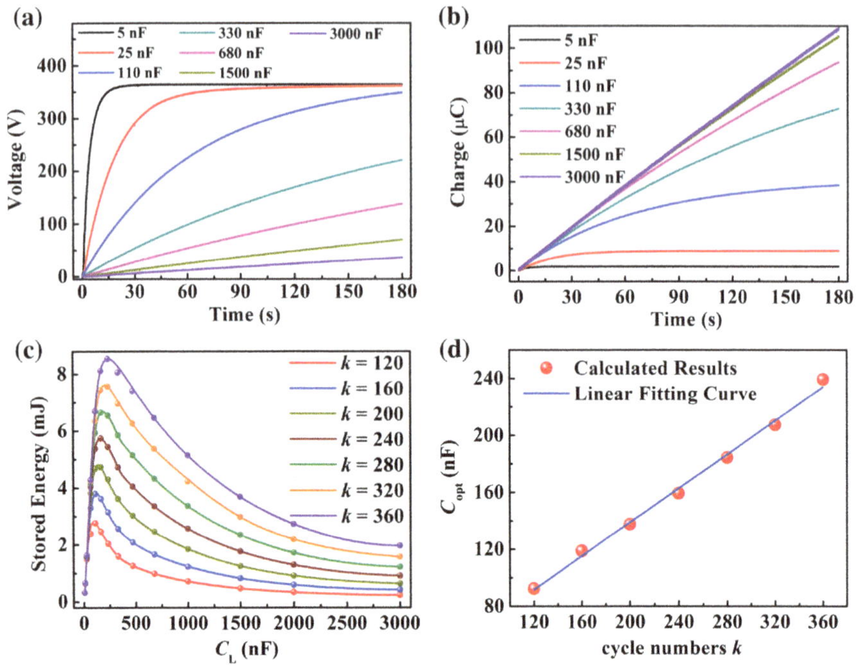


**Fig. 6.10** Experimental measured multi-step charging profiles. **a** Measured Voltage-time relationship at different load capacitances. **b** Measured stored charge-time relationship at different load capacitances. **c** Measured stored energy-load capacitance relationship at different time. **d** Extracted optimum capacitance with time. Reproduced with permission from IEEE [8]

## 6.6 Summary

In this chapter, we have carefully addressed the theoretical basis and fundamental physics of the TENG. The TENG has intrinsic capacitive behavior since it is based on a conjugation of contact electrification and electrostatic induction. Its governing equation is its *V-Q-x* relationship and its first-order equivalent model is a series of a voltage source and a capacitor. We have developed theoretical calculation method for TENGs and studied their load characteristics. When TENGs are connected with resistive loads, a three-working-region behavior is shown due to the impedance match mechanism. An optimum resistance is observed to maximize the TENG output on the load. When TENGs are utilized to charge a capacitor through bridge rectifier, it has a saturation charging behavior and is equivalent to first-order RC charging. An optimum capacitive load is also observed to maximize the energy storage. Our systematic analysis provides guidance to the design of nanogenerator based systems for the maximum utilization of output power.

# References


1. S.M. Niu, Z.L. Wang, Theoretical systems of triboelectric nanogenerators. Nano Energy **14**, 161–192 (2015)
2. S.M. Niu, Y. Liu, S.H. Wang, L. Lin, Y.S. Zhou, Y.F. Hu, Z.L. Wang, Theory of sliding-mode triboelectric nanogenerators. Adv. Mater. **25**(43), 6184–6193 (2013)
3. S.M. Niu, S.H. Wang, L. Lin, Y. Liu, Y.S. Zhou, Y.F. Hu, Z.L. Wang, Theoretical study of contact-mode triboelectric nanogenerators as an effective power source. Energy Environ. Sci. **6** (12), 3576–3583 (2013)
4. S.M. Niu, Y. Liu, S.H. Wang, L. Lin, Y.S. Zhou, Y.F. Hu, Z.L. Wang, Theoretical investigation and structural optimization of single-electrode triboelectric nanogenerators. Adv. Funct. Mater. **24**(22), 3332–3340 (2014)
5. S.M. Niu, Y.S. Zhou, S.H. Wang, Y. Liu, L. Lin, Y. Bando, Z.L. Wang, Simulation method for Optimizing the performance of an integrated triboelectric nanogenerator energy harvesting system. Nano Energy **8**, 150–156 (2014)
6. J. Stoer, R. Bulirsch, Introduction to numerical analysis. 3rd ed. (Springer, New York, 2002). p. xv, 744 p
7. S.M. Niu, Y. Liu, X.Y. Chen, S.H. Wang, Y.S. Zhou, L. Lin, Y.N. Xie, Z.L. Wang, Theory of freestanding triboelectric-layer-based nanogenerators. Nano Energy **12**, 760–774 (2015)
8. S.M. Niu, Y. Liu, Y.S. Zhou, S.H. Wang, L. Lin, Z.L. Wang, Optimization of triboelectric nanogenerator charging systems for efficient energy harvesting and storage. IEEE Tran. Electron. Dev. **62**(2), 641–647 (2015)

# Chapter 7
# Figure-of-Merits for Quantifying Triboelectric Nanogenerators

To improve the performance and broaden the applications of TENG, numerous efforts have been made with focus on both enhancement of the surface charge density $\sigma$ [1, 2] and development of new structures/modes [3–7]. However, without a common standard it becomes difficult to evaluate the performance of a TENG. In comparison to its counter parts, standards have been established for alternative power generators, such as Carnot efficiency for heat engines [8, 9] and pyroelectric nanogenerators [10, 11], ZT factor for thermoelectric materials [12, 13], and energy conversion efficiency for solar cells [14, 15]. As described in previous chapters, four basic modes of TENG have been developed, including vertical contact-separation (CS) mode, lateral sliding (LS) mode, single-electrode (SE) mode, and freestanding triboelectric-layer (FT) mode. Each mode has its own structure and choice of materials as well as specific mechanical triggering configurations. Taking the CS mode as an example, it is triggered by a vertical periodic driving force that causes a repeated contact-separation of two dissimilar materials that have coated electrodes on the top and bottom surfaces, respectively. As for the LS mode, it is triggered by a lateral sliding motion between two dissimilar materials in parallel. To evaluate and compare the performance of the TENGs in different structures/modes, a universal standard has to be introduced to quantify the performance of the TENG regardless its operation mode.

Zi et al. proposed a standard method to quantitatively evaluate TENG's performance from both structure's and the materials' points of view [16]. Starting from the plot of built-up voltage $V$—total transferred charges $Q$, the TENG operation cycle with maximized energy output is firstly proposed. Based on this cycle, a performance figure-of-merit (FOM) is proposed for TENG, which consists of a structural FOM related to the design of the TENG and a material FOM as the square of the surface charge density. To characterize and compare different structures of TENGs, the structural FOM for each configuration of TENG is derived and simulated. A standard method is proposed to quantify the material FOM. The proposed standards will set the foundation for the further applications and industrialization of the TENGs.



Z.L. Wang et al., *Triboelectric Nanogenerators*,
Green Energy and Technology, DOI 10.1007/978-3-319-40039-6_7

## 7.1 Operation Cycles of Triboelectric Nanogenerators

### *7.1.1* V-Q *Plot and Its Characteristics*

The fundamental working principle of the TENG is a conjugation of triboelectrification and electrostatic induction. For a basic triboelectric nanogenerator, there is at least one pair of triboelectric layers (possibly two pairs for FT mode and one pair for other modes) that are facing each other for creating opposite triboelectric charges via physical contacts. Besides the triboelectric layers, there are two electrodes (for SE mode, the ground is considered as the second/reference electrode [17]) that are carefully insulated from each other, which are required for the free electrons to be transferred between them through an external load. Triggered by the external mechanical force, there is a periodical relative motion between the triboelectric layers that breaks the balanced distribution of electrostatic charges; as a result, free electrons are driven to flow between the electrodes in order to build a new equilibrium. Therefore, the governing equations of TENG can be developed based on the relationship among the transferred charges between the electrodes $Q$, the built-up voltage $V$, and the relative displacement $x$ between the triboelectric layers. Usually, we define the status while one pair of the triboelectric layers are fully contacting with each other as $x = 0$, and $x = x_{max}$ is defined at the farthest achievable displacement. (Please be noticed that actually the $x_{max}$ can be picked at any displacement larger than 0. In application, the $x_{max}$ is fixed inside a range which depends on the structural design.) The detailed definitions of $x$ for different structures of TENG are illustrated in Fig. 7.3. We use the most commonly utilized minimum achievable charge reference state (MACRS) [18], so both the absolute short-circuit transferred charges $Q_{SC}(x)$ and the absolute open-circuit voltage $V_{OC}(x)$ at $x = 0$ position are set to be 0. The definitions of the displacement $x$ and the two electrodes for an LS mode TENG are illustrated in Fig. 7.1a. The maxima of $Q_{SC,max}$ and $V_{OC,max}$ are expected to be reached at $x = x_{max}$ for these basic modes of TENG.

For a continuous periodic mechanical motion, the electrical output signal from the TENG is also periodically time-dependent. In such a case, the average output power $\bar{P}$, which is related to the load resistance, is used to determine the merits of the TENG. Given a certain period of time $T$, the output energy per cycle $E$ can be derived as:

$$E = \bar{P}T = \int_0^T VI\mathrm{d}t \int_{t=0}^{t=T} V\mathrm{d}Q = \oint V\mathrm{d}Q \tag{7.1}$$

Therefore, the electrostatic states and the energy output of TENG can be represented by the plot of built-up voltage $V$ against the transferred charges $Q$.

Based on the *V-Q* curves, different operation cycles were developed to quantitatively describe the working status of TENG. Currently reported cycles include

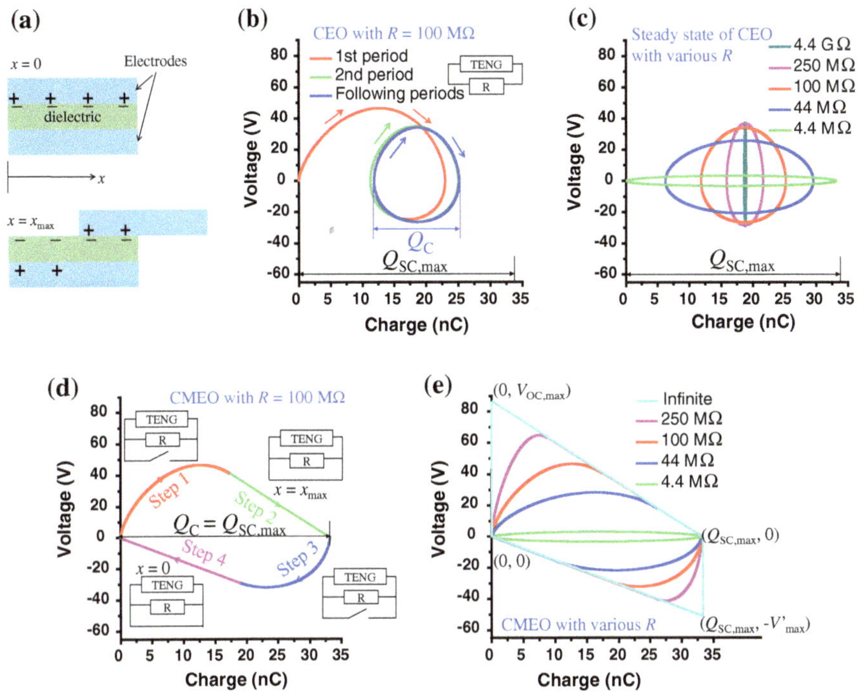


**Fig. 7.1** Operation cycles of TENG. **a** Schematic diagram of the LS mode TENG with displacement $x = 0$ and $x = x_{\max}$. **b** The cycle for energy output (*CEO*) with load resistance $R = 100$ MΩ. The total cycling charge $Q_C$ was marked and the inset shows the operation circuit. **c** The steady-state of CEO with various load resistances. **d** The cycle for maximized energy output (*CMEO*) with load resistance $R = 100$ MΩ, with the maximum total cycling charge $Q_C = Q_{SC,\max}$. The *insets* show the corresponding status of the switch in circuits during different steps. **e** The CMEO with various load resistances. The vertices of the CMEO with infinite load resistances are marked. Reproduced with permission from Nature Publication Group [16]

that of energy output and maximized energy output for TENG connected with a load resistance [16], and the direct charging cycle and the designed charging cycle for TENG to charge a battery or capacitor [19]. In following sections, we mainly focused on the former two since they are closely related to the figure-of-merits as proposed [16].

### *7.1.2 Cycle of Energy Output*

The *V-Q* plot for an LS mode TENG was firstly simulated by finite element method (FEM) operated under external load resistance of 100 MΩ, starting from $(Q, V) = (0, 0)$. The parameters of this LS mode TENG was shown in Table 7.1. From the *V-Q* plot, we noticed that the operation of the TENG will go to its

**Table 7.1** Parameters used for simulating the lateral sliding (LS) mode TENG

| | |
|---|---|
| Dielectric effective thickness $d_0 = \sum d_i/\varepsilon_{ri}$ | 90.8 μm |
| Triboelectrification area $A$ (Length $l$ × Width $w$) | 8.5 cm × 7.9 cm |
| Maximum displacement $x_{max}$ | 3.5 cm |
| Surface charge density $\sigma$ | 12 μC m$^{-2}$ |

steady state after only a few periods (Fig. 7.1b), and thus we can directly focus on the output of the steady-state operation. Because the steady-state output signal of TENG is periodic in responding to the mechanical triggering, the *V-Q* plot should be a closed loop. As indicated by Eq. (7.1), the output energy per cycle *E* can be calculated as the encircled area of the closed loop in the *V-Q* curve. The steady-state *V-Q* plots for this LS mode TENG were also simulated by FEM under various external loads, as shown in Fig. 7.1c. From the encircled areas of these *V-Q* curves, we noticed that the output energy per cycle *E* can be optimized by applying a matched load resistance [20]. The cycles given here can be named as "cycles for energy output" (CEO). For each CEO, the difference between the maximum and the minimum transferred charges in its steady-state is defined as the total cycling charge $Q_C$, as marked in Fig. 7.1b.

### *7.1.3 Cycle of Maximized Energy Output*

From Fig. 7.1b, c, it is noticed that for each CEO, the total cycling charge $Q_C$ was always less than the maximum transferred charges $Q_{SC,max}$, especially for cycles with large external load resistances. If we could maximize the $Q_C$ to be $Q_{SC,max}$ for these cycles, the output energy per cycle *E* would be further enhanced. Having noticed that the $Q_C = Q_{SC,max}$ occurs at short-circuit condition, following repeated steps are designed to achieve instantaneous short-circuit conditions during operations, with the use of a switch in parallel with the external load (as shown in Fig. 7.1d): step 1, triboelectric layers displace relatively from $x = 0$ to $x = x_{max}$ at switch off; step 2, turn the switch on to enable $Q = Q_{SC,max}$, and then turn the switch off; step 3, triboelectric layers displace relatively from $x = x_{max}$ to $x = 0$ at switch off; step 4, turn the switch on to enable $Q = 0$, and then turn the switch off. Therefore, the maximized total cycling charge $Q_C = Q_{SC,max}$ was enabled by the instantaneous short-circuit conditions in steps 2 and 4, as controlled by the switch. The simulation results with respect to various external load resistances were plotted as Fig. 7.1e. These cycles can be named as "cycles for maximized energy output" (CMEO). Clearly, benefited from the maximized total cycling charges, the output energy per cycle of the CMEO was always higher than that of the CEO with the same load resistance *R*, as observed in the encircled areas in Fig. 7.1c, e.

We noticed that for the CMEO, higher output energy per cycle can be achieved with larger external load resistance $R$ (Fig. 7.1e). So the maximized output energy per cycle can be achieved at $R = +\infty$, which is the open-circuit condition. We simulated this maximized output energy by simply removing the external load and operating the remaining part of the circuit the same as that for the steps in the CMEO case as stated above. The corresponding $V$-$Q$ curve was plotted as the CMEO with infinite load resistance as shown in Fig. 7.1e. This CMEO has a trapezoid shape, the vertices of which are determined by the maximum short-circuit transferred charge $Q_{SC,max}$, the maximum open-circuit voltage $V_{OC,max}$, and the maximum achievable absolute voltage $V'_{max}$ at $Q = Q_{SC,max}$. It can be easily proven that the $V$-$Q$ plots for all kinds of the TENG operations are limited inside the four edges of this trapezoid as following.

We only consider the charge $Q$ with $0 \le Q \le Q_{SC,max}$ since the motion part only operates between $x = 0$ and $x = x_{max}$. $0 \le Q \le Q_{SC,max}$ represents two of the four boundary lines.

At arbitrary displacement $x$, the capacitance was fixed at $C(x)$. Therefore, as the electrical potential superposition of the open-circuit voltage and the voltage drop due to the charge transfer, the total voltage was derived as:

$$V = -\frac{Q}{C(x)} + V_{OC}(x) \tag{7.2}$$

Here we defined a variable $V'(x)$, which is the absolute voltage value when $Q = Q_{SC,max}$ at displacement $x$. So from the definitions, at fixed $x$, $(Q, V)$ plots (Eq. (7.2)) should include points of $(Q_{SC}(x), 0)$, $(0, V_{OC}(x))$ and $(Q_{SC,max}, -V'(x))$. By putting these points into Eq. (7.2) we can get:

$$V'(x) = \left| -\frac{Q_{SC,max}}{C(x)} + V_{OC}(x) \right| = \frac{Q_{SC,max}}{C(x)} - V_{OC}(x) \tag{7.3a}$$

$$0 = -\frac{Q_{SC}(x)}{C(x)} + V_{OC}(x) \tag{7.3b}$$

Then the relationship among $Q_{SC}(x)$, $V_{OC}(x)$, $C(x)$ and $V'(x)$ can be shown as:

$$C(x) = \frac{Q_{SC}(x)}{V_{OC}(x)} = \frac{Q_{SC,max} - Q_{SC}(x)}{V'(x)} \tag{7.4}$$

If we replace $C(x)$ in Eq. (7.2) by Eq. (7.4), we can get:

$$\begin{cases} \dfrac{V}{V_{oc}(x)} + \dfrac{Q}{Q_{sc}(x)} = 1 \quad \text{(a)} \\ -\dfrac{V}{V'(x)} + \dfrac{Q_{sc,max} - Q}{Q_{sc,max} - Q_{sc}(x)} = 1 \quad \text{(b)} \end{cases} \tag{7.5}$$

$V_{OC}(x)$ and $Q_{SC}(x)$ always increase with the increase of $x$. Therefore, at $x = 0$, $V_{OC}(x)$ and $Q_{SC}(x)$ should achieve their minimum value $V_{OC}(0) = 0$ and $Q_{SC}(0) = 0$; and at $x = x_{max}$, $V_{OC}(x)$ and $Q_{SC}(x)$ should achieve their maximum value $V_{OC}(x_{max}) = V_{OC,max}$ and $Q_{SC}(x_{max}) = Q_{SC,max}$. Therefore, $0 \leq Q_{SC}(x) \leq Q_{SC,max}$, $0 \leq V_{OC}(x) \leq V_{OC,max}$.

We noticed if we define a new displacement $x' = x_{max} - x$, and redefine the MACRS in this new coordinate system as $V_{OC}'(x') = 0$ and $Q_{SC}'(x') = 0$ at $x' = 0$. There would be $Q_{SC}'(x') = Q_{SC,max} - Q_{SC}(x)$ and $C'(x') = C(x)$ from the definitions. Therefore,

$$V'_{OC}(x') = \frac{Q'_{SC}(x')}{C'(x')} = \frac{Q_{SC,max} - Q_{SC}(x)}{C(x)} = V'(x) \tag{7.6}$$

So as the open-circuit voltage at this new charge reference state, $V_{OC}'(x') = V'(x)$ should achieve the minimum value 0 at $x' = 0$ and the maximum value $V'_{max}$ at $x' = x_{max}$. Then there would be $0 \leq V'(x) \leq V'_{max}$.

So for arbitrary $(Q, V)$ with displacement $x$ and satisfying Eq. (7.5):
When $V \geq 0$, $\frac{V}{V'_{max}} + \frac{Q}{Q_{SC,max}} \geq 0$ is always valid, and besides,

$$\frac{V}{V_{OC,max}} + \frac{Q}{Q_{SC,max}} \leq \frac{V}{V_{OC}(x)} + \frac{Q}{Q_{SC}(x)} = 1;$$

When $V < 0$, $\frac{V}{V_{OC,max}} + \frac{Q}{Q_{SC,max}} < \frac{Q}{Q_{SC,max}} \leq 1$ is always valid, and besides,

$$\frac{V}{V'_{max}} + \frac{Q}{Q_{SC,max}} = \frac{V}{V'_{max}} - \frac{Q_{SC,max} - Q}{Q_{SC,max}} + 1 \geq \frac{V}{V'(x)} - \frac{Q_{SC,max} - Q}{Q_{SC,max} - Q_{SC}(x)} + 1 = 0.$$

Therefore, for any $(Q, V)$ in any states of the TENG, $\begin{cases} \frac{V}{V_{OC,max}} + \frac{Q}{Q_{SC,max}} \leq 1 \\ \frac{V}{V'_{max}} + \frac{Q}{Q_{SC,max}} \geq 0 \end{cases}$,

consequently:

$$\begin{cases} V \leq V_{OC,max} - V_{OC,max}\frac{Q}{Q_{SC,max}} & \text{(a)} \\ -V \leq V'_{max}\frac{Q}{Q_{SC,max}} & \text{(b)} \end{cases} \tag{7.7}$$

Equation (7.7) represent the other two of the four boundary lines. Then the four boundary lines of the $V$-$Q$ curves were fixed.

Thus, with the largest encircled area, this cycle has the largest possible output energy per cycle, which can be calculated using the following equation:

$$E_{\mathrm{m}} = \frac{1}{2} Q_{\mathrm{SC,max}} \left( V_{\mathrm{OC,max}} + V'_{\mathrm{max}} \right) \quad (7.8)$$

$E_{\mathrm{m}}$ is demonstrated to be the largest possible output energy per cycle: For TENG under arbitrary operations, energy per cycle $E = \oint V\mathrm{d}Q$, we can divide the route into two parts:

Part 1: when $V \geq 0$, $\mathrm{d}Q/\mathrm{d}t = I = V/R \geq 0$, so $Q$ will always increase at $V \geq 0$ (assume it is from $Q_{1\min}$ to $Q_{1\max}$);
Part 2: when $V < 0$, $\mathrm{d}Q/\mathrm{d}t = I = V/R < 0$, so $Q$ will always decrease at $V < 0$ (assume it is from $Q_{2\max}$ to $Q_{2\min}$);

So: $E = \oint V\mathrm{d}Q$

$$= (\text{Part 1}: V \geq 0) \int_{Q_{1\min}}^{Q_{1\max}} V\mathrm{d}Q + (\text{Part 2}: V < 0) \int_{Q_{2\max}}^{Q_{2\min}} V\mathrm{d}Q$$

$$= (\text{Part 1}: V \geq 0) \int_{Q_{1\min}}^{Q_{1\max}} V\mathrm{d}Q + (\text{Part 2}: V < 0) \int_{Q_{2\min}}^{Q_{2\max}} (-V)\mathrm{d}Q$$

$$\leq (\text{Part 1}) \int_{Q_{1\min}}^{Q_{1\max}} \left( V_{\mathrm{OC,max}} - V_{\mathrm{OC,max}} \frac{Q}{Q_{\mathrm{SC,max}}} \right) \mathrm{d}Q + (\text{Part 2}) \int_{Q_{2\min}}^{Q_{2\max}} \left( V'_{\mathrm{max}} \frac{Q}{Q_{\mathrm{SC,max}}} \right) \mathrm{d}Q$$

due to Eq. (7.7)

$$\leq \int_{0}^{Q_{\mathrm{SC,max}}} \left( V_{\mathrm{OC,max}} - V_{\mathrm{OC,max}} \frac{Q}{Q_{\mathrm{SC,max}}} + V'_{\mathrm{max}} \frac{Q}{Q_{\mathrm{SC,max}}} \right) \mathrm{d}Q$$

due to $0 \leq Q \leq Q_{\mathrm{SC,max}}$

$$= \frac{1}{2} Q_{\mathrm{SC,max}} \left( V_{\mathrm{OC,max}} + V'_{\mathrm{max}} \right) = E_{\mathrm{m}}$$

### 7.1.4 Experimental Realization of the Operation Cycles

Here we also demonstrated that the CEO and the CMEO can be easily achieved experimentally. An LS mode TENG with parameters close to that in Table 7.1 was fabricated. An aluminum (Al) foil was used as one electrode and the motion part. The copper (Cu) layer deposited on the fluorinated ethylene propylene (FEP) film was utilized as the other electrode. The Al foil and the FEP film are the

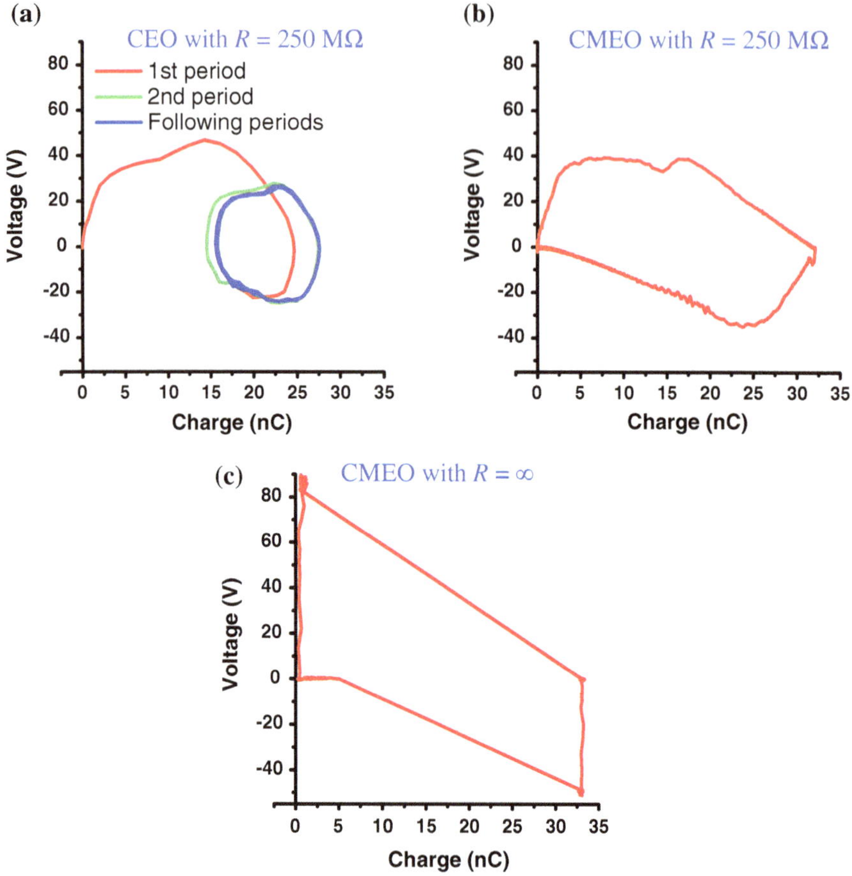


**Fig. 7.2** Experimental results of the *V-Q* plot of **a** the CEO with external load resistance of 250 MΩ. **b** The CMEO with external load resistance of 250 MΩ. **c** The CMEO with infinite external load resistance. Reproduced with permission from Nature Publication Group [16]

pair of triboelectric layers. Firstly, the *V-Q* plot of CEO was demonstrated by using an external load resistance of 250 MΩ, as shown in Fig. 7.2a. We noticed that the steady-state of the CEO was achieved only after a few periods of operation. And then, controlled by a switch in parallel, the *V-Q* plot of CMEO with external load resistances of 250 MΩ and infinite were also demonstrated, as shown in Fig. 7.2b, c, respectively. All of the features of the three *V-Q* plots are very consistent with the simulated results as shown in Fig. 7.1. We found that the trends of the characteristics with the varied resistances of these plots are very consistent with that of the simulated results. We can easily conclude from the encircled areas that the output energy per cycle of the CMEO with infinite external load resistance was the highest one in that of these cycles. Each output energy per cycle was calculated using Eq. (7.1), as shown in Table 7.2. These results demonstrated and further confirmed

**Table 7.2** Output energy per cycle for the three cycles operated in the LS mode TENG

| Cycle type | Output energy per cycle (μJ) |
|---|---|
| CMEO with $R = +\infty$ | 1.99 |
| CMEO with $R = 250\ \text{M}\Omega$ | 1.48 |
| CEO with $R = 250\ \text{M}\Omega$ | 0.47 |

that the experimental-achievable CMEO can harvest the maximized output energy per cycle. In fact, the instantaneous discharging TENG with a switch triggered by the motion of TENG, which was actually operated in CMEO, has been demonstrated to have gigantic enhanced instantaneous power compared to the traditional TENG operated in CEO [21].

## 7.2 Figure-of-Merits of Triboelectric Nanogenerators

For the TENG operating in CMEO with infinite load resistance, the period $T$ includes two parts of time. One part is from the relative motion in TENG, and the other part is from the discharging process in short-circuit condition. With the minimal resistance in short-circuit condition, the second part of $T$ can be short enough so that it can be omitted (less than 0.4 ms experimentally as reported in [21]). Thus, the average output power $\bar{P}$ at CMEO should satisfy:

$$\bar{P} = \frac{E_{\mathrm{m}}}{T} \approx \frac{E_{\mathrm{m}}}{2\frac{x_{\max}}{\bar{v}}} = \frac{\bar{v}}{2}\frac{E_{\mathrm{m}}}{x_{\max}} \tag{7.9}$$

where $\bar{v}$ is the average velocity value of the relative motion in TENG, which depends on the input mechanical motions. In this equation, $E_{\mathrm{m}}/x_{\max}$ is the only term that depends on the characteristics of the TENG itself.

On the other hand, the energy conversion efficiency of the TENG can be expressed as (at CMEO with $R = +\infty$):

$$\eta = \frac{E_{\mathrm{out}}}{E_{\mathrm{in}}} = \frac{E_{\text{output per cycle}}}{E_{\text{output per cycle}} + E_{\text{dissipation per cycle}}} = 1/\left[1 + 1/\left(\frac{E_{\mathrm{m}}}{2\bar{F}x_{\max}}\right)\right] \tag{7.10}$$

Here, $\bar{F}$ stands for the average dissipative force during the operation of the TENG. This force can be frictional force, air resistance force or others.

Therefore, from both Eqs. (7.9) and (7.10) we conclude that the term $E_{\mathrm{m}}/x_{\max}$ determines both the average power and the energy conversion efficiency from the characteristics of TENG itself. We also notice that as indicated in Eq. (7.8), $E_{\mathrm{m}}$ contains $Q_{\mathrm{SC,max}}$ that is proportional to the triboelectrification area $A$. Therefore, to exclude the effect of the TENG size on the output energy, the area $A$ should be placed in denominator of this term. Due to the reasons stated above, we confirm that the term $E_{\mathrm{m}}/Ax_{\max}$ determines the merits of TENG.

We notice that in Eq. (7.8), $Q_{SC,max}$, $V_{OC,max}$ and $V'_{max}$ are all proportional to the surface charge density $\sigma$. Therefore, $E_m$ is proportional to the square of the surface charge density $\sigma$. Then we can define a dimensionless structural FOM ($FOM_S$) of TENG, as the factor only depends on the structural parameters and $x_{max}$:

$$\mathrm{FOM_S} = \frac{2\varepsilon_0}{\sigma^2}\frac{E_m}{Ax_{max}} \tag{7.11}$$

Here, $\varepsilon_0$ is the permittivity of the vacuum. This structural FOM represents the merit of the TENG from the structural design. And then the performance FOM ($FOM_P$) of TENG can be defined as:

$$\mathrm{FOM_P} = \mathrm{FOM_S} \cdot \sigma^2 = 2\varepsilon_0 \frac{E_m}{Ax_{max}} \tag{7.12}$$

in which the $\sigma^2$ can be the material FOM ($FOM_M$), which is the only component related to the material properties. The performance FOM can be considered as the universal standard to evaluate varieties of TENGs, since it is directly proportional to the greatest possible average output power and related to the highest achievable energy conversion efficiency, regardless of the mode and the size of the TENG.

## 7.3 Structural Figure-of-Merit: Calculation and Simulation

Currently, four basic modes of TENGs have been developed for various applications [22, 23]. To compare these TENGs from the structural design's point of view, the structural FOM of CS mode, LS mode, single-electrode contact (SEC) structure in SE mode, sliding freestanding triboelectric-layer (SFT) and contact freestanding triboelectric-layer (CFT) structures in FT mode TENGs, were calculated by their analytical formulas and simulated by FEM, with the same surface charge density $\sigma$ and area $A$.

The detailed simulation method is described in following: We assume that each TENG takes surface charge density $\sigma$. All the electrodes have the size of area ($A$) = length ($l$) × width ($w$). The sliding direction in LS and SFT structures is along the length direction. The maximum displacement of the motion part is $x_{max}$, and the effective thickness of the dielectric layer is defined as $d_0 = \Sigma_i d_i/\varepsilon_{ri}$ in which $d_i$ and $\varepsilon_{ri}$ represent the thickness and dielectric constant of each dielectric layer, respectively. The capacitances between different electrodes and triboelectric surfaces are marked in Fig. 7.3, and $C_{Total}(x)$ is the total capacitance, which is the capacitance between two electrodes. The parameters for each structure of TENG were listed in Fig. 7.3 and Table 7.3.

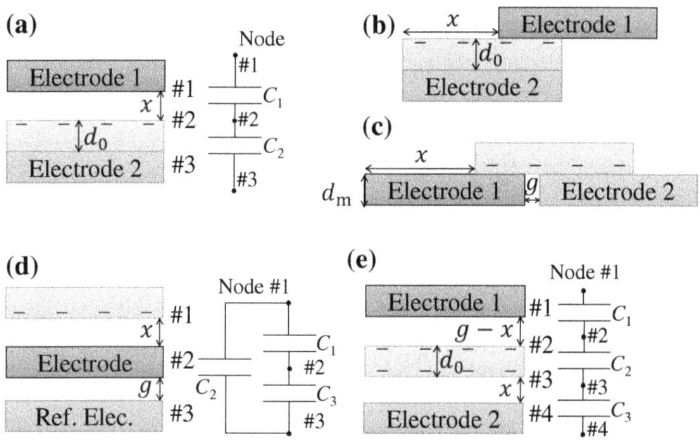


**Fig. 7.3** Illustration of symbols in 5 structures of TENG. **a** Vertical contact-separation (CS) mode TENG with capacitance $C_1(x)$ between electrode 1 (node #1) and triboelectric surface of the dielectric layer (node #2), and capacitance $C_2$ between node #2 and electrode 2 (node #3). **b** Lateral sliding (LS) mode TENG. **c** Sliding freestanding triboelectric-layer (SFT) structure TENG. **d** Single-electrode contact (SEC) structure TENG with capacitance $C_1(x)$ between triboelectric surface of the dielectric layer (node #1) and the electrode (node #2), capacitance $C_2(x)$ between node #1 and the reference electrode (node #3), and capacitance $C_3(x)$ between nodes #2 and #3. These capacitances are "virtual" capacitances as described in Ref. [1]. **e** Contact freestanding triboelectric-layer (CFT) structure TENG with capacitance $C_1(x)$ between electrode 1 (node #1) and the top triboelectric surface of the dielectric layer (node #2), capacitance $C_2$ between node #2 and the bottom triboelectric surface of the dielectric layer (node #3), and capacitance $C_3(x)$ between nodes #3 and electrode 2 (node #4). Reproduced with permission from Nature Publication Group [16]

**Table 7.3** Parameters used in simulation for different structures of TENG

| Structure | | CS | LS | SEC | SFT | CFT |
|---|---|---|---|---|---|---|
| Dielectric | | Effective thickness $d_0 = \sum d_i/\varepsilon_{ri} = 50\ \mu m$ | | | | |
| Air | | Relative dielectric constant $\varepsilon_r = 1$ | | | | |
| Electrodes | Size | Area $A$ = Length $l \times$ Width $w = 0.1$ m × 0.1 m = 0.01 m$^2$ | | | | |
| | Thickness | $d_m = 1 \mu m$ | | | | |
| Triboelectric surface charge density | | $\sigma = 10\ \mu C\ m^{-2}$ | | | | |
| Displacement $x_{max}$ | | 0∼0.2 m | 0∼0.1 m | 0∼1 m | 0∼0.101 m | 0∼0.01 m |
| (Ref. to Fig. 7.4) Gap g between electrodes | | – | – | 1 cm | 1 mm | 1 cm |

The analytical formulas of $V_{OC,max}$ and $Q_{SC,max}$ in CS, LS, SEC and CFT structures have been given by previous papers [17, 18, 20, 24]. For these structures, the $V'_{max}$ can also be derived by the models in these papers. The LS structure with $x_{max} > 0.95l$ and SFT structure were not calculated due to the lack of analytical formulas. That is because in the both cases, the capacitances $C_{Total}$ are between two fully misaligned parallel plates, which currently don't have appropriate analytical formulas. And besides, the analytical formulas of $Q_{SC}(x)$ cannot be derived, either, since they are strongly related to the side effects.

The formulas of $V_{OC,max}$, $Q_{SC,max}$ and $V'_{max}$ used were listed below (Please refer to the previous section for the meaning of symbols):

CS:

$$V_{OC,max} = \frac{\sigma A}{C_1(x_{max})} \tag{7.13}$$

$$Q_{SC,max} = V_{OC,max} C_{Total} = \sigma A \frac{C_2}{C_1(x_{max}) + C_2} \tag{7.14}$$

$$V'_{max} = \frac{Q_{SC,max}}{C_2} = \frac{\sigma A}{C_1(x_{max}) + C_2} \tag{7.15}$$

LS (only for $0 \leq x_{max} \leq 0.95l$):

$$V_{OC,max} = \frac{\sigma x_{max}}{\varepsilon_0(1 - x_{max})} d_0 \tag{7.16}$$

$$Q_{SC,max} = \sigma w x_{max} \tag{7.17}$$

$$V'_{max} = \frac{\sigma x_{max}}{\varepsilon_0 l} d_0 \tag{7.18}$$

SEC:

$$V_{OC,max} = V'_{max} = \sigma wl \frac{C_2(x_{max})}{C_1(x_{max})C_2(x_{max}) + C_2(x_{max})C_3(x_{max}) + C_3(x_{max})C_1(x_{max})} \tag{7.19}$$

$$Q_{SC,max} = \sigma wl \frac{C_2(x_{max})}{C_1(x_{max}) + C_2(x_{max})} \tag{7.20}$$

CFT:

$$Q_{SC,max} = \sigma wl \left[ \left( \frac{1}{C_2(x_{max})} + \frac{2}{C_3(x_{max})} \right) C_{Total}(x_{max}) - \left( \frac{1}{C_2(0)} + \frac{2}{C_3(0)} \right) C_{Total}(0) \right] \tag{7.21}$$

$$V_{OC,max} = \frac{Q_{SC,max}}{C_{Total}(x_{max})} \tag{7.22}$$

$$V'_{max} = \frac{Q_{SC,max}}{C_{Total}(0)} \tag{7.23}$$

The capacitances listed above were calculated by considering 1-side (1S) and 2-side (2S) side effects of the non-ideal parallel-plate capacitor. The equations are listed below [25]:

$$C_{1S} = \varepsilon_0 \varepsilon_r \left\{ \frac{lw}{d} + \frac{l}{\pi}\left[1 + \ln\left(1 + 2\pi\frac{w}{d} + \ln\left(1 + 2\pi\frac{w}{d}\right)\right)\right]\right\} \tag{7.24}$$

$$C_{2S} = \varepsilon_0 \varepsilon_r \left\{ \frac{lw}{d} + \frac{l}{\pi}\left[1 + \ln\left(1 + 2\pi\frac{w}{d} + \ln\left(1 + 2\pi\frac{w}{d}\right)\right)\right] + \frac{w}{\pi}\left[1 + \ln\left(1 + 2\pi\frac{l}{d} + \ln\left(1 + 2\pi\frac{l}{d}\right)\right)\right]\right\} \tag{7.25}$$

Here, $l$, $w$ and $d$ represent length, width and gap distance of the parallel-plate capacitor.

And then, the largest possible output energy per cycle $E_m$ was calculated by Eq. (7.8) and the structural FOM with respect to different $x_{max}$ was calculated by Eq. (7.11).

The calculated structural FOMs for the five structures with respect to varied maximum displacement $x_{max}$ were plotted and compared in Fig. 7.4. For CS, SEC and CFT structures, the non-ideal parallel capacitances considering both one-side (1S) and two-side (2S) edge effects were used for calculations by analytical formulas, [25] while in the FEM simulation only one-side (1S) edge effect was considered. The calculated results from the analytical formulas fit the FEM simulation results very well. We can extract the maximum value of the structural FOM ($FOM_{S,max}$) from FEM simulations as a standard criterion for evaluating the structures. As shown in Fig. 7.4f and Table 7.4, we found for $FOM_{S,max}$:

CFT > CS > SFT > LS > SEC

Here we can make several general conclusions. Firstly, the performance of the paired-electrode TENGs is dramatically better than that of single-electrode TENGs with the same size and materials, due to the limited transferred charges and suppressed built-in voltages in single-electrode TENG, which has been reported previously [17]. Secondly, the performance of TENG triggered by contact-separation action is better than that of TENG triggered by sliding (assuming the triboelectric charge densities created by contact-separation and sliding are the same), because to achieve the same $V_{OC,max}$, a much higher $x_{max}$ of TENG triggered by sliding is required compared with that of TENG triggered by contact-separation, resulting in a smaller $FOM_{S,max}$. Finally, the freestanding configuration enhances the performance by dramatically decreasing the capacitance between electrodes. The high $FOM_{S,max}$ of CFT structure is also benefited by the large transferred charges induced by the double-side triboelectrification of the middle dielectric layer during operation (as shown in Fig. 7.3e).

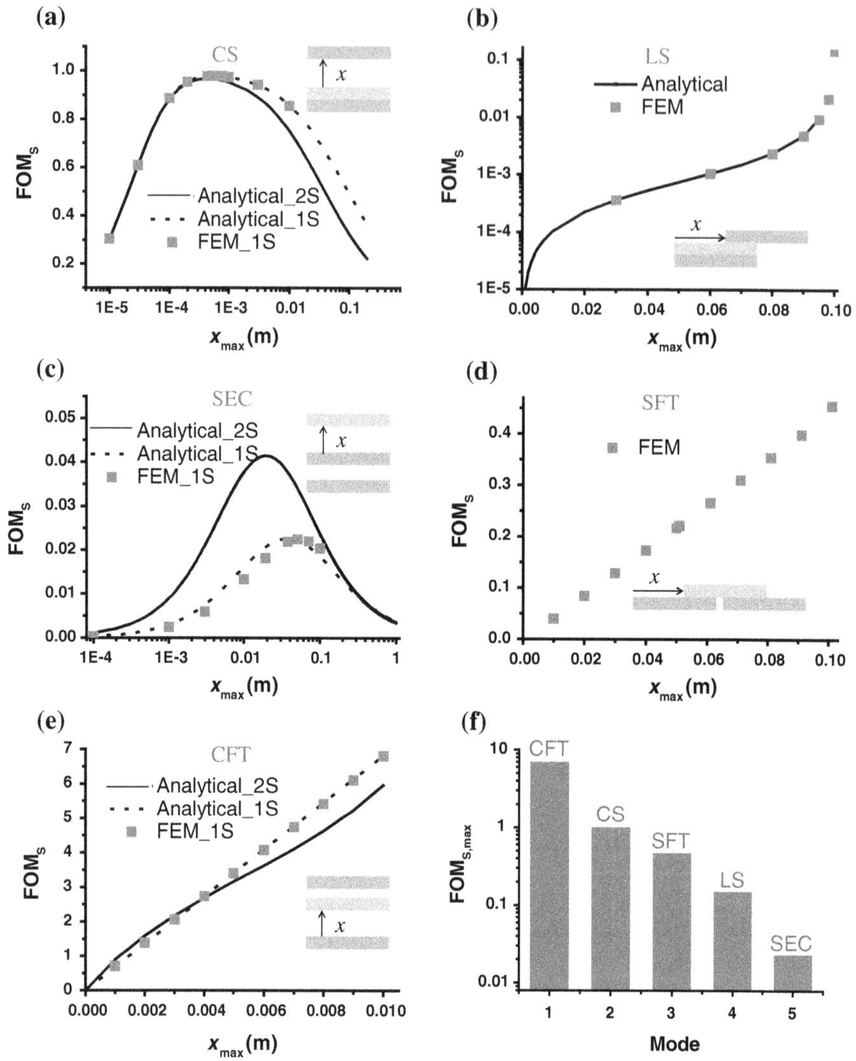


**Fig. 7.4** $FOM_S$ versus $x_{max}$ for different TENG structures. **a–e** $FOM_S$ for vertical contact-separation (*CS*) mode, lateral sliding (*LS*) mode, single-electrode contact (*SEC*) structure, sliding freestanding triboelectric-layer (*SFT*) structure and contact freestanding triboelectric-layer (*CFT*) structure calculated by analytical formulas and FEM simulation. The insets show the corresponding schematic diagrams of the corresponding structures. 1S and 2S stand for calculations and simulations considering 1-side and 2-side side effects of the non-ideal parallel-plate capacitor. **f** The maximum structural FOM ($FOM_{S,max}$) of different structures extracted from FEM simulations. Reproduced with permission from Nature Publication Group [16]

**Table 7.4** The simulated maximum structural FOM ($FOM_{S,max}$)

| Structure | $FOM_{S,max}$ |
|---|---|
| CFT | 6.81 |
| CS | 0.98 |
| SFT | 0.45 |
| LS | 0.15 |
| SEC | 0.022 |

## 7.4 Measurement of Material Figure-of-Merit

The triboelectric surface charge density $\sigma$ is the only material-related parameter in performance FOM as shown in Eq. (7.12). The optimization of $\sigma$ would enhance the performance of TENG significantly since the performance FOM is proportional to $\sigma^2$. This surface charge density was determined by the triboelectric performance of the materials in contact. Since the existing triboelectric series is a qualitative measure of the materials' triboelectric performances [22], a quantitative measuring matrix is required for TENGs.

Current studies of the triboelectric performance are influenced by the low contact intimacy induced by the nanometer-to-micrometer-level surface roughness in solid materials [26–28], Therefore, the measured surface charge density by solid-solid contact was not able to reach its highest possible value. To overcome this limitation of the solid surface and to build the measuring matrix for triboelectric performance of materials, we can utilize liquid metals such as liquid gallium (Ga, melting point: 29.8 °C), galinstan (a eutectic alloy consisting of 68.5 wt% gallium, 21.5 wt% indium and 10 wt% tin, melting point: −19 °C), and mercury (melting point: −38.8 °C) as one triboelectrification material. It is expected that, by using these liquid metals, the contact intimacy could be greatly enhanced since the liquid metals are shape-adaptive to the solid surfaces [29, 30].

### 7.4.1 *Measurement of Triboelectric Surface Charge Density*

Tang et al. develop the liquid-metal-based triboelectric nanogenerator (LM-TENG) using mercury for surface charge density measurement [28]. The LM-TENG is composed of two parts, as schemed in Fig. 7.5a, b depicts the working principle of that step by step. The detailed structure and the working process have been described in Sect. 3.3.6 in Chap. 3. By using this method and assuming the short-circuit charge transfer is the total triboelectric charge created in the surface, the charge densities of different materials respect to mercury are measured as shown in Fig. 7.6b.

Zi et al. reported charge density measurement by liquid gallium and gallium alloy [16]. Here the non-toxic liquid gallium and galinstan were utilized in experiments. To prevent them from being oxidized that would happen very fast for

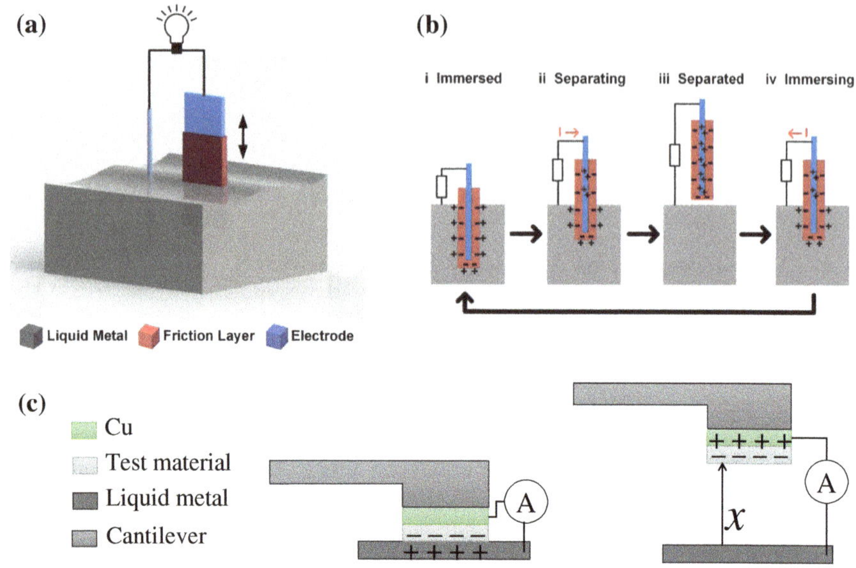


**Fig. 7.5** Methods of triboelectric surface charge measurement. **a** Schematic illustration showing the first method of measurement setup. **b** Step-by-step illustration showing the working principle of the first method of measurement. **c** The structure and the working principle of the second method of measurement. Reproduced with permission from Wiley [28] and Nature Publication Group [16]

all the liquid metals containing gallium [29–32], the experiments were performed in a glove box with argon environment and the fixed condition of room temperature, 1 atm and 0.006 % relative humidity (RH).

The setup of the measurement experiment is shown in Fig. 7.5c. Firstly the tested material was deposited with Cu thin film as one electrode, and the liquid galinstan (from rotometals) or liquid gallium (from VWR) was used as the other electrode. When the material surface was contacted with the liquid metal surface, the charges were created on both the material and liquid metal surfaces by triboelectrification effect. And then, after the material along with the Cu electrode was lifted above the liquid metal surface for a height much larger (over 100 times) than the thickness of the materials, the charges induced by triboelectrification would be approximately fully transferred to the Cu electrode in short-circuit condition. The transferred charges were measured as the triboelectric charges, and then the charge density was calculated by the charges over the area of the triboelectrification surface. To make comparisons, solid gallium was also utilized to replace liquid metals as one electrode for charge density measurements. The good repeatability of charge transfer was observed during multiple contact-separation process. The tested materials included FEP, Kapton, polarized polyvinylidene fluoride (PVDF), polyethylene (PE), nature rubber, and cellulose. All of the measurement results are shown in Fig. 7.6a, c. When the tested material is more negative than the liquid

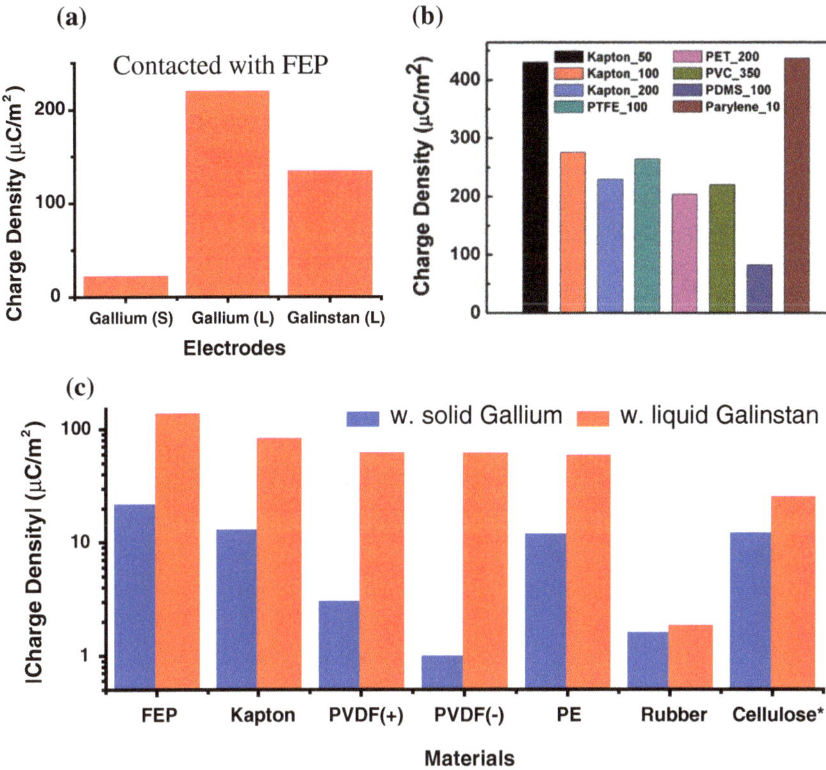


**Fig. 7.6** The standard $\sigma$ measured by liquid metal as one electrode. **a** The charge density measured by contacting FEP with solid gallium, liquid gallium and galinstan. **b** The charge density measured by contacting different materials with mercury. **c** The absolute charge density measured by contacting different materials with solid gallium and liquid galinstan. Among them, cellulose was measured to be more positive than galinstan (as marked by *). Reproduced with permission from Wiley [28] and Nature Publication Group [16]

metal, the measured charge density is recorded as a positive value; when the material is more positive than liquid metal, the measured charge density is recorded as a negative value (such as cellulose).

### 7.4.2 *Quantified Triboelectric Series Based on Normalized Charge Density and Dimensionless Material Figure-of-Merit*

As we noticed from the results, the measured charge densities by contacting tested materials with a liquid metal are always larger than that by contacting the material with the corresponding solid metal. This is because with liquid metals used as one contact, the contact intimacy can be greatly enhanced. In fact, by using liquid mercury as the contact, even higher surface charge densities can be achieved [28]. The varied charge densities measured by contacting the same material with different liquid metals might be due to the different capabilities of the liquid metals to absorb the electrons. The triboelectric order as derived from our measurement is: (most negative) FEP—Kapton—PVDF—PE—Nature Rubber—Galinstan—Cellulose (most positive), which are consistent with existing triboelectric series as reported [22]. Triboelectric performance of each material can be standardly quantified using the triboelectric surface charge density with respect to a certain liquid metal. For example, the triboelectric performance of FEP can be quantified as $\sigma_{\mathrm{FEP/galinstan}} = 133.24\ \mu\mathrm{C\ m}^{-2}$ with respect to galinstan, or $\sigma_{\mathrm{FEP/Ga(L)}} = 218.64\ \mu\mathrm{C\ m}^{-2}$ with respect to liquid gallium. FEP (Teflon) is usually considered as the most triboelectric negative material; galinstan should be about in the middle position of the triboelectric series since it is close to nature rubber and it is more negative than cellulose. Therefore, we can consider the surface charge densities while contacting FEP with galinstan as the reference triboelectric charge densities. Then the normalized triboelectric charge density $\sigma_N$ and dimensionless material figure-of-merit $\mathrm{FOM_{DM}}$ for triboelectrification (with respect to the charge density of FEP contacting with galinstan) can be defined as:

$$\mathrm{FOM_{DM}} = (\sigma_N)^2 = \frac{\sigma^2_{\mathrm{Material/Galinstan}}}{\sigma^2_{\mathrm{FEP/Galinstan}}} \tag{7.26}$$

Then the normalized triboelectric charge densities and dimensionless material FOM of measured materials are listed in Table 7.5. For a certain material, if $\sigma_N < 0$, then the material is more positive than the reference liquid metal; if $0 < \sigma_N < 1$, then this material is more negative than the reference liquid metal and more positive than FEP; if $\sigma_N > 1$, then this material is more negative than FEP.

**Table 7.5** $\sigma_N$ and $\mathrm{FOM_{DM}}$ of different materials. These results are respect to the charge density of contacting FEP with galinstan

| Materials | Normalized triboelectric charge density | Dimensionless material figure-of-merit | Position in triboelectric series |
|---|---|---|---|
| FEP | 1 | 1 | Most— |
| Kapton | 0.60 | 0.36 | |
| PVDF | 0.45 | 0.20 | |
| PE | 0.43 | 0.18 | |
| Nature Rubber | 0.0135 | 0.000183 | |
| Galinstan | 0 | 0 | About middle |
| Cellulose | −0.185 | 0.0342 | + |

## 7.5 Summary

In this chapter, we have developed methods for standardized evaluations on the performance of TENGs. Starting from the built-up voltage $V$—transferred charge $Q$ plot, the CMEO with infinite load resistance was derived to have the maximized output energy per cycle, which represents the maximum energy production of TENG, similar to the Carnot cycle in heat engines. Based on the maximum output energy per cycle, and considering both the maximized energy conversion efficiency and the maximized average output power, the performance figure-of-merit (FOM) was derived to evaluate each TENG design, composed by a structural FOM and a material FOM. The structural FOMs for different structures of TENGs were simulated by analytical formulas and FEM, respectively, showing the maximum value of structural FOM for each TENG structure. The standard evaluation of the material FOM was also demonstrated by measuring triboelectric surface charge density via contacting the materials with liquid metals, and then the normalized triboelectric charge density and dimensionless material FOM were defined and derived for various materials. The standards and evaluation methods provide here set the foundation for the further applications and industrialization of TENG technology.

## References

1. G. Zhu, C. Pan, W. Guo, C.-Y. Chen, Y. Zhou, R. Yu, Z.L. Wang, Triboelectric-generator-driven pulse electrodeposition for micropatterning. Nano Lett. **12**(9), 4960–4965 (2012)
2. X.S. Meng, G. Zhu, Z.L. Wang, Robust thin-film generator based on segmented contact-electrification for harvesting wind energy. ACS Appl. Mater. Interfaces **6**(11), 8011–8016 (2014)
3. G. Zhu, J. Chen, Y. Liu, P. Bai, Y.S. Zhou, Q. Jing, C. Pan, Z.L. Wang, Linear-grating triboelectric generator based on sliding electrification. Nano Lett. **13**(5), 2282–2289 (2013)
4. S. Wang, L. Lin, Y. Xie, Q. Jing, S. Niu, Z.L. Wang, Sliding-triboelectric nanogenerators based on in-plane charge-separation mechanism. Nano Lett. **13**(5), 2226–2233 (2013)
5. G. Zhu, J. Chen, T. Zhang, Q. Jing, Z.L. Wang, Radial-arrayed rotary electrification for high performance triboelectric generator. Nat. Commun. **5**, 3426 (2014)
6. Y. Yang, H. Zhang, J. Chen, Q. Jing, Y.S. Zhou, X. Wen, Z.L. Wang, Single-electrode-based sliding triboelectric nanogenerator for self-powered displacement vector sensor system. ACS Nano **7**(8), 7342–7351 (2013)
7. S. Wang, S. Niu, J. Yang, L. Lin, Z.L. Wang, Quantitative measurements of vibration amplitude using a contact-mode freestanding triboelectric nanogenerator. ACS Nano **8**(12), 12004–12013 (2014)
8. F.L. Curzon, B. Ahlborn, Efficiency of a Carnot engine at maximum power output. Am. J. Phys. **43**(1), 22–24 (1975)
9. N. Giordano, *College Physics: Reasoning and Relationships* (Cengage Learning, 2009)
10. G. Sebald, E. Lefeuvre, D. Guyomar, Pyroelectric energy conversion: optimization principles. IEEE Trans. Ultrason. Ferroelectr. Freq. Control **55**(3), 538–551 (2008)

11. S.P. Alpay, J. Mantese, S. Trolier-McKinstry, Q. Zhang, R.W. Whatmore, Next-generation electrocaloric and pyroelectric materials for solid-state electrothermal energy interconversion. MRS Bull. **39**(12), 1099–1111 (2014)
12. T.M. Tritt, M.A. Subramanian, Thermoelectric materials, phenomena, and applications: a bird's eye view. MRS Bull. **31**(03), 188–198 (2006)
13. D.M. Rowe, *CRC Handbook of Thermoelectrics* (Taylor & Francis, 2010)
14. M.A. Green, *Solar cells: Operating Principles, Technology, and System Applications* (Prentice-Hall, 1982)
15. J. Nelson, *The Physics of Solar Cells.* (Imperial College Press, 2003)
16. Y. Zi, S. Niu, J. Wang, Z. Wen, W. Tang, Z.L. Wang, Standards and figure-of-merits for quantifying the performance of triboelectric nanogenerators. Nat. Commun. **6**, 8376 (2015)
17. S. Niu, Y. Liu, S. Wang, L. Lin, Y.S. Zhou, Y. Hu, Z.L. Wang, Theoretical investigation and structural optimization of single-electrode triboelectric nanogenerators. Adv. Funct. Mater. **24** (22), 3332–3340 (2014)
18. S. Niu, Y. Liu, X. Chen, S. Wang, Y.S. Zhou, L. Lin, Y. Xie, Z.L. Wang, Theory of freestanding triboelectric-layer-based nanogenerators. Nano Energy **12**, 760–774 (2015)
19. Y. Zi, J. Wang, S. Wang, S. Li, Z. Wen, H. Guo, Z.L. Wang, Effective energy storage from a triboelectric nanogenerator. Nat. Commun. **7**, 10987 (2016)
20. S. Niu, Y. Liu, S. Wang, L. Lin, Y.S. Zhou, Y. Hu, Z.L. Wang, Theory of sliding-mode triboelectric nanogenerators. Adv. Mater. **25**(43), 6184–6193 (2013)
21. G. Cheng, Z.-H. Lin, L. Lin, Z.-L. Du, Z.L. Wang, Pulsed nanogenerator with huge instantaneous output power density. ACS Nano **7**(8), 7383–7391 (2013)
22. Z.L. Wang, Triboelectric nanogenerators as new energy technology for self-powered systems and as active mechanical and chemical sensors. ACS Nano **7**(11), 9533–9557 (2013)
23. Z.L. Wang, Triboelectric nanogenerators as new energy technology and self-powered sensors —Principles, problems and perspectives. Faraday Discuss. **176**, 447–458 (2014)
24. S. Niu, S. Wang, L. Lin, Y. Liu, Y.S. Zhou, Y. Hu, Z.L. Wang, Theoretical study of contact-mode triboelectric nanogenerators as an effective power source. Energy Environ. Sci. **6**(12), 3576–3583 (2013)
25. Y. Li, Y.H. Li, Q.X. Li, Y.Y. Zi, Computation of electrostatic forces with edge effects for nonparallel comb-actuators. J Tsinghua Univ (Sci & Tech) **43**(8), 1024–1026, 1030 (2003)
26. H.T. Baytekin, A.Z. Patashinski, M. Branicki, B. Baytekin, S. Soh, B.A. Grzybowski, The mosaic of surface charge in contact electrification. Science **333**(6040), 308–312 (2011)
27. T.A.L. Burgo, T.R.D. Ducati, K.R. Francisco, K.J. Clinckspoor, F. Galembeck, S.E. Galembeck, Triboelectricity: macroscopic charge patterns formed by self-arraying ions on polymer surfaces. Langmuir **28**(19), 7407–7416 (2012)
28. W. Tang, T. Jiang, F.R. Fan, A.F. Yu, C. Zhang, X. Cao, Z.L. Wang, Liquid-metal electrode for high-performance triboelectric nanogenerator at an instantaneous energy conversion efficiency of 70.6 %. Adv. Funct. Mater. **25**(24), 3718–3725 (2015)
29. R.C. Chiechi, E.A. Weiss, M.D. Dickey, G.M. Whitesides, Eutectic Gallium-Indium (EGaIn): a moldable liquid metal for electrical characterization of self-assembled monolayers. Angew. Chem. Int. Ed. **47**(1), 142–144 (2008)
30. M.D. Dickey, R.C. Chiechi, R.J. Larsen, E.A. Weiss, D.A. Weitz, G.M. Whitesides, Eutectic Gallium-Indium (EGaIn): a liquid metal alloy for the formation of stable structures in microchannels at room temperature. Adv. Funct. Mater. **18**(7), 1097–1104 (2008)
31. Q. Xu, N. Oudalov, Q. Guo, H.M. Jaeger, E. Brown, Effect of oxidation on the mechanical properties of liquid gallium and eutectic gallium-indium. Phys. Fluids (1994-present) **24**(6), 063101 (2012)
32. L. Tingyi, P. Sen, K. Chang-Jin, Characterization of liquid-metal Galinstan for droplet applications. In *2010 IEEE 23rd International Conference on Micro Electro Mechanical Systems (MEMS)* (IEEE, Wanchai, Hong Kong, 2010), pp. 560–563

# Part II
# Applications as a Sustainable Power Source

# Chapter 8
# Harvesting Body Motion Energy

## 8.1 Integrated Structure Triboelectric Nanogenerators

Contact-separation (CS) mode TENG, based on the universally known contact electrification effect, has been proven to be a cost-effective and robust approach for harvesting ambient environmental energy [1–17]. Periodic contact and separation between two materials with different charge affinities act as a charge pump to alternately drive induced electrons to flow between the electrodes through an external load. A key challenge to TENG is its relatively low output current. Yang et al. [1] presented a rationally designed TENG with integrated rhombic gridding as an effective solution to this problem. The newly designed TENG greatly improved the total current output as well as the vibration-to-electric energy conversion efficiency owing to its structurally multiplied unit cells connected in parallel. With the hybridization of both contact and sliding electrification among polytetrafluoroethylene (PTFE) nanowire arrays and aluminum nanopores, the open-circuit voltage ($V_{OC}$) and short-circuit current density ($I_{SC}$) of the TENG reach up to 428 V and 36.6 mA/m$^2$, respectively, with peak power density of 30.7 W/m$^2$. On the basis of this result, a self-powered backpack was developed. Under the circumstance of natural human walking with loads of 2.0 kg, the power generated by one unit cell is high enough to simultaneously light up more than 40 commercial LEDs, which unambiguously demonstrated the capability of the integrated rhombic gridding based TENG acting as a sustainable power source for mobile electronics.

The structure of integrated rhombic gridding based TENG is shown in Fig. 8.1a, in which, the total number of unit cells in one TENG can be expressed as:

$$N_{\mathrm{Total}} = 2n^2 \tag{8.1}$$

where $n$ is the number of unit cells along the edge length. The plastic sheets of polyethylene terephthalate (PET) with a thickness of 600 μm are utilized. Each PET sheet is cut half through and then locked into each other to form the framework of

 
Z.L. Wang et al., *Triboelectric Nanogenerators*,
Green Energy and Technology, DOI 10.1007/978-3-319-40039-6_8

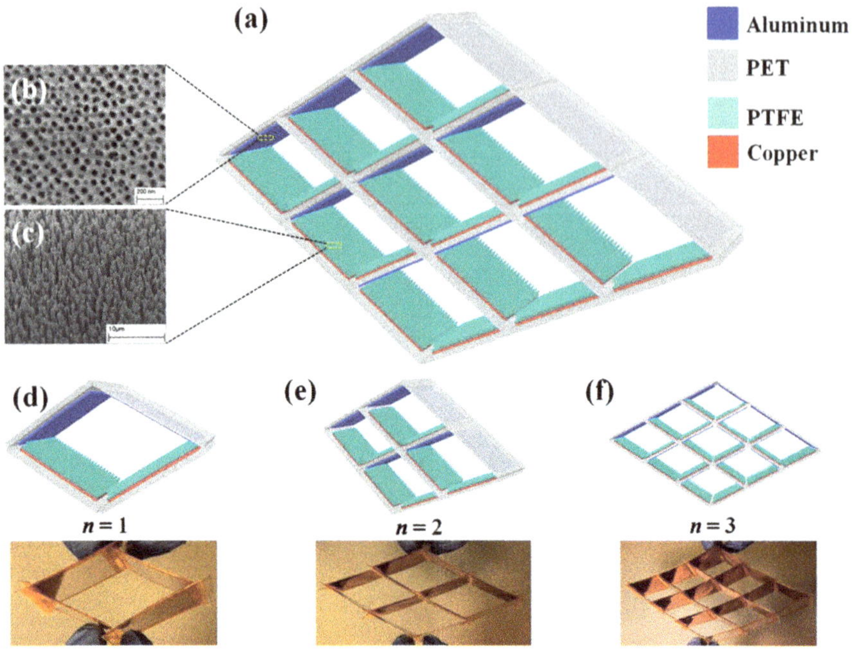


**Fig. 8.1** Integrated rhombic gridding based triboelectric nanogenerator. **a** Sketch of a typical TENG with n = 3. **b** SEM image of nanopores on aluminum electrode. **c** SEM image of PTFE nanowires. **d–f** Sketch and corresponding photograph of integrated rhombic gridding based TENG with n = 1, 2, 3, respectively. Reproduced with permission from American Chemical Society [1].

the TENG. On one side of the PET substrate, an aluminum thin film with nanoporous modification plays dual roles as a contact electrode and a contact surface. A scanning electron microscopy (SEM) image of the aluminum nanopores is shown in Fig. 8.1b. A layer of PTFE film with nanowire arrays was adhered onto the other side of the PET substrate with deposited copper thin film as back electrode. An SEM image of PTFE nanowire arrays is presented in Fig. 8.1c. As demonstrated in Fig. 8.1d–f, there are 2, 8, and 18 unit cells in the TENGs (sketch and corresponding photograph of real devices) with $n$ = 1, 2, and 3, respectively. This integrated rhombic gridding structure, with unit cells electrically connecting in parallel, is capable of improving the current output with multifold enhancement. Furthermore, the structural coupling of aluminum nanopores and PTFE nanowires arrays dramatically increases the effective contact area of the TENG.

TENG generally has two working modes: contact-separation mode and lateral sliding mode. The currently designed TENG works on the basis of a hybridization of the two modes. Because of the coupling of the triboelectric effect and electrostatic induction effect, a periodic contact and separation between two materials with opposite triboelectric polarities alternatingly drives induced electrons through an external circuit. According to the triboelectric series, a list of materials based on their tendency to gain or lose charges, electrons are injected from an aluminum electrode into PTFE at the original position, resulting in negative charges at the surface of the PTFE film and positive charges at the aluminum electrode. Once a periodic external force acts on the TENG, a separation of the two plates forms with an included angle of $\theta$, producing an electric potential difference between the contact electrode and back electrode. According to the Gauss Theorem, the electric potential deference of the two electrodes can be expressed as

$$V_{\rm A} - V_{\rm b} = \left(\int_A^C \frac{\sigma - \Delta\sigma}{2\varepsilon_0} + \frac{\sigma}{2\varepsilon_0} - \frac{\Delta\sigma}{2\varepsilon_0}\right)\mathrm{d}_l + \int_C^B \left(\frac{\sigma - \Delta\sigma}{2\varepsilon_0\varepsilon_r} + \frac{\sigma}{2\varepsilon_0\varepsilon_r} - \frac{\Delta\sigma}{2\varepsilon_0\varepsilon_r}\right)\mathrm{d}_l \tag{8.2}$$

where $V_{\rm A}$ and $V_{\rm B}$ are the electric potential of the contact electrode and back electrode, respectively. $\sigma$ is the triboelectric charge density, $\Delta\sigma$ is the induced charge density, $\varepsilon_0$ and $\varepsilon_{\rm r}$ is the vacuum permittivity and the relative permittivity of PTFE, respectively. If the device is in the open-circuit state, there are no transferring induced charges, that is $\Delta\sigma = 0$. So, the open circuit voltage can be calculated as:

$$V_{\rm oc} = \frac{2\sigma l \sin\frac{\theta}{2}}{\varepsilon_0} \tag{8.3}$$

where $l$ is the edge length of one rhombic unit cell. If the device is in the short-circuit state, the electric potential deference of the two electrodes becomes zero, $V_{\rm A} - V_{\rm B} = 0$. The transferred induced charge density of the TENG can be shown as

$$\Delta\sigma = \frac{\sigma}{1 + \frac{d_1}{2\varepsilon_r l \sin(\theta/2)}} \tag{8.4}$$

where $d_1$ is the thickness of PTFE. So, the short-circuit current can be expressed as

$$I_{\rm sc} = \frac{\sigma s d_1 v}{\varepsilon_{\rm r}[2l\sin(\theta/2) + d_1/\varepsilon_{\rm r}]} \tag{8.5}$$

where $v$ is the relative velocity of the two contact plates whose motion direction determines the flowing direction of the induced charge, thus, the polarity of the short circuit current; $s$ is the effective contact area. Such a raise of the electric potential difference drives electrons from the back electrode to the contact electrode through the external circuit, screening the positive triboelectric charges on the back electrode. When the included angle reaches its maximum $\theta_{\rm max}$, the positive triboelectric charges are nearly completely neutralized by the inductive electrons. Once the external force is applied on the TENG again, the decreasing of electric potential difference drives the induced electrons to move in a reversed direction. The TENG acts as an electron-pump that drives electrons back and forth flowing between two electrodes, producing an alternating current in the external circuit.

On the basis of Eq. (8.5), various factors, such as the charge density $\sigma$, the effective contact area $s$, and the relative velocity $v$ of the contact or separation of two plates, determine the short circuit current of TENG. Also it can be tuned from following aspects: (i) Selection of contact electrification materials. The triboelectric charge density $\sigma$ is largely determined by the charge affinity difference of two contact materials. Of all materials, PTFE is so far the most electronegative one with a charge affinity of –190 nC/J. Moreover, being electropositive, the charge affinity of aluminum is about 10–30 nC/J. So large charge affinity difference assures large amount of triboelectric charges and thus a superior electric output of the device. (ii) Surface modification by nanomaterials. The nanomaterials-based surface modification can greatly enhance the effective contact area $s$. (iii) Relative velocity of surface contact and separation. The periodical contact and separation of the contact surfaces acts as an electron pump to drive the induced charges back and forth, producing alternating current in the external circuit.

To enhance the total current output of the TENG, all of the unit cells are electrically connected in parallel. Under the fixed triggering frequencies and amplitude, the electric output measurement was performed on the integrated rhombic gridding based TENG with $n = 1, 2, 3$. The effective contact area of TENG is $2n^2 \times 4.6$ cm $\times$ 4.6 cm, which produces an open-circuit voltages ($V_{OC}$) 445 V at $n = 1$,439 V at $n = 2$, and 428 V at $n = 3$. The voltage output is almost constant for all the measurements because all of the rhombic unit cells are electrically connected in parallel. A positive voltage peak is generated due to the immediate charge separation at the departure of aluminum from PTFE. Since the electrons cannot flow back to screen the induced electric potential difference between the two electrodes under the open-circuit condition, the voltage holds at a plateau until the next contact emerges. Meanwhile, the peak values of the short-circuit current ($I_{SC}$) reach up to 245 μA at $n = 1$,801 μA at $n = 2$, and 1.395 mA at $n = 3$. In addition, the output current has an alternating behavior with asymmetrical amplitudes, with the larger peaks corresponding to the process in which the two contact surfaces move toward each other, while the smaller ones are generated as the two surfaces move apart. According to Eq. (8.5), the faster approach is expected to produce larger current peaks than the slower separation.

The current enhancement factor $\alpha$ is a function of the number of unit cells along the edge length, $\alpha = bn^2$. The fitting result renders the coefficient $b$ a value of 1.66. Considering the non-ideal experimental factors, such as humidity, particle contaminations in the air, the imperfection from the device fabrication process, and the difficulty of synchronizing of all the device units, which may potentially have negative impact on the actual output, the experimental result of the enhancement factor is considerably approaching the ideal value of $2n^2$, revealing that the integrated rhombic gridding structure can effectively enhance the total current output. The accumulative induced charges increase with $n$, which reaches up to 142.68 μC within 2.75 s when $n = 3$, further indicating that the integrated rhombic gridding structure can dramatically enhance the electric output of TENG. It is noteworthy that the accumulative induced charges are the sum of all the back-and-forth induced charges in the entire process of the TENG working as a "charge pump". Consequently, it is a monotonically increasing function of time throughout the experimental time window.

To prove the capability of the integrated rhombic gridding based TENG as a sustainable power source, a backpack was developed to harvest vibration energy from natural human walking. As indicated in Fig. 8.2a, four acrylics plates were built into a supporting shelf with a size of 5 cm × 7.5 cm × 20 cm, which bridged the backpack and its two straps using four springs and two long screw shanks. The integrated rhombic gridding based TENG with $n = 1$ is sandwiched between two acrylic sheets and the photograph of a real backpack is shown in Fig. 8.2b. A total of forty commercial LED bulbs were assembled in series on a piece of electric board (Fig. 8.2c), electrically connected to the newly designed backpack. When a person walks naturally carrying the designed backpack with a total weight of 2.0 kg, the power harvested from the body vibration is high enough to simultaneously light all the 40 LEDs (see Fig. 8.2d).

In summary, a novel integrated rhombic gridding based TENG was demonstrated. This innovative structure provides $2n^2$ unit cells electrically connected in parallel, which is able to greatly enhance the current output as well as the vibration-to-electric energy conversion efficiency. Also, the surface modification by PTFE nanowire-arrays and aluminum nanopores renders a hybridization of both contact-separation and lateral sliding electrification modes for effectively improving

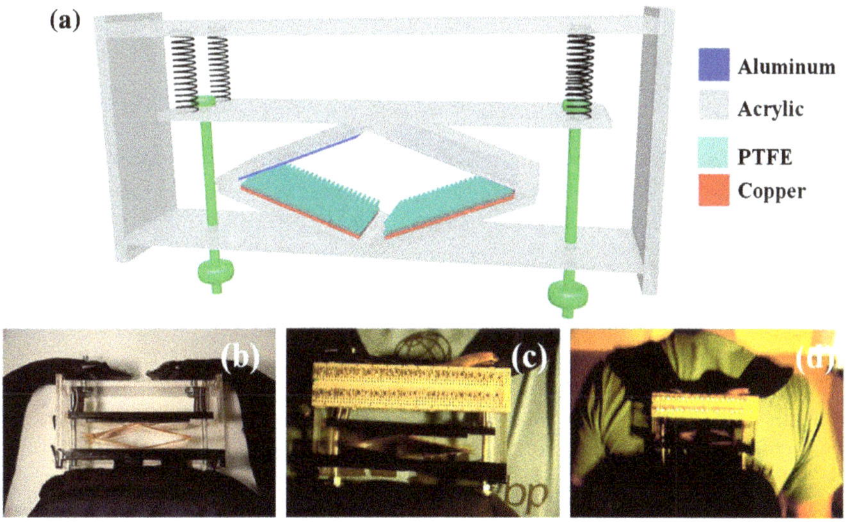


**Fig. 8.2** **a**, **b** Sketch and photograph of a self-powered backpack which has been developed on the basis of the integrated rhombic gridding based TENG. **c** Photograph of the backpack on the shoulder with human standing still. **d** Photograph of the backpack under normal human walking. Forty commercial LED bulbs were lit simultaneously. Reproduced with permission from American Chemical Society [1].

the electric output. The newly designed TENG with $n = 3$ produces an open-circuit voltage up to 428 V, and a short-circuit current of 1.395 mA with a peak power density of 30.7 $W/m^2$. The electric output is related to $n$ and increases manifoldly as a function of $2n^2$. Moreover, based on the TENG, a self-powered backpack was developed. When a person walks naturally carrying the designed backpack with a total weight of 2.0 kg, the power harvested from the body vibration is high enough to simultaneously light all the 40 LEDs. The newly designed TENG provides an innovative approach to effectively enhance the device current output and thus it is capable of harvesting vibration energy from natural human walking, which can have a range of applications for extending the lifetime of a battery as well as the possibility of replacing a battery for building self-powered systems.

## 8.2 Fabric Based Triboelectric Nanogenerators

Human body is a rich source of mechanical energy and the fabrics on it can be directly integrated with nanogenerators for energy harvesting. At the same time, one important developing trend of electronics is towards wearable and even implantable. Therefore, a paradigm shift idea is to fabricate fabric based TENG for directly driving wearable and implantable electronics on human body. The self-powered system based on this integration can provide plenty of functions including health monitoring, biomedical surgery, and sensing on others. Therefore, a series of efforts have been devoted to develop fiber-based and fabric-based TENG [5–11].

### *8.2.1 Fiber Based Triboelectric Nanogenerators*

The first fiber-based triboelectric nanogenerator (FBG) was introduced by Zhong et al. [5]. This FBG is composed by two modified-cotton threads entangled to each other: carbon nanotube (CNT) coated cotton thread (CCT), and polytetrafluoroethylene (PTFE) and carbon nanotube coated cotton thread (PCCT). This work demonstrated the first proof-of-concept fabrics woven by FBG to harvest energy from biomechanical motions in human body to power wearable electronics and medical applications.

The fabrication process of the FBGs is shown in Fig. 8.3a. After treated by ethanol flame (to eliminate redundant fibers) and nitric acid solution (to increase the hydrophilicity), the cotton threads were coated by homemade CNT ink through simple and cost-effective "dipping and drying" method to make them conductive [18]. The surface of the cotton thread covered by CNTs with diameter of ~240 μm was shown in scanning electron microscopy (SEM) in Fig. 8.3a (the left column). As tested, the final CCTs have shown a great flexibility, conductivity of ~0.644 kΩ/cm, and maximum tensile stress of ~180 MPa. PCCTs were made by PTFE coating on CCTs through "dipping and drying" method and then an

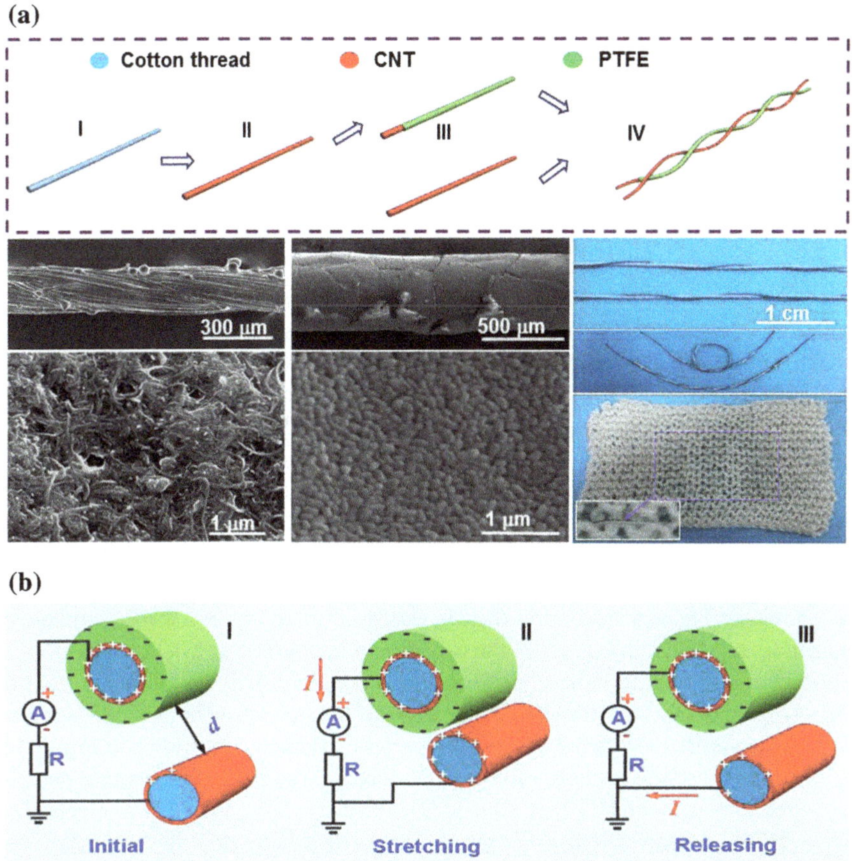


**Fig. 8.3** **a** Schematic diagram illustrating the fabricating process of an fiber-based generator (FBG) with SEM images of a carbon nanotube coated cotton thread (CCT) with low and high magnification, respectively. SEM images of polytetrafluoroethylene (PTFE) and carbon nanotube coated cotton thread (PCCT) with low and high magnification, respectively. Digital photography of FBGs with linear shape, with curved shape, and woven into fabric are also shown. **b** The working mechanism of the FBG. Reproduced with permission from American Chemical Society [5].

annealing process to enhance the adhesion. The surface of PCCTs with minor cracks and diameter of ~500 μm was shown in Fig. 8.3a (the middle column). The cracks might be due to the stress process, which can enhance the flexibility of PCCTs. The zoomed in SEM picture of the PCCT surface also shows oval-like nanoparticles with diameters of <200 nm. As an electret material, the ability of PTFE surface to retain electrostatic charges for 20 days was demonstrated. At last, one CCT and one PCCT was entangled to double-helical structure and then woven to form fabrics.

The working mechanism of FBG was illustrated in Fig. 8.3b, when the CNT layers of two kinds of threads were connected together. After contact between surfaces of PCCT and CCT, due to different ability to attract electrons, net negative charges appear on the PTFE surface, and positive charges are distributed on the two CNT layers (Fig. 8.3b-I). Then during stretching (Fig. 8.3b-II) and releasing (Fig. 8.3b-III), the contact-separation between two kinds of threads break original electrostatic balance periodically, which makes charges flow through external circuits to rebalance the electrostatic status. This process is similar to that of the common contact-separation mode TENG. Therefore, alternating current (AC) was produced between the threads.

To evaluate the performance of the fabricated device, a FBG with length of ~9.0 cm and 8 helix turns were stretched and released periodically under controlled frequencies and strains. It was demonstrated that the peak current (with an external load of 80 MΩ) increase with the strain, from 3.98 nA at 0.54 % to 11.22 nA at 2.15 %, with frequency of 5 Hz. The peak current also increase with the frequency, while the total transferred charge keeps the same at ~0.16 nC (at the strain of 2.15 %).

To demonstrate the FBG as a flexible power source for electronics, 8 FBGs were woven into a fabric, and then it was sew on a lab coat as a "power shirt" (Fig. 8.4c). The generated output current was shown in Fig. 8.4a (black) under periodical shaking. This output current was firstly rectified through a full-wave bridge rectifier (as shown in Fig. 8.4a red), and then it was used to charge a 2.2 μF commercial capacitor. As shown in Fig. 8.4b, the capacitor was charged to 2.4 V in about 27 s, and then it was used to light a light-emitting diode (LED), as shown in the inset of Fig. 8.4b. It is reported this FBG was also used to power a homemade wireless body temperature monitoring system, which was schematically shown in Fig. 8.4d.

In summary of this section, FBG was fabricated through dipping-coating process on cotton threads, with PTFE as triboelectric dielectric layer and CNT as electrodes. This FBG is demonstrated to output current, which increase with frequency and strain. This first proof-of-concept fiber-based TENG was demonstrated to charge a capacitor, light an LED and power a wireless body temperature sensor system, which shows the possibility to build a self-powered smart cloth system for wearable electronics and biomedical applications.

### 8.2.2 *Textile Based Triboelectric Nanogenerators*

Pu et al. designed a wearable power unit by integrating a whole-textile TENG-cloth and a flexible lithium-ion battery (LIB) belt [6]. Common flexible but insulating polyester fabrics were transformed into conductive with an electroless plating of a conformal Ni film, which were then utilized both as electrodes in the TENG-cloth and as current collectors in the LIB belt. The TENG-cloth demonstrated the capability of converting the mechanical energy of various human motions into electricity when being worn at different positions on the human body; the LIB belt showed decent electrochemical performances even being severely folded at 180° for 30 times. Furthermore, the LIB belt was charged by the TENG-cloth for 3 cycles,

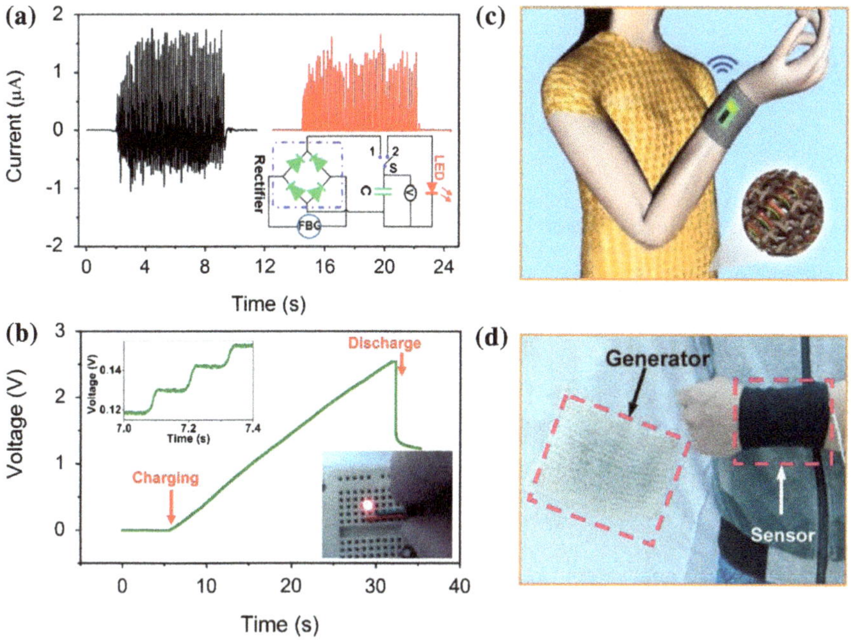


**Fig. 8.4** Electricity generation of the "power shirt" to power wireless body temperature sensor system. **a** Output current of the "power shirt" with (*black curve*) and without (*red curve*) rectification when the lab coat was being shaken. The inset depicts the equivalent loop circuit for storing the electrical energy produced by the "power shirt" and lighting the LED. **b** Voltage charging curve of a 2.2 μF commercial capacitor by the "power shirt". The upper inset shows an enlarged plot during charging. The lower inset depicts digital photography of a lighted LED powered by the charged capacitor. **c** Schematic diagram and **d** digital photography of a wireless body temperature monitor system triggered by the power clothes. Reproduced with permission from American Chemical Society [5].

and powered a heartbeat meter strap capable of remote communication with a smart phone, verifying the viability of the whole-wearable and self-charging power unit for future wearable smart electronics.

As the conductive electrode is one of the key components in a TENG, it is crucial to develop a soft and wearable conductor for wearable TENG Common soft polyester fabric was selected as the starting substrate, and was consecutively coated with conductive Ni film (Ni-cloth) and insulating parylene film (parylene-cloth), as schematically illustrated in Fig. 8.5a. Due to the difficulty in handling individual polyester microwire, belt-type Ni-cloth, and parylene-cloth (5-mm wide) were used as the building block (see Fig. 8.5a, b), and finally woven into a 5 cm × 5 cm TENG-cloth (see Fig. 8.5a, c). All the Ni-cloth belts were connected together as one electrode of the TENG, while all the parylene-cloth belts were connected as the other. After the Ni-coating, white polyester cloth changes into silver; while parylene layer is relatively transparent (see Fig. 8.5b). The woven TENG-cloth textile retains the mechanical flexibility, air-breathability, water-washability, and thus

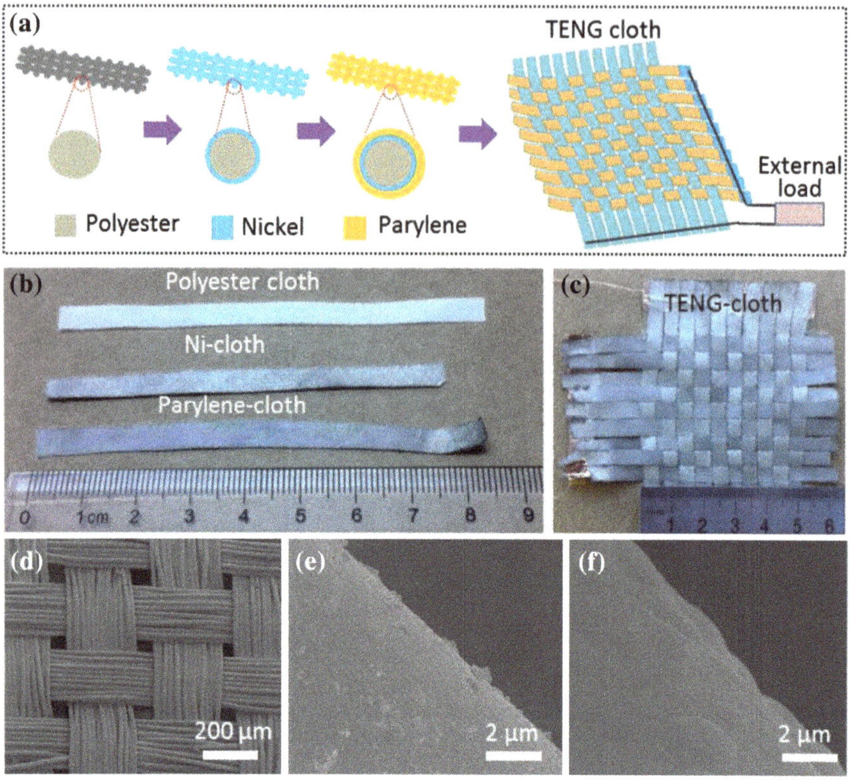


**Fig. 8.5** The fabrication of triboelectric nanogenerator cloth (TENG-cloth). **a** Schematic illustration of the fabrication of TENG-cloth. **b** An optical image of pristine polyester cloth, Ni-cloth and parylene-cloth. The Ni-cloth is coated with conductive Ni coating by an electroless deposition; the parylene-cloth has an insulating parylene film on top of the Ni coating. **c** An optical image of the TENG-cloth fabricated by weaving Ni-cloth and parylene-cloth. SEM images of **d** pristine polyester cloth, **e** Ni-cloth, and **f** parylene-cloth. Reproduced with permission from Wiley [6].

comfortability of the original polyester cloth. Figure 8.5d shows the SEM image of the pristine polyester cloth with a woven structure of polyester microwires. After the electroless coating of Ni, the surface of the polyester cloth was coated with a conformal Ni layer with nanoscale roughness (see Fig. 8.5e). The resistance of a 0.5 cm × 10 cm Ni-cloth is measured to be several ohms, comparable with typical metal foils and more conductive than reported CNT or graphene papers [18, 19]. Meanwhile, this coating method is low-cost and suitable for scale-up, and the polyester cloth can be easily coated with arbitrary size and shape in a short time and at low temperature. In contrast, the expensive synthesis of CNT or graphene by chemical vapor deposition at very high temperature and extra efforts in dispersing them in solvents make them difficult for large-scale application. Figure 8.5f shows a relatively smooth layer of parylene film covering the underneath Ni film after the chemical vapor deposition of parylene. It should be noted that coating methods used for Ni and parylene both produced conformal layers that cover the wavy surface of the textile substrate.

In order to elucidate the working principle, the output characteristics of the fabricated TENG-cloth at a TENG-cloth–TENG-cloth contact–separation (CS) mode were characterized. The releasing and pressing of two pieces of TENG-cloths can be simplified into the contact–separation between a parylene-cloth wire and a Ni-cloth wire, since all the parylene-cloth wires and Ni-cloth wires are respectively connected together as two electrodes. Contact electrification, i.e., electron transfer from the Ni film that tends to lose electrons, to the parylene film that tends to gain electrons, occurs once two TENG-cloths are brought into close contact. Then, as the external force is released and the two charged surfaces get separated, the corresponding charge separation will create an electric potential difference (EPD) between the two electrodes:

$$U_{\mathrm{EPD}} = -\sigma d/\varepsilon_0 \tag{8.6}$$

where $\sigma$ is the triboelectric charge density, $\varepsilon_0$ is the vacuum permittivity, and $d$ is the distance between two electrodes. When the TENG is being released, the open-circuit voltage ($V_{\mathrm{OC}}$) increases as the EPD is increased, while $V_{\mathrm{OC}}$ decreases when the TENG is pressed. When the TENG is shortened, the EPD will drive the flow of electrons between two electrodes through the external circuit, creating two current pulses ($I_{\mathrm{SC}}$) with opposite polarities when being released and pressed. The maximum $V_{\mathrm{OC}}$ and $I_{\mathrm{SC}}$ is 50 V and 4 μA, respectively, which is dramatically improved compared with previously reported textile nanogenerators made of CNT and Ag wires [5, 11]. With a rectifier, the AC output current could be tuned into DC output profile.

Figure 8.6 demonstrates the self-charging power unit by integrating the TENG-cloth as energy generation device and the LIB belt as energy storage device. A prototype of the power system is shown in Fig. 8.6a, where a heartbeat meter strap worn at the chest is powered by a LIB belt, which can be further charged or compensated by a TENG-cloth worn under the arm. The heartbeat meter has Bluetooth function, so that a remote smart phone can record and analyze the physiological information of human during sports or daily life. Figure 8.6b shows the rear side of the heartbeat strap, confirming that its original rigid coin cell is replaced by the wearable LIB belt. The equivalent electric circuit of the whole system is shown in Fig. 8.6c. This power unit can be easily modified for many other wearable smart electronics, such as smart watches and smart glasses.

To further verify the capability and stability of the wearable TENG-cloth in charging LIB belt, a LIB belt was charged by the TENG-cloth and subsequently discharged galvanostatically (GD) at 1 μA constant current for 3 cycles, as shown in Fig. 8.6d. A linear motor was utilized to mimic the TENG-cloth–TENG-cloth CS mode motion of human at 0.7 Hz low frequency during the charge process. The voltage of the LIB belt increases rapidly to the operational voltage (≈1.9 V) upon the charging by the TENG-cloth. The charging time for the first, second, and third cycle is 4, 9, and 14 h, respectively. The corresponding discharge capacities are 1.3,

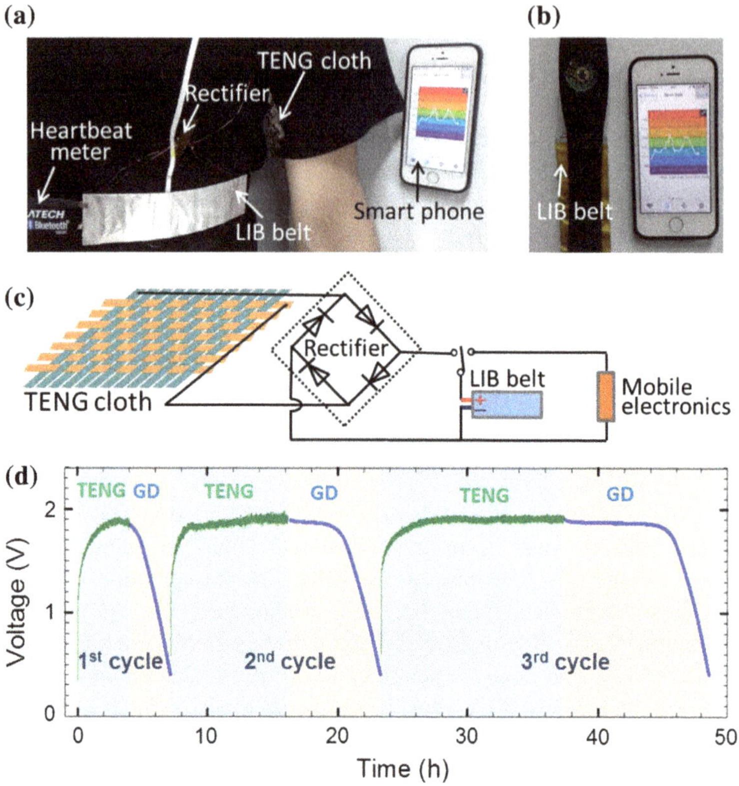


**Fig. 8.6** Self-charging power unit. **a** An optical image of the integrated self-charging power system that harvests mechanical energy of human motion with TENG-cloth, stores the energy with LIB belt, and then powers heartbeat meter strap, which has remote communication with a smart phone. **b** The rear side of heartbeat meter, indicating that its original coin cell was replaced by the LIB belt. **c** The equivalent electrical circuit of the self-charging power system. **d** The voltage profi les of the LIB belt charged by the TENG-cloth and galvanostatically discharged (GD) at 1 μA for 3 cycles. Reproduced with permission from Wiley [6].

2.8, and 4.4 mA h $m^{-2}$, respectively, when being normalized by the area of the TENG-cloth. During sports, the motion frequency of human body is actually much larger than 1 Hz, and thus higher energy generation can be expected. This energy is enough to power many small healthcare electronics (e.g. heartbeat meter, pedometer, pulse meter, body temperature meter), making the self-charging and wearable power unit possible to diminish or even eliminate the battery charge by a bulky household battery charger.

In summary, a self-charging power unit has been developed for smart electronics by integrating a textile TENG-cloth as an energy harvester and a flexible LIB belt as energy storage. The whole-textile TENG-cloth can be worn at many different positions of the human body (e.g., under the foot, under the arm, and at a joint), due to its versatility in scavenging energies of various modes of human motions.

Excellent flexibility and stable electrochemical performances of the LIB belt was achieved even when the LIB belt was completely folded at 180° for 30 times, owing to the utilization of conductive Ni-cloth textile as current collector in both cathode and anode, and the use of Al pouch as package. The electric energy converted from the daily human motion by the TENG-cloth was demonstrated to be able to charge the LIB belt, which then was able to power a heartbeat meter with remote communication with a smart phone. The stability of the power unit was further confirmed by charging the LIB with TENG-cloth and subsequently galvanostatic discharging for 3 cycles. The viability of the wearable and self-charging power unit provides potential solution to pave the way for the progress of wearable/flexible electronics in the near future.

### 8.2.3 Fiber Based Hybrid Nanogenerators

Li et al. design a fully integrated TENG and piezoelectric nanogenerator (PENG) on the carbon fiber cloth which contains numerous carbon fibers [7]. The fiber based hybrid nanogenerator (FBHNG) is very flexible and can be utilized to fabricate smart cloth for the purpose of collecting the mechanical/kinetic energy from ambient environment. A novel, low-cost 3D fabrication method is demonstrated to fabricate FBHNG consisting of central ZnO nanorods (NRs) on carbon fibers and outer foldable double side cooper coated nylon film as tribo- and piezo-electrode. Coaxial functional fiber with PENG inside and TENG outside were fully integrated on a carbon fiber cloth. PENG is the supplement for TENG which can also generate power and act as strain sensor under the motion mode without TENG. 3D coaxial structure ensured that the TENG and PENG have the same action axis and can improve the effective impact area greatly. ZnO NRs with the length of 10 μm were successfully grown around the conductive carbon fibers by hydrothermal growth approach. The output current and voltage of the TENG and the PENG were characterized by a mechanical motor. The measured instantaneous power density of 12 contact array points for TENG and PENG is 42.6 and 10.2 mW/m$^2$, respectively. The instantaneous power density and current density for FBHNG could multiply with the increasing of contact array points and area. The rectified outputs have been applied to charge commercial capacitors and drive light-emitting diodes which can be stored in a capacitor and used in a self-powered alert system. In addition, the FBHNG can be used as a self-powered sensor to detect human elbow motion.

The fabrication process of the FBHNG is schematically depicted in Fig. 8.7a. Basically, the FBHNG is a 3D coaxial and fully integrated device containing a PENG unit (Output 2) in the core and a TENG unit (Output 1) in the shell. Before the fabrication process, the carbon fiber cloth was cleaned sequentially with acetone, ethanol, and deionized water. First, a 250 nm ZnO thin film was sputtered on the carbon fiber as the seed layer (i). Then, the ZnO NRs were grown on the ZnO seed layer through the hydrothermal method (ii) [20]. Typically, the growth solution contained equal volume (50 mM) of zinc nitrate hexahydrate,

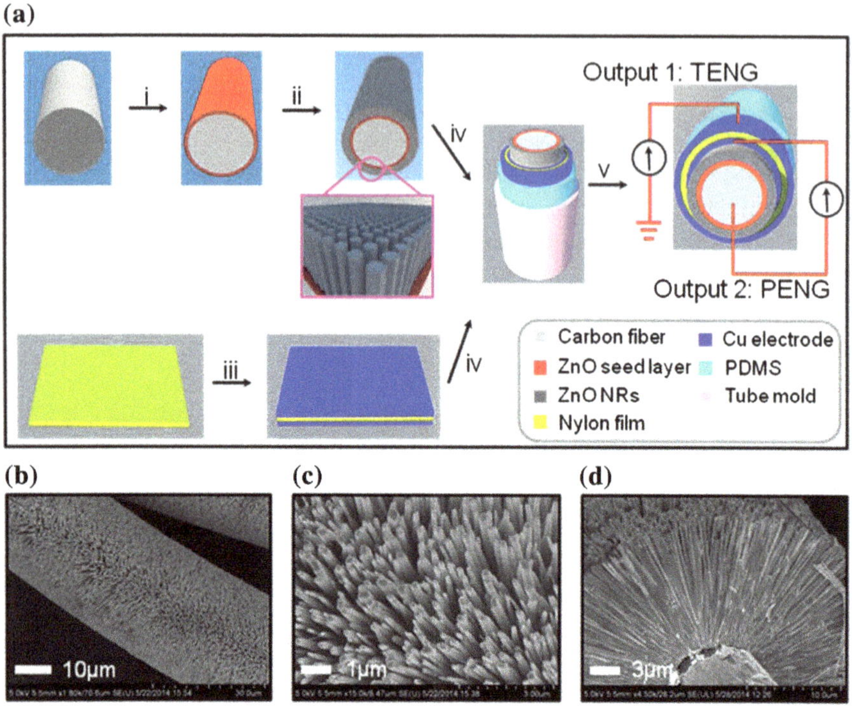


**Fig. 8.7** Fully integrated 3D fabrication method for FBHNG. **a** Fabrication process step for 3D Coaxial FBHNG. (*i*) Depositing of ZnO seed layer by PVD; (*ii*) growing of ZnO NRs around carbon fiber by hydrothermal method; (*iii*) depositing of double side copper electrode on nylon film; (*iv*) assembling the carbon fibers and nylon film into tube mold by PDMS; (*v*) peeling of FBHNG from the tube mold. SEM images for the whole fiber with ZnO NRs (**b**), top view of ZnO NRs (**c**), and cross-sectional view of ZnO NRs (**d**). Reproduced with permission from American Chemical Society [7].

hexamethylenetetramine (HMTA) (50 mM) and ammonia–water (7 mM) in deionized water and the reaction temperature was fixed at 85 °C for 12 h. As shown in SEM image (Fig. 8.7b–d), ZnO NRs are uniformly distributed on each fiber and the length of the ZnO NRs is 10 μm. In the other way, metal thin films of Ti/Cu (100 nm) were sputtered on both sides of a nylon film as the conducting electrodes (iii). The inner Ti/Cu thin film acted as the top electrode of PENG and the outer Ti/Cu thin film was the output electrode (Output 1) of TENG. The Ti/Cu-coated nylon film was wrapped around the carbon fiber and packaged in a tube mold with a diameter of 3 mm by polydimethylsiloxane (PDMS) (iv). The elastomer and the cross-linker of PDMS were mixed in a 10:1 weight ratio. After the mixture was degassed for 2 h, the carbon fibers and the Ti/Cu-coated nylon film were immersed in the mixture and degassed again to fully fill the tube. After the solution was further heated at 85 °C for 1 h, the coaxial FBHNG with TENG outside and PENG inside was obtained. Before the test of the prepared FBHNG, the mold needs to be peeled.

The working mechanisms of the FBHNG can be explained separately as a TENG and a PENG. For optimization of the output of the TENG, the nylon fiber is chosen as the other contact material to deform the FBHNG. When the nylon fiber contacts the PDMS layer, the electrons on the nylon fiber surface will transfer to the PDMS layer and create charged surfaces of the PDMS layer and the nylon fiber (i). Once the nylon fiber leaves, a negative electric potential difference between the PDMS layer and the ground is established (ii). The electrons will flow from the outer Cu electrode to the ground through the external load and finally reach equilibrium (iii). This contributes to the positive pulse output current of TENG from Output 1. As the nylon fiber approaches the PDMS layer, a positive electric potential difference between the PDMS layer and the ground is formed. This will cause the electrons flow from the outer Cu electrode to the ground and contribute to the negative pulse output current of TENG from Output 1 (iv), until achieving another equilibrium (i). At the same time, when the FBHNG is deformed by the nylon fiber, the inner PENG also starts to work because of the coaxial structure. The ZnO NRs on the carbon fiber are compressively strained and a positive electric potential difference is formed between the inner Cu electrode and the conductive carbon fiber (i). This will force the electrons to flow from the inner Cu electrode to the carbon fiber through the external load and contribute to the positive pulse output current of PENG from Output 2, until reaching equilibrium (ii). As the nylon fiber leaves and the compressive strain is released, the piezoelectric potential will be diminished. The electrons accumulated at the carbon fiber will flow back to the inner Cu electrode, generating the negative pulse output current of PENG from Output 2 (iii), finally reaching another equilibrium (iv). The continuous output of the FBHNG from Output 1 (TENG) and Output 2 (PENG) can be provided once the nylon fiber repeatedly deforms and separates from the FBHNG.

To demonstrate the working mode of the FBHNG, a mechanical motor is used as the applied force to ensure the contact of fibers. The nylon fiber is attached on the motor and will periodically contact and separate from the FBHNG. The short-circuit current ($I_{SC}$) and open-circuit voltage ($V_{OC}$) were measured to characterize the performance of the FBHNG. The effective contact area between the FBHNG and the nylon fiber is assumed to be 9 mm$^2$. Upon the impacting frequency of 2 Hz, the generated current from Output 1 is 20 nA and the generated current for Output 2 is 5.5 nA. The generated voltage from Output 1 can reach 1.5 V, while the generated voltage from Output 2 has a lower value of 0.5 V. The generated output from TENG (Output 1) is higher than that from PENG (Output 2).

To address the potential applications of the FBHNG, the FBHNG is demonstrated to be weaved with cotton fiber to become a power cloth or power bag. The generated current from Output 1 of 5 weaved FBHNGs can reach 70 nA (Fig. 8.8a), while the generated current from Output 2 of 5 weaved FBHNGs is 4 nA (Fig. 8.8b). The generated outputs then connect to a full-wave rectifying bridge to transform the AC outputs to a same direction and charge a 2.2 μF capacitor (Fig. 8.8c). By applying commercial LEDs as an indicator, a self-powered

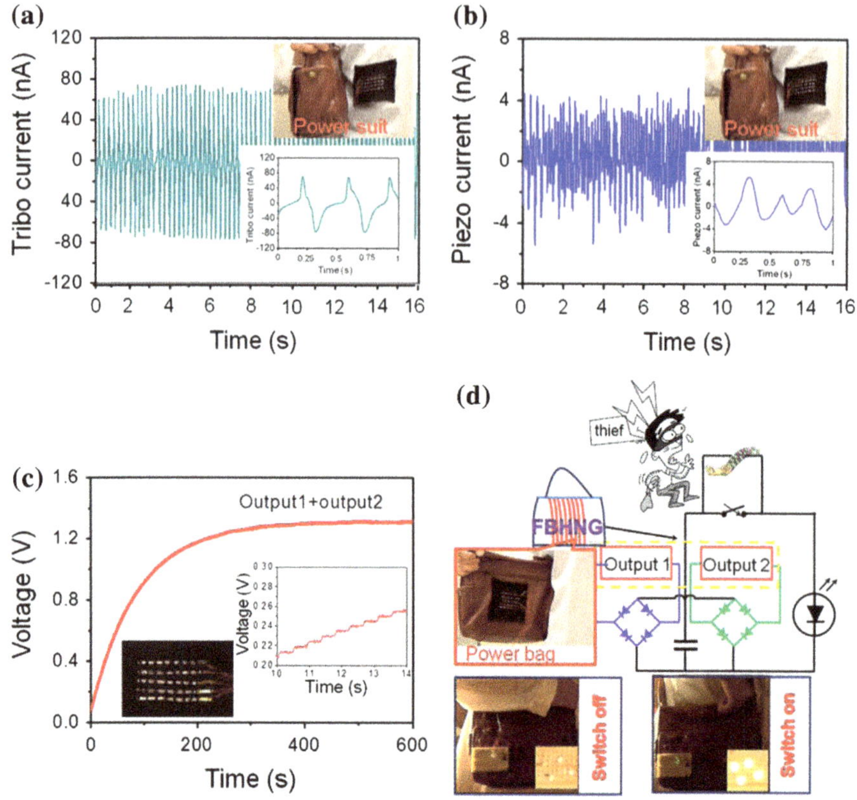


**Fig. 8.8** Generated current from Output 1 (**a**) and Output 2 (**b**) of power suit with FBHNG weaved in under the impact of human motion. **c** Charging curves of 2.2 μF capacitors by the rectified outputs from Output 1 and Output 2 of hybrid NGs. **d** The photograph of the rectified outputs from Output 1 and Output 2 to charge a capacitor and being used in bag burglar proof system. Reproduced with permission from American Chemical Society [7].

alert system is designed (Fig. 8.8d). Once the bag is opened, the sliding of zipper changes the switch to an "on" status and the LEDs will be turned on. This can remind the owner to be careful of the bag. The FBHNG can also function as a self-powered strain sensor for monitoring health condition.

In summary, a newly designed FBHNG is developed by integrating the TENG and PENG on coaxial carbon fiber cloth. The FBHNG can be more efficient at harvesting the mechanical energy as compared to the individual TENG or PENG. The presence of PENG in the hybrid structure is a supplement for TENG which can still generate output when the TENG is not working in certain situations. The generated instantaneous power density for TENG and PENG can achieve 42.6 and 10.2 mW/m$^2$, respectively. The total charge transfer, output current, and instantaneous power density can be largely enhanced by increasing the area and number of

FBHNG, which shows the potential in producing the power cloth and suit. The combination of commercial LEDs with FBHNG indicates that it could be used in a self-powered alert system. Besides, the FBHNG can also function as a self-powered strain sensor for monitoring health condition.

## 8.3 Paper Based Triboelectric Nanogenerators

As an important invention in human civilization and a useful tool until modern society, paper has been used for thousands of years due to its light-weight, cost-effective, and environment-friendly. Previously, paper has been applied in disposable and flexible electronics [21, 22]. At the same time, TENG has broad choice of materials including papers, which makes a paper-based TENG fabrication possible for powering paper-based electronics. The whole paper-based self-powered system is expected to be achieved as a low-cost, recyclable, and flexible system for multiple applications [12, 13, 23].

### *8.3.1 A Single Paper Based Triboelectric Nanogenerators*

The paper-based TENG (pNG) was firstly reported by Zhong et al. [12], which opens a new field of research on paper applications through triboelectrification and electrostatic induction. The fabrication process of pNG is shown in Fig. 8.9a. Firstly, the papers are deposited with silver (Ag) film as the electrode material, with SEM picture shown in Fig. 8.9b. The successful deposition of Ag is confirmed by the energy dispersive X-ray spectrum (EDX). And then, a piece of 3 cm × 2 cm paper (with Ag film) is spin coated with PTFE above the Ag film. The cross-sectional view confirms the PTFE layer on paper in Fig. 8.9c, d shows the top view surface of the PTFE-Ag-paper. At last, the PTFE-Ag-paper is integrated with Ag-paper by facing the PTFE side on the Ag side, forming a CS mode TENG. The two parallel sides of the top paper are fixed by kapton tapes to form an arch structure with an effective triboelectric area of 2.5 cm × 2 cm.

If we consider the pNG as an ideal parallel-board capacitor by ignoring the edge effects, the equations in Chap. 2 can be directly utilized for CS mode TENG. Assuming uniform distribution of the charges, the equilibrium surface charge density on Ag-paper electrode is given below:

$$\sigma_2 = -\frac{\sigma d_1}{d\varepsilon_r + d_1} \tag{8.7}$$

where $d_1$ is the thickness of the PTFE, $\varepsilon_r$ is the dielectric constant of PTFE ($\sim 1.93$), and $d$ is the gap distance between the PTFE surface and the Ag layer of

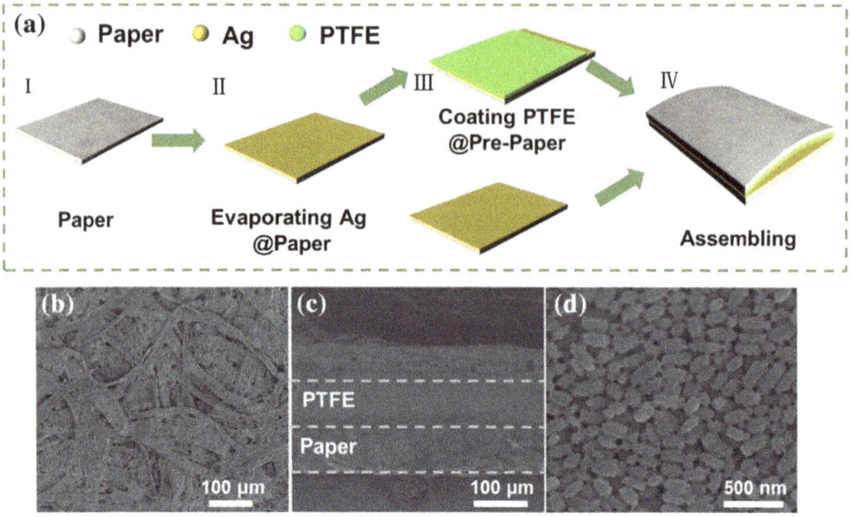


**Fig. 8.9** **a** A schematic diagram illustrating the process of fabricating an arch-structured pNG. SEM images of **b** the paper coated with a Ag layer, **c** the cross-sectional and **d** the top view of the Ag–paper after spin-coating with PTFE. Reproduced with permission from Royal Society of Chemistry [12].

the Ag–paper. The net negative charge on PTFE should be able to be retained for a long time. Therefore, under periodical press and release process, the distance $d$ varies that breaks the electrostatic balance. And to rebalance the electrostatic status, electrons flow through external circuit to form AC output.

The output performance of the pNG is studied and shown in Fig. 8.10. The voltage is found to increase with the external load resistance, while the current decreases with that, which are ~7 V, ~10 μA at 1 MΩ and ~230 V, ~1.2 μA at 200 MΩ. The maximum output power density is measured as ~90.6 μW $cm^{-2}$ at load resistance of ~40 MΩ. This pNG is demonstrated to drive LEDs, with the schematic diagram of the circuit shown in Fig. 8.10d. The pNG is pasted on a page of a book, and the page turning is used as the mechanical energy source. The current (~4.6 μA) and voltage (~2.4 V) are simultaneously measured, as shown in Fig. 8.10a, b. At the same time, a liquid crystal display (LCD) can also be lighted by this pNG.

In summary of this section, a light-weight and flexible paper-based TENG is fabricated, with a maximum power density of ~90.6 μW $cm^{-2}$ demonstrated. This pNG is used to drive LEDs and an LCD with the energy source of turning a page. This pNG will enable the significant potential applications of paper-based self-powered system for energy harvesting and storage, active sensing, monitoring, and even security purposes.

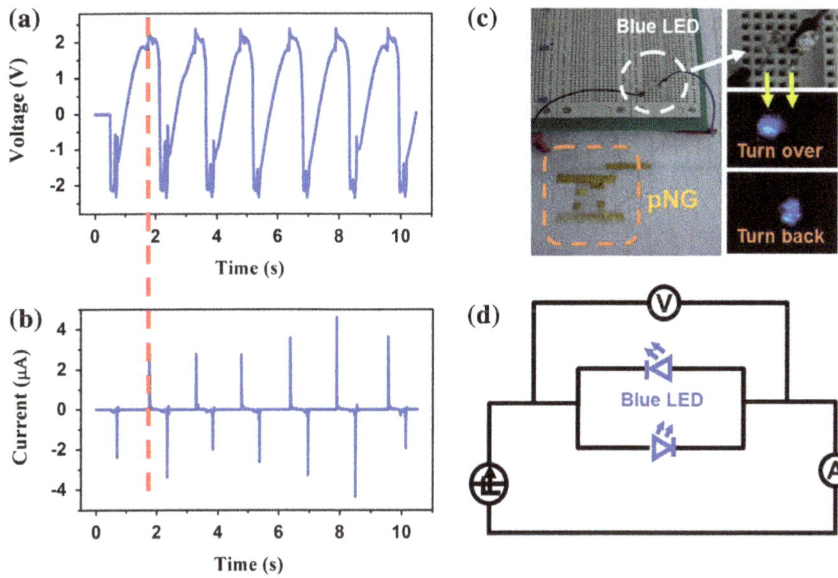


**Fig. 8.10** The performance of a pNG attached to a page of a book. **a** The voltage across and **b** the current through the two blue LEDs connected in antiparallel driven by turning a book page. **c** The digital photograph of the prototype of the energy harvesting circuit and LED display. **d** A schematic diagram of the circuit. Reproduced with permission from Royal Society of Chemistry [12].

### *8.3.2 A Paper Origami Based Triboelectric Nanogenerators*

Yang et al. develop a new type of paper-based TENG with origami configurations for harvesting mechanical energy [13]. The origami approach provides a facile path to fabricate a stacked TENG without expanding the area or complicating the fabrication process. A TENG with slinky shape is developed by connecting seven units in parallel. The output short-circuit current ($I_{SC}$) and open-circuit voltage ($V_{OC}$) of the slinky TENG can reach 2 μA and 20 V, respectively. A corresponding instantaneous power density of 0.14 W $m^{-2}$ is obtained at a load resistance of 400 MΩ. The generated electrical output has been demonstrated to directly light up commercial LEDs. The slinky TENG with the advantage of harvesting energy from various kinds of human motions, such as stretching, lifting, and twisting. Furthermore, the slinky TENG can serve as a self-powered pressure sensor to differentiate the weight difference of coins. The doodlebug TENG is also demonstrated to expand the application scope of the origami approach. These findings will pave the way for future energy harvesting and sensor design, especially for the development of green electronics.

The configuration of the slinky TENG is schematically illustrated in Fig. 8.11a. It is clearly shown that the slinky TENG is composed of three parts, including a paper substrate, polytetrafluoroethylene (PTFE) thin film, and aluminum

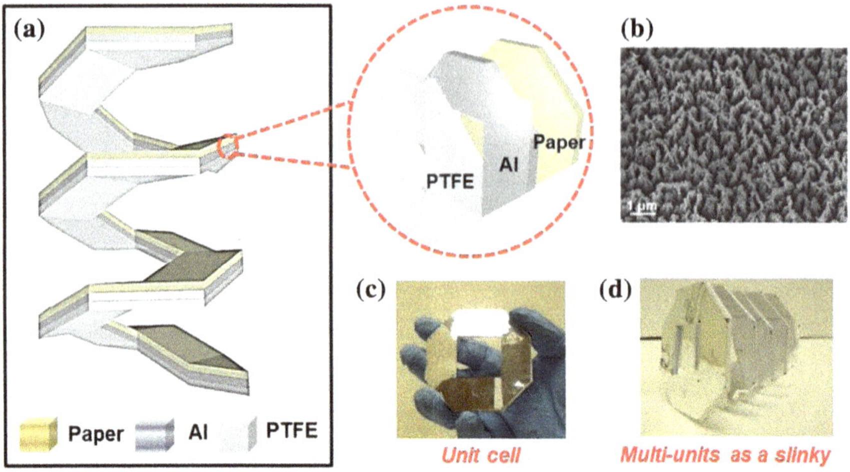


**Fig. 8.11** Structure and photographs of a slinky TENG. **a** Schematic diagram of a slinky TENG. **b** SEM image of the etched PTFE surface with nanowire-like structure. **c** Photograph of the unit cell in slinky TENG. **d** Photograph of multiunits in forming a slinky structure. Reproduced with permission from American Chemical Society [13].

foil. Figure 8.11b shows a scanning electron microscopy (SEM) image of the PTFE thin film surface. A uniformly distributed nanowire-like structure can be observed with a dimension about ~700 nm in length and ~70 nm in width. It has been theoretically reported that the charges on the PTFE surface can be retained for years, which guarantees the sustainable active status of the device within the period of the measurements and applications. Figure 8.11 c and d shows the photographs of real devices, including a single unit cell and multiple units in parallel connection.

The working mechanism of the slinky TENG is operated in single-electrode mode. At the original state (without external force), paper and the PTFE thin film are separated due to the stiffness of the paper substrate, and there is no electric output. By applying an external force on the slinky TENG, paper and the PTFE thin film are contacted, which contribute to the charged surfaces of paper and the PTFE thin film. Since PTFE is more triboelectrically negative than paper in the triboelectric series, electrons will transfer from the paper surface to the surface of the PTFE thin film. After these two surfaces are separated, a potential difference will be formed between the paper and the PTFE thin film, causing electrons to flow from the Al foil to the ground and achieve equilibrium. This is the half-cycle of slinky TENG operation.

Once the slinky TENG is pressed again, it will generate another potential difference, causing electrons to flow from the ground to the Al foil and finally achieve a new equilibrium. This is the other half-cycle of the slinky TENG operation. If the slinky TENG is operated periodically, the electric output will be continuously provided.

To form a slinky TENG, each octagonally shaped unit can be easily connected via paper joints that are produced during origami folding without any extra wiring processes. The Al electrodes are separated and attached into each unit cell independently. $V_{OC}$ and $I_{SC}$ were measured to characterize the performance of the slinky TENG. The obtained $I_{SC}$ and $V_{OC}$ are 0.3 μA and 20 V, respectively. To further understand the dependence of electrical outputs on the number of units in a parallel connection, rectified $I_{SC}$ of different number of units was measured. It should be noted that the output current is approximately linearly proportional to the number of connection units. When seven units are all set in a parallel connection, a rectified $I_{SC}$ of 2 μA can thus be achieved. The transferred charge ($\Delta Q$) of a different number of units is also shown in the inset. The maximum amount of transferred charge is 0.55 μC for seven units. In contrast, the $V_{OC}$ of the entire TENG is not the sum but the equivalent to that in the individual unit. The rectified $I_{SC}$ and $V_{OC}$ of different connection series were investigated. One can see that the $V_{OC}$ is almost identical regardless of the connection series, in contrast to the rectified $I_{SC}$. From the above-mentioned results, it is expected to enhance the electric output by incorporating more units while keeping the contact area of the TENG constant. Moreover, the stability of the electrode is one of the key issues that affect output characteristics of TENGs. The resistance distributions of Al foil on a paper substrate were measured after repeated pressing and releasing cycling tests. It is clear that the resistance distributions of Al foil remain stable with only negligible changes after 60,000 cycling tests. The results indicate excellent reliability of the Al foil on a paper substrate.

To understand the effect of inducing origami configurations to TENGs, the slinky TENG is compared with a real metal slinky, which can realize various kinds of human motions, such as stretching, lifting, and twisting (Fig. 8.12a). Figure 8.12b, c shows the typical electric outputs of the slinky TENG in harvesting various human motions, including $I_{SC}$ and $V_{OC}$. A monotonical decrease observed in electrical outputs from stretching to twisting is due to the fact that the triboelectric outputs strongly depend on the distance of the gap created during mechanical motions. The results indicate that the slinky TENG can effectively harvest ambient mechanical energy from various directions in contrast to the conventional TENG.

To further assess the applicability of the origami configuration in TENG technology, another famous structure inside the scope of origami is implemented, which is named doodlebug TENG. Figure 8.12d, e displays the electric outputs of the doodlebug TENG, where the $I_{SC}$ and $V_{OC}$ are approximately 0.15 μA and 30 V, respectively. Figure 8.12f shows the maximum output power density of 0.9 mW $m^{-2}$ at a load resistance of 200 MΩ. A photograph of the as-fabricated device based on doodlebug configuration is also shown as the inset. The results further point out the feasibility of utilizing diverse origami configurations in future TENG technology.

In summary, a low-cost, light-weight, and environmentally friendly paper-based TENG with origami configurations is demonstrated for harvesting ambient mechanical energy and serving as a self-powered pressure sensor. The dependence of

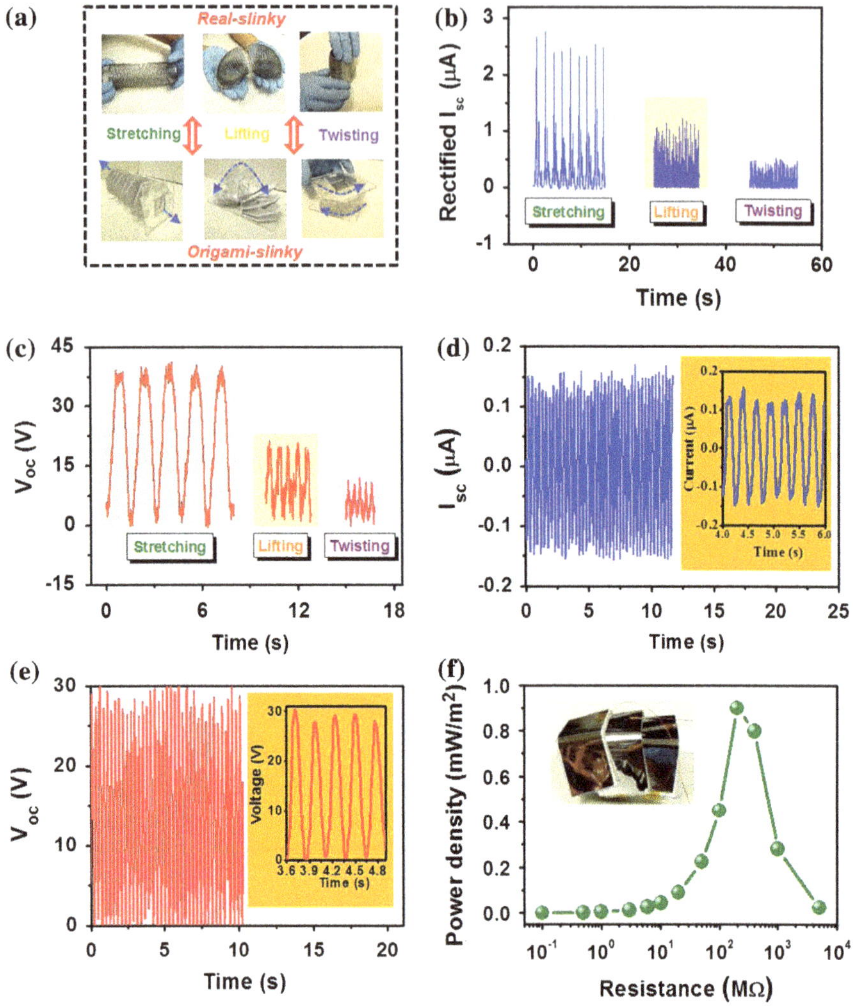


**Fig. 8.12** Mechanical energy harvester. **a** Realization of slinky motions by slinky TENG. **b** Rectified $I_{sc}$ and **c** $V_{oc}$ from various human motions of the slinky TENG. The inset images demonstrate the corresponding motions, where the *arrows* indicate the direction of applied force. **d** $I_{sc}$ and **e** $V_{oc}$ of the doodlebug TENG. The inset shows five enlarged cycles within the electrical output measurement. **f** Relationship between instantaneous power density and resistance of the external loads of the doodlebug TENG. The maximum power density reaches 0.9 mW $m^{-2}$ when the external load is 200 MΩ. The inset image shows the photograph of the as-fabricated TENG. Reproduced with permission from American Chemical Society [13].

electric outputs on the number of connection units and contact areas has been systematically studied. This prototype slinky TENG is able to provide an open-circuit voltage of 20 V and a short-circuit current of 2 μA. The corresponding peak power density is approximately 0.14 W $m^{-2}$, which is capable of lighting four LEDs simultaneously. Moreover, the slinky TENG can harvest energy from various kinds of human motions, indicating its multifunctionality in comparison with the conventional TENG. Furthermore, the slinky TENG can act as a self-powered pressure sensor to sense the difference in mass of standard coins. This study opens up the possibility of utilizing origami configuration into TENG design, extending its application scope, especially for paper-based portable power sources and pressure sensors.

## 8.4 Human Skin Based Single-Electrode Mode Triboelectric Nanogenerators

All the previously reported biomechanical energy harvesting techniques are based on external devices attached to the human body. An interesting question is what if the human body itself can be used as a power-generating material so that energy can be generated in the touching process. According to the triboelectric series, human skin as one of the triboelectric materials can be used to fabricate this kind of device by utilizing TENG techniques. However, TENGs usually need metal electrodes deposited on the triboelectric materials for effective electrical output, which cannot be realized on human skin. Developing single-electrode-based TENGs can solve this problem and enable us to harvest biomechanical energy from human skin as well as building new touch pad technology.

Yang et al. present a human skin based TENG that is based on periodic contact/separation between a human skin patch and a polydimethylsiloxane (PDMS) film with a surface of micropyramid structures on an ITO electrode. Due to the coupling between the triboelectric effect and the electrostatic effect, the periodic change of distance between two surfaces results in charge transfer between the ITO electrode and the ground, thus driving the flow of electrons across an external load. The fabricated TENG can produce an open-circuit voltage up to $\sim$1000 V and a short-circuit current density of 8 $mA/m^2$, which can be used to drive tens of green light-emitting diodes (LEDs) instantaneously. This work is important progress toward the practical applications of skin-based biomechanical energy harvesting techniques.

The fabricated TENG consists of a human skin patch and a PDMS film on an ITO electrode. Figure 8.13a illustrates a photograph of the PDMS film, where the microstructured surface was fabricated using a Si micropyramid template. Figure 8.13b shows a scanning electron microscope (SEM) image of the fabricated PDMS film. The surface is uniformly covered by micropyramid structures with sizes less than 5 μm. The micropyramid structures were further confirmed by the cross-sectional SEM image of the PDMS film, as shown in Fig. 8.13c. The corresponding thickness of the flexible PDMS film is about 400 μm. The micropyramid structure of the PDMS film can be used to induce a larger triboelectric charge density to enhance the output performance of TENG. Additionally this micropyramid structure can also change some other physical properties such as the wettability of the polymer surface. Figure 8.13d shows that

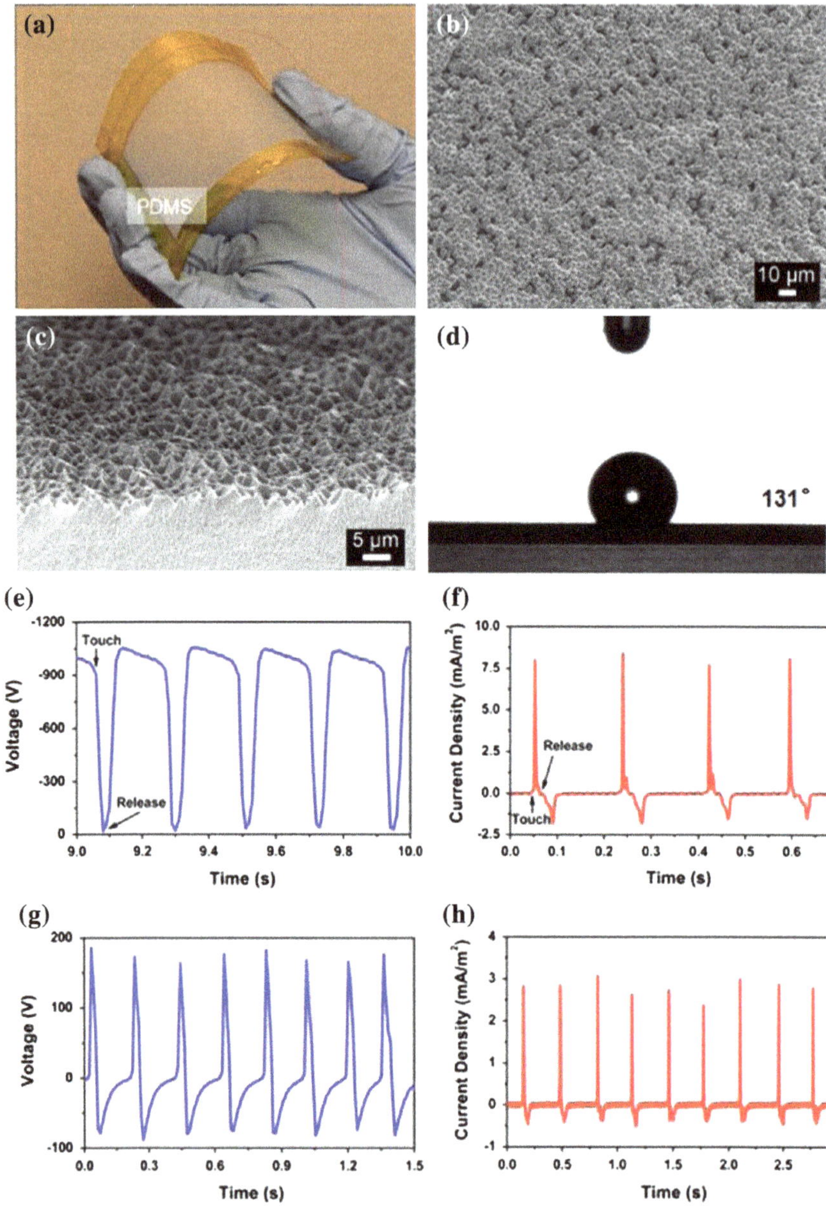


**Fig. 8.13** **a** Photograph of the fabricated PDMS film on the ITO electrode. **b** SEM image of the PDMS surface with micropyramid structures. **c** Tilted SEM image of the PDMS surface. **d** Photograph of the contact angle between the PDMS and water. **e** Measured open-circuit voltage of the device up to 1000 V. **f** Measured short-circuit current density of the TENG. **g**, **h** Measured output voltage (**g**) and current density (**h**) of the TENG with the used loading resistance of 100 MΩ. Reproduced with permission from American Chemical Society [14].

the contact angle between the PDMS polymer of the microstructured surface and water (131°) is obviously larger than that (113°) of the PDMS polymer without microstructure. The increase of the contact angles is beneficial since water droplets can easily roll off the surfaces of the device to take with them dirt and contaminants, realizing a self-cleaning function [24].

Figure 8.13e shows the measured open-circuit voltages of the fabricated TENG by utilizing an electrometer with very large input resistance, where the voltage can reach ~1000 V during fast contact/separation between the skin and PDMS film. The obtained negative voltages are associated with the negative triboelectric charges on the surface of the PDMS film. The corresponding short-circuit current density of the device reached 8 $mA/m^2$, as shown in Fig. 8.13f. Figure 8.13 g and h present the corresponding output voltage and the current density when a load resistor of 100 MΩ was connected between the ITO electrode and ground, respectively. The device produced an output voltage peak of 180 V with a current density peak of 2.8 $mA/m^2$, resulting in an output power density of 500 $mW/m^2$. The obtained output performance of the device with a micropyramid-structured surface is much larger than that without the microstructures. To demonstrate that the produced energy can be used as an effective power source, the PDMS film on the ITO electrode was attached to the back of a human hand, which was then connected to tens of green LEDs. Upon a fast swing of the hand, these green LEDs can be lighted up by the produced energy.

The working principle of the fabricated TENG is by the coupling of contact electrification and electrostatic induction. In the original position, the surfaces of the skin and PDMS fully contact each other, resulting in charge transfer between them. According to the triboelectric series, electrons were injected from the skin to the PDMS since PDMS is more triboelectrically negative than skin, which is the contact electrification process. The produced triboelectric charges with opposite polarities are fully balanced/screened, leading to no electron flow in the external circuit. Once a relative separation between PDMS and skin occurs, these triboelectric charges cannot be compensated. The negative charges on the surface of the PDMS can induce positive charges on the ITO electrode, driving free electrons to flow from the ITO electrode to the ground. This electrostatic induction process can give an output voltage/current signal. When negative triboelectric charges on the PDMS are fully screened from the induced positive charges on the ITO electrode by increasing the separation distance between the PDMS and skin, no output signals can be observed. Moreover, when the skin was reverted to approach the PDMS, the induced positive charges on the ITO electrode decrease and the electrons flow from the ground to the ITO electrode until the skin and PDMS fully contact each other again, resulting in a reversed output voltage/current signal. This is a full cycle of the electricity generation process for the TENG.

In summary, a human skin based TENG has been developed, consisting of a skin patch, a PDMS film with a micropyramid-structured surface, and an ITO electrode connected with a ground across an external load. The mechanism of TENG is based on the charge transfer between the ITO electrode and the ground by utilizing the contact/separation between the skin patch and the PDMS film. By converting the biomechanical energy into electricity, the TENG produces an open-circuit voltage up to ~1000 V, a short-circuit current density of 8 $mA/m^2$, and a power density of 500 $mW/m^2$ on a load resistance of 100 MΩ, which can be used to directly light up tens of green LEDs. This work pushes forward a significant step toward the practical applications of skin-based biomechanical energy harvesting techniques and self-powered touch pad technology.

## 8.5 Sliding Freestanding-Triboelectric-Layer Mode Triboelectric Nanogenerators

Mostof the existing TENGs necessitate the attachment of electrodes and lead wires onto moving triboelectric layers, which largely limits the versatile applications of TENGs especially when targeted motions are irregular and from arbitrary objects. Recently, sliding freestanding triboelectric-layer (SFT) nanogenerator, as a new fundamental mode, has been demonstrated to resolve the above issue by making it possible to harvest energy from independent and freely moving targets [16]. In such structure, one electrode-free triboelectric layer alternatively approaches two stationary electrodes by sliding motion, resulting in the flow of free electrons across the external load as driven by the periodical change of induced potential difference. With this unique mechanism, the device can even work without direct contact between the triboelectric surfaces, which is called non-contact operation mode. However, this type of device requires a relative large sliding displacement (the whole length of one electrode) to realize an effective electricity generation, which is unfavorable for its achievable power output and energy scavenging in practical use. Since mechanical vibration in our environment is always irregular and varying in amplitude and frequency. To solve this problem, advanced structures including rotational [17, 25] and grating [15] SFT structures have been developed to harvest body motion energy effectively.

Xie et al. report a linear grating-structured freestanding triboelectric-layer nanogenerator (GF-TENG) with largely-elevated output, high efficiency, and broad range of applications. [15] This new design is composed of a grating-segmented triboelectric layer and two pairs of interdigitated electrodes, all in the same periodicity. Both of the theoretical analysis and experimental study indicated significant enhancement in the amount of collectable charges, current density, and output frequency through introducing finer grating segments. As driven by a linear motor at an acceleration of 30 $m/s^2$, the GF-TENG with 16 grating segments delivered an open circuit voltage ($V_{OC}$) of ~135 V and a short-circuit current density ($J_{SC}$) of ~9 $mA/m^2$, corresponding to a maximum power density of ~1.2 $W/m^2$. Moreover, the device working in non-contact mode (that is free of sliding friction) exhibited supreme stability for a long period and achieved extremely-high total conversion efficiency of 85 % at low operation frequency. With these unique features, the GF-TENG was utilized to generate electricity from a wide range of ambient mechanical motion, such as sliding of human hand, acceleration or

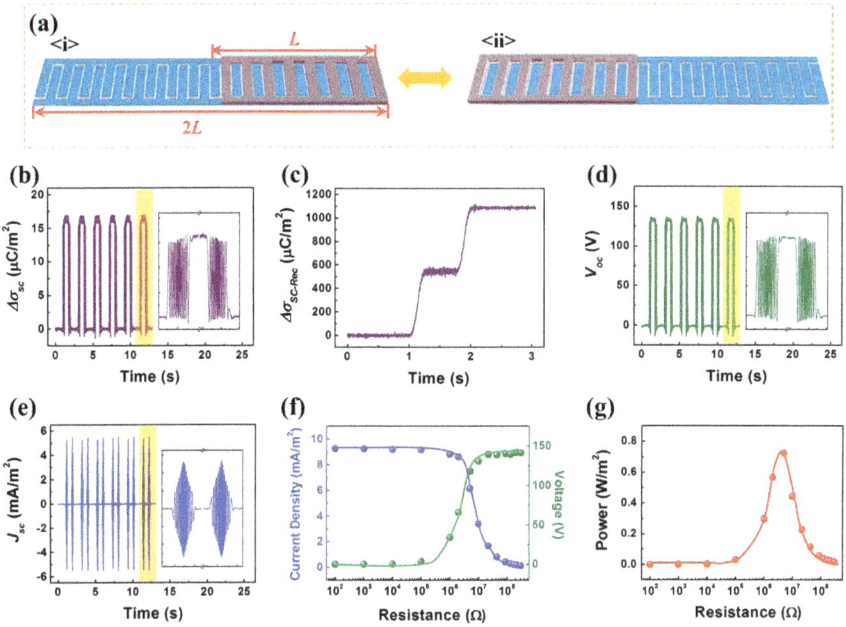


**Fig. 8.14** Electrical measurement results of the extended GF-TENGs. **a** The initial and final states of one-way sliding for a typical extended GF-TENG. **b–g** The measured results for the extended GF-TENG with 16 segments: **b** The $\Delta\sigma_{SC}$; **c** The $\Delta\sigma_{SC\text{-}Rec}$; **d** The $V_{OC}$; **e** The $J_{SC}$; **f** The dependence of the output voltage, current density, and **g** power density on the resistance of the external load. Reproduced with permission from Wiley [15].

deceleration of vehicles, and people walking. All these demonstrations reveal a great potential of GF-TENG as a high-efficiency device for practically-applicable mechanical energy harvesting.

The structure, fabrication process and basic output performances of this device have been described in Sect. 5.3.1. In the previous test where the FEP layer and the electrode groups have the same number of units, the total effective area decreases while the FEP slides outwards, resulting in a damped oscillation behavior of $\Delta\sigma_{SC}$. In order to make the electricity generation more effective through maintaining the same effective area during the sliding of FEP, an extended GF-TENG in which electrode part has the doubled number of units compared with the FEP layer is fabricated. In this design, the FEP will always slides within the range of the electrodes (Fig. 8.14a). Figure 8.14 b-e present the output performance of the extended GF-TENG with 16 grating segments. It can be observed that each unit step of sliding transfers almost the same amount of charges ($\Delta\sigma_{SC}$ = 16.5 $\mu C/m^2$) between the two electrodes without any decay during an entire unidirectional sliding process (Fig. 8.14b). If these transferred charges are accumulated, $\Delta\sigma_{SC\text{-}Rec}$ can reach 1100 $\mu C/m^2$ in the process (Fig. 8.14c), which is nearly twice as large as that from the previous non-extended structure. A similar undamped oscillation behavior) is observed for $V_{OC}$, although the amplitude (135 V) becomes smaller than the original one (Fig. 8.14d). The decrease can be ascribed to the increase of the total capacitance between two electrodes from the doubled number of electrode fingers. As for the current, when the acceleration is kept at 10 m/s$^2$, the peak value of $J_{SC}$ increases to 5.5 mA/m$^2$ due to the enhanced amount of the transferred charges. When the sliding movement has a larger speed, the magnitude of $J_{SC}$ will be further increased. As a power source in practical use, the TENG will be connected with external loads. Therefore, the actual voltages and current densities on varied load resistances were measured and the results are summarized in Fig. 8.14f. The current density (at the acceleration of 30 m/s$^2$) drops at larger external resistances, while the voltage across the load exhibits a reversed tendency. Accordingly, the power density that the device provides firstly rises at low resistance region and then declines at high resistance, showing a maximum value of 0.72 W/m$^2$ at the load resistance of 5 MΩ (Fig. 8.14g). As described previously in Sect. 5.3.1, as assisted by four springs, this design can scavenge the mechanical energy from people's walking with the total energy conversion efficiency up to 85 % as shown in Fig. 5.15c–e.

In summary, a new design of triboelectric nanogenerator has been demonstrated, which consists of a freestanding triboelectric-layer with grating segments and two stationary electrodes with interdigitated patterns. With grating units sweep across electrode fingers, multiple alternating currents are generated between the two electrodes due to electrostatic induction in both contact and non-contact mode. According to the theoretical and experimental investigation, the GF-TENG with finer grating units exhibits improved performance in the respect of the amount of collectable charges, current density, and output frequency. When the device is operated in non-contact mode, it shows an excellent stability and high total conversion efficiency up to 85 % at low operation frequency. On basis of these advantages, the GF-TENG is capable of harvesting energy from a variety of ambient sources, such as sliding of human hand, acceleration of vehicles, and people walking. This work represents a significant progress of TENG and exhibits huge potential of the GF-TENG as a high-efficiency energy harvester for practical applications.

## 8.6 Summary

Body motion is the richest source of energy for powering wearable, personal and portable electronics. To effectively harvest body motion energy using TENG, a number of approaches have been developed: the integrated TENGs; fabric based and paper based TENGs using flexible materials; human skin based TENG; and the SFT structure TENGs for high-efficient energy harvesting. The fundamental mechanisms, structure designs and performances of these technologies have been presented in this chapter with considering the low frequency and variable amplitude of the mechanical triggering. These methods not only can effectively harvest body motion energy, but also provide significant routes to achieve self-powered wearable devices and systems.

## References


1. W. Yang, J. Chen, G. Zhu, J. Yang, P. Bai, Y. Su, Q. Jing, X. Cao, Z.L. Wang, Harvesting energy from the natural vibration of human walking. ACS Nano **7**(12), 11317–11324 (2013)
2. P. Bai, G. Zhu, Z.-H. Lin, Q. Jing, J. Chen, G. Zhang, J. Ma, Z.L. Wang, Integrated multilayered triboelectric nanogenerator for harvesting biomechanical energy from human motions. ACS Nano **7**(4), 3713–3719 (2013)
3. T.-C. Hou, Y. Yang, H. Zhang, J. Chen, L.-J. Chen, Z.L. Wang, Triboelectric nanogenerator built inside shoe insole for harvesting walking energy. Nano Energy **2**(5), 856–862 (2013)
4. G. Zhu, P. Bai, J. Chen, Z.L. Wang, Power-generating shoe insole based on triboelectric nanogenerators for self-powered consumer electronics. Nano Energy **2**(5), 688–692 (2013)
5. J. Zhong, Y. Zhang, Q. Zhong, Q. Hu, B. Hu, Z.L. Wang, J. Zhou, Fiber-based generator for wearable electronics and mobile medication. ACS Nano **8**(6), 6273–6280 (2014)
6. X. Pu, L. Li, H. Song, C. Du, Z. Zhao, C. Jiang, G. Cao, W. Hu, Z.L. Wang, A self-charging power unit by integration of a textile triboelectric nanogenerator and a flexible lithium-ion battery for wearable electronics. Adv. Mater. **27**(15), 2472–2478 (2015)
7. X. Li, Z.-H. Lin, G. Cheng, X. Wen, Y. Liu, S. Niu, Z.L. Wang, 3D fiber-based hybrid nanogenerator for energy harvesting and as a self-powered pressure sensor. ACS Nano **8**(10), 10674–10681 (2014)
8. K.N. Kim, J. Chun, J.W. Kim, K.Y. Lee, J.-U. Park, S.-W. Kim, Z.L. Wang, J.M. Baik, Highly stretchable 2D fabrics for wearable triboelectric nanogenerator under harsh environments. ACS Nano **9**(6), 6394–6400 (2015)
9. S. Lee, W. Ko, Y. Oh, J. Lee, G. Baek, Y. Lee, J. Sohn, S. Cha, J. Kim, J. Park, Triboelectric energy harvester based on wearable textile platforms employing various surface morphologies. *Nano Energy* **12**(410–418 (2015)
10. W. Seung, M.K. Gupta, K.Y. Lee, K.-S. Shin, J.-H. Lee, T.Y. Kim, S. Kim, J. Lin, J.H. Kim, S.-W. Kim, Nanopatterned textile-based wearable triboelectric nanogenerator. ACS Nano **9**(4), 3501–3509 (2015)
11. T. Zhou, C. Zhang, C.B. Han, F.R. Fan, W. Tang, Z.L. Wang, Woven structured triboelectric nanogenerator for wearable devices. ACS Appl. Mater. Interfaces **6**(16), 14695–14701 (2014)
12. Q. Zhong, J. Zhong, B. Hu, Q. Hu, J. Zhou, Z.L. Wang, A paper-based nanogenerator as a power source and active sensor. Energy Environ. Sci. **6**(6), 1779–1784 (2013)
13. P.-K. Yang, Z.-H. Lin, K.C. Pradel, L. Lin, X. Li, X. Wen, J.-H. He, Z.L. Wang, Paper-based origami triboelectric nanogenerators and self-powered pressure sensors. ACS Nano **9**(1), 901–907 (2015)
14. Y. Yang, H. Zhang, Z.-H. Lin, Y.S. Zhou, Q. Jing, Y. Su, J. Yang, J. Chen, C. Hu, Z.L. Wang, Human skin based triboelectric nanogenerators for harvesting biomechanical energy and as selfpowered active tactile sensor system. ACS Nano **7**(10), 9213–9222 (2013)
15. Y. Xie, S. Wang, S. Niu, L. Lin, Q. Jing, J. Yang, Z. Wu, Z.L. Wang, Grating-structured freestanding triboelectric-layer nanogenerator for harvesting mechanical energy at 85 % total conversion efficiency. Adv. Mater. **26**(38), 6599–6607 (2014)
16. S. Wang, Y. Xie, S. Niu, L. Lin, Z.L. Wang, Freestanding triboelectric-layer-based nanogenerators for harvesting energy from a moving object or human motion in contact and non-contact modes. Adv. Mater. **26**(18), 2818–2824 (2014)

17. G. Zhu, J. Chen, T. Zhang, Q. Jing, Z.L. Wang, Radial-arrayed rotary electrification for high performance triboelectric generator. Nat. Commun. **5**, 3426 (2014)
18. L. Hu, M. Pasta, F.L. Mantia, L. Cui, S. Jeong, H.D. Deshazer, J.W. Choi, S.M. Han, Y. Cui, Stretchable, porous, and conductive energy textiles. Nano Lett. **10**(2), 708–714 (2010)
19. L. Hu, J.W. Choi, Y. Yang, S. Jeong, F. La Mantia, L.-F. Cui, Y. Cui, Highly conductive paper for energy-storage devices. Proc. Natl. Acad. Sci. **106**(51), 21490–21494 (2009)
20. Q. Liao, M. Mohr, X. Zhang, Z. Zhang, Y. Zhang, H.-J. Fecht, Carbon fiber-ZnO nanowire hybrid structures for flexible and adaptable strain sensors. Nanoscale **5**(24), 12350–12355 (2013)
21. M. Berggren, D. Nilsson, N.D. Robinson, Organic materials for printed electronics. Nat. Mater. **6**(1), 3–5 (2007)
22. R. Martins, A. Nathan, R. Barros, L. Pereira, P. Barquinha, N. Correia, R. Costa, A. Ahnood, I. Ferreira, E. Fortunato, Complementary metal oxide semiconductor technology with and on paper. Adv. Mater. **23**(39), 4491–4496 (2011)
23. K.-H. Kim, K.Y. Lee, J.-S. Seo, B. Kumar, S.-W. Kim, Paper-based piezoelectric nanogenerators with high thermal stability. Small **7**(18), 2577–2580 (2011)
24. A. Tuteja, W. Choi, M. Ma, J.M. Mabry, S.A. Mazzella, G.C. Rutledge, G.H. McKinley, R.E. Cohen, Designing superoleophobic surfaces. Science **318**(5856), 1618–1622 (2007)
25. L. Lin, S. Wang, S. Niu, C. Liu, Y. Xie, Z.L. Wang, Noncontact free-rotating disk triboelectric nanogenerator as a sustainable energy harvester and self-powered mechanical sensor. ACS Appl. Mater. Interfaces **6**(4), 3031–3038 (2014)

# Chapter 9
# Harvesting Vibration Energy

## 9.1 Vibration Energy Harvesting with Basic Operation Modes

### *9.1.1 Vertical Contact-Separation Mode*

Superior to other technologies of its kind [1–9], a harmonic-resonator-based TENG is the first reported TENG that can harness random and tiny ambient vibration as a sustainable power source, which can effectively respond to input vibration frequency from 2 to 200 Hz with a considerably wide working bandwidth of 13.4 Hz in low frequency range [10]. The harmonic-resonator-based TENG works on vertical contact-separation mode, and it has similar structure with the spring-assisted separation structure discussed in Sect. 2.3.3, which is also schematically shown in Fig. 9.1a. A photograph of an as-fabricated TENG is shown in Fig. 9.1b. On the upper substrate, aluminum thin film with nanoporous surface plays dual roles of an electrode and a contact surface. Scanning electron microscopy (SEM) image of nanopores on the aluminum is presented in Fig. 9.1c. The average diameter of aluminum nanopores are 57 ± 5 nm and a pore depth of 0.8 ± 0.2 μm with a distribution density of 212 per $\mu m^2$. A layer of polytetrafluoroethylene (PTFE) film was adhered to the lower substrate with deposited copper as another electrode. PTFE nanowires arrays were created on the exposed PTFE surface by a top-down method through reactive ion etching [11]. SEM image of the PTFE nanowires is displayed in the Fig. 9.1d. The average diameter of PTFE nanowires is 54 ± 3 nm with an average length of 1.5 ± 0.5 μm. The whole fabrication process for the TENG is sketched in Fig. 9.1e.

To investigate the TENG's performance of harvesting vibration energy, an electrodynamic shaker (Labworks Inc.) that provides sinusoidal wave was employed as a vibration source with tunable frequency and amplitude. The lower substrate of the TENG was anchored on the shaker, leaving the upper part free-standing. At a fixed vibration amplitude, the reliance of electric output on the input vibration frequency is presented in Fig. 9.2a, b. The electric output can be



Z.L. Wang et al., *Triboelectric Nanogenerators*,
Green Energy and Technology, DOI 10.1007/978-3-319-40039-6_9

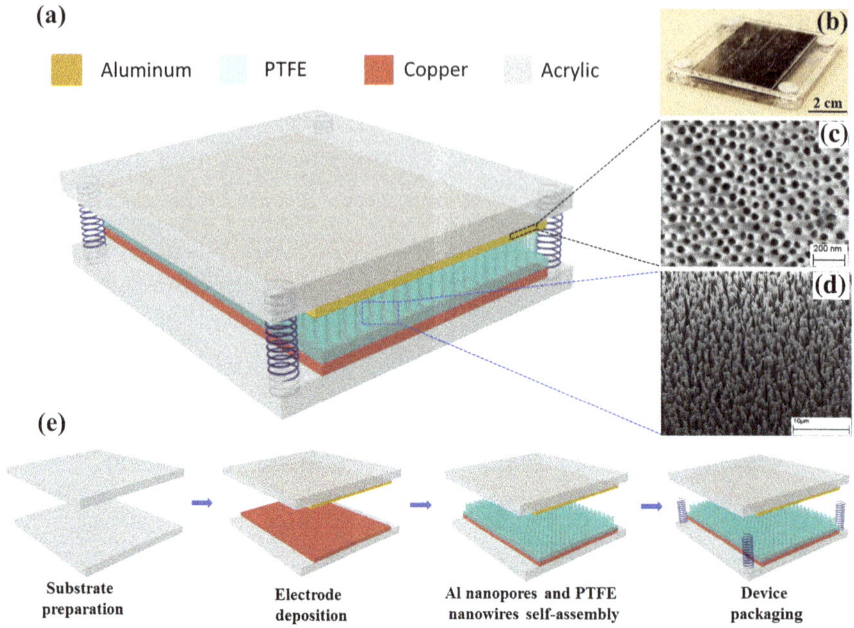


**Fig. 9.1** Harmonic resonator based triboelectric nanogenerator. **a** Sketch and **b** photograph of a typical harmonic resonator based TENG. **c** SEM image of nanopores on aluminum electrode. **d** SEM image of PTFE nanowires. **e** Process flow for fabricating the Harmonic Resonator Based TENG. Reproduced with permission from Wiley [10]

measured with broad input frequency varying from 2 to 200 Hz. Compared to state-of-the-art vibration energy harvesters that are based on nonlinear and topology variation, it has a considerably wider working bandwidth 13.4 Hz.

Experimentally, both the open-circuit voltage ($V_{OC}$) and the short-circuit current ($I_{SC}$) are maximized at the vibration frequency of 14.5 Hz with maximum values of 287.4 V and 76.8 μA, respectively, indicating that 14.5 Hz is the resonance frequency of the TENG. Theoretically, for a single degree-of-freedom vibration system, the natural frequency is given by

$$f_0 = \frac{1}{2\pi}\sqrt{\frac{4k}{m_0}} \tag{9.1}$$

where $f_0$ is the natural frequency, $m_0$ is the mass of the upper substrate plus the aluminum film, and $k$ is the stiffness coefficient of each spring. For the TENG, $m_0$ is 56.8 g, and spring stiffness coefficient is 112 N/m. Submitting the values into Eq. (9.1), we can obtain the natural frequency $f_0$ of 14.1 Hz, which is well consistent with the experimental results.

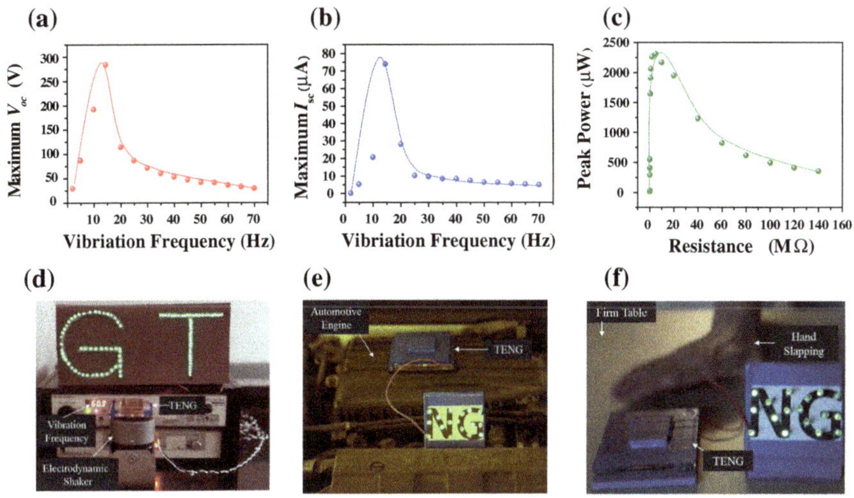


**Fig. 9.2** Electrical measurement and demonstration of the harmonic resonator based TENG as a sustainable power source. **a** Open-circuit voltage ($V_{OC}$) as a function of vibration frequency. The curve is the fitted result. **b** Short-circuit current ($I_{SC}$) as a function of vibration frequency. The curve is a fitted result. **c** Dependence of the peak power output on the resistance of the external load, indicating maximum power output when R = 5 MΩ. The curve is fitted result. **d** Photograph that shows TENG is working on an electrodynamic shaker under the vibration frequency of 60 Hz, which is the United States national grid frequency. About 200 LEDs are being lighted up simultaneously without strobe perceived by a naked eye. **e** Photograph that shows TENG is working on an automotive engine. When the car starts up, the "NG" is being lighted up simultaneously. **f** Photograph that shows TENG works when a human hands slapping a firm table. Due to the table vibration, about 20 LEDs are being lighted up simultaneously. Reproduced with permission from Wiley [10]

By measuring the output votlage across various load resistances, the maximum instantaneous peak power $\left(V^2_{peak}/R\right)$ is obtained at a load resistance of 5 MΩ, corresponding to a peak power density of 726.1 mW/m$^2$ (Fig. 9.2c).

To prove the capability of the harmonic resonator based TENG as a sustainable power source, it is installed in three scenarios for collecting vibration energies from various sources to power up electronic devices. As shown in Fig. 9.2d–f, the TENG could be excited by an electrodynamic shaker (60 Hz), an automotive engine, and a human-palm-initiated table, respectively, to instantaneously drive series of light-emitting diodes (LEDs). These demonstrated applications prove that the TENG is sensitive to small ambient vibrations, making it suitable to a wide range of circumstances for energy-harvesting purpose, e.g. on highways, bridges, and tunnels.

## 9.1.2 Contact Single-Electrode Mode

A contact single-electrode mode TENG with a wave-structure Cu-Kapton-Cu sandwiched between two flat nanostructured PTFE films was reported to harvest energy from ambient vibration [12]. A set of metal rods (with a diameter of 1/4 in.), as shown in Fig. 9.3a, was used to assist with the fabrication of the wave-structured core. PTFE films were processed with inductively coupled plasma (ICP) etching to produce the nanostructures shown in Fig. 9.3b, which would largely enhance contact electrification. The device structure is schematically shown in Fig. 9.3c, accompanied by a magnified schematic in Fig. 9.3d and a picture of a real device in Fig. 9.3e. The length and width of the TENG here is 10 and 5 cm, respectively. The size of the device could be easily scaled up for large-scale applications.

Similarly, the output voltage and current was measured across various load resistances (Fig. 9.4a) to determine the effective output power of the TENG. Instantaneous peak power density, calculated by $P_d = I^2R/S$, is plotted according to load resistance in Fig. 9.4b, and the highest peak power density of 0.4 W/m$^2$ was obtained at the load resistance of 5 MΩ, which is high enough to lit up series of 104 LEDs driven by both a mechanical shaker or human stepping, as displayed in Fig. 9.4c–f.

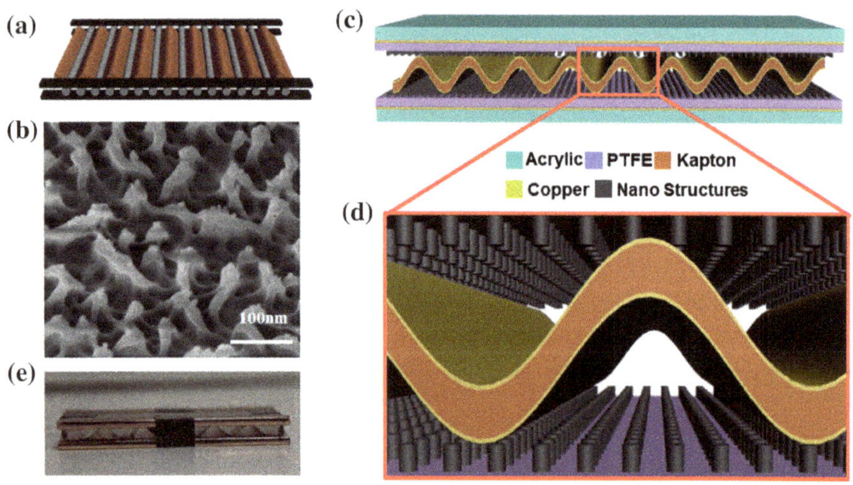


**Fig. 9.3** **a** Schematic of the method to fabricate wavy Kapton films. **b** SEM image of the ICP-processed PTFE film surface. **c** Schematic of the device structure. **d** Magnified schematic of the device, showing that the wavy core is in periodical contact with the nanostructures on the PTFE films. **e** Photograph of an as-fabricated TENG device before packaging. Reproduced with permission from American Chemical Society [12]

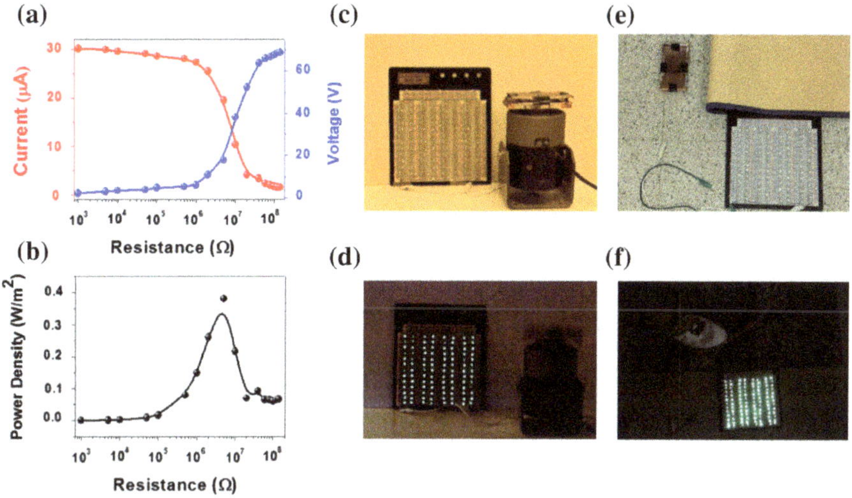


**Fig. 9.4** **a** At the vibration frequency of 100 Hz, output peak voltage and peak current dependence on load resistance. **b** At the vibration frequency of 100 Hz, output peak power density dependence on load resistance. **c** Photograph of the TENG mounted on the shaker, connected to 104 green LEDs. **d** Photograph of the 104 green LEDs lit up simultaneously by the TENG triggered by the shaker. **e** Photograph of the TENG to be covered by carpet, connected to 104 green LEDs. **f** Photograph of the 104 green LEDs lit up simultaneously by the TENG triggered by footsteps. Reproduced with permission from American Chemical Society [12]

### *9.1.3 Contact Freestanding-Triboelectric-Layer Mode*

In addition, a contact-mode freestanding triboelectric nanogenerator (CF-TENG) was reported to harvest ambient vibration energy [13]. And its output performance under the vibration amplitude of 1.5 cm and frequency of 15 Hz was described in Sect. 5.2.1 and Fig. 5.10. Owing to the linearity in the electricity generation behavior, the physical contact between the triboelectrically-charged resonator and the electrodes are not mandatory for the effective mechanical energy conversion. When the vibration amplitude of the shaker is gradually tuned down to a much smaller value, these LEDs still can be lit up by the CF-TENG. Similarly with the sliding-based freestanding-TENG, the CF-TENG's capability of non-contact operation also brings about the advantages of improved energy conversion efficiency and better stability.

# 9.2 Vibration Energy Harvesting with Advanced Structural Designs

## 9.2.1 *Multi-directional Vibration Energy Harvesting*

Vibrations in our living environments are generally distributed over a wide frequency spectrum and exhibit multiple motion directions over time, which renders most of the current vibration energy harvesters impractical to fully scavenge energy from all directions. With this regard, a three-dimensional triboelectric nanogenerator (3D-TENG) was developed, which can work in hybridization of both the vertical contact-separation mode and the in-plane sliding mode [14]. The innovative design facilitates harvesting random vibration energy in multiple directions over a wide bandwidth.

The 3D-TENG has a multilayer structure with circular acrylic as supporting substrates, as schematically shown in Fig. 9.5a. The cylindroid core of the 3D-TENG lies at the center of the acrylic substrate with a bottom diameter of 3 cm. On the top of the core, an iron mass is mobile and suspended by three identical springs with an included angle of 120° between each other. The designed structural symmetry assures that the whole system has a constant resonant frequency at arbitrary in-plane directions. A layer of polytetrafluoroethylene (PTFE) film as one contact surface was adhered onto the bottom side of circular iron mass with deposited copper thin film as the back electrode. Attached to the bottom acrylic substrate, aluminum thin film with nanopores modification plays dual roles of a contact electrode and the other contact surface. The scanning electron microscopy (SEM) image of aluminum nanopores are shown in Fig. 9.5b. Figure 9.5c is a photograph of the real 3D-TENG device.

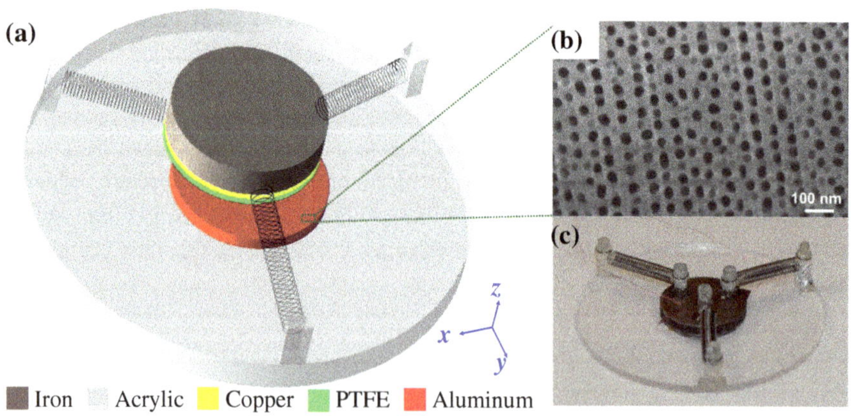


**Fig. 9.5** Three-dimensional triboelectric nanogenerator. **a** Schematic of a 3D-TENG. **b** SEM image of nanopores on aluminum electrode. **c** A photograph of the fabricated 3D-TENG. Reproduced with permission from Wiley [14]

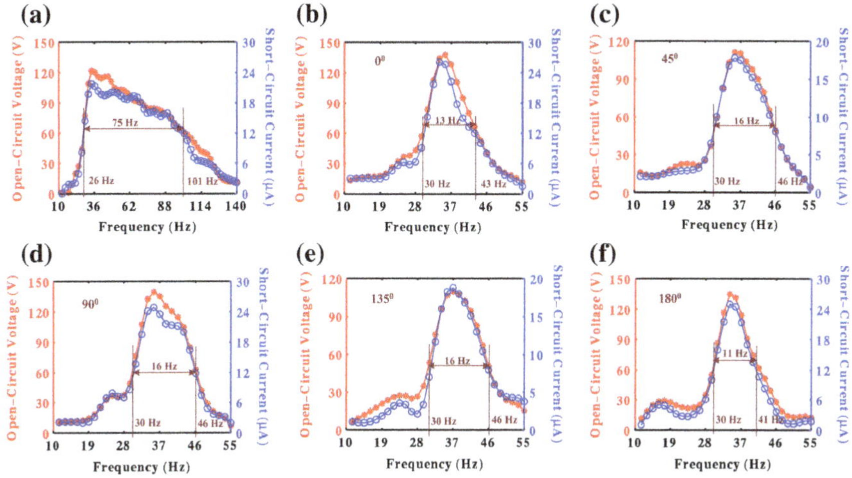


**Fig. 9.6** Frequency responses of the 3D-TENG. **a** Frequency response of 3D-TENG under out-of-plane excitation. **b–f** Frequency responses of 3D-TENG at in-plane excitation angles of 0°, 45°, 90°, 135° and 180°, respectively. Reproduced with permission from Wiley [14]

To operate, an electrodynamic shaker is still used as an external vibration source with controlled amplitude and acceleration. The supporting acrylic substrate of the 3D-TENG is anchored on the shaker table to investigate the relationship between the electrical outputs and input frequency when it works in a vertical contact-separation mode. As shown in Fig. 9.6a, both the voltage and current present a rapid increasing with the increase of frequency from 10 to 36 Hz. The maximum values of the voltage and current respectively reach 123 V and 21 μA at the frequency of 36 Hz. Then, the voltage and current both gradually decrease to their minima as the frequency increases from 36 to 140 Hz. The larger separation at 36 Hz would cause a larger contacting force when the two objects impact together. Since the aluminum surface was patterned with nanopores, a larger impact will greatly improve the effective contact area between the two surfaces, and thus the total electric output. In addition, if the half peak voltage point is adopted as the criteria of the working bandwidth, 3D-TENG shows an extremely wide working bandwidth of up to 75 Hz in a low vibration frequency range under the out-of-plane excitation.

When it comes to the in-plane sliding mode, the electric outputs were measured with the input excitation frequencies in a range from 10 to 55 Hz at different in-plane excitation angles (0°, 45°, 90°, 135°, 180°), as shown in Fig. 9.6b–f. The results proved that the 3D-TENG is capable of harvesting vibration energy from all directions. Considering the non-ideal experimental factors, such as the variation of the spring stiffness factors and the deviation of the fabricated structure symmetry, the 3D-TENG has almost the same responses to the external vibrations at arbitrary in-plane directions. Furthermore, both the voltage and current are maximized at the frequency of 36 Hz at all excitation angles. And slightly nonlinear behaviors are observed in all of the output responses mainly owing to the non-linear topology structure of the spring vibration system. In addition, the 3D-TENG also shows a considerably wide working bandwidth of 14.4 Hz in a low vibration frequency range under the in-plane excitation. The practicability of harvesting energy from multiple directions with considerably wide working bandwidth enables the 3D-TENG have tremendous applications of building up self-powered systems by harvesting ambient vibration energy.

To prove the capability of the 3D-TENG as a sustainable power source, three sets of practical applications were demonstrated. First, as shown in Fig. 9.7a, the 3D-TENG works on a national grid transmission line which can effectively harvest wind or rain droplet induced line vibration energy. Forty serial-connected commercial LEDs were lighted up due to the line oscillating. Figure 9.7b plots the output voltages of the 3D-TENG under different line swing amplitudes. It can be seen that the voltages present an obvious increasing tendency with the increased swing amplitudes. This is because larger swing amplitude of the line will contribute to a larger out-of-plane separation of the two contact surface in the 3D-TENG, leading to higher output voltage.

Secondly, as shown in Fig. 9.7c, the 3D-TENG is mounted on a human trail leg to harvest the vibration energy from human walking. This human motion induced

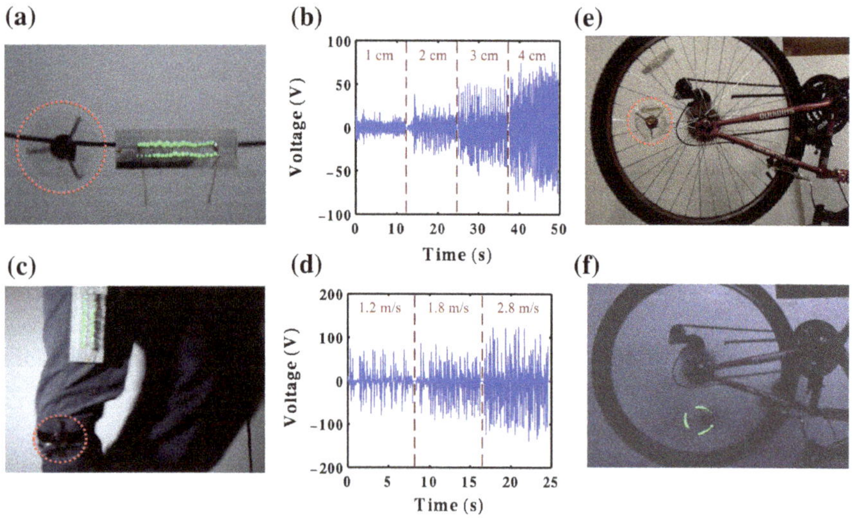


**Fig. 9.7** Demonstration of the 3D-TENG (marked in *red circles*) as a sustainable power source and self-powered active sensors. **a** The photograph of line vibration energy harvesting. Forty commercial LED bulbs were lighted up simultaneously. **b** The output voltage of the 3D-TENG under different line swing amplitudes. **c** The photograph of human walking energy harvesting. Forty commercial LED bulbs were lighted up simultaneously. **d** The output voltages at different walking speeds. **e** The photograph of 3D-TENG mounted on a still bicycle wheel. **f** Thirty commercial LED bulbs were lighted up simultaneously under the rotation of the bicycle wheel. Reproduced with permission from Wiley [14]

vibration can be used as an external excitation to the 3D-TENG and 40 serial-connected commercial LEDs were lighted up. Figure 9.7d plots the output voltages under different walk speeds, which indicates that a faster speed leads to a larger voltage output. A third practical application is shown in Fig. 9.7e. The 3D-TENG was anchored on a bicycle wheel to harvest the rotation energy. And 30 serial-connected commercial LEDs were lighted up during the wheel rotating (Fig. 9.7f). This demonstration proves that the 3D-TENG not only can harvest wheel rotation energy but also can be developed as self-powered sensing system to monitor the tire-pressure of automobiles, airplanes and other wheel-based vehicles.

In a word, the 3D-TENG can effectively harvest ambient vibration energy in out-of-plane direction with extremely wide working bandwidth up to 75 Hz and in arbitrary in-plane directions with prepotent bandwidth of 14.4 Hz in low vibration frequency range. This superior capacity enable the 3D-TENG to harvest energy from various vibration sources, such as wind, rain droplets, human activity, and automobiles, demonstrating its broad and versatile applications.

### 9.2.2 *Multi-layer Structural Design*

The applications of a single-layer triboelectric nanogenerator for vibration energy harvesting may be challenged by its low output current, and a possible solution to improve the output current is multiple-layer integration, and the most important point is to synchronize the outputs of all the TENGs so that the instantaneous output power can be maximized, as is presented in the following work [15]. The 3D-TENG has a multilayered structure with acrylic as supporting substrates, as schematically shown in Fig. 9.8a. A photograph of an as-fabricated TENG is shown in Fig. 9.8b, in which, the total number of the unit cells can be expressed as:

$$N_{total} = 4n \tag{9.2}$$

where $n$ is the number of pinned fingers of a TENG. Eight identical springs were employed to bridge the moveable and pinned fingers. All fingers were made from acrylic sheets with a thickness of 3 mm and parallel to each other with identical but tuneable gap distance. The thickness of acrylic sheets is large enough to prevent the mutual charges influences among the unit cells. Moreover, all the unit cells are electrically connected in parallel. Aluminium thin films were deposited onto both sides of the pinned fingers, which played dual roles of a contact electrode and a contact surface. A layer of polytetrafluoroethylene (PTFE) film was adhered to the both sides of the movable fingers with deposited copper as another electrode. PTFE nanowires arrays were created on the exposed PTFE surface by a top-down method through reactive ion etching. SEM image of the PTFE nanowires is displayed in Fig. 9.8c. Additionally, to promote the triboelectrification and to increase the effective contact area between the two contacting surfaces, PTFE nanowires are created as surface modifications, which greatly increases the triboelectric charges

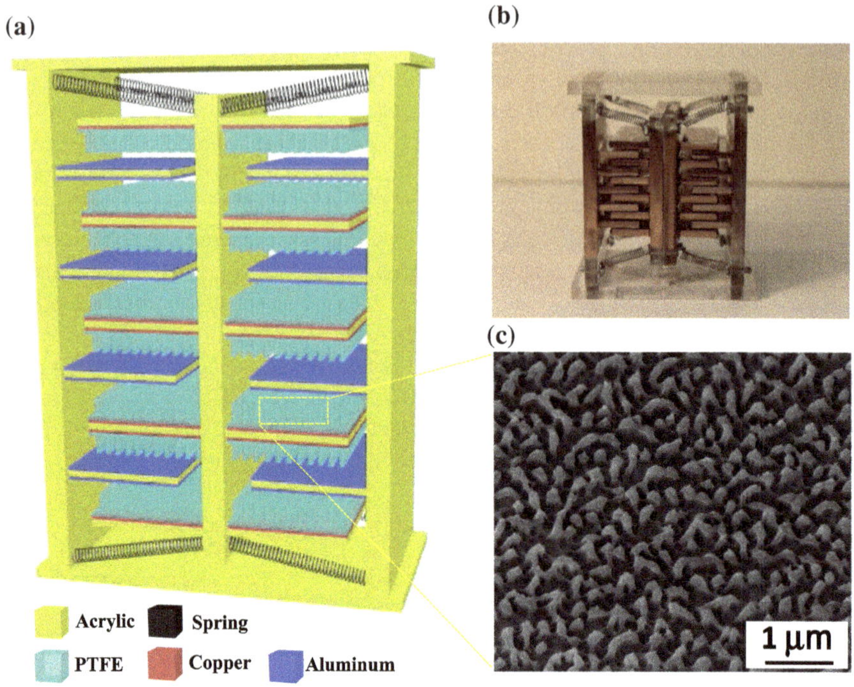


**Fig. 9.8** Three-dimensional triboelectric nanogenerator. **a** Schematic of a 3D-TENG. **b** SEM image of nanopores on aluminum electrode. **c** A photograph of the fabricated 3D-TENG. Reproduced with permission from Wiley [15]

and thus the overall electrical output. Theoretically, the short-circuit current of the 3D-TENG can be expressed as:

$$I_{SC} = \frac{\sigma s d_0 v}{\varepsilon_r [d + \frac{d_0}{\varepsilon_r}]^2} \tag{9.3}$$

where $d_0$ is the thickness of PTFE, $\varepsilon_r$ is the relative permittivity of PTFE, $s$ is the effective contact area and $v$ is the relative velocity at which they will contact, whose motion direction determines the flowing direction of the induced charge, and thus the direction of the short circuit current. $n$ is the number of pinned fingers, $\Delta S$ is the effective contact area and $\Delta d$ is the gap distance between the two adjacent fingers.

Resistors were utilized as external loads to further investigate the output power of the 3D-TENG at its resonance frequency. As displayed in Fig. 9.9a, the current amplitude drops with increasing load resistance owing to the Ohmic loss, while the voltage follows a reverse trend. Consequently, the instantaneous peak power ($P = I^2_{\mathrm{peak}} R$) is maximized at a load resistance of 2 MΩ, corresponding to a peak power density ($P_{\mathrm{d}} = I^2_{\mathrm{peak}} R / \Delta S$) of 104.6 W m$^{-2}$ (Fig. 9.9b). To prove the capability of

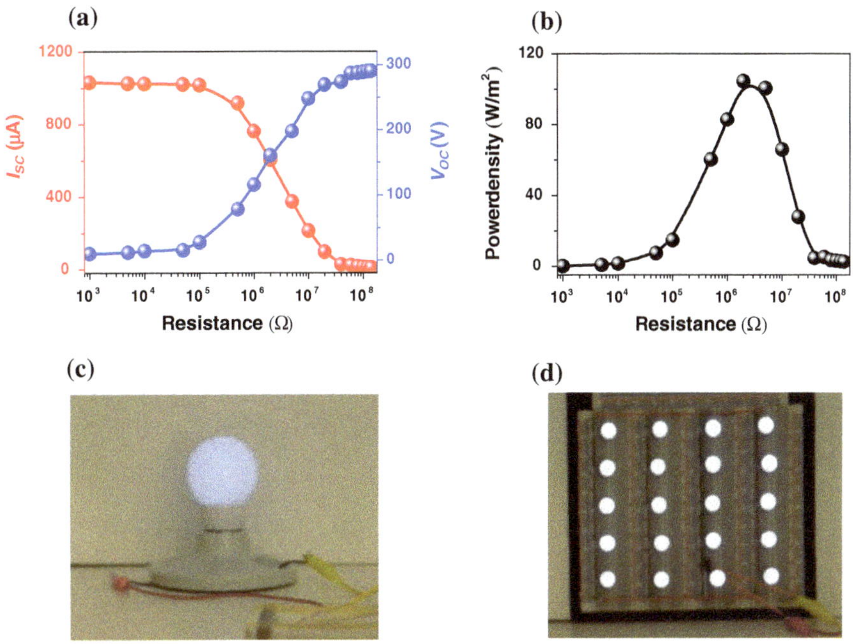


**Fig. 9.9** Demonstration of the 3D-TENG as a direct power source. **a** Dependence of the voltage and current output on the external load resistance for the TENG with n = 5. The points represent peak values of electric signals while the lines are the fitted results. **b** Dependence of the peak power output on the resistance of the external load for the TENG with n = 5, indicating maximum power output at R = 2 MΩ. The curve is a fitted result. **c** Photograph of a white G16 globe light that is directly powered by the TENG (Input frequency 40 Hz). **d** Photograph of 20 spot lights (0.6 W ea.) connected in series that are lighted up simultaneously and continuously without observable stroboflash by the TENG (Input frequency 40 Hz). Reproduced with permission from Wiley [15]

the 3D-TENG as a sustainable power source, two sets of practical applications were demonstrated. First, as shown in Fig. 9.9c, the TENG was excited by an electrodynamic shaker at a vibration frequency of 40 Hz, lighting up a white G16 globe light. Inset is a photograph when a white G16 globe light is off. Meanwhile, a total of 20 spot lights (0.6 W each) connected in series were lighted up simultaneously and continuously without observable flashing (Fig. 9.9d).

### 9.2.3 *Liquid-Metal Based Structural Design*

Solid materials based TENG may be suffering from a poor physical contact due to the roughness of contact surfaces, which would lead to a reduction of the energy conversion efficiency of TENG for vibration energy harvesting. In this regard, a liquid-metal-based TENG (LM-TENG) was developed for an intimate contact between the metal and the dielectric layer, and an instantaneous energy conversion efficiency up to 70.6 % was demonstrated, as discussed in Chap. 3, Sect. 3.3.6 [16].

To demonstrate the capability of the LM-TENG as a power source, it is directly connected to hundreds of light-emitting diode (LED) bulbs. Driven by the hand motions, even very gentle, the LM-TENG can power up 150 LEDs. Additionally, the liquid is normally easy-fluctuating, which suggests an inherent technique of vibration energy harvesting. The configuration of the device is illustrated in Fig. 9.10a. An acrylic box with 4 cm * 4 cm * 1.5 cm in size is fabricated to contain and seal the liquid metal, while ten slices are mounted in the box's lid, with their bottom ends suspended over the liquid level. Figure 9.10b shows the testing results of the device under various vibration frequencies with the amplitude of 1.2 mm. The maximum output current and voltage are 5.8 μA and 145 V, obtained at 10 Hz. With a bridge rectifier, the output charge without external load reaches 100 μC in 65 s (red line in Fig. 9.10c).

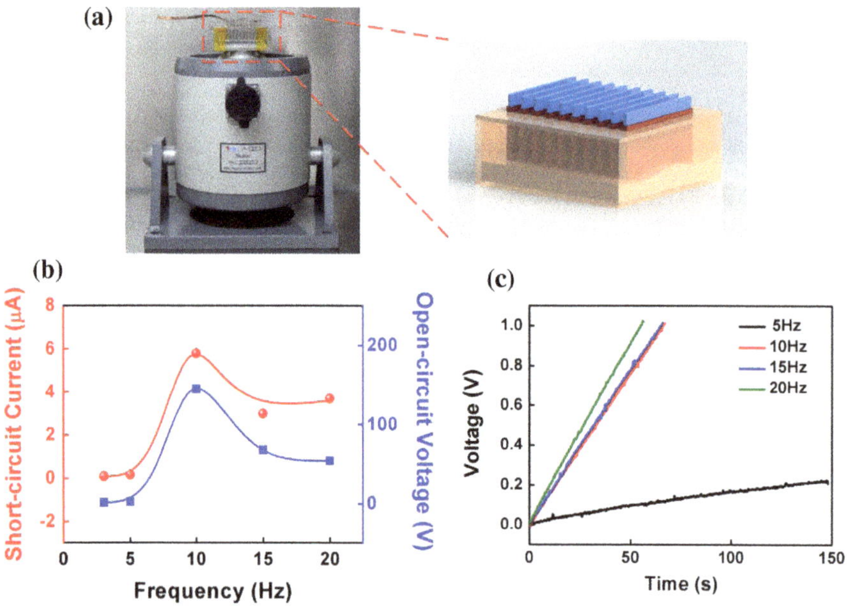


**Fig. 9.10** LM-TENG's application in the vibration energy harvesting. **a** Schematic illustrations of the device and the testing setup. **b** The output current and voltage as a function of the vibration frequency. **c** The charging curves of a 100 μF capacitor under various vibration frequencies. Reproduced with permission from Wiley [16].

## 9.3 Sound Wave Energy Harvesting

Acoustic wave is also a specific form of vibration. As a clean, ubiquitous and sustainable energy source, acoustic waves, such as various sound noises from any living activities, airports, construction sites and traffic, are one of the wasted energies that are abundant in our daily life. Sound energy is usually taken as unwanted noise that is polluting our living environment. Since acoustic energy is usually low in power density, effective technologies are lacking for harvesting such energy. Here we introduce the TENG as a potential approach for harvesting acoustic wave energy.

### *9.3.1 Organic Film Based TENG*

The TENG was firstly demonstrated to harvest ambient acoustic energy by fabricating an organic thin-film based TENG [17]. Relying on a Helmholtz cavity with a size-tunable narrow neck on its back, the core of the nanogenerator is in a circular shape and embedded as the flexible front plate of the cavity, as schematically shown in Fig. 9.11a. For a better illustration, a cross-sectional view of the core is shown in Fig. 9.11b, with a multilayered structure. Aluminum thin film with nanoporous surface plays dual roles as an electrode and a contact surface. A scanning electron microscopy (SEM) image of nanopores on the aluminum is presented in Fig. 9.11c.

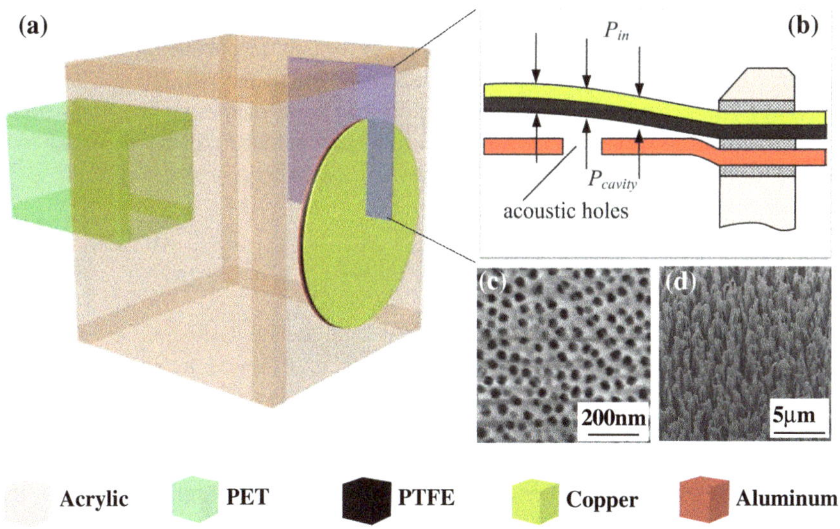


**Fig. 9.11** Structural design of the nanogenerator. **a** A sketch and **b** Cross sectional view of the nanogenerator. **c** An SEM image of nanopores on aluminum electrode. **d** An SEM image of PTFE nanowires. Reproduced with permission from American Chemical Society [17]

The nanopores were uniformly distributed on the surface of aluminum with an average diameter of 57 ± 5 nm and a density of 210 per $\mu m^2$. A layer of polytetrafluoroethylene (PTFE) film with deposited copper was employed as another electrode. An SEM image of the PTFE nanowires is displayed in Fig. 9.11d. The average diameter of PTFE nanowires is 54 ± 3 nm with an average length of 1.1 ± 0.4 μm. To investigate the performance of the proposed nanogenerator in harvesting acoustic energy, a loudspeaker (Fostex Inc.) that provides sinusoidal sound waves was used as an acoustic source with tunable frequency and amplitude. A sound level meter (Extech Inc.) with 2 dB accuracy and 0.1 dB resolution is used to measure the incident acoustic pressure. The meter is located near the harvester at a distance far less than the acoustic wavelength. Displacements in the radial direction are prescribed at the outer circular boundary of the PTFE film to intentionally introduce the initial in-plane pre-stress in order to tune the resonant frequency of the film, thus render it acting as an elastic element to oscillate under acoustic wave excitation. Under pre-stresses of 2.8, 5.6 and 8.4 kPa and a constant incident acoustic pressure of 110 $dB_{SPL}$, the reliance of open-circuit voltages ($V_{OC}$) and short-circuit currents ($I_{SC}$) on the input acoustic frequency is presented in Fig. 9.12a, b, respectively. At the resonant frequencies, the $V_{oc}$ under different pre-stresses are respectively 16.50, 60.50 and 7.50 V, respectively. While $I_{sc}$ are 5.50, 15.10 and 2.10 μA, respectively. The output voltage shows an enhancement of 3.7 times for the device with a pre-stress of 5.6 kPa than that with the pre-stress of 2.8 kPa, and also an enhancement of 8.1 times is achieved compared to the device with pre-stress of 8.4 kPa. The results show that a suitable initial pre-stress can optimize the oscillation coupling between the air trapped in the cavity and PTFE thin film, and thus the electrical output.

Regarding the optimized design, namely, the device with pre-stress of 5.6 kPa, experimentally, both $V_{OC}$ and $I_{SC}$ are maximized at the acoustic frequency of 240 Hz, indicating that 240 Hz is the resonant frequency of the as-fabricated generator. Theoretically, if the dimensions of device are smaller than or comparable to the incident acoustic wavelength, its dynamic behavior can be modeled as a lumped system. And its resonant frequency can be expressed as:

$$f_R = \frac{c}{2\pi}\sqrt{\frac{S}{L'V}} \tag{9.4}$$

where $S$ and $L'$ are respectively the cross-sectional area and effective length of the neck. $V$ is the cavity volume and $c$ is the speed of sound in air (343 m $s^{-1}$). Submitting the values into Eq. (9.4), the natural frequency was calculated as 238.8 Hz, which is consistent with the experimental result. Besides the pre-stress showed a marked impact on the electric output of the device, aimed at further improving the electric output, a serial of acoustic holes were punched through the aluminum electrode, which acted as communicating vessels to integrate the air gap between two contact surfaces with the air in the cavity. The open ratio (OR) is defined as the area ratio of all acoustic holes' area to the surface area of the aluminum electrode, which largely influences the damping of the air. A larger value

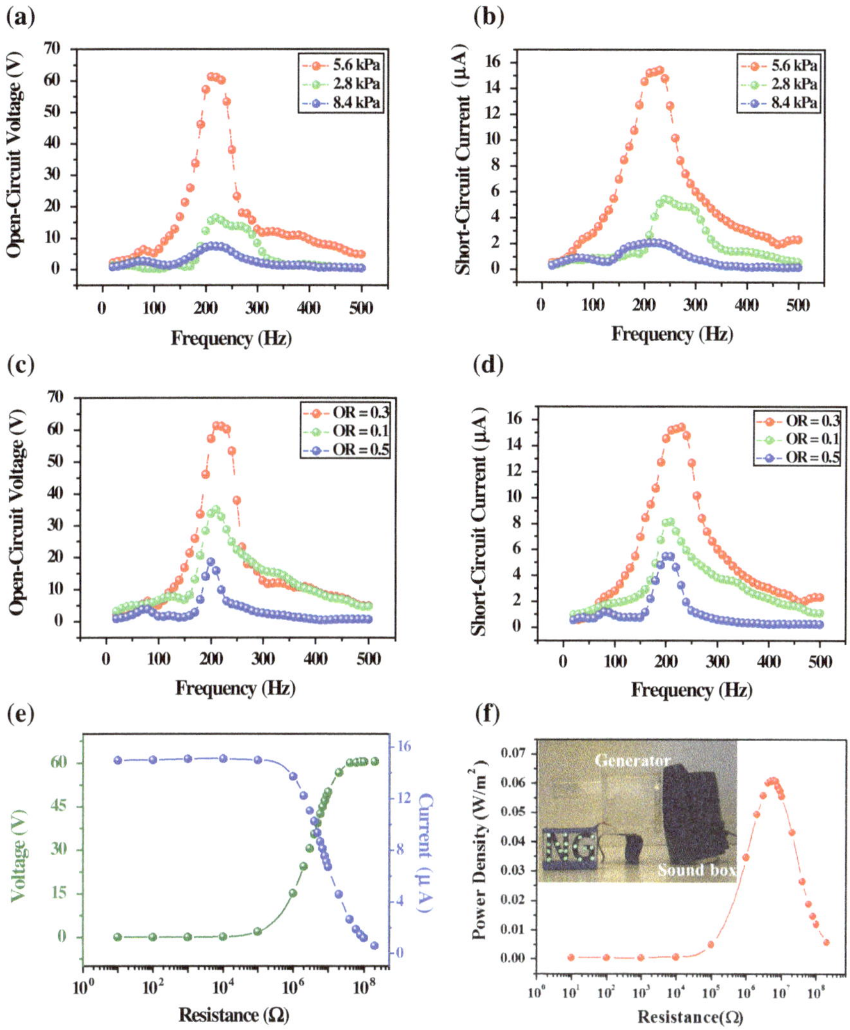


**Fig. 9.12** Performance evaluation of a nanogenerator. **a** Open-circuit voltages ($V_{OC}$) and **b** Short-circuit currents ($I_{SC}$) as a function of acoustic frequency with different pre-stresses of 2.8, 5.6 and 8.4 kPa. The curves are the fitted results. **c** $V_{OC}$ and **d** $I_{SC}$ as a function of acoustic frequency with different open ratios (ORs) of 0.1, 0.3 and 0.5. The curves are the fitted results. **e** Dependence of the voltage and current output on the external load resistance. The points represent peak value of electric signals while the lines are the fitted results. **f** Dependence of the peak power output on the resistance of the external load, indicating maximum power output at R = 6 MΩ. The curve is fitted result. Inset is a photograph that shows nanogenerator works under the excitation from a sound box. Due to the external sound, 17 LEDs are being lighted up simultaneously. Reproduced with permission from American Chemical Society [17]

of OR results in a higher flow velocity and lower damping, which will contribute to a larger deformation of the PTFE film under the same acoustic pressure excitation. Under a constant acoustic pressure of 110 $dB_{SPL}$, the $V_{OC}$ and $I_{SC}$ of the generator with different ORs of 0.1, 0.3 and 0.5 were respectively measured, as shown in Fig. 9.12c, d. At the resonant frequencies, the $V_{OC}$ for open ratio of 0.1, 0.3 and 0.5, are respectively 35.20, 60.50 and 18.70 V, and the corresponding $I_{SC}$ are 8.20, 15.10 and 5.50 μA. The results indicate that an enhancement of 1.7 times of the electric output is obtained for the device with OR value of 0.3, compared to that of 0.1, and also an enhancement of 3.2 times is achieved compared to that of 0.5. This experimental observation is resulted from a tradeoff between the PTFE deformation and the effective contact area. A larger OR leads to a larger deformation of the PTFE thin film, and thus a higher electric output. However, increased OR will reduce the effective contact area for triboelectrification, and thus a lower electrical output. Consequently, an optimum OR is needed to maximize the electrical output.

Resistors were utilized as external loads to further investigate the useful output power of the as-fabricated nanogenerator at the resonant frequency. As displayed in Fig. 9.12e, the voltage amplitudes increase with increasing load resistance, while the current follows a reverse trend owing to the Ohmic loss. As a result, the instantaneous peak power $(V^2/R)_{peak}$ is maximized at a load resistance of 6 MΩ, corresponding to a peak power density of 60.2 mW $m^{-2}$ (Fig. 9.12f). To prove the capability of the proposed nanogenerator as a sustainable power source, powered by a nanogenerator with the cavity dimensions of 8 cm × 8 cm × 8 cm under acoustic pressure of 110 $dB_{SPL}$ at the resonant frequency of 240 Hz, 17 commercial LED bulbs were lighted up simultaneously, as demonstrated in the inset of Fig. 9.12f.

In a word, TENG was first developed to harvest acoustic energy in this work. Rationally designed structure, coupled with nanomaterial modification, the as-fabricated TENG enables superior performance in harvesting ambient acoustic energy. It produced a maximum electric power density of 60.2 mW $m^{-2}$, which directly lighted up 17 commercial LEDs simultaneously. This work presents an adaptable, cost-effective and fundamentally new approach for ambient acoustic energy harvesting with potential applications in infrastructure monitoring, sensor networks, military surveillance and environmental noise reduction.

### *9.3.2 Rollable Paper Based TENG*

The TENG introduced in the last section has a bulky structure due to the requirement of a resonance cavity, which greatly limited its wide-range application in acoustic energy harvesting. A further improved performance of the acoustic energy harvesting is on the basis of an ultrathin, rollable, and paper-based TENG, which innovatively employs arrays of micro holes for acoustic response enhancement [18].

An ultrathin TENG has a multilayered structure composed of thin film materials that are vertically laminated. A layer of multiholed paper forms the structural backbone of the TENG, which was coated with copper acting as an electrification layer that generates triboelectric charges upon contacting with a thin polytetrafluoroethylene (PTFE) membrane, as schematically shown in Fig. 9.13a. Papers were selected as the structural materials owing to its flexibility, lightweight, good machinability, low cost as well as biodegradability. And, instead of papers, other expensive organic thin films can be chosen as well. For purpose of enhancing a broad-band acoustic response, holes with diameters of 400 μm were evenly punched and distributed on the paper substrate. A photograph of the multi-hole paper electrode is demonstrated in bottom-left of Fig. 9.13a. In order to enhance the triboelectrification, the polymer nanowires array was purposely created onto the PTFE membrane, as the SEM image shown in bottom-right of Fig. 9.13a. Figure 9.13b is a photograph of an as-fabricated paper-based TENG with a thickness less than 125 μm.

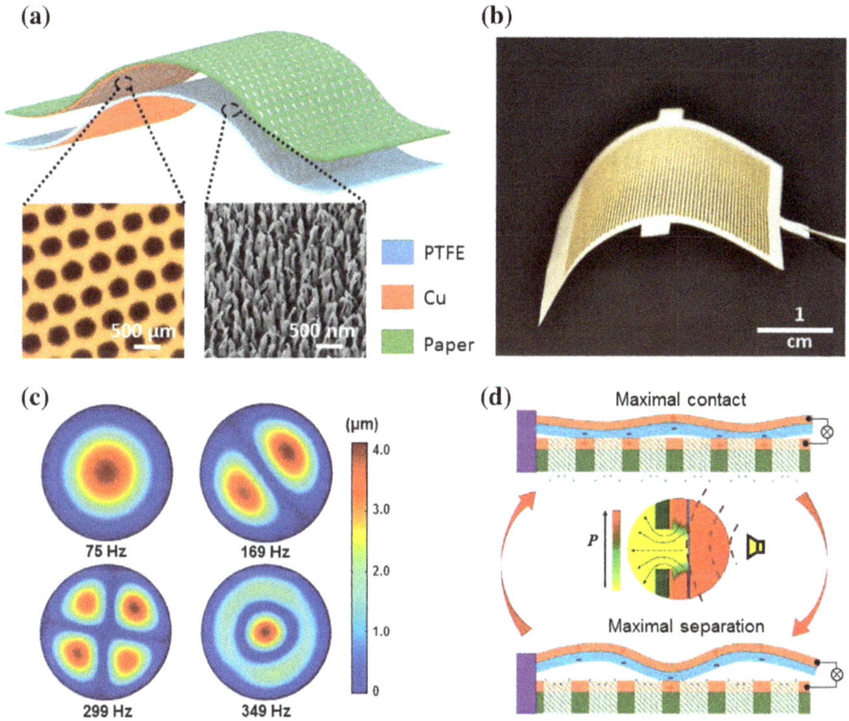


**Fig. 9.13** Structural design and operating principle of the ultrathin paper-based TENG. **a** Schematic illustrations of the paper thin TENG. *Below-left* A photograph of the multi-hole paper electrode. *Below-right* A SEM image of the PTFE polymer nanowires. **b** A photograph of an as-fabricated paper thin TENG. The scale bar is 1 cm. **c** The ANSYS software was employed to characterize the PTFE membrane vibrations under various sound frequencies. **d** An illustration to interpret the sound wave induced PTFE membrane mechanical vibration. Reproduced with permission from American Chemical Society [18]

The working principle of the thin-paper-based triboelectric nanogenerator can be elucidated from two aspects, namely, sound induced membrane vibration and vibration induced electricity generation. On one hand, when an external sound wave is continuously incident onto the thin-paper-based TENG, the flexible PTFE membrane would vibrate driven by the periodical air-pressure difference. This vibrational mechanism is distinctly different from the traditional acoustic energy harvester that relies on a resonator cavity, for which the mechanical vibration is attributed to the alternative air compression-expansion within the cavity. An ANSYS software is employed to simulate the sound wave induced PTFE membrane vibration under various frequencies, as shown in Fig. 9.13c, assuming that the acoustic pressure difference of 20 Pa is uniformly distributed over the 0.025 mm-thick PTFE membrane with Young's modulus of 440 MPa. As it can be observed from the simulation results that the deformation regions and magnitudes of the PTFE membrane are highly related to the external sound excitation frequencies, which can be attributed to different resonant frequencies under different vibration modes. On the other hand, the vibration induced electricity generation is attributed to a coupling effect between contact electrification and electrostatic induction, similar with what we have discussed in Chap. 2. A cycle of electricity generation process under external pressure is schematically depicted in Fig. 9.13d. At its maximum contact state, sound wave induced contact between PTFE and copper will generate electrical charges. The two materials have a different affinity for electrons, with the PTFE attracting electrons from copper, resulting in positive triboelectric charges on the copper side and negative ones on the PTFE side [10, 19–21]. Subsequently, the acoustic pressure will separate the PTFE from copper. As a result, an inner dipole moment between the two contact surfaces is consequently altered, which drives free electrons to flow from the copper electrode on the PTFE membrane to the multi-hole paper electrode till the maximum separation state is reached. And the free electrons will flow in a reverse direction in the process from maximum separation state towards maximum contact state, which completes a full cycle of electricity generation process.

Resistors were utilized as external loads to further investigate the output power of the structurally optimized paper thin TENG at acoustic frequency of 250 Hz with an acoustic pressure of 114 $dB_{SPL}$. As displayed in Fig. 9.14a, the voltage amplitudes increase with increasing load resistance, while the current follows a reverse trend owing to the Ohmic loss. As a result, the instantaneous peak power is maximized at a load resistance of 800 KΩ, corresponding to a peak power density of 121 $mW/m^2$ (volume power density of 968 $W/m^3$), as shown in Fig. 9.14b. Furthermore, the dependence of the electrical output on the incident sound pressures was also investigated and a direct proportional function was experimentally observed, as indicated in Fig. 9.14c. Holding a collection of compelling features, including thin paper based, rollable, broad working bandwidth, independence of resonator, the proposed nanogenerator demonstrated its unique power in the field of acoustic energy harvesting. And it was demonstrated to recycle acoustic energy from a commercial cell phone when playing music, human talking on the phone. As shown in Fig. 9.14d, the recycled acoustic energy from a cell phone is capable of

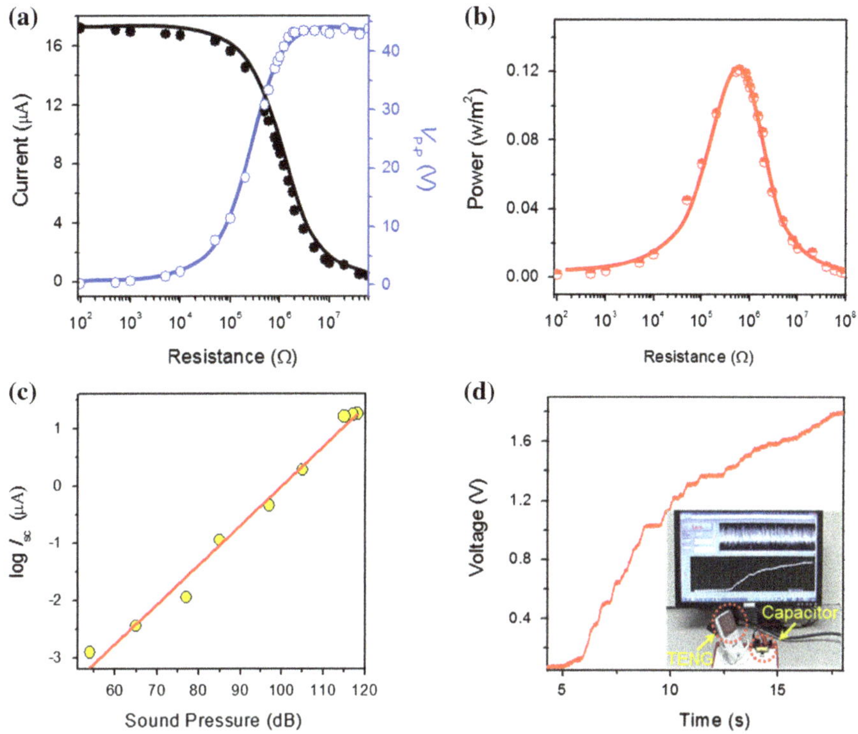


**Fig. 9.14** Performance evaluation of the paper thin triboelectric nanogenerator as a sustainable power source. **a** Dependence of the peak-to-peak voltage ($V_{p\text{-}p}$) and current output on the external load resistance. **b** Dependence of the peak power output on the resistance of the external load, indicating the maximum power output at R = 800 KΩ. **c** Dependence of the electrical output on the input sound pressures. **d** Recycling the acoustic energy from the cell phone via charging a 2 μF capacitor. Inset is a photograph that shows a commercial cell phone equipped with a paper thin triboelectric nanogenerator for capacitor charging. Reproduced with permission from American Chemical Society [18]

charging a commercial capacitor up to 1.8 V in about 12 s. Especially, when the paper thin triboelectric nanogenerator was installed on a wall or glass window, it can still recycle the environmental noise for electricity generation.

## 9.4 Summary

Vibrational energy harvesting is a traditional and growing research field. Since the first report of TENG for mechanical energy harvesting in 2012, various efforts have been devoted to apply the TENG as a novel approach for ambient vibration energy harvesting as well as self-powered vibration sensing. This chapter presents a number of approaches that have been developed for harvesting vibration energy with TENGs, especially sound wave energy, which is rather unique for TENG it is rather challenging for other techniques. To introduce the triboelectric effect into the field of vibration energy harvesting, a harmonic-resonator-based TENG was firstly reported. Then, a 3-D TENG was developed to harvest energy from both the vertical and lateral directions, and it was also capable of energy harvesting from a broader frequency spectrum. To improve the energy conversion efficiency as well as current output, a three-dimensional (3D) stack integrated multilayered TENG was developed to synchronize the outputs from all the parallel-connected TENG units. In addition, a liquid-metal-based TENG (LM-TENG) was developed for an intimate contact between the metal and the dielectric layer, and an instantaneous energy conversion efficiency up to 70.6 % was demonstrated.

Lastly, TENGs were also demonstrated to harvest energy from acoustic wave, which is also a specific form of vibration. The TENG was firstly demonstrated to harvest ambient acoustic energy by fabricating an organic thin-film based TENG. However, the organic thin-film based TENG has a bulky structure due to the requirement of a resonance cavity, which greatly limited its wide-range application, such as harvesting acoustic energy from a commercial cell phone when playing music or from human talking on the phone. And a further improvement of the acoustic energy harvesting is on the basis of an ultrathin, rollable, and paper-based TENG, which innovatively employs arrays of micro holes for acoustic response enhancement.

## References

1. Z.L. Wang, J. Song, Piezoelectric nanogenerators based on zinc oxide nanowire arrays. Science **312**(5771), 242–246 (2006)
2. Y. Qin, X. Wang, Z.L. Wang, Microfibre-nanowire hybrid structure for energy scavenging. Nature **451**, 809–813 (2008)
3. X. Wang, J. Song, J. Liu, Z.L. Wang, Direct-current nanogenerator driven by ultrasonic waves. Science **316**(5821), 102–105 (2007)
4. S. Xu, Y. Qin, C. Xu, Y. Wei, R. Yang, Z.L. Wang, Self-powered nanowire devices. Nat. Nanotechnol. **5**, 366–373 (2010)
5. R. Yang, Y. Qin, L. Dai, Z.L. Wang, Power generation with laterally packaged piezoelectric fine wires. Nat. Nanotechnol. **4**, 34–39 (2009)
6. X. Bai, Y. Wen, J. Yang, P. Li, J. Qiu, Y. Zhu, A magnetoelectric energy harvester with the magnetic coupling to enhance the output performance. J. Appl. Phys. **111**, 07A938 (2012)
7. P.D. Mitcheson, P. Miao, B.H. Stark, E.M. Yeatman, A.S. Holmes, T.C. Green, MEMS electrostatic micropower generator for low frequency operation. Sens. Actuator A-Phys. **115** (2), 523–529 (2004)
8. W. Lei, F.G. Yuan, Vibration energy harvesting by magnetostrictive material. Smart Mater. Struct. **17**, 045009 (2008)
9. H. Kulah, K. Najafi, Energy scavenging from low-frequency vibrations by using frequency up-conversion for wireless sensor applications. IEEE Sens. J. **8**(3), 261–268 (2008)

10. J. Chen, G. Zhu, W. Yang, Q. Jing, P. Bai, Y. Yang, T.-C. Hou, Z.L. Wang, Harmonicresonator-based triboelectric nanogenerator as a sustainable power source and a self-powered active vibration sensor. Adv. Mater. **25**(42), 6094–6099 (2013)
11. H. Fang, W. Wu, J. Song, Z.L. Wang, Controlled growth of aligned polymer nanowires. J. Phys. Chem. C **113**(38), 16571–16574 (2009)
12. X. Wen, W. Yang, Q. Jing, Z.L. Wang, Harvesting broadband kinetic impact energy from mechanical triggering/vibration and water waves. ACS Nano **8**(7), 7405–7412 (2014)
13. S. Wang, S. Niu, J. Yang, L. Lin, Z.L. Wang, Quantitative measurements of vibration amplitude using a contact-mode freestanding triboelectric nanogenerator. ACS Nano **8**(12), 12004–12013 (2014)
14. J. Yang, J. Chen, Y. Yang, H. Zhang, W. Yang, P. Bai, Y. Su, Z.L. Wang, Broadband vibrational energy harvesting based on a triboelectric nanogenerator. Adv. Energy Mater. **4**(6), 1301322 (2014)
15. W. Yang, J. Chen, Q. Jing, J. Yang, X. Wen, Y. Su, G. Zhu, P. Bai, Z.L. Wang, 3D stack integrated triboelectric nanogenerator for harvesting vibration energy. Adv. Funct. Mater. **24** (26), 4090–4096 (2014)
16. W. Tang, T. Jiang, F.R. Fan, A.F. Yu, C. Zhang, X. Cao, Z.L. Wang, Liquid metal electrode for high performance triboelectric nanogenerator at an instantaneous energy conversion efficiency of 70.6 %. Adv. Funct. Mater. **25**(24), 3718–3725 (2015)
17. J. Yang, J. Chen, Y. Liu, W. Yang, Y. Su, Z.L. Wang, Triboelectrification-based organic film nanogenerator for acoustic energy harvesting and self-powered active acoustic sensing. ACS Nano **8**(3), 2649–2657 (2014)
18. X. Fan, J. Chen, J. Yang, P. Bai, Z. Li, Z.L. Wang, Ultrathin, rollable, paper-based triboelectric nanogenerator for acoustic energy harvesting and self-powered sound recording. ACS Nano **9**(4), 4236–4243 (2015)
19. G. Zhu, J. Chen, T. Zhang, Q. Jing, Z.L. Wang, Radial-arrayed rotary electrification for high performance triboelectric generator. Nat. Commun. **5**, 3426 (2014)
20. G. Zhu, J. Chen, Y. Liu, P. Bai, Y.S. Zhou, Q. Jing, C. Pan, Z.L. Wang, Linear-grating triboelectric generator based on sliding electrification. Nano Lett. **13**(5), 2282–2289 (2013)
21. P. Bai, G. Zhu, Y. Liu, J. Chen, Q. Jing, W. Yang, J. Ma, G. Zhang, Z.L. Wang, Cylindrical rotating triboelectric nanogenerator. ACS Nano **7**(7), 6361–6366 (2013)

# Chapter 10
# Harvesting Wind Energy

Wind energy, as a key mechanical energy offered by nature, has been regarded as one of the most important renewable and green energy sources under the threat of the global warming and energy crisis [1]. The conventional approach of generating electricity from wind relies on the structure of wind turbines and the principle of electromagnetic induction. However, there are several drawbacks of this kind of device, including large size and weight, high cost of installation, difficulty of being driven under low wind speeds, and thus low efficiency, which limits its usability especially for the weak wind in our daily living environment such as around houses and in the city [2]. Recently, piezoelectric windmills employing piezoelectric bimorph transducer structures have been reported with low start-up speed and small sizes [3]. However, the structures are still complex and the outputs are relatively low. In this regard, developing innovative concepts and strategies is of great necessity for extending the applications of low-magnitude wind in our living environment. In this chapter, TENG is applied to harvest wind energy based on rotational structures [4–9] or flutter-driven structures [10–18].

## 10.1 Wind Energy Harvesting Based on Rotational Structure

For the traditional wind energy harvesting through electromagnetic generator, wind turbines or wind cups are used to convert wind blow to rotational mechanical motion. Therefore, the rotational TENG structures were firstly considered for harvesting wind energy [4]. In them, the rotational sliding freestanding mode TENG is widely used for varieties of applications due to its high output performance [6–9]. As reported, there are also other types of TENG applied on the wind energy harvesting [4, 5].



Z.L. Wang et al., *Triboelectric Nanogenerators*,
Green Energy and Technology, DOI 10.1007/978-3-319-40039-6_10

### 10.1.1 Rotational Sliding Freestanding-Triboelectric-Layer Mode Wind-Driven Triboelectric Nanogenerators

Li et al. demonstrated a disk-based rotational sliding freestanding-triboelectric-layer (SFT) mode TENG that uses the wind force to simultaneously harvest wind energy and control the automatic transition between the contact and non-contact states to achieve an unprecedented robustness [6]. Wind energy harvesting is demonstrated through this structure. And through the automatic transition, TENGs can work in the non-contact state for the majority of their working processes to minimize the surface wear, and they can also automatically transit into contact state intermittently for the replenishment of surface charges without interrupting the operation.

In the structural design of this TENG (Fig. 10.1a), there are two major parts: the rotational inner acrylic barrel (~1″ in diameter) that connects with the freestanding rotor of the disk TENG, and the stationary outer barrel (~3″ in diameter) that connects with the stator of the TENG. Two bearings are used to link the two parts and enable the relative rotation. The entire framework is supported by a group of acrylic sheets. The rotor in the TENG is made of a uniformly-distributed-four-segmented acrylic sheet (~3″ in diameter), with its surface laminated with a layer of fluorinated ethylene propylene (FEP) thin film (~50 μm in thickness) of the same shape. In order to further improve the charge density from the tribo-electrification, the contacting surface of FEP film is treated by inductive coupling plasma (ICP) to create nanostructures. The stator is made of an eight evenly-segmented acrylic sheet (~3″ in diameter and the segments are in the same shape and size as the ones on the rotor), with each segment laminated with one piece of aluminum foil (~50 μm in thickness) on the surface, which are connected into two groups of electrode to be the stator in a freestanding disk TENG. The automatic transition between the two states is achieved by creating two opposite forces on the rotor disk. The force pulling it apart from the stator's surface comes from a fan of four rectangular blades (~6 × 2.5 × 0.0625″ in size) that is attached to the rotor through a small cylinder. In order to create this axial force as well as a tangential force to drive the rotation, each blade is tilted for 35°. On the other hand, the force to push the rotor to the stator comes from one tensile spring with the original tensile deformation $\Delta x_0 = 17$ mm inside the inner barrel ($k = 25.375$ N/m as the spring constant), which connects the bottom of the inner barrel with the rotor disk. An acrylic stopping ring surrounding the cylinder is utilized to restrict the maximum vertical separation distance ($\Delta x_{max} = 3$ mm) between the freestanding rotor and the stator (the inset in Fig. 10.1a). There are four feet extending out from the stopping ring to connect the inner barrel. There are two functions of these feet: (1) to drive the inner barrel to spin together with the rotor disk because the spring cannot be twisted; (2) to create a restriction for the separation distance between the rotor and the stator in the non-contact state (the insets in Fig. 10.1a show the amplified images of this part of the TENG in both contact state and maximum non-contact state).

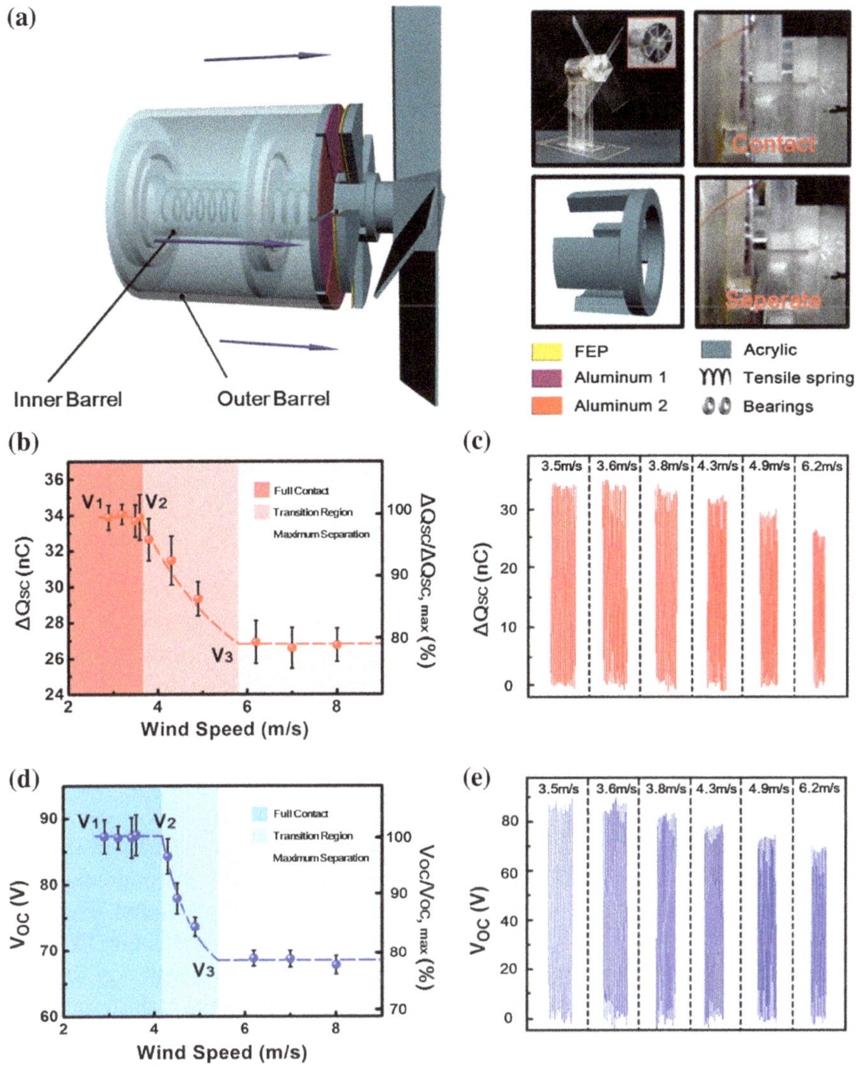


**Fig. 10.1** Device structure and electrical output performance of the wind-driven TENG. **a** Schematic of the basic structure of the wind-driven TENG. The *insets* are the images of the whole structure as well as the disk TENG, the structure of the stopping ring, the amplified images of the TENG in contact state and noncontact state, as well as the figure legend. **b** and **d** $\Delta Q_{SC}$ and $V_{OC}$ of the TENG under different wind speeds, respectively. **c** and **e** Measured profiles of $\Delta Q_{SC}$ and $V_{OC}$ in the state transition region. Reproduced with permission from American Chemical Society [6]

Under the uniform wind (with the speed $v_{wind}$) blowing from the barrel side, the force on the tilted rectangle blades will have two perpendicular components: one along the tangential direction ($F(X_0)$) and one along the axial direction ($F(Y_0)$). They can be expressed by the following two equations:

$$F(X_0) = \frac{1}{2} n \times \rho_{air} \times v_{wind}^2 \times a \times b \times \left(\cos^2\theta \sin\theta - \sin^2\theta \cos\theta\right)^2 \quad (10.1)$$

$$F(Y_0) = \frac{1}{2} n \times \rho_{air} \times v_{wind}^2 \times a \times b \times \left(\cos^3\theta + \sin^3\theta\right)^2 \quad (10.2)$$

where $n$, $a$, $b$ are the number, length and width of the rectangle fan blades, $\rho_{air}$ is the density of the surrounding air, and $\theta$ is the tilted angle ($\sim 35°$) of each blade. The separation between freestanding FEP and electrodes is determined by the equilibrium between the $F(Y_0)$ and the tensile force $f(v_{wind})$ in the spring, which can be expressed as:

$$F(Y_0) = f(v_{wind}) = k(\Delta x_0 + \Delta x) \quad (10.3)$$

where $\Delta x_0$ is the original deformation of the spring for the contact state between rotor and the stator in the beginning, and $\Delta x$ is the extra tensile deformation caused by the vertical separation. According to the conditions of vertical separation between the two disk parts in this TENG, the wind speed can be divided into three regions, with the boundary wind speed expressed as $v_1$, $v_2$ and $v_3$. In the first region (Region I, $v_1 < v_{wind} < v_2$), $F(X_0)$ is larger than the friction force caused by original tensile deformation of the spring. At the same time, $F(Y_0)$ is smaller than the tensile force from the original deformation of the spring. As a result, there is no vertical separation between the two disks and the TENG is working in full contact state. In the second region (Region II, $v_2 < v_{wind} < v_3$), $F(Y_0)$ exceeds the original tensile force from the spring, but smaller than the tensile force under maximum separation distance ($\Delta x_{max}$) created by the stopping ring. As a result, the freestanding FEP disk is pushed to get separated from the electrode disk, with the separation distance smaller than the maximum value ($0 < \Delta x < \Delta x_{max}$). This is the transition region between the contact working state (Region I) and the non-contact working state with the maximum separation (Region III). In the third region (Region III, $v_{wind} > v_3$), $F(Y_0)$ is larger than tensile force under maximum separation, so that the separation reaches at its maximum ($\sim 3$ mm restricted by the stopping ring) and the TENG begins to work in the non-contact state with constant dielectric-electrode separation distance. In Region I, the magnitude of the short-circuit charge transfer ($\Delta Q_{SC}$) should equal to the amount of triboelectric charges on the FEP. When the wind speed keeps increasing in Region II, the vertical dielectric-electrode separation distance gradually gets larger, so that $\Delta Q_{SC}$ will gradually decrease. Then, in Region III, because the separation distance stays constant at its maximum value, $\Delta Q_{SC}$ should remain unchanged in minimum under different wind speeds.

The experimental characterization of the automatic transition in this wind-driven TENG has been performed. Under a series of different wind speeds, 10 sets of short-circuit charge transferred ($\Delta Q_{SC}$) and open-circuit voltage ($V_{OC}$) were measured (Fig. 10.1b–e). As shown in Fig. 10.1b, the relationship between $\Delta Q_{SC}$ and $v_{wind}$ is similar to the theoretical curve predicted. The measured $\Delta Q_{SC}$–$v_{wind}$ curve can be divided into three regions. In the first region with $v_{wind}$ between $v_1$ (2.8 m/s) and $v_2$ (3.6 m/s), the $\Delta Q_{SC}$ remains at about a constant value of 34 nC due to the full contact between the rotor and the stator. In the second region with $v_{wind}$ between $v_2$ and $v_3$ (3.6–5.8 m/s), there is an evident decrease of $\Delta Q_{SC}$ from 34 to 27 nC with the increase of $v_{wind}$. This decrease is due to gradual separation of the two disks. Thus, it is the transition region. The profiles of the measured $\Delta Q_{SC}$ in this region are shown in Fig. 10.1c. Subsequently, when the wind speed increases to $v_3$ (5.8 m/s), $\Delta Q_{SC}$ reaches another constant value of 27 nC, which is because that the separation distance has reached the maximum value. This is the third region. From the full contact state to the non-contact state with maximum separation of the TENG, $\Delta Q_{SC}$ reduces by only ~21 %, which demonstrates a good electric output. As for the open-circuit voltage ($V_{OC}$), it has the following relationship with the $\Delta Q_{SC}$:

$$V_{OC} = \frac{\Delta Q_{SC}}{C} \tag{10.4}$$

where $C$ is the capacitance between the two groups of stationary electrodes, which remains constant during the operation. Therefore, $V_{OC}$ should have a similar response to the wind speed as $\Delta Q_{SC}$, which is shown by the measurement results in Fig. 10.1d. Under $v_{wind}$ from $v_1$ (2.8 m/s) to $v_2$ (4.1 m/s), there is the first constant value of $V_{OC}$ (~88 V), representing the contact state of the TENG. The transition region from full contact to maximum separation appears with $v_{wind}$ of $v_2$ to $v_3$ (4.1–5.4 m/s), in which $V_{OC}$ decreases from ~88 to ~69 V. The profiles of the measured $V_{OC}$ in this region are shown in Fig. 10.1e. Finally there is the second constant $V_{OC}$ of ~69 V when the freestanding rotor contacts the stopping ring, which happens when $v_{wind}$ exceeds $v_3$ (5.4 m/s). It can be found that the decrease of $V_{OC}$ due to state transition is in the same degree with $\Delta Q_{SC}$, and the boundary velocities for the three regions of the $\Delta Q_{SC}$ and the $V_{OC}$ has a good correspondence.

In order to demonstrate the supreme robustness and ultra-long lifetime of the above wind-driven TENG brought by the automatic transition between the non-contact state and the contact state, long-term stability tests for both this wind-driven TENG and a TENG purely working in contact state as the control group have been carried out. For this wind-driven TENG, wind blow was in the speed of 10.8 m/s to ensure the non-contact working state. In reality, the natural wind always has a fluctuating speed, so the entire TENG can get to the contact state occasionally to replenish the triboelectric charges. In order to simulate this behavior in the experiment, the blower to generate wind was stopped after every 10,000 rotation cycles for charge replenishment through triboelectrification in low wind speed. From the SEM images of the FEP on the wind-driving TENG before and

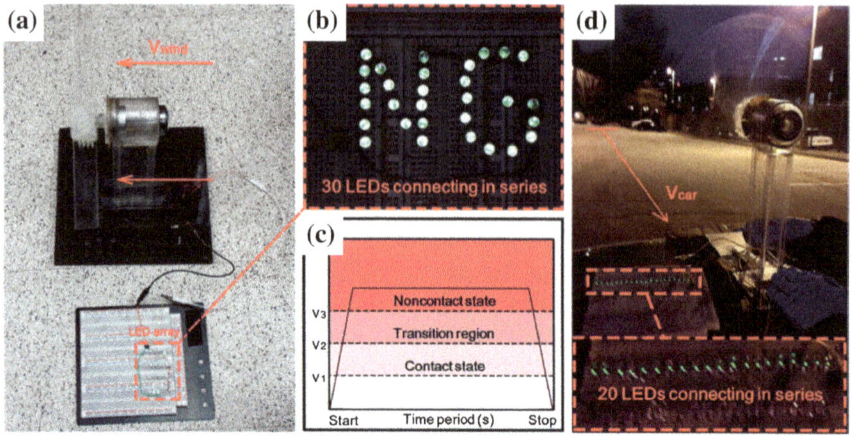


**Fig. 10.2** Applications of the wind-driven TENG to harvest energy from the ambient environment. **a** and **b** Images of the wind-driven-TENG to light up 30 LEDs connected in series with wind at a speed of 10.8 m/s. **c** Simplified *v-t* curve of a car (from *start* to *stop*). **d** Image of the wind-driven TENG to harvest wind energy generated by a moving car ($v_{car}$ = 10 m/s) to light up 20 LEDs connected in series. Reproduced with permission from American Chemical Society [6]

after 120,000 rotation cycles, the surface nanostructures are preserved in a good condition, showing the minimized surface wear even after the extended operation period. After 120,000 rotation cycles, nearly 95 % of the maximum (original) $\Delta Q_{SC}$ is preserved in the wind-driving TENG. As a comparison, a disk-based TENG operated in the contact state was assembled with the same size as the wind-driven TENG. From the SEM image, the nanostructures on the FEP have been almost completely wiped out due to constant wear after only 24,000 cycles. Moreover, the $\Delta Q_{SC}$ is only 60 % of the original value. Thus, the automatic transition between the contact state and the non-contact state leads to significant improvement of the TENG's robustness and lifetime.

Due to its outstanding long-term stability, the application of the wind-driving TENG can be largely expanded (Fig. 10.2). As demonstrated in Fig. 10.2a, b, the electricity generated by this TENG in the non-contact state under a wind speed of 10.8 m/s is enough to light up 30 LEDs in series without surface wear. The moving objects (such as vehicles) are important resources for generating wind with various wind speed depending on the moving speed of the object. For example, the motion of a car can be simplified into three different stages (Fig. 10.2c): acceleration, uniform motion and deceleration. When the car is just started or nearly stopped, there will be charge replenishment in the contact state in low speed. While the car is running in a high speed, the TENG will work in the non-contact state without surface wear. In this way, automatic transition between the two states will be realized, which will largely enhance the long-term stability of the wind-driving TENG. As shown in Fig. 10.2d, 20 LEDs in series has been turned on by this TENG on a moving car in the speed of 10 m/s.

In summary, a SFT mode wind-driven TENG with ability of automatic transition between contact state and non-contact state has been demonstrated. The TENG has shown promising output performance with extraordinary long-term stability due to automatic transition. This TENG has been demonstrated to be an ideal energy harvester for wind energy.

### *10.1.2 Other Rotational Structures for Wind Energy Harvesting*

Xie et al. developed a rotary structured triboelectric nanogenerator (R-TENG) for scavenging weak wind energy [4]. Under the wind flow, the wind-cup structure will be driven to rotate and thus the soft and flexible polytetrafluoroethylene (PTFE) film based rotor blade will sweep across the Al sheets based stators consecutively, so that a repeating process of contact-sliding-separation-contact between the two charged surfaces can be achieved by hybridizing the two modes. On the basis of this new design, an open-circuit voltage ($V_{OC}$) of 250 V and short-circuit current ($I_{SC}$) of 0.25 mA have been reached, corresponding to a maximum power output of 62.5 mW, which is capable of either driving hundreds of electronic devices (such as commercial LEDs) instantaneously or efficiently charging energy storage units. This work is the first one to use TENG technology to effectively harvest wind energy, which is an important progress in the practical applications of nanogenerators and also exhibits enormous potential as the future wind power technology.

The structure of the R-TENG springs from the conventional wind cup structure, which includes a framework, a shaft, a flexible rotor blade, two stators, as illustrated in Fig. 10.3a. The framework for supporting the entire structure consists of two Acrylic rectangles that are joined together perpendicularly at the common central axes, where the metallic shaft rod is positioned through two bearings. A flexible and soft rotor blade structurally made of a polyester (PET) film ($\sim$125 μm in thickness) is connected to the shaft, and thus can rotate with the shaft. Around the circumference of the device, two Al covered plates as the symmetric stators stretch out from the framework toward the shaft direction. On top of the shaft, the wind cup setup is mounted to convert the wind flow into the rotation of the shaft and the flexible rotor, during which the rotor blade will periodically sweep across the stators with small resistance. In this process, a consecutive face-to-face contact (in area of 2.5 cm × 6.4 cm) and separation between the rotor and the stators will be enabled, which can serve as the basic TENG-based process for generating electricity. According to the triboelectric series, a PTFE film was chosen to adhere at the end of the PET blade as a triboelectric layer to get into contact with Al that acts not only as the counter triboelectric layer, but also as an electrode. For the other electrode, a layer of metal film was deposited between the PTFE and PET to enable the transfer of induced charges. To enhance the surface roughness and therefore improve the triboelectric charge density, the PTFE film was dry-etched through inductive

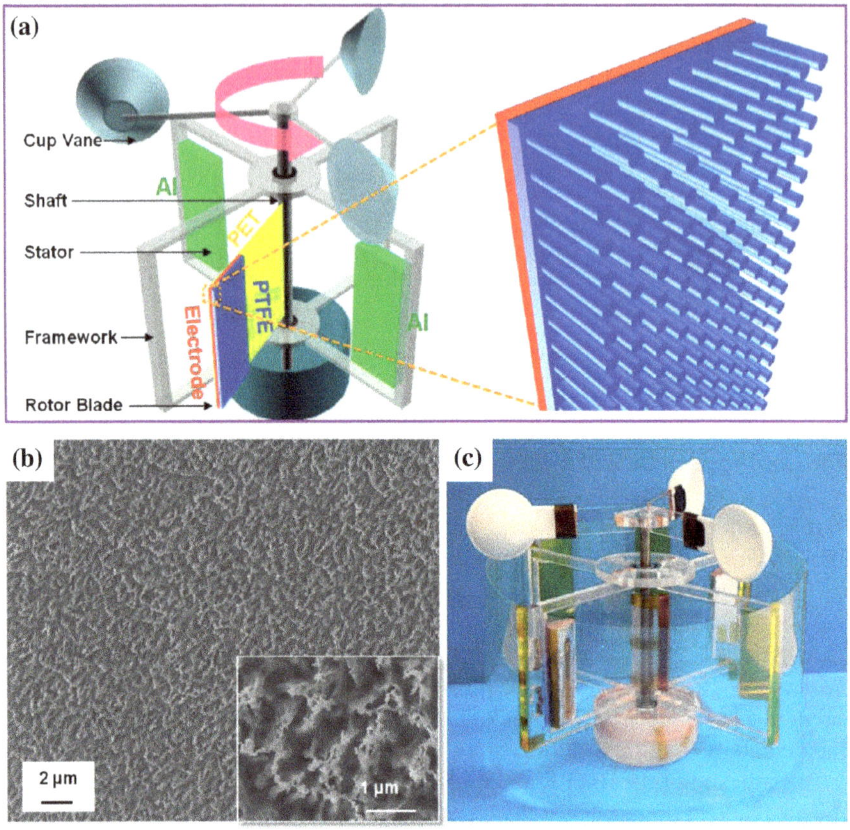


**Fig. 10.3** Device structure of the rotary triboelectric nanogenerator (R-TENG). **a** The schematic diagram showing the structural design of the R-TENG, with the enlarged picture showing the nanowire-like structures on the surface of PTFE. **b** The SEM image of the PTFE surface with etched nanowire-like structures; the *inset* is an SEM image of the nanowires. **c** A photograph of the fabricated R-TENG. Reproduced with permission from American Chemical Society [4]

coupled plasma (ICP) reactive ion etching to prepare nanowire-like structures on the surface (with the magnified scheme in Fig. 10.3a). As shown in the scanning electron microscopy (SEM) images, after a 40 s etching, the nanowire-like structures were uniformly distributed on the surface of PTFE, with an average length of ~700 nm (Fig. 10.3b). The photograph of an as-fabricated R-TENG device is shown in Fig. 10.3c. This device is light in weight, compact in size, cheap in cost, and robust in operation.

The electricity generation of the R-TENG is a hybridization of contact-sliding-separation-contact processes. In the original state where the rotor blade is stationary and the triboelectric layers are separated from each other, there is no tribo-charges generated on the surfaces. When the wind blows on the wind cups, the rotor blade will be driven to rotate around the shaft, which will bring the PTFE film into full contact with the Al on either one of the stators. Owing to the different tendencies to gain or lose electrons, the triboelectric effect will enable the generation of surface charges at the contact area due to relative sliding, leaving the PTFE with net negative charges and the Al with positive charges. At this very moment, because the opposite charges on the contact surfaces are in equal densities and locate at the same plane, there is little electric potential difference generated in the space. As the polymeric rotor blade continues to rotate, the flexible blade will be bent in order to sweep across the rigid stator. Because of the strong electrostatic attraction between the two tribo-charged surfaces, the PTFE film has the tendency to keep the intimate contact with the Al stator, so that most of the bending happens on the PET portion of the rotor blade. The PTFE plate will be guided to slide outward across the Al surface, leading to a continuous decrease in the overlapping area of the two tribo-charged surfaces and thus the in-plane charge separation. The lateral dipole moment in parallel to the sliding surface will generate a higher potential on the Al surface, thus drives a current flow in the external load from the Al to the electrode of PTFE to offset the tribo-charge-induced potential. This process will last until the PTFE fully slides out of the Al surface, and the total transferred charges will equal the amount of the triboelectric charges on each surface. This is the working mechanism of the TENG in the sliding mode, which can help to generate higher triboelectric charge densities because the sliding provides much more friction than the contact mode, and it is more effective for triboelectrification. Subsequently, the rotor blade will continue to rotate toward the other stator, with the attached electrode having positive charges with equal density of the negative tribo-charges on the PTFE surfaces. When the rotor blade approaches the other Al blade, the two surfaces will get into contact momentarily in vertical direction. An electric potential difference pointing from the electrode on the PTFE to the Al sheet will be generated, driving a reversal current flow in order to reach the electrostatic equilibrium where all of the positive charges on the PTFE electrode have transferred back to the Al stator. This is the working mechanism of the TENG in the contact-separation mode. At this point, a cycle is completed. Thus, the entire electricity generation process of the R-TENG hybridizes the in-plane contact-sliding mode and the separation-contact mode in vertical direction as a full cycle process and generates a pair of alternating currents: one lower-magnitude but wider peak corresponding to the in-plane charge separation and one sharper but narrower peak corresponding to the vertical charge recontact.

The electrical output measurement of the R-TENG was carried out under a wind speed of $\sim 15$ m/s. In the process of separation-contacting-sliding-separation, the open-circuit voltage ($V_{OC}$) jumped from zero to a peak value of $\sim 250$ V and to zero again (Fig. 10.4a), with the positive probe connecting to the electrode of the PTFE film. The transferred charges ($\Delta Q$) driven by this potential difference were also measured, as illustrated in Fig. 10.4b, and the charges with a maximum amount of 140 nC flew back and forth between the two electrodes when the two tribo-charged surfaces were contacted and separated with each other. Consequently, the transfer of the charges produced an alternating-current (AC) output with a peak short-circuit current ($I_{SC}$) of 250 μA corresponding to the vertical contact process

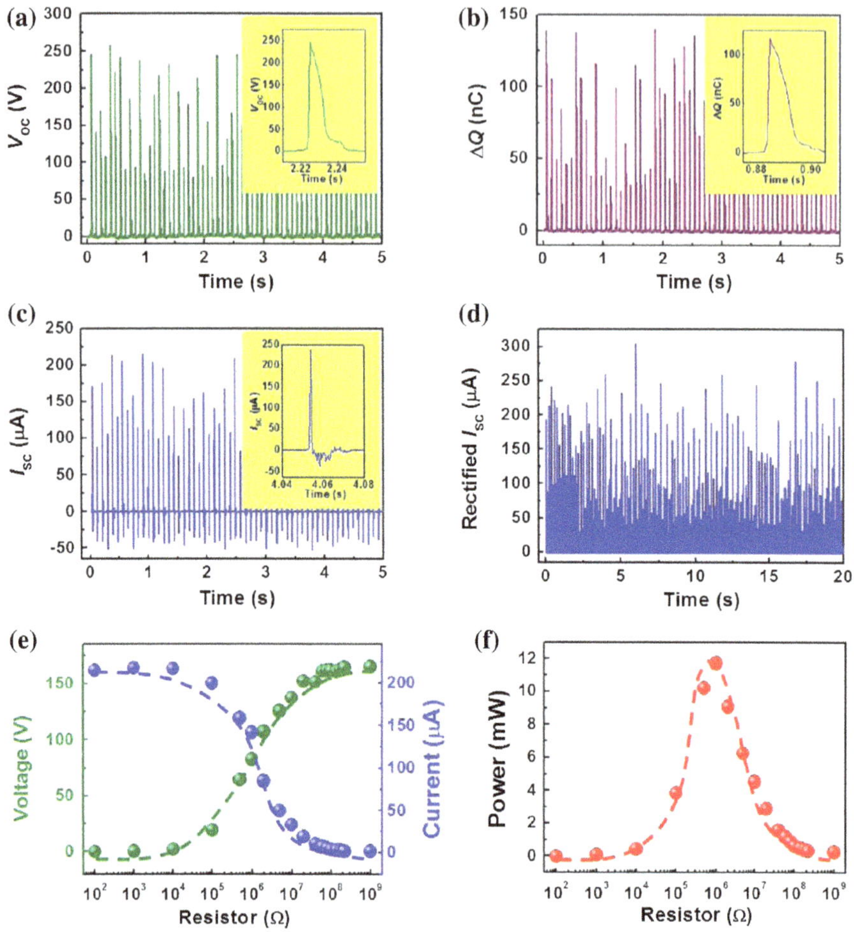


**Fig. 10.4** Performance of the R-TENG driven by the wind flow. **a** The open-circuit voltage ($V_{OC}$), **b** the transferred charges ($\Delta Q$), **c** the $I_{SC}$, and **d** the rectified $I_{SC}$ under the wind speed of 15 m/s (in the scale of 7 BF). **e** and **f** The dependence of (**e**) the output voltage (*green*) and current (*blue*) and (**f**) the power (*red*) on the resistance of the external load. Reproduced with permission from American Chemical Society [4]

(Fig. 10.4c). The amount of the transferred charges can also be obtained through the integration of each current peak over time. The total charges $\Delta Q$ generated from the sharp peak corresponding to the vertical-contact process equal those from the wider peak from sliding-separation process. From the enlarged profiles of the $V_{OC}$, $\Delta Q$ and $I_{SC}$ shown in the insets of Fig. 10.4a–c, the differences in the characteristics of the contacting mode and the sliding mode are clearly reflected: the steep increases of the $V_{OC}$ and $\Delta Q$ with a sharp and higher $I_{SC}$ peak for the vertical contact; the gradual decreases of the $V_{OC}$ and $\Delta Q$ with a wider and lower $I_{SC}$ peak for the sliding separation. This AC output can be rectified to a DC output by a full-way rectifying bridge, as shown in Fig. 10.4d. After the rectification, the current output retained the same magnitude. Besides the $V_{OC}$ and $I_{SC}$, the relationship between the effective output power and the resistance of external load is another important characteristic of the R-TENG. The voltage and current outputs on a series of different resistors have been experimentally studied. As depicted in Fig. 10.4e, the current drops with the increase of the external resistance, while the voltage across the load follows a reversed tendency. Consequently, the instantaneous power on the load reaches a maximum value of ~12 mW when driving the load with a resistance of ~1 MΩ (Fig. 10.4f).

In summary, an innovative type of triboelectric nanogenerator based on the traditional vertical wind-cup structure is demonstrated for effectively harvesting wind energy, especially weak-wind available outdoors. By hybridizing the two basic modes for the first time, the R-TENG generated an open-circuit voltage of 250 V and a short-circuit current of 0.25 mA with a maximum power of 62.5 mW at the wind speed of 15 m/s, which is capable to either light up hundreds of commercial LEDs directly or efficiently charge energy storage units. It is the first time that triboelectric nanogenerators are combined with traditional wind power technology, which shows a number of advantages over the electromagnetic induction based technology.

## 10.2 Wind Energy Harvesting Based on Flutter-Driven Triboelectrification

A non-rotational TENG structure, which is based on flutter-driven triboelectrification, is firstly reported by Yang et al. [10] and has been rapidly developed to be an effective wind energy harvester [11–18]. This structure is based on the contact-separation between the flutter and the electrodes on the both sides, which does not rely on the wind cups or wind turbines. Moreover, this structure is able to recognize the direction of the wind, which can be used as self-powered wind-vector sensor system [10]. These research open a novel strategy to effective harvest wind energy.

### *10.2.1 The First Flutter-Driven TENG for Wind Energy Harvesting*

The first work of this type TENG for wind energy harvesting is reported by Yang et al. [10] The as-fabricated wind energy harvester is based on periodic contact/separation between two aluminum (Al) foils and a fluorinated ethylene–propylene (FEP) film with a surface of nanowire structures. By utilizing the coupling between the triboelectric effect and the electrostatic effect, the periodic change of distance between the two Al foils and the FEP film can induce charge transfer between the Al electrode and the ground, resulting in the flow of free electrons in the external circuit as an alternating current. The integrated TENGs with a size of 2.5 cm × 2.5 cm × 22 cm deliver an output voltage of up to 100 V and an output current of 1.6 μA, which can be used to directly drive tens of commercial light-emitting diodes (LEDs). This work is important progress toward the practical applications of wind energy harvesting and the related self-powered sensor systems.

The fabricated TENG consists of two Al foils and a FEP film in a cuboid acrylic tube, as schematically illustrated in Fig. 10.5a. The Al foil plays dual roles as a triboelectric surface and as an electrode. One side of the FEP film was fixed at the middle of the end surface of the tube, leaving the other side free-standing. The distance between the two Al foils and the FEP film can be changed periodically due to the wind-induced vibration of the FEP film, resulting in an output voltage/current across an external load. Figure 10.5b, c are the side view and front view of a typical as-fabricated TENG, respectively, indicating that the device has the dimensions 2.5 cm × 2.5 cm × 5.5 cm. The device electrode E1 or E2 is connected through an external resistance of 100 MΩ to the ground. To increase the triboelectric charge density, FEP nanowires are created on the surface of the FEP film via an inductively coupled plasma method, as shown in Fig. 10.5d.

To investigate the TENG's performance for harvesting wind energy, a compressive air injector was employed to stimulate a wind source with uniform and tunable flow speed. The measured output voltage and current of the fabricated TENG under a wind speed of about 10 m/s are measured, where the voltage and current are about 5 V and 0.5 μA under a loading resistance of 100 MΩ, respectively. Figure 10.5e presents the rectified output voltage and current of the TENG. The rectified output current of the TENG is still about 0.5 μA, while the output voltage can be increased to about 20 V, which is associated with the parasitic capacitance in the bridge rectification circuit under the high working frequency and the large loading resistance. Due to the charging effect of the parasitic capacitance, the output voltage of the TENG can be increased in this study. To increase the output power of the device, four TENGs were integrated in one acrylic tube. The top and bottom Al electrodes were connected with two bridge rectification circuits, respectively. Then the two bridge rectification circuits were connected in parallel. The measured output voltage and current of the integrated TENGs can be increased to 100 V and 1.6 μA, respectively, corresponding to an output power of 0.16 mW. The increase of the total output is due to the increase of the current flow through the external loading resistor. The output signals of the TENGs were rectified and then were connected in parallel. The total output performance can be enhanced further by connecting more TENG units. To demonstrate that the produced energy can be utilized as an effective power source, one TENG with a size of 2.5 cm × 2.5 cm × 5.5 cm was fixed on a car. The device was then connected with a bridge rectification circuit and a 9-LED flashlight. The photograph in Fig. 10.5g indicates that the TENG can harvest the wind energy induced by the movement of a car and light up the flashlight. Figure 10.5h illustrates that the produced energy of the four integrated TENGs can be used to directly drive 40 green LEDs.

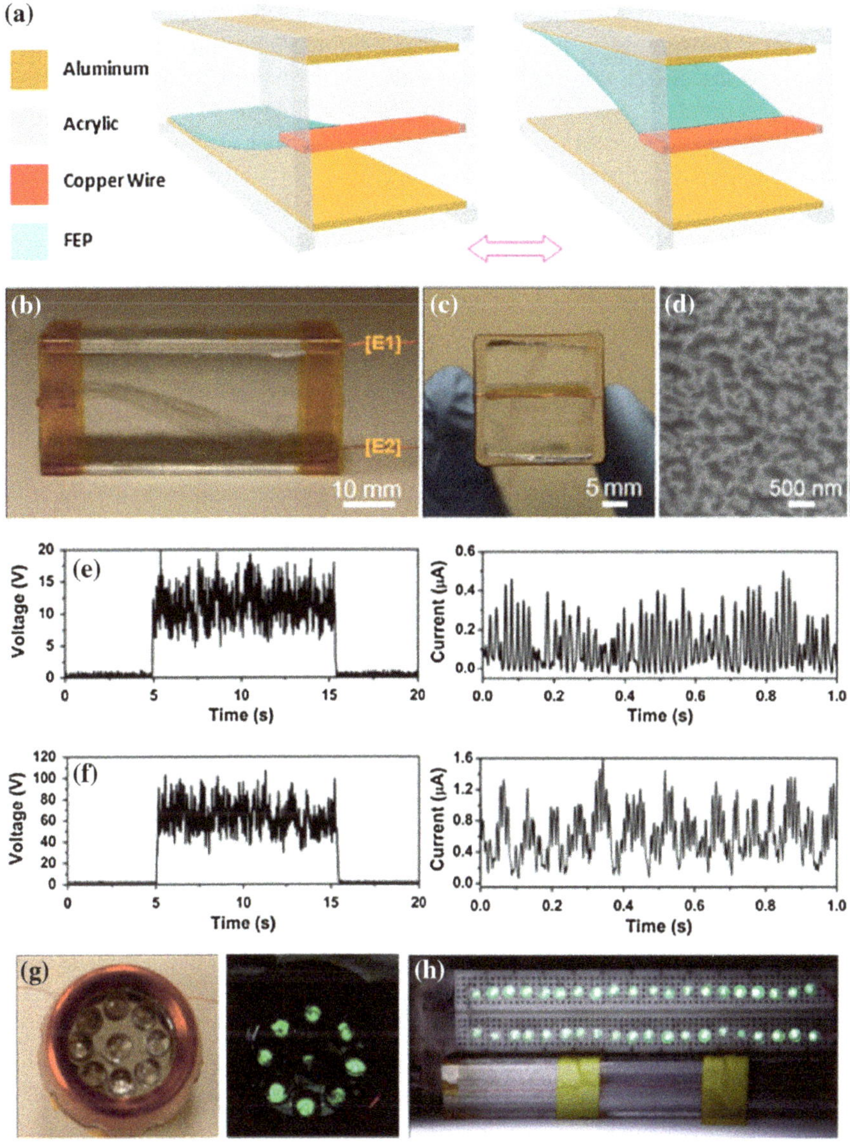


**Fig. 10.5** **a** Schematic diagram of the TENG. **b** and **c** Photographs of the TENG. **d** SEM image of the etched FEP film surface. **e** and **f** Rectified output voltage and current of one (**e**) and four (**f**) TENGs under a loading resistance of 100 MΩ. **g** Photographs of a 9-LED flashlight before and after being driven. **h** Photograph of the 40 green LEDs driven by the integrated TENGs. Reproduced with permission from American Chemical Society [10]

A cycle of electricity generation of the fabricated TENG is stated below. At the initial state, the surfaces of the bottom Al foil and the FEP film fully contact each other, inducing charge transfer between them. According to the triboelectric series, the electrons are injected from Al to FEP, resulting in positive and negative triboelectric charges on the Al and FEP surfaces, respectively. The produced triboelectric charges with opposite polarities are fully balanced; no electron flow occurs in the external circuit. Once a wind flow separates the FEP film from the bottom Al, these triboelectric charges cannot be compensated. The negative ones on the FEP film will drive free electrons to flow through the external circuits, where the movement directions of electrons are opposite at the top and bottom Al electrodes. This electrostatic induction process can give the output voltage/current signals until the negative triboelectric charges on the FEP film are fully screened from the induced positive charges as well as the contact-induced triboelectric positive charges. No output voltage/current signals can be observed at this state. When the FEP film was moved back from the top Al electrode to the bottom Al electrode, the negative triboelectric charges on the FEP film can induce the electrons to flow from the ground to the top Al electrode due to the increase of the relative separation between the top Al electrode and the FEP film. Moreover, the electrons can be driven to flow from the bottom Al electrode to the ground due to the decrease of the distance between the FEP and the bottom Al electrode. When the FEP film is in contact with the bottom Al electrode, the negative triboelectric charges on the two surfaces of the FEP film are fully screened by the induced positive charges, resulting in no electron flow in the external circuit. When the FEP film was moved again from the bottom Al electrode to the top Al electrode, the opposite voltage/current signals can be observed in the external circuit. Therefore, when the wind flow vibrates the FEP film up and down, the TENG acts as an electron pump that drives electrons back and forth between the Al electrodes and the ground, producing alternating current in the external circuit. Two circuits were used to connect the top and bottom Al electrodes, respectively. When the FEP film moved between the two Al electrodes, the output signals at both the top and bottom electrodes can be obtained by using the two circuits. The output signals were then rectified and connected in parallel, producing the largest electric output.

Meanwhile, due to the diversity of contact electrification, an additional double-electrode-based TENG has also been developed for wind energy harvesting in this study. The working mechanism of the TENG is based on the contact/separation between two triboelectric material layers to drive the electrostatic induction charge flow in the external circuit. Figure 10.6a presents a schematic diagram of the TENG, which consists of a copper electrode, a polytetrafluoroethylene (PTFE) film, and an Al electrode on a Kapton film as the wind-sensitive unit. As illustrated in Fig. 10.6b, the photograph of the TENG indicates that the device has the dimensions 2.5 cm × 0.6 cm × 5 cm. The Al, Kapton, and the bottom acrylic can also form a single-electrode-based TENG. However, the short distance (<5 mm) between the Kapton film and the bottom acrylic can result in a much smaller output performance than that of the top double-electrode-based TENG. Thus the bottom single-electrode-based TENG for

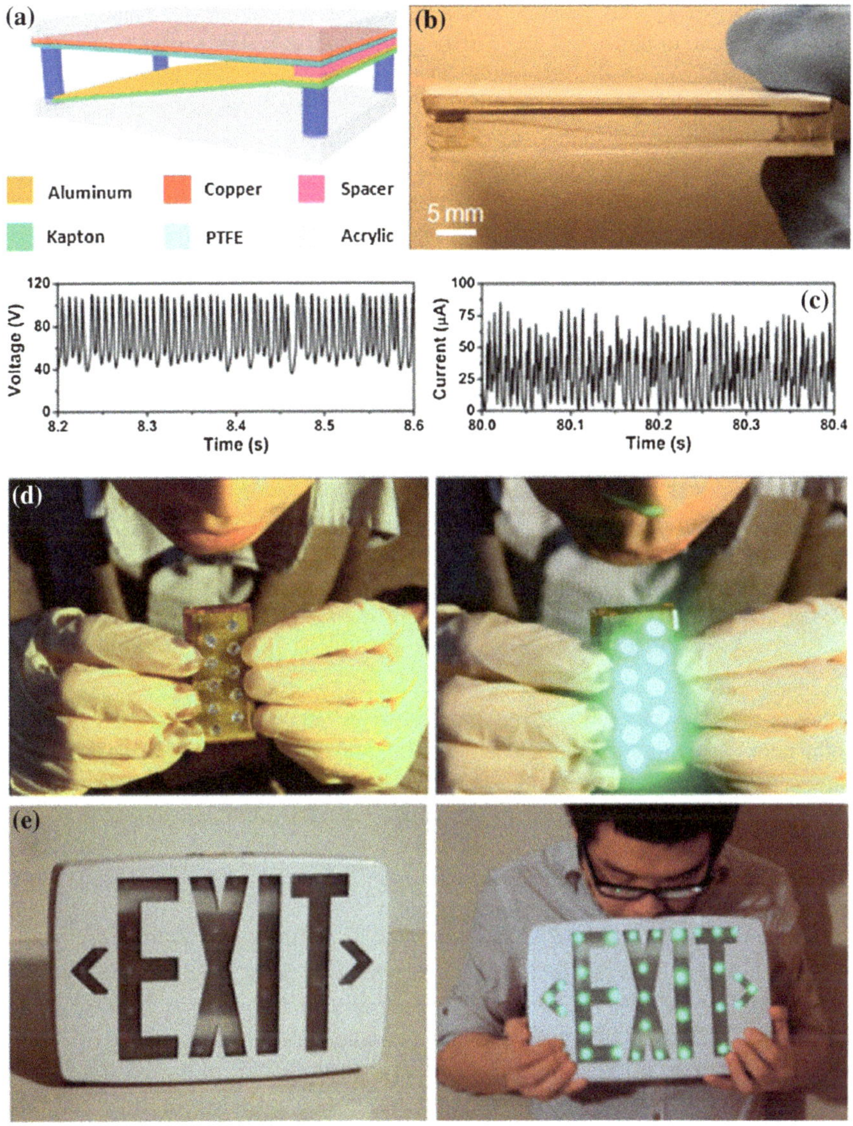


**Fig. 10.6** **a** Schematic diagram of the designed double-electrode-based TENG. **b** Photograph of the TENG. **c** Measured open-circuit voltage and the short-circuit current of the TENG under a wind speed of 10 m/s. **d** and **e** Photographs show that the TENG was used to harvest the human mouth blowing induced wind energy to drive 10 LEDs (**d**) and an exit sign (**e**). Reproduced with permission from American Chemical Society [10]

this case is not considered. The TENG has an open-circuit voltage up to 110 V under a wind speed of about 10 m/s. As depicted in Fig. 10.6c, the enlarged voltage–time curve reveals that the output voltage changes between 40 and 110 V, which is associated with the high working frequency of the device. The measured largest short-circuit current of the TENG is about 80 μA. The produced electricity energy can be used to directly drive 10 LEDs by scavenging the human mouth blowing induced wind energy, as displayed in Fig. 10.6d. An exit sign can also be driven by the produced energy, as shown in Fig. 10.6e.

In summary, a TENG, consisting of an FEP film and two Al foils connected with a ground through an external load has been demonstrated. The mechanism of the TENG is based on the charge transfer between the Al electrode and the ground by utilizing the wind-induced vibration of the FEP film between the two Al electrodes. The integrated TENGs produce an output voltage up to 100 V, an output current of 1.6 μA, and an output power of 0.16 mW under a load of 100 MΩ, which can be used to light up tens of commercial LEDs. An additional double-electrode-based TENG is also demonstrated to harvest wind energy from a human blowing, and the produced electricity is enough to power an exit sign. This invention may push forward a significant step toward the practical applications of wind energy harvesting techniques and self-powered sensor systems.

### *10.2.2 Elasto-Aerodynamics-Driven Triboelectric Nanogenerator*

Wang et al. report an elasto-aerodynamics-driven TENG characterized by underlining aero-elastic flutter phenomena [12]. Rational design of the TENG achieves an unprecedented power density of 9 $kW/m^3$. The modular design of the reported TENG enables the good output power capability. The ultrahigh output power of TENG allows for direct powering of a body temperature sensor by simply harvesting human breath-induced air-flow energy. This research marks a step toward self-powered healthcare monitoring systems.

Figure 10.7a illustrates the schematic diagram of the fabricated TENG, which includes a Kapton film with two Cu electrodes at both sides and two polytetrafluorothylene (PTFE) films with Cu electrodes at the top and bottom acrylic substrates. The Kapton film was fixed at the middle of the device by using two bolts, where two narrow air gaps were created between the PTFE films and the Kapton film. Figure 10.7b, c display photographs of the TENG with the different view angles, where the device has the inner dimensions of $100 \times 10 \times 1.6$ $mm^3$. The air flows into the gaps of the device and induces the vibration of the Kapton film, which can drive the working of the TENG. As depicted in Fig. 10.7d, the dynamic process of the Kapton film under the wind speed of about 15 m/s includes three typical states. A periodic contact and separation between the Cu electrode on the Kapton film and the PTFE film under the air-flow condition can be clearly observed by using a high speed camera.

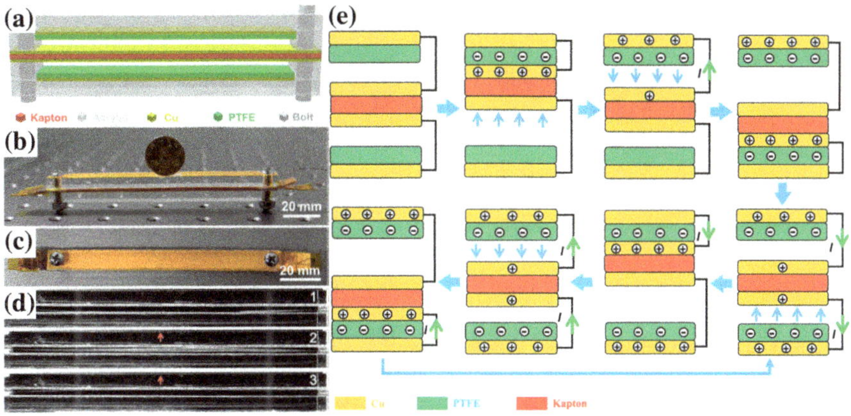


**Fig. 10.7** **a** Schematic diagram of the fabricated TENG. **b** The cross-sectional view of the fabricated TENG. **c** Top view of the fabricated TENG. **d** Photographs of TENG in motion captured by using a high-speed camera. **e** Schematic diagrams of the electricity generation process of TENG. Reproduced with permission from American Chemical Society [12]

The working of the TENG is based on a periodic contact and separation between the Cu electrode on the Kapton film and the PTFE film, where the coupling between triboelectrification and electrostatic induction results in alternating flow of electrons between electrodes. As illustrated in Fig. 10.7e, the electricity generation process includes several different states. At the initial state, the Cu electrodes on the Kapton film and the PTFE films are separated by the air gaps, where there is no current/voltage output for the TENG. When the Cu electrode on the Kapton film vibrates up to get in touch with the top PTFE film under the air-flow through the TENG, the negative and positive triboelectric charges can be created on the surfaces of PTFE film and Cu electrode, respectively. When the Kapton film vibrates down to approach the bottom PTFE film, the triboelectric charges can keep on the surface of top PTFE film, while the charges on the top two Cu electrodes can flow when there is an external circuit due to the electrostatic induction of the charged PTFE film. In the vibration process of the Kapton film, both the top and bottom PTFE films have the negative triboelectric charges by rubbing the Cu electrodes. Under the strong electrostatic attraction, both the top and bottom TENGs can deliver the current/voltage signals under the air-flow condition.

Output voltage and current of TENGs in Fig. 10.8 were measured under a wind speed of about 15 m/s. As shown in Fig. 10.8a, b, the top TENG delivers an alternating output signals with a short-circuit current of about 80 μA and an output voltage of about 240 V under a loading resistance of 10 MΩ. The bottom TENG produces a short-circuit current of about 60 μA and an output voltage of 220 V, as displayed in Fig. 10.8c, d. Figure 10.8e presents the measured output voltage under the different loading resistance, exhibiting a gradual increase with increasing the loading resistance, where the corresponding largest power density of the top TENG is about 3 kW/m$^3$ under the loading resistance of 2.3 MΩ. Figure 10.8f

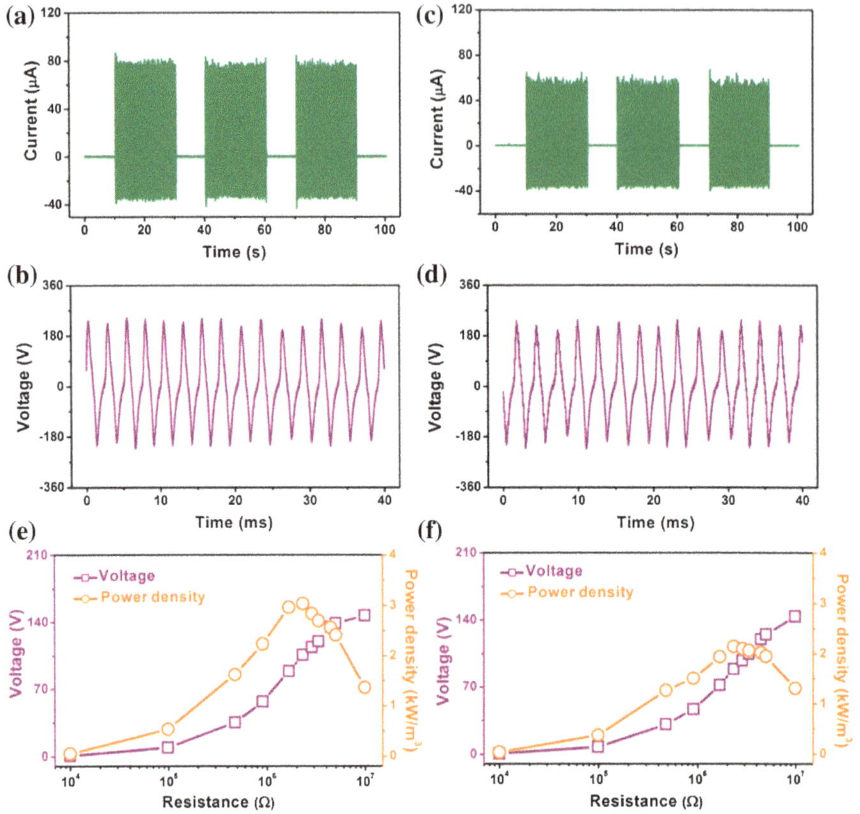


**Fig. 10.8** **a** Measured short-circuit current of TENG 1. **b** Measured output voltage of the TENG 1 under a loading resistance of 10 MΩ. **c** Short-circuit current of the TENG 2. **d** Measured output voltage of TENG 2 under a loading resistance of 10 MΩ. **e** Dependence of the output voltage and the power density for TENG 1 on the external loading resistance. **f** Dependence of the output voltage and the power density for TENG 2 on the external loading resistance. Reproduced with permission from American Chemical Society [12]

displays a similar change in the measured output voltage of the bottom TENG with increasing loading resistance, where the bottom TENG has the largest power density of about 2 kW/m$^3$. The difference of the two TENGs in one device is associated with the minor unevenness of the flexible Kapton film that can result in the different distances between the Kapton film and the top/bottom PTFE film in the vibration process.

As depicted in Fig. 10.9a–c, the maximum deformation of the Kapton film at different inlet wind speeds was simulated, indicating that the Kapton film is mainly working at bending mode. Moreover, the corresponding pressure and the wind flow velocity distributions in the TENG under the wind speed of 15 m/s are illustrated in Fig. 10.9, panels d and e, respectively. The film deformation values are similar when the wind speed exceeds 10 m/s, suggesting that the Kapton film works at a

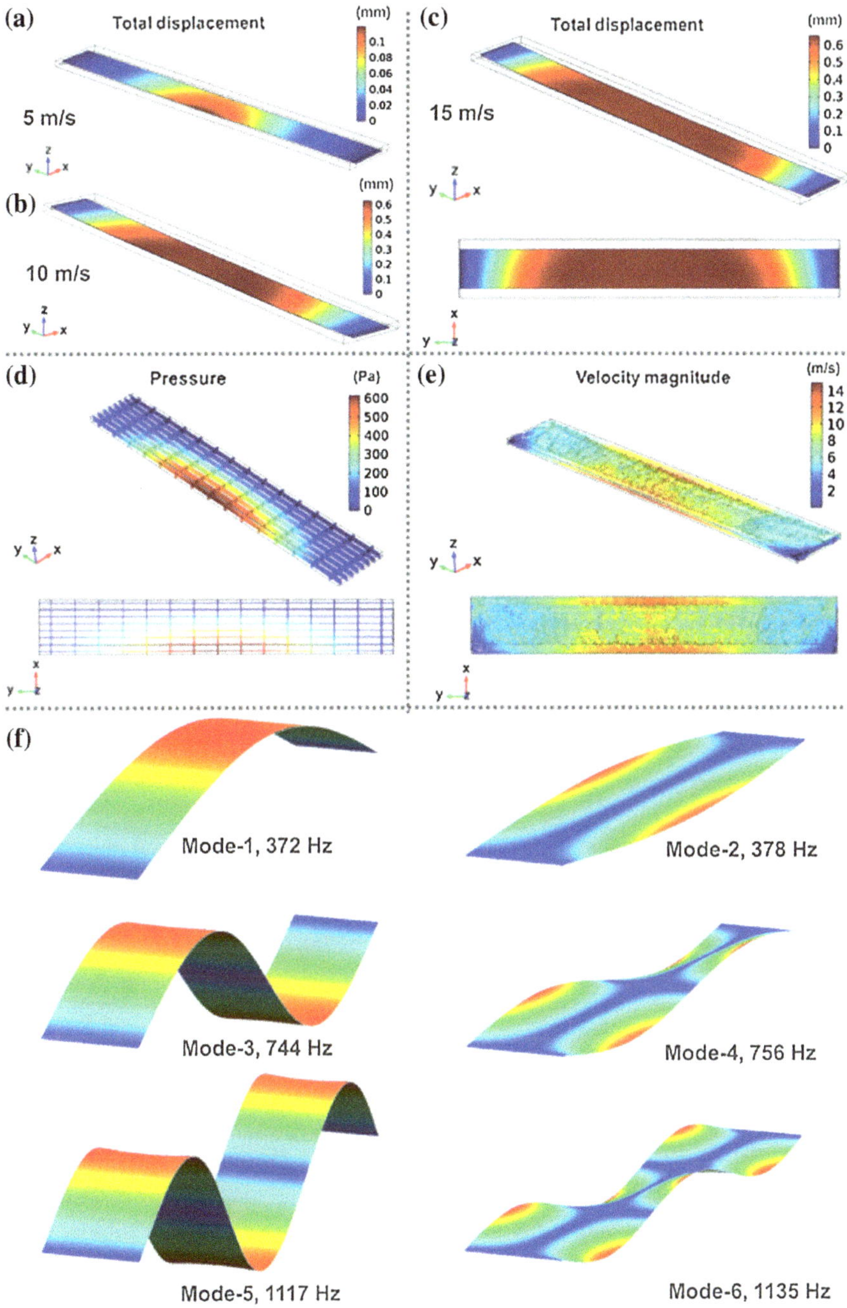
(a)
Total displacement
(mm)
0.1
0.08
0.06
0.04
0.02
0
5 m/s
(b)
(mm)
0.6
0.5
0.4
0.3
0.2
0.1
0
10 m/s
(c)
Total displacement
(mm)
0.6
0.5
0.4
0.3
0.2
0.1
0
15 m/s
(d)
Pressure
(Pa)
600
500
400
300
200
100
0
(e)
Velocity magnitude
(m/s)
14
12
10
8
6
4
2
(f)
Mode-1, 372 Hz
Mode-2, 378 Hz
Mode-3, 744 Hz
Mode-4, 756 Hz
Mode-5, 1117 Hz
Mode-6, 1135 Hz

◀ **Fig. 10.9** **a–c** Total displacement distributions of the Kapton film at the air-flow rate of 5 m/s (**a**), 10 m/s (**b**), and 15 m/s (**c**). **d** Pressure distribution of the Kapton film. **e** Velocity magnitude distribution of the Kapton film. **f** Different vibration modes of the Kapton film. Reproduced with permission from American Chemical Society [12]

stable state (up and down bending vibration) once the inlet wind speed approaches a specific value. On the other hands, a more interesting phenomenon can be observed that at the low inlet wind speed (5 m/s), the leading edge area of the thin film deforms larger than that of the trailing edge, resulting in the twist (torsion) motion of the film but not the bending motion. The deformation in this condition is 5 times smaller than that of the stable state at large wind speed. Normally, such small amplitude twist motion contributes less to the electric charge generation, whereas it produces attack angle of wind, which is the requirement condition of flutter occurrence. Once the inlet wind speed approaches the critical wind velocity of the Kapton thin film, the damping ratio of the twist motion will be decreased to zero and thus start the fluttering of the Kapton film. The flutter phenomenon enables the vibrated film extracting energy from the free wind flow when there is phase difference between film structure instantaneous aerodynamic force and elastic deformation. It will not occur at relative low air flow speed, while the large constant amplitude vibration occurs when it reaches critical wind velocity. This critical wind velocity can be determined by the frequency ratio of first order bending and twist mode and the respective damping ratio. Since the flutter related vibration is periodical and neutrally stable at the critical wind speed or above, effective contact between the Kapton thin film and the top/bottom device surface will maintain in a huge wind speed range. The calculation results are completely consistent with the actual deformations in Fig. 10.7d.

Flutter effect means that the air-flow induces the structure vibrating at its natural frequency. The natural frequencies and the mode shapes of first six order vibration mode were simulated by Coventorware. As illustrated in Fig. 10.9f, the first and second order mode shape are bending and twisting (torsion) at the frequency of 372 and 378 Hz, respectively, where the two vibration modes are consistent with the actual vibrations of the Kapton film that were observed by using a high-speed camera. The necessary condition of the occurrence of the flutter is that the frequency of the same order twist mode is closely above that of the bending mode and the wind flow speed should approach or above the value of the critical wind velocity as well. It is clearly observed that the first two mode are in the range of 370–380 Hz, which aligns well with experimentally results in Fig. 10.8c, d.

To demonstrate the capability of the TENG as a power source, tens of TENGs in parallel using the rectification circuits were connected to a globe light without using a Li-on battery. Under a wind speed of 15 m/s, the white globe light can be directly driven by the TENGs (Fig. 10.10a). Moreover, the TENGs-driven globe light can provide sufficient light intensity of about 43 lx for reading printed text in complete darkness, as presented in Fig. 10.10b.

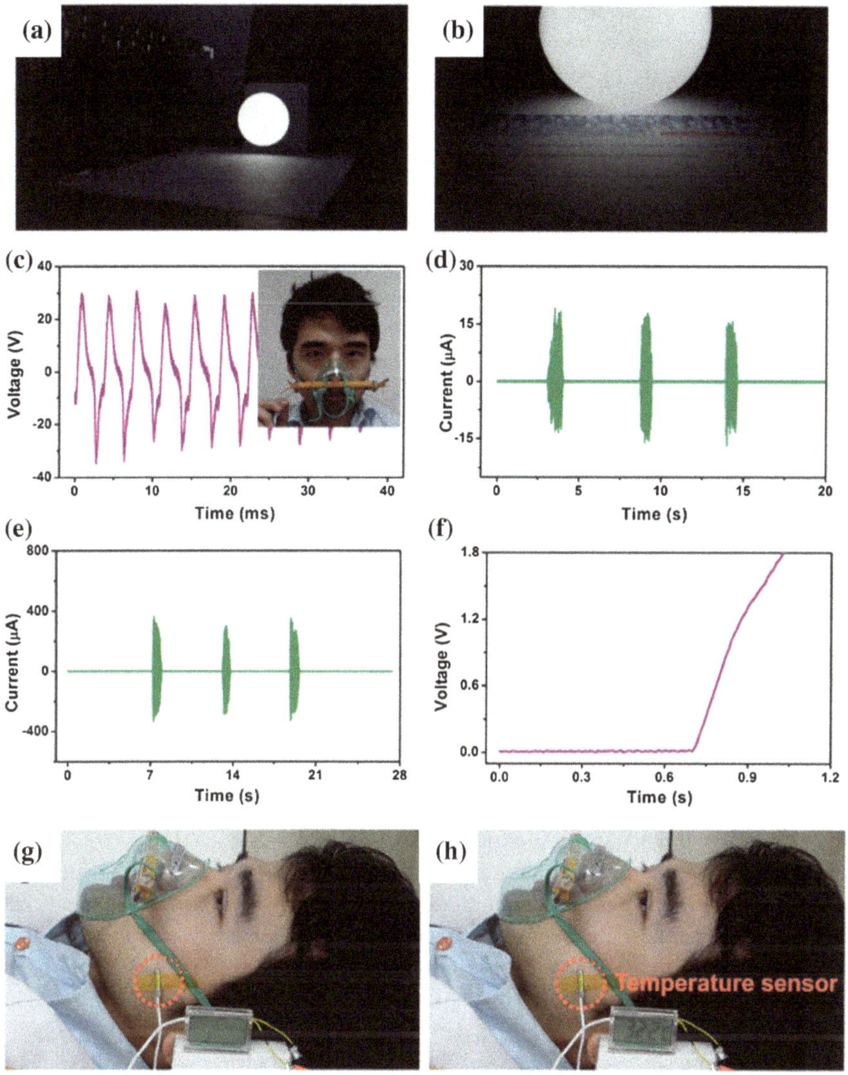


**Fig. 10.10** **a** Photograph of a spot light powered by 10 TENGs connected in parallel. **b** Photograph of a printed text on a paper illuminated by the TENGs driven spot light in complete darkness. **c** and **d** Measured output voltage (**c**) and output current (**d**) of the TENG for scavenging a human nose breathing induced air-flow energy. **e** Measured output current of TENG after using a transformer. **f** Measured voltage of a 10 μF capacitor charged by the TENG for scavenging a human nose breathing induced air-flow energy. **g** and **h** Photographs of using the TENG to scavenge human breath induced air-flow energy to sustainably power a temperature sensor system. Reproduced with permission from American Chemical Society [12]

To demonstrate that the TENG has the potential applications in self-powered healthcare monitoring, a human nose breath induced low air-flow can be used to drive the working of TENG, as shown in the inset of Fig. 10.10c. As illustrated in Fig. 10.10c, d, the corresponding output current and voltage can be up to 30 V and 18 μA under a loading resistance of 2.3 MΩ, respectively, resulting in an output power of 0.75 mW. As shown in Fig. 10.10e, the output current of the TENG after using a transformer can be increased to 350 μA under a loading resistance of 4 kΩ, where the corresponding output power is about 0.49 mW. A capacitor of 10 μF can be charged from 0 to 1.8 V in 0.33 s by using the human nose breath-driven TENG, as depicted in Fig. 10.10f. A self-powered temperature sensor system for monitoring the temperature of human body has been realized by integrating the TENG near human nose. The temperature sensor can sustainably work by using the human nose blowing induced low air-flow driven TENG, as displayed in Fig. 10.10g, h.

As compared with the conventional wind generation technology, the invented TENG in this study has the following advantages: (1) The conventional wind generator uses the electromagnetic effect to extract wind power, which usually requires a high cost for the device materials and installation. The TENG has very low investment cost by using the cheap organic film materials and conductive electrodes. (2) Due to the large volume, the conventional wind generators are often located in remote locations, far from cities where the transmission lines must be built to transfer the electricity from wind farm to the city. In comparison, the TENG has a very small volume so that it is suitable for TENG installation in the city such as on the house roofs and windows, where the produced electricity can be directly used by the customers. (3) The volume power density of conventional wind generator is smaller than 500 W/m$^3$, while the TENG has the much larger volume power density of 9 kW/m$^3$ under a wind speed of 15 m/s.

In summary, an elasto-aerodynamics-driven TENG is demonstrated for scavenging air-flow energy, where the flutter effect was employed to induce the periodic contact and separation between two triboelectric materials to realize the output of the device. By designing the different dimensions, the output performance of TENG has been optimized, where the maximum output power density of about 9 kW/m$^3$ has been achieved under a device working size of $125 \times 10 \times 1.6$ mm$^3$. Potential large-scale air-flow energy can also be demonstrated by parallel connecting more TENGs for powering some electric devices such as a globe light. The TENG has the potential applications for achieving large-scale wind energy harvesting at low cost and self-powered sensor systems.

## 10.3 Summary

Wind energy as an alternative to fossil fuels is plentiful, renewable, widely distributed, clean, and produces no greenhouse gas emissions during operation. TENGs with low cost, simple fabrication process, and good scalability, exhibit great potential as a wind energy harvester through both rotational structures and flutter-based structures. As compared with traditional wind energy harvesting through electromagnetic effect, TENG has shown advantages of light weight and flexibility. Additionally, TENG can also effectively harvest low speed wind energy, which is hard for traditional electromagnetic generators. As a miniaturized, modularized, and efficient wind energy harvesting solution, TENG will work as a distributed power supply for powering electronics and achieving self-powered system.

## References


1. G.M. Joselin Herbert, S. Iniyan, E. Sreevalsan, S. Rajapandian, A review of wind energy technologies. Renew. Sustain. Energy Rev. **11**(6), 1117–1145 (2007)
2. T. Ackermann, L. Söder, Wind energy technology and current status: a review. Renew. Sustain. Energy Rev. **4**(4), 315–374 (2000)
3. P. Shashank, C. Chih-Ta, F. Darren, Z. Jeff, Piezoelectric windmill: a novel solution to remote sensing. Jpn. J. Appl. Phys. **44**(1L), L104 (2005)
4. Y. Xie, S. Wang, L. Lin, Q. Jing, Z.-H. Lin, S. Niu, Z. Wu, Z.L. Wang, Rotary triboelectric nanogenerator based on a hybridized mechanism for harvesting wind energy. ACS Nano **7**(8), 7119–7125 (2013)
5. Y. Wu, X. Zhong, X. Wang, Y. Yang, Z. Wang, Hybrid energy cell for simultaneously harvesting wind, solar, and chemical energies. Nano Res. **7**(11), 1631–1639 (2014)
6. S. Li, S. Wang, Y. Zi, Z. Wen, L. Lin, G. Zhang, Z.L. Wang, Largely improving the robustness and lifetime of triboelectric nanogenerators through automatic transition between contact and noncontact working states. ACS Nano **9**(7), 7479–7487 (2015)
7. S. Chen, C. Gao, W. Tang, H. Zhu, Y. Han, Q. Jiang, T. Li, X. Cao, Z. Wang, Self-powered cleaning of air pollution by wind driven triboelectric nanogenerator. Nano Energy **14**, 217–225 (2015)
8. Z. Wen, J. Chen, M.-H. Yeh, H. Guo, Z. Li, X. Fan, T. Zhang, L. Zhu, Z.L. Wang, Blow-driven triboelectric nanogenerator as an active alcohol breath analyzer. Nano Energy **16**, 38–46 (2015)
9. G. Zhu, J. Chen, T. Zhang, Q. Jing, Z.L. Wang, Radial-arrayed rotary electrification for high performance triboelectric generator. Nat. Commun. **5**, 3426 (2014)
10. Y. Yang, G. Zhu, H. Zhang, J. Chen, X. Zhong, Z.-H. Lin, Y. Su, P. Bai, X. Wen, Z.L. Wang, Triboelectric nanogenerator for harvesting wind energy and as self-powered wind vector sensor system. ACS Nano **7**(10), 9461–9468 (2013)
11. J. Bae, J. Lee, S. Kim, J. Ha, B.-S. Lee, Y. Park, C. Choong, J.-B. Kim, Z.L. Wang, H.-Y. Kim, J.-J. Park, U.I. Chung, Flutter-driven triboelectrification for harvesting wind energy. Nat. Commun. **5**, 4929 (2014)
12. S. Wang, X. Mu, X. Wang, A.Y. Gu, Z.L. Wang, Y. Yang, Elasto-aerodynamics-driven triboelectric nanogenerator for scavenging air-flow energy. ACS Nano **9**(10), 9554–9563 (2015)
13. X. Wang, S. Wang, Y. Yang, Z.L. Wang, Hybridized electromagnetic-triboelectric nanogenerator for scavenging air-flow energy to sustainably power temperature sensors. ACS Nano **9**(4), 4553–4562 (2015)
14. H. Guo, J. Chen, L. Tian, Q. Leng, Y. Xi, C. Hu, Airflow-induced triboelectric nanogenerator as a self-powered sensor for detecting humidity and airflow rate. ACS Appl. Mater. Interfaces **6**(19), 17184–17189 (2014)

15. X.S. Meng, G. Zhu, Z.L. Wang, Robust thin-film generator based on segmented contact-electrification for harvesting wind energy. ACS Appl. Mater. Interfaces **6**(11), 8011–8016 (2014)
16. M. Taghavi, A. Sadeghi, B. Mazzolai, L. Beccai, V. Mattoli, Triboelectric-based harvesting of gas flow energy and powerless sensing applications. Appl. Surf. Sci. **323**, 82–87 (2014)
17. H. Guo, X. He, J. Zhong, Q. Zhong, Q. Leng, C. Hu, J. Chen, L. Tian, Y. Xi, J. Zhou, A nanogenerator for harvesting airflow energy and light energy. J. Mater. Chem. A **2**(7), 2079–2087 (2014)
18. S. Wang, X. Mu, Y. Yang, C. Sun, A.Y. Gu, Z.L. Wang, Flow-driven triboelectric generator for directly powering a wireless sensor node. Adv. Mater. **27**(2), 240–248 (2015)

# Chapter 11
# Harvesting Large-Scale Blue Energy

Widely distributed across the globe, hydrokinetic energy is one of the richest energy sources for large-scope applications. The state-of-the-art wave energy harvesting relies mainly on electromagnetic generator (EMG). However, this traditional approach is likely to have possible limitations towards large-scale blue energy harvesting. First, the electromagnetic generator is usually heavy and has a large mass density owing to the presence of magnets and metal coils, so it cannot naturally float on the surface of the water unless supported by a floater or a buoy platform. In ocean, the most dynamic energy presents at the surface of water. Secondly, the EMG is effective to catch the power from flowing streams at a certain direction, so that most of the oscillatory wave motion cannot be effectively harvested. Thirdly, EMG is ineffective for harvesting low frequency mechanical energy due to the very low output voltage (<3 Hz), so that it is not efficient for harvesting water wave energy in ocean. Lastly, the fabrication of EMG requires high-quality materials, so that it may not be cost-effective for applications in a large surface area. Therefore, a lightweight, low-cost approach that can harvest energy from not only flowing streams but also the waves is greatly desirable as a key to solving the above problems. This chapter present a network based on TENGs as an approach toward large-scale blue energy.

## 11.1 TENG for Water Wave Energy Harvesting

### *11.1.1 Liquid-Solid Electrification-Based TENG*

Featured as being simple, reliable, cost-effective, triboelectric nanogenerators could provide a fundamentally new route toward large-scale wave energy harvesting [1–5]. And certain efforts were devoted to this direction in the past 2 years [6–10]. On one hand, resulting from the novel working mechanism, the water itself can be one



Z.L. Wang et al., *Triboelectric Nanogenerators*,
Green Energy and Technology, DOI 10.1007/978-3-319-40039-6_11

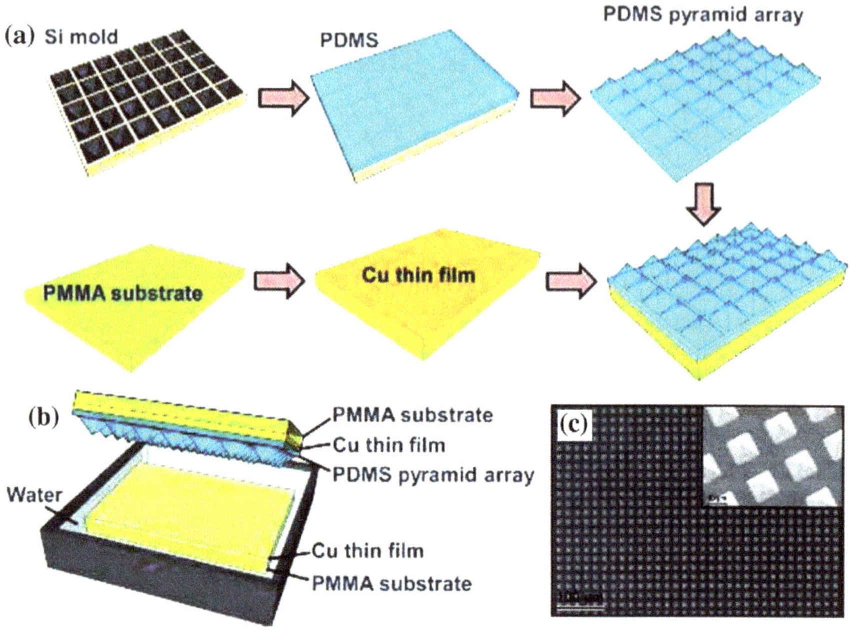


**Fig. 11.1** **a** Fabrication process of patterned PDMS pyramid array and related device. **b** Schematic diagram of the water-based TENG. **c** SEM image of patterned PDMS pyramid array. *Inset* is an SEM image with higher magnification. Reproduced with permission from Wiley [7]

triboelectric material to harvest wave energy when it interacts with insulating polymer films.

A first example was using the PDMS to interact with water as the two triboelectric layers, which forms a water-TENG device [7]. Figure 11.1 shows the fabrication process of the TENG and how the water contact electrification is included in the action unit. The TENG fabrication starts from the design of a PDMS film with patterned pyramid array (Fig. 11.1a). The Si wafer mold was made by photolithography and then etched by a dry etching process. Liquid PDMS elastomer and cross-linker were mixed, degassed, and uniformly spin-coated on the Si wafer mold. After thermal incubation, a uniform PDMS film with a patterned pyramid array was formed. For the other part of the construct, thin films of Cu (100 nm) were deposited on two poly methyl methacrylate (PMMA) substrates by a RF magnetron sputtering deposition system. PMMA is selected as the substrate material because it provides a flat surface, light weight, and high strength [11–13]. The PDMS film with the pyramid array pattern was peeled off the Si wafer mold and then placed on one of the Cu thin-film-deposited PMMA substrates with uncured PDMS mixture on top. Finally, the structure was incubated again to obtain a fully developed device for use in the next step. To investigate the contact electrification between water and the PDMS film, the second Cu thin-film-deposited

PMMA substrate was placed on the bottom of an insulating tank (Fig. 11.1b), acting as the conducting electrode for the water. The dimensions of the tank were 11 cm × 7 cm. After the tank was filled with water, the device consisting of the patterned PDMS pyramid array was controlled by a linear motor to periodically make contact with the water and then separate. The depth of water layer is 2 cm. During the contact electrification process, the PDMS film only made contact with the water surface and did not touch the Cu thin-film deposited PMMA substrate on the bottom of the tank. Figure 11.1c displays scanning electron microscopy (SEM) image of the patterned PDMS pyramid array, showing uniform microstructures distributed across the whole area. The thickness of PDMS film is nearly 140 μm. These microstructures not only act as steady gaps between two contact materials in a strain-free condition, but also enable the contact area of two materials to be enlarged, which will increase the electrical output of the water–TENG. High-magnification SEM image (inset of Fig. 11.1c) further reveals that the size of single pyramid is 10 μm, and each pyramid has a perfect geometric structure and a sharp tip.

To characterize the device output performance, the electrical output of the water–TENG was obtained through the contact electrification between the patterned PDMS pyramid array and deionized water. The TENG was mechanically triggered by a linear motor that provides dynamic impact with controlled force at a frequency of 2 Hz. $V_{oc}$ and $J_{sc}$ were measured separately to characterize the electrical performance of the water–TENG. From Fig. 11.2a, the $V_{oc}$ of the water-TENG surged from 0 to 82 V upon removing the patterned PDMS pyramid array from the deionized water surface to a separation distance of 1.5 cm. The generated voltage held as a plateau because in an open circuit condition the electrons cannot flow to screen the electric potential difference between patterned PDMS pyramid array and the deionized water (inset of Fig. 11.2a). When the PDMS pyramid array moving back into contact with deionized water again, the $V_{oc}$ fell back to 0. In the short-circuit case, the charges transferred between Cu electrode 1 and Cu electrode 2 in the external circuit during the contact and separation between the PDMS film and the water surface instantaneously generated positive and negative currents. The peak value of the $J_{sc}$ reaches 1.05 $mAm^{-2}$ (Fig. 11.2b). The integration of positive and negative current peaks represents the transferred charges in releasing and pressing movements. Both the generated $V_{oc}$ and $J_{sc}$ exhibit corresponding reversed signs when the connection polarity to the electrometer is switched, indicating that the measured signals were generated by the water-TENG. The patterned pyramid array is effective for enhancing the output. Furthermore, the AC output could be transformed to a pulse output in the same direction simply by using full-wave rectifying bridge (Fig. 11.2c) and stored in a capacitor of 33 μF (Fig. 11.2d). With such a power output generated from water-contact electrification, 60 commercial LEDs can be instantaneously driven by a single water–TENG device (Fig. 11.2e, f).

A second example of the liquid-solid electrification enabled TENG was consisted of a layer of the nano-structured PTFE contacting with water [6]. Figure 11.3a shows a scanning electron microscopy (SEM) image of the PTFE thin film. The surface of the PTFE thin film is composed of irregular microstructures,

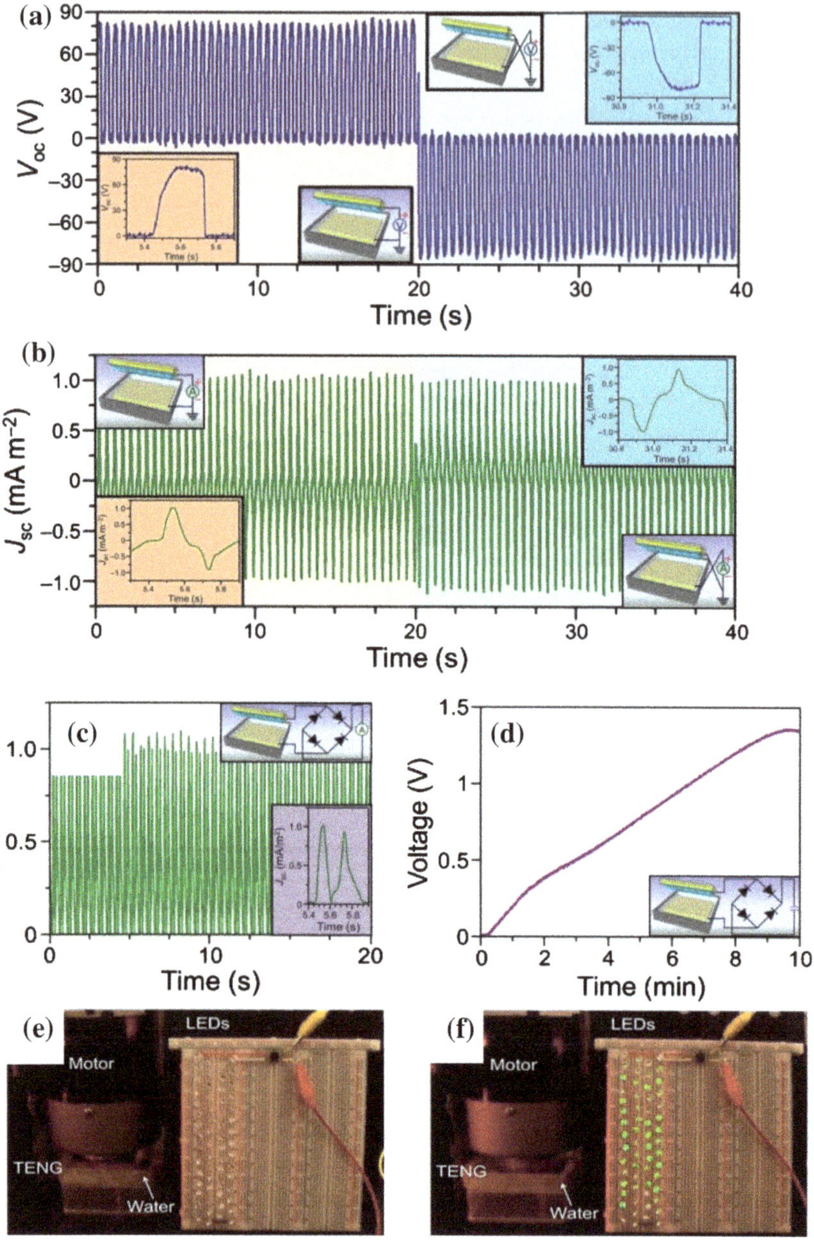
(a)
$V_{oc}$ (V)
90
60
30
0
−30
−60
−90
0
10
20
30
40
Time (s)
(b)
$J_{sc}$ (mA m$^{-2}$)
1.0
0.5
0
−0.5
−1.0
Time (s)
(c)
$J_{sc}$ (mA m$^{-2}$)
Time (s)
(d)
Voltage (V)
1.5
1
0.5
0
2
4
6
8
10
Time (min)
(e)
LEDs
Motor
TENG
Water
(f)
LEDs
Motor
TENG
Water

◀ **Fig. 11.2** **a** $V_{OC}$ and **b** $J_{SC}$ of the water-TENG under a constant external force with a frequency of 2 Hz. Measurements operated both at the forward connection, shown at the left-hand side, and the reversed connection, shown at the right-hand side. **c** The AC output transformed into a pulse output in the same direction by a full-wave rectifying bridge. **d** The rectified output used to charge a capacitor. The insets are the magnified output curve in one cycle (**a**–**c**) and a diagram of the corresponding connection polarity (**a**–**d**). **e** Photograph of a setup in which the water-TENG acts as a direct power source for 60 commercial *green* LED lamps and **f** when the patterned PDMS pyramid array is in contact with the water, the LEDs light up. Reproduced with permission from Wiley [7]

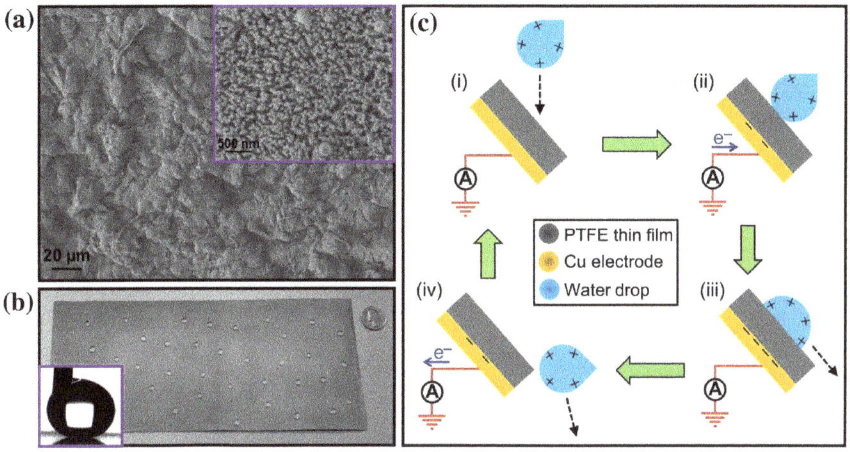


**Fig. 11.3** **a** SEM image of the prepared PTFE thin film with hierarchical micro-/nanostructures. *Inset* is an SEM image with higher magnification. **b** Advancing contact angle of the PTFE thin film. **c** Working mechanism of the water-TENG when the generated triboelectricity is dominated by the contact electrification process with the air/pipes. Reproduced with permission from Wiley [6]

which is further covered by high-density nanorods. The mean length and diameter of the nanorods determined from SEM images are around 360 and 45 nm, respectively. It was demonstrated that large-sized PTFE thin films can be prepared with a uniform superhydrophobic property (Fig. 11.3b). This is very important for the application of this concept to harvest the water-related energy in ambient environment. The hydrophobicity of the PTFE thin film was investigated by measuring the contact angle of water drop on it (Inset of Fig. 11.3b). Because the hierarchical micro-/nanostructures contain trapped air, it will reduce the actual contact area between the surface and water drop, making the surface superhydrophobic. The average measured contact angle of the PTFE thin film is 169°. The operation of the water-TENG will be explained as single-electrode mode. Figure 11.3c shows before the contact with the water-TENG, the water drop already contains triboelectric charges on its surface because of the friction with air/pipes. Previous studies have shown that when a water drop is falling from the sky or flowing through an insulating tube, triboelectricity will be generated and contribute to the charged surface of water drop [14]. For example, a surface charge density of 4.5 μC $m^{-2}$ is measured from each water drop pipetted from a PTFE coating tip. It is worth noting that the charge on the water drop could be positive or negative depending the counterpart the water drop interacts with.

To demonstrate that this design can be used to harvest the water-related energy in the environment, flowing tap water from a kitchen faucet is first evaluated. The flow rate was set at around 30 mL/s and the distance between the water-TENG and faucet was 20 cm. This demonstration could be important because it represents not only in the future we can collect the wastewater energy from the environment in daily life but also we are moving forward about discovering and facilitating new renewable energy source. To highlight the potential application, the generated output now will be divided by the water-TENG dimensions (4 cm × 4 cm). Figure 11.4a shows the output current density of the water-TENG. It is noticed that the maximum output current density can reach 1.5 μA $cm^{-2}$ and the signal varies due to the unstable flow of tap water from the faucet. The output signal drops to zero after we turn off the faucet, revealing the output is generated from the faucet

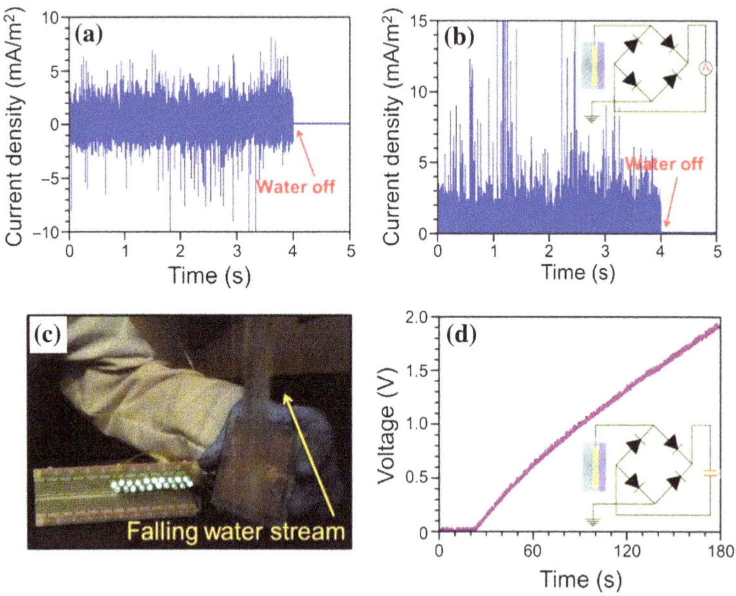


**Fig. 11.4** **a** Output current density of the water-TENG generated from flowing tap water. **b** The AC output transformed to unidirectional pulse output by a full-wave rectifying bridge. **c** Photograph of the water-TENG used as a direct power source to light up 20 LEDs. **d** The rectified output used to charge a commercial capacitor of 33 F. *Insets* of (**b**) and (**d**) are the sketches of the corresponding circuit connection polarities. Reproduced with permission from Wiley [6]

flowing water. The AC output could be transformed to unidirectional pulse output by attaching a full-wave rectifying bridge (Fig. 11.4b). The rectified output can be utilized to drive 20 commercial LEDs instantaneously (Fig. 11.4c) and charge a 33-μF capacitor (Fig. 11.4d). The LEDs are lighted up as soon as the flowing water contacts with the water-TENG and totally go out when the faucet was turned off. In the cases of flowing water or spray water drop like raindrop, it is necessary to consider there could be more water remaining on the surface of the water-TENG during the energy harvesting process. Here it was noting that under those conditions, the water-TENG can still generate output is because of the amount of water on the water-TENG surface is not constant. Like the water flowing from a kitchen faucet is not fully continuous. The water flow has some fluctuation and causes the amount of water on the water-TENG surface to be constantly changing. Besides, the superhydrophobic property of the water-TENG also enables the change of water amount on the surface.

A third example of the liquid-solid electrification enabled TENGs was consisted of advanced grating structure, as it is sketched in Fig. 11.5a [9]. On one side of a fluorinated ethylene propylene (FEP) thin film, two parallel strip-shaped electrodes are fabricated, which are discrete with a fine gap in between. The other side of the FEP layer, modified to intentionally create patterned surface roughness at the nanoscale (Fig. 11.5b), directly interacts with ambient water to generate surface triboelectric charges at the liquid-solid interface. The vertically aligned nanowires, ~100 nm in diameter and ~2 μm in length, are for promoting the output power of the TENG by enhancing the contact area with water. The FEP is chosen for a number of reasons. Foremost, it is the most negative material that is commercially available with respect to triboelectric polarity, which enables superior surface charging ability compared with other materials [15–17]. In addition, other

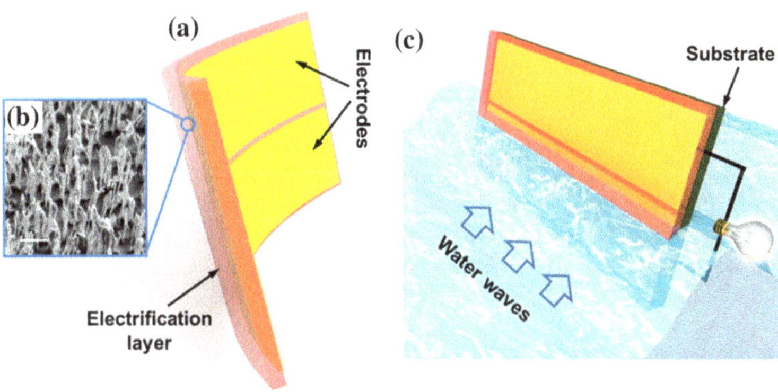


**Fig. 11.5** Structural design of the LSEG. **a** Schematic of the bent electrification layer with two electrodes on one side. **b** SEM image (tilting angle of 30°) of the polymer nanowires on the opposite side of the polymer layer. The scale bar is 1 μm. **c** Schematic of a substrate-supported LSEG positioned in water waves. The *up*-and-*down* movement of the surrounding water body induces electricity generated between the two electrodes. Reproduced with permission from American Chemical Society [9]

advantageous features, including heat resistance, radiation stability, and chemical inertness, make the FEP a desirable and durable material for underwater operation. The fabricated layer is then applied onto a substrate (Fig. 11.5c) for quantitative measurement.

Instead of being driven by an electric motor for quantitative measurement, an integrated TENG (I-TENG) with a scaled-up design was further tested in a normal environment where energy from ambient water motions was harvested. The I-TENG consists of six strip-shaped electrodes with each having lateral dimensions of 20 cm by 6 mm. Then, a total of five basic units were formed by any pair of adjacent electrodes. The electric output of each pair was first rectified through an electric bridge and then constructively superimposed through a parallel connection.

In the first demonstration, the I-TENG was vertically fixed in a water container with a tilting angle of 30° away from the vertical position. The water surface leveled with the lower edge of the device. The container was then gently tilted back and forth to create traveling waves that washed the I-TENG in a reciprocating way, similar to the circumstance along the seashore and river banks. Pulses of DC current are produced, as shown in Fig. 11.6b. A zoomed-in view of a single current pulse in

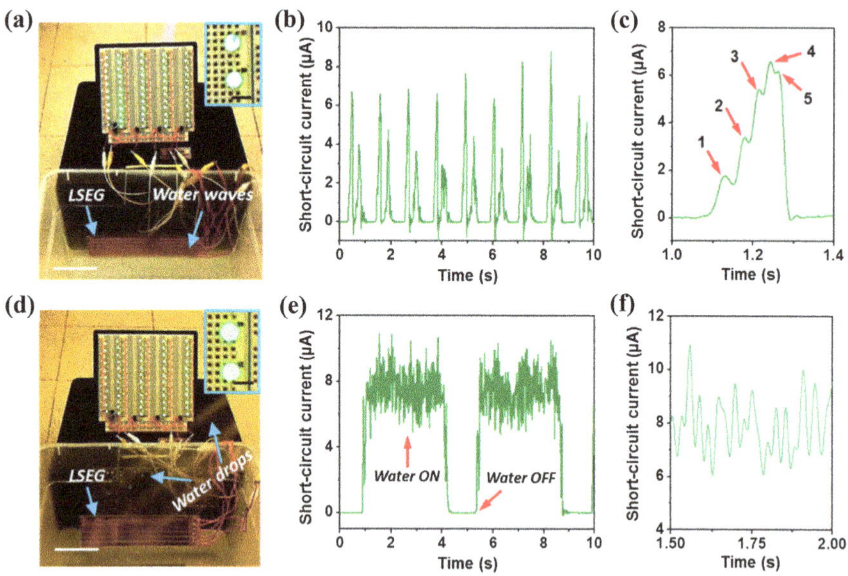


**Fig. 11.6** Performance evaluation of the LSEG in harvesting energy from ambient water motions. **a** Demonstration of an integrated LSEG with an array of electrode in powering tens of LED bulbs by harnessing water waves. The scale bar is 8 cm. *Inset* Zoom-in view of two lit LEDs. **b** Short-circuit current of the integrated LSEG when it interacts with water waves at a frequency of ~0.7 Hz. **c** Zoom-in view of a large current peak. **d** Demonstration of the integrated LSEG in powering tens of LED bulbs by harnessing water drops. The scale bar is 8 cm. *Inset* Zoom-in view of two lit LEDs. **e** Short-circuit current of the integrated LSEG when it interacts with falling water drops. **f** Zoom-in view of the current signal when water drops are shed upon the device. Reproduced with permission from American Chemical Society [9]

Fig. 11.6c reveals multiple peaks. Since the five units interact with the water sequentially instead of simultaneously, they result in five current pulses that merge together but with timing mismatch. The generated electricity could light up tens of LED bulbs, proving the capability of the TENG as a power source for electronics (Fig. 11.6a). This demonstration clearly indicates that our device can be potentially used along the seaside/riverside area. Besides, accompanied with a buoy, the TENG can also be possibly deployed offshore on the ocean surface. In light of the operating principle that requires only changing of the contact area with water, we further demonstrated energy harvesting from falling water or rain drops. The I-TENG was still braced with the same tilting angle (Fig. 11.6d). A sprinkler head was used to spray water onto the device. As the water accumulated, droplets slid across the I-TENG surface perpendicular to the electrode array. Since a moving droplet could momentarily cover part of the device, electricity was generated as the droplet crossed boundaries between adjacent electrodes. With numerous droplets, a large number of current pulses merged together, leading to an apparent continuous DC output (Fig. 11.6e, f). Larger water droplets tend to create more fluctuation in the output current but do not greatly change the overall performance. The electric output was again sufficient to power the LED bulbs (Fig. 11.6d). This demonstration further expands the applicability of the TENG, providing a new means of energy harvesting from a largely unexploited energy source, namely, natural precipitation.

### *11.1.2 TENG Based Hydrokinetics Energy Harvesting*

A multi-layer integration of the disk TENGs was developed to harvest the hydrokinetics energy [18]. And the multi-layered disk TENG is mainly composed of two groups: rotors with Al film as the attached triboelectric layers, and stators with the purposely chosen polytetra-fluoroethylene (PTFE) film as the other triboelectric layers. They are coaxial with the input rotation motion, as schematically depicted in Fig. 11.7a. For achieving flat surfaces, acrylic sheets were used as template substrates for both of them. At first, they were processed by laser cutting to form eight-sector structures with through-holes in the center for the shaft connection, which were of the same shape as the attached triboelectric layers. For the rotor templates, the center-hole is of D-shape, strictly matching the D-shaft we used, so that they can be led by the shaft to rotate with little angular variations (Fig. 11.7b). Both sides of a rotor substrate were deposited with Al films by e-beam evaporation, which serve as one side in a disk TENG unit. As for the stators, the center-hole is in round-shape, with the diameter a bit larger than the D-shape shaft. In this way, the stators will not rotate with the spinning shaft. On their surfaces, 50-μm-thick PTFE films with aluminum electrodes on the back side were securely attached. For the four-layer integration as depicted in Fig. 11.7, the stator placed in the middle was double-side coated with PTFE. The multiple layered structure was obtained by stacking three stators and two rotors in an alternating sequence, with the D-shape

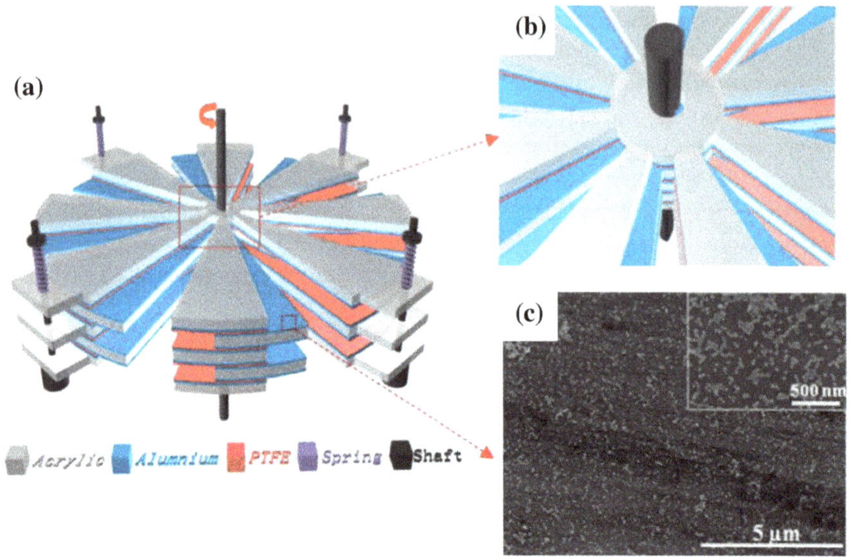


**Fig. 11.7** Device structure of the multi-layered disk triboelectric nanogenerator (TENG). **a** The schematic diagram showing the structure design of the multi-layered disk TENG. **b** The enlarged picture showing the D-shape shaft going through all the center holes. **c** The SEM image of the silver nanoparticles coated on the aluminum electrode of the rotor part, the inset is the SEM image in higher magnification. Reproduced with permission from Elsevier [18]

shaft going through all the center-holes. Each pair of PTFE film and aluminum film next to each other constitutes a disk-TENG unit. The stators were fixed together by 4 screws going through the holes in their four corners, with one screw fastened to a bracket to ensure their stationary during operation. In order to maintain intimate surface contact between adjacent tribo-layers, low-stiffness springs were installed on these screws to apply a well-controlled gentle pressing force between the plates. When the D-shape shaft is connected to a rotating object, e.g. a rotary motor in the experimental measurement, all the Al layers on the rotors will be driven to rotate with the same phase, while the PTFE layers on the stators will stay still. In this manner, the relative rotation between the adjacent layers enables periodic contact-separation cycles of the opposite triboelectric charges, and thus generates electricity in each layer. When all of the four layers of the multi-layered disk TENG are connected in parallel, the current with the same phase can be added up, contributing to a multiplied power output. It is worth noting that, in order to enhance the surface roughness and promote the triboelectrification intensity, the aluminum films on the rotor part were coated with silver nanoparticles. Figure 11.7c shows the SEM images of the assembled silver nanoparticles on Al film, which uniformly distribute across the whole surface with an average diameter of 50 nm. In this experimental demonstration, the projected surface area of the TENG device is 34.2 $cm^2$.

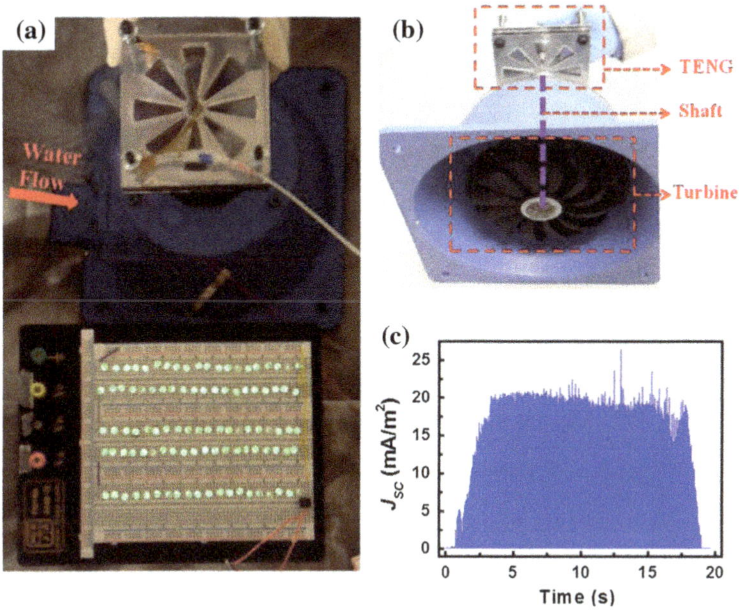


**Fig. 11.8** Application of the four-layer disk TENG for harvesting energy from water flow. **a** The picture of the multi-layered disk TENG coupled with a water turbine for harvesting water flow energy and lighting up 100 commercial LEDs. **b** The picture of the energy harvesting system showing the detailed structural design. **c** The real-time measurement of the current signal of the system. Reproduced with permission from Elsevier [18]

Practically, irregular flow motions of fluid, such as water flow or wind, can be effectively transformed to rotary motion of the shaft and hence enable the electricity generation of the device. As an example, the nanogenerator was demonstrated as a practically applicable structure for harvesting the energy of water flow. Through connecting the shaft with a water turbine, the water flowing through the turbine can effectively drive the rotation of the disk TENG (Fig. 11.8b). As shown in Fig. 11.8a, when the water flow that comes from an ordinary household faucet was turned on, the rotors of the multi-layered disk TENG were driven to spin against the stators. In this manner, the mechanical energy of the water flow was effectively converted into electricity, which is capable of instantaneously lighting up 100 commercial LEDs. Through a real-time measurement of the current signal (Fig. 11.8c), the short-circuit current generated by the water flow can be as high as 20 mA/m$^2$. From this demonstration, the similar methodology can be expanded to the applications of harvesting energies from ocean wave and wind power, which will make the multi-layered TENG an effective approach to harvesting large-scale mechanical energies.

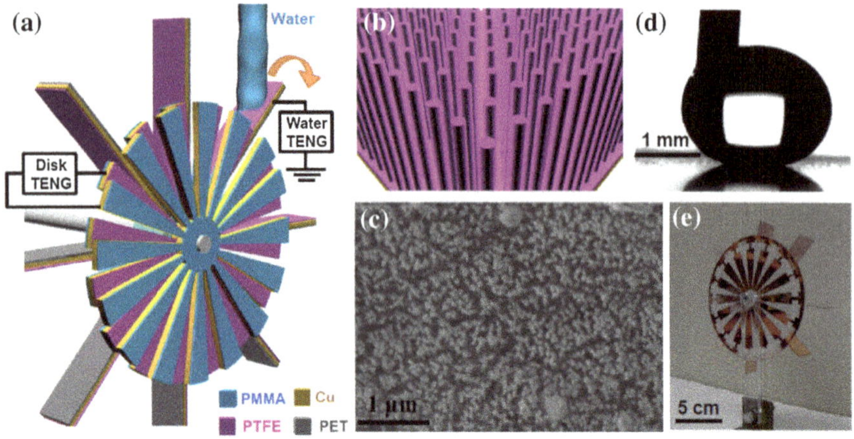


**Fig. 11.9** **a** The structure diagram of the hybridized TENG. The structure diagram (**b**), the SEM image (**c**), and the contact angle (**d**) of the PTFE thin film with nanostructure. **e** The photography of the prepared hybridized TENG. Reproduced with permission from American Chemical Society [19]

### *11.1.3 Dual Mode TENG for Electrostatic and Mechanical Energies Harvesting*

Flowing water contains not only mechanical kinetic energy, but also the electrostatic energy owing to the triboelectric charges caused by its contact with surrounding media such as air. In this regards, a water wheel hybridized TENG, composed of a water-TENG part and a disk-TENG part, has been developed for simultaneously harvesting the two types of energies from the tap water flowing from a household faucet [19]. The structure diagram of the hybridized TENG with a water wheel structure is shown in Fig. 11.9a, which consists of two different types of TENGs: one is the water-TENG and the other is the disk-TENG. The water TENG is composed of 8 wheel blades and operates as a single-electrode-based TENG to harvest the electrostatic energy from flowing water, in which the electrical channel is formed between the Cu electrode on each blade and the ground. The disk-TENG is composed by two disks, a back disk and a front disk, and each disk has 16 segments. The water wheel structure is assembled using a smooth metal rod as a rotation axis through the centers of the wheel and the two disks. The back disk is connected with the wheel, and the wheel blades are also used to gather the flowing water force to drive the rotation of the wheel and the back disk. The front disk is fixed on the rotation axis and keeps motionless even during the whole process. The disk-TENG is used to harvest the mechanical energy of the water, in which the electrical channel is formed between the Cu electrodes of the two disks. For fabricating the blades, a 100-nm thin Cu film was deposited on a square polyethylene terephthalate (PET) sheet to serve as the electrode layer, and then a

PTFE film with nanostructures fabricated by using porous anodic Al oxide (AAO) as template was attached onto the Cu electrode using a double-sided tape. The Cu electrodes of the 8 blades were electrically connected together. Two poly (methyl methacrylate) (PMMA) sheets were first processed by laser cutting to form the 16-segment structured circle disks as the substrates of the disk TENG. For the back disk, a 100-nm thin Cu film was deposited on one PMMA substrate, and then a PTFE film with nanostructures was attached onto the Cu electrode. For the front disk, another 100-nm thin Cu film was deposited on the other PMMA substrate and directly used as the contact material. The PTFE layer of the back disk and the Cu layer of the front disk were brought to a face-to-face intimate contact to form a sliding TENG operating at rotation disk mode.

The surface of PTFE film on the wheel blades of water-TENG and the back disk of disk-TENG is fully covered by high-density PTFE nanorods (Fig. 11.9b). The mean length and diameter of the PTFE nanorods are 360 and 45 nm from the SEM image displayed in Fig. 11.9c. The contact angle of the PTFE film is 162° (Fig. 11.9d), which represents the surface of the PTFE film is superhydrophobic. The high-density nanorods covered on the surface will contain trapped air and reduce the actual contact area between the surface and water droplets, which causes the superhydrophobic ability of the PTFE film. The superhydrophobic ability on the surface of the wheel blades is a critical factor for the performance of the water-TENG, which will be discussed later. In addition, the nanostructures on the PTFE surface will also increase the effective contact area of the two tribo-surfaces and enhance the electrical output of the disk-TENG. The photograph of the prepared hybridized TENG is shown in Fig. 11.9e.

To investigate the output power of the water wheel TENG at an external load resistor, the output voltage and output current of the water-TENG and the disk-TENG at various load resistances ranging from 100 Ω to 1 GΩ were measured, and the corresponding output power value at each resistance was calculated using $P = IV$. The dependences of open-circuit voltage($V_{oc}$) and short-circuit current ($I_{sc}$) on the flowing water rates for the water-TENG are shown in Fig. 11.10a, in which $V_{oc}$ values nearly keep constant around 72 V for various flow rates, and the $I_{sc}$ value increases from 4.3 to 12.9 μA as the flowing water rate increases from 7 to 54 mL/s. The increase of the $I_{sc}$ is caused by more tribo-charges carried in the flowing water in a unit time at higher flowing rate. The dependences of $V_{oc}$ and $I_{sc}$ on the flowing water rate for the disk-TENG are shown in Fig. 11.10b, in which $V_{oc}$ values nearly keep constant around 101 V for various flowing rates, and the $I_{sc}$ value nearly increases linearly with the flowing water rate. For TENG operating at the rotation disk mode, the $I_{sc}$ values will increase linearly with the rotation speed of the disk.

For practical applications in driving LEDs or charging a capacitor, the electrical outputs of the water-TENG and the disk-TENG are connected in parallel after they are converted to pulse output in the same directions by using two full-wave rectifying bridges, as shown in the inset of Fig. 11.10c. As the hybridized TENG is driven by water at a flow rate of 54 mL/s, 20 commercial green LEDs are lighted up, as shown in Fig. 11.10c. The hybridized TENG is also demonstrated to charge a capacitor of 4.7 μF at a flowing water rate of 54 mL/s, and the voltage curves of the

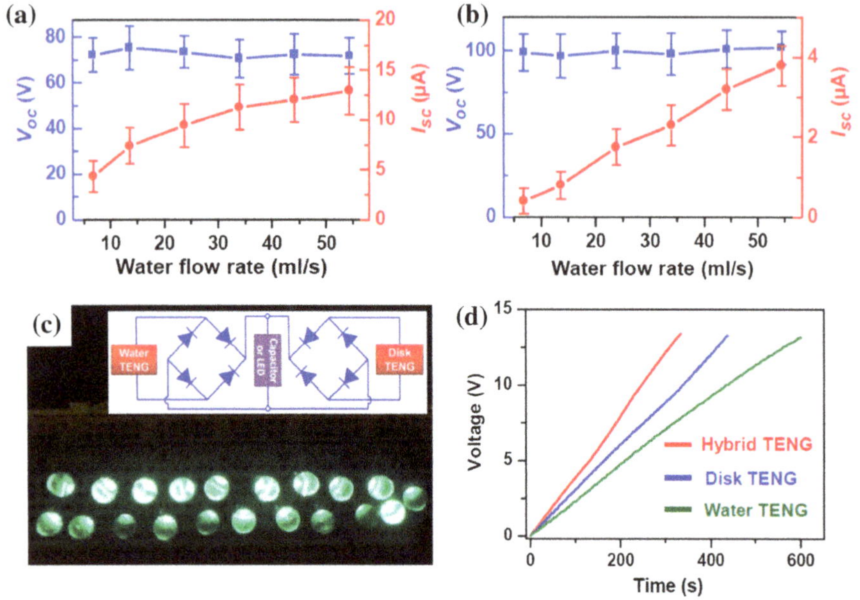


**Fig. 11.10** Performance evaluation of the hybrid TENG for water energy harvesting. The dependences of the $V_{OC}$ and $I_{SC}$ values on the flowing water rate of the **a** water-TENG and the **b** disk TENG. **c** The photography of 20 commercial LED bulbs driven by the hybridized TENG at a flowing water rate of 54 mL/s, and inset shows the diagram of the rectifying circuit. **d** The measured voltage of a 4.7 μF capacitor charged by the hybridized TENG and a single part of the hybridized TENG at a flowing water rate of 54 mL/s. Reproduced with permission from American Chemical Society [19]

capacitor are shown in Fig. 11.10d. When the capacitor is charged by the rectified output of water-TENG and the rectified output of disk-TENG, respectively, it takes about 596 and 435 s to charge the capacitor to a voltage of 13 V, respectively. When using the total hybridized TENG, it takes 326 s to charge the capacitor to a voltage of 13 V, which obviously demonstrates the effectiveness of the hybridized TENG as a novel approach to harvest multiple types of energies from the ambient environment.

### 11.1.4 *Fully Enclosed TENG for Water Wave Energy Harvesting*

The energy provided by water has a much less dependence on season, day or night, weather, and/or temperature. Since the humid environment can largely weaken the triboelectric effect, the TENG with an open structure cannot work under harsh conditions with the presence of water. In this regard, a rolling structured,

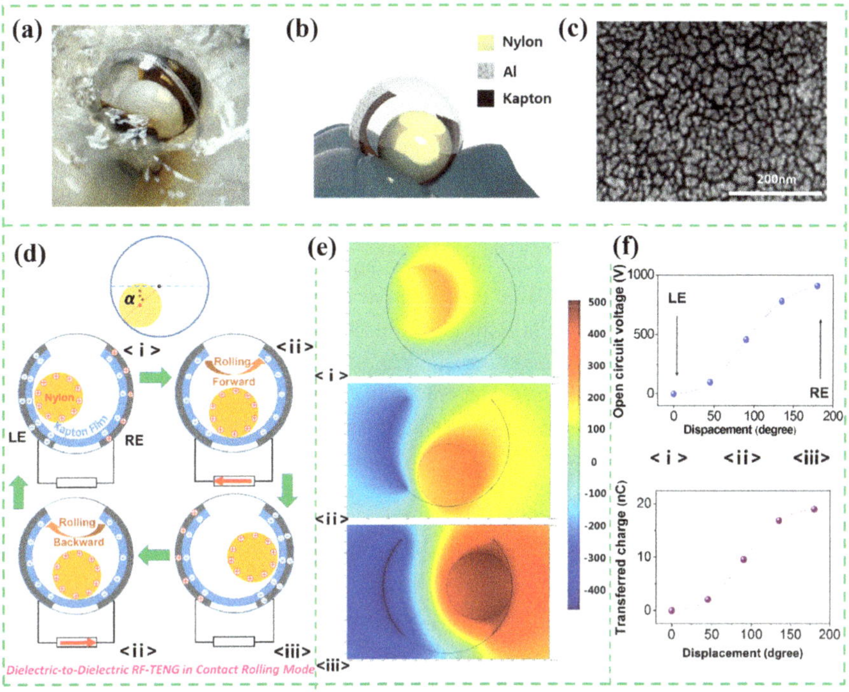


**Fig. 11.11** Device structure, basic operations and working principles of the freestanding-triboelectric-layer based nanogenerator (RF-TENG) with a rolling nylon ball enclosed. **a** Photograph of a rocking nanogenerators floating on water. **b** Schematic diagrams of freestanding-structured design. **c** SEM image of nanorod structure on the Kapton surface. **d** Schematic working principle of a dielectric-to-dielectric RF-TENG in rolling mode. **e** The simulated potential distributions at three different rolling displacements (α). **f** The simulated open-circuit voltages ($V_{OC}$) and transferred change ($Q_{SC}$) at different rolling displacements. Reproduced with permission from Wiley [20]

freestanding-triboelectric-layer based nanogenerator (RF-TENG) was demonstrated to harvest energy from low-frequency wave movement in harsh environment [20].

An image of the fabricated RF-TENG floating on water is shown in Fig. 11.11a. Figure 11.11b shows the schematic diagram of the freestanding structured design that consists of one rolling ball and two stationary electrodes. When driven by a wave vibration, the freestanding ball can roll back and forth between the two electrodes, thus providing alternating current to the external load. To maximize the electrifi cation during contact, a polyhexamethylene adipamide (Nylon 6/6) ball and a polyimide (Kapton) film are chosen as the materials for the triboelectric layers. To enhance the electric output of the TENG, nanowire arrays are fabricated on the surface of the Kapton film (Fig. 11.11c) that provides a large contact area to generate more triboelectric charges on the surface. The area for each electrode is 15 cm$^2$ and the diameter of the outside shell sphere is 6 cm. To more accurately simulate a realistic wave environment, all of the tested RFTENGs in the experiment are driven by an actual water wave system equipped with a pump. The operating principle of the proposed RF-TENG is based on the conjugation of the triboelectric effect and electrostatic induction. When the freestanding Nylon ball rolls on the top of the Kapton film, equal amounts of charges generate on the top surface of the Kapton and on the surface of the Nylon ball (Fig. 11.11d–i). When the Nylon ball rolls from the left-hand electrode (LE) toward the right-hand electrode (RE), the negative charges flow from the LE to the RE via the external load to screen the local field of the non-mobile negative charges on the dielectric due to the electrostatic induction process (Fig. 11.11d–ii). When the ball reaches the overlapping position of the RE (Fig. 11.11d–iii), all of the negative charges will then be driven to the RE. Subsequently, a backward rolling of the Nylon ball from the RE to the LE should drive the flow of the negative charges back from the right electrode to the left electrode, forming a reverse current in the load. A theoretical model of the RF-TENG has been established to simulate the change of the open-circuit voltage ($V_{OC}$) between the two electrodes when the displacement angle α changes due to external vibration (Fig. 11.11d, e). The displacement angle $\alpha$ is the angle between the horizontal level line and the line through the centers of the two balls. Using COMSOL software, we employ the finite element method (FEM) to simulate the open-circuit voltage ($V_{OC}$) and short-circuit transferred charge ($Q_{SC}$) for freestanding structures under a minimum achievable charge reference state. At the initial state, the Nylon ball is sitting on top of the LE (the displacement angle $\alpha = 0°$) and the simulated electrical potential difference between the two electrodes is zero. When the Nylon ball with positive surface charges rolls to the middle position ($\alpha = 90°$), a potential difference ($V_{OC}$) of ≈456 V is generated between the two electrodes under the open-circuit condition (Fig. 11.11e–ii). Finally, when the Nylon ball reaches the overlapping position of the RE ($\alpha = 180°$), $V_{OC}$ will increase to ≈903 V, as shown in Fig. 11.11e–iii. If there is an external load connected between the two electrodes, this potential difference will drive electrons to flow between the electrodes. Based on this fundamental relationship, a large open-circuit voltage can be generated from the RF-TENG due to the relatively small inherent capacitance. Via FEM, short-circuit transferred charges $Q_{SC}$ have a similar shape to the open-circuit voltage profile (shown in Fig. 11.11f), which derives from the nearly constant internal capacitance due to the stationary electrodes in the RF-TENG.

This type of TENG can have a wide range of applications in harvesting water wave energy. In an initial demonstration, 70 green LEDs are connected in series with the TENG to show its direct application in driving electronic devices, as shown in Fig. 11.12a. As shown in Fig. 11.12b, the TENG has sufficient power to drive dozens of green LEDs by harvesting the wave energy. Moreover, these devices can also be employed with energy storage devices to act as power sources for commercial electronic sensors. In a second demonstration, a series of ELDCs (each of 100 mF) is utilized along with the TENG and bridge rectifiers to form an entire self-powered system, as shown in Fig. 11.12c. The charging performance of this series of EDLCs is shown in Fig. 11.12d. These EDLCs can be charged from 0 V to a working voltage of 2.4 V within several hours by continuously oscillating

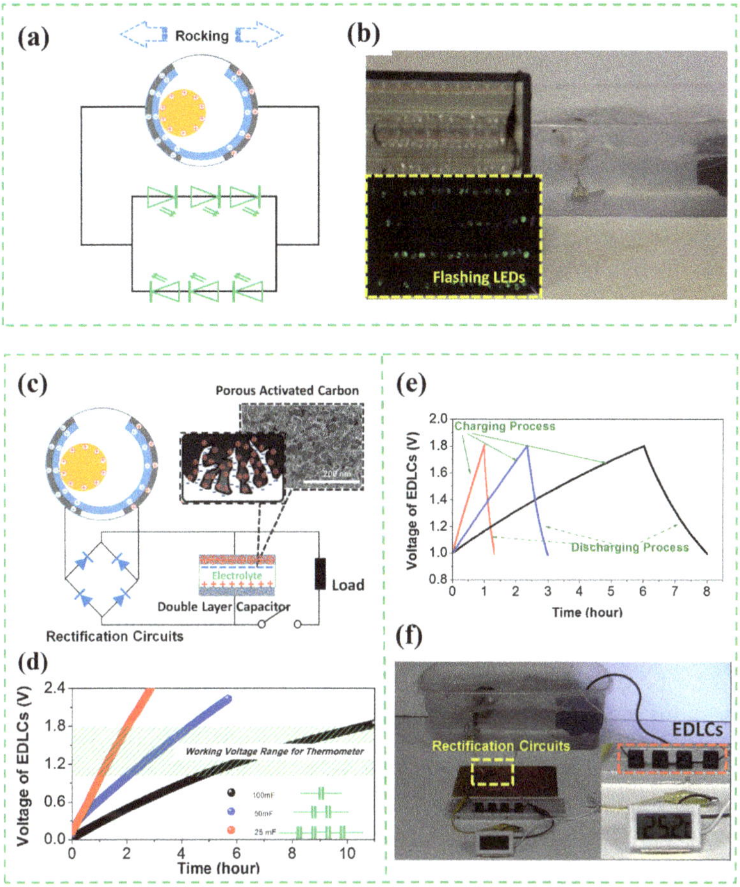


**Fig. 11.12** Applications of RF-TENGs for wave energy harvesting. **a** Schematic diagram of the connection between the TENG and 70 *green* LEDs. **b** Optical images of 70 *green* LEDs being driven by the rocking TENG in a wave system. **c** Schematic diagrams of the self-power system with TENG, rectification circuits, EDLC and Load. **d** The charging *curve* for one, two and four EDLC capacitors in series, respectively. **e** Charging curve of EDLCs in series by using a RF-TENG and discharging curve of EDLCs with a thermometer as load. **f** Photographs of the self-power system with rectification circuits, EDLCs and thermometer

the RF-TENGs in the waves. A commercialized thermometer can be easily powered by this self-powered system; the working process of this self-powered temperature sensor system is shown in Fig. 11.12e. As shown by the upward line representing the charging process, the voltage of the EDLCs reaches 1.8 V when the RF-TENG is driven to charge four EDLCs in series for a period of 60 min. At this point, the switch is closed and the thermometer begins operating; the voltage of the EDLCs then begins to decrease, as shown by the line for the discharging process. As shown in Fig. 11.12f, the temperature of the water can be recorded and visualized through the LCD display for a period of more than 20 min before the thermometer automatically turns off once the voltage of the EDLCs (25 mF in total) drops to 1.0 V; the EDLCs then begin to recharge. For this self-powered system with four EDLCs in series, the percentages of charging and working times are 73.5 and 26.5 %, respectively. The above energy supply processes for temperature sensors with specific power consumption requirements are very representative. Most sensors in unattended water monitoring systems adopt an intermittent operating mode to gather or send data and demand discontinuous high power consumption. The above concept and design can offer a feasible power solution to the long-term, wide-area, in-situ, and near real-time monitoring of water parameters, and may be the best power choice, particularly in closed environments such as sealed pipelines.

## 11.2 Network of TENGs for Blue Energy Harvesting

Relying on surface contact electrification effect between the conventional polymers and very thin layer of metal as electrodes, a TENG-Network (TENG-NW) structure was proposed towards large-scale blue energy harvesting [10]. Featured as being extremely light-weight, low-cost, high anticorrosion to the marine environment, the networks of TENGs are capable of floating on the surface of water for wave energy harvesting. And by using the collision of a rolling ball caused contact and separation, the TENG converts the slow, random and high-force all-directional oscillatory motions into electricity.

For wave energy harvesting, four basic units, vertically anchored, wall a metal ball to form a TENG, as the photograph shown in Fig. 11.13a. Triggered by the water wave motion, the ball at the center of the single unit will collide with the walls, namely, the standing basic units. To develop a TENG-NW, thousands of TENGs will be electrically connected and wove into a network, as the sketch showing in Fig. 11.13b.

To deduce and demonstrate the law of the TENG-NW for blue energy harvesting, a further study was taken to investigate the output performance on the unit numbers. With this regards, four units were fabricated and wove into a small-scale network. And the seesawing motion of the tank with water inside will generate the water wave for experimental measurement. The output characteristics regarding the scale of the TENG-NW were demonstrated in Fig. 11.13c–f. The output current and voltage of the TENG-NW with unit number n = 1, 2, 3, 4 were respectively demonstrated in Fig. 11.13c, d. From the evolution of the output signals regarding the increasing unit numbers, certain trends can be derived for the TENG-NW. Firstly, current amplitudes are drastically increased with elevated unit numbers. The average current amplitude at n = 1 is about 50.44 μA, which is greatly increased to 301.95 μA at n = 4, as indicated in Fig. 11.13c. Secondly, the peak density of current output is also obviously increased when the unit number increased from 1 to 4. Thirdly, the peak density of voltage signals is still an increasing function of the unit number. Here the average voltage peak amplitudes hold almost constant with elevated unit numbers, which is attributed to the electrically parallel-connected units. Based on the

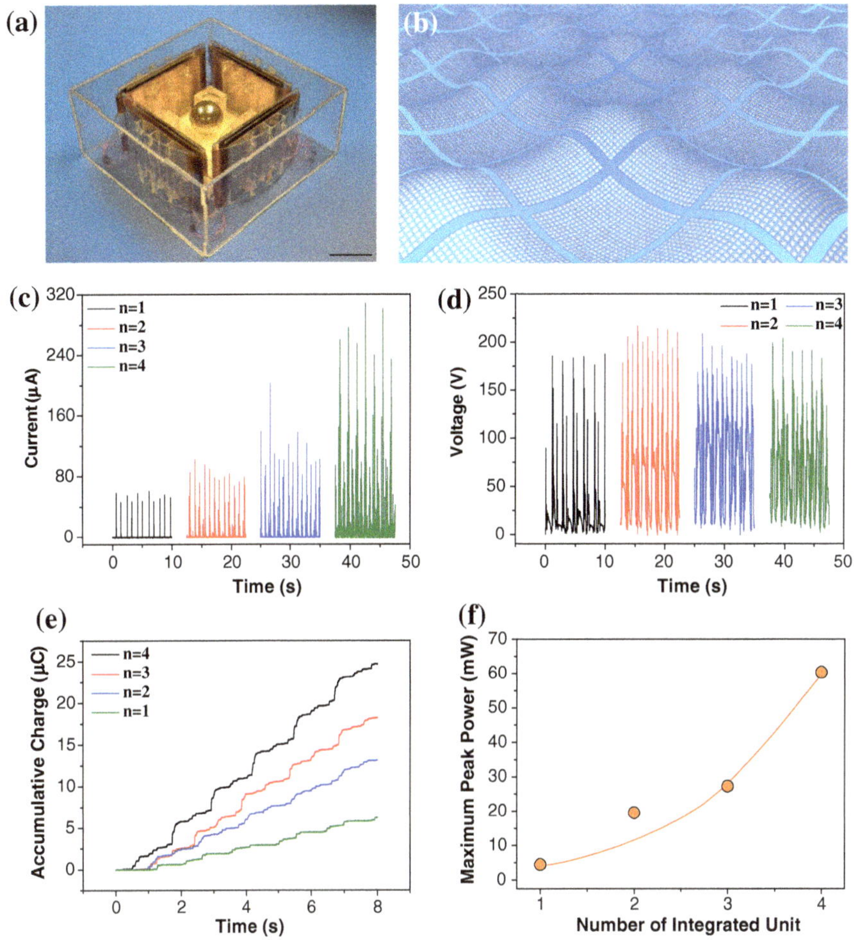


**Fig. 11.13** TENG-NW and its electrical output characterization for water wave energy harvesting. **a** Photograph of an as-fabricated single unit of the TENG-NW. The scale bar is 5 cm. **b** Schematic illustration of the TENG-NW that is consisted of thousands of single units. **c** Rectified short circuit current of the TENG-NWs with unit number n = 1, 2, 3, 4. **d** Open circuit voltage of the TENG-NWs with unit number n = 1, 2, 3, 4. **e** Accumulative induced charges generated by the TENG-NWs with unit number n = 1, 2, 3, 4. **f** Dependence of the peak power output on the resistance of the unit numbers of TENGs. Here, all of the units are electrically connected in parallel. Reproduced with permission from American Chemical Society

above observations, it can be inferred that a quasi-direct/direct output signal could be obtained if thousands of units working together as a TENG-NW. And also the output power frequency is totally controllable and tunable by configuration design of the TENG-NW. With a diode bridge, the total accumulative induced charges can also be measured, as show in Fig. 11.13e. And a directly proportional relationship was found between the unit number and the charging accumulation rate. This is because more units in the TENG-NW means more collisions are launched in a unit period of time, thus faster of triboelectric charges generation, and higher charging accumulation rate is expected. Additionally, further effort was committed to investigate the dependence of the peak power output on the resistance of the TENG-NW unit numbers. As demonstrated in Fig. 11.13f, the peak output power of the TENG-NW is exponentially increased with the elevated unit numbers.

To extrapolate the capability of the TENG-NW for large scale blue energy harvesting, the TENG-NW with a single unit was considered. The generated average power $E_0$ in a single current peak can be calculated as:

$$E_0 = \alpha \cdot I_{sc} \cdot V_{oc} \cdot \Delta t \tag{11.1}$$

where $I_{sc}$ and $V_{oc}$ are the average short-circuit current and open-circuit voltage for the single unit, which are 50 μA and 180 V, respectively. $\Delta t$ is the peak width of the short-circuit current with a value of 0.0184 s. And $\alpha$ is a factor in a range of 0–1. For a rough estimation, assuming $\alpha = 0.5$. Submit all the value into Eq. (11.1), $E_0$ is calculated to be 82.8 μJ.

Regarding a single unit of the TENG-NW, one collision will generate two current peaks with identical $I_{sc} \cdot \Delta t$, consequently, the generated average power $E_c$ in one collision can be estimated as:

$$E_c = 2\,E_0 \tag{11.2}$$

And the generated power per second per unit volume $E_{cv}$ can be expressed as:

$$E_{cv} = f \cdot \beta \cdot \frac{E_c}{V_o} \tag{11.3}$$

where $\beta$ is the volume ratio of TENG-NW, since all of the units in the TENG-NW is not close packing. Here, $\beta$ is designed to be 0.6. $V_o$ as the effective volume of a single unit.

According to the experimental design, $V_o$ = 6 cm • 12 cm • 12 cm = 864 cm$^3$. $f$ is the ball collision frequency, assuming an average collision frequency is 2 Hz for the ocean wave. Thus, $E_{cv}$ is estimated to be 0.23 J/m$^3$.

Features as extremely light-weight, high anticorrosion to the marine environment, it is reasonable to construct the TENG-NW with a depth of 5 m in the ocean, then, the generated energy in a water area of 1 km$^2$ per s could be estimated as:

$$E = 5\,\text{m} \cdot 1\,\text{km}^2 \cdot E_{cv} \tag{11.4}$$

Submit Eq. (11.3) into Eq. (11.4), $E$ is thus calculated to be 1.15 MJ.

Consequently, an average power output of 1.15 MW was expected in a water area of 1 km square for the reported TENG-NW. To demonstrate its feasibility as a practical power source, a small scale TENG-NW was developed with four units connected in parallel. As displayed in Fig. 11.14a, a small TENG-NW with 4 units

**Fig. 11.14** Demonstration of a small scale TENG-NW as a sustainable power source. **a** Photograph of a small scale TENG-NW with 4 units in a swimming pool. The scale bar is 10 cm. **b** Photograph shows the TENG-NW works and is capable of realizing a self-powered SOS system for ocean emergency. The scale bar is 10 cm. **c** Schematic illustration of the configuration of the proposed TENG-NW for practical applications. Reproduced with permission from American Chemical Society

is floating on the water surface of a home swimming pool. As a light wind passed by, the aroused gentle wave can start to drive the TENG-NW, which is capable of realizing a self-powered SOS system for ocean emergency, as shown in Fig. 11.14b. Figure 11.14c is a schematic illustration of the configuration of the TENG-NW for practical applications. Here, a multi-layer electrical connection is proposed. In a foot layer, thousands of single units are electrically connected in parallel to form a community, which could effectively enhance the output current. In the upper layer, thousands of communities could be electrically connected in series to obtain an enhanced output voltage. As a result, both the current and voltage, amplitude and peak density, will be greatly boosted up to a high level for practical applications. Compared with current technologies for wave energy harvesting, the TENG-NW holds unprecedented advantages towards large-scale ocean wave energy harvesting. Firstly, TENG-NW is suitable to harvest wave energy in a wide range of wave motions, from subtle to strong, transverse wave to lateral wave. Furthermore, not like the electromagnetic effect based wave energy harvesting, which mainly relies on the undercurrents, the TENGs will show great potential in harvesting energy from both the undercurrent and the surface waves. In addition, most of the wave motions are multidirectional, the TENG-NW, with a rationally designed structure, renders an innovative and effective approach to fully utilize the wave motion from all-directions. And it can not only be applied in the epicontinental sea, but also can easily be implemented in almost all of the water area.

Secondly, since the TENGs are mostly made from polymer materials without magnets, the load of the total device is expected to be decreased greatly compared to current electromagnetic generators, which are made from heavy materials (such as metals). Moreover, relying on the surface charging effect, only small amount of materials are needed [21–23]. The TENG-NW is thus cost-effective as well as light-weight, which makes it possible high anticorrosion to the marine environment and floating on the water surface for wave energy harvesting, which will greatly eliminate the needs of building poles or towers for holding traditions electromagnetic generators for wave energy harvesting.

Thirdly, with the network architecture and packaging of individual TENG, the failure rate of the structure could be greatly decreased compared to the state-of-art technologies. And besides, in a case if any of a single unit in the TENG-NW was broken, the entire network can tolerate the failed one and the working status of the rest of the TENGs will not be affected.

In a word, with a distinctive working mechanism and rational designed device structures, the TENG-NW will distinguish itself in the field of energy harvesting with unique applicability, especially playing a complementary role in current development of wave energy harvesting.

## 11.3 Summary

By using TENGs for blue energy harvesting holds unique advantages. Firstly, TENG is suitable to harvest wave energy in a wide range of wave motions, from subtle to strong. Compared with current electromagnetic effect based wave energy harvesting, which mainly replies on the undercurrents, the TENGs will show great potential in harvesting energy from both the undercurrent and the surface waves, especially with considering the low frequency of water wave.

Secondly, since the TENGs are mostly made from polymer materials without magnets, the load of the total device is expected to be decreased greatly compared to current machines made from heavy materials (such as metals). The total load reduction is expected to be more than 70 % by simply comparing the densities of common polymers and metals. Also, the total cost could thus be decreased greatly. The use of such light TENG eliminate the needs of building poles or towers for holding traditions EMGs.

Thirdly, with the network architecture and packaging of individual TENG, the failure rate of the structure could be greatly decreased compared to the state-of-art technologies. And besides, in a case if any of a single unit in the TENG network was broken, the entire network can tolerate the failure one and the working status of the rest of the TENGs will not be affected.

In summary, the development of the TENG as an innovative technology for marine and hydrokinectic energy harvesting would have great impact on the development of sustainable energy. It is anticipated to be the most effective approach for harvesting kinetic wavy energy from sea water in the vicinity of the surface, which, however, cannot be effectively done using traditional EMG. The TENG network harvests energy from multi-dimensional waves and flowing streams at the same time, maximizing the utilization of ocean energy. This approach is high performance and it can be applied to any water area of any depth without building towers. At the same time, the cost of the energy production could be reduced greatly by utilizing the cost-effective materials such as polymers. TENG, as a fundamentally new and promising energy technology, shows great potential to harvest "blue" energy with unique advantages. The TENG network is a new paradigm for sea water energy harvesting with unprecedented performance.

# References


1. Z.L. Wang, Triboelectric nanogenerators as new energy technology and self-powered sensors —principles, problems and perspectives. Farad. Discuss. **176**, 447–458 (2014)
2. Z.L. Wang, Triboelectric nanogenerators as new energy technology for self-powered systems and as active mechanical and chemical sensors. ACS Nano **7**(11), 9533–9557 (2013)
3. Z.L. Wang, J. Chen, L. Lin, Progress in triboelectric nanogenertors as new energy technology and self-powered sensors. Energy Environ. Sci. **8**, 2250–2282 (2015)
4. S. Wang, L. Lin, Z.L. Wang, Triboelectric nanogenerators as self-powered active sensors. Nano Energy **11**, 436–462 (2015)
5. G. Zhu, B. Peng, J. Chen, Q. Jing, Z.L. Wang, Triboelectric nanogenerators as a new energy technology: from fundamentals, devices, to applications. Nano Energy **14**, 126–138 (2015)
6. Z.-H. Lin, G. Cheng, S. Lee, K.C. Pradel, Z.L. Wang, Harvesting water drop energy by a sequential contact-electrification and electrostatic-induction process. Adv. Mater. **26**(27), 4690–4696 (2014)
7. Z.-H. Lin, G. Cheng, L. Lin, S. Lee, Z.L. Wang, Water–solid surface contact electrification and its use for harvesting liquid-wave energy. Angew. Chem. Int. Ed. **125**(48), 12777–12781 (2013)
8. Z.-H. Lin, G. Cheng, W. Wu, K.C. Pradel, Z.L. Wang, Dual-mode triboelectric nanogenerator for harvesting water energy and as a self-powered ethanol nanosensor. ACS Nano **8**(6), 6440–6448 (2014)
9. G. Zhu, Y. Su, P. Bai, J. Chen, Q. Jing, W. Yang, Z.L. Wang, Harvesting water wave energy by asymmetric screening of electrostatic charges on a nanostructured hydrophobic thin-film surface. ACS Nano **8**(6), 6031–6037 (2014)
10. J. Chen, J. Yang, Z. Li, X. Fan, Y. Zi, Q. Jing, H. Guo, Z. Wen, K.C. Pradel, S. Niu, Z.L. Wang, Networks of triboelectric nanogenerators for harvesting water wave energy: a potential approach toward blue energy. ACS Nano **9**(3), 3324–3331 (2015)
11. W. Yang, J. Chen, G. Zhu, J. Yang, P. Bai, Y. Su, Q. Jing, X. Cao, Z.L. Wang, Harvesting energy from the natural vibration of human walking. ACS Nano **7**(12), 11317–11324 (2013)
12. W. Yang, J. Chen, Q. Jing, J. Yang, X. Wen, Y. Su, G. Zhu, P. Bai, Z.L. Wang, 3D Stack integrated triboelectric nanogenerator for harvesting vibration energy. Adv. Funct. Mater. **24**(26), 4090–4096 (2014)
13. J. Yang, J. Chen, Y. Liu, W. Yang, Y. Su, Z.L. Wang, Triboelectrification-based organic film nanogenerator for acoustic energy harvesting and self-powered active acoustic sensing. ACS Nano **8**(3), 2649–2657 (2014)
14. Y. Su, G. Zhu, W. Yang, J. Yang, J. Chen, Q. Jing, Z. Wu, Y. Jiang, Z.L. Wang, Triboelectric sensor for self-powered tracking of object motion inside tubing. ACS Nano **8**(4), 3843–3850 (2014)
15. J. Chen, G. Zhu, J. Yang, Q. Jing, P. Bai, W. Yang, X. Qi, Y. Su, Z.L. Wang, Personalized keystroke dynamics for self-powered human–machine interfacing. ACS Nano **9**(1), 105–116 (2015)
16. J. Yang, J. Chen, Y. Su, Q. Jing, Z. Li, F. Yi, X. Wen, Z. Wang, Z.L. Wang, Eardrum-inspired active sensors for self-powered cardiovascular system characterization and throat-attached anti-interference voice recognition. Adv. Mater. **27**(8), 1316–1326 (2015)
17. G. Zhu, Y.S. Zhou, P. Bai, X.S. Meng, Q. Jing, J. Chen, Z.L. Wang, A shape-adaptive thin-film-based approach for 50 % high-efficiency energy generation through micro-grating sliding electrification. Adv. Mater. **26**(23), 3788–3796 (2014)
18. Y. Xie, S. Wang, S. Niu, L. Lin, Q. Jing, Y. Su, Z. Wu, Z.L. Wang, Multi-layered disk triboelectric nanogenerator for harvesting hydropower. Nano Energy **6**, 129–136 (2014)
19. G. Cheng, Z.-H. Lin, Z.-L. Du, Z.L. Wang, Simultaneously harvesting electrostatic and mechanical energies from flowing water by a hybridized triboelectric nanogenerator. ACS Nano **8**(2), 1932–1939 (2014)

20. X. Wang, S. Niu, Y. Yin, F. Yi, Z. You, Z.L. Wang, Triboelectric nanogenerator based on fully enclosed rolling spherical structure for harvesting low-frequency water wave energy. Adv. Energy Mater. **5**, 1501467 (2015)
21. G. Zhu, W. Yang, T. Zhang, Q. Jing, J. Chen, Y.S. Zhou, P. Bai, Z.L. Wang, Self-powered, ultrasensitive, flexible tactile sensors based on contact electrification. Nano Lett. **14**(6), 3208–3213 (2014)
22. G. Zhu, J. Chen, T. Zhang, Q. Jing, Z.L. Wang, Radial-arrayed rotary electrification for high performance triboelectric generator. Nat Commun. **5**, 3426 (2014)
23. S. Wang, L. Lin, Z.L. Wang, Nanoscale triboelectric-effect-enabled energy conversion for sustainably powering portable electronics. Nano Lett. **12**(12), 6339–6346 (2012)

# Chapter 12
# Hybrid Cell Composed of Triboelectric Nanogenerator

Our living environment has abundant energy resources in a variety of forms, such as mechanical energy (including vibration, sonic wave, wind and hydraulic), thermal energy, solar energy, and electrochemical energy. Harvesting these types of energies is of critical importance for long-term energy needs and sustainable development of the world. Over the years, rationally designed materials and technologies have been developed for converting these energies into electricity. Mechanical energy harvesting has been achieved by triboelectric nanogenerators (TENG), piezoelectric nanogenerators (PENG), and traditionally, electromagnetic generators. Thermal energy can be harvested by thermoelectric devices and pyroelectric nanogenerators. Solar and electrochemical energies are mainly harvested by solar cells and electrochemical cells, respectively. Innovative approaches have to be developed for conjunctional harvesting of multiple types of energies using an integrated structure/material so that the energy resources can be effectively and complimentarily utilized whenever and wherever one or all of them are available. This is the case for portable/wearable electronics as well as micro grid, because the environment the device will be deployed varies from time to time and from location to location. Therefore, hybrid cells have been developed that are designed for simultaneously harvesting different types of energies through two or more approaches. The multiple energy harvesting approaches can work simultaneously or individually, and they can be integrated in parallel or in series for optimizing the output current or voltage, respectively. This chapter is to demonstrate the innovative approach of hybrid cell based on TENG for developing integrated technologies for effectively scavenging available energies in our environment around the clock.



Z.L. Wang et al., *Triboelectric Nanogenerators*,
Green Energy and Technology, DOI 10.1007/978-3-319-40039-6_12

## 12.1 AC–AC Hybrid Cell

### *12.1.1 Hybrid Electromagnetic and Triboelectric Nanogenerator*

Since the electromagnetic induction effect was discovered by Michael Faraday in 1831, the electromagnetic generator (EMG) has been invented as the most important mean for power generation. In terms of TENG, it can provide not only micro-scale energy for powering micro/nanosystems, but also macro-scale energy for powering portable electronics and household appliances, which is likely to be a parallel or possibly equivalently important technology as the traditional EMG for power generation. And besides, TENG is demonstrated to be irreplaceable and unbeatable for harvesting low-frequency mechanical energy in cases such as human motion and water wave, [1] which makes it complement with EMG for effectively harvesting broadband-frequency mechanical energy.

By integrating TENG and EMG, a hybridized mechanical energy harvesting technology has been used to largely increase the total output power of the energy harvesting device, and the core idea of this technology is based on how to extract more electricity from one mechanical motion. In this regard, different designs of hybrid generators have been developed to fulfill this goal in various forms [2–8].

#### 12.1.1.1 Theoretical Comparison and Conjunction Operations of Electromagnetic Generator and Triboelectric Nanogenerator

Before we go into details of device architectures of the TENG-EMG hybrid cell, it is essential to give a comprehensive theoretical comparison of these two power generators, considering their similar output profile but distinct working principles. As reported by Zhang et al., systematic theoretical analysis and comparison of EMG and TENG from their working mechanisms and governing equations are given [2]. Based on the two different electricity generating principles, a hybrid generator is designed and two hybrid operation modes in parallel and series are presented and characterized. It has also established the basis of the TENG as a new energy technology that could be equivalently important as the EMG for general power application even at a large scale.

The EMG is based on the fundamental principle of the electromagnetic induction phenomenon, which is schematically illustrated in Fig. 12.1a. When a conductor stick with an external electric circuit moves across the magnetic induction lines, the induced electrodynamic potential is generated across the conductor stick:

$$E = B \cdot l \cdot v \tag{12.1}$$

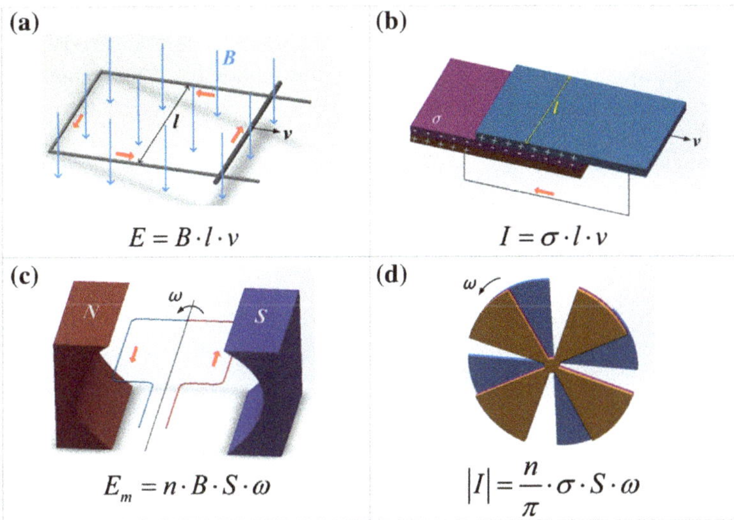


(e)

| Symbol | Definition |
|---|---|
| $E$ | induced electrodynamic potential |
| $E_m$ | peak value of $E$ |
| $B$ | magnetic flux density |
| $l$ | length of wire |
| $v$ | velocity of movement |
| $n$ | turns of coils |
| $S$ | coil area |
| $\omega$ | angular velocity |
| $\frac{\Delta\Phi}{\Delta t}$ | change rate of magnetic flux |

(f)

| Symbol | Definition |
|---|---|
| $I$ | induced current |
| $\|I\|$ | absolute value of $I$ |
| $\sigma$ | electric charge density |
| $l$ | width of friction surface |
| $v$ | velocity of movement |
| $n$ | segments number |
| $S$ | friction area |
| $\omega$ | angular velocity |
| $\frac{\Delta Q}{\Delta t}$ | transfer rate of electric charge |

**Fig. 12.1** Theoretical comparison of EMG and TENG. **a**, **c**, **e** Schematic diagram, fundamental principles and governing equation and parameters of the electromagnetic generator (EMG). **b**, **d**, **f** Schematic diagram, fundamental principles and governing equation and parameters of the triboelectric nanogenertor (TENG). Reproduced with permission from Wiley [2]

where $B$ is the magnetic flux density, $l$ is the length of the conductor stick and $v$ is the velocity of the conductor stick cutting the magnetic induction lines.

Unlike the EMG, the TENG is based on the triboelectric effect and the electrostatic induction phenomenon to convert mechanical energy into electricity. The triboelectric effect is a contact-induced electrification in which a material becomes electrically charged after it is contacted with a different material. The polarity of the charges to be carried by a material depends on its relative electron affinity in comparison with its counterpart. The electrostatic induction is a redistribution of electrons between two electrons with respect to the change of electric field generated by periodic contact and separation of the electrostatic charges. As discussed in Part I, four fundamental operation modes of the TENG have been developed: vertical contact-separation mode, lateral sliding mode, single-electrode mode and freestanding-triboelectric-layer mode. All of the modes have the same electricity generation process including the charge generation and separation process.

Here, for convenient comparison with the EMG, we pick up the lateral sliding mode for its theoretical analysis, which is schematically illustrated in Fig. 12.1b. The fundamental theory and working principle of the slidng mode TENG has been extensively discussed in Sect. 3.2.1. It should be noted that, given a constant sliding velocity, the short-circuit current of the TENG could be expressed as the following equation:

$$I_{SC} = \frac{\sigma \cdot \Delta S}{\Delta t} = \sigma \cdot l \cdot v \tag{12.2}$$

where $\sigma$ is the triboelectric charge density of the friction surface, $l$ is the width of the friction surface perpendicular to the sliding direction and $v$ is the sliding velocity of the top metal.

Compared with the two governing equations, there are similar expressions and symmetric effects of them, which indicate that the TENG can generate the induced current as a current source while the EMG can generate the induced electrodynamic potential as a voltage source. The output current and voltage are both dependent on three physical variables, respectively. The short-circuit current of the TENG depends on the triboelectric charge density, the width of the friction surface perpendicular to the sliding direction, and the sliding velocity; on the other hand, the electrodynamic potential of the EMG depends on the magnetic flux density, the length of the conductor stick, and the cutting velocity of the conductor stick. Sequentially, the three pair of parameters refer to the intrinsic properties of materials, the device geometry, and the mechanical motion, which demonstrates similarity of the two type of power generators.

When it comes to rotation motion, the rotating counterparts of both generators are also investigated. The schematic diagram of the rotating EMG is reviewed and illustrated in Fig. 12.1c. The rotating EMG consists of a pair of magnets and a group of coils located in the magnetic field. When the coil rotates around the central axis, the magnetic flux through the coils is varying, which will generate an induced electrodynamic potential across the coils. The output voltage of the rotating EMG is alternating as a sinusoidal wave, of which the peak value can be represented as:

$$E_m = n \cdot B \cdot S \cdot \omega \tag{12.3}$$

where $n$ is the turns of coils, $S$ is the area of a single turn of the coil and $\omega$ is the rotational angular velocity of the coil.

Figure 12.1d demonstrates a schematic diagram of the segmentally patterned disk-shaped rotating TENG, in which a periodic overlapping and separation process of the two groups of sectors on the two concentric and closely contacted disks is achieved by relative rotation. The theoretical analysis, working principle, and output performance has been extensively discussed in Sect. 3.2.3. The absolute value of the short-circuit current of the rotating TENG can be represented as:

$$|I| = \frac{\sigma \cdot \Delta S}{\Delta t} = \sigma \cdot \frac{n \cdot r^2 \cdot \Delta \theta}{2\Delta t} = \frac{n}{\pi} \cdot \sigma \cdot S \cdot \omega \tag{12.4}$$

where $n$ is the segments number of the disk, $S$ is the friction area of the disk (half area of the whole circle) and $\omega$ is the rotational angular velocity of the top metal.

The rotating EMG and TENG both have an alternating output, which also have the similar and symmetric expressions in Eqs. (12.3) and (12.4). For the different output waves of sinusoidal and square in theory, the expressions have a difference of $\pi$ in coefficients. The short-circuit current of the TENG and electrodynamic potential of the EMG are both dependent on four physical variables, respectively. Similarly, these four pair of variables are affiliated with intrinsic properties of materials, device structure, device geometry, and extrinsic mechanical motions. There is one more variable of device structure component presented here, as compared with the sliding-mode structures, because the degree of device complexity is higher for the rotating-driven generator.

In general, for the EMG, the general governing equation is reviewed and represented as:

$$E = n \cdot \frac{\Delta \Phi}{\Delta t} \tag{12.5}$$

where $n$ is the turns of coils, and $\Delta\phi/\Delta t$ is the change rate of magnetic flux in each coil. For the TENG, no matter the operation mode is straight-line sliding, rotational sliding or other modes, the induced current is generated due to the transfer of electric charge between two objects. Therefore, the output current of the TENG can be described in a general governing equation as:

$$I = n \cdot \frac{\Delta Q}{\Delta t} \tag{12.6}$$

where $n$ is the segments number which also represents the number of electric charge round-trips between two objects in a unit time, and $\Delta Q/\Delta t$ is the transfer rate of electric charge in each segment. Compared with the two general governing equations, we can see that the TENG supplies induced current due to the transferring electric charge, while the EMG supplies induced voltage due to the changing magnetic flux.

The fundamental principles, rotational structures, governing equations and significances of symbols of the EMG and TENG are illustrated and described all together in Fig. 12.1 for comparison. The results of the analysis and comparison in above indicate that the TENG has a comparative and symmetric relationship with the traditional EMG in theory, which can be considered as a current source while the EMG is considered as a voltage source.

**Table 12.1** Parameters design of EMG and TENG

| EMG | | TENG | |
|---|---|---|---|
| Turns of coils | $n_1 = 12$ | Segments number | $n_2 = 4$ |
| Magnetic flux density | $B = 0.05$ T | Triboelectric charge density | $\sigma = 70\ \mu C/m^2$ |
| Coil area | $S_1 = 25.3\ cm^2$ | Friction area | $S_2 = 12.4\ cm^2$ |
| Angular velocity | $\omega = 20\ \pi$ rad/s | Angular velocity | $\omega = 20\ \pi$ rad/s |
| Induced electrodynamic potential | $E_m = 95.3$ mV | Induced current | $I = 6.94\ \mu A$ |

Reproduced with permission from Wiley [2]

Besides the theoretical comparison, the output characteristics of the rotating EMG and TENG are also compared with two sets of parameters as detailed in Table 12.1. The rotating TENG is designed with four sectors structure and the empirical triboelectric charge density of 70 μC/m$^2$, while the rotating EMG is designed with twelve turns of a coil and the magnetic flux density of 0.05 T. The friction area of the TENG is 12.4 cm$^2$ and the coil area of the EMG is 25.3 cm$^2$. The disk surface of the TENG is two times of the friction area that is approximately the coil area of the EMG. Both of them have the same angular velocity of 20 π rad/s. According to Eqs. (12.3) and (12.4), the induced peak voltage of the EMG is 95.3 mV and the induced current of the TENG is 6.94 μA in theoretical calculation.

The electric brushes are used for the rotating EMG to convert the AC output into the direct current (DC) output and the output characteristics of the DC-EMG are systematically studied at different external load resistances. The output current decreases with the increasing resistance while the output voltage shows the reverse trend, but both the current and voltage tend to saturate when the resistance is considerably large. The measured open circuit output voltage is 67.6 mV, which accords well with the theoretical peak value of the sinusoidal induced voltage. The maximum output power of 102.6 μW is received when the external resistance is 12.3 Ω. The experimental results indicate that the optimum impedance of the EMG is 12.3 Ω as the EMG has a small output impedance. When the external resistive load is considerably larger than the output impedance, the EMG is equivalent to be considered as a voltage source.

The output of the rotating TENG is also converted into the DC output as a constant current wave by the electric brushes [9] and systematically characterized at different external load resistances. As for the EMG, the output current decreases with the increasing resistance while the output voltage shows the reverse trend, but both the current and voltage curves are smooth when the resistance is small. The measured short-circuit output current is 6.96 μA, which accords with the theoretical value of the induced current well. The output power increases at a lower resistance region and then decreases at a higher resistance region. The maximum output power of 140.4 μW is obtained when the external resistance is 13.8 MΩ. Therefore, the TENG can be equivalent to be considered as a current source with a large output

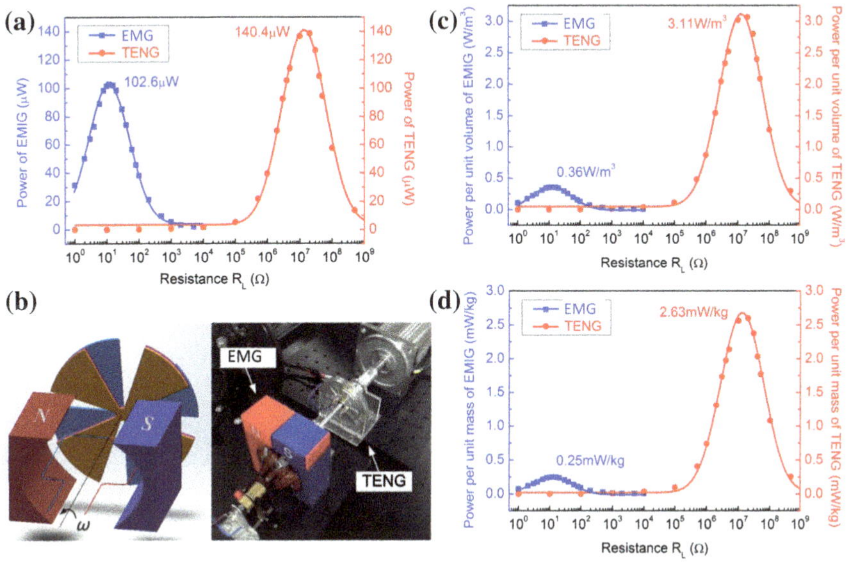


**Fig. 12.2** Output power comparison of the rotating DC-EMG and DC-TENG in the hybrid generator. **a** The output power of the rotating DC-EMG and DC-TENG with different external resistive load. **b** The schematic diagram and photo of the hybrid generator. **c** The output power per unit volume of the rotating DC-EMG and DC-TENG with different external resistive load. The maximum power per unit volume is 0.36 W/m$^3$ for the DC-EMG and 3.11 W/m$^3$ for the DC-TENG. **d** The output power per unit mass of the rotating DC-EMG and DC-TENG with different external resistive load. The maximum power per unit mass is 0.25 mW/kg for the DC-EMG and 2.63 mW/kg for the DC-TENG. Reproduced with permission from Wiley [2]

impedance when the external load resistance is considerably smaller than the output impedance.

The output power of the rotating DC-EMG and DC-TENG are compared in Fig. 12.2a. The matching impedances for receiving the maximum output powers from them are gigantically different in magnitude and the measured maximum output power of the rotating DC-TENG is slightly larger than the DC-EMG. The output power per unit volume and mass of the two generators are also measured and plotted in Fig. 12.2c, d. Since the EMG has two thick and heavy magnets but the TENG has two supported organic glass plates with smaller volume and lower weight, the maximum power per unit volume is 0.36 W/m$^3$ for the EMG and 3.11 W/m$^3$ for the TENG, as well as the maximum power per unit mass is 0.25 mW/kg for the EMG and 2.63 mW/kg for the TENG. Obviously, considering the size and weight, the TENG has significantly larger output power and more advantages over the EMG.

The measured characteristics including the electric output, output impedance, maximum power, volume and mass of the EMG and TENG are summarized all together in Table 12.2 for comparison, which quantitatively supports the theoretical

**Table 12.2** Characteristics comparison of EMG and TENG

| EMG | | TENG | |
|---|---|---|---|
| Open-circuit voltage | $U_S = 67.6$ mV | Short-circuit current | $I_S = 6.96$ μA |
| Internal resistance | $R_{01} = 12.3$ Ω | Internal resistance | $R_{02} = 13.8$ MΩ |
| Maximum power | $P_{m1} = 102.6$ μW | Maximum power | $P_{m2} = 140.4$ μW |
| Volume | $V_1 = 288$ cm$^3$ | Volume | $V_2 = 45.2$ cm$^3$ |
| Mass | $M_1 = 412$ g | Mass | $M_2 = 53.3$ g |
| Maximum power per unit volume | $P_{V1} = 0.36$ W/m$^3$ | Maximum power per unit volume | $P_{V2} = 3.11$ W/m$^3$ |
| Maximum power per unit mass | $P_{M1} = 0.25$ mW/kg | Maximum power per unit mass | $P_{M2} = 2.63$ mW/kg |

Reproduced with permission from Wiley [2]

comparison and well validated that the TENG is a distinct but likely to be a parallelly important technology as the traditional EMG for power generation.

Based on the two different electricity generating principles, a hybrid generator of the rotating DC-EMG and DC-TENG is designed, which is schematically and practically illustrated in Fig. 12.2b. The two generators have a common rotational axis and the input mechanical energy can simultaneously drive both generators under the same angular velocity. The output characteristics of the hybrid generator are measured and analyzed in different conjunction operation modes of the EMG and TENG.

An ideal voltage source has zero internal resistance and the output voltage is independent of the external load, while an ideal current source has infinite internal resistance and the output current is independent of the external load. In the circuit, if an ideal voltage source is connected with an ideal current source in parallel, the output voltage and current on the external load is dictated by the voltage source. On the contrary, if an ideal voltage source is connected with an ideal current source in series, the output voltage and current on the external load is dictated by the current source.

The output characteristics of the rotating DC-EMG and DC-TENG in parallel and series connection are measured and analyzed. Figure 12.3a shows the output current and voltage by varying the load resistance, from 1 Ω to 10 kΩ. The output power is also plotted as a function of external resistance in Fig. 12.3b. The measured saturation value of the effective output voltage is 67.5 mV and the maximum power 103.1 μW is received when the external resistance is 12.3 Ω. The measured curve and values are nearly identical as those for the rotating DC-EMG as shown in Fig. 12.2a, indicating that the parallel connection of the EMG and TENG is nearly equivalent to the EMG for the external load. Likewise, Fig. 12.3c shows the resistance dependence of the output current and voltage when the two are connected in series. The output power is plotted as a function of external resistance in Fig. 12.3d. The measured short-circuit output current is 6.94 μA and the maximum power 140.4 μW is received when the external resistance is 13.8 MΩ. The measured curve and values are almost the same as those for the rotating DC-TENG as

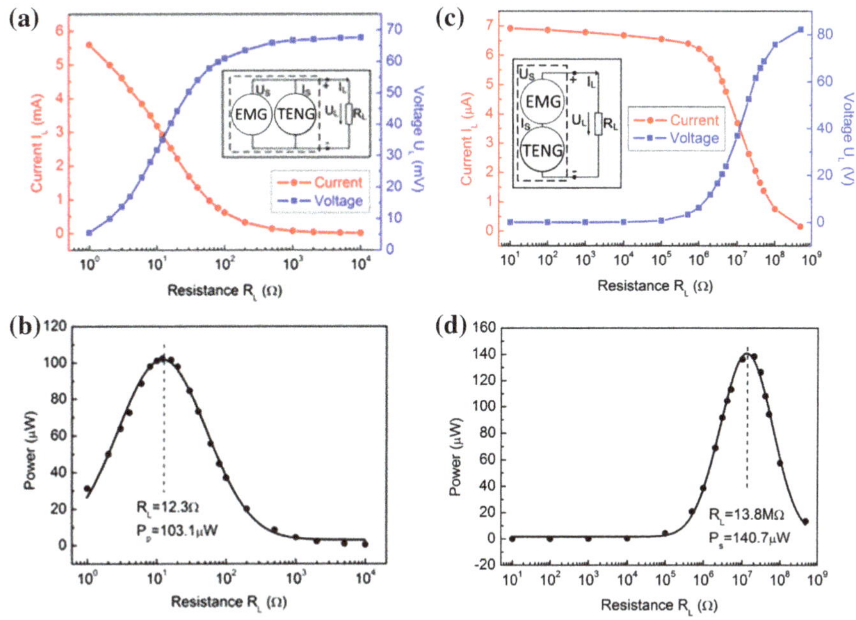


**Fig. 12.3** Output characteristics with different external resistive load of the rotating DC-EMG and DC-TENG in **a**, **b** parallel and **c**, **d** serial connection. **a** and **c** The relationship between the output voltage/current and the resistance of the external load. The *insets* are the circuit diagram of the rotating DC-EMG and DC-TENG in parallel and serial connection with external resistive load, respectively. **b** and **d** The relationship between the output power and the resistance of the external load. The maximum power of 103.1 μW is received when the external resistance is 12.3 Ω in parallel connection, and the maximum power of 140.7 μW is received when the external resistance is 13.8 MΩ in serial connection. Reproduced with permission from Wiley [2]

shown in Fig. 12.2a, which indicate that a serial connection of the EMG and TENG is nearly equivalent to the TENG for the external load. The experimental results further validate the characteristics of the TENG as a current source while the EMG as a voltage source.

In summary, the analysis and comparison in details of the theoretical models, working mechanisms and governing equations of the EMG and TENG for harvesting mechanical energy are performed, which has a comparative and symmetric relationship with each other. The TENG is a current source with a large output impedance, and the EMG is equivalent to a voltage source with a small output impedance. A hybrid generator based on the two different electricity generating principles is presented and characterized in different circuit connections, which are in good agreement with the theoretical analysis. The theoretical comparison and experimental validations is a milestone for in-depth analysis and comparison between the TENG and the traditional EMG, and establishes the basis of using the TENG as a new energy technology that could be equivalently important as the EMG for general power application.

#### 12.1.1.2 Rotating-Disk-Based Hybridized Electromagnetic-Triboelectric Nanogenerator

Zhong et al. demonstrated a rotating-disk-based hybridized nanogenerator that integrated an EMG and a TENG for simultaneously scavenging biomechanical energy from one rotating motion as a mobile power source [5]. The hybridized nanogenerator with a planar structure and a diameter of 14 cm can produce output powers of 17 mW for TENG and 50 mW for EMG through scavenging human-hand-induced biomechanical energy. The hybridized nanogenerator has a much better charging performance than that of individual energy harvesting unit (EMG or TENG) for charging a capacitor of 6600 μF. A commercial globe light can be sustainably powered by using a capacitor charged by the hybridized nanogenerator in outdoor environment, which gives an illumination intensity of 1700 lx. The hybridized-nanogenerator-based mobile power source has the potential applications in sustainably driving some mobile electronic devices.

Figure 12.4a illustrates a schematic diagram of a hybridized electromagnetic-triboelectric nanogenerator, which consists of an EMG including the top and bottom layers (1 and 5) and a TENG including the middle layers (2, 3, 4) with the planar structures, where the rotator and the stator are composed of layers 1–2 and layers 3–5, respectively. For the rotator of EMG (layer 1) at the top, six magnets uniformly distribute in an acrylic substrate with the magnetic poles in alternating arrangement manner. For the stator of EMG (layer 5) at the bottom, the corresponding six groups of coils were integrated on a substrate in series connection at the same positions of the magnets. The rotator of the TENG consists of radially-arrayed sectors separated by equal-degree intervals in between, where each sector unit has a central angle of 1° (layer 2). The stator of TENG includes a layer of triboelectric material (polyamide film, layer 3) and a layer of complementary-patterned electrode networks that are separated by trenches in between (layer 4). Figure 12.4b displays the corresponding photographs of each layer.

Under a rotation rate of 200 r/min, the output voltage and current of TENG can reach about 75 V and 0.33 mA, respectively, as depicted in Fig. 12.5a, b. Under the same rotation condition, the output voltage and current of EMG are about 0.62 V and 57.8 mA, respectively, as displayed in Fig. 12.5 c, d. By analysis of the output voltage/current curves, the corresponding working frequencies of TENG and EMG are about 600 and 10 Hz, respectively. Figure 12.5e, f display the resistance dependence of both the output current and the corresponding output power of the TENG and EMG, respectively. The output currents of both the TENG and EMG decrease with increasing the loading resistance, while the instantaneous powers of both the two generators increase in the initial stage and then decrease under the larger loading resistances. The largest instantaneous powers of TENG and EMG are about 8.6 mW (powers per unit mass/volume: 119 μW/g and 261 W/m$^3$) at a loading resistance of 0.2 MΩ and 8.4 mW (powers per unit mass/volume: 24 μW/g and 56 W/m$^3$) at a loading resistance of 12 Ω, respectively. To exactly compare the output performances of the TENG, EMG, and hybridized nanogenerator, Fig. 12.5g, h show the generated electrical energy and the output current signals of

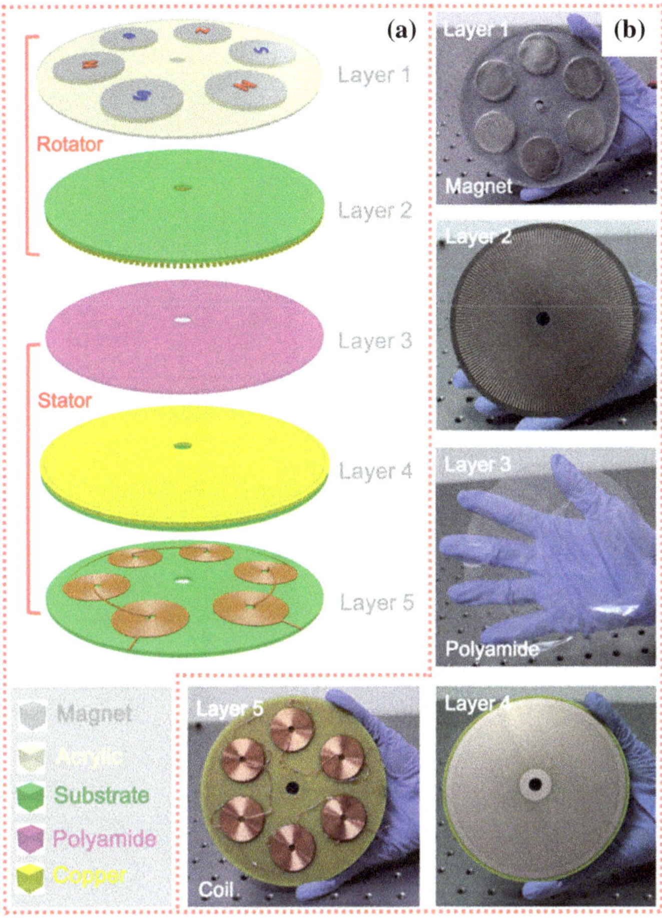


**Fig. 12.4** **a** Schematic diagram of the designed hybridized nanogenerator. **b** Photographs of the hybridized nanogenerator. Reproduced with permission from Elsevier [5]

TENG and EMG at a loading resistances of 0.2 MΩ and 12 Ω, respectively. Here, the generated electric energy $E_{\text{electricity}}$ can be expressed as:

$$E_{\text{electricity}} = \int_{t_1}^{t_2} I^2 \cdot R \cdot dt \tag{12.7}$$

where $I$ is the output current of the nanogenerator, and $R$ is the loading resistance. Under the same time interval between $t_1$ and $t_2$ (100 ms), the produced electrical energies of TENG and EMG are the same (0.4 mJ), suggesting that the efficiency for converting the mechanical energy into electricity can be enhanced 100 % by using the hybridized nanogenerator technology as compared with the individual energy harvesting unit (TENG or EMG).

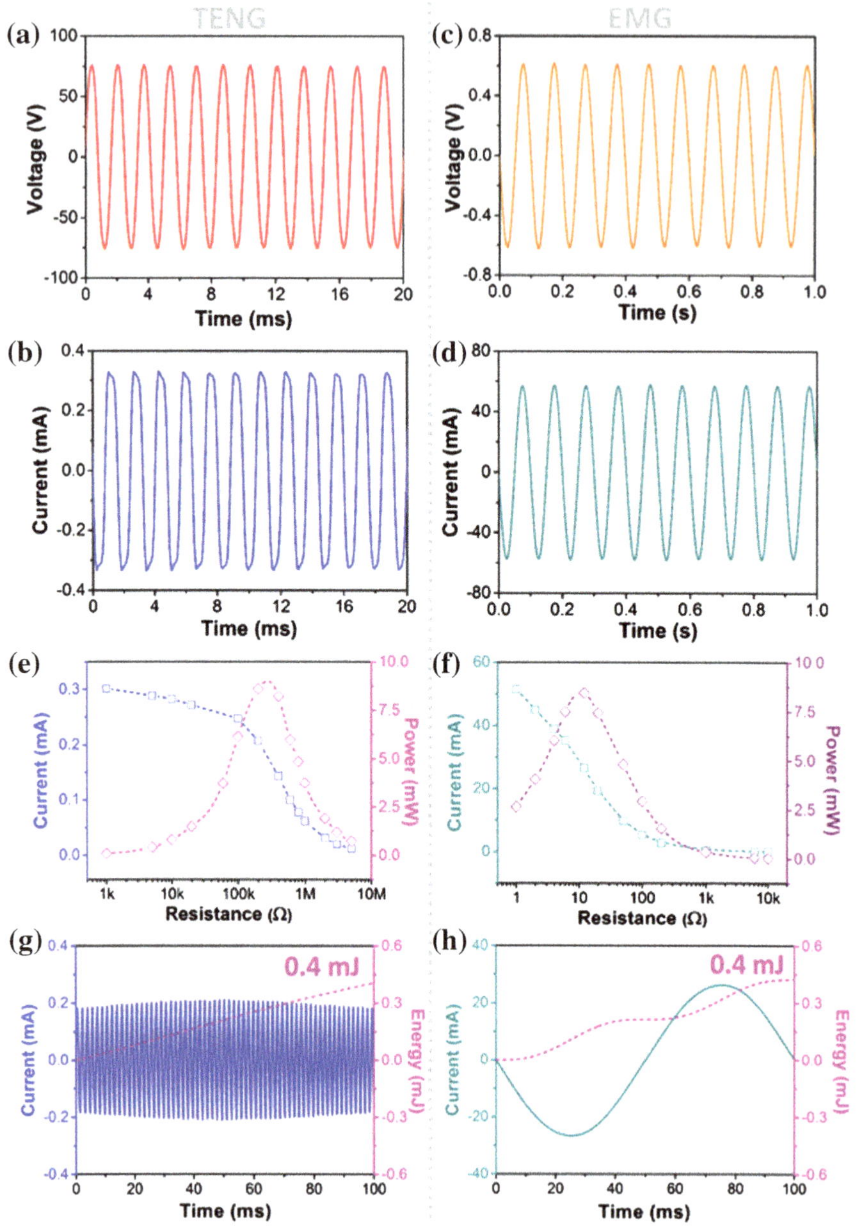


**Fig. 12.5** **a**, **b** Open-circuit voltage (**a**) and short-circuit current (**b**) of TENG under a rotation rate of 200 r/min. **c**, **d** Open-circuit voltage (**c**) and short-circuit current (**d**) of EMG under a rotation rate of 200 r/min. **e**, **f** The relationships between the output current/power and the loading resistance of TENG (**e**) and EMG (**f**). **g**, **h** The generated electrical energy of TENG (**g**) and EMG (**h**) in 100 ms. Reproduced with permission from Elsevier [5]

Figure 12.6a depicts that the different generators were used to charge a capacitor of 6600 μF under the rotation of human hand, clearly showing that the hybridized nanogenerator has much better charging performance than that of individual energy harvesting unit (TENG or EMG). To demonstrate the capability of the hybridized nanogenerator as a mobile power source, the TENG and the EMG were connected to two groups of LEDs (TENG for top 20 LEDs, EMG for bottom 20 LEDs), respectively. As illustrated in Fig. 12.6b, these LEDs can be lighted up at the same time under the human hand driven rotation motion of hybridized nanogenerator. Figure 12.6c shows that a white globe light can be continually lighted up by the hybridized nanogenerator, where a capacitor of 6600 μF was utilized to connect both the nanogenerator and the white globe light. The corresponding intensity of illumination is up to 1700 lx, as displayed in Fig. 12.6d. The obtained illumination was sufficient for reading the printed texts in complete darkness condition (Fig. 12.6e). The mobile characteristic of the hybridized nanogenerator was also demonstrated by using the nanogenerator to sustainably power a globe light in the wild, as depicted in Fig. 12.6f.

In summary, a rotating-disk-based hybridized nanogenerator including an EMG and a TENG for simultaneously scavenging biomechanical energy from human hand induced rotating motions as mobile power source has been demonstrated. Under a rotating rate of 200 r/min, the EMG and TENG can produce an output powers of about 8.4 mW at a loading resistance of 12 Ω and 8.6 mW at a loading resistance of 0.2 MΩ, respectively, where the generated electrical energy of the hybridized nanogenerator is more than double of that generated by individual energy harvesting units. At the best matched loads for each, the EMG and TENG can produce an output power density per unit mass/volume of 24 μW/g and 56 $W/m^3$, and 119 μW/g and 261 $W/m^3$, respectively, indicating the unique advantages that can be offered by TENG owing to its light weight, low-cost and the potential for large-scope application as EMG. By charging a capacitor, the hybridized nanogenerator exhibited better charging performance than that of the individual energy harvesting units. A commercial globe light can be sustainably lighted up by scavenging the biomechanical energy from human hand induced rotating motions, which has an illumination intensity up to 1700 lx. This invention of the hybridized nanogenerator may push forward a significant step toward the practical applications of biomechanical energy harvesting units as a mobile power source in the wild.

### *12.1.2 Hybrid Triboelectric-Piezoelectric/Pyroelectric Nanogenerator*

In additional to TENG and EMG, piezoelectric nanogenerators (PENG) have drawn a lot of attention due to its capability to harvest mechanical energy through the deformation of piezoelectric materials, including ZnO [10], polyvinylidene fluoride

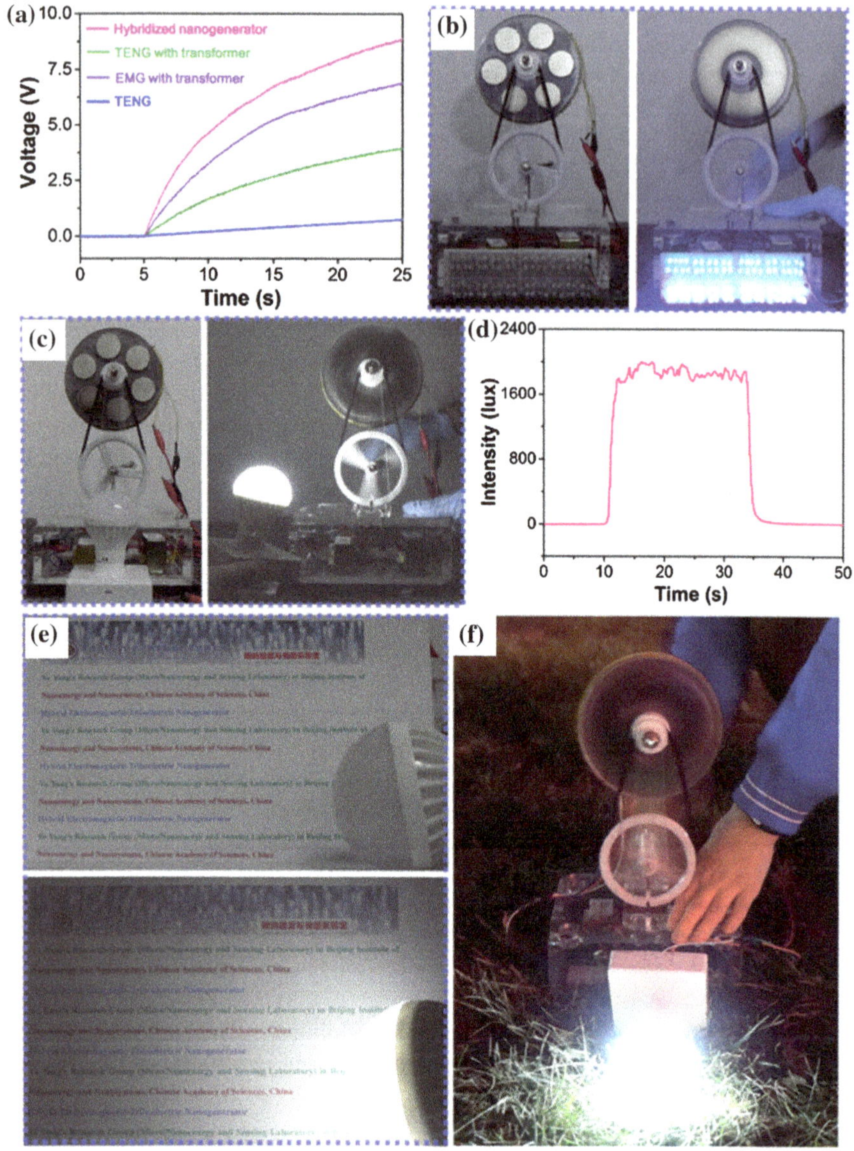


**Fig. 12.6** **a** Measured voltages of a 6600 μF capacitor charged by the TENG, TENG or EMG with transformer, and the hybridized nanogenerator (EMG and TENG with transformer in parallel) driven by human hand rotation. **b** Photographs of 40 LEDs lighted up by the hybridized nanogenerator. The top and bottom 20 LEDs were connected to the TENG and EMG, respectively. **c** Photographs of a globe light that can be directly powered by the hybridized nanogenerator with a capacitor of 6600 μF. **d** The illumination intensity of the globe light in (**c**). **e** Photographs of the hybridized nanogenerator driven globe light that can be used for providing the illumination to read the printed text in complete darkness condition. **f** Photograph of a globe light that can be directly powered by the human hand rotation driven hybridized nanogenerator in the wild. Reproduced with permission from Elsevier [5]

(PVDF) [11], PZT [12], $BaTiO_3$ [13], and PMN-PT [14]. During operation of the PENG, the friction or contact between the electrodes/piezoelectric material and substrate is inevitable. Thus the triboelectrification effect during the friction or contact can be combined with electrostatic induction to drive a TENG. Through novel designs, TENG can be hybridized with PENG to supply more power to drive electronics and to achieve self-powered active sensors. Therefore, hybrid triboelectric-piezoelectric NG have been developed effectively to harvest mechanical energies [15–18].

Pyroelectric effect can be used to harvest the thermal energy from the time-dependent temperature fluctuations [19, 20]. Since the mechanical and thermal energies are not always available at the same time in the environment, it is necessary to develop a hybrid energy cell technology to simultaneously/individually harvest both energies by using an integrated device. Triboelectric-pyroelectric hybrid energy cell has been reported by Yang et al. for degradation of methyl orange by self-powered electrocatalytic oxidation [21].

At the same time, the energy conversion efficiency of the TENG may still be limited by the energy loss during the power generation process. Take the sliding TENG as an example, substantial amount of energy input will be wasted through friction induced heat dissipation, especially for the high-frequency sliding motion; besides, the work done by the indispensable normal force applied on the contact surface will result in additional energy loss. In this regard, high-efficient nanogenerators have been developed based on non-contact approach [3, 22] to minimize the energy loss in the friction; from another point of view, it is also reasonable to fully scavenge the lost energy by hybridizing supplementary power generators. Here, pyroelectric nanogenerators can be employed for harvesting the thermal energy from the friction induced temperature fluctuation, and piezoelectric nanogenerators can harvest the mechanical energy from the small deformation of materials caused by the normal force. Triboelectric-pyroelectric-piezoelectric hybrid cell has been reported by Zi et al. to achieve high-efficiency mechanical energy harvesting [23].

#### 12.1.2.1 r-Shaped Hybrid Triboelectric-Piezoelectric Nanogenerator

Han et al. present the first r-shaped hybrid triboelectric-piezoelectric NG to generate electricity through mechanical input [15]. Both piezoelectric and triboelectric outputs are generated through the hybrid NG, with the micro/nanostructures fabricated on the surface to enhance the performance. A schematic diagram and the photograph of the r-shape hybrid NG are shown in Fig. 12.7a. This r-shape hybrid NG consists of a PENG (the upper bent part) and a TENG (the lower straight part). The bent part includes a polarized PVDF film with thickness of 100 μm, surface-nanostructured Al electrodes on the both sides of the PVDF film, and a polyethylene terephthalate (PET) film which is bent to be an arc-shape with length of 4 cm and diameter of 6.5 cm. The straight part consists of the micro/nano dual-scale (MNDS) polydimethylsiloxane (PDMS), PET/ITO electrode, and the

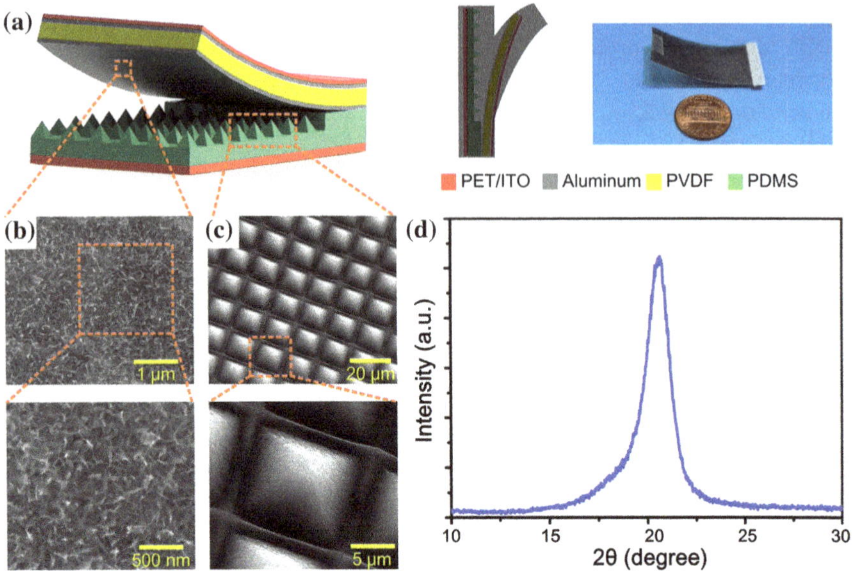


**Fig. 12.7** **a** Schematic diagram and the photo of the fabricated r-shaped hybrid NG. **b** and **c** SEM images of **b** the Al electrode with nanostructures and **c** MNDS PDMS. **d** XRD spectrum of the polarized PVDF. Reproduced with permission from American Chemical Society [15]

surface-nanostructured Al electrode in the bottom. One Al electrode in PENG is shared by the TENG. The r-shape structure is formed by fixing the two parts together at one end. The nanostructure in the Al surface and the MNDS surface of PDMS are shown in Fig. 12.7b, c, respectively. The X-ray diffraction (XRD) pattern of the PVDF film indicates the dominant β-phase, which is crucial for the PENG, as shown in Fig. 12.7d.

The working mechanism of the TENG has been illustrated in Chap. 2 for the contact-separation mode. The output of the PENG can be explained by the variation of the polarization density in PVDF. When the force is applied on the PVDF film, the polarization density is changed by the corresponding strain variation and the current flows from the one electrode to the other electrode to balance the extra polarization density in the short-circuit condition. When the force and hence the strain is released, then the current flows back to balance the charge induced by the recovered polarization density in the short-circuit condition. The outputs from TENG and PENG can be achieved simultaneously during the operation.

Through a vibration system with a frequency of 5 Hz, the output performance of the hybrid NG is investigated. With a 100 MΩ probe, the output voltage of the hybrid NG is measured as 52.8 V for PENG (Fig. 12.8a) and 240 V for TENG (Fig. 12.8b); however, the half-wave width for PENG is much higher than that for TENG. With a 10 kΩ external load resistance, the output currents are measured as 166 μA for PENG (Fig. 12.8c) and 27.2 μA for TENG (Fig. 12.8d), respectively. Compared with TENG, PENG has a smaller output voltage but a larger output

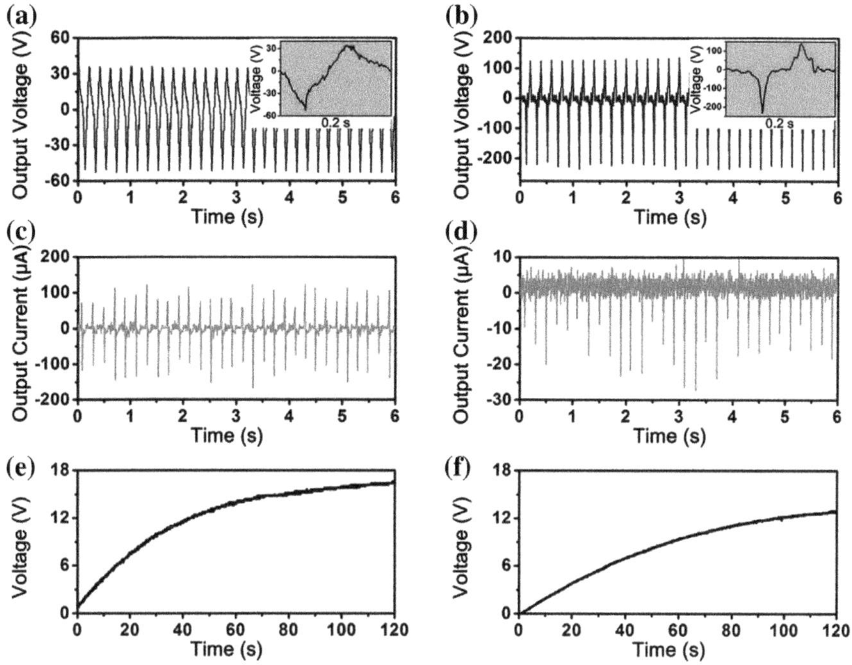


**Fig. 12.8** Electrical output performance including output voltages of **a** PENG and **b** TENG, output current of **c** PENG and **d** TENG, and the voltage of a 1 μF capacitor as charged by **e** PENG and **f** TENG. The *insets* in **a** and **b** show the enlarged voltages in one cycle. Reproduced with permission from American Chemical Society [15]

current, which is caused by the relatively lower output impedance. To double confirm that, the outputs of the hybrid NG with connections in series and parallel are measured, respectively. With the connection in series, the hybrid output voltage reaches 328 V, which is more similar to the output voltage of the TENG. With the connection in parallel, the hybrid output voltage is only about 58.6 V, which is close to that of PENG. Therefore, similar to the comparison between EMG and TENG in Sect. 12.1.1.1, the output impedance of the PENG is concluded to be smaller than that of TENG.

The instantaneous power densities of PENG and TENG are measured to be 1.10 and 0.82 mW/cm$^2$, respectively. The generated outputs are also used to charge a 1 μF capacitor through full-wave rectifiers. For PENG and TENG, the voltage reaches 16.6 V (Fig. 12.8e) and 13.0 V (Fig. 12.8f) in 120 s, respectively.

To demonstrate practical applications of the hybrid NG, a 10 Hz external force is applied and the hybrid NG can work continuously for over 6000 cycles. The outputs from PENG and TENG are firstly rectified to instantaneously drive commercial light-emitting diodes (LEDs), with the circuit as shown in Fig. 12.9a. The rectified output voltages of PENG and TENG are shown in Fig. 12.9b. The green and white LEDs are lighted by PENG and TENG through a gentle finger typing,

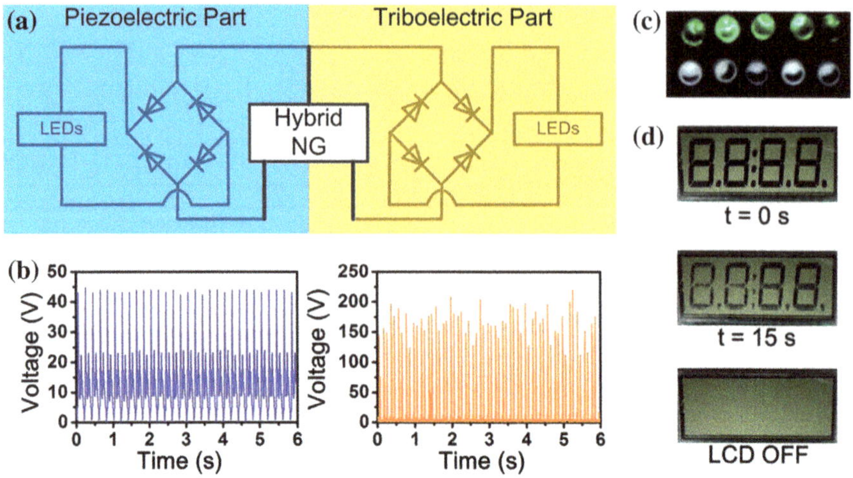


**Fig. 12.9** **a** The schematic diagram of the circuit of the hybrid NG for lighting up LED bulbs. **b** The rectified output voltage from PENG (*left*) and TENG (*right*). **c** The LED bulbs as lighted by PENG (*upper green*) and TENG (*lower white*). **d** The LCD screen as charged by one cycle of the hybrid NG operation at 0 s (*top*) and 15 s (*middle*). The bottom one shows the off state of the LCD screen. Reproduced with permission from American Chemical Society [15]

respectively, as shown in Fig. 12.9c. And besides, a 4-bit liquid crystal display (LCD) screen is lighted continuously for tens of seconds by only one cycle of power generation (Fig. 12.9d). And even after 15 s, the numbers displayed can still be clearly observed. Thus the hybrid NG which can be easily integrated to surfaces is demonstrated to be driven by finger typing to achieve power generation and active sensing.

In summary, the r-shaped hybrid NG consisting of a PENG and a TENG is developed to enhance the output. With a 5 Hz periodic vibrating force, the output voltage, the current density and the volume power density of PENG and TENG reach 52.8 and 240 V, 20.75 and 3.40 $\mu A/cm^2$, 10.95 and 2.04 $mW/cm^3$, respectively. The outputs of the hybrid NG are demonstrated to power LEDs and LCDs, and charge the capacitors, which shows potentials towards broader applications in our daily life.

#### 12.1.2.2 Hybrid Triboelectric-Pyroelectric Energy Cell

Yang et al. demonstrated a hybrid energy cell, which was used to simultaneously/individually harvest the mechanical and thermal energies for a self-powered electrodegradation of methyl orange (MO) [21]. A TENG was fabricated at the top by using a flexible polydimethysiloxane (PDMS) nanowire array with the diameters of about 200 nm to harvest the mechanical energy, and a lead zirconate titanate (PZT) film-based pyroelectric nanogenerator was fabricated at the

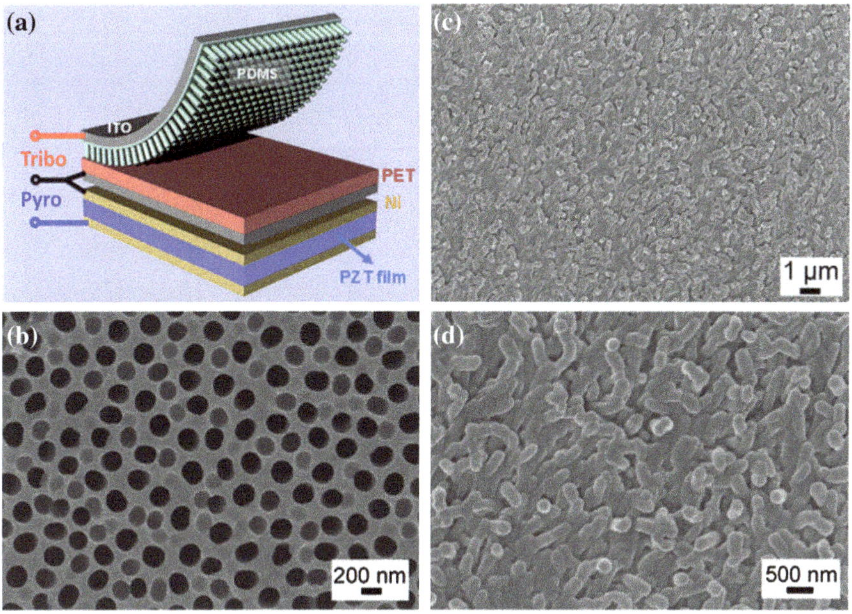


**Fig. 12.10** **a** Schematic diagram of the fabricated hybrid energy cell. **b** SEM image of the fabricated AAO template. **c** SEM image of the PDMS nanowire array. **d** Enlarged SEM image of the PDMS nanowire array. Reproduced with permission from American Chemical Society [21]

bottom to harvest thermal energy. These energies produced by the NGs can be stored in a Li-ion battery or directly used for the degradation of MO by the self-powered electrocatalytic oxidation.

Figure 12.10a shows a schematic diagram of the fabricated hybrid energy cell. In this study, a home-made anodic aluminum oxide (AAO) template is used to fabricate the PDMS nanowire array. The top TENG consists of a transparent indium tin oxide (ITO) film electrode, the PDMS nanowire array layer of about 200 μm, and a polyester (PET) film of 127 μm. Figure 12.10b shows a scanning electron microscopy (SEM) image of the fabricated AAO template, revealing that the diameters of the holes are about 200 nm. The depth of the holes is about 500 nm. Figure 12.10c shows a SEM image of the obtained PDMS nanowire array, where the distribution of the nanowires is uniform. The enlarged SEM image indicates that the diameters of the PDMS nanowires are about 200 nm, as shown in Fig. 12.10d. Some nanowires are not vertical on the substrate, which is associated with the attraction forces among the PDMS nanowires when they were pulled out from the AAO template. The bottom pyroelectric NG consists of two Ni electrodes and a PZT film with the thickness of about 1.5 mm.

Figures 12.11 a and b show the output performance of the fabricated TENG under the forward and reversed connections, where the output voltage is up to 12 V and the output current is about 0.2 μA (current density of 0.17 $mA/m^2$). The mechanism of

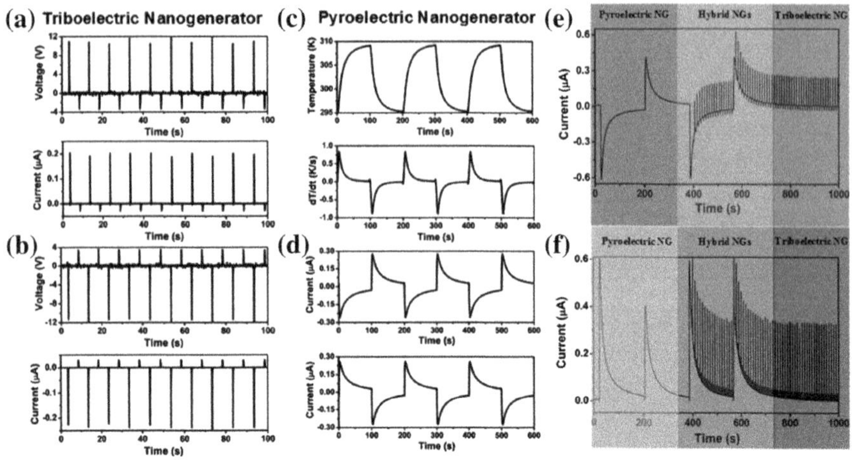


**Fig. 12.11** **a**, **b** The output voltage and current of the TENG under the forward connection (**a**) and the reversed connection (**b**) to the measurement system. **c** Cyclic change in temperature of the pyroelectric NG and the corresponding differential curve. **d** Measured output current of the pyroelectric NG under the forward and reversed connections. **e** Output current of the hybrid energy cell, where the TENG and pyroelectric NG can work simultaneously and individually. **f** Output current of the hybrid energy cell after the output signals were rectified by a full-wave bridge circuit. Reproduced with permission from American Chemical Society [21]

the TENG has been discussed in Chap 2. The output voltage $V_{OC(TENG)}$ of TENG can be given by

$$V_{OC(TENG)} = \frac{\sigma d}{\varepsilon_0} \tag{12.8}$$

where $\sigma$ is the triboelectric charge density, $d$ is the interlayer distance, and $\varepsilon_0$ is the vacuum permittivity. According to Eq. (12.8), the output voltage $V_{OC(TENG)}$ will increase with increasing the interlayer distance $d$.

The output performance of the pyroelectric NG was measured by varying the temperature in the vicinity of the device from 295 to 309 K, as shown in Fig. 12.11c. The peak value of the temperature changing rate is about 0.8 K/s. The output current pulse of the pyroelectric NG is about 0.3 μA (current density of 0.33 $mA/m^2$) under the forward and reversed connections, as shown in Fig. 12.11d. The corresponding output voltage pulse is about 16 V. The output voltage $V_{OC(Pyro)}$ of the pyroelectric NG can be expressed as

$$V_{OC(Pyro)} = \frac{pD\Delta T}{(\varepsilon_r - 1)\varepsilon_0} \tag{12.9}$$

where $p$ is pyroelectric coefficient, $D$ is the thickness of the device, $\Delta T$ is the change in temperature, and $\varepsilon_r$ is the relative dielectric constant of the sample. According to the Eq. (12.9), the output voltage of pyroelectric NG will increase with increasing the change in temperature of the device.

Figure 12.11e shows the output current of the hybrid energy cell, where the two NGs were in parallel connection. It can be seen that the TENG and pyroelectric NG can work simultaneously and individually to harvest mechanical and thermal energies, respectively. Since both the TENG and pyroelectric NG have the alternating electric (AC) output signals, the total output current was sometimes weakened rather than enhanced (see the time from 400 to 550 s in Fig. 12.11e). To solve this problem, these output signals need to be rectified by the full-wave bridge circuits. Figure 12.11f shows the output current of the hybrid energy cell after the output signals were rectified. It can be seen that all of the output current pulses are positive and the output current of the hybrid cell is always larger than that of the individual NGs. The application of this NG in self-powered system for degradation of MO is described in Sect. 13.2.1.

In summary, a hybrid energy cell, which can simultaneously/individually harvest the mechanical and thermal energies for self-powered degradation of MO, has been demonstrated. By using the flexible PDMS nanowire array, the fabricated TENG at the top was used to harvest the mechanical energy, and a PZT film-based pyroelectric NG at the bottom was used to harvest the thermal energy. The fabricated hybrid energy cells have the potential electrochemical applications in the self-powered electrodeposition, pollutant degradation, corrosion protection, and water splitting.

#### 12.1.2.3 Triboelectric-Pyroelectric-Piezoelectric Hybrid Cell for High-Efficiency Energy Harvesting

Zi et al. reported a triboelectric-pyroelectric-piezoelectric hybrid cell composed of a sliding mode TENG and a pyroelectric-piezoelectric nanogenerator (PPENG) with outstanding performances [23]. By harvesting the majority of the mechanical energy during the sliding motion, the TENG with a size of 63.5 $cm^2$ has produced an area power density of 0.15 $W/m^2$ at a sliding frequency of 4.41 Hz. The PPENG is able to harvest the thermal energy of the friction induced heat and the mechanical energy induced by the normal force. The hybrid cell has been demonstrated to extend the illumination time of an LED and charge a supercapacitor at a charging rate about twice of that charged by the TENG alone. These outstanding performances demonstrate the hybrid cell as a promising energy harvester and self-powered sensor for future applications.

As shown in Fig. 12.12a, the hybrid cell has a multi-layered planar structure composed of a sliding-mode TENG on the top and a PPENG on the bottom. In the TENG, a layer of aluminum (Al) foil was used as the sliding part and a piece of polytetrafluoroethylene (PTFE) film deposited with copper (Cu) electrode was utilized as the static part. For the PPENG, a piece of 110-μm-thick polarized polyvinylidene fluoride (PVDF) deposited with Cu electrodes on both sides was attached underneath the TENG. A Kapton film was laminated between the bottom

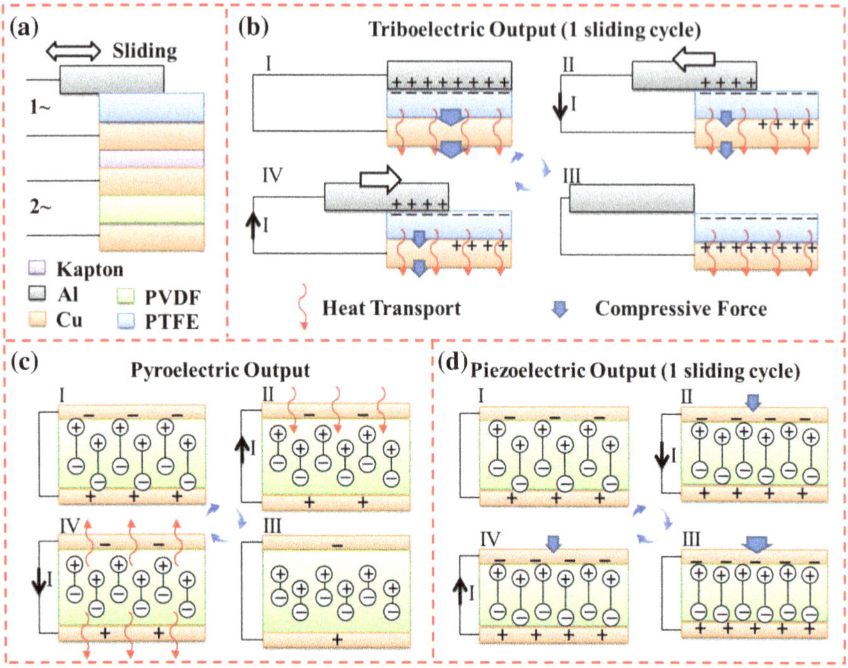


**Fig. 12.12** The structure and working mechanism of the hybrid cell. **a** The structure of the hybrid cell, where 1 is the output from triboelectric nanogenerator (TENG) and 2 is the output from pyroelectric-piezoelectric nanogenerator (PPENG); **b** the working mechanism of the output in TENG; **c** and **d** the working mechanism of the pyroelectric and the piezoelectric outputs in PPENG, respectively. Reproduced with permission from Wiley [23]

electrode of the TENG and the top electrode of the PPENG as an electronically insulating but thermally conductive layer. The heights of both the sliding and static parts were adjusted to ensure that the Al foil and the PTFE film contact firmly and keep aligned during the sliding. The motion of the Al foil was controlled by a linear motor to slide on the PTFE surface periodically with a period of 0.2267 s and a displacement of 80 mm. The total frictional area was determined to be 63.5 $cm^2$. A thermocouple was placed in the center of the frictional area underneath the bottom side of the PPENG to record the temperature variation.

The operation and theoretical study of the sliding mode TENG have been reported previously. As shown in Fig. 12.12b, the working principle of the TENG is based on the coupling of triboelectrification and electrostatic induction. Specifically, upon the contact/friction between the two parts of the TENG, the Al layer is positively charged, while the surface of PTFE gets negatively charged due to their distinct surface polarities. The net negative charges on the PTFE are immobile and tend to preserve for a long period of time considering the insulating nature of the electret polymer material. Once the Al foil is sliding away from the aligned position, the separation of triboelectric charges on both surfaces leads to potential difference between the Al and Cu electrodes. This potential difference is represented as the output voltage in the open-circuit condition, and serves as the driving force for the charge flow in the short-circuit condition (Fig. 12.12b–II), until the Al foil fully slides away from the top of the PTFE film (Fig. 12.12b–III). When the Al foil is moving back, the open-circuit potential difference drops and the electrons flow back in the short-circuit condition to reach equilibrium (Fig. 12.12b–IV). Eventually the Al foil goes back to the aligned position to complete one sliding circle (Fig. 12.12b–I).

The operation of the PPENG highly depends on the performance of the polarized PVDF film. Figure 12.12c, d explain the pyroelectric and piezoelectric outputs of the PPENG, respectively. In the device structure, the PVDF in PPENG is placed with the polarization direction upward (Fig. 12.12c–I, d–I). During the periodical sliding motion, the underlying PVDF film undergoes a normal force and receives the heat induced by the friction. With the increase of the temperature due to the friction induced heat, the dipole moments decrease and the volume expands, both of which reduce the polarization density. To balance the extra charges owing to the reduction of the polarization density, a negative potential difference from the top to the bottom electrodes is created across the PVDF, and a current is driven to flow from the bottom electrode to the top electrode of the PVDF (Fig. 12.12c–II) in the short-circuit condition. After the sliding motion stops (Fig. 12.12c–III), the heat dissipation from the PVDF film leads to the increase of the polarization density, which reduces the absolute value of the potential difference in the open-circuit condition, and drives the current flow in the reverse direction in the short-circuit condition (Fig. 12.12c–IV). This open-circuit voltage variation and this short-circuit current owing to the pyroelectric effect can be derived as [24, 25]:

$$\Delta V = \frac{A\Delta\sigma}{C} = \frac{A}{A'}\frac{\Delta\sigma d}{\varepsilon_r\varepsilon_0} = \frac{A}{A'}\frac{d}{\varepsilon_r\varepsilon_0}p\Delta T \tag{12.10}$$

$$I = \frac{\mathrm{d}(\sigma A)}{\mathrm{d}t} = pA\frac{\mathrm{d}T}{\mathrm{d}t} \text{ or } \Delta Q = \int I\mathrm{d}t = pA\Delta T \tag{12.11}$$

Here $d$, $A'$ and $C$ are the thickness, the total area and the capacitance of the PVDF in PPENG, respectively. $A$ is the frictional area, which is considered to be the thermally affected area in PVDF. $p$ and $\varepsilon_r$ are the pyroelectric coefficient and the relative dielectric constant of the PVDF, respectively, and $\varepsilon_0$ is the permittivity of the vacuum. $\Delta V$, $\Delta Q$, $\Delta T$ and $I$ represent the voltage variation, the transferred charge, the temperature variation and the current, respectively. Since the polarization direction is perpendicular to the film plane, we can consider the charge density variation as $\Delta\sigma = \Delta P = p\Delta T$ [26].

When the sliding part moves in, the normal compressing force is applied on the PVDF film, resulting in a negative strain and reduced volume of the PVDF film. Thus, the polarization density is enhanced and the current flows from the top electrode to the bottom electrode to balance the extra polarization density (Fig. 12.12d–II) in the short-circuit condition. In the aligned position of the TENG, the largest normal force is achieved (Fig. 12.12d–III). When the sliding part slides out, the applied force decreases and then the current flows back to balance the charge induced by the release of the strain (Fig. 12.12d–IV) in the short-circuit condition. The open-circuit voltage variation and the short-circuit transferred charge owing to the piezoelectric effect have been derived to be proportional to the applied normal force [27]. Compared to the pyroelectric effect, the piezoelectric effect is considered to be more complex since all of the 3-dimensional strains can affect the polarization density.

The output performance of the TENG was measured while the Al foil was sliding periodically. For the open-circuit voltage as shown in Fig. 12.13a, it oscillated with a peak-to-peak value of 1132.6 V. In the short-circuit condition, the current density was measured to be an AC output with a peak value of 1.45 mA/m$^2$ (9.2 μA in the current peak) as shown in Fig. 12.13d. Both the voltage and the current were measured to have the same period as that of the sliding motion. These results are comparable to previously reported values [28, 29]. To determine the optimized load resistance and the maximum power density outputs, the external loads with different resistances were applied. As the load resistance increased, the

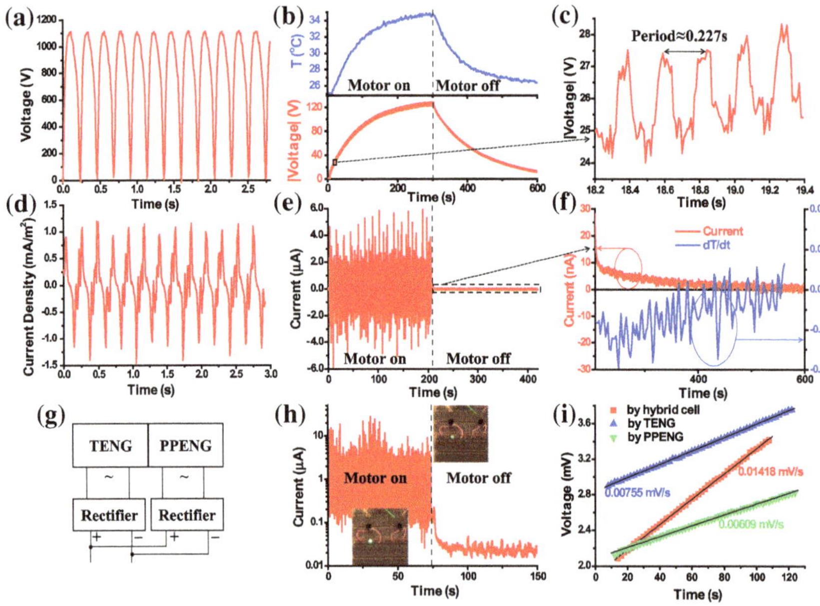


**Fig. 12.13** Electrical outputs of the hybrid cell. **a–c** The open-circuit voltage of **a** the TENG, **b** and **c** the PPENG; **d–f** The short-circuit current density of **d** the TENG, **e** and **f** the PPENG. **g** The circuits to hybridize the outputs from TENG and PPENG; **h** The hybridized output current, with the LED lighted by the hybridized output when the motor was on and after the motor was off, as shown in *insets*; **i** The voltages of the supercapacitor charged by the hybridized output, the rectified TENG and PPENG outputs of the hybrid cell when the motor was on. Reproduced with permission from Wiley [23]

amplitudes of the output current peaks were observed to drop and the voltage peak values were observed to be enhanced. Then the maximum power density was determined as 146.2 $mW/m^2$ at the matched load of 140 MΩ.

The output performances of the PPENG were measured in the entire range of the sliding motion (motor on) and after the sliding motion stopped (motor off). The open-circuit voltage variation between the top and the bottom electrodes was firstly measured as shown in Fig. 12.13b. When the motor was on, the voltage became negative, and the voltage decreasing rate was high initially and then approached to zero gradually. After the motor was off, the voltage increased back to zero gradually. To clarify the origin of this output, the temperature variation of the PPENG in the entire process was measured, also as shown in Fig. 12.13b. We noticed that the main part of the voltage decrease is proportional to the temperature increase, which was consistent with the linear relationship between $\Delta V$ and $\Delta T$ in Eq. (12.10) and the fact of the negative pyroelectric coefficient $p$ [30]. This result indicated that the majority of the open-circuit voltage variation was due to the pyroelectric effect from the temperature variation, as a result of the friction induced heat. It is also observed when the motor was on, there were periodical oscillations in the voltage (Fig. 12.13c); as a comparison, after the motor was off, the voltage became relatively smooth without periodical oscillations. Thus, these oscillations were determined to come from the piezoelectric effect induced by the periodically applied normal force. The period of the voltage oscillations was measured as about 0.227 s, which was consistent with the period of the sliding motion.

The short-circuit current was measured as an AC output when the motor was on (Fig. 12.13e) and a decreasing unidirectional output after the motor was off (Fig. 12.13e, f). Due to Eq. (12.11), the pyroelectric short-circuit current was proportional to the changing rate of the temperature $\mathrm{d}T/\mathrm{d}t$. The temperature kept increasing when the motor was on and decreasing after the motor was off, as a result, the pyroelectric part of the current should be a negative unidirectional output when the motor was on and a positive unidirectional output after the motor was off, respectively. At the same time, the normal force was applied periodically only when the motor was on, resulting in an AC output as the piezoelectric part of the current. It was reasonable to determine the majority of the measured AC output (larger than $10^{-6}$ A) as the piezoelectric current, since the normal force changed frequently (in the period of 0.2267 s), resulting in a large $\mathrm{d}F/\mathrm{d}t$ as well as a larger piezoelectric current than the pyroelectric current (estimated as only about $10^{-8}$ A level). After the motor was off, there was only unidirectional current existed decaying from about $3 \times 10^{-8}$ A. To verify it was the pyroelectric current, the temperature variation is measured and the changing rate of the temperature $\mathrm{d}T/\mathrm{d}t$ is calculated, as plotted in Fig. 12.13f. Even though there was a large noise level in the $\mathrm{d}T/\mathrm{d}t$ curve, the negative $\mathrm{d}T/\mathrm{d}t$ was still observed to be approaching zero, which was consistent with the variation of the current. Therefore, this measured unidirectional current was confirmed as the pyroelectric current.

It can be found from the above measurement results that the PPENG can convey a high short-circuit current from the piezoelectric effect, and a high open-circuit voltage from the pyroelectric effect. Hence, the overall PPENG output with an external load resistance was hybridized by AC and unidirectional outputs. With large load resistances (≥60 MΩ), the output profile of the PPENG was measured to be close to unidirectional output from the pyroelectric effect, even though oscillations from the piezoelectric effect could still be observed while the motor was on. With small load resistances (<60 MΩ), the output profile was measured to be close to the AC output from the piezoelectric effect. The average power densities of the PPENG and the TENG subject to various load resistances were calculated, and then the ratio between the average power densities of the PPENG and the TENG was calculated as the enhancement of the energy-harvesting efficiency of the hybrid cell compared with that of the TENG only. This efficiency enhancement was found to be larger than 10 % subject to small load resistances (≤100 kΩ), and it could reach up to 26.2 % with the load resistance of 1 kΩ. By using Eq. (12.10) and the integral form of Eq. (12.11), the pyroelectric coefficient p was calculated as $-1.80 \pm 0.07$ nC/(cm$^2$ K).

By using rectifiers, both the output currents from the TENG and PPENG could be converted to be unidirectional. The output performances from both the TENG and PPENG were hybridized by connecting the two rectified output currents in parallel (Fig. 12.13g). The hybridized output current was measured, as shown in Fig. 12.13h, which was up to $10^{-5}$ A level when the motor was on. This hybridized output was utilized to power up an LED bulb first. When the motor was on, the LED was turned on with a strong illumination intensity (left inset of Fig. 12.13h); as soon as the motor was off, the LED turned off gradually since the pyroelectric output current enabled the illumination in a lower intensity (right inset of Fig. 12.13h). This phenomenon was not observed in the LED driven by only the rectified TENG output. It was also demonstrated that this hybridized output could charge a supercapacitor with a high capacitance ($C$) of 41.9 mF. The voltage across the supercapacitor during charging was plotted in Fig. 12.13i. As comparison, the voltages across the supercapacitor charged by only the rectified TENG and PPENG outputs were also plotted, respectively. It was observed that the supercapacitor could be charged faster by the hybridized output than that by only one individual output. The charging rate by the hybridized output was about the sum of the charging rates of the rectified TENG and PPENG outputs. The average value of the charging current could be calculated by the following equation:

$$\bar{I} = \frac{\mathrm{d}Q}{\mathrm{d}t} = C\frac{\mathrm{d}V}{\mathrm{d}t} \tag{12.12}$$

when the motor was on, the average charging currents of the hybrid cell, the rectified TENG and the rectified PPENG outputs were calculated as about 0.59, 0.32 and 0.26 μA, respectively.

In summary, a hybrid cell composed of a triboelectric nanogenerator (TENG) and a pyroelectric-piezoelectric nanogenerator (PPENG) to efficiently harvest energy during the friction motion has been developed. For the TENG, high oscillating voltage and current outputs with a maximum power density of 146.2 mW/m$^2$ were achieved at the sliding frequency of 4.41 Hz. For the PPENG, during the period when motor was on, increasing DC output voltage in open-circuit condition and oscillating AC output current in short-circuit condition were observed, and the voltage variation was proportional to the temperature increase; after the motor was off, decaying DC pyroelectric output was observed. With the PPENG integrated on the TENG, the mechanical energy harvesting efficiency enhancement could be up to 26.2 %. The hybridized output was demonstrated for instantaneously driving the LED with extended illumination time and charging a supercapacitor in the charging rate about twice of that charged by the TENG only. These outstanding performances of the hybrid cell make it a perfect candidate for both power generators and active sensors, which is a substantial advancement in nanogenerators and will push forward its practical applications to a new paradigm.

## 12.2 AC–DC Hybrid Cell

### *12.2.1 Hybrid Solar Cell and Triboelectric Nanogenerator*

Solar cell is so far the commercial technology for large scale harvesting of solar energy. In practical applications, the surface of a Si solar cell is usually covered by a layer of transparent material for protecting the device from corrosion, contamination and mechanical damage. When a solar cell is exposed to nature, raindrops and wind are existing sources of mechanical energy. An interesting question is that if the surface layer for protecting the solar cell can be used for harvesting this kind of mechanical energy. A triboelectric nanogenerator (TENG) can be fabricated using flexible transparent polymer materials, which can be a part for the TENG, but also as a surface protection layer for the solar cell, so that a hybrid energy cell technology may be fabricated to simultaneously/individually harvest multimode energies whichever is available. Therefore, to harvest solar energy and mechanical energy simultaneously, multiple hybrid energy cells have been designed [31–34].

#### 12.2.1.1 Hybrid Solar Cell and Triboelectric Nanogenerator to Harvest Solar and Mechanical Energies

Yang et al. demonstrated a hybrid energy cell that is built directly based on a micropyramid Si solar cell for simultaneously harvesting both solar and mechanical energy [31]. A flexible transparent polydimethysiloxane (PDMS) nanowire array-based TENG was fabricated at the top for not only harvesting the mechanical energy, but also serving as the protection layer for Si solar cell.

A schematic diagram of the fabricated hybrid energy cell is shown in Fig. 12.14a. In this study, the PDMS nanowire array was fabricated by using a home-made anodic aluminum oxide (AAO) template, where the diameter and depth of the holes are about 200 and 500 nm, respectively. The top TENG is based on the

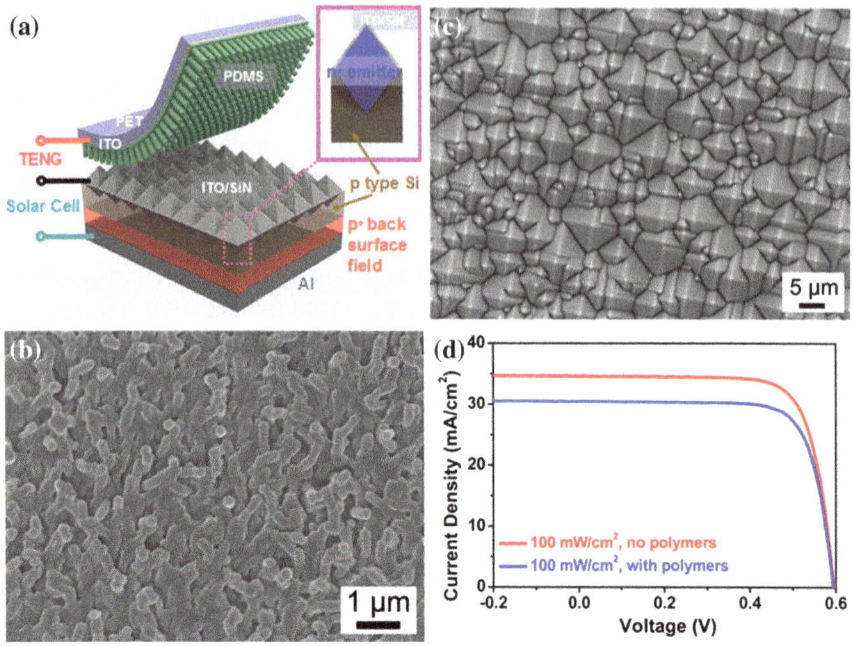


**Fig. 12.14** **a** Schematic diagram of the fabricated hybrid energy cell. **b** SEM image of the PDMS nanowire array. **c** SEM image of the fabricated Si pyramids. **d** *J-V curves* of the Si solar cell covered with and without PET-PDMS polymers under the light illumination intensity of 100 mW/cm$^2$. Reproduced with permission from American Chemical Society [31]

contact/separation between the transparent PDMS nanowire array and the indium tin oxide (ITO) film. Figure 12.14b shows a scanning electron microscopy (SEM) image of the obtained PDMS nanowire array, revealing that the diameters are about 200 nm. The bottom Si micropyramid solar cell consists of an Al film electrode, a p$^+$ back surface field layer, a p type Si layer, n$^+$ emitter layer, SiN film, Ag grids and the ITO film electrode. The pyramid surface is designed to enhance the light absorption, which is also used here as an effective means for enhancing the triboelectrification process in TENG. The TENG and the Si solar cell have a co-electrode of ITO film on the Si pyramids. Figure 12.14c shows a SEM image of the Si pyramids, indicating that the sizes of Si pyramids are from 1 to 10 μm.

Figure 12.14d shows the output performance of the fabricated Si pyramid solar cell. Under the light illumination intensity of 100 mW/cm$^2$, the open-circuit voltage and the short-circuit current density of the device are about 0.6 V and 35 mA/cm$^2$, respectively. The corresponding energy conversion efficiency is up to about 16 %, which is much larger than the reported Si nanowire hetero-junction solar cells. The output performance of the Si solar cell covered by the PDMS-PET polymers is also measured, as shown in Fig. 12.14d. Under the same light illumination intensity, the conversion efficiency of the solar cell covered with the polymers was found to decrease from 16 to 14 %.

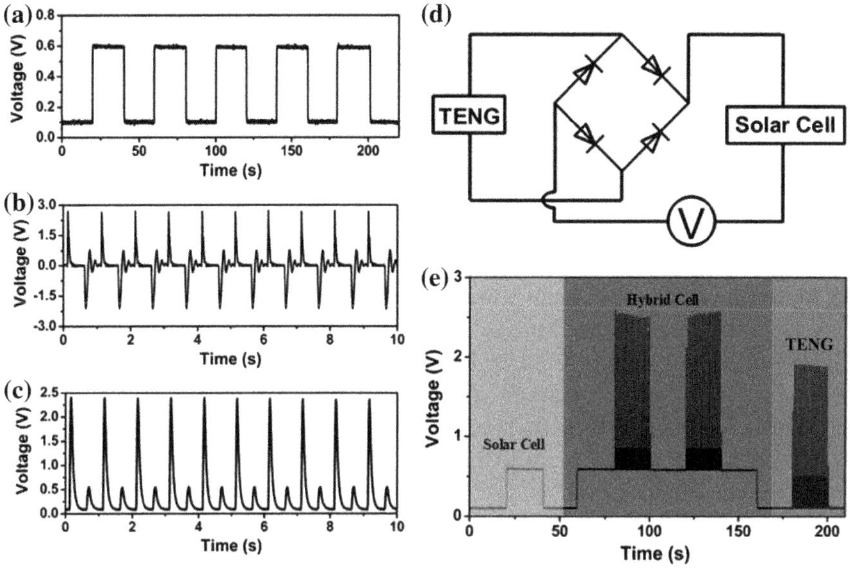


**Fig. 12.15** **a** Output voltage of the fabricated Si solar cell. **b** Output voltage of the fabricated TENG. **c** Output voltage of the TENG after the output signals were rectified. **d** Schematic diagram of the connection between the Si solar cell and TENG. **e** Output voltage of the hybrid solar cell and TENG (after rectification) for harvesting solar and mechanical energies. Reproduced with permission from American Chemical Society [31]

Figure 12.15a shows that the output voltage of a Si solar cell device is about 0.6 V, where the corresponding output current is up to 18 mA. By controlling the periodic contact and separation between the PDMS nanowire array and the ITO layer on the Si pyramids, the output voltage pulse of the fabricated TENG is about 3 V under the forward connection, as shown in Fig. 12.15b. It can be found that the TENG gives an alternating current (AC) output signal, which does not match with the direct current (DC) output signal of the Si solar cell. A bridge rectification circuit was used to convert the AC to DC signals. The obtained rectified output voltage pulse of the device is about 2.5 V, as shown in Fig. 12.15c. Figure 12.15d shows a schematic diagram of the integrated hybrid energy cell, where the solar cell and the rectified TENG device were connected in series, which can ensure that there is always a voltage/current output when either the solar or mechanical energy is available. Figure 12.15e shows the output voltage of the hybrid energy cell, where the output signals of TENG were rectified. It can be clearly seen that the solar cell and the TENG can work simultaneously/individually to harvest the solar and mechanical energies, respectively.

The mechanism of the TENG is to use the triboelectric effect induced electrostatic charges to drive the electrons to flow in the external circuit once the charged layers are separated by a small distance. In this study, when the PDMS is in contact with the ITO film, the surface charges will be transferred from ITO to PDMS due to the electrostatic electrification, resulting in the observed output current and voltage. The pyramid structure of the electrode largely increases the friction between the PDMS and the ITO film on the silicon surface. The output voltage $V_{OC}$ of the TENG can be expressed by

$$V_{OC} = \frac{\sigma_0 d_S}{\varepsilon_0} - \frac{\Delta\sigma}{\varepsilon_0}\left(d_S + \frac{d_{PDMS}}{\varepsilon_{r(PDMS)}}\right) \tag{12.13}$$

where $\sigma_0$ is the triboelectric charge density, $\Delta\sigma$ is the transferred charges between the two electrodes, $d_S$ is the interlayer distance, $d_{PDMS}$ is the thickness of the PDMS film, $\varepsilon_0$ is the vacuum permittivity, and $\varepsilon_{r(PDMS)}$ is the relative permittivity of PDMS. According to Eq. (12.13), the output voltage $V_{OC}$ can be increased by increasing the interlayer distance $d_S$. By increasing the interlayer distance and integrating 5 Si solar cells in series connection, the output voltage of the hybrid cell can be increased to 12 V, which is much larger than that in Fig. 12.15e. The corresponding output current is about 17.4 mA.

In summary, the first hybrid energy cell that consists of a solar cell and a TENG, which can be used to simultaneously/individually harvest solar and mechanical energies has been developed. The bottom solar cell device was designed by using a Si micropyramid $n^+$-p-$p^+$ junction, which can give an output voltage of 0.6 V and an output current density of 35 mA/cm$^2$ under the light illumination intensity of 100 mW/cm$^2$. The top TENG is based on the contact/separation between the PDMS nanowire array and ITO film on the Si pyramids.

#### 12.2.1.2 Hybridized Power Panel to Simultaneously Generate Electricity from Sunlight, Rain Drops and Wind Around the Clock

Zheng et al. reported a multifunctional solar cell panel by hybridization with a transparent dual-mode TENG [34]. Consequently, it can not only harvest energy from sunlight, but also from light wind and natural rain drops. Given no compromise of the solar energy harvesting, the presented hybrid cell holds a greatly expanded working time and provides good output compensation except for solar energy. In a windy day, the hybrid cell, harvesting energy from ambient wind, can deliver an average output of 8 mW m$^{-2}$ at a wind speed of 2.7 m s$^{-1}$. In a rainy day, harvesting energy from the dripping rain drops, the hybrid cell can achieve an average output of 86 Mw m$^{-2}$ at a dripping rate of 13.6 ml s$^{-1}$. These functions render that the hybrid cell plays an important and compensative role in the current field of solar energy harvesting. Given a collection of advantages, such as being cost-effective, easy fabrication, greatly expanded working time without comprising the original solar energy harvesting, the presented hybrid cell is a unique and practical step towards a high efficiency ambient green energy harvesting with a lower production cost per watt.

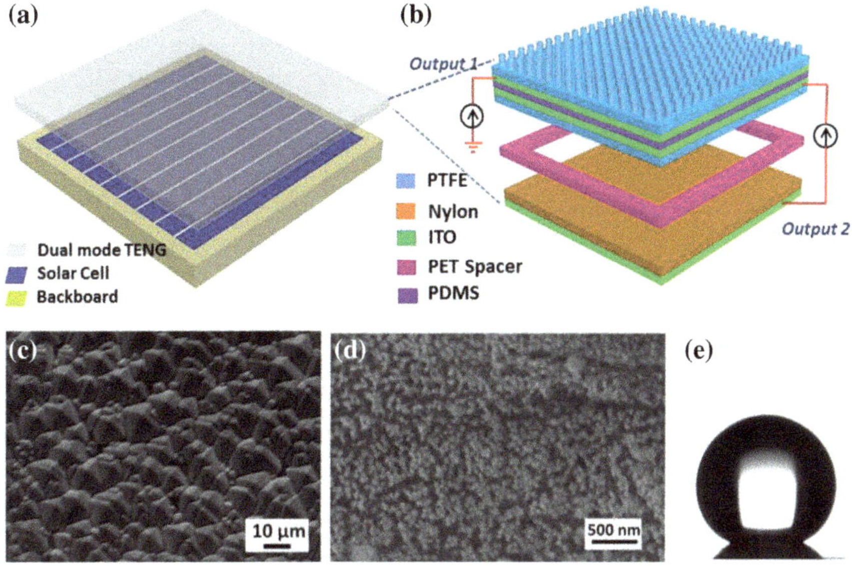


**Fig. 12.16** **a** Schematic diagram of the as-fabricated hybridized power panel. **b** An enlarged view to illustrate the dual-mode TENG in detail. A PET spacer was used to maintain the gap distance between the nether PTFE film and nylon film, integrating all the parts into a dual-mode TENG. The electricity generated from water TENG and contact TENG can be collected through the output 1 and output 2 of the dual-mode TENG. **c** SEM image of the fabricated silicon-based solar cell surface with micro/nanostructured pyramids taken from a tilted view of 20°. **d** SEM image of the PTFE thin film with hierarchical micro/nanostructures. **e** The apparent contact angle of the PTFE thin film, indicating the film is superhydrophobic. Reproduced with permission from Wiley [34]

The reported hybridized power panel is mainly consisted of three parts: an acrylic backboard, a silicon-based solar cell and a dual-mode TENG, as schematically depicted in Fig. 12.16a. As the bottom part, acrylic is selected to be the backbone of the whole device owing to its decent strength, light weight, good machinability, and low cost. The middle part is a silicon-based solar cell consists of an Al film electrode, a $p^+$ back field layer, a p-type Si bulk layer, a $n^+$ emitter layer, SiN film, Ag grids, and an ITO film electrode. Micro pyramidal morphology was created on the surface of the Si-based solar cell to enhance light absorption and suppress light reflectance so as to improve the light-electricity conversion efficiency of the solar cell. The top part is a dual-mode TENG which is an integration of a water-TENG and a contact-TENG using a spacer of PET, and the specific corresponding structure and electric output of this dual-mode TENG is schematically shown in Fig. 12.16b. A scanning electron microscopy (SEM) image of the Si solar cell surface is exhibited in Fig. 12.16c, indicating that the surface is micro pyramidal with a size range of 1–10 μm. The superhydrophobic PTFE thin film was fabricated by using a homemade porous anodic Al oxide (AAO) template. Figure 12.16d shows an SEM image of the prepared PTFE film, where the surface of the PTFE film is composed of high-density nanorods, which will contain trapped air to reduce the actual contact area between the surface and water, contributing to the superhydrophobicity. An apparent contact angle of 155° of the fabricated PTFE film is shown in Fig. 12.16e, confirming that the surface is superhydrophobic (>150°).

As tested, the fabricated dual-mode TENG is slightly more transparent than the conventional glass that is commercially used in industry in the visible light region. The J-V curves of fabricated Si solar cell integrated with the dual-mode TENG and conventional glass are also measured respectively. Under a solar light irradiation (100 mW $cm^{-2}$), the short-circuit current density ($J_{SC}$) of the hybrid cell is slightly higher than that of the traditional solar cell panel, revealing that the dual-mode TENG will not affect the function of the solar cell compared with the traditional ones. Moreover, the superhydrophobicity of PTFE thin film in water TENG can also render a self-cleaning property of the solar cell surface, which could keep the surface from dust in the air and always provide a transparent protection layer, maintaining the standard working efficiency of the hybridized power panel.

The fundamental working principle of dual-mode TENG is based on a sequential process of contact-electrification and electrostatic induction, which can be depicted separately as a water-TENG (output 1) and a contact-TENG (output 2). To apply this technique in real life and achieve high energy conversion efficiency, the water-TENG and contact-TENG are operated in single-electrode and dual-electrode mode, respectively.

It is essential to investigate the performance of dual-mode TENG for harvesting energy from natural rain drop and ambient wind. Here the water drops sprayed from a shower which is connected to a household faucet are used to simulate rain drops. The distance between the outlet of water drops and the dual-mode TENG surface is around 40 cm, and the water dripping rate was controlled at a rate of 13.6 ml $s^{-1}$. The incident angle between water dripping direction and TENG surface was set at 45° [33]. The values of $J_{SC}$ and $V_{OC}$ of water-TENG and contact-TENG are 3.5 and 9.8 mA $m^{-2}$, 17.5 and 27.2 V, respectively. Due to the fact that when countless water drops fall randomly on the PTFE film surface, not all the drops can reach or leave the PTFE film simultaneously, thus the curves of output voltage and current exhibit random fluctuating characteristics. Under the same experimental conditions, when a hydrophilic nylon film is used to replace the nanostructured PTFE film, the output $J_{sc}$ of water TENG decreased sharply, indicating that the superhydrophobic surface of PTFE film plays a critical role in prompting the removal of the water drops from TENG surface and can enhance the electric output of water TENG.

On the other hand, when the dual-mode TENG is blew by the natural wind, it will also cause the random contact and separation of the PTFE film with nylon film, which then generate continuous electrical output. To reveal the relationship between the output of wind contact-TENG and wind speed numerically and perform this experiment in an easy way, a wind faucet in lab is used to simulate the natural wind and fan the dual-mode TENG, where the wind speed can be controlled by a rotary knob and can be tested by an anemometer. Under a speed of 2.7 m $s^{-1}$ which is a very usual wind speed in our daily life, the basic output $J_{SC}$ and $V_{OC}$

values of wind contact TENG are 3.1 mA $m^{-2}$ and 10.7 V, respectively. With the increase of wind speed, the output of wind contact TENG will also increase.

By measuring the output performance at various load resistances, the optimum output powers of the water-TENG, contact-TENG, and wind-TENG are determined as 6, 86, and 8 mW $m^{-2}$ at a resistance of 5, 0.2, and 1 MΩ, respectively.

The reported hybridized power panel can harvest solar, rain and wind energy simultaneously or individually when either of them is available in the ambient environment, and this mechanism could serve as a complementary backup when solar energy may not be available. In sunny or cloudy days, the as-fabricated hybrid cell can harvest solar and wind energy simultaneously. The output voltage of the Si solar cell under a light intensity of 100 mW $cm^{-2}$ is around 0.6 V, while the output voltage of hybrid cell driven by sunlight and wind at speed of 2.7 m $s^{-1}$ reaches nearly 11 V, where the solar cell and wind contact TENG are connected in series after rectified by a full wave bridge circuit. The output current density of wind contact TENG is 3.1 mA $m^{-2}$, while the $J_{SC}$ of hybrid cell where the solar cell and wind contact TENG is connected in parallel is 25.9 mA $cm^{-2}$, respectively. Though the $J_{SC}$ of the wind-TENG is smaller than that of the solar cell, it could function as necessary supplement especially considering the variations of weather conditions and much higher output voltage of the TENG.

In rainy days, due to the dimness of sunlight, the functionality of solar cell is greatly depressed. But there are countless raindrops, and more importantly, the raindrops carry energies, i.e. the electrostatic energy gained from the contact with air/particles and the mechanical energy gained from gravitational force when falling in sky. In such circumstance, though the output of solar cell part is relatively small, the hybridized power panel can harvest both the raindrop electrostatic and mechanical energy using a dual-mode TENG, which is very meaningful for natural energy harvesting, especially in raining seasons. In this experiment, water drops sprayed from a household shower are used to measure the output voltage and output current of the water TENG, water contact TENG and dual-mode TENG driven by water drops at a dripping rate of 13.6 ml $s^{-1}$. Both the output voltage and output current density of the water contact TENG is much higher than the water TENG itself, and this is due to a fact that the traditional contact electrification between two different dry materials is much stronger than water-solid triboelectric effect. By integrating the two outputs together, the hybridized power panel provides much higher output voltage and current density than either of the two working modes of the dual-mode TENG, indicating a higher efficiency in harvesting rain drop energy.

For practical applications in driving LEDs or charging a capacitor, the electrical outputs of the water TENG and the water contact TENG are connected in parallel through a full-wave bridge. As the dual-mode TENG is driven by water drops at a dripping rate of 20 ml $s^{-1}$, 10 commercial green LEDs are lighted up by the water TENG and 20 LEDs are lighted up by water contact TENG, respectively, as shown in Fig. 12.17a. Combined with solar cell, 50 LEDs can be lighted up by the hybridized power panel driven by water drops under room light conditions. The dual-mode TENG is also demonstrated to charge a capacitor of 3.3 μF at a water dripping rate of 20 ml $s^{-1}$. As shown in Fig. 12.17b, it takes 59 s and 6.5 s to

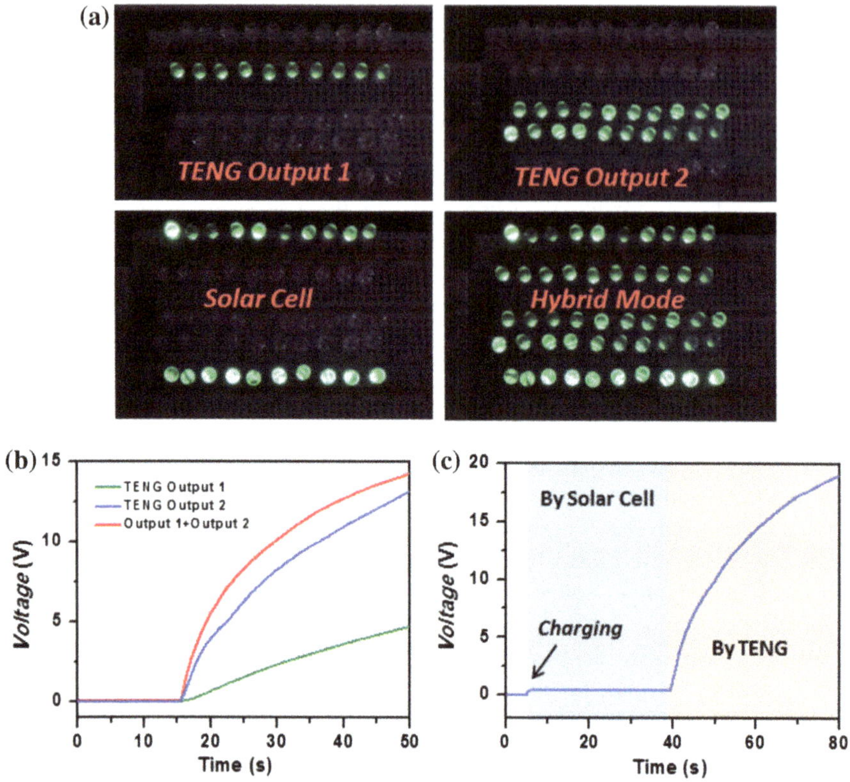


**Fig. 12.17** **a** The photograph of 50 commercial LED bulbs driven by the hybridized power panel (10 LEDs lighted by water TENG, 20 LEDs by water contact TENG at a dripping rate of 20 mL $s^{-1,}$ and 20 LEDs by solar cell). **b** The measured voltage curves of a 3.3 μF capacitor charged by water TENG, water contact TENG, and dual-mode TENG, separately. **c** The measured voltage curve of a 3.3 μF capacitor charged by the solar cell and dual-mode TENG impacted by water drops consequently. Reproduced with permission from Wiley [34]

charge the capacitor to a voltage of 7 V by the water TENG and water contact TENG, respectively. The output power of contact TENG is much more efficient compared with the single water TENG under the same experimental conditions. When using the dual-mode TENG, it takes 4.5 s to charge the capacitor to a voltage of 7 V. As for the comparison of the output voltage performance between solar cell and dual-mode TENG, the fabricated solar cell under a light illumination of 100 mW $cm^{-2}$ was first used to charge the capacitor to a maximum of 0.6 V due to its upper voltage limit. Then the rectified dual-mode TENG was used to charge the same capacitor, it took 39 s to charge the capacitor to a voltage of 19 V and the corresponding curves are displayed in Fig. 12.17c. The comparison results indicate that the hybridized power panel can compensate the lower voltage output of the solar cell itself.

In summary, a hybridized power panel to generate electricity from solar, rain drops and wind is reported. By hybridizing a transparent dual-mode TENG with a common solar cell, the hybridized power panel can simultaneously/individually harvest solar, rain drop and wind energy in various weather conditions around the clock. Without compromising the output performance of the original solar cell itself, the presented hybrid cell can respectively deliver an average output of 86 mW $m^{-2}$ from the dripping water drops at a rate of 13.6 ml $s^{-1}$, and an average output of 8 mW $m^{-2}$ at a wind speed of 2.7 m $s^{-1}$. This creative hybridization greatly expanded the device working time of the commonly-used solar cell, playing an important and compensative role in the current field of solar energy harvesting. Given its exceptional properties of being cost-effective, easy fabrication, greatly expanded working time, unique applicability resulting from distinctive device structure, the hybridized power panel is a practical approach for higher efficiency ambient energy harvesting with lower production cost per watt. And the justified concepts and demonstrations in this work can be immediately and extensively adopted in the real industrial field of solar energy harvesting, and come into effect of improving the way of our living.

### *12.2.2 Hybrid Thermoelectric Cell and Triboelectric Nanogenerator*

Thermal energy can be harvested using thermoelectric cells, which are based on Seebeck effect [35]. However, these energies are not always available at the same time, depending on weather, day/night, and physical locations. Hybrid energy cells have been developed to simultaneously or individually harvest thermal and mechanical energies by designing an integrated system, which consists of two kinds of energy harvesters. Yang et al. demonstrated a hybrid energy cell that consists of a polyamide (PA)-perfluoroalkyoxy (PFA) polymer film-based TENG, a $Bi_2Te_3$-based thermoelectric cell, and a silicon (Si)-based solar cell, which can be used to simultaneously/individually harvest mechanical, thermal, and/or solar energies [36]. Here we only talk about the hybridization of thermoelectric cell and triboelectric nanogenerator.

The fabricated TENG includes a PA film (thickness of 25 μm), a PFA film (thickness of 25 μm), and two Cu electrodes. A homemade force loading system was used to apply the periodically compressive forces on the TENG, where the contact/separation between the PA and PFA films can result in the output of the device. The bottom $Bi_2Te_3$-based thermoelectric cell was used to give the output voltage/current under the temperature difference across the device, which was induced by using a heated source. A temperature sensor was used to detect the change in temperature of the device.

A schematic diagram of a fabricated hybrid energy cell is sketched in Fig. 12.18a. A TENG was fabricated on the top for harvesting mechanical energy.

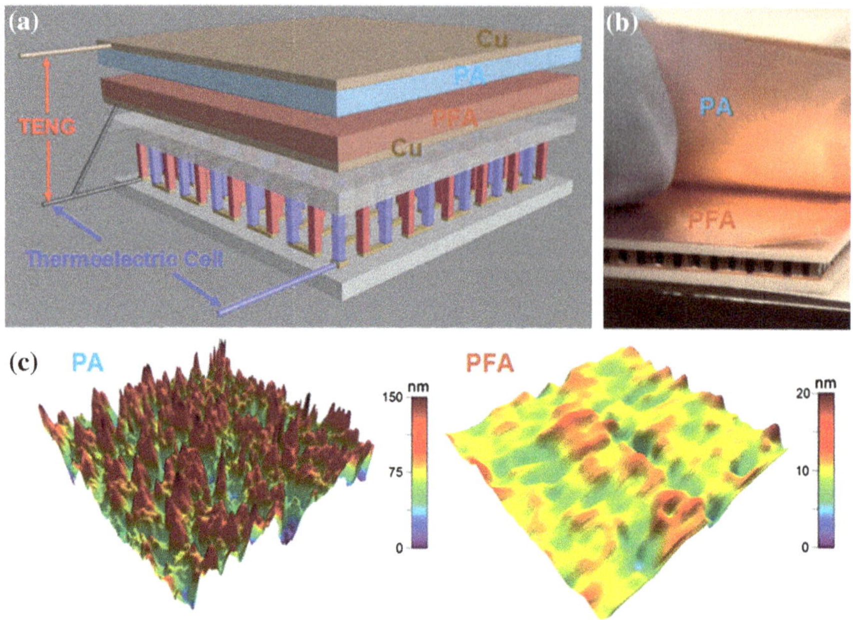


**Fig. 12.18** **a** Schematic diagram of the fabricated hybrid energy cell including a TENG and a thermoelectric cell. **b** Photograph of the hybrid energy cell. **c** AFM images of the PA and PFA films. Reproduced with permission from Royal Society of Chemistry [36]

In this study, the PA and PFA polymer films at the opposite ends of the triboelectric series were purposely chosen as the triboelectric layers for the effective electrification during the contact/separation process. On each polymer film, a layer of metal Cu electrode was fabricated on the external surface. A periodic loading force was applied on the top of the device to induce the contact and separation between the two polymer films. As illustrated in Fig. 12.18a, a $Bi_2Te_3$-based thermoelectric cell was designed at the bottom for harvesting thermal energy. A thermal source was used at the bottom of the device to create a temperature difference across the device. Figure 12.18b shows the photograph of the fabricated device, where an $Al_2O_3$ layer with the thickness of 1 mm was placed between the thermoelectric cell and the TENG. Figure 12.18c shows atomic force microscopy (AFM) images of the triboelectric PA and PFA films with an effective area of 30 μm × 30 μm. The surface of PA film is rougher than that of the PFA film. The height of the aligned nanorod structures on the PA film surface is about 150 nm, which can enhance the triboelectric charge density on the surface.

Figure 12.19a, b show the output performance of the fabricated TENG under the forward and reversed connections, respectively. The corresponding output voltage and current are about 110 V and 60 μA (the current density of 2.4 $\mu A/cm^2$), respectively. The open-circuit voltage $V_{OC}$ of TENG can be given by

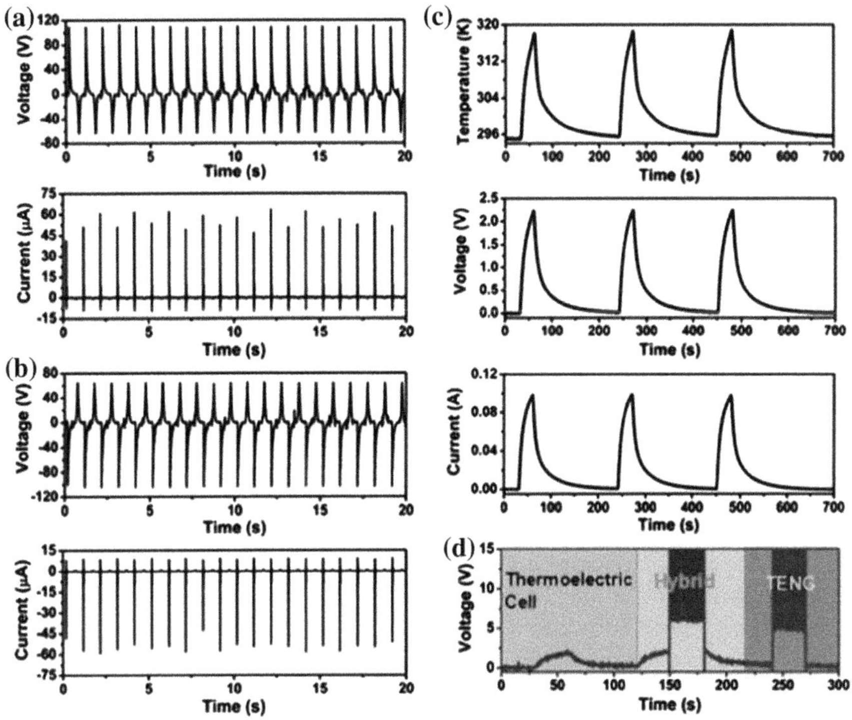


**Fig. 12.19** **a** and **b** Output voltage and current of the fabricated TENG under the forward (**a**) and reversed (**b**) connections to the measurement system. **c** Change in temperature at the bottom of the thermoelectric cell, and output voltage and current of the thermoelectric cell. **d** Enlarged output voltage of the hybrid thermoelectric cell and TENG (after rectification) for harvesting thermal and mechanical energies. Reproduced with permission from Royal Society of Chemistry [36]

$$V_{OC} = \frac{\sigma d}{\varepsilon_0} \tag{12.14}$$

where $\sigma$ is the triboelectric charge density, $d$ is the interlayer distance, and $\varepsilon_0$ is the permittivity in vacuum. According to Eq. (12.14), the output voltage $V_{OC}$ can be increased by increasing the interlayer distance $d$.

Figure 12.19c shows the output performance of the thermoelectric cell. For a temperature increase from 295 K at the top to 318 K at the bottom of the device, the largest output voltage and current of the device are about 2.2 V and 0.1 A, respectively. As compared with the TENG, the fabricated thermoelectric cell has a much smaller output voltage but much larger output current. Since the TENG gives alternating current (AC) output signals, which do not match with the direct current (DC) output signals of the thermoelectric cell, a bridge rectification circuit was used to convert the AC to DC signals. Figure 12.19d shows the enlarged output voltage of the integrated hybrid energy cell, where the solar cell and the rectified TENG were connected in series. To clearly observe the small output voltage of the thermoelectric cell, the maximum output voltage was cut off at 15 V. The total output voltage of the hybrid energy cell is obviously larger than that of the thermoelectric cell. It can also be seen that the hybrid energy cell can simultaneously/individually harvest mechanical and thermal energies.

### *12.2.3 Hybrid Electrochemical Cell and Triboelectric Nanogenerator*

For the fabrication of any electricity generators, two groups of materials are indispensable. The first group is the action materials that exhibit optoelectronic, piezoelectric, thermal electric and/or pyroelectric properties for converting various forms of energy into electricity. The second group is the electrode materials for charge transport. The mechanism of an electrochemical cell (EC) for converting chemical energy into electricity is usually based on spontaneous oxidation-reduction reactions at two different metal electrodes, respectively, in the electrolyte solution to drive electrons to flow in the external circuit. A triboelectric nanogenerator (TENG) has been extensively reported to harvest mechanical energy from irregular mechanical vibrations. Usually, each type of energy converter has its own electrodes. An interesting question is if the electrode materials for the two distinctly different devices can be shared. Moreover, the ECs are fabricated using a low-cost and a simple method of immersing two different metal materials in an electrolyte such as seawater. To avoid the possible corrosion, contamination and mechanical damage, these devices are usually covered by a protection layer, which may be used as a component to fabricate another energy harvesting unit, such as TENG.

Yang et al. demonstrate the first integration of the EC and the TENG as a hybrid energy cell for simultaneously/individually harvesting the chemical and mechanical energies [37]. By using the spontaneous oxidation-reduction reactions, a Cu/NaCl solution/Al-based EC was designed to harvest the chemical energy, where the Cu and Al films were used as both the electrode materials and the energy harvesting materials. A polydimethysiloxane (PDMS) film was covered on the Al electrode of the EC to make a TENG for harvesting mechanical energy, where the PDMS film was used as not only the triboelectric material for the TENG, but also the protection layer of the EC. The output performance of the TENG can be largely enhanced by embedding of $BaTiO_3$ nanoparticles into the PDMS matrix film. The hybrid energy cell has been used for driving biosensors and lighting up to 30 green LEDs.

Figure 12.20a shows a schematic diagram of the fabricated hybrid energy cell. An Al film was used as a common electrode for both the TENG and the EC. The EC was fixed on a wood structure that was located on the surface of the NaCl solution. The top TENG consists of a Cu film electrode, a PDMS film, and the Al film. In this study, the PDMS film with a micropyramid surface was fabricated

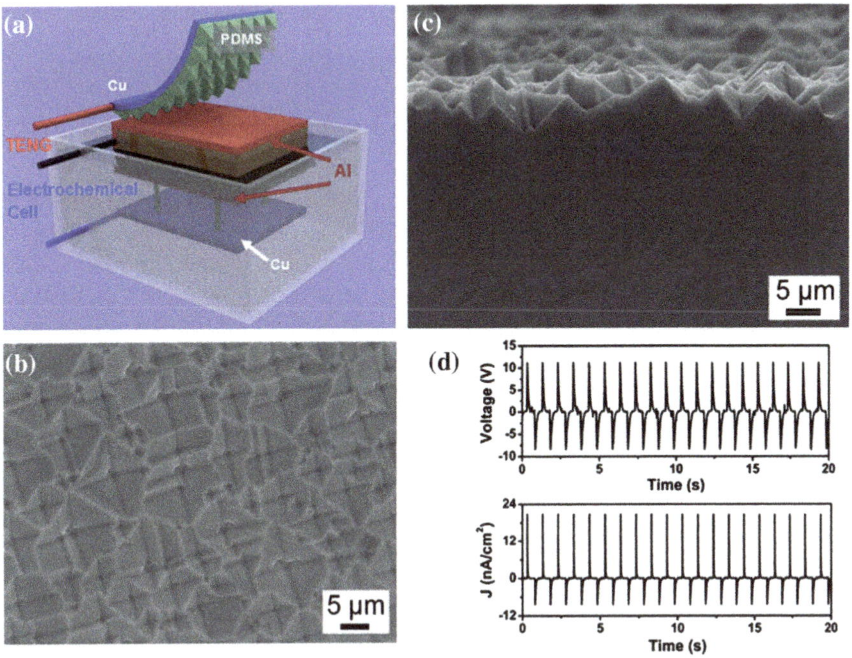


**Fig. 12.20** **a** Schematic diagram of the fabricated hybrid energy cell. **b** SEM image of the obtained PDMS film with the micropyramid surface. **c** Cross-sectional SEM image of the PDMS film. **d** The output open circuit voltage and the output short circuit current density of the fabricated TENG. Reproduced with permission from Royal Society of Chemistry [37]

using a Si micropyramid template, which was used to enhance the electrification process in the tribo-contacting for the TENG. Figure 12.20b shows a scanning electron microscopy (SEM) image of the fabricated PDMS film, revealing that the surface of the film has a micropyramid structure, which was confirmed further by the cross-sectional SEM image of the PDMS film, as shown in Fig. 12.20c. The thickness of the PDMS film is about 350 μm. The bottom EC consists of the Al film, the NaCl water solution, and the Cu film (Fig. 12.20a).

The mechanical energy applied to the device was harvested using the TENG. Figure 12.20d shows the output performance of the fabricated TENG, where the open circuit voltage and the short circuit current density are 11 V and 21.2 nA/cm$^2$, respectively. In this study, the mechanism of the TENG is based on the electrostatic charges generated in the contact-separation process between the PDMS film and the Al film due to the triboelectric effect. When the PDMS is in contact with the Al film, the surface charges will be transferred from Al film to PDMS due to the different triboelectric coefficients. A separation of the Al film from PDMS produces an open circuit voltage, which drives the electrons to flow in the external circuit in order to create an opposite voltage for compensating that from the triboelectric charges, resulting in the observed output signals. The output open circuit voltage $V_{OC}$ of TENG can be expressed by

$$V_{OC} = \frac{(-\sigma_0 - \sigma_1) \cdot d_S}{\varepsilon_0} - \frac{\sigma_1 \cdot d_{PDMS}}{\varepsilon_0 \varepsilon_r} \tag{12.15}$$

where $\sigma_0$ is the triboelectric charge density, $\sigma_1$ is the transferred charge density on Cu electrode on the PDMS film, $d_S$ is the distance between the two electrodes, $d_{PDMS}$ is the thickness of the PDMS film, $\varepsilon_0$ is the vacuum permittivity, and $\varepsilon_r$ is the relative permittivity of the PDMS layer. It is noticed that the $\sigma_0$ and $\sigma_1$ have the negative and positive signs, respectively.

According to the Maxwell model, the relative permittivity $\varepsilon_{r(composite)}$ of the $BaTiO_3$-PDMS composite material can be given by

$$\varepsilon_{r(composite)} = \varepsilon_{r(PDMS)} \left[ 1 + \frac{3\phi_f}{\frac{\varepsilon_{r(BTO)}/\varepsilon_{r(PDMS)}+2}{\varepsilon_{r(BTO)}/\varepsilon_{r(PDMS)}-2} - \varphi_f} \right] \tag{12.16}$$

where $\varepsilon_{r(PDMS)}$ is the relative permittivity of the PDMS, $\varepsilon_{r(BTO)}$ is the relative permittivity of the $BaTiO_3$ nanoparticles, and $\phi_f$ is the volume ratio of $BaTiO_3$ nanoparticles in the composite material. According to Eq. (12.16), we can obtain the relative permittivity $\varepsilon_{r(composite)}$ of the $BaTiO_3$-PDMS composite film. It can be seen that the relative permittivity $\varepsilon_{r(composite)}$ can increase from 2.6 to 6.5 with increasing the volume ratio of $BaTiO_3$ nanoparticles from 0 % to 33.3 %. According to Eq. (12.16), the first term is independent of $\varepsilon_{r(composite)}$, but the second term is associated with the $\varepsilon_{r(composite)}$. The increase of the relative permittivity $\varepsilon_{r(composite)}$ can effectively increase the output voltage of the TENG. According to Eqs. (12.15) and (12.16), the ratio $\sigma_1/(-\sigma_0)$ between the total transferred charge density $\sigma_1$ and the triboelectric charge density $\sigma_0$ at the equilibrium state can be given by

$$\frac{\sigma_1}{-\sigma_0} = \frac{d_s}{d_s + d_{PDMS}/\varepsilon_{r(composite)}} \tag{12.17}$$

According to Eq. (12.17), the obtained ratio $\sigma_1/(-\sigma_0)$ can increase with increasing the relative permittivity $\varepsilon_{r(composite)}$ of the $BaTiO_3$-PDMS composite material. Under the same triboelectric charge density, the increase of the total transferred charge density $\sigma_1$ can result in the increase of the output current density, which is consistent with the measured output current density.

The chemical energy in the electrolyte was harvested using the EC. Figure 12.21a, b show the output voltage and the output current of the fabricated Cu/NaCl solution/Al-based EC under the forward and reversed connections to the measurement system, respectively. The output voltage is about 0.55 V and there is an obvious output current peak of about 3 mA, which is associated with the charge-discharge process of the capacitance induced by the ON/OFF switch in the

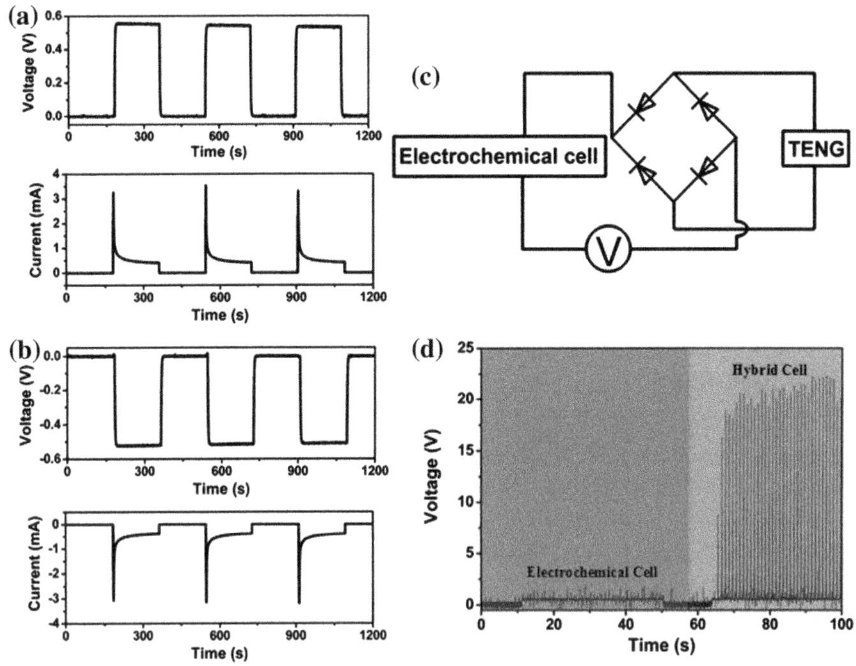


**Fig. 12.21** The output open-circuit voltage and short-circuit current of the fabricated EC under the forward (**a**) and reversed (**b**) connections to the measurement system. **c** Schematic diagram of the connection between the TENG and the EC. **d** Measured output open-circuit voltage of the hybrid TENG (after rectification) and EC. Reproduced with permission from Royal Society of Chemistry [37]

circuit, where the stable output current is about 0.4 mA (the output current density of 13.3 μA/cm$^2$). The mechanism of the fabricated EC is proposed as follows:

$$2\mathrm{Al} - 6e^- \rightarrow 2\mathrm{Al}^{3+}(\text{cathode}) \tag{12.18}$$

$$6\mathrm{H_2O} + 6\mathrm{e}^- \rightarrow 6\mathrm{OH}^- + 3\mathrm{H_2}(\text{anode}) \tag{12.19}$$

where the oxidation and reduction reactions occur at the cathode and anode, respectively. The spontaneous oxidation-reduction reactions in the NaCl solution can drive electrons to flow in the external circuit, resulting in the observed output voltage and current in Fig. 12.21a, b. Figure 12.21c shows a schematic diagram of the hybrid energy cell including the TENG and the EC, where the TENG was connected with a bridge rectification circuit for converting alternating current to direct current. Figure 12.21d shows the output voltage of the hybrid energy cell. It can be seen that all of the output voltage signals for the hybrid cell are positive and are larger than that of the individual energy harvesters.

In summary, a hybrid energy cell is demonstrated, which can simultaneously harvest mechanical and chemical energies for driving biosensors and personal electronics. A TENG was fabricated based on the contact-separation between a PDMS layer and the Al electrode layer to harvest mechanical energy. The output performance of the TENG can be increased by adding $BaTiO_3$ nanoparticles in the PDMS matrix to enhance the dielectric property of the PDMS layer. By using the spontaneous oxidation-reduction reactions at two different metal electrodes in the electrolyte solution, a Cu/NaCl solution/Al-based EC is fabricated to harvest chemical energy. Both the TENG and the EC share the Al electrode and the PDMS layer. The fabricated hybrid energy cells have the potential applications in some self-powered electrochemical reactions and driving some personal electronics.

## 12.3 Summary

Our living environment has an abundance of energies in forms of mechanical, light, thermal and chemical energies. Scavenging these types of energies is of critical importance for long-term energy needs and sustainable operation of portable/wearable electronics. However, these energies are not always available at the same time, which depends on the weather, working conditions, and location. The purpose for developing the hybrid cells is to simultaneously/individually scavenge these energies by using an integrated device, so that the devices/sensors can be continuously powered by using whatever energy that is available at their environment. The hybrid cells have the potential for fully utilizing the multimode energies in the environment under which the sensors/devices will be continuously and stably operating.

Hybridized with triboelectric nanogenerators, many kinds of hybrid cells for simultaneously harvesting mechanical energy along with multi-types of energies, has potential applications in self-powered electronics and some electrochemical reactions. The different energy harvesting units can work simultaneously or individually, and can be integrated in series or in parallel for enhancing the output voltage and current, respectively. In this chapter, the existing hybrid cells based on TENGs and other energy harvesters are described and summarized. These investigations have demonstrated many innovative approaches for developing integrated technologies for effectively harvesting available energies in our environment.

## References


1. Y. Zi, H. Guo, Z. Wen, M.-H. Yeh, C. Hu, Z.L. Wang, Harvesting low-frequency (<5 Hz) Irregular mechanical energy – a possible killer application of triboelectric nanogenerator. ACS Nano **10**(4), 4797–4805 (2016)
2. C. Zhang, W. Tang, C. Han, F. Fan, Z.L. Wang, Theoretical comparison, equivalent transformation, and conjunction operations of electromagnetic induction generator and triboelectric nanogenerator for harvesting mechanical energy. Adv. Mater. **26**(22), 3580–3591 (2014)

3. Y. Hu, J. Yang, S. Niu, W. Wu, Z.L. Wang, Hybridizing triboelectrification and electromagnetic induction effects for high-efficient mechanical energy harvesting. ACS Nano **8**(7), 7442–7450 (2014)
4. F. Feng-Ru, T. Wei, Y. Yan, L. Jianjun, Z. Chi, W. Zhong, Lin, Complementary power output characteristics of electromagnetic generators and triboelectric generators. Nanotechnology **25** (13), 135402 (2014)
5. S. Niu, Y. Liu, X. Chen, S. Wang, Y.S. Zhou, L. Lin, Y. Xie, Z.L. Wang, Theory of freestanding triboelectric-layer-based nanogenerators. Nano Energy **12**, 760–774 (2015)
6. K. Zhang, X. Wang, Y. Yang, Z.L. Wang, Hybridized electromagnetic-triboelectric nanogenerator for scavenging biomechanical energy for sustainably powering wearable electronics. ACS Nano **9**(4), 3521–3529 (2015)
7. H. Guo, Z. Wen, Y. Zi, M.-H. Yeh, J. Wang, L. Zhu, C. Hu, Z.L. Wang, A water-proof triboelectric-electromagnetic hybrid generator for energy harvesting in harsh environments. Adv Energy Mater (2015)
8. T. Quan, Y. Wu, Y. Yang, Hybrid electromagnetic–triboelectric nanogenerator for harvesting vibration energy. Nano Res. **8**(10), 3272–3280 (2015)
9. C. Zhang, T. Zhou, W. Tang, C. Han, L. Zhang, Z.L. Wang, Rotating-disk-based direct-current triboelectric nanogenerator. Adv Energy Mater **4**(9), n/a–n/a (2014)
10. Z.L. Wang, J. Song, piezoelectric nanogenerators based on Zinc Oxide nanowire arrays. Science **312**(5771), 242–246 (2006)
11. C. Chang, V.H. Tran, J. Wang, Y.-K. Fuh, L. Lin, Direct-Write piezoelectric polymeric nanogenerator with high energy conversion efficiency. Nano Lett. **10**(2), 726–731 (2010)
12. X. Chen, S. Xu, N. Yao, Y. Shi, 1.6 V nanogenerator for mechanical energy harvesting using PZT nanofibers. Nano Lett. **10**(6), 2133–2137 (2010)
13. K.-I. Park, M. Lee, Y. Liu, S. Moon, G.-T. Hwang, G. Zhu, J.E. Kim, S.O. Kim, D.K. Kim, Z.L. Wang, K.J. Lee, Flexible nanocomposite generator made of batio3 nanoparticles and graphitic carbons. Adv. Mater. **24**(22), 2999–3004 (2012)
14. S. Xu, Y.-W. Yeh, G. Poirier, M.C. McAlpine, R.A. Register, N. Yao, Flexible piezoelectric PMN-PT nanowire-based nanocomposite and device. Nano Lett. **13**(6), 2393–2398 (2013)
15. M. Han, X.-S. Zhang, B. Meng, W. Liu, W. Tang, X. Sun, W. Wang, H. Zhang, r-shaped hybrid nanogenerator with enhanced piezoelectricity. ACS Nano **7**(10), 8554–8560 (2013)
16. W.-S. Jung, M.-G. Kang, H.G. Moon, S.-H. Baek, S.-J. Yoon, Z.-L. Wang, S.-W. Kim, C.-Y. Kang, High output piezo/triboelectric hybrid generator. Sci. Rep **5**, 9309 (2015)
17. M. Han, X. Zhang, W. Liu, X. Sun, X. Peng, H. Zhang, Low-frequency wide-band hybrid energy harvester based on piezoelectric and triboelectric mechanism. Sci. Chin. Technol. Sci. **56**(8), 1835–1841 (2013)
18. C. Xue, J. Li, Q. Zhang, Z. Zhang, Z. Hai, L. Gao, R. Feng, J. Tang, J. Liu, W. Zhang, D. Sun, A novel arch-shape nanogenerator based on piezoelectric and triboelectric mechanism for mechanical energy harvesting. Nanomaterials **5**(1), 36 (2014)
19. Y. Yang, W. Guo, K.C. Pradel, G. Zhu, Y. Zhou, Y. Zhang, Y. Hu, L. Lin, Z.L. Wang, Pyroelectric nanogenerators for harvesting thermoelectric energy. Nano Lett. **12**(6), 2833–2838 (2012)
20. Y. Yang, J.H. Jung, B.K. Yun, F. Zhang, K.C. Pradel, W. Guo, Z.L. Wang, Flexible pyroelectric nanogenerators using a composite structure of lead-free KNbO3 nanowires. Adv. Mater. **24**(39), 5357–5362 (2012)
21. Y. Yang, H. Zhang, S. Lee, D. Kim, W. Hwang, Z.L. Wang, Hybrid energy cell for degradation of methyl orange by self-powered electrocatalytic oxidation. Nano Lett. **13**(2), 803–808 (2013)
22. J. Bae, J. Lee, S. Kim, J. Ha, B.-S. Lee, Y. Park, C. Choong, J.-B. Kim, Z.L. Wang, H.-Y. Kim, J.-J. Park, U.I. Chung, Flutter-driven triboelectrification for harvesting wind energy. Nat. Commun. **5,** 4929(2014)
23. Y. Zi, L. Lin, J. Wang, S. Wang, J. Chen, X. Fan, P.-K. Yang, F. Yi, Z.L. Wang, Triboelectric–pyroelectric–piezoelectric hybrid cell for high-efficiency energy-harvesting and self-powered sensing. Adv. Mater. **27**(14), 2340–2347 (2015)

24. K.L. Bing, T. Li, H.H. Hng, F. Boey, T. Zhang, S. Li, *Waste Energy Harvesting: Mechanical and Thermal Energies* (Springer, 2014)
25. Q. Leng, L. Chen, H. Guo, J. Liu, G. Liu, C. Hu, Y. Xi, Harvesting heat energy from hot/cold water with a pyroelectric generator. J. Mater. Chem. A **2**(30), 11940–11947 (2014)
26. Y. Yang, S. Wang, Y. Zhang, Z.L. Wang, Pyroelectric nanogenerators for driving wireless sensors. Nano Lett. **12**(12), 6408–6413 (2012)
27. K. Lefki, G.J.M. Dormans, Measurement of piezoelectric coefficients of ferroelectric thin films. J. Appl. Phys. **76**(3), 1764–1767 (1994)
28. S. Niu, Y. Liu, S. Wang, L. Lin, Y.S. Zhou, Y. Hu, Z.L. Wang, Theory of sliding-mode triboelectric nanogenerators. Adv. Mater. **25**(43), 6184–6193 (2013)
29. G. Zhu, J. Chen, Y. Liu, P. Bai, Y.S. Zhou, Q. Jing, C. Pan, Z.L. Wang, Linear-grating triboelectric generator based on sliding electrification. Nano Lett. **13**(5), 2282–2289 (2013)
30. R.G. Kepler, R.A. Anderson, Piezoelectricity and pyroelectricity in polyvinylidene fluoride. J. Appl. Phys. **49**(8), 4490–4494 (1978)
31. Y. Yang, H. Zhang, Y. Liu, Z.-H. Lin, S. Lee, Z. Lin, C.P. Wong, Z.L. Wang, Silicon-based hybrid energy cell for self-powered electrodegradation and personal electronics. ACS Nano **7** (3), 2808–2813 (2013)
32. H. Guo, X. He, J. Zhong, Q. Zhong, Q. Leng, C. Hu, J. Chen, L. Tian, Y. Xi, J. Zhou, A nanogenerator for harvesting airflow energy and light energy. J. Mater. Chem. A **2**(7), 2079–2087 (2014)
33. L. Zheng, Z.-H. Lin, G. Cheng, W. Wu, X. Wen, S. Lee, Z.L. Wang, Silicon-based hybrid cell for harvesting solar energy and raindrop electrostatic energy. Nano Energy **9**, 291–300 (2014)
34. L. Zheng, G. Cheng, J. Chen, L. Lin, J. Wang, Y. Liu, H. Li, Z.L. Wang, A hybridized power panel to simultaneously generate electricity from sunlight, raindrops, and wind around the clock. Adv. Energy Mater. n/a–n/a (2015)
35. W. Liu, X. Yan, G. Chen, Z. Ren, Recent advances in thermoelectric nanocomposites. Nano Energy **1**(1), 42–56 (2012)
36. Y. Yang, H. Zhang, Z.-H. Lin, Y. Liu, J. Chen, Z. Lin, Y.S. Zhou, C.P. Wong, Z.L. Wang, A hybrid energy cell for self-powered water splitting. Energy Environ. Sci. **6**(8), 2429–2434 (2013)
37. Y. Yang, H. Zhang, J. Chen, S. Lee, T.-C. Hou, Z.L. Wang, Simultaneously harvesting mechanical and chemical energies by a hybrid cell for self-powered biosensors and personal electronics. Energy Environ. Sci. **6**(6), 1744–1749 (2013)

# Chapter 13
# Applications in Self-powered Systems and Processes

## 13.1 Integration of TENG with Energy Storage Units for Sustainably Driving Portable Electronics

TENG has intrinsic capacitive behavior and its equivalent circuit model is a voltage source in series with an inherent capacitor. Since capacitor cannot conduct direct current (DC), TENGs are fundamentally alternating current (AC) power sources. So it can only directly drive AC loads, like light-emitting diodes (LEDs) and light bulbs, but cannot serve as a direct power source for commercial electronics. If a TENG wants to be utilized to provide energy for commercial electronics, an energy storage device such as a battery or a large storage capacitor is necessary to be integrated to form a self-powered system [1] to serve a reservoir for the generated charges and output them in a regulated manner. Besides, a power management circuit is also necessary to convert the AC current from TENG to DC current to charge the energy storage elements.

### *13.1.1 Direct Integration and Transformer Integration*

The simplest power management circuit is a bridge rectifier and this type of integration is called direct integration. A typical example is developed by Wang et al. [2]. They demonstrate the first flexible self-charging power unit (SCPU) that is capable of simultaneously harvesting and storing ambient mechanical energy, by integrating a TENG-based mechanical energy harvester with a Li-ion battery (LIB) based energy storage. In this SCPU, the Li-ion battery portion can be directly charged by mechanical motions. The integration was realized through developing a flexible LIB on an arch-shaped TENG structure [3]. When the surrounding mechanical energy is applied onto the SCPU, an alternating-current in response to the external triggering is generated. After rectification, the electrical



Z.L. Wang et al., *Triboelectric Nanogenerators*,
Green Energy and Technology, DOI 10.1007/978-3-319-40039-6_13

energy can be stored in the LIB, which can be fully charged in 11 h. With this SCPU, a new working mode for a power source—the "sustainable mode" was demonstrated, in which the environmental mechanical energy is scavenged to charge the battery while the battery keeps driving an external load as a DC source. In this mode, the demonstrated SCPU can provide a continuous and sustainable DC current of 2 μA at a stable voltage of 1.55 V for as long as there is mechanical motion/agitation. It can be used to continuously drive a UV sensor [4] for an extended period of time. The LIB in the SCPU serves not only as energy storage, but also as a power regulator and management for the entire system by utilizing the stable electrode-potential difference [5, 6]. Since mechanical motion is available almost everywhere and at any time to replenish the energy draining in the battery, a large-capacity battery with a high density may not be necessary to ensure a long operating lifetime. This could introduce a new alternative to the current research trend of solely seeking high density batteries.

The integration of a LIB and a TENG in the SCPU was achieved through the construction of a flexible LIB, which was made in an arch shape for creating the gap necessary for the TENG structure, [3] as illustrated in Fig. 13.1a. In order to make the battery flexible, a polymeric shell is designed for the packaging and sealing, to replace the conventional ridge coin cell. The supporting substrate of this polymeric shell is a bent Kapton film (3 cm × 4 cm) with purposely-introduced thermal stress through the deposition of a $SiO_2$ film at high temperature [3]. The three kernel components of the LIB: the $TiO_2$ nanowires anode, the polyethylene (PE) separator and the $LiFePO_4$/active carbon/binder mixture cathode on the Al current collector, were laminated on top of the convex surface (the surface deposited with $SiO_2$) of this Kapton substrate. This kernel structure with an effective area of ~1.5 cm × 1.5 cm was immersed in the electrolyte. In order to enhance the flexibility of the battery from the anode side, the $TiO_2$ nanowires with an anatase crystal structure were hydrothermally grown on a piece of soft carbon cloth that also serves as the current collector, as shown by the scanning electron microscopy (SEM) image displayed in Fig. 13.1b. From the picture of the $TiO_2$/carbon cloth composite shown in the inset (bottom) of Fig. 13.1b, this anode layer exhibits excellent flexibility. SEM images with higher magnifications are shown at the top inset of Fig. 13.1b (the morphology of a single fiber from the carbon cloth after the growth of $TiO_2$ nanowire networks), and Fig. 13.1c, from which it is shown that every single fiber in the carbon cloth is uniformly covered by $TiO_2$ nanowire networks with an average length of ~1 μm. The battery was isolated from air by covering with a layer of thin Kapton film and sealing with polydimethylsiloxane (PDMS) and epoxy at the sides. For the hybridization with the TENG, this LIB's supporting Kapton substrate was deposited with Al film on the back and bonded with another bent Kapton film of the same size in a face-to-face manner. This forms an arch-shaped TENG structure based on the triboelectrification between the top Al layer and the bottom Kapton that has the corresponding electrode (Al film) deposited on its convex surface. With this hybridized SCPU structure, the LIB can be charged through harnessing ambient mechanical energy, only by rectifying the TENG-generated alternating-current pulses to DC with a full-wave bridge rectifier.

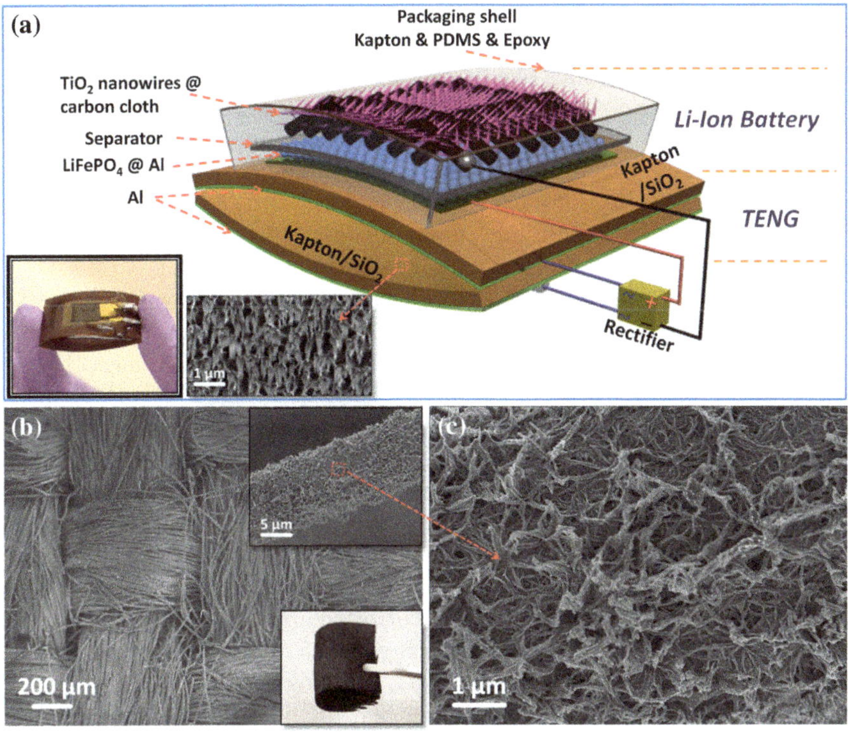


**Fig. 13.1** Structure design of a flexible self-charging power unit (SCPU). **a** Schematic diagram showing the detailed structure of the SCPU, the *inset* on the *left* is a photograph of a typical SCPU device, and the *inset* on the *right* is an SEM image of the Kapton nanorod-structure on the concave surface of the *bottom* Kapton, with 30°-tilted view. **b** SEM image of the carbon cloth grown with $TiO_2$ nanowire networks, the *top inset* is the enlarged SEM image of a single fiber in the fabric, and the *bottom inset* is a photograph of the $TiO_2$-covered carbon cloth as the anode, showing its good flexibility. **c** SEM image with higher magnification showing the detailed morphology of the $TiO_2$ nanowire networks on the carbon cloth. Reproduced with permission from American Chemical Society [2]

A photo of a typical SCPU device is shown in the inset (left) of Fig. 13.1a, clearly revealing its flexibility and arch-shaped structure.

When external mechanical vibration is applied onto the flexible SCPU to periodically flatten it, the two surfaces across the gap will get into contact from time to time. Coupling of the triboelectric effect with the electrostatic induction will generate a flow of current in the external circuit, which will be stored in the LIB in the same device. Since the electrode potential is not affected by the state of charge (SOC) within the wide-range discharging plateau, the irregular and fluctuating current generated by the TENG due to the change of the surround mechanical motion will not cause an obvious change in the output voltage of the LIB. This stable constant voltage on the load should be able to be maintained for an infinitely long time in principle, as long as the averaged current provided by the TENG over time is about the same as the current consumed by the load, so that the SOC of the LIB does not shift out of the voltage plateau. In this proposed integration mode of mechanical energy harvesting and energy storage, the consumed energy is generated from the surrounding mechanical motion, thus the amount of energy stored in the LIB will not change very much although the load is being driven continuously. Therefore, the battery works not only as an energy storage unit, but more importantly as a power regulator that converts the irregular, unstable and time-dependent input from the TENG (due to the varying ambient condition), to a constant DC output. This is the reason that the SCPU can be a stable DC power source.

The performance of the SCPU is first characterized by the outputs of the TENG and the LIB as individual units (Fig. 13.2). When the arch-shaped structure was periodically flattened under a frequency of 8 Hz, an open-circuit voltage ($V_{OC}$) of ~220 V was generated between the two electrodes of the TENG part (Fig. 13.2a), which is the potential difference induced by the separation of opposite tribo-charges on the Al and Kapton films. Under a short-circuit condition, this voltage drives ~110 nC of charges ($\Delta Q$) to flow between the two electrodes in each half deformation cycle (as shown in the measured charge transfer in Fig. 13.2b). Since the voltage of the LIB is much smaller than the TENG open-circuit voltage, this $\Delta Q$ value should theoretically equal to the amount of tribo-charges on each triboelectric surface (with an area of ~12 $cm^2$), corresponding to a charge density of ~91.6 $\mu C/m^2$. This back-and-forth charge transfer produced a pair of AC short-circuit current ($I_{SC}$) peaks in each cycle. As shown in Fig. 13.2c, the positive peak with a larger magnitude of ~150 μA came from the pressing motion, which had a higher deformation rate compared to the elastic releasing process. But these two asymmetric peaks per cycle carry the same amount of charges due to charge conservation (Fig. 13.2b). This AC output can be rectified to DC by a low-loss full wave bridge rectifier (Fig. 13.2d), which generates ~220 nC electricity per cycle to be stored in the LIB.

The capability of the flexible LIB in the SCPU for charge storage was evaluated using deep charging/discharging galvanostatic test between 0.7 and 2.7 V, at a current of 2 μA, as for conventional batteries. As indicated by the voltage profile in the 6 consecutive cycles (Fig. 13.2e), the LIB gave clear voltage plateaus in both charging and discharging processes. The discharging plateau is ~1.53 V, which is the real working voltage of the LIB as a power source. The LIB also has a stable capacity around 17.5 μAh, revealing its expected performance as an energy storage unit.

When the TENG part and the LIB part work conjunctionally as a full unit, the SCPU will work as a self-sufficient power source. Conventionally, the charging and discharging of an energy storage unit is in two separated and consecutive steps. This working mode for the SCPU is called the "standby-active mode", as shown schematically in the circuit diagrams in Fig. 13.3a. The LIB is connected to the TENG through the rectifier but disconnected from the load (with S1 closed and S2 open), during which it is only storing the electrical energy generated by the TENG part. As demonstrated in Fig. 13.3b, the LIB was charged to its full capacity in ~11 h by the TENG under a deformation frequency of 9 Hz, with the voltage

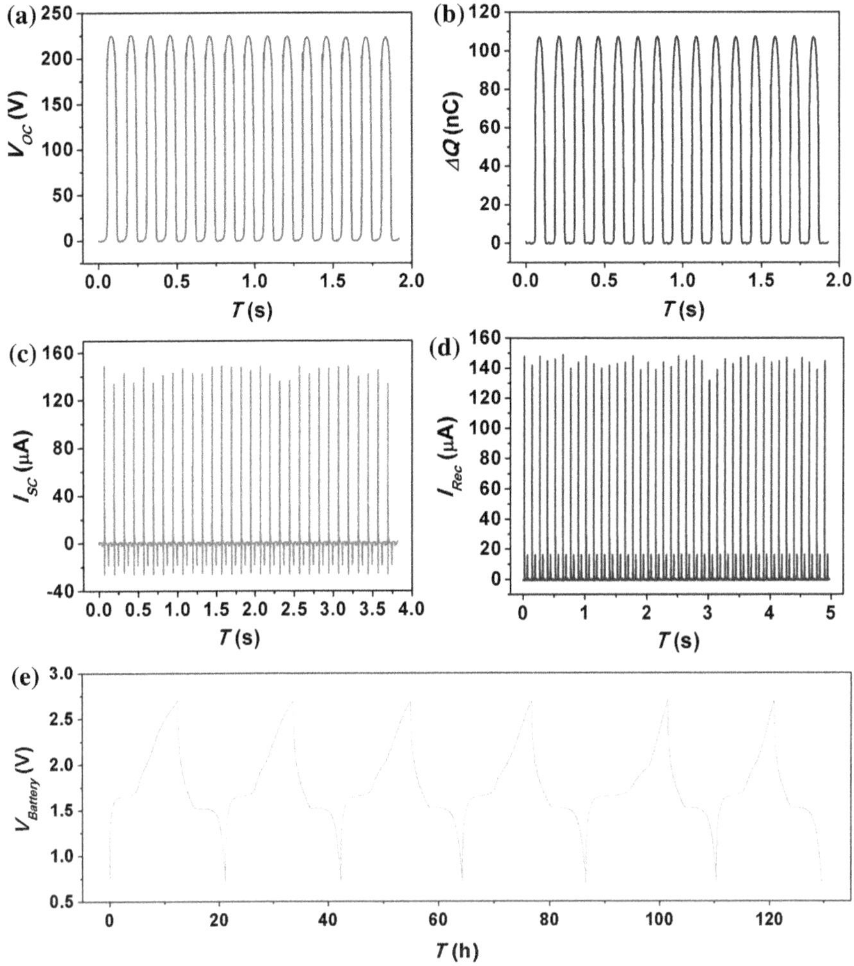


**Fig. 13.2** Characterizing the individual performances of the TENG part and the LIB part in the SCPU. **a** The open-circuit voltage ($V_{OC}$), **b** the amount of charges transferred ($\Delta Q$), **c** the short-circuit current ($I_{SC}$) and **d** the rectified current ($I_{Rec}$) generated by the TENG part when the SCPU is periodically flattened by external mechanical agitation under the frequency of 8 Hz. **e** The voltage profile of the LIB part during the deep charging/discharging galvanostatic test with a current of 2 μA. Reproduced with permission from American Chemical Society [2]

($V_{Battery}$) increasing from 0.7 to 2.5 V crossing the charging plateau at ~1.62 V. This is the "standby mode" during which the SCPU is charged. Next, the fully-charged SCPU will be used as a battery to drive a device by switching the connection with S1 open and S2 closed. As shown in Fig. 13.3b, the fully charged LIB could provide a constant current of 2 μA to the external load for ~10 h (~3 h at the discharging plateau, which is the actual working time for this LIB as a constant-voltage power source). This is the "active mode" of the SCPU during which it is used to drive a device. In practical applications, the SCPU switches between the "standby mode" and the "active mode", which correspond to the charging and discharging of the LIB, respectively. This operational integration methodology of energy harvesting and storage is particularly suitable for applications where the current required by certain electronic devices is much larger than the average current (rather than the peak current) generated by the TENG, and the electronic devices do not need to work continuously. This meets the requirement of some sensors for intermittent detections.

Alternatively, in the cases that the electronic devices need to work constantly, the SCPU can work in a "sustainable mode", as proposed in Fig. 13.3c. In this mode, the LIB is being charged by the TENG while it is simultaneously powering an external load. As long as the averaged current converted from the surrounding mechanical energy is approximately the same as the current required by the load, the energy stored in the LIB will remain at a stable level so that the voltage can remain at its discharging plateau "forever" due to the nature of electrochemical reactions. As shown in Fig. 13.3d, when the SCPU was being triggered by a pressing motion at 9 Hz (the averaged current produced can be estimated as $\sim 1.98$ μA), the LIB part provided a constant current of 2 μA at the voltage of $\sim 1.55$ V for more than 40 h! Thus, in this proposed "sustainable mode", the SCPU could work as an independent, self-sufficient and sustainable power source for providing a constant voltage.

In a lot of applications, a constant voltage from the power source is needed for the operation of the electronic systems. The demonstrated SCPU under the "sustainable mode" can serve as a non-stop power source for these applications. A ZnO-nanowire-based UV sensor is utilized to demonstrate this unique capability of the SCPU. In the sensor, the nanowire is bonded onto the substrate at its two ends with silver paste, forming a metal-semiconductor-metal structure (the left inset of Fig. 13.4a). A Schottky contact is formed so that the I–V has a rectifying behavior. In the first case (Fig. 13.4a), the fully charged LIB portion of the SCPU was used to power this UV sensor as shown in the inset (right) of Fig. 13.4a, which is the "active mode" of the SCPU illustrated in Fig. 13.3a. When the UV light is off, with an input voltage of $\sim 1.53$ V from the battery at its discharging plateau, the measured current through the sensor was around 1.55 μA. When UV light is turned on, the current jumped to a level of $\sim 2.95$ μA. The UV light was turned on and off periodically, and the voltage of the battery remained stable during the first few cycles. However, after $\sim 3.5$ h, the voltage started to drop, which indicated that the battery was running out of charge. As a result, the current levels for the UV-off and UV-on states both decreased to lower levels, and thus could not be an effective indicator for the presence of UV-illumination anymore since it cannot be judged using the calibrated current, which means that this battery has to be replaced after about 4 h usage.

This problem can be solved using the SCPU in the proposed "sustainable mode". As shown in Fig. 13.4b, when the LIB was being charged by the TENG under a deformation frequency of 9 Hz while driving the UV sensor, the output voltage of the battery can stay at the plateau of around 1.55 V with only small fluctuation for

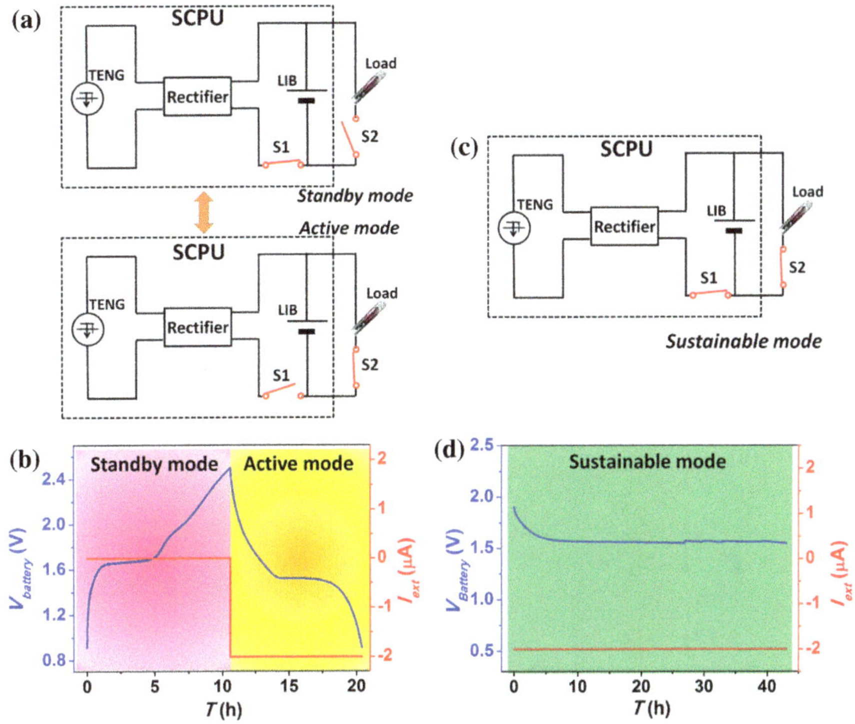


**Fig. 13.3** Performance of the SCPU as an integrated DC power source. **a** Electrical circuit diagram showing the "standby-active mode" for the integration of the TENG part and the LIB part. The portion enclosed by the dotted square is the SCPU. **b** The voltage profile showing the LIB part fully charged by the TENG part (the "standby mode") and then discharging under a current of 2 μA (the "active mode"). **c** The electrical circuit diagram showing the newly-proposed "sustainable mode", in which the LIB part is driving the external load while it is being charged by the TENG part. **d** In this sustainable mode, the SCPU provided a 2-μA DC current with a constant voltage of 1.53 V for more than 40 h, when the SCPU was deformed under a frequency of 9 Hz. Reproduced with permission from American Chemical Society [2]

over 12 h, during which the UV light was periodically turned on and off for more than 10 times without obvious decay in current level, which can clearly convey the information of the UV illumination to the calibrated system. Therefore, it has been undoubtedly demonstrated that with the "hybridization" of the mechanical energy harvesting and energy storage in the SCPU, an electronic system could be truly self-powered continuously and sustainable.

The above shown direct integration has many advantages, such as the simple integration technique. However, the most severe problem for this direct integration is its ultralow charging efficiency from the impedance mismatch between the energy harvesting and storage devices. As discussed in Chap. 6, triboelectric nanogenerators have inherent very high impedance while the typical energy storage element like the LIBs and supercapacitors have very low impedance. Therefore, this huge

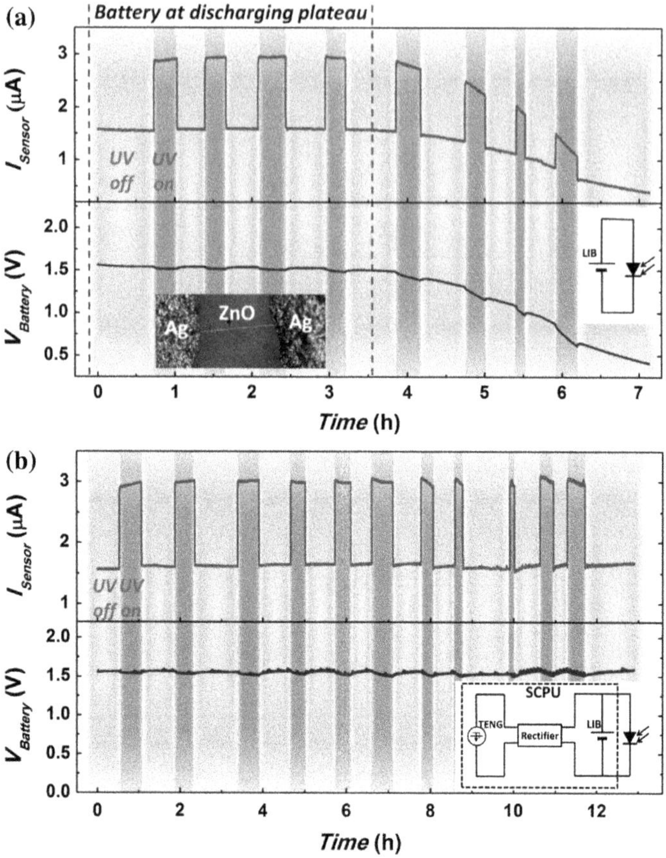


**Fig. 13.4** The SCPU as a sustainable power source for driving a UV sensor. **a** Operation of the UV sensor when solely driven by the LIB part in the SCPU, which was recorded from the point that the battery just entered its discharging plateau. The *left inset* is an optical microscope image of the ZnO-nanowire-based UV sensor in this demonstration, and the *right inset* is the circuit diagram showing the power source. **b** The operation of the UV sensor continuously driven by the SCPU in the "sustainable mode" for ~13 h. The *inset* is the circuit diagram showing the UV sensor powered by the SCPU in the "sustainable mode". In both of (**a**) and (**b**), the *top curve* (*red*) is the current though the sensor, reflecting its response to the UV light, and the *bottom curve* (*blue*) is the real time voltage of LIB part. Reproduced with permission from American Chemical Society [2]

impedance mismatch usually results in an ultralow charging efficiency that is typically less than 1 %. Much power was lost during the charging process and this is the reason why the provided DC power is only about 3 μW even under very tense (~10 Hz) motor-triggered mechanical motion in this direct integration example. This kind of power level (3 μW) is still far away from any practical applications and the entire system cannot be exclusively driven by the harvested energy if the data measurement and processing systems as well as data transmission components are included.

Many efforts have been devoted to solve the impedance match problem and improved the charging efficiency. Zhu et al. tried to utilize a transformer as an impedance match network and showed some progress [7]. However, since TENGs have high internal impedance and pulsed signal output, all practical transformers suffer a huge loss in power conversion, as dictated by their limited quality factor, insufficient primary inductance, poor low-frequency response, and finite narrow bandwidth. Transformer only shows improvement of charging efficiency when the output signal frequency is high enough. Under other conditions, the charging efficiency when transformer is utilized is even lower than direct charging.

### 13.1.2 Through Power Management Circuit Board

From the above section, it is noticed that the impedance mismatch induces huge energy loss through integration with an energy storage unit so a proper power management is critical for any practical TENG application. Finally, that problem was solved by Niu et al. by presenting the first genuine self-powered system to meet mW requirement of personal electronics [8]. The system includes a TENG, a power management circuit, and a low leakage energy storage device. In that work, Niu et al. perform system-level optimization to ensure the collaborative work of all the system components. Especially a power management circuit is designed to solve the impedance mismatch problem, which can achieve 90 % board efficiency and 60 % total efficiency, about two orders of magnitude improvement compared to direct charging. Driven by palm tapping, the power unit can provide a continuous DC electricity of 1.044 mW on average power in a regulated and managed manner that can be universally applied as a standard power source for continuously driving numerous conventional electronics, such as a thermometer, a heart rate monitor (electrocardiograph/ECG system), a pedometer, a wearable electronic watch, a scientific calculator, and a wireless radio-frequency communication system. This work demonstrates the first power unit that utilizes widely accessible biomechanical energy source to sustainably drive a broad range of commercial mobile and wearable electronic devices.

This integrated human-motion driven self-powered system hierarchy is shown in Fig. 13.5a, including a human-motion powered self-charging power unit and functional circuits for different applications. The TENG scavenges the human biomechanical energy and converts it to AC electricity. Then this AC electricity is converted to a DC electric output that matches to the input of the energy storage unit by the power management circuit. Finally, the energy storage unit provides DC electricity to the whole functional circuits, including sensors, data processors, displays, and wireless transmitters. The signals from sensors and data processors are either visualized on a liquid-crystal display (LCD) or sent out remotely through a wireless transmitter.

The TENG is first designed and optimized to efficiently harvest human biomechanical energy. A multilayered attached-electrode contact-mode TENG is

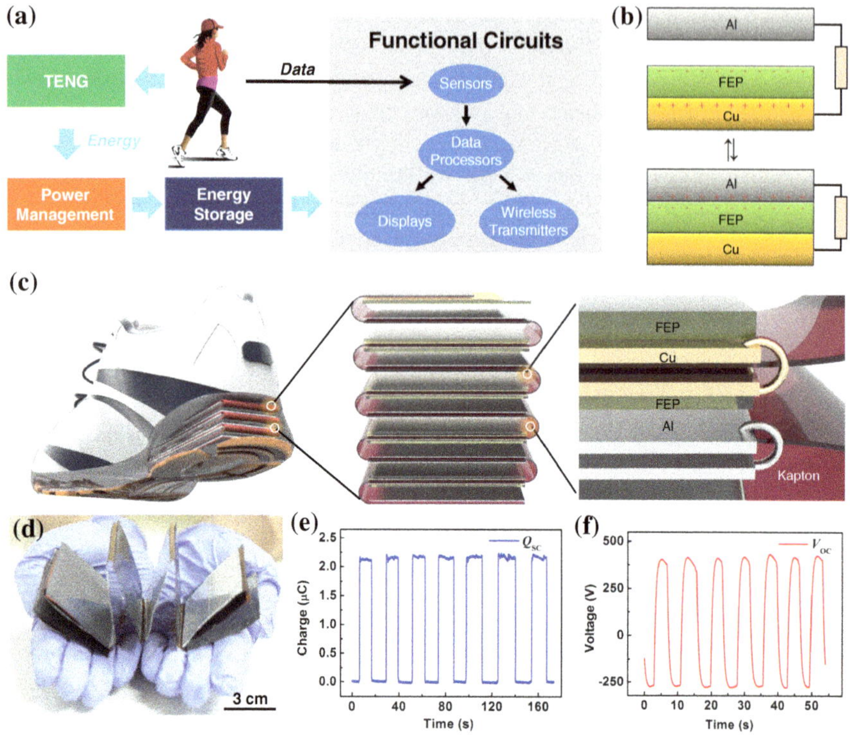


**Fig. 13.5** Structure design of a completely self-powered system that is driven exclusively by human biomechanical energy. **a** System diagram of a TENG-based self-powered system. **b** Working mechanism of an attached-electrode contact-mode TENG. **c** Structure of the designed multilayer TENG. **d** Photo of an as-fabricated TENG. **e–f**, **e** Short-circuit current output and **f** open-circuit voltage output of the as-fabricated TENG. Reproduced with permission from Nature Publishing Group [8]

chosen to effectively collect the energy from human walking and running. The basic working principle of attached-electrode contact-mode TENGs is shown in Fig. 13.5b, as discussed in Chap. 2. As shown in Fig. 13.5c, a Kapton film is shaped into a zigzag structure with 10–15 layers. For each layer, a thin Aluminum foil and FEP (Fluorinated ethylene propylene) layer are utilized as the triboelectric materials. The Al foil also serves as one electrode. Copper is evaporated at the backside of the FEP layer as the other electrode. As shown in Fig. 13.5d, the as-fabricated TENG has small volume and light weight (5.7 cm × 5.2 cm × 1. 6 cm/29.9 g for a 10-layer TENG and 5.7 cm × 5.2 cm × 2.4 cm/43.6 g for a 15-layer TENG). Embedded in the shoe insoles, a human walking can drive this TENG to generate about 2.2 μC short-circuit transferred charge and about 700 V voltage output, as shown in Fig. 13.5e, f.

From theoretical understanding of TENG charging characteristics discussed in Chap. 6, Niu et al. designed the following charging strategy for maximized energy storage efficiency. First, a small temporary capacitor ($C_{temp}$) is charged by the TENG from 0 V. Once its voltage reaches $V_{opt}$ (impedance match condition is reached), the energy stored in $C_{temp}$ begins to be transferred to the final energy storage unit (a large capacitor or a battery) to maximize the theoretical total efficiency $\eta_{total}$. When the energy transfer finishes, the voltage of $C_{temp}$ drops back to close to 0, and then $C_{temp}$ is recharged by the TENG to reach $V_{opt}$ again. With this optimized charging cycle, $\eta_{total}$ can theoretically reach 75 %.

This theoretical charging cycle can be realized by a two-stage power management circuit, as shown in Fig. 13.6a. At the first stage, a temporary capacitor $C_{temp}$ is charged by a TENG through a bridge rectifier. The second stage is for energy transfer from $C_{temp}$ to the final energy storage unit. Since transferring electrostatic energy directly from a small capacitor to a large capacitor (or a battery) results in huge energy loss, two automatic electronic switches (controlled by a logic control unit, the power of both switches and their logic control unit is supplied from the final energy storage unit) and a coupled inductor are utilized in the second stage [9]. The detailed operation process is shown below to achieve efficient energy transfer in this stage. First, both the switches $J_1$ and $J_2$ are open to avoid interference of the charging process of $C_{temp}$. When $V_{temp}$ reaches $V_{opt}$, the electronic switch $J_1$ closes. As a result, the energy starts to transfer from $C_{temp}$ to inductor $L_1$ and $V_{temp}$ starts to drop. When the energy is thoroughly transferred to $L_1$, the switch $J_1$ opens and $J_2$ closes. As a consequence, the current of $L_1$ falls to 0 instantaneously. Besides, since the total magnetic flux linkage in the coupled inductance cannot change abruptly, the current of $L_2$ will suddenly rise up, corresponding to the energy transfer from $L_1$ to $L_2$. Finally, the energy stored in $L_2$ will automatically transfer to the final energy storage unit because of the closure of $J_2$. When the energy stored in $L_2$ is thoroughly sent out, $J_2$ is open again and another charging cycle begins.

The operation of such a power management system is shown in Fig. 13.6c–g. Figure 13.6c shows the voltage profiles of both temporary capacitor ($V_{temp}$) and final storage capacitor ($V_{store}$) when one 15-layer TENG and a 1 mF Al-electrolyte capacitor were utilized as the energy harvester and the energy storage unit ($C_{store}$), respectively. While the TENG is driven by palm tapping, $C_{temp}$ is charged by the TENG and discharged through the energy transfer network, resulting in an oscillation of $V_{temp}$ between 230 and 0 V. During the drop phase of $V_{temp}$, $V_{store}$ is raised up by the transferred energy from $C_{temp}$. Note that if there was no mechanical energy input (from 5.4 to 5.7 s in Fig. 13.6c), both $V_{temp}$ and $V_{store}$ will decrease slowly because of the system leakage current and the power consumption of the electronic switches. The performance of the energy transfer network can be evaluated by the board efficiency ($\eta_{board}$), which is defined as the ratio of the total energy stored in $C_{store}$ to the total energy transferred out from $C_{temp}$. From the data shown in Fig. 13.6c, the total energy sent out from $C_{temp}$ is calculated as 9.160 mJ, while the total energy stored in $C_{store}$ is calculated as 8.243 mJ. So this energy transfer network has $\eta_{board}$ = 90.0 %, which is hugely improved in comparison to that of one-step direct energy transfer process.

To measure the DC power delivered to the load by the system as driven by palm tapping, a load resistor $R_L$ is connected in parallel with the storage capacitor. The

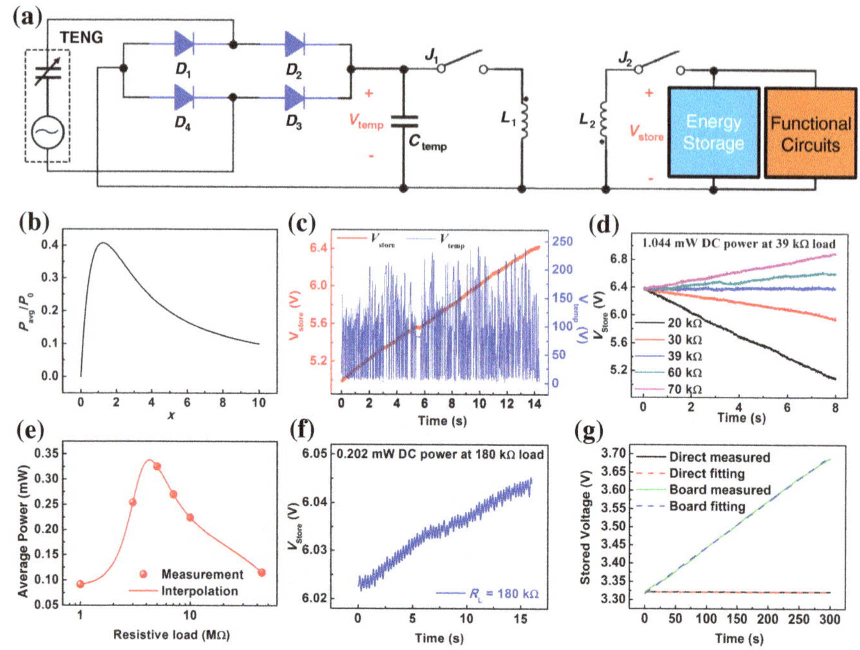


**Fig. 13.6** Design of the power management part for converting a "random" AC output from TENG to a regulated and managed power for directly driving electronics. **a** Circuit diagram of the power management circuit. **b** Theoretical calculation of the optimized charging time to demonstrate the design of the power management circuit. **c** Board efficiency measurement results. **d** Measurement of the maximum DC power of this system driven by human biomechanical energy. **e–f** Total efficiency measurement results. **e** Measurement of the AC harvested power from a resistor. **f** Measurement of the DC harvested power from the power management circuit. **g** Comparison of the charging current between direct charging and board charging. Reproduced with permission from Nature Publishing Group [8]

storage capacitor $C_{store}$ is charged by two 15-layer as fabricated TENGs and then provides DC power for $R_L$. As shown in Fig. 13.6d, when $R_L$ is high, the power consumption of $R_L$ is lower than the power provided by TENG, so $V_{store}$ has a positive slope with time. As the load resistance $R_L$ decreases, the load power consumption increases and the charging slope of $V_{store}$ decreases. When the charging slope reduces to 0 (39 kΩ in Fig. 13.6d), the power delivered from palm tapping is the same as the power consumed on the load, which is calculated as 1.044 mW (7.34 $Wm^{-3}$). When $R_L$ continues to decrease (20 kΩ), the power delivered from palm tapping is not enough to compensate the energy consumption on the load, and the charging slope becomes negative.

The most important parameter of the power management circuit is the total efficiency $\eta_{total}$, which is defined as the ratio of the maximum DC power stored into the storage unit to the maximum AC power delivered to a resistive load. To measure $\eta_{total}$, first the maximum AC power delivered to a resistive load can be extracted by the TENG resistance matching measurement. As shown in Fig. 13.6e, the maximum AC energy generated by TENG is 0.3384 mW at an optimum load resistance of 4.26 MΩ. Second, the maximum DC power delivered through the power management board can be measured using the method shown above, which is 0.2023 mW (Fig. 13.6f) under the same mechanical triggering (from an electric motor). Therefore, $\eta_{\text{total}}$ is calculated as 59.8 %.

Compared with direct charging, the power management board shows a huge enhancement of the charging efficiency. In another supercapacitor charging experiment, the net supercapacitor charging current is 13.82 nA through direct charging and 15.14 μA through the designed power management board, which is improved by as high as 1096 times! Note that a transformer shows even worse performance than direct charging. Under the same experimental condition, the charging current of a supercapacitor cannot even compensate the leakage current when a 10:1 transformer is utilized as the impedance match network.

This human-motion charged power unit have wide applications in realizing self-powered human-activity sensors, which makes them self-sufficient without any external power sources. Several applications have been demonstrated by utilizing this self-charged unit to sustainably drive various commercial electronic systems (Fig. 13.7a). In the first demo, this human-motion charged power unit was connected to a commercial temperature sensor (Fig. 13.7b). This temperature sensor utilizes a thermal couple to sense the external environment temperature. Then the sensing analog signal is digitalized through an analog-digital converter (ADC). Finally, the digital signal is shown on a LCD display to fulfill the visualization of the output. Even under very gentle 1.6 Hz palm tapping (palm tapping is utilized mainly for easier filming purpose, other kinds of human-motion, such as foot tapping is also suitable for the application), this human-motion charged power unit could supply enough power to maintain continuous operation of the temperature sensor. There were not any external power sources/batteries inside to power any parts of the whole functionalized system.

This kind of power unit can also serve as power source for human health monitoring. As a typical example, a self-powered electrocardiograph system (ECG system, Fig. 13.7c) was realized. The bioelectricity signal across human left-hand and right-hand was first collected through the ECG signal collector (two conducting Al rods), and then amplified and filtered through an ECG signal amplifier (AD8232, Analog devices). After that, an ADC was utilized to convert the analog signal to digital signal, which was then captured by a digital counter. Finally the output of the digital counter was shown on the LCD display to visualize the signal. Besides, a buck converter was utilized to solve the supply voltage mismatch between the ECG signal amplifier (2–3.3 V) and the LCD display (>4.5 V) [9]. The number on the LCD display increased by 1 every time the heartbeat. At the same time, the storage capacitor voltage maintained a positive slope, which showed that the power provided from palm tapping was high enough to power the whole ECG system.

This kind of mechanical power source is also capable of being utilized in various fitness applications. For example, a self-powered pedometer system was realized (Fig. 13.7d). In this system, the mechanical sensing part was realized by another

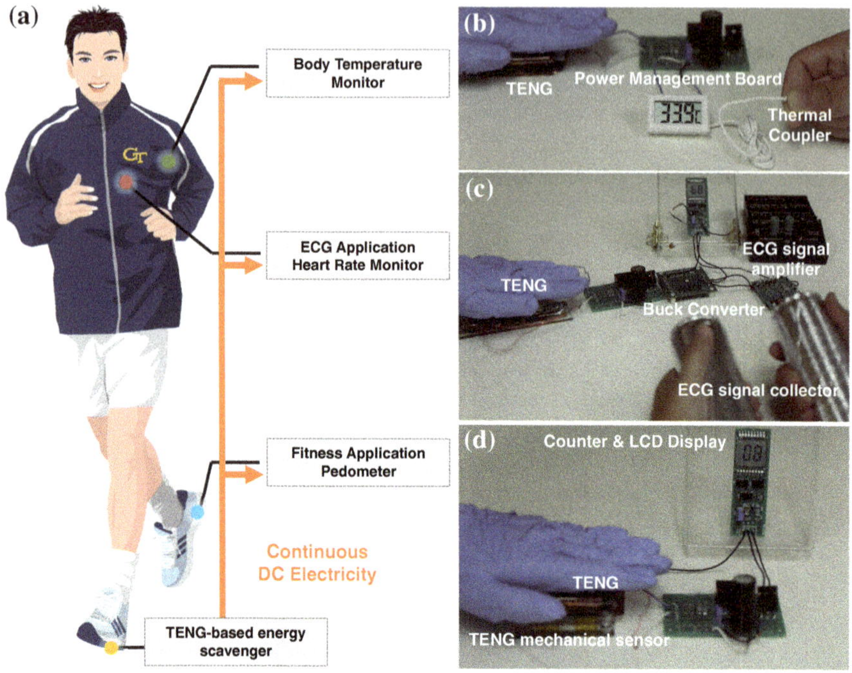


**Fig. 13.7** Application of this self-charging unit in self-powered wearable human activity sensors. **a** System configuration of self-powered human activity sensors. **b** Demonstration of a self-powered temperature sensor. **c** Demonstration of a self-powered heart-rate monitor (ECG) system. **d** Demonstration of a self-powered pedometer. Reproduced with permission from Nature Publishing Group [8]

attached-electrode TENG. When pressure was applied, this attached-electrode TENG could generate a voltage peak, which made this mechanical sensor consume zero electric power. Similar to the above heart rate signal, this voltage peak signal can be captured by an ADC and a digital counter and then visualized on the LCD display. The number on the LCD display increased by 1 every time pressure was applied on the mechanical sensor. Besides, while the pedometer maintained its normal function, $V_{\mathrm{store}}$ increased quickly from the initial value of 5.02–5.67 V. This fast increase is from the ultralow system power consumption due to the utilization of the self-powered TENG mechanical sensor.

Besides the self-powered human-activity sensors, this kind of mechanical power unit also has broad applications for other personal electronics. First, its application in wearable electronics is demonstrated through a self-powered wearable watch with mathematical calculation function (Fig. 13.8a). A very slow and gentle palm tapping (1 Hz) could drive this wearable watch with full functionality: accurately recording time and calculating square root as an example. Under even irregular palm tapping (have irregular interval between the palming), this wearable watch still remains its normal function as long as the average power provided by human

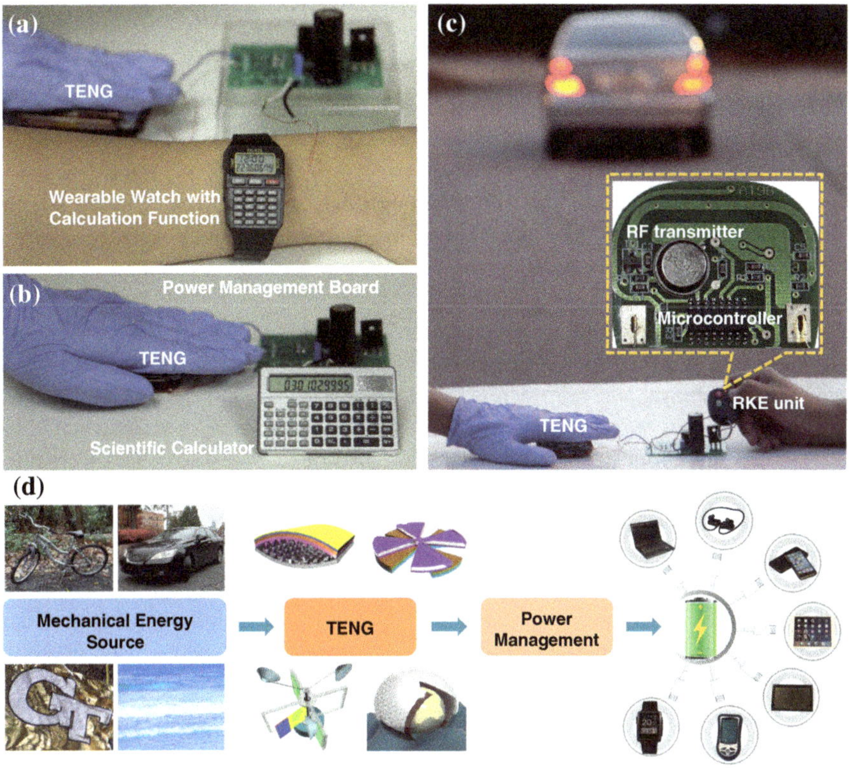


**Fig. 13.8** Application of the self-charging unit in self-powered data processing and transmission system. **a** Demonstration of a self-powered wearable watch and calculator. **b** Demonstration of a self-powered scientific calculator. **c** Demonstration of a self-powered remote keyless entry system. **d** Extension of this TENG-based self-charging unit for various applications as a universal adaptable power source. Reproduced with permission from Nature Publishing Group [8]

motion is larger than its power consumption. In addition, more complicated data processing work can also be finished with the power provided by palm tapping. For example, a scientific calculator is a commonly-used complicated data-processing system, containing clock generators, registers, read-only memories, arithmetic logic units, and a LCD display (Fig. 13.8b). Such an advanced scientific calculator could be easily powered by the mechanical energy from palm tapping. All of the advanced calculation, such as trigonometric function, exponent function and logarithm function could be realized without existence of any batteries. Finally, this power unit is applicable in wireless communication system as well. As a typical demo, a remote keyless entry (RKE) module was connected with this power unit and utilized to wirelessly control a car that was about 50 m away. The RKE module is a complicated radio-frequency system, mainly containing a microcontroller and a radio-frequency transmitter (Fig. 13.8c). Once the button on the RKE module is pressed, the microcontroller processes and encrypts the signal and transfers the encrypted 64–128 bit data to the radio-frequency transmitter. Then the 433.92 MHz radio-frequency transmitter can send out the code remotely with a data transfer rate about 2–20 kHz. This power unit could charge the storage capacitor from 5.9 to 6.4 V in about 3–5 s. Once the storage capacitor's voltage reached 6.4 V, this RKE unit could functionalize and send out the encrypted signal and the storage capacitor's voltage went back to about 5.9–6.2 V. After receiving this radio-frequency signal, the receiver inside the car could process, decode, and respond to this command (unlocking its door and lighting up its low beam light), clearly showing the success of remote wireless transmission. Then the storage capacitor was charged by this power unit to reach 6.4 V again.

The demos presented above have covered all of the fundamental parts of mobile and wearable systems, including sensors, microcontrollers, memories, arithmetic logic units, displays, and even wireless transmitters, which have broad applications in personal sensor systems and internet of things. Besides human biomechanical energy, this kind of designs and concepts can also be extended to other mechanical energy sources, such as rolling wheels, moving cars and trains, blowing wind, and surging waves, by applying various available TENG designs. If high frequency mechanical agitations are utilized to drastically improve the TENG output power, such system will have a potential to serve as a universal standard power source for sustainably driving more complicated electronic system, including smart watches, cellphones, navigation system, tablets, personal computers, and sensor nodes in internet of things (Fig. 13.8d).

## 13.2 TENG for Self-powered Electrochemical Applications

The purpose of developing self-powered system is to use nanogenerators (NG) to power small electronic devices and other applications [10–12]. Currently, there are two possible methods to utilize the random-pulsed AC output energy from NG: the first is through a power management circuit, a rectifier and an energy storage unit (batteries/capacitors) to convert output from NG to be a DC output with suitable voltage and current which is required by most of electronic devices and some applications, as stated in above sections; the second is to directly apply the pulsed output on some small electronic devices (such as LED, LCD) and electrochemical applications (such as electrodeposition [13], electrodegradation [14–16], water-splitting [17, 18], and etc. [19–25]). The electrochemical application is a typical one that can be directly driven by the rectified pulsed output from NG. In NG, TENG with high output power and infinite internal resistance has shown a great promise. Consequently, a series of self-powered electrochemical systems based on TENG have been developed.

### 13.2.1 *Self-powered Electrochemical Degradation*

The purpose of electrochemical degradation is to remove pollutions in the environment such as dyes. Usually, wastewater from textile manufacturing contains dyes, which must be removed before it can be discharged. The purification of dyeing wastewater by electrochemical techniques has been proven to be a very effective method, where it has the simple equipment, easy operation, lower temperature requirements and no sludge formation [26, 27]. Although the feasibility of electrochemical degradation of dyes [such as methyl orange (MO), rhodamine B] for wastewater treatment has been extensively reported, [28, 29] all these electrochemical investigations need external power sources, which largely limit the development of this technique. To solve this problem, the NGs may be used to achieve self-powered electrochemical degradation of dyeing wastewater by harvesting energy from environment. Usually, the mechanism of the electrodegradation of MO is based on an electrocatalytic oxidation by hydroxyl radical that is generated on the anode [30].

Here, Yang et al. demonstrated the first hybrid energy cell, which was used to simultaneously/individually harvest the mechanical and thermal energies for a self-powered electrodegradation of MO [14]. A TENG was fabricated at the top by using a flexible polydimethysiloxane (PDMS) nanowire array with the diameters of about 200 nm to harvest the mechanical energy, and a lead zirconate titanate (PZT) film-based pyroelectric nanogenerator was fabricated at the bottom to harvest thermal energy. These energies produced by the NGs can be stored in a Li-ion battery or directly used for the degradation of MO by the self-powered electrocatalytic oxidation.

For the output performance of this hybrid energy cell, please refer to Sect. 12.1.2.2. To illustrate the potential applications of the hybrid energy cell, the energies produced by the fabricated NGs were demonstrated to be used for the electrodegradation of MO. Figure 13.9a shows the schematic diagram of the self-powered degradation of MO by electrocatalytic oxidation. There are two methods to be used for the electrodegradation of MO. When the point "1" was connected to the point "2", the energies can be stored in a Li-ion battery and then was used for the electrodegradation of MO. The second method is that the fabricated NGs were directly used for electrodegradation of MO when the point "1" was connected to the point "3". In this study, the Pt electrode was used as the anode for electrocatalysis and NaCl as electrolyte was added into MO solution to improve the conductivity. The mechanism of the electrodegradation of MO is proposed as follows:

$$2H_2O \rightarrow 2OH^* + 2H^+ + 2e^- \quad (13.1)$$

$$Cl^- + H_2O \rightarrow ClO^- + 2H^+ + 2e^- \quad (13.2)$$

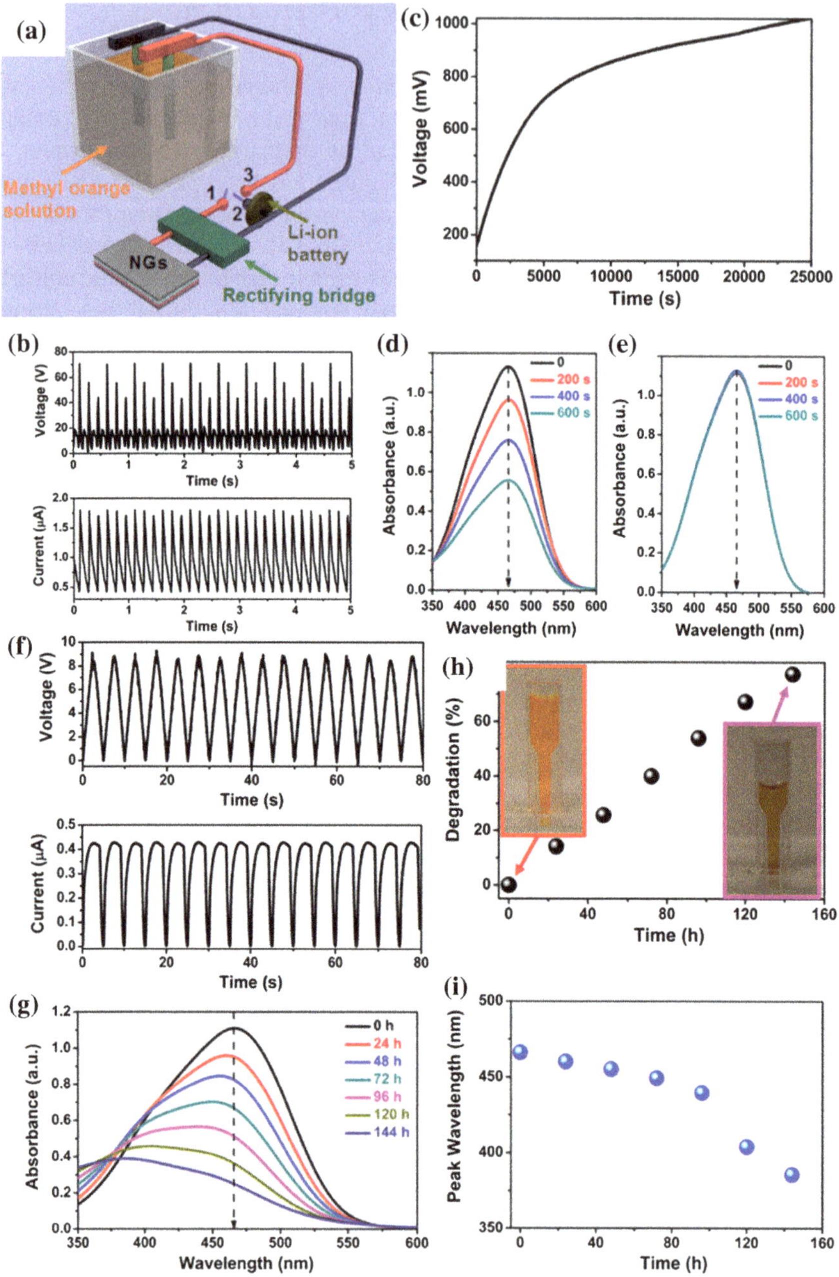
(a)
Methyl orange solution
1
2
3
Li-ion battery
NGs
Rectifying bridge
(c)
Voltage (mV)
Time (s)
(b)
Voltage (V)
Time (s)
Current (μA)
Time (s)
(d)
0
200 s
400 s
600 s
Absorbance (a.u.)
Wavelength (nm)
(e)
0
200 s
400 s
600 s
Absorbance (a.u.)
Wavelength (nm)
(f)
Voltage (V)
Time (s)
Current (μA)
Time (s)
(h)
Degradation (%)
Time (h)
(g)
0 h
24 h
48 h
72 h
96 h
120 h
144 h
Absorbance (a.u.)
Wavelength (nm)
(i)
Peak Wavelength (nm)
Time (h)

◀ **Fig. 13.9** **a** Schematic diagram of the self-powered electrodegradation of MO. **b** Measured output voltage and current of the TENG. **c** Charging curve of a Li-ion battery by using the energy in part b. **d**, **e** Absorption spectra of the MO solution with (**d**) and without (**e**) a Li-ion battery of about 2 V under the same time intervals. **f** Output voltage and current of the PENG. **g** Absorption spectra of the MO solution under the some time intervals. **h** Plot of degradation percentage versus the electrodegradation time. The *inset* shows the optical images of the MO solution before and after the electrodegradation. *i* Corresponding peak wavelength in b under the different electrodegradation time intervals. Reproduced with permission from American Chemical Society [14]

According to Eqs. (13.1) and (13.2), $H_2O$ and $Cl^-$ were adsorbed to Pt electrode (anode) by the discharge reaction of anode, resulting in the formation of super-oxidative hydroxyl radical $OH^*$ and $ClO^-$ due to the electrocatalysis of Pt [31]. Then the oxidative species can degrade MO into the small molecules like $CO_2$, $H_2O$, etc. The electrodegradation of MO which proceeds via series of parallel and consecutive reactions is accomplished simultaneously by the redox processes. Also, a small amount of organic pollutants can be oxidized directly on the surface of Pt electrode.

Here, the TENG was chosen as a model to demonstrate that the energy from NGs can be stored in a Li-ion battery and then was used for the electrodegradation of MO. Figure 13.9b shows that the output voltage and current of the TENG were increased to 70 V and 1.8 μA (current density of 0.28 $mA/m^2$), respectively. The structure of the TENG in Fig. 13.9b was the same as the one for Fig. 12.10a, except the device size and the interlayer in Eq. (12.8) were increased, resulting in the increase of both the output voltage and current. A Li-ion battery can be charged by the TENG from about 0.18–1 V, as shown in Fig. 13.9c. Figure 13.9d shows that the UV-visible absorption spectra of the MO solution for electrodegradation by using a charged Li-ion battery of about 2 V at the same time intervals. The characteristic absorption peak of MO, which is centered at 461 nm, was selected for monitoring the electrodegradation process. It can be seen that the characteristic absorption peak intensity of MO decreases with increasing the electrodegradation time, indicating the decrease of the MO concentration.

In order to confirm that the decrease of MO absorption peak intensity is due to the electrodegradation further, a control experiment was finished, where no batteries or other power sources were used. Under this condition, there is no change in the absorption spectra of the MO solution at the same time intervals in Fig. 13.9d, as shown in Fig. 13.9e. The dynamic electrodegradation process of the MO solution was also shown in a movie file in the supporting file. It can be clearly seen that the electrodegradation process starts at the Pt anode electrode, where the color of the MO solution was changed from yellow to red and then to colorless. The change in color of the MO solution between red and yellow can be expressed as [32].

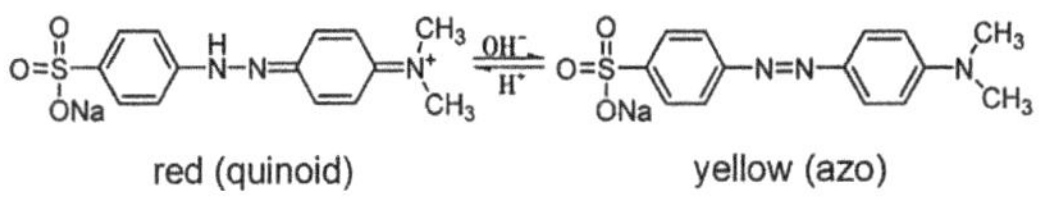


(13.3)

The existing form of MO is mainly $ArSO_3^-$ anion, which has the color of yellow. According to Eqs. (13.1) and (13.2), there are many $H^+$ anions produced on the anode in the electrodegradation process, which can result in the change in color from yellow to red according to Eq. (13.3). The change in color of MO is consistent with the proposed mechanism of the electrodegradation of MO in the Eqs. (13.1–13.3). The optical images of the MO solution before and after degradation show an obvious difference, indicating that the electrodegradation of MO is effective.

Without Li-ion batteries as the storage units, the pyroelectric nanogenerator was demonstrated as a model directly used for the electrodegradation of MO. Figure 13.9f shows the output performance of the pyroelectric nanogenerator, where the output voltage and current are about 9 V and 0.4 μA, respectively. Figure 13.9g shows that the UV-visible absorption spectra of the MO solution for electrodegradation by using the pyroelectric nanogenerator at the same time intervals. It can be seen that the characteristic absorption peak intensity of MO decreases with increasing the degradation time. The degradation percentage of MO is up to 80 % after 144 h, as shown in Fig. 13.9h. The optical images of the MO solution before and after degradation are shown in the inset in Fig. 13.9h. The color of MO solution at 144 h was obviously shallowed as compared with that before the degradation, indicating that the MO has been effectively degraded by using the pyroelectric nanogenerator. Figure 13.9i shows the change in maximum absorption peak of MO under the different degradation time, where the peak wavelength has been moved to the lower wavelength (a shift of about 80 nm). It is associated with the higher voltage (9 V) produced by the pyroelectric nanogenerator than that of the Li-ion battery (2 V). Usually, the certain characteristic peak corresponds to the certain chemical groups. Under the different voltages, the different intermediates can be induced in the electrocatalytic oxidation process, resulting in the shift of the absorption peak [33].

In summary, a hybrid energy cell, which can simultaneously/individually harvest the mechanical and thermal energies, was demonstrated for self-powered degradation of MO. The mechanical energy produced by the TENG was stored in a Li-ion battery, which can be used for degradation of MO by electrocatalytic oxidation. Moreover, the pyroelectric nanogenerator was directly used for degradation of MO, where the degradation percentage is up to 80 % after 144 h. The fabricated hybrid energy cells have the potential electrochemical applications in the self-powered electrodeposition, pollutant degradation, corrosion protection, and water splitting.

### *13.2.2 Self-powered Water Splitting*

Water splitting is utilized as a regenerative life supporting system and is part of the energy conversion process in nature. Its manipulation is often accomplished through the application of electrical potential by an external power supply [34, 35].

In order to develop a cost-effective and environment friendly technology, many researchers have focused on this field. Photocatalytic splitting of water into $H_2$ and $O_2$ using a catalyst and the solar energy is one of the ideal approaches [35, 36]. Among these technologies, the energy source for water splitting includes solar, chemical, and thermal energy; whereas, the mechanical energy is rarely involved, [17] which is, however, of major interest, due to its universal availability.

Recently, the invention of TENG has provided an effective approach to convert ambient mechanical energy into electricity. The working principle of the TENG is based on the combination of contact electrification and electrostatic induction. Contact-induced charge transfer between two tribomaterials with opposite polarity results in a potential difference when the two materials are separated. This potential difference will drive the electrons/ions in the external circuit to flow, which could be used for water splitting.

Tang et al. developed a water splitting system that is fabricated by coupling a TENG and a water splitting unit [18]. When the TENG's spinning speed is 600 rpm, the hydrogen producing rate in 30 % (w.t.) potassium hydroxide (KOH) solution reaches $6.25 \times 10^{-3}$ mL $min^{-1}$. Particularly, when the KOH solution is replaced by pure water, the system is demonstrated to be even four times more effective for hydrogen generation than that driven by an electrochemical workstation at 10 V. Furthermore, the normal tap water flowing is successful to drive TENG for effectively splitting power, which shows the first fully self-powered water splitting system.

What the authors intend to do is to drive the water splitting system using the electricity harvested from the water itself, so that no external power source is needed for the water splitting. Figure 13.10a shows the experiment setup. The TENG was connected to a miniaturized water turbine, which was then driven by the flow of normal tap water at a flow rate of 3 L $min^{-1}$. Due to the decay of the spinning speed and the driven force given by the tap water, the TENG's output current and voltage decreased to some degree. Then, the electricity generated by water flow was used to split the water. In order to present the splitting process in a visible manner, a 1000 μF capacitor was employed as a charge reservoir. The circuit diagram is shown in Fig. 13.10b (the transformer is not used in this experiment). Electricity was first stored in the capacitor, and then used for the splitting in the KOH solution. The voltage of the capacitor was monitored and plotted in Fig. 13.10c. As we can see, after 60 min, the capacitor was charged from 1.7 to 4.5 V. When being connected with the water splitting system, it discharged rapidly from 4.5 to 1.8 V (Noting that 1.8 V could be the decomposition voltage of the KOH solution in experiments). Photographs of the splitting process are illustrated in Fig. 13.10d. Bubbles can be apparently seen in both electrodes, which are hydrogen in the cathode, and oxygen in the anode. The volume of generated $H_2$ during once discharging is found to be $0.3 \times 10^{-3}$ mL, which means the amount of transferred charges is about 2.57 mC and the consumed electrolytic energy is about 3.8 mJ. The amount of charges is consistent with that released by the capacitor, and the corresponding energy conversion efficiency is about 43.8 %.

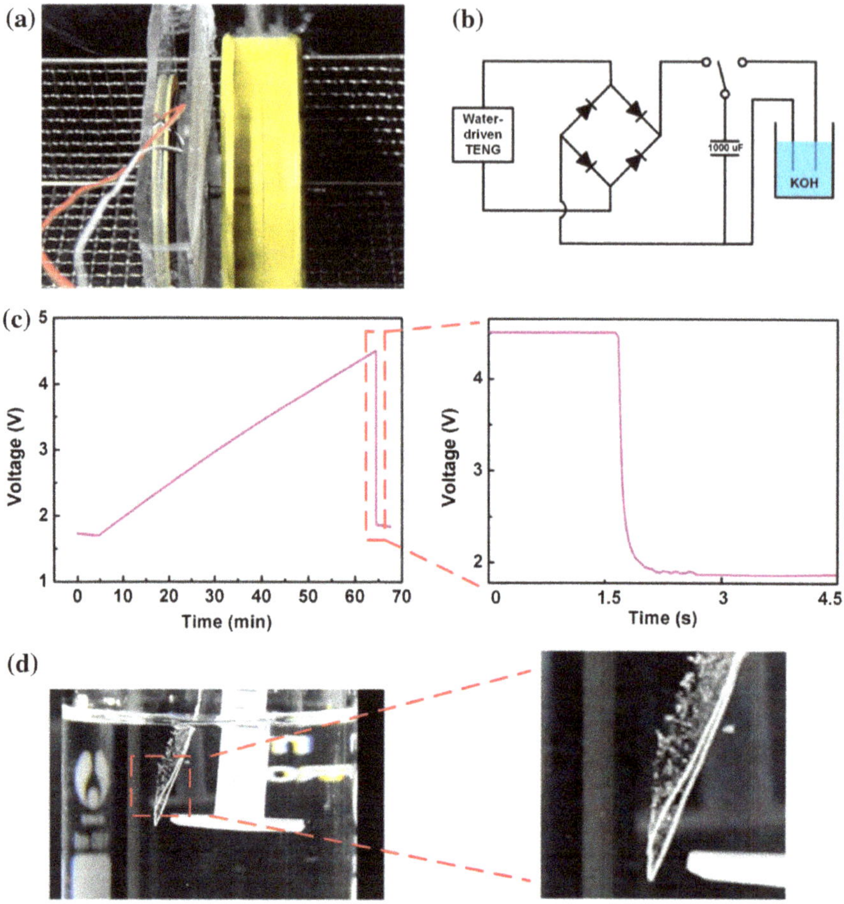


**Fig. 13.10** Self-powered water splitting. **a** The TENG is assembled with a miniature water turbine. **b** The circuit diagram of the self-powered water splitting system. **c** The storage capacitor's working *curve*. **d** Photographs of the self-powered splitting process, where the produced $O_2$ and $H_2$ bubbles are bursting out. This experiment proves the feasibility of fully self-powered water splitting process. Reproduced with permission from Wiley [18]

In summary, the TENG can be driven by the flow of normal tap water, demonstrating the fully self-powered water splitting system. Therefore, the TENG-driven water splitting is an effective approach for in situ hydrogen generation either for energy storage or chemical reactions without an external power source. It will surely initiate a research direction in this field of triboelectrolysis and possibly impose impact on energy science.

### 13.2.3 Self-powered Anticorrosion

Corrosion of metal surfaces in areas such as battle ships and historic relics costs billions of dollars worldwide each year. The corrosion mechanism of the metal surfaces is an electrochemical process that occurs in a thin electrolyte, and it is accelerated in rainy days. Among the protection methods, cathodic protection (CP) is widely used as a primary technology [37]. Generally, there are two typical CP systems: the first is the sacrificial anode cathodic protection (SACP) system, in which the metal is electrically coupled to a sacrificial anode; the second one is the impressed current cathodic protection (ICCP) system, in which a cathodic current is applied from an external power source that is connected between an 'inert' anode electrode and the metal to be protected. For the SACP system, more active metals are sacrificed to protect the specimen from corrosion, and what's more, the protected area is limited to the size of sacrificed metal [38]. For the ICCP system, external power sources are required to provide sufficient output, resulting in very high cost and environmental pollutions [39]. In recent years, other CP systems have been reported, such as photogenerated cathode protection, however, an ultraviolet light source is required in this system [40]. Since corrosion is a slow process, and anti-corrosion could also be a slow process, therefore, an ideal CP system should be a self-powered system that operates by harvesting available energy from environment (kinetic energy of water drops and wind) without supplying an additional power source.

Zhu et al. designed a self-powered CP system for metal surface protection by utilizing the energy harvested from the wind and rain drops using a flexible TENG. During the rainy season, metal surface can experience an accelerated corrosion. At this time, the self-powered CP system driven by TENG can protect the metal surface from corrosion. To the best of our knowledge, it is the first time that a self-powered CP system for metal surface protection is reported. This research demonstrates that the flexible TENG can produce highly efficient CP system for almost all of metallic structures. Considering the extremely-low cost, non-polluting and unique applicability, the as-established self-powered CP system can also be widely used in ocean engineering and industrial equipment protection.

In order to characterize the feasibility of the self-powered CP system, the electrical characteristics of the flexible TENG were investigated. The TENG is based on the contact electrification of polydimethylsiloxane (PDMS) and ITO. Before fabrication of TENG, PDMS was pretreated to form inverted pyramid structures. The output current is over 130 μA, and the voltage reaches about 500 V. CP experiments of metal surface were conducted with and without the CP of TENG. For the one powered with TENG, the cathode was the protected specimen that was immersed into the electrolyte and the anode was a carbon rod (Fig. 13.11a). The TENG was rectified and then connected to the whole system with the external stimulation at a frequency of 1 Hz. Corrosion data are compared and discussed in the following sections. Furthermore, when raining, TENG can harvest the mechanical energy of wind and rain drops, and then convert it into

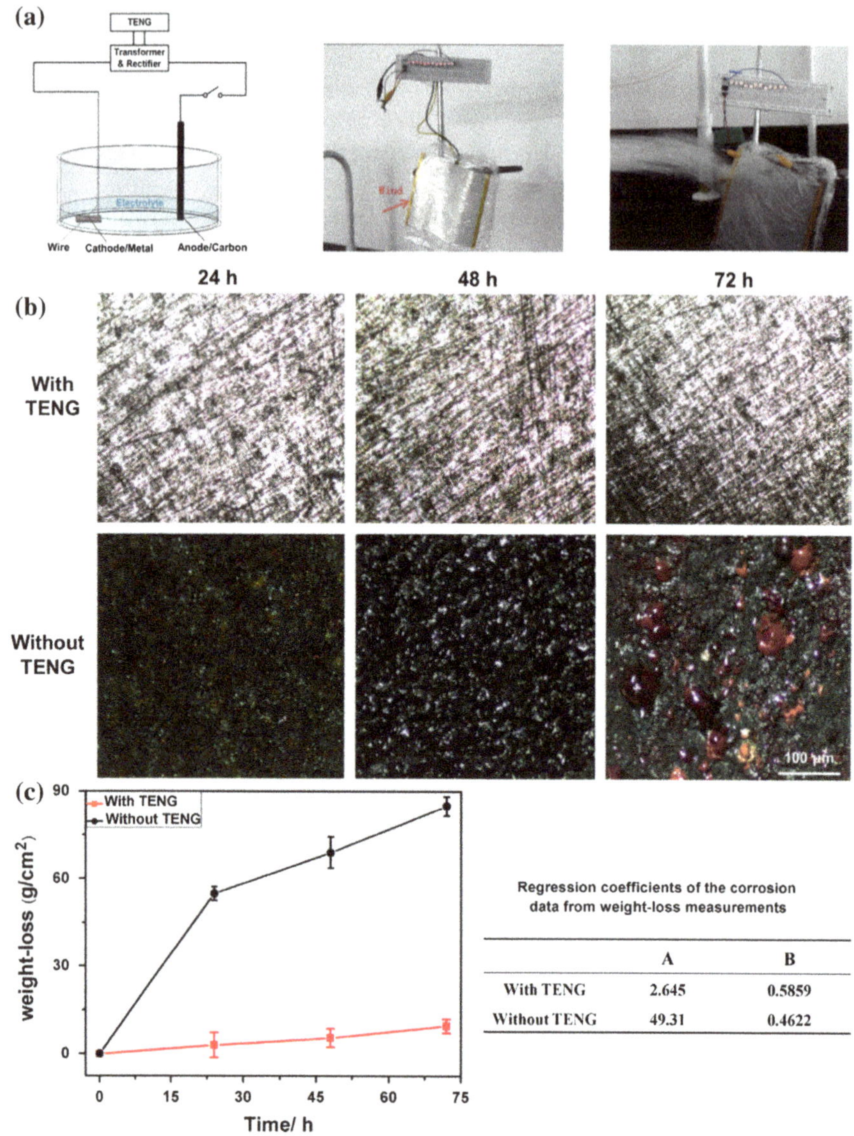


| | A | B |
|---|---|---|
| With TENG | 2.645 | 0.5859 |
| Without TENG | 49.31 | 0.4622 |

**Fig. 13.11** **a** Schematic diagram and photos of the circuits for self-powered anticorrotion. **b** Micrographs of the rusted specimens after accelerated corrosion. The specimens were separated into two groups; one group (*the tops*) were under CP driven by TENG while the other ones (*the bottoms*) were not. **c** Weight-loss *curves* of testing specimens after accelerated corrosion and the regression coefficients of the corrosion data from weight-loss measurements. Reproduced with permission from Elsevier [24]

electricity to provide the power for the CP. Since the TENG can be driven by human motions, wind and rain drops, it can power up multiple light-emitting diodes (LEDs) by harvesting the ambient mechanical energy. As shown in Fig. 13.11a, the multiple LEDs were powered up, which indicated the self-powered CP system is feasible.

Generally speaking, rusts can intuitively reflect the corrosion degree of metal surface. Then, the rusted specimens with and without CP system were studied under microscopy to compare their micro-morphology after accelerated corrosion test. As shown in Fig. 13.11b, for the specimens protected with CP system, there are several gray distributing areas on the surface and the gray distributing areas increase with corrosion time elapsed. However, no apparent corrosion was observed on these samples. On the other hand, for the specimen without the CP driven by TENG, pitting were observed clearly in the three pictures, and more big pits were found on the sample that was corroded in simulated electrolyte for 72 h. The macro-morphology analysis of specimen also showed the same corrosion state. The micro and macro-morphology analysis demonstrated that the CP system can protect the specimen from corrosion effectively. Further, the specimen without the CP driven by TENG corroded deeply as the corrosion time increased, however the specimens with the CP system have not been corroded obviously.

In order to compare the corrosion degree of specimens with and without the CP systems quantitatively, corrosion data of the specimens were measured by weight loss tests. These data were only recorded when the specimens had already been uniformly treated. The weight-loss data of the samples placed in the simulated electrolyte with and without the CP driven by the TENG are shown in Fig. 13.11c. It is observed that the corrosion rate of samples without the CP system is about 10 times higher than the one when the power from TENG was applied at the same corrosion times. Furthermore, the corrosion data were fitted by using the well-know bilogarithmic equation approximately: [41]

$$C = At^B \tag{13.4}$$

where $C$ is the weight loss (g), $t$ is the periodic test time (h), $A$ and $B$ are constants. It is clear that $A$ is equivalent to the weight loss when time is unity. $B$ reflects the change in weight loss tests with time. The table in Fig. 13.11c lists the regression coefficients of the corrosion data from the weight-loss test.

From Fig. 13.11c, it is clearly observed that the corrosion rate of samples without the CP is higher during the corrosion process in the test and increase notably as time elapses. However, corrosion rates induced by the sample with the CP driven by TENG do not increase as corrosion proceeds. It is well known that constant $B$ is roughly equal to 0.5 of the corrosion which is controlled by an ideal diffusion process; if the value of $B$ is smaller than 0.5, it means a decrease in diffusion coefficient and the rust layer becomes more protective. Thus the rust layer induced by the corrosion without the CP of TENG is more protective than that with TENG according to the value of $B$ as showed in Fig. 13.11c. The results also demonstrate that the CP powered from TENG is feasible in the protection for metal surface.

In summary, the first self-powered CP system for metal surface protection driven by flexible TENG using the energy provided by wind and rain drop in the environment has been developed. Systematic investigations were carried out to compare the corrosion degree of metal surface with and without the CP system. The results confirm that the metal surfaces with the CP driven by TENG show no apparent corrosion, while the others without the CP system corroded heavily in the same simulated electrolyte. The work demonstrates a highly simple fabrication, strong performance and cost-effect approach to protecting metal surface from corrosion. The new technique of self-powered CP system can be used in metal cultural relics, ocean engineering and transportation.

### *13.2.4 Self-powered Air Filtering*

With the growing threat of air pollution and the crisis of energy, the search for cost-effective, renewable and green methods to clean air is one of the most urgent challenges. Especially in China that is experiencing a fast development, people have long been tormented by polluted air, which is mainly due to Particulate Matter (PM), sulfur dioxide and nitrogen oxide ($NO_X$). In Beijing, during 2000–2014, the annual average concentration of PM10 is never lower than 100 $\mu g/cm^3$. Various health problems are believed to be associated with long-term exposure to polluted air, [42] such as respiratory diseases, rises in numbers of deaths from cardiovascular and respiratory disease among older people, and increases of lung cancer and cardiopulmonary mortality [43]. Therefore, it is essential to find a cost-effective way to resolve this issue in order to improve the quality of people's life.

Multiple approaches have been extensively demonstrated to remove dust and $SO_2$. For dust removal, methods such as simple filtration, adsorption with solid sorbents, [44] pulsed electron beam, reaction with negative ion [45] and electrostatic precipitation [46], have been reported. Among these de-dusting approaches, electrostatic precipitation has been widely used in industries and indoors [47]. But a major drawback of this approach is that it requires an external power even at a high voltage, making it expensive and not adequate for large-scale outdoor usage. Moreover, some of the methods rely on corona effect for electrically charging particulates, which could generate ozone and $NO_2$ as byproduct [48]. The filtration is very simple and effective, but it could be easily clogged by submicron solid aerosol particles [49]. As for $SO_2$ removal, there are also many methods such as electro-catalytic and metal catalytic oxidation [50]. However, electro-catalytic oxidation needs external power, and the metal catalytic oxidation relies on novel metals which are very expensive and are rather limited.

Chen et al. innovatively fabricated a very simple and practical system for self-powered clearing of toxic materials in the air by electrostatic precipitating of flying dust particles and oxidizing $SO_2$. Instead of utilizing an external electricity power, the electricity generated from natural wind with the use of R-TENG is utilized. The output voltage of the R-TENG is about 300 V, which is not too high to generate byproducts such as ozone and $NO_X$, [51] as produced in traditional electrostatic precipitation by using high voltage discharge. In addition, no noble metals are needed to electrochemically oxidize $SO_2$, since the generated current is high enough to electro-catalyze the oxidation of $SO_2$. Last but not the least, the self-powered $SO_2$ and flying dust removal system is cost-effective, versatile in size and easy to be applied, making it feasible for environmental cleaning and indoor air purification. It is also worth mentioning that the method of oxidizing $SO_2$ may have a great influence on sulfur acid production, and this principle could be used in many other electrochemical reduction or oxidation applications.

The multi-layered R-TENG is mainly composed of a rotator and a stator. The key components of the stator and rotator are top grating and bottom grating electrodes which are fabricated by the print circuit board (PCB) technology. Both of the electrodes are clusters of radially arrayed segments, made of copper, with radius of 70 mm and a central angle of 1°. The difference is that the top electrode has 180 sector unites separated by an interval of 1°; while the bottom electrode is composed of two complementary patterned electrode networks which are separated by fine trenches in between. Each group of patterned electrode networks is respectively connected at one end. The rotator rotates under the driving force from wind blowing; while the stator is fixed on a bracket. For the fabrication of rotator, the top grating was just needed to be immobilized onto the top substrate (made of PMMA). As to the stator, the bottom grating was firstly immobilized onto the bottom substrate, and then adhered with a Kapton film (thickness of 30 μm).

To characterize the optimized performance, the fabricated R-TENG was mechanically driven by a rotary motor with a rotating speed of 600 rpm. The open-circuit voltage ($V_{OC}$) was measured by a low-noise voltage preamplifier (Keithley 6514 System Electrometer) and the short-circuit current ($I_{SC}$) of the R-TENG was measured by a low-noise current preamplifier (Stanford Research SR570). The measured $I_{SC}$ and $V_{OC}$ are 3.4 mA and 320 V respectively. For supplying the air cleaning system, the alternating current (AC) output of the R-TENG should be rectified into direct current (DC) output. To indicate the capability of the R-TENG as a power source, it was connected to conventional light bulbs after rectification. With the wind gusting up to 8.5 m/s (5 BF in Beaufort wind force scale), the spot lights on the panel were all lighten up. To drive the systems using electricity harvested from wind energy, the R-TENG was connected to a miniaturized wind cup structure that was driven by air blower at a wind speed of 15.1 m/s (7 BF).

The oxidation of $SO_2$ and adsorption of flying dust were all performed in a closed transparent cubic chamber of 125 L with an air inlet, an air outlet and two needle-like copper wires across the wall at ambient temperature. Through wires, the R-TENG outside of the chamber was connected to two paralleled copper meshes inside. For the experiment of $SO_2$ oxidation, the chamber was filled with $SO_2$ beforehand. Then the copper meshes were connected through a sink of saturated $NaHSO_3$ solution (Fig. 13.12a). At given time intervals, the $SO_2$ inside the chamber

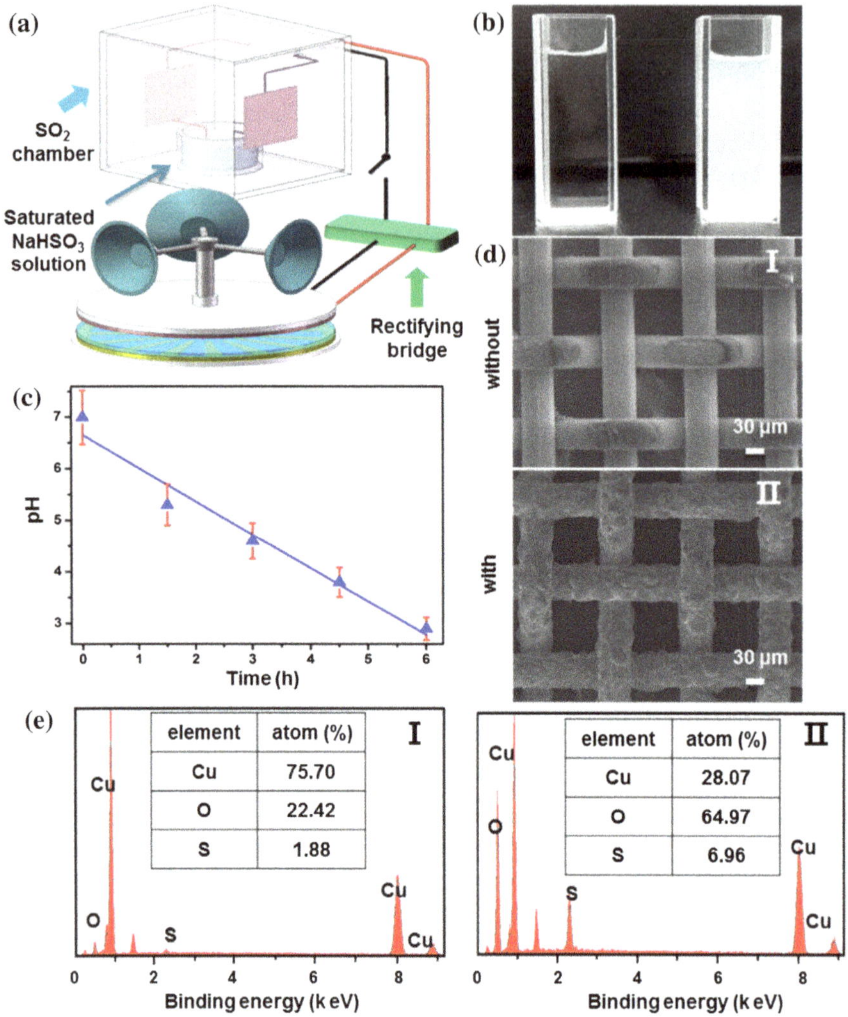


**Fig. 13.12** Self-powered oxidation of $SO_2$. **a** Schematic diagram of the system. **b** The optical images of the $BaCl_2$–HCl solution after addition of rinse solution of copper mesh without and with R-TENG. **c** pH of rinse solution of anodic copper mesh under a wind speed of 15.1 m/s (7BF). SEM image (**d**) and EDS spectra (**e**) of copper mesh without (*I*) and with (*II*) R-TENG. Wind speed, 15.1 m/s (7BF); working time, 6 h. Reproduced with permission from Elsevier [19]

was extracted by vacuum sample bags, and assayed by gas chromatography (GC) to determine the concentration change, which was then compared (for both setups with and without air cleaning system) to evaluate the air cleaning system's effectiveness. To verify that $SO_2$ is oxidized into sulfur acid in the chamber with R-TENG, the anodic copper mesh is characterized by scanning electron microscope (SEM), stereomicroscope and energy dispersive spectrometer (EDS). For the experiment of dust removal, the chamber was replenished with dust aerosol through the inlet by solid aerosol generator series SAG 410. The weight of the dust adsorbed to the meshes was measured with and without R-TENG in real time to determine adsorption efficiency.

A schematic diagram of the system for self-powered oxidation of $SO_2$ is sketched in Fig. 13.12a. The R-TENG captured wind energy with the assistance of wind cups, generating AC electricity. In order to oxidize $SO_2$ continuously, the generated AC electricity is conversed into DC electricity through a rectifying bridge. The saturated $NaHSO_3$ solution is connected with two copper meshes in between to circulate the current and eliminate the interference of $SO_2$ dissolving for the reason that saturated $NaHSO_3$ solution has certain conductivity and that $SO_2$ is almost insoluble in saturated $NaHSO_3$ solution. NaCl was added into the saturated $NaHSO_3$ solution till no dissolving to improve the conductivity. When the system works, $SO_2$ can be oxidized into sulfur acid on anode, as verified by Fig. 13.12b where a cloudy white precipitate appeared in the $BaCl_2$–HCl solution after addition of rinse solution of copper mesh with R-TENG, while no changes occurred in the same situation without R-TENG. Because of the generation of sulfur acid, the pH of anodic copper mesh's rinse solution decreased (Fig. 13.12c) and the corrosion of copper mesh in $SO_2$ atmosphere accelerated (Fig. 13.12d by SEM) compared to the system without R-TENG, thus proving the assumption. The increase of sulfur and oxygen elements in the EDS result (Fig. 13.12e) of the copper mesh with R-TENG compared to the system without R-TENG further demonstrate the assumption from another point of view.

Figure 13.13a illustrates the schematic diagram of the system for self-powered adsorption of flying dust, among which electricity supplying and electrode components are the same as Fig. 13.12a for $SO_2$ removing. Figure 13.13b implies that the voltage remains a saturated value of 300 V when wind speed rises. This result is due to the fact that the contact area of the tribo-electrifaction surfaces has not been changed.

In order to confirm the increase of dust absorption is due to the electrostatic precipitation, a comparative experiment was performed. As we can see from Fig. 13.13c, the weight increase of the dust in the condition without R-TENG at the same time intervals is much lower than that with R-TENG where the weight increase of dust goes obviously with working time. After 5 h, when the wind speed is 15.1 m/s (7BF), 1 g copper mesh could adsorb 0.327 g dust, proving the feasibility of this de-dusting system.

Besides time, distance between two copper meshes is another influence factor for dust removal. As implied in Fig. 13.13d, the weight increase goes up with the reduction of the distance. It is because that electric field is inversely proportional to distance. With the decrease of distance between two collecting plates, electric field between two copper meshes rises correspondingly, and the Coulombic force exerted on flying dust rises consequently, resulting in the enhance of weight increase on copper mesh, which will be studied in future.

In summary, a novel self-powered air cleaning system that is capable of effectively removing $SO_2$ and flying dust particles in the air has been introduced. By

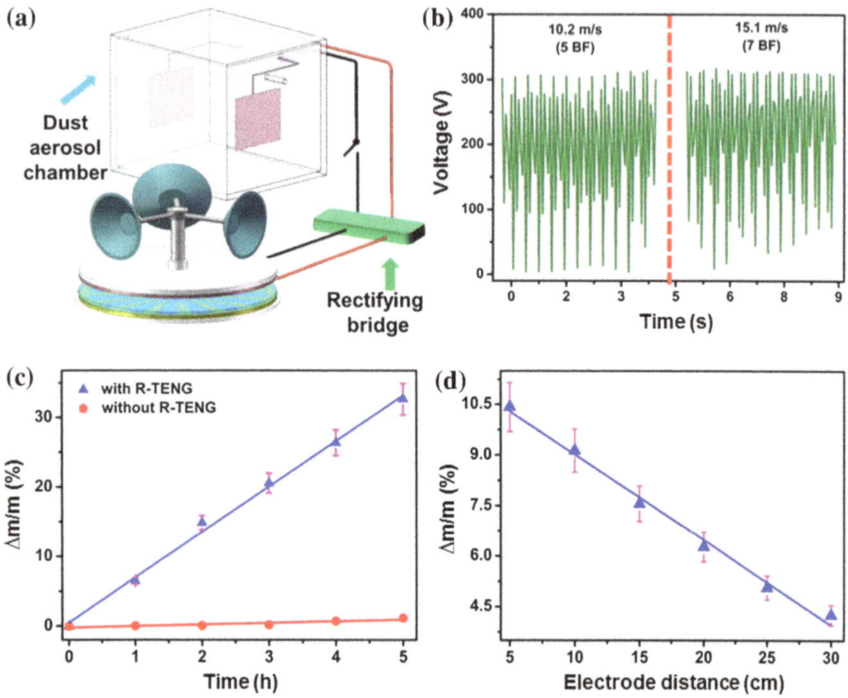


**Fig. 13.13** Self-powered precipitation of dust particles. **a** Schematic diagram of the system. **b** Rectified voltage of R-TENG connected with copper meshes under different wind speeds from 10.2 m/s (5BF) to 15.1 m/s (7BF). **c** Weight increase of copper mesh with and without R-TENG. Distance of collecting plates, 15 cm. **d** Dust adsorption rate as a function of collecting plates' distances of 5, 10, 15, 20, 25, and 30 cm. Wind speed, 15.1 m/s (7BF). Reproduced with permission from Elsevier [19]

using grating electrodes as electrical performance enhancer, the as-developed R-TENG could achieve a high voltage of 320 V and a high power density of 3.4 mA. Through the fabricated R-TENG, wind energy was harvested, and then immediately used to oxidize the $SO_2$ and adsorb the flying dust in the air. The working principle of the air cleaning system has been elucidated through conjunction of schematic illustrations and experiments. Not only the air cleaning system makes full use of the wind energy, but also shows advantage in simplicity, low cost and effectiveness. Thus, this self-powered dust and $SO_2$ removing system holds great potential for cleaning and prevention of air pollutants. As most polluting molecules can be oxidized or reduced through electrochemical method, and most polluting ions could be absorbed through electrostatic interaction, this work may inspire the development of R-TENG toward other pollutants control, such as sewage treatment and formaldehyde removal, in the near future. This is a new direction in self-powered electrochemistry.

### 13.2.5 Self-powered Electrochemical Recovery

Industrial wastewater is produced as a by-product with various kinds of toxic chemical substances (e.g. acid cyanide, alkaline cleaning agent, and degreasing solvents) and heavy metals (e.g. Cu, Ni, and Cr). Most of the heavy metals in wastewater are extremely harmful when they are discharged directly to the environment without any treatment, for their high solubility in the aquatic environments high absorption by living organisms. Once these heavy metals enter into the food chain of human, large concentrations of heavy metals may accumulate in a human body and may further induce severe illnesses. Hence, providing an efficient industrial wastewater treatment for removing or collecting these heavy metals before discharging becomes important and necessary. Copper (Cu) is one of the most comment heavy metal ions in the wastewater used for PCB process. Several approaches have been proposed to remove or collect Cu in industrial wastewater, such as chemical precipitation, adsorption, and ion exchange [52]. One of the most effective ways for collecting Cu ions is electrodeposition process [53]. Cu ions were removed with high efficiency from a dilute industrial effluent in an electrochemical reactor with plate electrode. Moreover, extremely high quality (>99 %) of electrodeposited Cu will be collected on the surface of electrode without any further purification process. However, for realizing the above electrochemical process, additional DC based power supplies are indispensable to keep it working.

The energy harvested from water-flow by TENG is applied in self-powered electrochemical recovery system for collecting Cu ions in the wastewater by Yeh et al. [21]. With the above design, the TENG delivers an open-circuit voltage ($V_{OC}$) up to ~650 V and a short-circuit current ($I_{SC}$) of 280 μA under a rotation speed of 1,500 rpm. Ultra-robust characteristic can be fully realized by introducing the frictionless free standing TENG structure with rolling bearing for refilling its superficial charge. Moreover, several unit of rolling free standing TENG can be easily integrated by means of incorporating with comment mechanical gears to bring about the integrated bundle structure of rolling free standing TENG. By integrating with a transformer and a power management circuit, stable DC output (5 V, 4 mA) performance can be achieved by water-flow driven RF-TENG to practically realize the concept of self-powered electrochemical recovery system for collecting Cu ions. Moreover, several electrochemical processes in practical wastewater electrochemical treatments, such as electrocoagulation, electroflotation, electrochemical degradation, and electrochemical advanced oxidation process, can also be suitable for utilizing the proposed self-powered electrochemical processor. This work provides a significant progress of TENG and exhibits huge potential of the RF-TENG as a high efficiency energy harvester for practical applications.

It is worth to note that TENGs have been proposed for providing much more output power and other advantages than a conventional electromagnetic generators (EMGs), which implied that TENGs can be the potential power generators to replace some self-powered systems that are driven by traditional EMGs. In practical application, for example, the TENG can be easily coupled with a waterwheel or a water turbine via the coaxial design. Photograph of water wheel for driving RF-TENG is exhibited as Fig. 13.14a. Water flow energy can be effectively transformed by a water wheel into rotary motion of the shaft to drive the TENG device. Furthermore, self-powered electrochemical copper collecting system was achieved by means of incorporating with water-flow-driven TENG for harvesting hydropower from waste water flowing energy. To simulate the proposed system, a two-electrode electrochemical cell (Fig. 13.14b) that consists of a graphite rod as anode and a copper plate as cathode was utilized and $CuSO_4$ aqueous solution (2.5 g/L) was used as electrolyte to simulate waste water with $Cu^{2+}$ ions. According to the electroplating effect, the $Cu^{2+}$ ions reduction reaction occurred on the surface of cathodic electrode and could be explained by following equation:

$$Cu^{2+} + 2e^- \rightarrow Cu \qquad E^o = 0.3402\,V/NHE \tag{13.5}$$

A DC power supply with steady output potential is indispensable for driving this reaction efficiently and smoothly. To realize the above requirement, the output of the TENG was integrated with a conventional transformer and a power management circuit composed of a full-wave bridge rectifier, a voltage regulator and capacitors, as shown in the circuit diagram of Fig. 13.14c. With the help of the power transformer, the output current of the TENG was effectively boosted up from 250 μA to 8 mA, while the voltage was reduced from ~650 to 16 V, which means that the transformed efficiency is around ~80 % (Fig. 13.14d). The transformed output could still produce enough potential to initiate the electrochemical reaction, but the speed of the reaction was greatly improved due to drastically enhancement of the current. After that, as shown in Fig. 13.14e, a stable DC electric output of 5 V and 4 mA was delivered by incorporating it with another power management circuit. Figure 13.14f shows the real-time monitoring of the UV-Vis absorption spectra of the $Cu^{2+}$ ions in the wastewater obtained at different duration times. With increasing in duration time of electrochemical collecting process, the characteristic absorption peak intensity of $Cu^{2+}$ ions in the wastewater decreases evidently, indicating the effectiveness of the route for self-powered electrochemical collection. Also, pictures of wastewater with $Cu^{2+}$ ions at different duration times are also presented as an inset for depicting the changed colours of the $Cu^{2+}$ solution. The recovery rate of $Cu^{2+}$ ions can be further estimated exactly by utilizing the calibration curve, as shown in Fig. 13.14g. As expected, the deposition amount of copper increased slowly as time passed by and then reached to saturation after 48 h. The photograph of the original copper plate (left) and the one fully deposited with copper (right) are also attached as insets in Fig. 13.14g. An energy conversion efficiency of up to 80 % (practical collection weight: ~180 mg; theoretical collection weight: ~227.5 mg) for collecting $Cu^{2+}$ ions in the wastewater can be achieved by utilizing the proposed self-powered electrochemical collection system. The proposed TENG can be easily installed in current wastewater treatment systems as a power supplier for harvesting water flowing energies and further driving electrowinning process. Moreover, several electrochemical processes in practical wastewater electrochemical treatments, such as electrocoagulation, electroflotation,

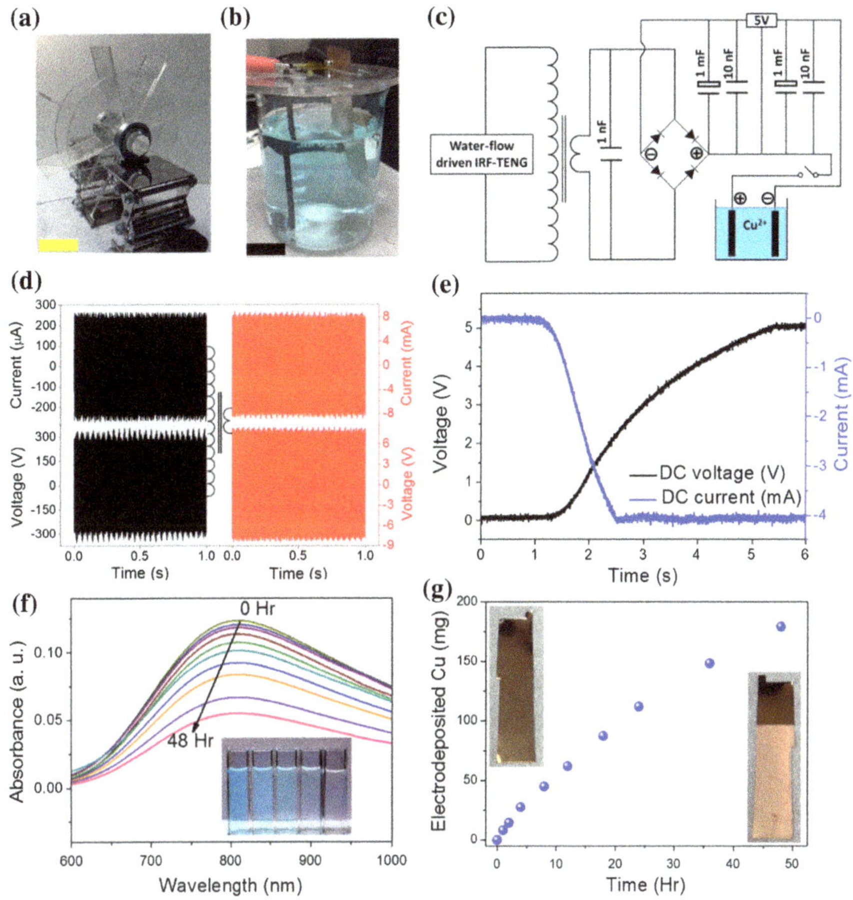


**Fig. 13.14** Demonstrations of the TENG for harvesting hydroenergy from water flow and its practical applications in self-powered electrochemical recovery system for collecting $Cu^{2+}$ ions. **a** Photograph of water wheel for driving TENG via harvesting water flow energy. The scale bar is 5 cm. **b** Photograph of two-electrode electrochemical system, which consisted of a graphite rod as anode and a copper plate as cathode; $CuSO_4$ aqua solution (2.5 g/L) was used as electrolyte. The scale bar is 2 cm. **c** Circuit diagram of self-powered electrochemical recovery system for collecting $Cu^{2+}$ ions, composed of the complete stable power-supplying system which integrated with a water-driven TENG and a power management circuit, and a two-electrochemical cell. **d** The performance of the $V_{OC}$ and the $I_{SC}$ for water-flow driven RF-TENG (*left hand side*) and corresponding electric performance by applying a transformer (*right hand side*). **e** Stable output DC voltage (5 V) and current (4 mA) were obtained by complete stable power-supplying system, which consists of a water-driven RF-TENG and a power management circuit. **f** UV-Vis absorption spectra of the $Cu^{2+}$ ions in the wasted water obtained at different duration times. Pictures of $Cu^{2+}$ solution at different duration times are also presented as *insets* for depicting the changed colors of the $Cu^{2+}$ solution at different times. **g** Corresponding weight of electrodeposited Cu as a function of operating time. The *inset* shows the photograph of flash Cu electrode (*left*) and after collected Cu one (*right*). Reproduced with permission from Wiley [21]

electrochemical degradation, and electrochemical advanced oxidation process etc., can be also applicable in the proposed self-powered electrochemical processor.

In summary, TENG is demonstrated to continuously harvest water-flow energy for driving the electrodeposition reaction and harvesting Cu metal in the wastewater through a power management system. A collection efficiency of up to 80 % for $Cu^{2+}$ ions in the wastewater can be achieved by utilizing the proposed self-powered electrochemical collection system. This research presents a substantial advancement in the practical applications of nanogenrators and self-powered electrochemical systems, which will initiate promissing improvement in related field.

### *13.2.6 Self-powered Electrochromic Device for Smart Window System*

Electrochromic devices (ECDs) are developed to provide reversible changes of their optical properties via the electrochemical redox reactions corresponding to an external electric field, [54] which are widely applied in displays, [55] switchable mirrors, [56] and electronic papers [57]. Specifically, some types of ECDs with a nonvolatile memory effect are ideal candidates for energy-saving smart windows on infrastructures and automobiles, since this type of ECD can maintain a certain coloring state with no additional power supply [58]. In this regard, the ECD can be integrated with a TENG as the power source instead of using batteries [25] to provide a constant voltage across the device as a self-powered smart window system. The concept of a "self-powered electrochromic window" can also be realized by integration with solar cells as the power sources to drive the electrochromic device [59]. However, there are unavoidable drawbacks, such as some part of ECD's surface area will be covered or sacrificed by solar cells or the transparency of the integrated device will be reduced since most of the solar cells are not fully transparent, which may limit its practical applications. Here, a new structure for a self-powered electrochromic window was proposed, which is composed of a transparent TENG to harvest natural mechanical energies to drive the ECD and can completely solve the problems mentioned above.

Here Yeh et al. established a fully integrated self-powered smart window composed of a dual-mode TENG and an ECD [22]. The entire device was a transparent, multilayered structure, which was compatible with the smart window structure. The ECD was composed of Prussian blue (PB) nanoparticles and zinc hexacyanoferrate (ZnHCF) nanocubes as the electrochromic material and the ion storage layer, respectively. By operating with a conventional electrochemical workstation, a maximum reversible change in transmittance ($\Delta T$) of 32.6 % could be achieved with respect to an external dc voltage. The TENG consisted of a multilayered structure with micropatterned polydimethylsiloxane (PDMS) thin films and transparent electrodes, which could be employed for harvesting the kinetic energy from wind impact and water droplets, with an optimized output power of 130 $mW/m^2$. By assembling the two devices on the same substrate, the self-powered smart window system was successfully realized with a transmittance change of up to 32.4 % with visualized color variations as well, which could compete with the results demonstrated by the electrochemical workstation. This work sheds light on motion-driven electrochemical reactions and paves the way for promising applications of TENGs, which will push forward the development of self-powered systems.

The multilayer structure of the self-powered smart window is schematically illustrated in Fig. 13.15a. On the top is a single-electrode TENG for scavenging the raindrop kinetic energy that is composed of a PDMS thin film attached to a conducting substrate. A micropatterned pyramid array structure was created on the surface of the PDMS thin film through photolithography and a template molding process, which plays an important role in improving the hydrophobic property and effective contact area of the surface, both aiming at enhancing the output performance of the single-electrode TENG. With a similar material system, a contact-mode TENG assembled by elastic springs was implemented right beneath the above-mentioned single-electrode-mode TENG. This structure was excellent for harvesting energy from wind impact, since the low-elastic-modulus springs could buffer the wind impact and brought the device back to its original shape when the wind stopped. Both the TENGs were stacked on the ECD with a common acrylic substrate to form a fully integrated self-powered system.

For the ECD, water-dispersible PB nanoparticles and ZnHCF nanocubes were synthesized as electrochromic and ion storage materials, respectively. The phase compositions of the as-synthesized powders were confirmed by X-ray diffraction, and the peaks for each of the XRD patterns can be assigned to $Fe_4[Fe(CN)_6]_3 \cdot xH_2O$ (JCPDS, PDF no. 73–0687) and $Zn_3[Fe(CN)_6]_2$ (JCPDS, PDF no. 38–0688), suggesting that both compounds were successfully synthesized and the residual reagents were completely removed. Uniform PB and ZnHCF films were successfully prepared by spin coating the corresponding dispersions on the FTO conducting glass. PB-coated and ZnHCF-coated FTO conducting glasses were employed as the working and counter electrode, respectively. The ECD was assembled by sandwiching the solid-state electrolyte (0.1 M $LiClO_4$/SN with 7 % $SiO_2$) between the PB and ZnHCF films.

The electrochromic performance of the integrated ECD driven by the dual-mode TENG was investigated, as shown in Fig. 13.15a. For simplicity purposes, only the wind-powered-TENG was involved in the powering of the smart window, and the implementation of the water-drop-TENG will add to the output current and enhance the overall performance of the integrated device. In addition, a mechanical shaker was used to simulate the wind impact at a constant frequency of 5 Hz. The intensity of the motor impact was set as 28.5 $kgf/m^2$, which was equivalent to a wind velocity of 16.1 m/s. Moreover, a full-wave bridge rectifier was employed to convert the generated AC pulses into DC signals, as shown in Fig. 13.15b, and the rectified dc current could generate 140 nC of charge transfer per cycle for driving the ECD. The $I_{SC}$ of 45 μA and the $V_{OC}$ of 140 V of the dual-mode TENG are measured. The effective output power of the TENG was evaluated by monitoring

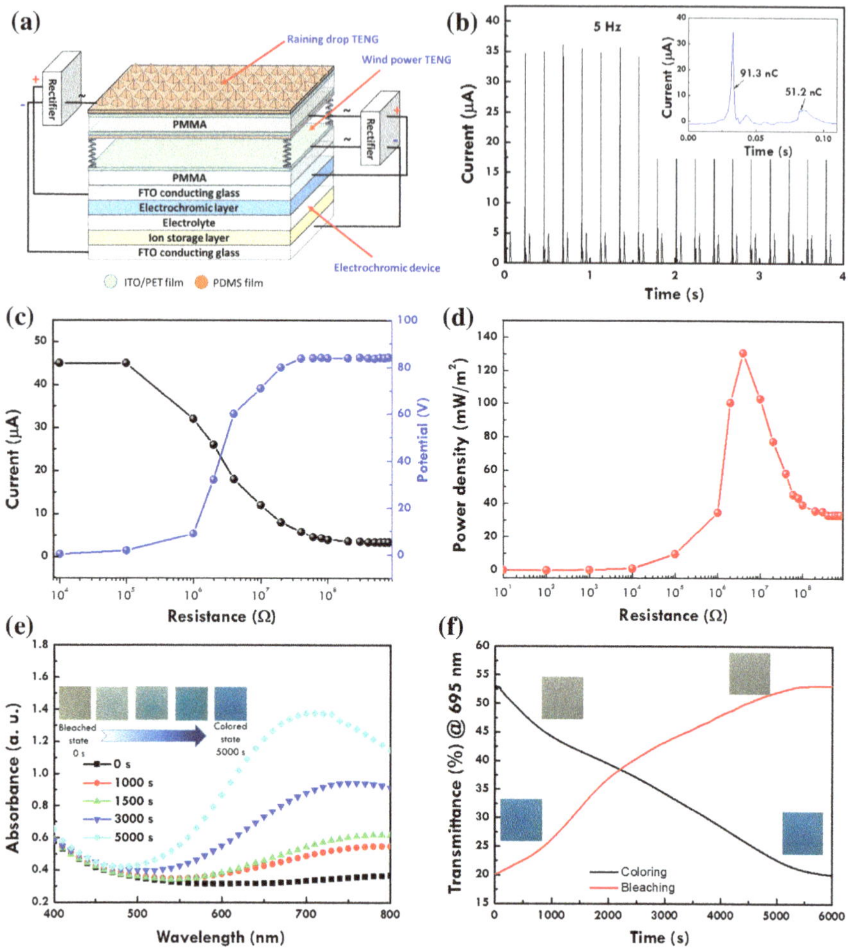


**Fig. 13.15** **a** Schematic diagram of the detailed structure of the self-powered smart window integrated with full-wave bridge rectifiers. **b** Output of the rectified $I_{SC}$ curve of the dual-mode TENG. Enlarged views of a single current peak and related current charge are also shown as *insets*. **c** Dependences of $I_{load}$ and $V_{load}$ of the dual-mode TENG at various load resistances ranging from 10 Ω to 1 GΩ. **d** Instantaneous P of the dual-mode TENG on the resistance of the external load. **e** UV absorption spectra of the self-powered ECD obtained at different duration times. Pictures of ECD at different duration times are also presented as *insets* for depicting the changed colors of the ECD at different times. **f** Transmittance change of the self-powered smart window in response to various duration times for bleaching and coloring. Reproduced with permission from American Chemical Society [22]

the voltage and current across variable load resistances ranging from 10 Ω to 1 GΩ (Fig. 13.15c), and an optimized power density of 130 mW/m$^2$ can be achieved with a load resistance of 4 MΩ (Fig. 13.15d). Figure 13.15e shows the UV/vis-absorbance change of the self-powered ECD driven by the shaker with various durations from 0 to 5000 s. Significant absorbance differences can be monitored by UV/vis spectroscopy, and photographs of the ECD at various durations are also presented as insets in Fig. 13.15e, showing the color switching of the ECD between colorless and dark blue driven by environmentally induced mechanical motions. Figure 13.15f shows the transmittance change of the self-powered ECD at a certain wavelength (695 nm) for bleaching and coloring driven by a dual-mode TENG. These results provide solid evidence to support that the self-powered smart window is successfully realized by an integrated device with an ECD and a TENG.

In summary, the concept of a motion-driven self-powered smart window system was realized by integrating an electrochromic device with a dual-mode triboelectric nanogenerator. The TENG can take advantage of environmental oscillations such as blowing wind and raindrops to deliver an output current of up to 45 μA, corresponding to an effective power density of 130 $mW/m^2$. Reversible electrochromic reactions can be driven directly by the TENG, and a maximum transmittance change of 32.4 % as well as visible color change can be obtained, which is comparable to the ECD driven by an electrochemical workstation. This research presents a substantial advancement in the practical applications of nanogenerators and self-powered systems, which will initiate promising improvements in self-powered flexible displays, wearable electronics, and energy efficient buildings.

## 13.3 TENG for Self-powered Biomedical Stimulation

### *13.3.1 In-Vivo Implanted TENG for Self-powered Pacemaker*

The applications of TENG under in-vivo are distinct from in vitro environments, which are more complicated and challenging. Some crucial problems need to be addressed. For example, the in-vivo environment is filled with body fluid, which will significantly affect the output performance of the TENG [60] if it leaks into the gap between the friction layers of TENG. So careful packaging is needed using biocompatible and soft materials. Secondly, the size and thickness of TENG need to be strictly controlled for the available space inside that is usually very narrow and irregular in shape. Another problem is that the in-vivo movement is so gentle and the amplitude is so small that the TENG must be sensitive enough to catch such small scale motion.

In-vivo biomechanical-energy harvesting using a TENG has been demonstrated by Zheng et al. for the first time [23]. An implantable TENG in a living rat has been developed to harvest energy from its periodic breathing. Then, the breathing generated energy was directly used to power a prototype pacemaker. This is a significant progress for fabricating self-powered implanted medical devices using TENG as a power source.

Powering medical devices directly with the energy harvested by TENG and forming a self-powered system are the main objectives of this work. In order to demonstrate the function of this self-powered system, a prototype pacemaker with adjustable parameters (frequency, width, and voltage of the stimulation pulse) is fabricated as a model (Fig. 13.16a, b). Here, the prototype based on a 555 timer integrated circuit (IC) is designed. It outputs a continuous stream of rectangular pulses at a specified frequency. Resistors ($R_1$, $R_2$, $R_3$) and capacitor $C_1$ co-determined the output performance of this prototype. Resistor $R_1$ was connected between the supply voltage and the discharge pin (pin 7) and another resistor ($R_2$) was connected between the discharge pin (pin 7) and the trigger (pin 2) and threshold (pin 6) pins that shared a common node. In order to achieve a duty cycle less than 50 %, two diodes ($D_1$ and $D_2$) is set up to make sure that the capacitor ($C_1$) can be charged approximately only through $R_1$, and discharged only through $R_2$. The capacitor $C_2$ connected between pin 5 and ground is to provide interference, and resistor $R_3$ is used to adjust the magnitude of the voltage of the output pulse (Fig. 13.16b).

For this prototype, the high time from each pulse can be given by:

$$T_\mathrm{h} = \ln(2) \cdot R_1 \cdot C_1 \tag{13.6}$$

And the low time from each pulse can be given by:

$$T_\mathrm{l} = \ln(2) \cdot R_2 \cdot C_1 \tag{13.7}$$

So, the frequency ($f$) of the pulse stream depends on the values of $T_\mathrm{h}$ and $T_\mathrm{l}$:

$$f = \frac{1}{T_\mathrm{h} + T_\mathrm{l}} = \frac{1}{\ln(2) \cdot C_1(R_1 + R_2)} \tag{13.8}$$

The pulse duty cycle ($q$) also can be achieved by:

$$q = \frac{R_1}{R_1 + R_2} \tag{13.9}$$

Basically, in the self-powered system, the AC output from TENG could be transformed to a pulse output in the same direction simply by using full-wave rectifying bridge and stored in a capacitor. To our knowledge, the commercial pacemaker worked within a voltage range between 1.8 and 2.8 V, when the voltage decreased to a certain level (Elected Replacement Index, ERI), the device has to be replaced as soon as possible. In experiments, the working range of the prototype was 2–3 V (Fig. 13.16d), and therefore the charging and discharging tests were carried out in this range. The capacitor can be charged from 2 to 3 V within 275 min (for a total of 13750 breathing cycles) by the TENG. The total charges ($C_\mathrm{T}$) can be calculated based on following equation:

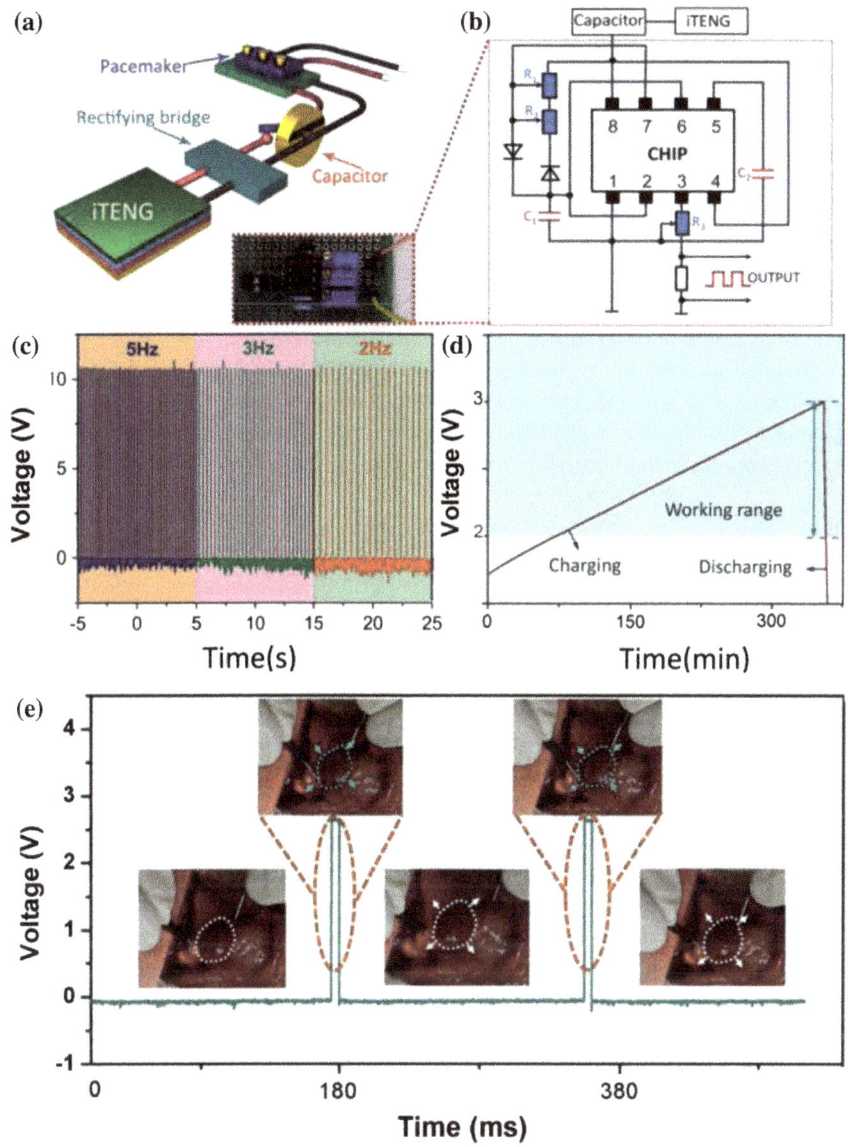


**Fig. 13.16** Scheme diagram of a self-powered prototype pacemaker. **a** Structure and photograph of the self-powered pacemaker. **b** Circuit diagram of prototype pacemaker ($R_1 \approx 0$–$50$ kΩ, $R_2 \approx 0$–$200$ kΩ, $R_3 \approx 0$–$100$ kΩ, $C_1 = 10$ μF, $C_2 = 0.01$ μF). **c** Stimulation pulse with different frequencies generated by self-powered pacemaker. **d** Charging curve of the capacitor by using implantable TENG. **e** Stimulating the heart of the rat by self-powered pacemaker (*insets* photographs of the heart beat of the rat regulated by self-powered pacemaker). Reproduced with permission from Wiley [23]

$$C_T = I_{TENG} \cdot W \cdot R_B \cdot T_C \tag{13.10}$$

where $I_{TENG}$ is the short-circuit current of TENG, $W$ is the pulse-width, $R_B$ is the breath rate and $T_C$ is the total charging time.

The output of a single stimulation pulse of the as-fabricated pacemaker prototype was about 25 μA ($I_P$), and the pulse width was set to 2 ms ($t_P$), which was comparable with the commercial pacemakers. Based on these parameters, it was theoretically calculated that the energy generated during 5 breaths can stimulate a single pulse in TENG (Eq. (13.11) below). Then, the charged capacitor was applied to drive the pacemaker prototype and stimulation pulses with different frequency were successfully demonstrated (Fig. 13.16c). Furthermore, the TENG powered pacemaker was used for regulating the heart rate of the rat (Fig. 13.16e). Briefly, stimulation pulse with different working frequency (2, 3 and 5 Hz) generated by the pacemaker was applied to the heart, and apparently, the heart rate was regulated to synchronized with the frequency of the stimulation pulse.

$$r = I_T t_T / I_P t_P \tag{13.11}$$

In summary, an in-vivo application of a TENG for harvesting biomechanical energy inside a living animal has been demonstrated for the first time. This new designed TENG successfully converted the mechanical energy from the rat's normal breathing into electricity with a power density up to 8.44 mW $m^{-2}$. Moreover, the harvested energy from in-vivo was demonstrated to drive an implantable medical device as a real application. The electricity from TENG was stored in a capacitor and successfully drove a pacemaker prototype to regulate the heart rate of a rat. This research shows a feasible approach to scavenge the biomechanical energy, such as heart beating, muscle stretching, blood flow, or even irregular vibration. This work presents a crucial step forward for lifetime implantable self-powered medical devices.

### 13.3.2 Implantable Self-powered Laser Cure System for Proliferation and Differentiation of Cells

Bone remodeling or orthodontic treatment is usually a long-term process [61, 62]. Currently, it is accepted that low-level laser therapy has a positive effect on the speed and quality of this process, [63] possibly because the laser's energy corresponds with the characteristic energy and absorption levels of the respiratory chain in mitochondria. Thus, it enhances the vitality actions of cells and accelerates regeneration of the damaged tissues [62]. To facilitate the applications of this technology, an attachable or implantable laser cure system is demanded. However, a normal battery would not work in this scenario because it is either too big in volume or uncomfortable for the patients.

A self-powered low-level laser cure (SPLC) system for osteogenesis was developed by Tang et al., which significantly accelerated the mouse embryonic osteoblasts' proliferation and differentiation [20]. Moreover, it was found that the system could also work under the drive of a living creature's motions, such as human walking or a mouse's breathing. This work shows great progress not only for TENGs' applications in portable or implantable medical devices but also for clinical therapy of bone remodeling and orthodontic treatment.

The SPLC system is composed of a flexible TENG and an infrared laser excitation unit. The fabrication of the TENG refers to the arch-shaped highly flexible device [64]. Polydimethylsiloxane (PDMS) and indium tin oxide (ITO) were utilized as the friction materials. However, to enhance the TENG's current output, a pyramid array patterned PDMS film was preferred instead of the plain film. It consists of a pyramid array with each side's length being 5 μm. The working principle of this TENG was previously reported and is based on the combination of contact electrification and electrostatic induction. The TENG output a short-circuit current of about 30 μA and an open-circuit voltage of 115 V. The transferred charge during one period was around 70 nC.

As for the biological experiment, murine calvarial preosteoblasts (MC3T3-E1) (ATCC CRL-2594) were chosen. It is a mouse calvaria-derived cell line that can provide a useful in vitro system to study the different stages during the osteoblast differentiation [65]. The cells were initially cultured with DMEM (Dulbecco's modified Eagle's medium, with 10 % fetal bovine serum and 1 % penicillin–streptomycin) in 5 % $CO_2$ at 37 °C, seeded in 96-well plates for 3000 cells/100 μL, and allocated in three groups: reference group—no laser treatment; laser-irradiated group driven by TENG (TENG-lasered group)—pulse laser irradiation driven by TENG (100 pulses/day); laser-irradiated group using a battery (battery-lasered group)—continuous laser irradiation powered by a battery (1 min/day), as shown in Fig. 13.17a. Before laser treatment, all of the cells were incubated for 24 h. Then, the laser treatment was performed, and cells were tested after another 24 h of incubation. During the incubating process, the whole plates with tinfoil was covered to avoid any ambient illumination.

For the infrared laser, it has a wavelength and power of 850 nm and 10 mW, with an irradiation diameter of 6 mm, which means the power density is about 35.4 mW/cm$^2$. The TENG-lasered irradiation was driven by the TENG with a capacitor as an electricity reservoir (Fig. 13.17b). The TENG was triggered by a motor with a displacement of ±0.5 mm and a frequency of 50 Hz. The infrared signals for the two laser groups are shown in Fig. 13.17c, d.

Since MC3T3-E1 becomes functional during the successive developmental stages, including the proliferation period, bone matrix formation, and mineralization stages, in the following experiments, a MTT [3-(4,5-dimethylthiazol-2-yl)-2,5-diphenyltetrazolium bromide] test is used to assess cell proliferation, ALP (alkaline phosphatase) level to detect early bone matrix formation, and Alizarin red S dye to characterize the mineralization.

To investigate the effect of infrared irradiation on MC3T3-E1, the cytoskeleton and nucleus with phalloidin and DAPI (Sigma-Aldrich) was stained to observe the

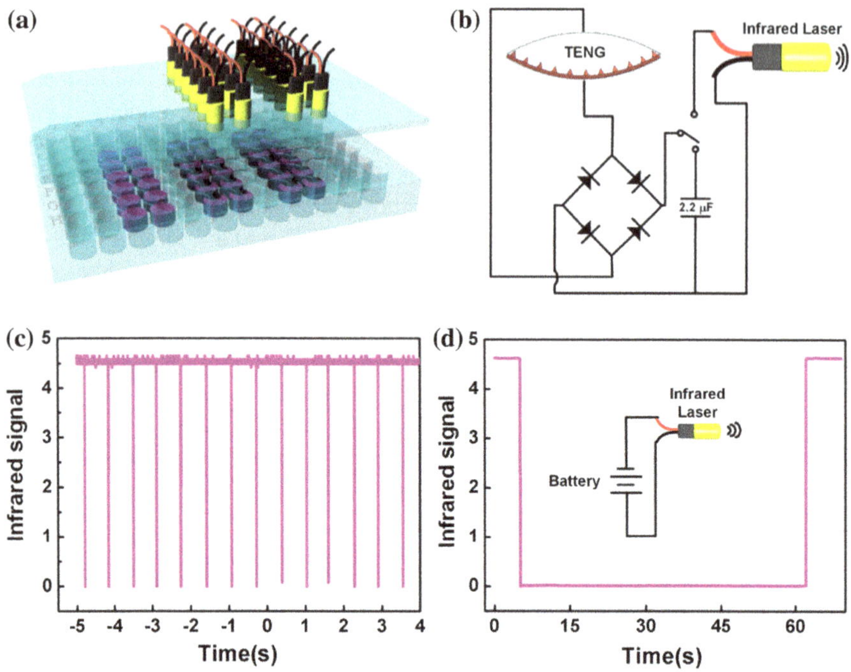


**Fig. 13.17** **a** Bioexperimental setup. **b** Schematic diagram of the SPLC system. **c** Infrared laser signal emitted by the SPLC system. **d** Infrared laser signal emitted by a battery. Reproduced with permission from American Chemical Society [20]

morphology and quantity of cells, as shown in Fig. 13.18. Phalloidin (red) specifically binds to cellular filamentous actin, which displayed the relative integrity of the cytoskeleton. There was no distortion of the cytoskeleton structure observed among the three groups, which implied that the cells in TENG-lasered (Fig. 13.18b, e, h, 1–3 days) and battery-lasered irradiation groups (Fig. 13.18c, f, i, 1–3 days) grew as normal as the reference group. Additionally, DAPI (blue) specifically binds to the DNA of the nucleus. The quantity of nucleus in all irradiated groups increased compared to that in the reference group (Fig. 13.18a, d, g, 1–3 days), which means the cells proliferated apparently under infrared irradiation.

After being plated, the cells entered a proliferative phase. Proliferation was evaluated at 24 h after laser irradiation and assessed using the MTT assay. After 10 μL of MTT (0.5 mg/mL in final concentration) was added per well, cells were incubated in the dark for another 4 h in 5 % $CO_2$ at 37 °C to induce the production of formazan crystals. One hundred microliters of dimethylsulfoxide was added to dissolve the aforementioned formazan crystals. Absorbance was read at 490 nm to evaluate the cell proliferation on an enzyme-linked immunosorbent assay (ELISA) reader. After treatment with infrared irradiation for 1 day, there was no significant difference observed in the reference, TENG-lasered, or battery-lasered irradiation groups. However, after irradiation by an infrared laser for 2 or 3 days, the cells in

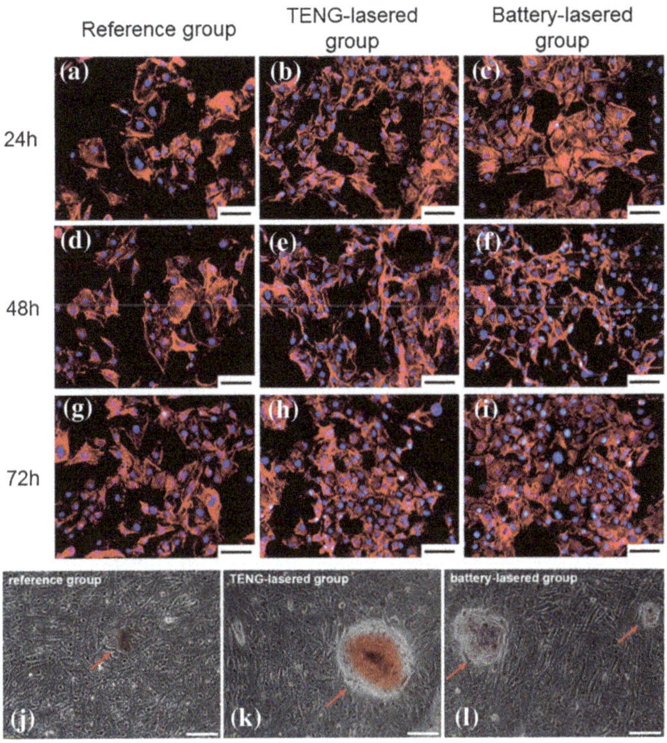


**Fig. 13.18** Quantity and morphology of the MC3T3-E1 cell with different processing methods, stained by phalloidin and DAPI. **a**, **d**, **g** Reference group cultured 24, 48, and 72 h, respectively. **b**, **e**, **h** TENG-lasered group, with the irradiation dose of 100 pulses/day, cultured 24, 48, and 72 h, respectively. **c**, **f**, **i** Battery-lasered group, with the irradiation dose of 1 min/day, cultured 24, 48, and 72 h, respectively. Scale bar is 100 μm. **j** MTT results show a promotion on the proliferation of MC3T3-E1 after different laser irradiation. The cell viability of the TENG-lasered group was 15 and 10 % higher than that of the reference group, respectively, after being irradiated for 2 and 3 days. Battery-lasered group was 66 and 32 % higher. **k** ALP level and **l** mineral deposition results implied that the laser irradiation could induce bone formation. The ALP level of the TENG-lasered group for 1, 3, and 5 day treatment was 9.2, 9.4, and 16.9 % higher than that of reference group, respectively. Battery-lasered group was 5.1, 15.0, and 21.7 % higher. The mineralized area in both TENG-lasered and battery-lasered groups was significantly larger than that in the reference group. The mineralized area is marked with a red arrow, which was stained by Alizarin red S. Scale bar is 100 μm. Reproduced with permission from American Chemical Society [20]

the irradiated groups increased statistically remarkably. Specifically, the cell viability of the TENG-lasered group was 15 and 10 % higher than that of the reference group for the 2 and 3 day samples, respectively. The cell viability of the battery-lasered group was 66 and 32 % higher than that of the reference group, respectively. The results indicated that the TENG-lasered and battery-lasered irradiation could promote cell proliferation.

The ALP level indicates the beginning of bone matrix synthesis and the maturity of the extracellular matrix of the MC3T3-E1 cell. The extracellular ALP level was measured with an ALP ELISA kit. This kit adopted a double sandwich method. The absorbance was read at 450 nm on an ELISA reader. The cell culture time was relatively longer than the proliferation test, and the irradiation times also increased. The ALP level of three groups was measured after 1, 3, and 5 days of irradiation. The TENG-lasered group was 9.2, 9.4, and 16.9 % higher than the reference group, respectively, and the battery-lasered group was 5.1, 15.0, and 21.7 % higher than the reference group. The ALP level implied that the bone matrix synthesis was activated by the infrared irradiation and the MC3T3-E1 cells had a more mature extracellular matrix after laser treatment.

Mineral deposition is a significant marker of bone formation in-vivo in MC3T3-E1 cells, which will begin after several days of induction. Here, 50 μg/mL ascorbic acid (AA) and 10 mM β-glycerophosphate was added into DMEM to induce mineral deposition of MC3T3-E1 cells. The Alizarin red S method was used to evaluate the mineral deposition of MC3T3-E1 cells at day 28 (Fig. 13.18j–l). After being stained by Alizarin red S, the optical images showed obvious mineral deposition in the infrared irradiation groups. Both the quantity of calcium nodules and mineralized area in both TENG-lasered and battery-lasered irradiation groups were obviously larger than those in the reference group (Fig. 13.18j–l).

In all, these experiments significantly demonstrated that the TENG-lasered treatment had a positive effect on osteoblasts' proliferation, differentiation, and bone formation. Compared to the continuous irradiation powered by a battery, the pulse irradiation powered by the SPLC system has a similar effect, consistent with previous work. Additionally, it is more convenient for portable or implantable clinical treatment.

In summary, a self-powered low-level laser cure system for osteogenesis was developed, based on the integration of a triboelectric nanogenerator and an infrared laser irradiation unit. It was demonstrated that the system enhanced MC3T3-E1 proliferation by 15 % after 2 days of laser irradiation and accelerated the cells' differentiation by 16.9 % after 5 days of irradiation. Osteoblast mineralization was also significantly increased by the SPLC system. This work shows a great progress not only for application of TENGs in portable or implantable medical devices but also for clinical therapy of bone remodeling and orthodontic treatment.

## 13.4 Summary

In this chapter, TENG has been demonstrated to form self-powered systems based on coupling with energy storage units and power management circuits. These self-powered systems have been applied on electronic devices, electrochemical reactions and biomedical stimulations. The self-powered systems proposed here provide an alternative solution other than battery to sustainably power functional devices. Since the energy is from ambient environment, the lifetime of the systems is expected to be far higher than battery-based devices. Self-powered electrochemistry is to use the energy harvested in the environment such as water flow using TENG to drive electrochemical processes, such as pollution degradation, water splitting, anticorrosion and air pollution filtering. Research in self-powered system will provide a paradigm shift solution for powering portable electronics and electrochemical/biomedical reactions.

## References


1. Z.L. Wang, J. Song, Piezoelectric nanogenerators Based on Zinc Oxide Nanowire arrays. Science **312**(5771), 242–246 (2006)
2. S.H. Wang, Z.H. Lin, S.M. Niu, L. Lin, Y.N. Xie, K.C. Pradel, Z.L. Wang, Motion charged battery as sustainable flexible-power-unit. ACS Nano **7**(12), 11263–11271 (2013)
3. S.H. Wang, L. Lin, Z.L. Wang, Nanoscale triboelectric-effect-enabled energy conversion for sustainably powering portable electronics. Nano Lett. **12**(12), 6339–6346 (2012)
4. J. Zhou, Y.D. Gu, Y.F. Hu, W.J. Mai, P.H. Yeh, G. Bao, A.K. Sood, D.L. Polla, Z.L. Wang, Gigantic enhancement in response and reset time of ZnO UV nanosensor by utilizing Schottky contact and surface functionalization. Appl. Phys. Lett. **94**(19), (2009)
5. J.M. Tarascon, M. Armand, Issues and challenges facing rechargeable lithium batteries. Nature **414**(6861), 359–367 (2001)
6. A.J. Bard, L.R. Faulkner, Electrochemical methods: fundamentals and applications. 2nd ed.; p xxi, 833 pages
7. G. Zhu, J. Chen, T.J. Zhang, Q.S. Jing, Z.L. Wang, Radial-arrayed rotary electrification for high performance triboelectric generator. Nat. Commun. **5**, 3426 (2014)
8. S.M. Niu, X.F. Wang, F. Yi, Y.S. Zhou, Z.L. Wang, A universal self-charging system driven by random biomechanical energy for sustainable operation of mobile electronics. Nat. Commun. **6**, 8975 (2015)
9. R.W. Erickson, D. Maksimović, *Fundamentals of Power Electronics*, 2nd ed. (Kluwer Academic, Norwell, MA, 2001), p. xxi, 883 p
10. S. Xu, Y. Qin, C. Xu, Y. Wei, R. Yang, Z.L. Wang, Self-powered nanowire devices. Nat. NANO **5**(5), 366–373 (2010)
11. Z.L. Wang, Toward self-powered sensor networks. Nano Today **5**(6), 512–514 (2010)
12. Y. Hu, Y. Zhang, C. Xu, L. Lin, R.L. Snyder, Z.L. Wang, Self-powered system with wireless data transmission. Nano Lett. **11**(6), 2572–2577 (2011)
13. G. Zhu, C. Pan, W. Guo, C.-Y. Chen, Y. Zhou, R. Yu, Z.L. Wang, Triboelectric-generator-driven pulse electrodeposition for micropatterning. Nano Lett. **12**(9), 4960–4965 (2012)
14. Y. Yang, H. Zhang, S. Lee, D. Kim, W. Hwang, Z.L. Wang, Hybrid energy cell for degradation of methyl orange by self-powered electrocatalytic oxidation. Nano Lett. **13**(2), 803–808 (2013)
15. Y. Yang, H. Zhang, Y. Liu, Z.-H. Lin, S. Lee, Z. Lin, C.P. Wong, Z.L. Wang, Silicon-based hybrid energy cell for self-powered electrodegradation and personal electronics. ACS Nano **7** (3), 2808–2813 (2013)
16. Y. Su, Y. Yang, H. Zhang, Y. Xie, Z. Wu, Y. Jiang, N. Fukata, Y. Bando, Z.L. Wang, Enhanced photodegradation of methyl orange with $TiO_2$ nanoparticles using a triboelectric nanogenerator. Nanotechnology **24**(29), 295401 (2013)
17. Y. Yang, H. Zhang, Z.-H. Lin, Y. Liu, J. Chen, Z. Lin, Y.S. Zhou, C.P. Wong, Z.L. Wang, A hybrid energy cell for self-powered water splitting. Energy Environ. Sci. **6**(8), 2429–2434 (2013)

18. W. Tang, Y. Han, C.B. Han, C.Z. Gao, X. Cao, Z.L. Wang, Self-Powered water splitting using flowing kinetic energy. Adv. Mater. **27**(2), 272–276 (2015)
19. S. Chen, C. Gao, W. Tang, H. Zhu, Y. Han, Q. Jiang, T. Li, X. Cao, Z. Wang, Self-powered cleaning of air pollution by wind driven triboelectric nanogenerator. Nano Energy **14**, 217–225 (2015)
20. W. Tang, J. Tian, Q. Zheng, L. Yan, J. Wang, Z. Li, Z.L. Wang, Implantable self-powered low-level laser cure system for mouse embryonic osteoblasts' proliferation and differentiation. ACS Nano **9**(8), 7867–7873 (2015)
21. M.-H. Yeh, H. Guo, L. Lin, Z. Wen, Z. Li, C. Hu, Z.L. Wang, Rolling friction enhanced free-standing triboelectric nanogenerators and their applications in self-powered electrochemical recovery systems. Adv. Funct. Mater. n/a–n/a (2015)
22. M.-H. Yeh, L. Lin, P.-K. Yang, Z.L. Wang, Motion-driven electrochromic reactions for self-powered smart window system. ACS Nano **9**(5), 4757–4765 (2015)
23. Q. Zheng, B. Shi, F. Fan, X. Wang, L. Yan, W. Yuan, S. Wang, H. Liu, Z. Li, Z.L. Wang, In-vivo powering of pacemaker by breathing-driven implanted triboelectric nanogenerator. Adv. Mater. **26**(33), 5851–5856 (2014)
24. H.R. Zhu, W. Tang, C.Z. Gao, Y. Han, T. Li, X. Cao, Z.L. Wang, Self-powered metal surface anti-corrosion protection using energy harvested from rain drops and wind. Nano Energy **14**, 193–200 (2015)
25. X. Yang, G. Zhu, S. Wang, R. Zhang, L. Lin, W. Wu, Z.L. Wang, A self-powered electrochromic device driven by a nanogenerator. Energy Environ. Sci. **5**(11), 9462–9466 (2012)
26. J.S. Do, M.L. Chen, Decolourization of dye-containing solutions by electrocoagulation. J. Appl. Electrochem. **24**(8), 785–790 (1994)
27. H. Ma, B. Wang, X. Luo, Studies on degradation of Methyl Orange wastewater by combined electrochemical process. J. Hazard. Mater. **149**(2), 492–498 (2007)
28. M. Panizza, G. Cerisola, Electrochemical Oxidation as a final treatment of synthetic tannery wastewater. Environ. Sci. Technol. **38**(20), 5470–5475 (2004)
29. N. Mohan, N. Balasubramanian, C.A. Basha, Electrochemical oxidation of textile wastewater and its reuse. J. Hazard. Mater. **147**(1–2), 644–651 (2007)
30. M. Panizza, G. Cerisola, Electrocatalytic materials for the electrochemical oxidation of synthetic dyes. Appl. Catal. B **75**(1–2), 95–101 (2007)
31. L.-C. Chiang, J.-E. Chang, T.-C. Wen, Indirect oxidation effect in electrochemical oxidation treatment of landfill leachate. Water Res. **29**(2), 671–678 (1995)
32. P. Niu, J. Hao, Fabrication of Titanium Dioxide and Tungstophosphate Nanocomposite Films and Their Photocatalytic Degradation for Methyl Orange. Langmuir **27**(22), 13590–13597 (2011)
33. S. Xu, Y. Zhu, L. Jiang, Y. Dan, Visible light induced photocatalytic degradation of Methyl Orange by Polythiophene/$TiO_2$ composite particles. Water Air Soil Pollut. **213**(1–4), 151–159 (2010)
34. E. Rasten, G. Hagen, R. Tunold, Electrocatalysis in water electrolysis with solid polymer electrolyte. Electrochim. Acta **48**(25–26), 3945–3952 (2003)
35. D.L. Stojić, M.P. Marčeta, S.P. Sovilj, Š.S. Miljanić, Hydrogen generation from water electrolysis—possibilities of energy saving. J. Power Sources **118**(1–2), 315–319 (2003)
36. Z. Zou, J. Ye, K. Sayama, H. Arakawa, Direct splitting of water under visible light irradiation with an oxide semiconductor photocatalyst. Nature **414**(6864), 625–627 (2001)
37. M.M.S. Cheung, C. Cao, Application of cathodic protection for controlling macrocell corrosion in chloride contaminated RC structures. Constr. Build. Mater. **45**, 199–207 (2013)
38. J.-G. Kim, J.-H. Joo, S.-J. Koo, Development of high-driving potential and high-efficiency Mg-based sacrificial anodes for cathodic protection. J. Mater. Sci. Lett. **19**(6), 477–479 (2000)
39. J. Xu, W. Yao, Current distribution in reinforced concrete cathodic protection system with conductive mortar overlay anode. Constr. Build. Mater. **23**(6), 2220–2226 (2009)
40. J. Yuan, S. Tsujikawa, Characterization of Sol-Gel-Derived $TiO_2$ coatings and their photoeffects on copper substrates. J. Electrochem. Soc. **142**(10), 3444–3450 (1995)

41. L. Veleva, P. Castro, G. Hernandez-Duque, M. Schorr, The corrosion performance of steel and reinforced concrete in a tropical humid climate. A Rev. (1998)
42. B. Brunekreef, S.T. Holgate, Air pollution and health. Lancet **360**(9341), 1233–1242 (2002)
43. I.C.A. Pope, R.T. Burnett, M.J. Thun, E.E. Calle, D. Krewski, K. Ito, G.D. Thurston, Lung cancer, cardiopulmonary mortality, and long-term exposure to fine particulate air pollution. J. Am. Med. Assoc. **287**(9), 1132–1141 (2002)
44. S. Gao, N. Nakagawa, K. Kato, M. Inomata, F. Tsuchiya, Simultaneous SO2/NOx removal by a powder-particle fluidized bed. Catal. Today **29**(1–4), 165–169 (1996)
45. C.C. Wu, G.W.M. Lee, S. Yang, K.-P. Yu, C.L. Lou, Influence of air humidity and the distance from the source on negative air ion concentration in indoor air. Sci. Total Environ. **370**(1), 245–253 (2006)
46. R. Hackam, H. Aklyama, Air pollution control by electrical discharges. IEEE Trans. Dielectr. Electr. Insul. **7**(5), 654–683 (2000)
47. A. Mizuno, Electrostatic precipitation. IEEE Trans. Dielectr. Electr. Insul. **7**(5), 615–624 (2000)
48. D.K. Brandvold, P. Martinez, D. Dogruel, Polarity dependence of N2O formation from corona discharge. Atmos. Environ. (1967) **23**(9), 1881–1883 (1989)
49. V.A. Kirsh, I.B. Stechkina, Kinetics of the clogging and optimization of prefilters in a two-stage air cleaning system. Theor. Found. Chem. Eng. **44**(1), 76–85 (2010)
50. D. Zhang, M. Wen, S. Zhang, P. Liu, W. Zhu, G. Li, H. Li, Au nanoparticles enhanced rutile $TiO_2$ nanorod bundles with high visible-light photocatalytic performance for NO oxidation. Appl. Catal. B: Environ. **147**, 610–616 (2014)
51. L. Dascalescu, R. Morar, A. Iuga, A. Samuila, V. Neamtu, I. Suarasan, Charging of particulates in the corona field of roll-type electroseparators. J. Phys. D Appl. Phys. **27**(6), 1242 (1994)
52. F. Fu, Q. Wang, Removal of heavy metal ions from wastewaters: a review. J. Environ. Manage. **92**(3), 407–418 (2011)
53. G. Chen, Electrochemical technologies in wastewater treatment. Sep. Purif. Technol. **38**(1), 11–41 (2004)
54. D.R. Rosseinsky, R.J. Mortimer, Electrochromic systems and the prospects for devices. Adv. Mater. **13**(11), 783–793 (2001)
55. R.J. Mortimer, A.L. Dyer, J.R. Reynolds, Electrochromic organic and polymeric materials for display applications. Displays **27**(1), 2–18 (2006)
56. T.J. Richardson, New electrochromic mirror systems. Solid State Ionics **165**(1–4), 305–308 (2003)
57. P. Andersson, D. Nilsson, P.O. Svensson, M. Chen, A. Malmström, T. Remonen, T. Kugler, M. Berggren, Active Matrix Displays Based on All-Organic Electrochemical Smart Pixels Printed on Paper. Adv. Mater. **14**(20), 1460–1464 (2002)
58. L. Su, H. Wang, Z. Lu, All-solid-state electrochromic window of prussian blue and electrodeposited WO3 film with poly(ethylene oxide) gel electrolyte. Mater. Chem. Phys. **56** (3), 266–270 (1998)
59. A.L. Dyer, R.H. Bulloch, Y. Zhou, B. Kippelen, J.R. Reynolds, F. Zhang, A vertically integrated solar-powered electrochromic window for energy efficient buildings. Adv. Mater. **26**(28), 4895–4900 (2014)
60. H.L. Zhang, Y. Yang, Y.J. Su, J. Chen, C.G. Hu, Z.K. Wu, Y. Liu, C.P. Wong, Y. Bando, Z.L. Wang, Triboelectric nanogenerator as self-powered active sensors for detecting liquid/gaseous water/ethanol. Nano Energy **2**(5), 693–701 (2013)
61. H. Pretel, R.F.Z. Lizarelli, L.T.O. Ramalho, Effect of low-level laser therapy on bone repair: Histological study in rats. Lasers Surg. Med. **39**(10), 788–796 (2007)
62. M. Jawad, A. Husein, M. Alam, R. Hassan, R. Shaari, Overview of non-invasive factors (low level laser and low intensity pulsed ultrasound) accelerating tooth movement during orthodontic treatment. Lasers Med. Sci. **29**(1), 367–372 (2014)

63. E.J. Luger, S. Rochkind, Y. Wollman, G. Kogan, S. Dekel, Effect of low-power laser irradiation on the mechanical properties of bone fracture healing in rats. Lasers Surg. Med. **22** (2), 97–102 (1998)
64. F.R. Fan, J. Luo, W. Tang, C. Li, C. Zhang, Z. Tian, Z.L. Wang, Highly transparent and flexible triboelectric nanogenerators: performance improvements and fundamental mechanisms. J. Mater Chem A **2**(33), 13219–13225 (2014)
65. J.-Y. Choi, B.-H. Lee, K.-B. Song, R.-W. Park, I.-S. Kim, K.-Y. Sohn, J.-S. Jo, H.-M. Ryoo, Expression patterns of bone-related proteins during osteoblastic differentiation in MC3T3-E1 cells. J. Cell. Biochem. **61**(4), 609–618 (1996)

# Part III
# Applications as Self-Powered Active Sensors

# Chapter 14
# Self-powered Sensing for Human-Machine Interface

In Part A of this book, we have presented the basic principle and fundamental working modes of triboelectric nanogenerators (TENGs) [1–3]. In part B, we show the applications of self-powered systems by integrating TENGs with various functional electronic devices. In response to an external mechanical impact, the TENG is able to produce electricity output based on triboelectrification and electrostatic induction. The amplitude and frequency of the output signal are dependent on those of the mechanical stimuli. In this regard, the information of the external mechanical input (like pressure, motion, vibration, etc.) could be obtained by analyzing the generated electric output signals. Additionally, the amount of surface triboelectric charges is a key factor responsible for the overall output performance of the TENG, and the environmental parameters that may affect the charge density of triboelectric surfaces could be detected by measuring the output voltage and current of the TENG. The TENGs will then be enabled to serve as self-powered active environmental/chemical sensors to detect solution concentration of certain species, intensity of UV illumination, moisture, and many more. The focus of this chapter is about TENGs as self-powered active sensors.

## 14.1 Self-powered Pressure/Touch Sensor

The first straightforward application of TENG-based active sensor is self-powered pressure/touch sensor [1–3]. A few works about self-powered pressure/touch sensor have been completed based on vertical contact-separation mode and single-electrode mode [4–8]. Moreover, an ultrasensitive pressure sensor has been developed by incorporating both triboelectric effect and conventional resistance variation [9]. To achieve tactile imaging function, multiple triboelectric pressure sensor units were integrated into an array for monitoring the pressure distribution over a certain area [5, 10]. Finally, as a practical application, single-electrode triboelectric sensor was integrated into a smart keyboard to realize self-powered

 
Z.L. Wang et al., *Triboelectric Nanogenerators*,
Green Energy and Technology, DOI 10.1007/978-3-319-40039-6_14

touch sensing and even biometric authentication purposes, with great potential towards self-powered human-machine interfacing [11].

### *14.1.1 Contact-Separation Mode Self-powered Pressure Sensor*

As the first example, we will introduce a flexible triboelectric active sensor (TEAS) with excellent performance and low processing cost [5]. The working principle of the TEAS is similar to that of TENGs at vertical contact-separation mode. Both static and dynamic pressure sensing were accomplished by the same device with different measurement approaches. Specifically, the open-circuit voltage as well as the amount of the transferred charge density was employed for static pressure detection, while the pulse-like short-circuit current peak was used for dynamic pressure monitoring. Owing to this active sensing principle, a supreme performance of pressure detection was achieved on the TEAS, including high sensitivity of 0.31 $kPa^{-1}$, fast response/relaxation time of $<5$ ms, long-term stability/reliability of 30,000 cycles, as well as a low detection limit of 2.1 Pa. This work is an unprecedented progress towards the practical application of nanogenerators, the realization of self-powered system, and further advancement of flexible electronic devices.

The device structure of the TEAS is schematically depicted in Fig. 14.1a. Based on the metal-insulator triboelectric pairs, the fabrication of the TEAS device could be divided into two parts. In the first part, the polydimethylsiloxane (PDMS) membrane was modified with micro-patterned pyramid structures through photolithography and template method. A scanning electron microscopy (SEM) image of its surface morphology is shown in Fig. 14.1b. The pyramid micro-structures were uniformly distributed on the inner surface of PDMS, with the size of $\sim 10$ μm. The processed PDMS membrane was deposited with gold electrode on its outer surface. In the second part, a piece of aluminum was assembled with the composite of silver nanowires and nanoparticles that were prepared by a seed-mediated approach [12], as shown in Fig. 14.1c. The micro/nano structures on both inner surfaces were introduced to enhance the triboelectric effect and the effective contact area between the two surfaces, which would help to improve the pressure response of the TEAS. The two layers were bonded by non-conductive tapes and both electrodes were connected to electrical leads. Due to the elastic resilience of the materials, the device naturally forms an arch shape, with an air gap presented for effective charge separation (Inset of Fig. 14.1a). The size of the TEAS is 2 cm $\times$ 2 cm, which of course can be much reduced in size.

The typical electrical responses of the TEAS under external agitations by a mechanical shaker are shown in Fig. 14.1d–f. With a 2 Hz cyclic pressing force applied on the TEAS, it produces an open-circuit voltage ($V_{OC}$) of $\sim 75$ V, transferred charge density ($\Delta\sigma$) of $\sim 45$ $\mu C/m^2$, and short-circuit current density

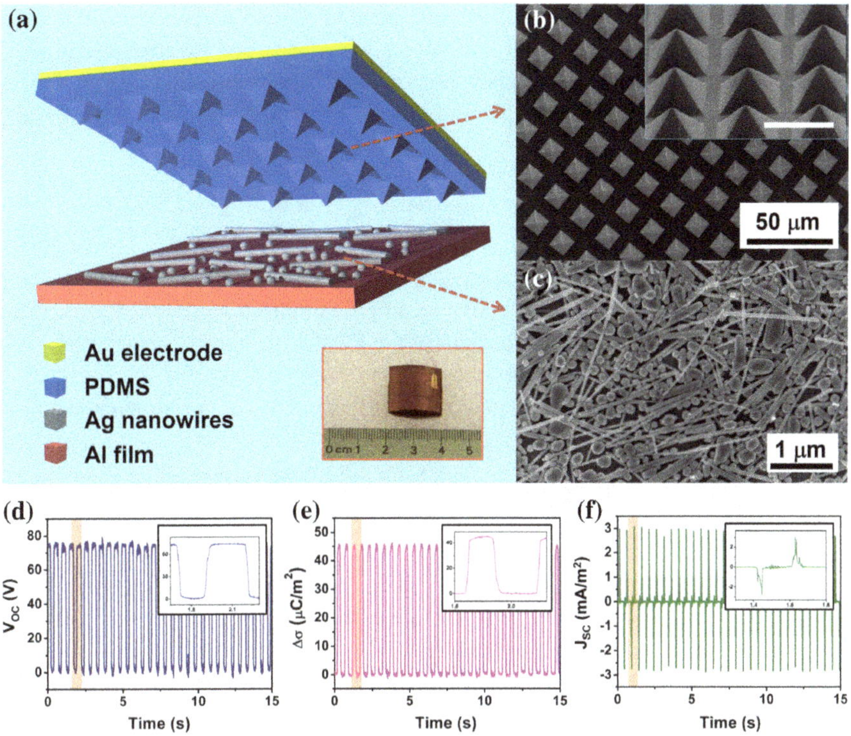


**Fig. 14.1** The basic structure and typical electrical response of the triboelectric active sensor applied with a cyclic pressure. **a** The schematic illustration of the TEAS device with both inner surfaces modified by micropattern PDMS structures and Ag nanowires/nanoparticles composite. The *inset* is a photograph of the TEAS device. **b** The SEM image of the pyramidstructured micropatterns fabricated on the inner surface of the PDMS membrane. The *inset* is a high magnification SEM image of the micropatterns, and the *inset* scale bar is 10 μm. **c** The SEM image of the Ag nanowires/nanoparticles composite assembled on the inner surface of the Al foil. **d**, **e** The measured typical electrical response including **d** the open-circuit voltage, **e** the transferred charge density, and **f** the short-circuit current density of the TEAS upon a cyclic pressure. Reproduced with permission from American Chemical Society [5]

($J_{SC}$) of ~3 mA/m$^2$. The working principle of the TEAS is based on the coupling effect of contact electrification and electrostatic induction, as we discussed in Chap. 2. In a contact-separation TENG (attached-electrode contact-mode TENG), we have already shown that the open-circuit voltage has linear relationship with the separation distance:

$$V_{oc} = \frac{\sigma \cdot d}{\varepsilon_0} \tag{14.1}$$

$\varepsilon_0$ is the permittivity in vacuum and $\sigma$ is the triboelectric charge density, which is considered as constant in this work. It should be noted that the $V_{OC}$ starts decreasing when the external pressure is applied (with smaller $d$), and reverts to its maximum level when the pressure is fully removed ($d = d_0$). As for the short-circuit (SC) condition, we have shown that the amount of the transferred charge density ($\Delta\sigma$) as a result of pressure change is also a function of $d$, which is given by:

$$\Delta\sigma = \frac{Q_0 - Q}{S} = \frac{\sigma \cdot d_0}{\frac{d_{PDMS}}{\varepsilon_{r,PDMS}} + d_0} - \frac{\sigma \cdot d}{\frac{d_{PDMS}}{\varepsilon_{r,PDMS}} + d} \tag{14.2}$$

$Q$ and $Q_0$ are the amount of induced charges on one electrode with and without pressure, respectively, $S$ is the effective area of electrodes, $d_{PDMS}$ and $\varepsilon_{r,PDMS}$ is the thickness and relative dielectric constant of the PDMS membrane, respectively, and $d_0$ is the maximum gap distance without pressure. Apparently, the $\Delta\sigma$ increases as $d$ is reduced, but not with linear relationship. Hence, the $J_{SC}$ could be deducted as

$$J_{sc} = \frac{d(\Delta\sigma)}{dt} = -\frac{dQ}{S \cdot dt} = \frac{\sigma \cdot \frac{d_{PDMS}}{\varepsilon_{r,PDMS}}}{\left(\frac{d_{PDMS}}{\varepsilon_{r,PDMS}} + d\right)^2} \cdot \frac{\delta d}{\delta t} \tag{14.3}$$

It can be found that the magnitude of the $J_{SC}$ is not only dependent on the gap distance but also affected by the loading rate of the external pressure ($\delta d/\delta t$). This expectation was confirmed in the measurement with higher frequency of the applied pressure (6 Hz), and the $J_{SC}$ displays a gigantic increase up to $\sim$40 $mAm^{-2}$. Therefore, the output open-circuit signal from the TEAS is a static signal that is only dictated by the gap distance $d$, which is a measure of the magnitude of the applied pressure/force. The output current is a dynamic signal that is determined by the speed at which the gap was squeezed, thus, it gives the information about the rate of the pressure/force being applied.

In order to correlate the external pressure with the $V_{OC}$, the material resilience of the TEAS was considered as the spring-entangled structure using the Hooker's Law:

$$p = \frac{k \cdot x}{S} = \frac{k \cdot (d_0 - d)}{S} \tag{14.4}$$

$p$ is the pressure and $k$ represents the elastic property of the materials in the TEAS. Combining Eqs. (14.1) and (14.4), we can get

$$\frac{V_{oc,0} - V_{oc}}{V_{oc,0}} = \frac{d_0 - d}{d_0} = \frac{S}{k \cdot d_0} \cdot p \tag{14.5}$$

Hence the relative variation of the $V_{OC}$ should be expected to show directly linear relationship with the applied pressure, and thus it is a reliable parameter for measuring the magnitude of the external pressure applied onto the TEAS, while the $J_{SC}$ is able to provide the dynamic information of the applied pressure (i.e. loading rate). Combining the analysis on the measurement of both the $V_{OC}$ and the $J_{SC}$, the TEAS device is capable of giving the detailed information about the ambient mechanical stimulations. On this basis, we designed and conducted a series of measurements to characterize the performance of the TEAS, as discussed below.

In the measurements, the TEAS was attached onto a flat plate and fixed on the end of a force gauge, and cyclic external pressures were applied by a mechanical linear motor. The data for both the output parameters (the $V_{OC}$, $\Delta\sigma$, and $J_{SC}$) and the real-time force/pressure were recorded and analyzed in the same computer. Figure 14.2a is the real-time measurement result of $V_{OC}$ under series of different pressures. The reference level of voltage was selected in the way that the $V_{OC}$ approached 0 with the maximum pressure when the two surfaces came into full contact. It could be observed that the $V_{OC}$ stays at the maximum level ($V_0 = 60$ V) when there is no pressure, and decreases to a lower level once the external pressure is applied. The voltage variation ($V_0 - V$) increases with increasing pressure, and the relationship between relative voltage variation $(V_0 - V)/V_0$ and magnitude of pressure is plotted in Fig. 14.2b (the subscript "OC" is omitted here for convenient expression). The plot can be divided into two regions based on difference in sensitivity. In the low pressure region (Region I), the sensitivity of the TEAS is 0.31 $kPa^{-1}$, which is much higher than that of the high pressure region (0.01 $kPa^{-1}$ in Region II). The separated fittings of the data in both regions display good linear behavior ($R_I^2 = 0.965$, and $R_{II}^2 = 0.985$), which reveals the validity of the device as a reliable pressure sensor. The reason for different sensitivities can be explained based on the previous analysis. In Region I, the pressure increase results in drastic gap distance change as estimated in Eq. (14.4), introducing larger voltage variations; while it is quite probable that the gap distance has already been close to 0 from the turning point in Region II, and the increasing pressure only works for enhancing the effective contact area between the two layers. In another word, the $V_{OC}$ tends to saturate to zero in Region II, and the applied pressure has little effect on the voltage variation.

To verify the theoretical expectation that the $V_{OC}$ merely depends on the magnitude of the applied pressure rather than the loading rate, the voltage response of the TEAS was tested at a constant pressure of ~2 kPa but of different loading rates that were controlled by the accelerations of the linear motor. As shown in Fig. 14.2c, d, as the accelerations vary from 2.5 to 30 $m/s^2$, the relative voltage variations almost remain constant; though a minor increasing trend is present, possibly resulting from minor pressure variations. The stability of the TEAS device was examined through continuously loading and unloading a constant pressure (~2.5 kPa) for over 30,000 cycles, as exhibited in Fig. 14.2e. The voltage response was measured after every 10,000 cycles, during which 300 cycles were recorded and displayed in Fig. 14.2e. The voltage response only shows a minor decay of ~6.7 % after 30,000 cycles, through which the stability of the TEAS is confirmed. Moreover, in Fig. 14.2f, the real-time pressure measurement and voltage response was displayed in the same plot and both plots match perfectly. Hence, the voltage response is capable of revealing the detailed information of the loading process of an external pressure. From the comparison of plots for the loading and unloading process of the external pressure, it is estimated that both the response time and the relaxation time are <5 ms, demonstrating the prompt response of the

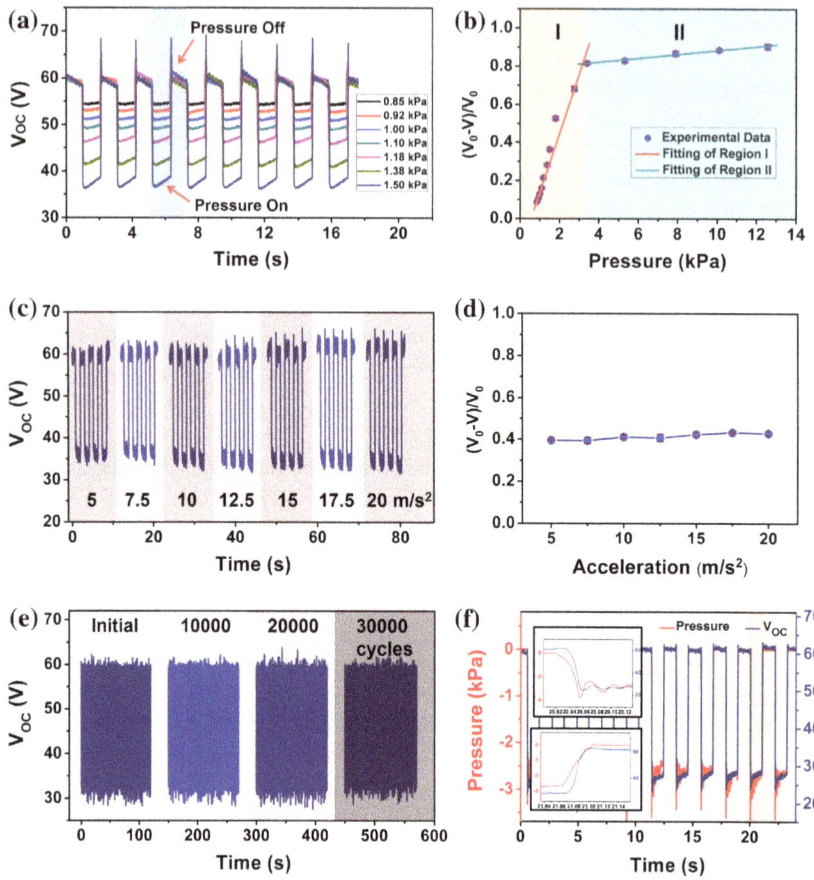


**Fig. 14.2** The static pressure sensing of the TEAS through measurement of the open-circuit voltage. **a** The real-time measurement of the $V_{OC}$ with cyclic and variable pressures applied on the TEAS device. The *highlighted region* indicates one cycle of the pressure loading and unloading process. **b** The summarized relationship and linear fitting between the relative variations of voltage and the pressure applied on the TEAS device. **c** The measurement of the $V_{OC}$ with a cyclic and constant pressure (2 kPa) applied on the TEAS with variable accelerations. **d** The summarized relationship between the relative voltage variations and accelerations. **e** The stability test for static pressure sensing of the TEAS with continuous loading and unloading a cyclic pressure (2.5 kPa) for 30,000 cycles. The $V_{OC}$ was measured after each 10,000 cycles, and 300 cycles of data were recorded and presented in the plot. **f** The comparison of the real-time measurement of both the applied pressure (*red curve*) and the $V_{OC}$ (*blue curve*). The *top inset* and *bottom inset* are the enlarged view of the loading and unloading process in the highlighted cycle, respectively. The response/relaxation times are <5 ms from the time interval of the two curves. Reproduced with permission from American Chemical Society [5]

nanogenerator-based active pressure sensor and proving its superior advantage to conventional sensors. Furthermore, beside the $V_{OC}$, the $\Delta\sigma$ was also a reliable parameter that could be used for static pressure sensing. In this aspect, the similar measurements were then carried out with variable pressures loaded on the TEAS device, and the data of the $\Delta\sigma$ was recorded and displayed in the same way. Different from the $V_{OC}$, the $\Delta\sigma$ is 0 without pressure and increases as the applied pressure is elevated. Similarly, the relationship of the $\Delta\sigma$ with increasing pressure also has two different sensitivity regions: a high sensitivity region at lower pressures and a drastically decreased sensitivity at higher pressures. The acceleration dependence at constant loading pressure was measured, and the amount of $\Delta\sigma$ also keeps constant with variable accelerations of the linear motor, though small variation exists. The stability of this type of measurement was confirmed by continuously loading and unloading a constant pressure of 2.5 kPa for 30,000 cycles.

Different from the $V_{OC}$ and the $\Delta\sigma$, the $J_{SC}$ is not only dependent on the applied force/pressure, but also related to the loading rate of the external force, as discussed in Eq. (14.3) above. Furthermore, the output profile of the $J_{SC}$ is a pulse-like current peak, which enables its application for dynamic pressure detection. Based on this expectation, both the effect of the force/pressure magnitude and the acceleration were tested on the TEAS for peak current measurement. The measured rectified $J_{SC}$ with variable pressures under the same loading rate were displayed in Fig. 14.3a, showing a clearly increasing trend with elevated pressures. The full data summarized in Fig. 14.3b shows the reversibility of the current response. The loading pressure was first increased from low pressure region (<1 kPa) to a high pressure region ($\sim$9 kPa), and then decreased to the original magnitude. The resulted current response in the decreasing-pressure curve does not exhibit much deviation in reference to the original curve, thus demonstrating the reliability of the current measurement. Similarly, the current response with a constant pressure (0.75 kPa) and various accelerations was measured and shown in Fig. 14.3c, d. The peak values of the output $J_{SC}$ also have positive correlation with the acceleration of the applied pressure. Both the relationship between the peak value of the $J_{SC}$ and the applied pressure/acceleration could be explained using Eq. (14.3), where the loading rate $\delta d/\delta t$ is positively related to the acceleration of the linear motor. Therefore, with increasing applied pressure or acceleration, the term $d$ will decrease or $\delta d/\delta t$ will increase, both leading to the enhancement of the absolute value of the $J_{SC}$.

Furthermore, the cycling stability test of the rectified $J_{SC}$ under continuous pressure triggering for 30,000 cycles was carried out in the same way as the test of the $V_{OC}$, and little decay could be found in the measurement results (Fig. 14.3e). In addition, the dynamic reliability of the TEAS was further confirmed by recording the peak current response with variable pressures after every 10,000 triggering cycles. The results shown in Fig. 14.3f also demonstrates that the measured $J_{SC}$-pressure relationship after continuous triggering is close to the original test. More impressively, the dynamic pressure sensing limit of the TEAS was examined by comparing the output signals of both a gentle pressure (0.125 kPa) and the corresponding noise level. The low-end detection limit ($p_0$) of the TEAS is then calculated by the following equation:

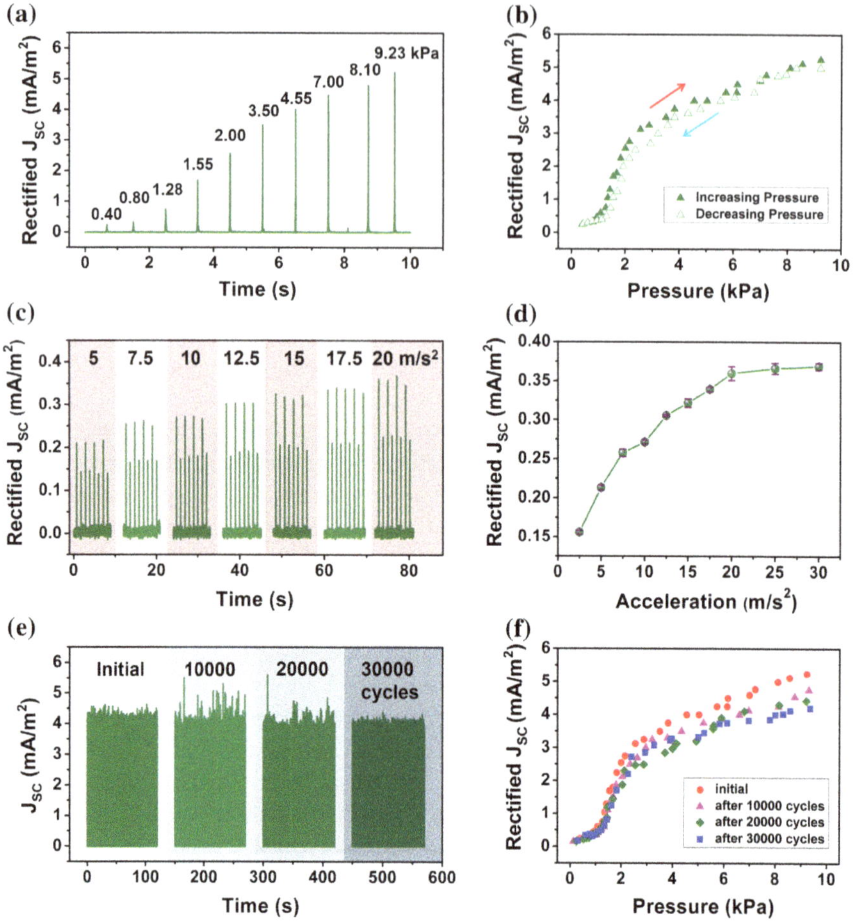


**Fig. 14.3** Dynamic pressure sensing of the TEAS through measurement of the short-circuit current density. **a** The measured rectified pulse-like JSC with variable pressures applied on the TEAS device. **b** The summarized data of the rectified JSC with variable pressures applied on the TEAS device and the reversibility test by increasing the pressure in the first test cycle and decreasing the pressure in the second test cycle. **c** The measured rectified JSC with a constant pressure (0.75 kPa) applied on the TEAS device but with variable accelerations. **d** The summarized data of the rectified JSC with a constant pressure but with variable accelerations ranging from 2.5 to 30 m/s$^2$. **e** The stability test for dynamic pressure sensing of the TEAS with continuous loading and unloading for 30,000 cycles. The JSC was measured after each 10,000 cycles, and 300 cycles of data were recorded and presented in the plot. **f** The reliability test of the dynamic pressure sensing. The measurement of the rectified JSC with variable pressures was carried out after each 10,000 cycles and compiled in the same plot. Reproduced with permission from American Chemical Society [5]

$$p_0 = \frac{p}{J_{sc}/J_{sc,noise}} = \frac{0.125}{0.15/0.0025}\,\text{kPa} = 2.1\,\text{Pa} \qquad (14.6)$$

the low-end detection limit of the TEAS is merely 2.1 Pa, elaborating the high sensitivity of the nanogenerator as an active pressure sensor. Above all, it could be clearly found that the high sensitivity range of the TEAS is confined within a low pressure region (<5 kPa), as both the voltage and current signal tends to get saturated when the pressure reaches a higher level. This performance might limit the practical applications of the TEAS, and the extension of higher detection range is highly desirable. In this regard, we designed a new type of TEAS with the original device entangled by a plastic bracket with elastic springs to enforce the whole structure. Based on this newly fabricated device, the linear and high sensitivity region has been expanded to ~40 kPa.

### 14.1.2 Single-Electrode Mode Self-powered Touch Sensor

In this work using single-electrode working mode, a TEAS is composed of thin-film materials that are vertically laminated (Fig. 14.4a) [6]. A layer of polyethylene terephthalate (PET) forms structural backbone of the TEAS, which is sandwiched by transparent indium tin oxide (ITO) layers as electrodes on both sides. On the top side, a layer of fluorinated ethylene propylene (FEP) is applied as an electrification layer that generates triboelectric charges upon contact with a foreign object. Surface modification on the FEP could be adopted to create vertically aligned polymer nanowires (PNWs) (Fig. 14.4b). They have an average diameter and length of

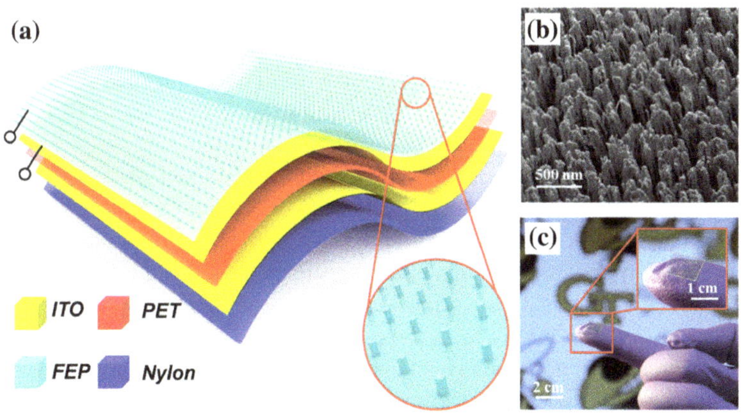


**Fig. 14.4** Structural design of the TEAS. **a** Schematic illustration of the TEAS. *Inset* Enlarged schematic of polymer nanowires on the top surface. **b** SEM image of polymer nanowires created by plasma dry etching. **c** Photograph of an as-fabricated TES. Reproduced with permission from American Chemical Society [6]

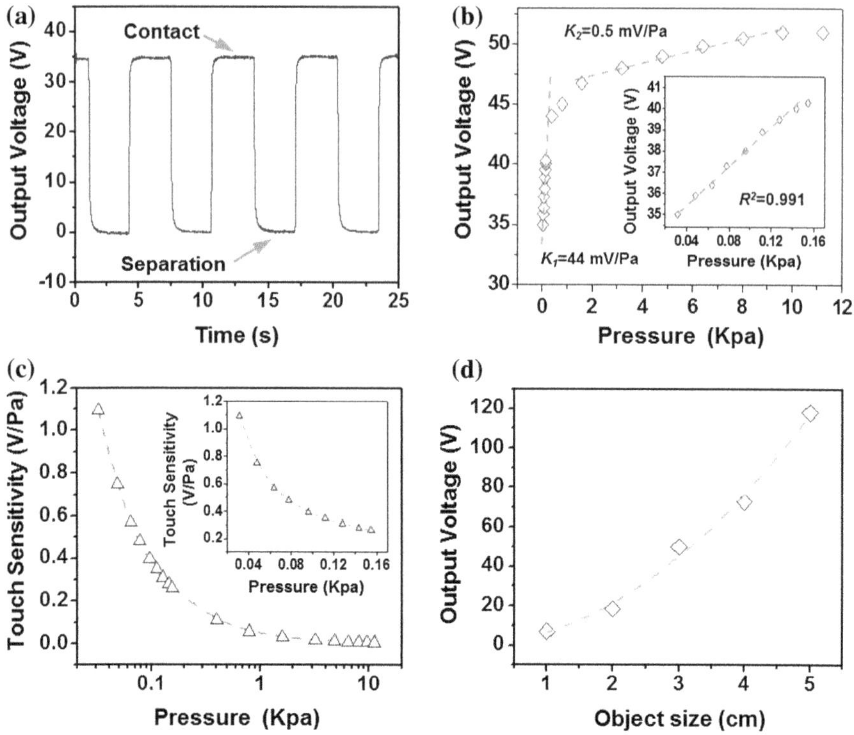


**Fig. 14.5** Characterization results of the TEAS. **a** Output voltage at a contact pressure of 0.03 kPa from a TES (5 cm in side length) when sensing a metal object (2.5 cm in side length). **b** Output voltage with increasing pressure. *Inset* Enlarged view in small-pressure region. **c** Touch sensitivity with increasing pressure. *Inset* Enlarged view in small-pressure region. **d** Output voltage with increasing the size of the metal object at a uniform pressure of 0.03 kPa. The TES has a fixed side length of 5 cm. Reproduced with permission from American Chemical Society [6]

150 nm and 1.5 μm, respectively, which play an important role in achieving high sensitivity for low pressure detection, as will be discussed in later sections. On the bottom side, a nylon film serves as a protection layer due to its outstanding mechanical and thermal properties. A photograph of an as-fabricated TEAS is shown in Fig. 14.4c. Made of conventional thin-film materials, the TEAS has an extremely low cost, which is advantageous for its practicability. The structural design, as well as the fabrication process, is compatible to possible large-scale manufacturing.

To characterize the response of the TEAS to a contact event, we used the output voltage defined as the difference of the open-circuit voltage with respect to that at the separation state. Consequently, the output voltage always has a positive value, which has a zero-baseline corresponding to the separation state. Here, a square-shaped TEAS having an edge length of 5 cm was utilized to detect a metal object with a planar dimension of 2.5 cm by 2.5 cm. Repetitive contact and separation were realized through a linear motor that provided a precisely controlled reciprocating motion. Details on measurement setup as well as sample preparation are presented in Methods. As shown in Fig. 14.5a, at a contact force of 20 mN (applied pressure of 0.03 kPa) the TEAS produces a uniform square-wave output voltage with the maximum amplitude of 35 V.

As the contact force/pressure increases, the output voltage rises and finally saturates at 50 V when the contact pressure approaches 10 kPa (Fig. 14.5b). This increasing behavior can be attributed to the increase of contact area. High pressure results in more contact area, which imparts higher density of surface charges to the TEAS. As a result, higher output voltage is obtained. It is noticed that the curve in Fig. 14.5b exhibits two distinct regions. In the extremely low-pressure region (<0.15 kPa), an exceptional pressure sensitivity of 44 mV/Pa is achieved with excellent linearity ($R^2$ = 0.991), corresponding to a sensitivity factor of 0.09 % $Pa^{-1}$. In the region beyond 2 kPa, the pressure sensitivity drops to 0.5 mV/Pa but still has good linearity ($R^2$ = 0.974). It is suggested that the two-region behavior is likely due to the enhancement effect resulting from the PNWs on the TEAS surface. Without the PNWs, the contact between two nominal flat surfaces is confined at certain points due to the surface asperities. In comparison, the high-aspect-ratio PNWs are likely to be readily bent and become adaptive to the morphology of the sensed object in a weak contact. Such a conformable interaction can result in largely enhanced increment of the real contact area in response to an increase of the pressure because it requires much less pressure to deform the PNWs than the bulk FEP film. As a result, the device with the PNWs exhibits a much more sensitive response in the low-pressure region (<0.15 kPa), as confirmed by a 7-fold enhancement in the pressure sensitivity for ultra-low pressure detection. It needs to be noted that a proper length of the PNWs is important for the sensitivity enhancement. On one hand, the enhancement was not observed for short PNWs due to possibly ineffective interaction with the contact object. On the other hand, excessively long PNWs (>3 μm) tend to fall down after a number of contacts, which reveals poor mechanical robustness.

To further characterize the ability of the TEAS in responding to a contact event, touch sensitivity is calculated and plotted in Fig. 14.5c. The touch sensitivity reaches an ultra-high value of 1.1 V/Pa, corresponding to a factor of 2.3 % $Pa^{-1}$. It indicates the superior applicability of the TEAS in detecting a touch event especially when the interaction is week. It is found that the PNW-based modification can also largely promote the touch sensitivity by as much as 150 % compared to the device without the modification. As discussed above, the PNWs can possibly increase the number of contact points by accommodating the surface asperities by the comfortable interaction. Compared to recent reports on plastic thin-film pressure sensors, our device demonstrates unprecedented sensitivity in extremely low-pressure region (<0.15 kPa) where interaction is at least two orders of magnitude weaker than a gentle finger touch.

The size is a major factor that determines the output voltage as well as the sensitivity of the TEAS. For a TEAS with a fixed size (5 cm in side length), a quadratic increase in output voltage is obtained with respect to side length of the sensed object that varies from 1 to 5 cm (Fig. 14.5d). Alternatively, the output voltage linearly scales with the object area. The reason is that a larger object imparts more triboelectric charges to the TEAS, which induces a higher potential in magnitude on the upper electrode given that the electrode has a fixed area. The proposed explanation is further supported by results that the output voltage is reversely proportional to the device area while the sensed object has a fixed size. If both the TEAS and the object scale altogether, the output voltage keeps stable. Therefore, an important guidance in achieving high sensitivity is that the TEAS needs to have a lateral dimension smaller than an object that is to be detected.

The large voltage signal induced by a contact event can be easily picked up by external readout circuits. For practical applications, a complete wireless sensing system was developed through integrating the TEAS with a signal-processing circuit. The system relies on the output voltage from the TEAS to trigger an IC timer that controls a wireless transmitter for remotely switching a siren between a panic state and a silence state. In the first demonstration shown in Fig. 14.6a, a TEAS was customized to a dimension of 1 cm by 1 cm. Once a human finger gently contacted with the TEAS, the output voltage (Fig. 14.6b) touched off the siren that produced a sharp alarm with flashing light (Fig. 14.6b). In the second demonstration, a square-shaped TEAS with side length of 10 cm was laid on the ground and imbedded underneath a piece of carpet (Fig. 14.6c). When a person stepped on top of the TEAS as walking by, an output voltage of 15 V shown in Fig. 14.6d was generated even though the shoe did not have an intimate contact with the TEAS. This was because the TEAS was loosely covered by the carpet, leaving a spatial gap between them. When the carpet was pressed down by the human foot, it fully touched the TEAS. Such a variance in contact area would give to an output voltage from the TEAS, which then triggered the siren (Fig. 14.6c). Moreover, the sensing system could be applied to other circumstances, such as a door handle (Fig. 14.6e). For this particular scenario, the TEAS was tailored to a long strip (5 cm by 3 mm) and fixed on the curved surface of the handle, exhibiting its ability in adapting to different shapes for flexible implementation. As demonstrated in Fig. 14.6e, the sensing system immediately started operation once a human hand grabbed the door handle. Through a minor revision that substitutes other functional electronics for the wireless transmitter in the signal-processing circuit, the sensing system could be adopted for more purposes. For example, it was demonstrated as a touch-enabled switch for a panel light, which is shown in Fig. 14.6f. Therefore, the TEAS along with the sensing system developed here has immediate applications in a variety of areas, including human-machine interface, automatic control, surveillance, remote operation, and security systems.

### 14.1.3 Dual-Mode Ultrasensitive Self-powered Pressure Sensor

In the above sections, we have discussed TENG-based self-powered pressure sensor. Most recently, this type of self-powered pressure sensor could also be integrated with conventional resistive pressure sensor into a hybrid device, which

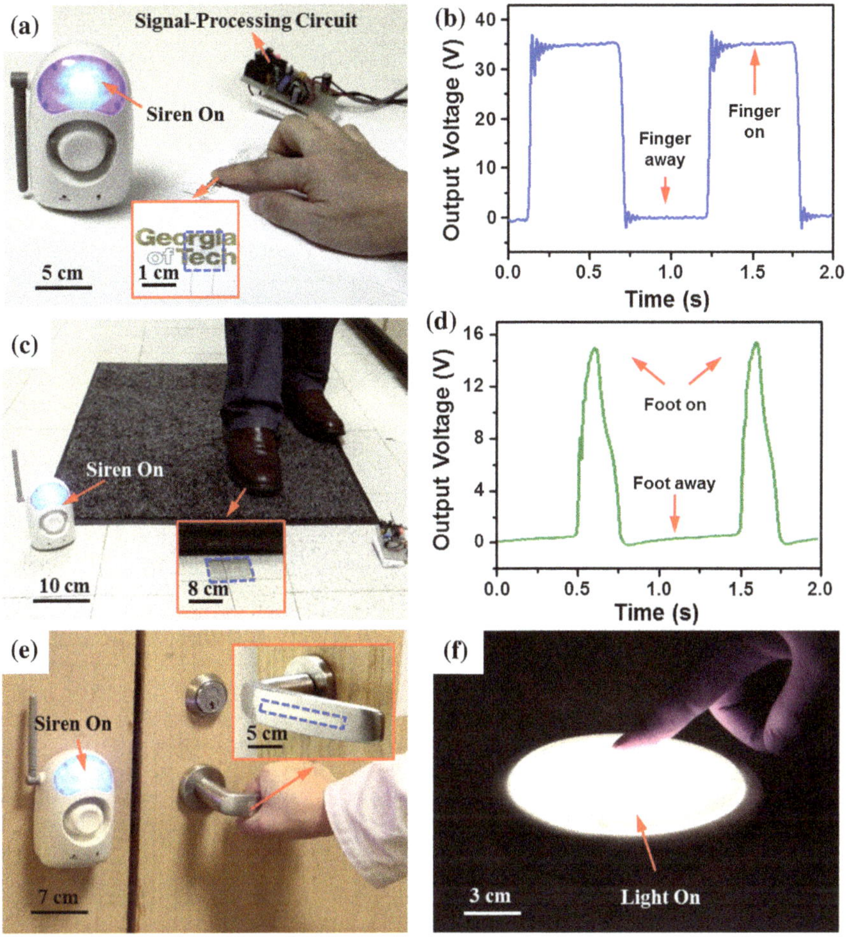


**Fig. 14.6** Complete sensing system based on the TEAS for practical applications. **a** Triggering a wireless alarm system by gentle finger tapping on a TEAS (1 cm in side length). *Inset* Enlarged view of the TES on a piece of paper. **b** Output voltage of the TEAS generated by finger tapping. **c** Triggering a wireless alarm system by stepping on *top* of a TEAS (10 cm in side length) that is embedded underneath a carpet. *Inset* Enlarged view of the TEAS on the floor. **d** Output voltage generated by stepping on the TEAS covered by the carpet. **e** Triggering a wireless alarm system by grabbing the TEAS (5 cm by 3 mm) that is applied on a door handle. *Inset* Enlarged view of the TEAS on the door handle. **f** Switching a panel light by finger tapping on the TEAS (1 cm in side length) that is applied on the surface of the light. Reproduced with permission from American Chemical Society [6]

incorporates both self-powered operation of TENG-based pressure sensor and the ultrahigh sensitivity exhibited by resistive pressure sensor [9]. The concept of this integrated self-powered sensing system is illustrated in Fig. 14.7a. First, the TENG can be used as self-powered active sensors. In the meantime, it harvests different kinds of energies from the environment and stores them in the energy storage unit.

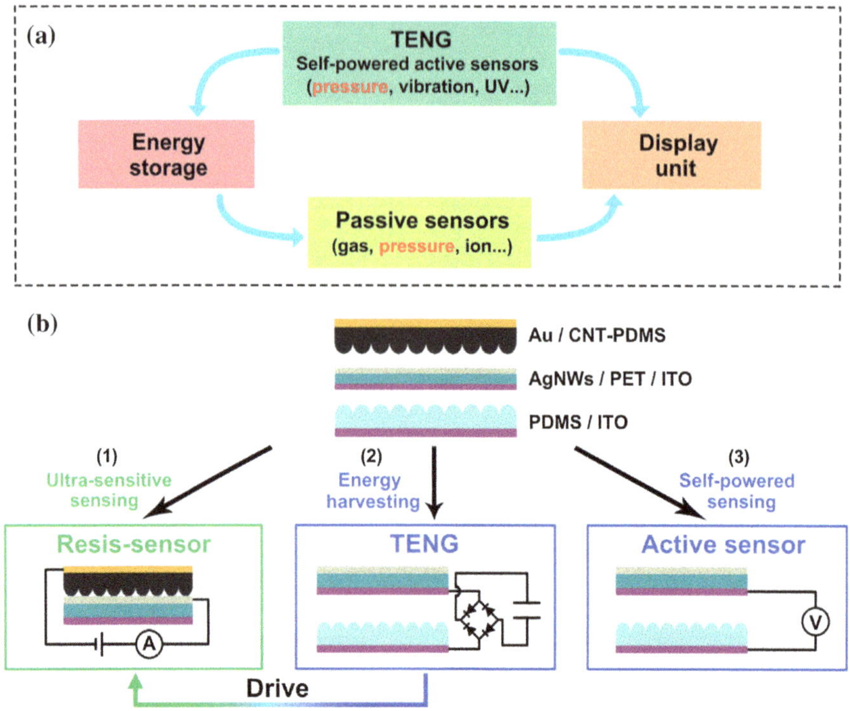


**Fig. 14.7** Schematic diagram of the integrated self-powered sensing system. **a** The integrated self-powered sensing system composed of four parts: TENG, energy storage unit, sensors and display unit. **b** A typical case of self-powered pressure sensing system through the combination of TENG and passive resistive pressure sensor together. Reproduced with permission from Elsevier [9]

Then the collected energy is used to drive the other passive sensor. These two kinds of sensors can work simultaneously and complementarily. The magnitude of detection parameters will be revealed in the display unit. Figure 14.7b shows a typical example of the self-powered sensing system through the integration of active triboelectric pressure sensor and passive resistive pressure sensor. The basic working principle of this system is composed of three modes: (1) the upper part is the ultrasensitive resistive pressure sensor (resis-sensor) with low operating voltage, which is suitable for continuous monitoring of pressures. But it can work only under the drive of a power source. (2) The lower part is a high-performance TENG for harvesting the environmental energy, and this generating power can be used to drive the resis-sensor. (3) The TENG can be used as a self-powered active pressure sensor, which is suitable to detect the pulsing pressure. Note that the silver nanowires (AgNWs) and ITO electrodes in the middle layer are connected with each other, constructing an integrated electrode, and these two parts share the middle electrode together, which effectively improves the operating facility of the device. With such a sandwich structure, the device can choose to work in proper mode, depending on different requirement.

The sensing mechanism of the resis-sensor is due to pressure-dependent change in effective contact area between wrinkled carbon nanotube (CNT)-PDMS and 3D structured AgNWs thin films. Compared to a bulk rigid planar mental, 3D structured AgNWs thin film has large specific surface area and excellent conductivity. Furthermore, the elasticity of PDMS facilitates the contact between these two films, imparting ultra-high sensitivity and reproducible sensing characteristics to our sensor. When applying an external pressure, the pressure-induced deformation increased the contact area of the wrinkled CNT-PDMS and 3D AgNWs thin films. This would cause the sensor to experience a decrease in resistance, leading to an increase in current in a constant voltage (0.1 V). When the external pressure was removed, the two films recovered to their original shapes. This leads to a decrease of contact area between the two films, therefore, causing a decrease of the current. Note that the sensor operating voltage of 0.1 V is sufficiently low to allow for powering by TENGs, thus enabling the integration of self-powered sensing system.

To measure the response of the resis-sensor, a home-made system containing a computer-controlled stepping motor and a force sensor has been designed. The resis-sensor was fixed on a flat plate under the force gauge, and cyclic external pressures were applied by a stepping motor. With such a system, external pressures and electrical signals can be recorded simultaneously. Figure 14.8a is a real-time measurement result of the relative change of current ($\Delta I/I_{off}$) under series of different pressures, and the relationship between $\Delta I/I_{off}$ and magnitude of pressure is plotted in Fig. 14.8b. When the applied pressure increased from 107 to 4500 Pa on the sensor, the measured current dramatically increased along with the increase of contact area between the CNT-PDMS wrinkles and AgNWs. The sensitivity S can be defined as:

$$S = \frac{\delta\left(\frac{\Delta I}{I_{off}}\right)}{\delta p} \tag{14.7}$$

$$\Delta I = I - I_{off} \tag{14.8}$$

where I is the current when applied pressure on the sensor, and $I_{off}$ is the current of sensor with no pressure, p is the applied pressure. Based on the difference in sensitivity, the plot can be divided into two regions. When the applied pressure was lowered than 800 Pa, the sensitivity (S) is 204.4 $kPa^{-1}$. To the best of our knowledge, this S in low-pressure regime (<800 Pa) is the highest reported for a flexible pressure sensor, which is significantly higher than those reported in previous literatures. Figure 14.8c shows the response of the sensor upon loading-unloading the leaf (20 mg) and petal (50 mg), indicating that the sensor is reliable in detecting minute pressure. The corresponding pressure of the leaf and petal were 0.2 and 0.5 Pa, respectively. Compared to other present flexible pressure sensors reported in the previous literatures, our resis-sensor shows a very low detection limit (0.2 Pa).

To further demonstrate the stability of our resis-sensor, a continuous pressure of 250 Pa at a frequency of 1 Hz was applied to the sensor, as shown in Fig. 14.8d.

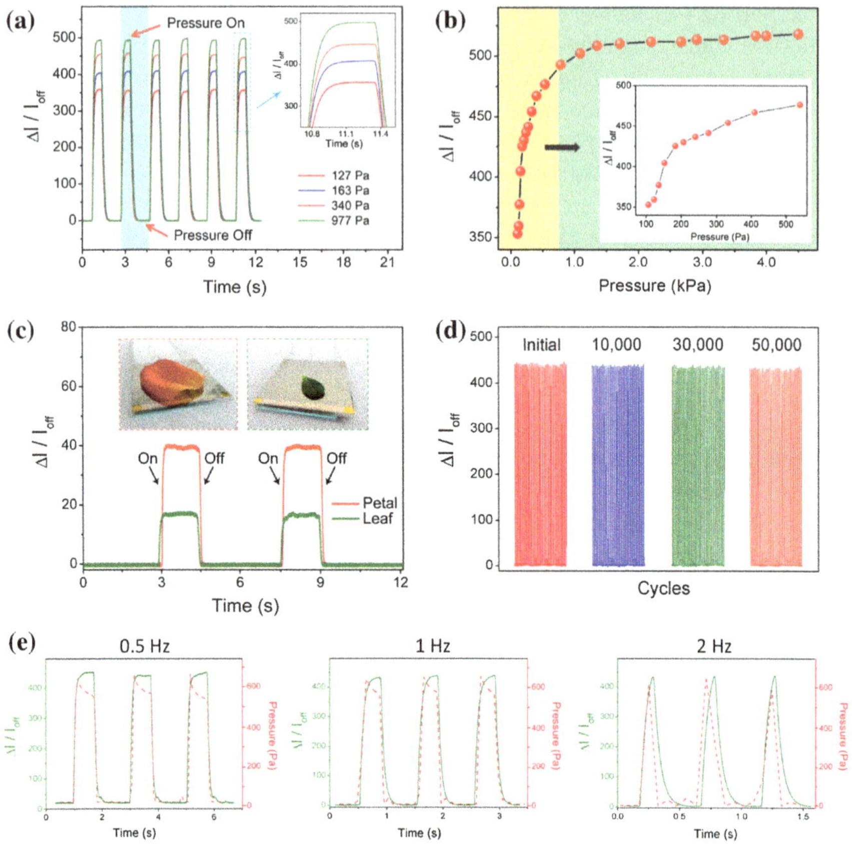


**Fig. 14.8** Electromechanical response of the resis-sensor. **a** The real-time measurement of the relative change of current with cyclic and variable pressures applied on the resis-sensor. The *highlighted region* indicates one cycle of the pressure loading and unloading process. **b** The summarized relationship and linear fitting between the relative variations of current and the pressure applied on the resis-sensor. **c** The resis-sensor is able to sense the application of very small pressure. Shown is the relative change of current on placing and removing a leaf (20 mg) and a petal (50 mg), respectively. **d** The stability test for pressure sensing of the resis-sensor with continuous loading and unloading a cyclic pressure (250 Pa) for 50,000 cycles, at a frequency of 1 Hz. **e** Plots showing frequency responses at the pressure of 250 Pa: pressure input frequency of 0.5, 1, 2 Hz. Reproduced with permission from Elsevier [9]

Note that the high signal-to-noise ratios (SNR) were well maintained and the current amplitude exhibited negligible changes after 50,000 loading-unloading cycles. The resis-sensor exhibits rapid response time of less than 0.07 s. The time-resolved response measurement shows that the current waves and the input pressure waves match well under a pressure of 250 Pa at the frequency of 0.5–1 Hz (Fig. 14.8e). When the frequency is up to 2 Hz, a tiny hysteresis can be observed. This might be attributed to the viscoelastic effects of the PDMS films.

As above, the TENG can be used as an active pressure sensor, and its sensing mechanism is based on coupling effect of contact electrification and electrostatic induction. When an external pressure is applied, the two triboelectric layers are brought into contact with each other. The open-circuit voltage ($V_{OC}$) will decrease linearly according to Eq. (14.1). Reversely, the $V_{OC}$ reverts to its maximum level when the pressure is fully removed. Considering the mechanism of this active sensor, it is suitable to detect the pulsing pressure.

Figure 14.9a is the measurement setup for the TENG as an active pressure sensor. In comparison with the resis-sensor, power supply is removed and only volt meter is needed due to the self-powered feature of the TENG. The real-time measurement result of voltage under various pressures is shown in Fig. 14.9b. When there is no pressure, the $V_{OC}$ stays at the maximum level ($V_{off}$ = 150 V). Once the external pressure is applied, the $V_{OC}$ decreases to a lower level. Figure 14.9c is the relationship between the relative variations of voltage ($\Delta V/V_{off}$) and the size of pressure ranging from 67 to 2556 Pa. Inset shows the favorable linear relationship of the sensor when the pressure is lower than 400 Pa, which reveals our sensor is reliable for pressure sensing. The sensitivity is defined by:

$$S = \frac{\delta\left(\frac{\Delta V}{V_{off}}\right)}{\delta p} \tag{14.9}$$

$$\Delta V = V_{off} - V \tag{14.10}$$

where $V_{off}$ is the voltage of sensor with no pressure, and V is the voltage when applied pressure on the sensor, p is the applied pressure. It can be found that the sensitivity of the sensor is 1.52 $kPa^{-1}$ in the low pressure region, which is much higher than that of high pressure region (0.11 $kPa^{-1}$). It is noteworthy that this sensitivity is apparently higher than other TENG-based pressure sensors reported previously [13–16]. This is because the electrical output performance of the TENG is greatly improved, using the wrinkled and flexible PDMS.

Furthermore, as shown in Fig. 14.9d, the sensor exhibits excellent mechanical robustness and stability under a pressure of 300 Pa at a frequency of 1 Hz, and it can still work properly after 50,000 cycles, with a high SNR. The response time of this sensor is less than 0.09 s. To examine the delay time of our sensor to external forces, the output voltage signals were compared with the pressure inputs at a frequency of 0.5–2 Hz (Fig. 14.9e). It can be found that the voltage waves were almost the same as the input pressure waves.

## 14.2 Self-powered Tactile Imaging

To achieve the function of tactile imaging or measuring the pressure distribution over a certain area, an array of multiple TEAS units is required to provide real-time and localized pressure information individually. In a typical work using vertical

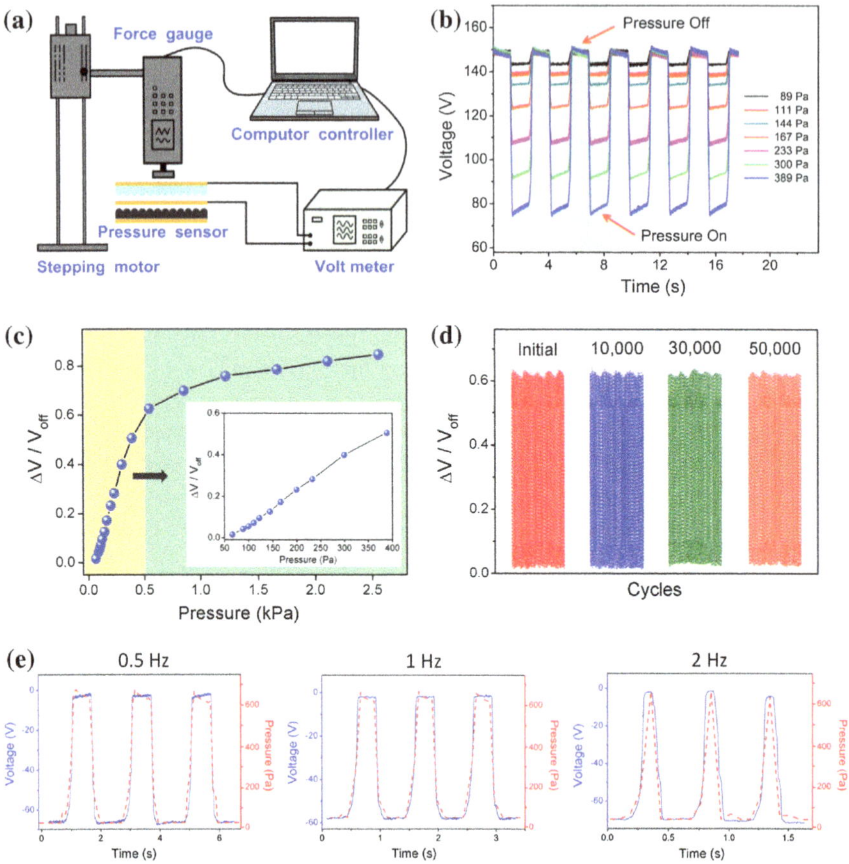


**Fig. 14.9** Electromechanical response of the TENG as an active pressure sensor. **a** Schematic illustration of the experimental setup of the active sensor. **b** The real-time measurement of the Voltage with cyclic and variable pressures applied on the active sensor. The *highlighted region* indicates one cycle of the pressure loading and unloading process. **c** The summarized relationship and linear fitting between the relative variations of voltage and the pressure applied on the active sensor. **d** The stability test for pressure sensing of the active sensor with continuous loading and unloading a cyclic pressure (300 Pa) for 50,000 cycles, at a frequency of 1 Hz. **e** Plots showing frequency responses at the pressure of 300 Pa: pressure input frequency of 0.5, 1, 2 Hz. Reproduced with permission from Elsevier [9]

contact-separation mode TEAS, the corresponding array device was fabricated on a flat plate with deposited aluminum thin film as a common electrode, as schematically illustrated in Fig. 14.10a [5]. The unit of the array device has the same size as the single TEAS device mentioned in Sect. 14.1.1. Each pixel was marked the letter/number coordinates from “A1” to “F6”. The response of the TEAS array with local pressure was measured through a multi-channel measurement system (National Instruments). The aluminum electrodes of all the TEAS units were connected to a common electrode on the plate with conductive paste, while the upper Au electrode of each unit was connected to the probe of each measurement channel. The measurement instrument is a voltage meter with an internal resistance of $\sim 10$ MΩ, and a corresponding voltage peak will be recorded as response of a local pressure on the corresponding TEAS unit in the matrix.

Based on this working principle, the tactile imaging capability of the TEAS matrix were demonstrated by loading pressure through pre-designed plastic architecture with the calligraphy of the letters "TENG", respectively. Before applying the pressure, the voltage outputs from all the pixels of the TEAS matrix were at the background level, as displayed in Fig. 14.10b. Figure 14.10c–f shows the two-dimensional contour plotting of the peak value of the voltage responses that were measured when external pressures were applied through each architecture, respectively. The highlighted color represents the area under pressing through each letter, as outlined by the white dash lines. These plots elaborate the spatial resolution of the TEAS matrix for distinguishably mapping the calligraphy of the applied pressure, and its potential applications such as personal signature recognition. In addition, to gain a more intuitive understanding of the "self-powered" pressure mapping functionality of the TEAS matrix, each nine units of the same array device were connected in parallel to power up a serially-connected array of LEDs showing "T", "E", "N", "G" characters, respectively. This demonstration proves that the TEAS matrix can work as an external power source and a sensor array simultaneously for a truly stand-alone self-powered system.

In another work by Yang et al., a single-electrode TENG can be utilized as a self-powered tactile sensor for monitoring the local touching actions of human fingers or hands [8]. Then, it was integrated as a $4 \times 4$ matrix of ITO electrode array, where each electrode with the size of 1 cm $\times$ 1 cm was connected with the ground through a load resistor of 10 MΩ. The fabricated matrix was completely covered by a layer of PDMS film as the sensitive unit to be touched. The device exhibits both the transparent and flexible features. When the PDMS film was touched by a human finger with the representative pressure of lower than 10 kPa, the corresponding output voltage signals were observed. The magnitude of the output voltage is proportional to the applied pressures in the experiments, where the corresponding pressure was measured by a commercial pressure sensor based on the piezoresistive effect.

Similarly, the output voltage signals of 16 devices were recorded in real-time as a mapping figure. By addressing and monitoring the positive output voltage signals in 16 channels of the tactile sensor system, the touching information of a human finger including the position and pressure can be attained by analysis of the measured mapping figures. Figure 14.11a presents a photograph of the device when the sixth and eleventh devices in the matrix were simultaneously touched. Two almost same pressures of about 4.9 kPa can be confirmed by the analysis of the obtained same output voltage signals in the corresponding mapping image, as shown in Fig. 14.11b. Figure 14.11c displays the response of the device to the localized pressure induced by two fingers when the matrix was rotated by 90°. The obtained mapping figure clearly reveals that the seventh and tenth devices were touched, where the produced pressures of about 7.0 kPa are larger than those in Fig. 14.11b. When the pressures were applied along the diagonal line of the matrix by using the

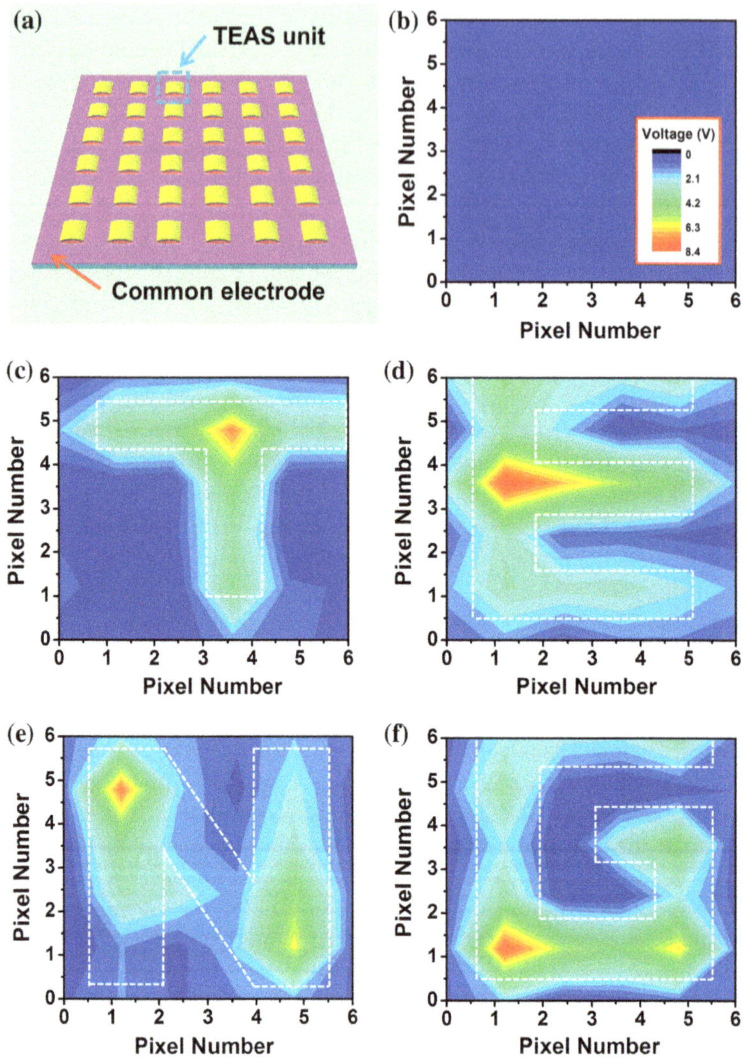


**Fig. 14.10** Self-powered pressure mapping based on the TEAS array. **a** A schematic illustration of the TEAS array device. **b** The background signal with no pressure applied on the TEAS matrix. The *inset* is color scaling of voltage for all the measurements in this figure. **c–f** The two-dimensional voltage contour plot from the multichannel measurement of the TEAS matrix with an external pressure uniformly and locally applied onto the device through architectures with calligraphy of "T", "E", "N", and "G", respectively. Reproduced with permission from American Chemical Society [5]

side surface of human hand, the distinctive changes in the output voltage signals can be observed, revealing an increase of the pressure from 3.2 to 6.0 kPa along the diagonal line, as illustrated in Fig. 14.11d. When the pressures were applied on all the 16 devices by using the human hand, the recorded mapping figure (Fig. 14.11e)

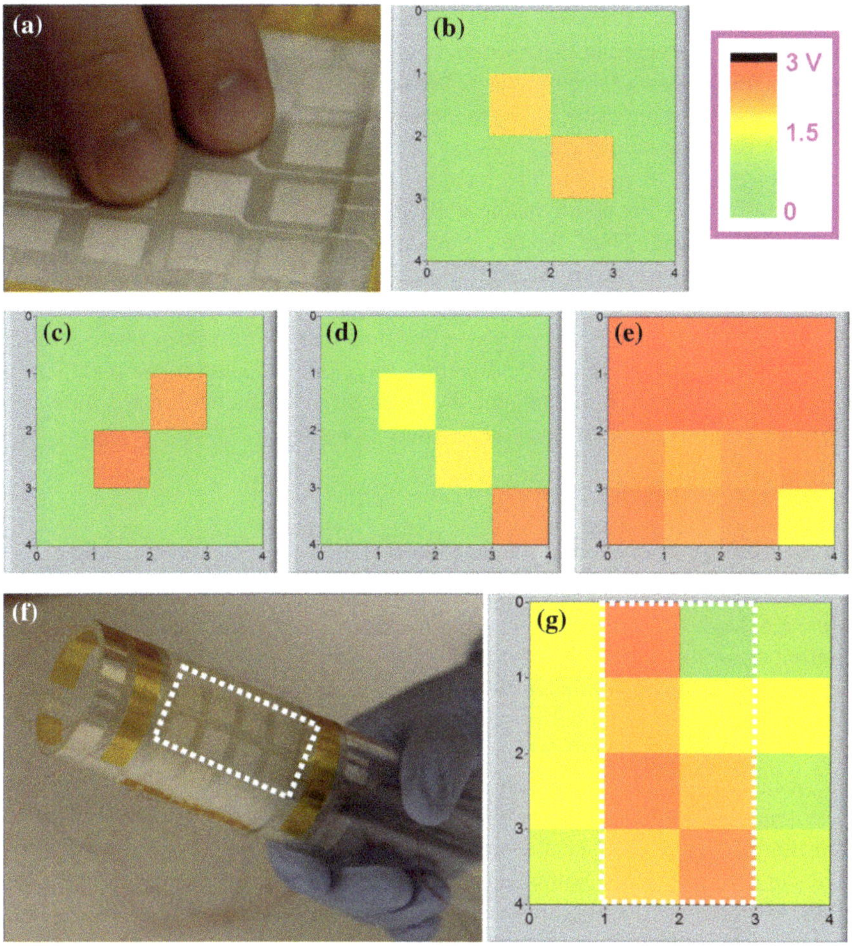


**Fig. 14.11** **a–c** Measured mapping figures under the different pressures. **d** Photograph of the matrix touched by two human fingers. **e** Measured mapping figure when the two devices in the matrix were touched in (**d**). **f–h** Measured mapping figures when two (**f**), three (**g**), and all devices (**h**) in the matrix were touched. **j** Photograph of the fabricated matrix attached to an acrylic tube. **k** Measured mapping figure when the device in **j** was touched by a human hand. Reproduced with permission from American Chemical Society [8]

shows that all the devices are functional and the pressures on the 1–8th devices are obviously larger than those on the 9–16th devices. The real-time detection of the touching actions on the flexible devices is a desirable feature for sensors embedded in robots or prosthetic devices. Figure 14.11f shows a photograph of the flexible matrix attached to a transparent acrylic tube. When the pressures were applied on the tube surface by using the human hand, the largest pressure of about 7.3 kPa occurs at the marked area (the white dashed line in Fig. 14.11f), which is consistent with the measured mapping figure, as shown in Fig. 14.11g.

The active pressure sensor and the integrated sensor array based on the triboelectric effect presented in these works have several advantages over conventional passive pressure sensors. First, the active sensor in this work is capable of both static pressure sensing using the open-circuit voltage and dynamic pressure sensing using the short-circuit current, while conventional sensors are usually incapable of dynamic sensing to provide the loading rate information. Second, the prompt response of both static and dynamic sensing enables the revealing of details about the loading pressure. Third, the detection limit of the TEAS for dynamic sensing is as low as 2.1 Pa, owing to the high output of the TENG. Last but not least, the active sensor array presented in this work has no power consumption and could even be combined with its energy harvesting functionality for self-powered pressure mapping. Future works in this field involve the miniaturization of the pixel size to achieve higher spatial resolution, and the integration of the TEAS matrix onto fully flexible substrate for shape-adaptive pressure imaging.

## 14.3 Self-powered Smart Keyboard

Although keyboard has been used for hundreds of years for advancing human civilization, studying human behavior by keystroke dynamics using smart keyboard remains a great challenge. Here, a self-powered, non-mechanical-punching intelligent keyboard (IKB) enabled by contact electrification between human fingers and keys is introduced, which converts mechanical stimuli applied onto the keyboard into local electronic signals without applying an external power [11]. The key functional element (KFE) of the IKB is composed of vertically-stacked transparent thin film materials. A layer of polyethylene terephthalate (PET) sits between two layers of ITO that are the bottom and the top electrodes. Then, a layer of fluorinated ethylene propylene (FEP) is applied onto the ITO surface as an electrification layer that generates triboelectric charges upon contact with human fingers, as schematically shown in Fig. 14.12a. FEP nanowires array was created on the exposed FEP surface by a top-down method through reactive ion etching. A scanning electron microscopy (SEM) image of vertically aligned FEP nanowires is displayed in Fig. 14.12b, which indicates that the average clustering diameter of FEP nanowires is 104 $\pm$ 21 nm with an average length of 0.8 $\pm$ 0.2 mm. The static contact angle of the FEP surface was measured by a sessile droplet method with a 2 ml water droplet. Inset of Fig. 14.12b is the contour of the resting droplet, which indicates a contact angle of 160° by Young-Laplace fitting. Meanwhile, a tilting base method was employed to investigate the surface sliding angle of the IKB. And a sliding angle of 7.2° $\pm$ 0.5° was experimentally observed. The super-hydrophobicity of the IKB surface assures its self-cleaning property, which can effectively keep the keyboard free of dirt and grime. A photograph of an as-fabricated KFE is shown in Fig. 14.12c, d, plainly demonstrating its flexibility, transparency and one-piece structure without mechanical punching. Figure 14.12e is a photograph of an IKB after assembling the KFE with a commercial keyboard.

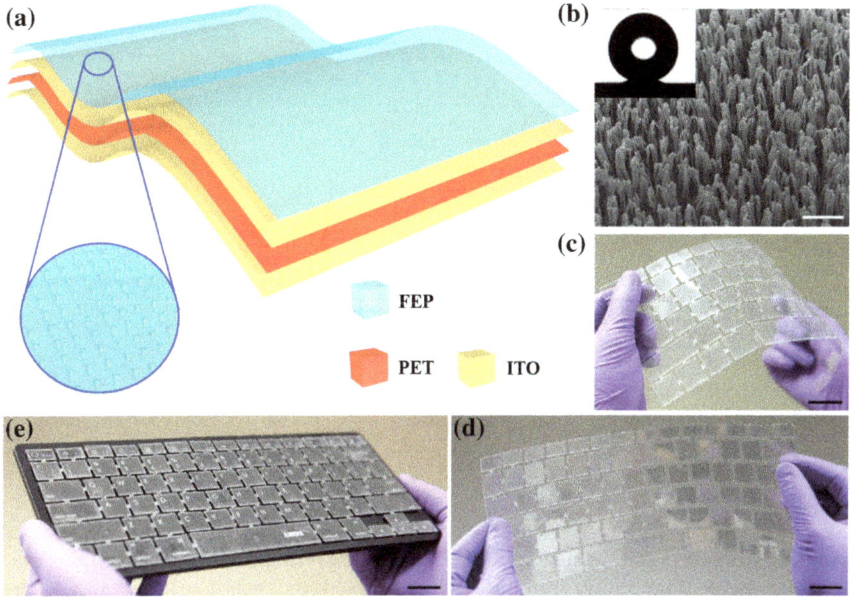


**Fig. 14.12** Structural design of the KFE of the intelligent keyboard. **a** Schematic illustrations of the KFE. *Inset* Enlarged schematic of FEP nanowires on the top surface. It is noted that these drawings do not scale. **b** SEM image of FEP nanowires. Inset is the contour of resting droplet for surface static contact angle measurement. The scale bar is 500 nm. **c**, **d** Photograph of a flexible and transparent KFE. The scale bars are 3 cm. **e** Photograph of an as-fabricated IKB, which is system integration of KFE and a commercial keyboard. The scale bars are 3 cm. Reproduced with permission from American Chemical Society [11]

Firstly, the keyboard was integrated with a signal-processing circuit to develop a complete wireless warning system. This customized circuit for the IKB holds two unique advantages. First, a designed trigger voltage threshold of 5 V enables the IKB to work stably even in a high electrical noise environment. Second, the alarm time is controllable and tunable by the signal processing circuit. Once a finger gently strokes a key, the generated voltage signal triggers an IC timer that controls a wireless transmitter and remotely switches a siren between a panic state and a silence state. The keys in the keyboards are categorized into seven regions according to their dimensions to compare their output voltage. The obtained results are shown in Fig. 14.13a. An increasing function between key dimensions and the output voltage is observed. The reason is that contact electrification is a surface charging effect, a larger amount of triboelectric charges will contribute to a higher voltage output. Figure 14.13b is an enlarged view of the output voltage when the smallest key is gently touched. Even this lowest voltage was capable of triggering the siren that produced a sharp alarm with flashing light (Fig. 14.13c).

Furthermore, a customized multi-channel data acquisition system was designed for the IKB to individually address the electric signal from each key (Fig. 14.13d).

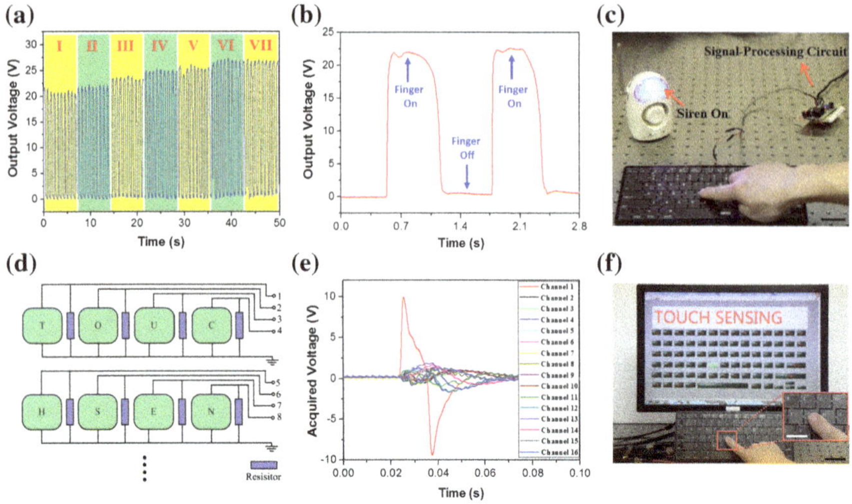


**Fig. 14.13** Intelligent keyboard as a self-securing system. **a** Measured output voltage is an increasing function of the key dimensions when a gentle keystroke was applied. **b** An enlarged view of obtained output voltage when the smallest key is gently touched. **c** Triggering a wireless alarm system by gentle finger tapping on the IKB. The scale bar is 5 cm. **d** Schematic diagram of keyboard based multi-channel data acquisition system for keystroke tracing and real-time recording. **e** The system acquired output voltage signals when the key "T" was stroked. **f** An photograph demonstrated the IKB for the real-time keystroke tracing and recording. A continuously typing string "TOUCH SENSING" was recorded in real time without uncomfortable delay. The scale bar is 5 cm. *Inset* An enlarged view of the key "G" was stroked. The scale bar is 2 cm. Reproduced with permission from American Chemical Society [11]

Consequently, the real-time tracing and recording during typing can be realized. Every channel was electrically but independently connected to a key in the keyboard as a functional unit, and each unit was connected to the ground through a 1 MW resistor. When a keystroke was initiated on a particular key, "T" for example, a peak output voltage up to 10 V was acquired from the corresponding channel, while signals from other keys were less than 2.5 V (Fig. 14.13e). Theoretically, the signals from untouched keys should be 0. However, a certain voltage signals were observed experimentally, which is mainly due to the coupling among all of the channels in the data acquisition system.

For recognition purpose, the maximum peak value of the output voltage from channel $i$, $V_{\mathrm{pi}}$, was extracted. Then the threshold voltage $V_{\mathrm{th}}$ can be analytically expressed by the following equation using Pauta Criterion Method.

$$V_{\mathrm{th}} = \frac{1}{n}\sum_{\mathrm{i}=1}^{n} V_{\mathrm{pi}} + \frac{3}{\sqrt{n}} \tag{14.11}$$

where $n$ is the total number of channels, and $i$ is integral from 1 to n. Once the extracted $V_{\mathrm{pi}}$ is higher than $V_{\mathrm{th}}$, the key corresponding to this particular channel is considered to be stroked. Figure 14.13f demonstrates the real-time keystroke tracing and recording. When a phrase "TOUCH SENSING" was continuously typed, it was simultaneously recorded and displayed without noticeable delay.

Typing patterns based on dimensional keystroke timing vector, lacking of uniqueness and permanence, largely hinder the practicality and acceptability of behavioral biometric of keystroke dynamics as effective identifier to current authentication system. The IKB also provides a superior route in creating accurate, unique, and permanent typing patterns for verification and identification purposes. Three sets of original typing patterns (in red curves) were respectively obtained for three typists Tom, Mike, and Alex, who independently typed the word "touch" for more than four times on the IKB in each accustomed manner. Each set of typing pattern corresponds to two subsets of characteristic signals (voltage and current), which are time-series data. These electric signals correlate to a variety of information, including the manner and rhythm of the keystroke, typing habit, finger size, individual bioelectricity, and applied typing force. Thus, they can not only characterize the keystroke timing, but also quantitatively record the concrete dynamic changes in the course of typing. Figure 14.14a, b represent a set of the typing pattern of Tom. For quantitative differentiation, Discreet Fourier Transformation (DFT) was performed to obtain frequency features of the electric signals. Wavelet transformation was carried out to simultaneously obtain both time domain and frequency domain features of the typing patterns. Based on DB4, the typing patterns in the form of electric signals can be expressed by the following formula *via* multiresolution analysis

$$f(\mathrm{t}) = S_4(t) + D_4(t) + D_3(t) + D_2(t) + D_1(t) \tag{14.12}$$

and

$$S_4 = \sum_{\mathrm{k}} s_{4,\mathrm{k}}\varphi_{4,\mathrm{k}}(t) \tag{14.13}$$

$$D_j = \sum_{\mathrm{k}} d_{\mathrm{j,k}}\psi_{\mathrm{j,k}}(t) \tag{14.14}$$

where $s_{4,k}$ and $d_{\mathrm{j,k}}$ are the wavelet coefficients, j = 1, 2, 3, 4… k is the number of translations of the wavelet for any given scale. $\varphi_{4,k}(t)$ and $\psi_{\mathrm{j,k}}(t)$ are respectively the father wavelets and mother wavelets. *f*(t) is either voltage or current signal of the typing patterns. $S_4$ is smooth. Wavelet details $D_4$, $D_3$, $D_2$ and $D_1$ represent a set of voltage or current components that provide representations of the original signals at different resolution levels.

On the basis of DB4 results, the corresponding higher resolution wavelet components of three individuals are significantly different from each other. Still, Pearson correlation coefficient was utilized to quantitatively measure the correlation between the wavelet components by the following equation

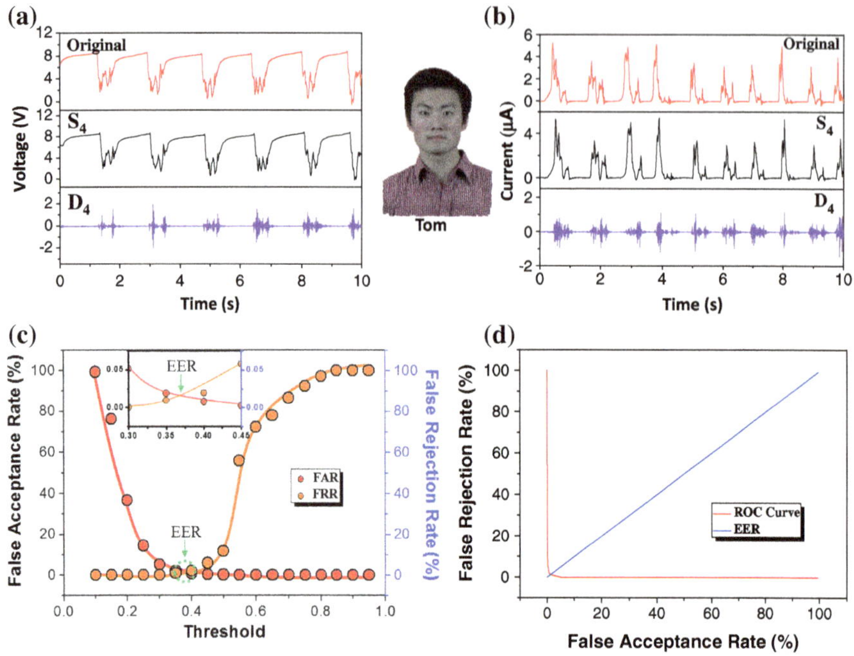


**Fig. 14.14** Personality identified keystroke dynamics using intelligent keyboard. Typing patterns **a** voltage and **b** current obtained when Tom was continuously typing the word "touch" more than four times into the computer via the IKB. $S_4$ and $D_4$ are the corresponding wavelet components after DB4 transformation. **c** The variation of FAR and FRR in relation to the threshold for the biometric authentication system using triboelectrification enabled keystroke dynamics. Inset is an enlarged view of the EER point, which indicates a remarkably low EER value of 1.34 % at the threshold of 0.39. **d** Receiver operating characteristic (ROC) curve of the biometric authentication system using triboelectrification enabled keystroke dynamics. The false rejection curve is plotted as a function of the false acceptance curve. Reproduced with permission from American Chemical Society [11]

$$\gamma = \frac{\sum_{\mathrm{i}=1}^{n}(t_{\mathrm{i}} - \bar{t})(x_{\mathrm{i}} - \bar{x})}{\sqrt{\sum_{\mathrm{i}=1}^{n}(t_{\mathrm{i}} - \bar{t})^2}\sqrt{\sum_{\mathrm{i}=1}^{n}(x_{\mathrm{i}} - \bar{x})^2}} \tag{14.15}$$

where $n$ is the length of compared date sequence, and $x$ is either the voltage or the current sequences after DB4. All of the Pearson correlation coefficients of the voltage or current components from each individual are larger than 0.85, suggesting superior permanence of the typing patterns, while other Pearson correlation coefficients are less than 0.45 that is indicative of excellent uniqueness among individuals. Therefore, the Pearson correlation coefficient can be harnessed to express the uniqueness and permanence among individual typing patterns. In the meanwhile, the correlation coefficient can also act as a threshold to separate the genuine typing patterns from the impostor typing patterns, if a verification system is established using behavioral biometric of keystroke dynamics.

To evaluate the performance of the triboelectrification enabled behavioral biometric of keystroke dynamics as an effective identifier, 104 participants were invited to independently type the word "touch" for more than four times on the IKB in each accustomed manner. All the participants are the randomly and diversely selected with an age span from 14 to 69 years old, male and female people from different countries.

Subsequently, a collection of 104 individual typing patterns were divided into 52 client patterns and 52 impostor patterns. In the whole testing set of 104 typing patterns, user template is composed of the 52 client patterns. And the performance of the authentication biometrics is characterized through two error rates: False Rejection Rate (FRR) and False Acceptance Rate (FAR). The FRR and FAR are variables depending on the classification threshold. By tuning the threshold, the FRR and FAR can typically be traded off against each other to achieve the preferable classification result. In this regard, the Pearson correlation coefficient was selected as the classification threshold to evaluate the behavioral biometric authentication system using triboelectrification enabled keystroke dynamics. The variation of FAR and FRR in relation to the threshold is shown in Fig. 14.14c. The FRR is increasing with the elevated threshold, while FAR follows a reverse trend. However, the FRR and FAR intersect at a certain threshold value, which indicates the EER point. The inset of Fig. 14.14c is an enlarged view of the EER point of the presented behavioral biometric authentication system. Compared with the state of the art biometric techniques, the presented biometric authentication system achieves a remarkably lower EER value of 1.34 % at the threshold of 0.37. In the meanwhile, the Receiver Operating Characteristic (ROC) curve is demonstrated in the Fig. 14.14d. In addition, to prove the capability of the IKB as an additional layer of stronger security to current authentication systems, a practical application was demonstrated. When four different individuals typing the password "touch" into the computer through IKB, only the genuine owner with matched typing pattern can access into the computer. Given its exceptional authentication ability, the IKB is able to identify the personal character of typing individuals, making it practical for developing a highly secured authentication system based on behavioral biometrics.

## 14.4 Summary

In this chapter, we discussed the first type of self-powered active sensor based on TENG, that is, the self-powered pressure/touch sensor. Both vertical contact-separation mode and single-electrode mode TENG have been employed as self-powered pressure/touch sensor to deliver high sensitivity (2.3 % $Pa^{-1}$), low detection limit ($\sim$0.2 Pa), and fast response time (<5 ms). In general, the pressure response contains a high sensitivity region at relatively low pressure, and a low sensitivity region at higher pressure. Possible explanations of this behavior include full closing of air gap or saturation of increasing contact area between the touch object and the device. It has been demonstrated that the open-circuit voltage could be employed for static pressure detection to obtain the amplitude of the applied pressure, while the short-circuit current mainly reveal the dynamic information of the applied pressure like loading rate or acceleration. Furthermore, the TENG-based active sensor has been integrated with a resistance-based passive sensor to form a dual-mode pressure sensor with both high sensitivity and self-powered operation. To achieve the goal of self-powered tactile imaging, multiple triboelectric active sensor units were integrated into a 6 × 6 array matrix with a common aluminum electrode at the bottom. Four different character-like objects were taken as templates to apply pressure onto the sensor matrix, and the distribution and confinement of the applied pressure could be displayed by the contour plot of the multi-channel measurement.

The basic idea has been further incorporated with a keyboard structure for self-powered human-machine interface. The polymer-coated smart keyboard has the function of recognizing personal typing patterns to achieve security authentication of personal electronics. Huge amount of data analysis further verifies the validity of this approach. This new concept of human-machine interface has numerous potential applications in next-generation smart electronics and wearable devices.

## References

1. Z.L. Wang, Triboelectric nanogenerators as new energy technology for self-powered systems and as active mechanical and chemical sensors. ACS Nano **7**(11), 9533–9557 (2013)
2. Z.L. Wang, Triboelectric nanogenerators as new energy technology and self-powered sensors —principles, problems and perspectives. Faraday Discuss. **176**, 447–458 (2014)
3. Z.L. Wang, J. Chen, L. Lin, Progress in triboelectric nanogenerators as a new energy technology and self-powered sensors. Energy Environ. Sci. **8**(8), 2250–2282 (2015)
4. F.R. Fan, L. Lin, G. Zhu, W.Z. Wu, R. Zhang, Z.L. Wang, Transparent triboelectric nanogenerators and self-powered pressure sensors based on micropatterned plastic films. Nano Lett. **12**(6), 3109–3114 (2012)
5. L. Lin, Y.N. Xie, S.H. Wang, W.Z. Wu, S.M. Niu, X.N. Wen, Z.L. Wang, Triboelectric active sensor array for self-powered static and dynamic pressure detection and tactile imaging. ACS Nano **7**(9), 8266–8274 (2013)
6. G. Zhu, W.Q. Yang, T.J. Zhang, Q.S. Jing, J. Chen, Y.S. Zhou, P. Bai, Z.L. Wang, Self-powered, ultrasensitive, flexible tactile sensors based on contact electrification. Nano Lett. **14**(6), 3208–3213 (2014)
7. P. Bai, G. Zhu, Q.S. Jing, J. Yang, J. Chen, Y.J. Su, J.S. Ma, G. Zhang, Z.L. Wang, Membrane-based self-powered triboelectric sensors for pressure change detection and its uses in security surveillance and healthcare monitoring. Adv. Funct. Mater. **24**(37), 5807–5813 (2014)
8. Y. Yang, H.L. Zhang, Z.H. Lin, Y.S. Zhou, Q.S. Jing, Y.J. Su, J. Yang, J. Chen, C.G. Hu, Z.L. Wang, Human skin based triboelectric nanogenerators for harvesting biomechanical energy and as self-powered active tactile sensor system. ACS Nano **7**(10), 9213–9222 (2013)

9. J.J. Luo, F.R. Fan, T. Zhou, W. Tang, F. Xue, Z.L. Wang, Ultrasensitive self-powered pressure sensing system. Extreme Mech. Lett. **2**, 28–36 (2015)
10. Y. Yang, H.L. Zhang, X.D. Zhong, F. Yi, R.M. Yu, Y. Zhang, Z.L. Wang, Electret film-enhanced triboelectric nanogenerator matrix for self-powered instantaneous tactile imaging. ACS Appl. Mater. Interfaces **6**(5), 3680–3688 (2014)
11. J. Chen, G. Zhu, J. Yang, Q.S. Jing, P. Bai, W.Q. Yang, X.W. Qi, Y.J. Su, Z.L. Wang, Personalized keystroke dynamics for self-powered human-machine interfacing. ACS Nano **9** (1), 105–116 (2015)
12. Y.G. Sun, B. Gates, B. Mayers, Y.N. Xia, Crystalline silver nanowires by soft solution processing. Nano Lett. **2**(2), 165–168 (2002)
13. C.L. Choong, M.B. Shim, B.S. Lee, S. Jeon, D.S. Ko, T.H. Kang, J. Bae, S.H. Lee, K.E. Byun, J. Im, Y.J. Jeong, C.E. Park, J.J. Park, U.I. Chung, Highly stretchable resistive pressure sensors using a conductive elastomeric composite on a micropyramid array. Adv. Mater. **26** (21), 3451–3458 (2014)
14. S. Gong, W. Schwalb, Y.W. Wang, Y. Chen, Y. Tang, J. Si, B. Shirinzadeh, W.L. Cheng, A wearable and highly sensitive pressure sensor with ultrathin gold nanowires. Nat. Commun. **5**, 3132 (2014)
15. C. Pang, G.Y. Lee, T.I. Kim, S.M. Kim, H.N. Kim, S.H. Ahn, K.Y. Suh, A flexible and highly sensitive strain-gauge sensor using reversible interlocking of nanofibres. Nat. Mater. **11**(9), 795–801 (2012)
16. W.Z. Wu, X.N. Wen, Z.L. Wang, Taxel-addressable matrix of vertical-nanowire piezotronic transistors for active and adaptive tactile imaging. Science **340**(6135), 952–957 (2013)

# Chapter 15
# Self-powered Sensing for Vibration and Biomedical Monitoring

## 15.1 Self-powered Vibration Sensor

The vertical contact-separation mode TENG generate electricity through the periodic contact and separation between the two triboelectric layers, which is generally enabled by the vertical-to-plane vibrational motion of at least one of the two plates in the TENG (see Chap. 2). Thus, through integrating the typical structure of the contact-mode TENG into an unstable mechanical structure that can self-balance under perturbation, the constructed device can not only harvest energy from vibrations [1–7], but also serve as a self-powered vibration sensor [8–10].

### *15.1.1 Position Tracking of the Vibration Source*

Among these structures, the 3D spiral structure is a typical example. It is a three-dimensional curve around an axis at a continuously reducing radius while moving parallel to the axis, which can be simply regarded as a conical shaped spring structure [10]. Loaded with a seismic mass, the suspended 3D spiral structure can oscillate as a response to external disturbance at a high sensitivity. A spiral TENG (STENG) that operates in the contact-mode was built at the front of the seismic mass in a spiral structure. The spiral structure behaves like a spring in response to an external disturbance. A schematic diagram of the STENG is shown in Fig. 15.1a. An acrylic sheet is chosen as the structural material. The working parts that generate the electricity are two round plates facing each other, which is marked in Fig. 15.1a with a red dashed line. One is attached at the bottom of the 3D spiral structure. The other one is centered and fixed on the upper surface of the cube's bottom. The height of the cube was adjusted to have these two plates be in slight contact when the spiral is relaxed and in equilibrium. Figure 15.1b is a scanning electron microscopy (SEM) image of the aluminum surface. There are

 
Z.L. Wang et al., *Triboelectric Nanogenerators*,
Green Energy and Technology, DOI 10.1007/978-3-319-40039-6_15

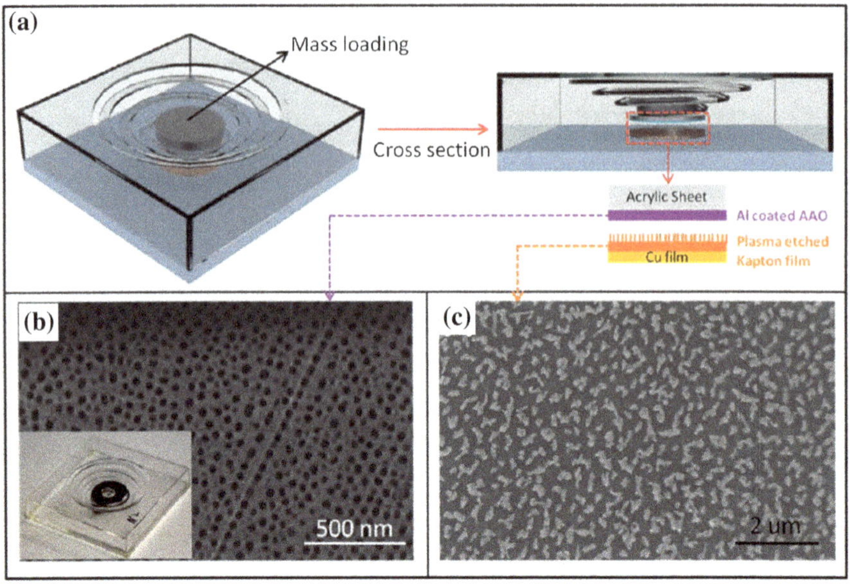


**Fig. 15.1** Triboelectric nanogenerator built on a suspended 3D spiral structure. **a** Schematic diagram of the device and its cross section view. SEM image of the surface morphology of **b** aluminum and **c** the Kapton film. The *inset* of **b** is a photo of the real device. Reproduced with permission from American Chemical Society [10]

nanopores uniformly distributed on the surface with an average diameter of 30 nm, which are able to increase the effective contact area for triboelectrification. The lower contact surface is a Kapton film. The surface of the Kapton film was modified to introduce aligned nanowire structures. Figure 15.1c is an SEM image of the obtained surface. The nanostructures on both the surfaces can increase the surface roughness and the effective surface area of the STENG for effective triboelectrification. When the STENG is working, the spiral oscillates in response to an external vibration source, so that the distance between the Al and Kapton film changes. During this process, the electricity is generated following the principle of contact-mode TENG.

The STENG can serve as self-powered/active sensors for vibration detection and monitoring. This means that whenever there is a vibration disturbance in the environment experienced by the STENG, the STENG will be triggered to generate an indicating electric signal. As in some situations vibrations are undesirable and harmful, in addition to vibration detection, the vibration source positioning is very important and of great practical value. Here, the STENGs were used to construct a multichannel active sensor system, and the in-plane vibration source positioning was realized. The vibration source positioning test was carried out on a wood board with dimensions of 1 in. × 24 in. × 48 in. Three STENGs were positioned on three corners of the board, as shown in Fig. 15.2a. A three-channel measurement was applied to acquire data from these three STENGs simultaneously.

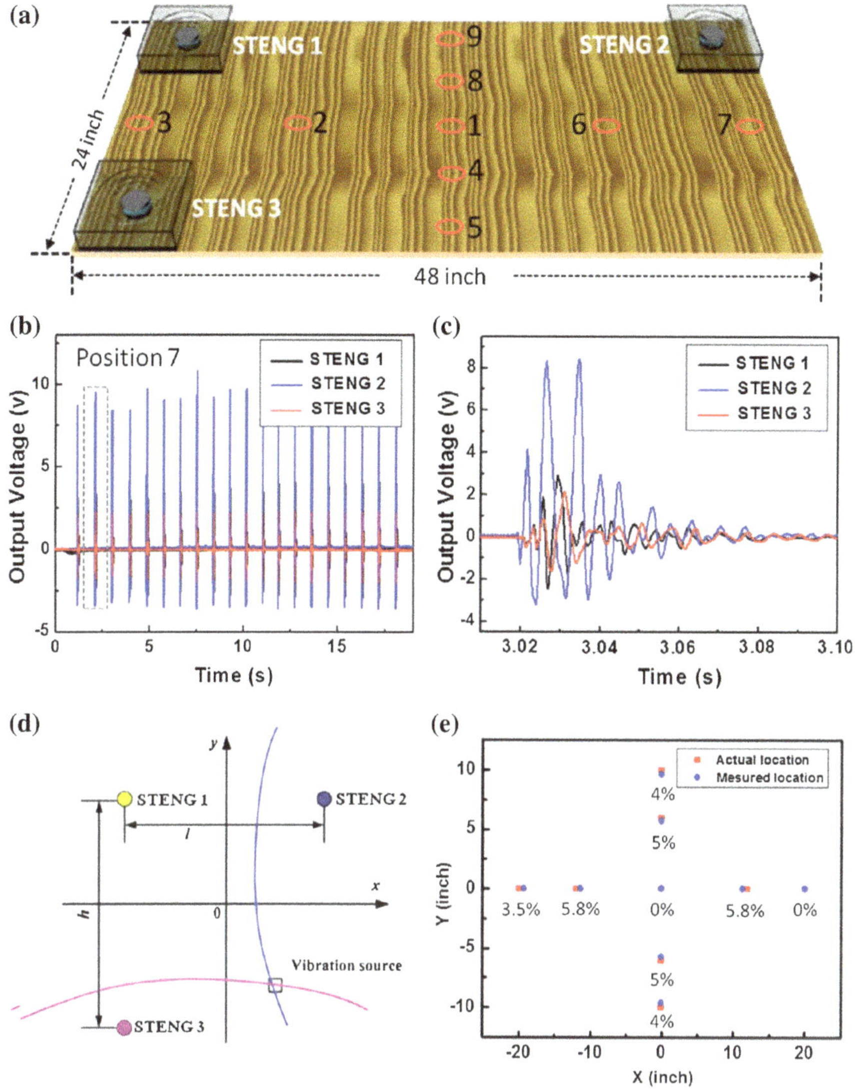


**Fig. 15.2** Vibration source location on a wood board. **a** Schematic diagram of the experimental setup shows three STENGs positioned at the three corners of a wood board. **b** Recorded signal patterns when the impact is located at position 7. **c** Enlarged view of one peak signal in **b**, which is indicated by a *dashed gray rectangle*. **d** Schematic diagram of the principle to locate the vibration source. **e** Comparison of the measured location and actual location of the vibration source and corresponding location error. Reproduced with permission from American Chemical Society [10]

A representative signal pattern is shown in Fig. 15.2b, which was acquired when the small hammer hit position 7. Because STENG 2 is the closest one to the vibration source, it obtains the highest signal level. For STENG 1 and STENG 3, the signal levels are similar due to the equal distance from the vibration source. Figure 15.2c is the enlarged view of one signal peak in Fig. 15.2b, which is indicated by a dashed gray rectangle. The vibration propagated and first arrived at STENG 2 and then arrived at STENG 1 and STENG 3 at the same time. Thus the first signal peak from STENG 2 is earlier than the two from STENG 1 and STENG 3. The principle to locate the vibration source is depicted in Fig. 15.2d. We can obtain the arrival time differences of the signal among the three STENGs. Combining the vibration propagation speed, we can get information on the distance difference from the vibration source to STENG 1 and STENG 2 (D1), as well as the distance difference from the vibration source to STENG 1 and STENG 3 (D2). The track of an object that maintains the same distance difference from two fixed points is a hyperbolic curve, as indicated by the blue and purple curves in Fig. 15.2d. The vibration source is located at the intersection of the two hyperbolic curves by solving a set of equations. The corresponding experimental results are shown in Fig. 15.2e. The position at the center of the board is accurately located without error, and the location error for other positions is less than 5.8 %. This error can be improved by increasing the sampling rate of the data acquisition equipment.

Besides the 3D-spiral-based TENG, several other structure designs of the contact-mode TENG have been developed for harvesting vibrational energy, such as the harmonic-resonator-based TENG, the 3-D TENG with broadband response, the 3-D stack integrated TENG, the triple-cantilever-based TENG, and the integrated rhombic gridding based TENG. They all can be utilized as the self-powered/active sensor to detect external vibrations.

### *15.1.2 Vibration Amplitude Measurement*

Generally, vibrations are characterized by two parameters: frequency and amplitude. The vibration frequency can be easily quantified by most vibration sensors, but a quantitative measurement of the amplitude is relatively more difficult because it often requires a linear relationship between the sensing signal and the amplitude. The TENG structures based on the vertical contact-separation mode is only applicable for measuring the frequency but not the amplitude. This is because that the electricity generation behavior of the vertical contact-separation mode is not linear to the separation distance, due to the nature of the changing capacitance between the two electrodes during its operation. Among the four fundamental modes for TENGs, the contact-mode freestanding-triboelectric-layer mode has stationary electrodes thus constant electrode capacitance. Based on this mode, a contact-separation-enabled freestanding-TENG (CF-TENG) has been proposed, in which the vibration of the freestanding triboelectric layer between the two electrodes periodically changes the induced potential difference between the two electrodes and thus generate electricity in external load [9].

The structural design of the vibration-enabled CF-TENG is introduced in the Sect. 5.2. When the CF-TENG with the resonator plate supported by springs is bonded to a vibration source, the resonator plate will be triggered to vibrate between the two electrodes. The linear characteristic gives the CF-TENG the supreme capability of quantitatively sensing the amplitude of the vibration. When the frequency was at the resonant frequency of the CF-TENG (Fig. 15.3a), both the peak-to-peak value of the $V_{OC}$ and the amplitude of the $I_{SC}$ increase almost linearly with the increase of the vibration amplitude, as long as it within the small range (below 1.5 mm, i.e. Region I shown in Fig. 15.3a). This response comes from the linear relationship between the agitated vibration amplitude of the resonator plate and the vibration amplitude of the entire CF-TENG. However, when the vibration amplitude is above ~1.5 mm, the resonator plate will start to get into contact with the two electrode plates. This results in the saturation behavior of the electrical outputs in Region II shown in Fig. 15.3a. Moreover, from Fig. 15.3b as the enlarged plot of the curves in Fig. 15.3a below the vibration amplitude of 0.3 mm, it can be found that both the $V_{OC}$ and the $I_{SC}$ have very good linear responses. When the vibration amplitude is only 3.5 μm, the peak-to-peak value of the $V_{OC}$

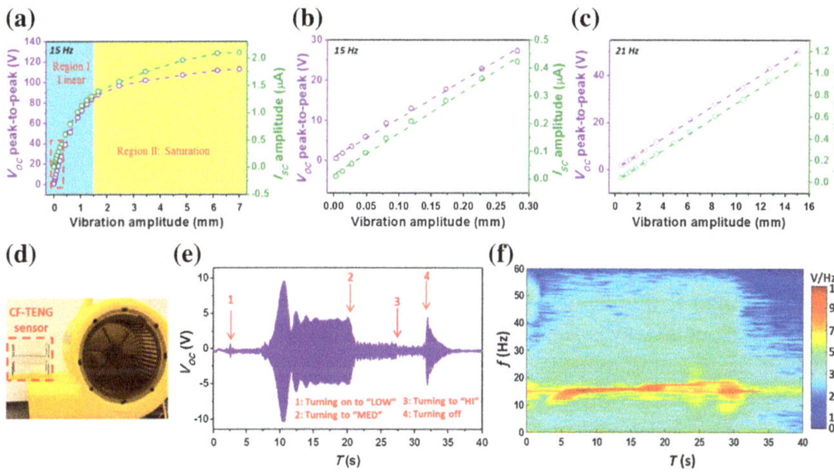


**Fig. 15.3** The contact-mode freestanding-TENG for self-powered vibration amplitude measurement. **a** $V_{OC}$ and $I_{SC}$ from the CF-TENG triggered by vibrations with different amplitudes, when the external vibration source is at the resonating frequency (15 Hz) of the CF-TENG. **b** Enlarged plot of panel with the vibration amplitude in the range below 0.3 mm. **c** $V_{OC}$ and $I_{SC}$ from the CF-TENG triggered by vibrations with different amplitudes, when the external vibration source is not at the resonating frequency of the CF-TENG. **d** Photograph showing the demonstration of using the CF-TENG to monitor the vibration of a wind blower during its operation. **e** $V_{OC}$ from the CF-TENG when the wind blower was sequentially switched to different speeds. **f** Short-time Fourier transform of the $V_{OC}$ signal shown in panel (**e**). Reproduced with permission from American Chemical Society [9]

and the amplitude of the $I_{SC}$ generated by the CF-TENG still have 0.54 V and 10 nA. Therefore, when the vibration source is at or close to the resonant frequency of the CF-TENG, it has an extremely high sensitivity for very subtle vibrations. In the other group of experiments with the frequency at 21 Hz, which is off the resonant frequency (Fig. 15.3c), when the vibration amplitude is increased up to 15 mm, both the $V_{OC}$ and the $I_{SC}$ follow very good linear behaviors. Thus, when the vibration source is not at the resonant frequency of the CF-TENG, this active sensor has the capability of quantitatively measuring the amplitude in a wide range.

As a demonstration of the CF-TENG's outstanding capability in vibration sensing, it was mounted onto a wind blower to monitor its vibration during the operation (Fig. 15.3d). When the wind blower was sequentially turned on from the "off" state to the "LOW" speed, then turned to the "MED" speed, then to the "HIGH" speed, and finally turned off, the $V_{OC}$ signal generated by the CF-TENG is shown in Fig. 15.3e. From the short-time Fourier transform (STFT) of the $V_{OC}$ signal, the instantaneous frequency distribution of the wind blower throughout this operation process can be obtained in Fig. 15.3f. In each stage with the same frequency, the change of the amplitude in the $V_{OC}$ signal directly reflects the change of the vibration amplitude of the wind-blower, as shown in Fig. 15.3e. Therefore, this demonstration convincingly shows the CF-TENG's supreme capability in quantitatively detecting both the amplitude and the frequency of any arbitrary vibration.

## 15.2 Self-powered Acoustic Sensor for Voice Recording

### *15.2.1 Helmholtz-Cavity-Based Acoustic Sensor*

Acoustic wave is a special form of vibration ubiquitously existing in the environment, which appears in a wide range of frequency and strength. The sensing of acoustic waves is very important not only for information technology, but also for the monitoring of equipment systems and environment [11]. With the possibility of converting sound waves into electricity, TENGs could provide an effective approach for realizing the self-powered acoustic sensing.

With a Helmholtz cavity, a triboelectric nanogenerator for acoustic energy harvesting was introduced in Sect. 9.3.1 With a superior capability in converting acoustic waves into electricity, this TENG can also act as an active self-powered acoustic sensor for different applications [12], such as to be utilized as a self-powered microphone. The natural frequencies of the TENG devices can be purposely tuned by the structural parameters configuration. Thus, we can design multiple TENGs and have their corresponding frequency bands overlap with each other to give a broadened working bandwidth. Here, as demonstrated in Fig. 15.4a, four nanogenerators, $NG_1$, $NG_2$, $NG_3$, and $NG_4$, respectively, with a series of dimensions corresponding to natural frequencies of 350, 650, 1100, and 1400 Hz, respectively, were employed to widen the overall working bandwidth from 10 to

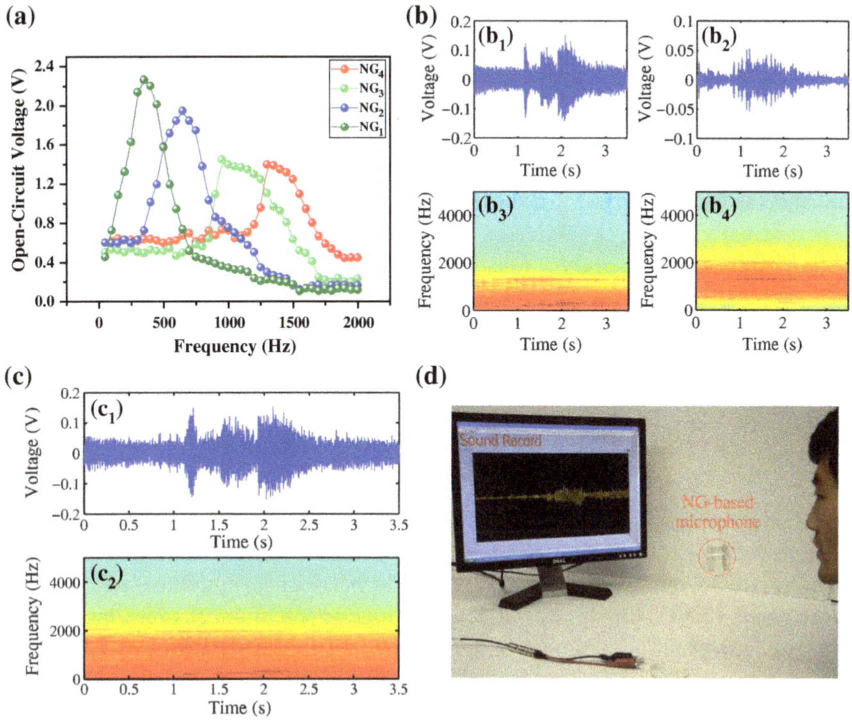


**Fig. 15.4** Demonstration of the organic film nanogenerator acting as a self-powered microphone. **a** Frequency responses from the nanogenerators array, which consists of four NGs with various designed natural frequencies, aimed to enhance the overall working bandwidth. **$b_1$**, **$b_2$** Sound waveforms of the signals acquired by $NG_1$ and $NG_2$, respectively; **$b_3$**, **$b_4$** Short-Time Fourier Transforms of the acquired signals by $NG_1$ and $NG_2$, respectively. **c** Sound waveform and corresponding Short-Time Fourier Transform of the signals acquired by the array of the NGs. **d** Photograph that shows a NG is working as a self-powered microphone for sound recording. Reproduced with permission from American Chemical Society [12]

1700 Hz, which enabled the superior performance of these multiple TENGs as a self-powered microphone for sound recording. Parts $b_1$ and $b_3$ of Fig. 15.4b are, respectively, the time domain waveforms of the recorded sounds from $NG_1$ and $NG_4$. Although $NG_1$ and $NG_4$ were triggered by the same sound source, the waveforms of the two apparently exhibit different characteristics, which is attributed to a different frequency response ranges of the two. With a natural frequency of 350 Hz, the waveform of $NG_1$ is smoother owing to its dominant response to lower frequency components from 10 to 600 Hz, as shown in Fig. 15.4$b_2$, which is the corresponding short-time Fourier transform (STFT), while the waveform of $NG_4$ with a natural frequency of 1400 Hz is rougher, due to its dominant response to the higher frequency components from 1100 to 1700 Hz, as demonstrated in Fig. 15.4$b_4$ of its corresponding STFT. In order to reconstruct the original sound, the acquired acoustic signals of the array are weighted according to the relative amount of information available from each source. The waveform of the reconstructed signal and its corresponding STFT are illustrated in parts $c_1$ and $c_2$, respectively, of Fig. 15.4c. The frequency components of the reconstructed signal cover all the frequencies ranging from 10 to 1700 Hz, as demonstrated in Fig. 15.4$c_2$. Figure 15.4d shows an as-fabricated TENG working as a self-powered microphone for sound recording. In another demonstration of the acoustic TENG for the sensing purpose, it acts an acoustic source localization sensor. Experimentally, three as-fabricated TENGs were arranged in an L shape and anchored in a 2D plane with dimensions of 2 m by 1.8 m. The results show that positioning within an average error circle of 7 cm in diameter is achieved based on multiple measurements.

### 15.2.2 Ultrathin Paper-Based Acoustic Sensor

The bulky structure of the Helmholtz resonance cavity greatly limited the wide-range application of the TENG as an acoustic sensor. A further improved performance of the acoustic sensor is on the basis of an ultrathin, rollable, and paper-based TENG, which innovatively employs arrays of micro holes for acoustic response [13]. And the structural design of the device was detailed introduced in the Sect. 9.3.2.

With a broad working bandwidth, the as-fabricated paper-thin TENG is also capable of acting as an active self-powered microphone for sound recording. As demonstrated in Fig. 15.5a, a low-cost ultra-thin self-powered microphone was developed, which can efficiently convert the human voice into electrical signals for the recording purpose. Compared to other existing technologies for acoustic energy harvesting, the paper based TENG distinguishes itself in many aspects and brings about a number of advantages, such as ultrathin, rollable, low-cost, environmentally friendly and extremely high volume power density. From structure point of view, the traditional acoustic energy harvester has a bulky structure due to the requirement of a resonance cavity. Superiorly, the presented ultrathin triboelectric nanogenerator innovatively employs a multi-hole structure on the paper electrode, which effectively gets rid of the traditional resonator for acoustic energy harvesting. The paper thin TENG achieves a volume power density of $\sim 1$ kW/m$^3$ at a sound pressure of 117 dB$_{SPL}$.

From the performance point of view, owing to the superior advantages of structurally ultrathin and flexible, the paper based TENG is rollable. Figure 15.5b, c are the schematic illustrations to show the measurement of the directional patterns of the flat and rolled paper thin triboelectric nanogenerator (the insets are the photographs of the as-fabricated devices). Figure 15.5d shows the corresponding shape dependent directional patterns of the TENG with flat and rolled structure, respectively. And a butterfly shaped directional pattern with mirror symmetry was

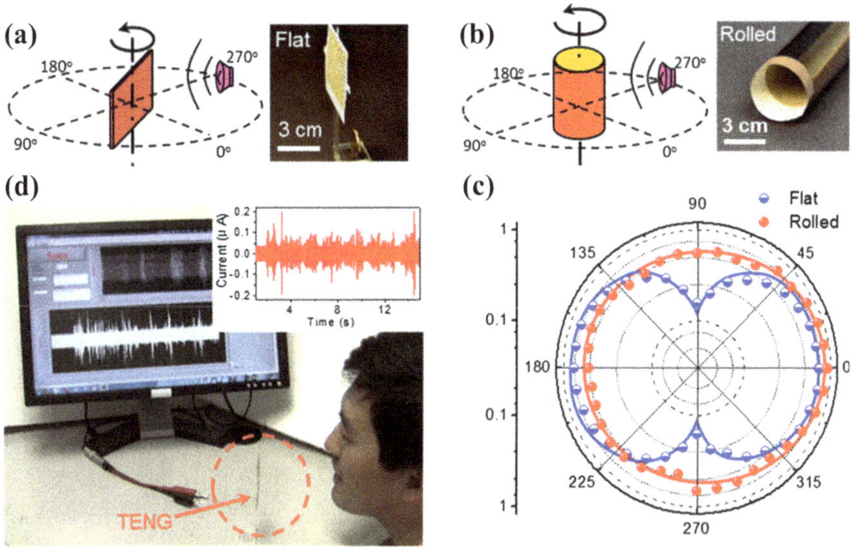


**Fig. 15.5** Demonstration of the paper thin triboelectric nanogenerator for self-powered sound recording. Schematic illustration to show the measurement of the directional patterns of the **a** flat and **b** rolled paper thin triboelectric nanogenerator. Right sides are the photographs of the as-fabricated devices. The scale bars are 2.5 cm. **c** Shape dependent directional patterns of the paper thin triboelectric nanogenerator. **d** Photograph that shows a paper thin triboelectric nanogenerator is working as a self-powered microphone for sound recording. Inset is the acquired electrical signals. Reproduced with permission from American Chemical Society [13]

observed for the flat type paper thin TENG. While the directional pattern of the rolled type is a highly symmetric circle, which indicates the output is independent of the sound wave incident direction. This is mainly attributed to the rigorous structural symmetry. Holding the advantage of directional independence, the rolled type TENG is suitable to a wide range of circumstances for either energy-harvesting or sensing purposes, such as theatric stage live recording, military surveillance, omnibearing acoustic energy harvesting.

From cost point of view, the ultrathin TENG is fabricated mainly based on low-cost, light-weight, and biodegradable paper materials with a simple structure. Besides, based on surface charging effect, the fabrication requires only very small amount of materials, which are conventional polymers or thin layer of metal as electrodes. Furthermore, the fabrication process of the paper-thin TENG is straightforward and compatible with possible large-scale manufacturing. Additionally, the backbone of paper thin TENG is made of commonly-used paper materials, which is biodegradable and greatly reduced the possible environmental costs. As a consequence, the paper-based TENG is extremely cost-effective, which is an unparalleled advantage compared to any other acoustic energy harvesting techniques.

## 15.3 Self-powered Biomedical Monitoring

### *15.3.1 Eardrum Inspired Bionic Membrane Sensor*

Measurements of human physiological or behavioral characteristics are the main designing functions of wearable and interactive electronics. A lightweight and self-powered technology is highly desired for simultaneously acquiring multi-functional characteristics. Here in this section, we will introduce a bionic membrane sensor (BMS) based on triboelectrification for self-powered physiological and behavioral measurements such as local internal body pressures [14].

A BMS has a multilayered structure with a thin layer of polyethylene terephthalate (PET) in an oval shape as the bottom supporting substrate, as schematically shown in Fig. 15.6a. Then, a layer of ITO coated Nylon thin film was laminated onto the PET substrate with the ITO acting as the back electrode and the Nylon functioning as one electrification layer. Mimicked a human eardrum, a layer of polytetrafluoroethylene (PTFE) tympanic membrane is tented outwards at the level of the tip of an umbo, which was centered at a Nylon layer with a size of Φ 0.6 mm × 0.4 mm. By peripherally being anchored, the tensed PTFE tympanic membrane, acting as another electrification layer, is extremely sensitive to external dynamic pressure over a wideband frequency range like human eardrum. The

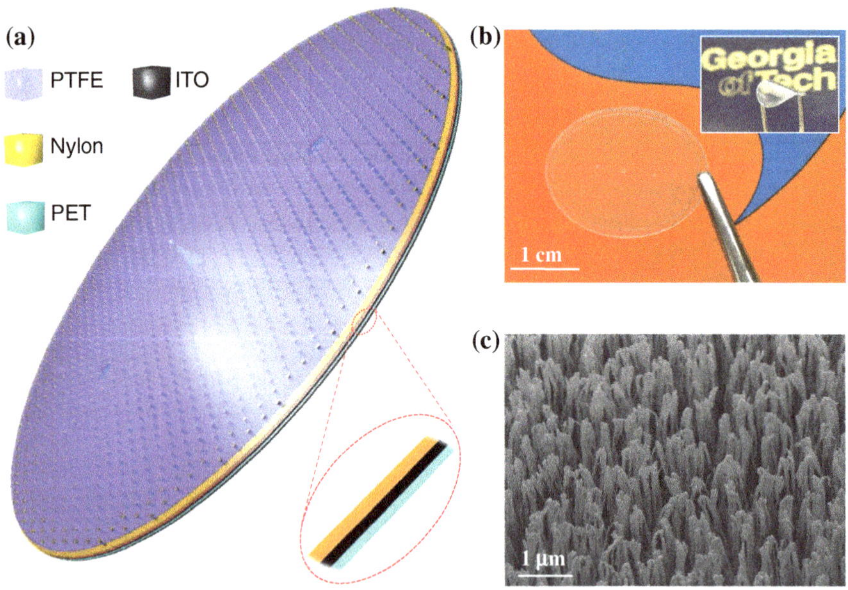


**Fig. 15.6** Structural design of the bionic membrane sensor. **a** Schematic illustrations of the bionic membrane sensor. **b** Photograph of an as-fabricated flexible and transparent bionic membrane sensor. The scale bar is 1.0 cm. **c** A SEM image of surface-etched PTFE nanowires. The scale bar is 1 μm. Reproduced with permission from Wiley [14]

formed slightly conical cavity between the PTFE and Nylon serves as an air spacer for the charge generation and transfer. And two circular acoustic holes with diameters of 0.5 mm are punched through the PET, ITO and Nylon three layers, acting as communicating vessels to integrate the conical cavity with the ambient air. A photograph of an as-fabricated flexible and transparent BMS is shown in Fig. 15.6b. In order to enhance the triboelectrification, vertically aligned polymer nanowires were created onto the PTFE surface for an intimate contact with Nylon. A scanning electron microscopy (SEM) image of PTFE nanowires is displayed in Fig. 15.6c, which indicates that the average clustering diameter of nanowires is 110 nm with an average length of 0.8 μm. The fabrication process of the BMS is straightforward and compatible with possible large-scale manufacturing.

#### 15.3.1.1 Self-powered Arterial Pulse Wave Measurement

The worldwide leading causes of death are from cardiovascular diseases. Intensive research is focused on early detection and prevention of such diseases, with a special emphasis on arterial compliance capability. Arterial pulse wave propagation in the arterial tree can provide essential information about the arterial physical situation. The BMS can provide a superior route to acquire the human arterial pulse wave in a self-powered and noninvasive manner. For a real-time monitoring of the arterial pulse, the BMS was attached onto the carotid, wrist and chest, just over the artery as it is usually done in arterial tonometry. The arterial pulse wave corresponds to the dynamic pressure change on the skin surface, which can be characterized by the electrical output of the attached BMS. Figure 15.7a, b respectively show the recorded real-time voltage and current outputs over several pulse periods when a BMS was placed over the carotid artery of a 30-year-old man. Insets are the enlarged views of one cycle for a detailed information delivery. The acquired voltage and current signals independently but complementarily give a comprehensive picture of the arterial physical situation.

On one hand, regarding the voltage signal, a typical characteristic pulse wave shape was obtained with three clearly distinguishable determinants for both the 30-year-old and 70-year-old men: Systolic peak ($P_S$), Point of inflection ($P_i$) and Dicrotic wave ($P_D$), which are known to be respectively resulted from the blood ejected from the left ventricle, the reflected pulse wave and the ejected blood back to the left ventricle. A comparison of the pressure pulse waves of a young man with that of an old man indicates apparent differences between the two, especially that $P_i$ occurs after the $P_S$ for the younger one, while the $P_i$ appears before the $P_S$ for the older one. The differences are significantly related to the physiological condition of human cardiovascular system, which can be quantified by two of the most commonly used parameters: the augmentation index $\mathrm{AIx}(\%) = \pm(P_s - P_i)/\mathrm{PP}$, and the reflection index $\mathrm{RI} = h/\Delta t$, where $h$ is the subject height, $\Delta t$ is the time delay between $P_S$ and $P_D$, while PP is the absolute pulse wave magnitude. In order to accurately evaluate the physiological conditions of the two peoples' cardiovascular systems, a time period of 10 min was taken to continuously record their atrial pulse

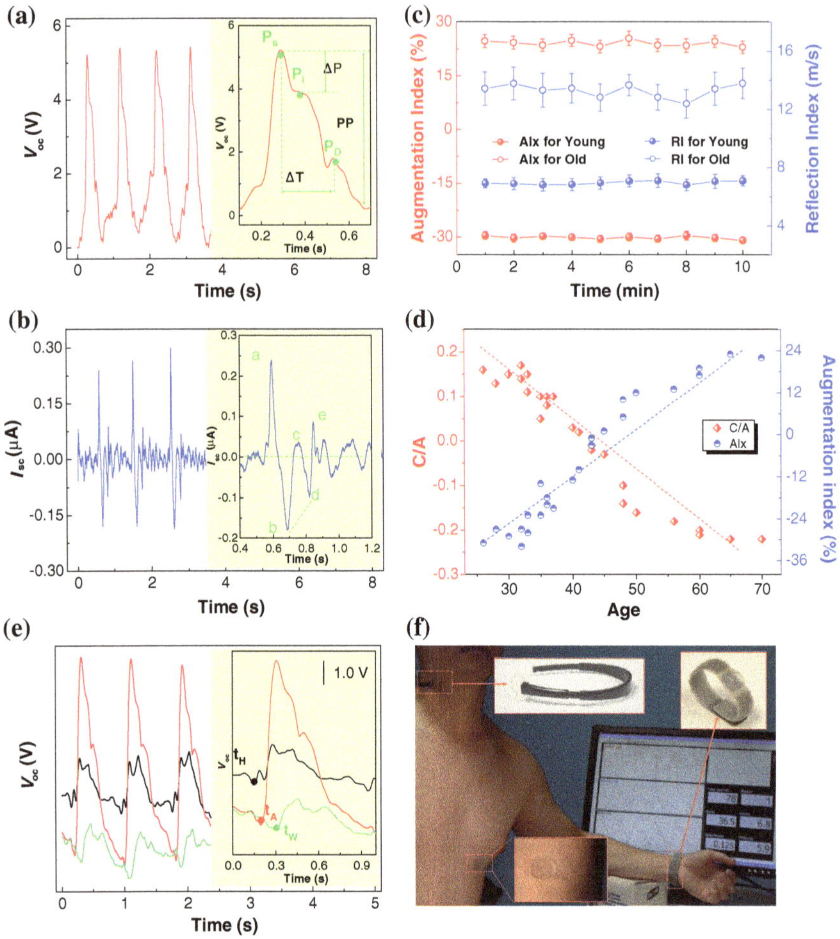


**Fig. 15.7** Bionic membrane sensor measured the arterial pulse wave for noninvasive self-powered human health monitoring. **a** The real-time voltage output when the sensor is placed over the carotid artery of a 30-year-old man. The *inset* is an enlarged view of one cycle, which is a graphic representation of the augmentation index (AIx), defined as $\text{AIx}(\%) = \pm(P_s - P_i)/\text{PP}$. PP is the absolute pulse amplitude. ΔT is the time delay between the systolic $P_S$ and diastolic peak $P_D$. And thus the reflection index (RI) of carotid artery can be obtained from subject height divided by the time delay. **b** The corresponding real-time current output. The *inset* is an enlarged view of one cycle of the current signal, whose waveform comprises of five main parts, labeled as A-wave to E-wave: initial positive (A-wave), early negative (B-wave), re-increasing (C-wave), late re-decreasing (D-wave) and diastolic positive (E-wave). **c** Quantitative comparison of the AIx and RI between the 30-year-old man and the 70-year-old one. **d** Relationship between the C/A ratio and AIx in the carotid arteries of 25 subjects age from 26 to 70 years' old. The C/A ratio decreased with age, while AIx followed a reverse trend. **e** The pulse waves acquired from three different sites of a participant: carotid artery, left wrist and the chest. The *inset* is an enlarged view of a cycle of output signals corresponding to three positions. And the *dots* indicate their start points of each cycle. **f** A photograph showing the bionic membrane sensors were directly attached to simultaneously monitor the pulse waves of the participant from his carotid artery, chest and wrist. Reproduced with permission from Wiley [14]

waves. Based on the acquired waveforms, a statistical result of the AIx and RI was respectively obtained for the young and old, as shown in Fig. 15.7c. For the younger one, average values of −30 % and 6.8 m/s were respectively obtained for the parameters AIx and RI, while for the older one, values of 24.8 % and 13.3 m/s were respectively obtained. A distinct difference of AIx and RI between the two subjects is mainly attributed to that, for the young man with better arterial compliance capability, the pulse wave is spread through the arteries at a lower velocity and also the reflected wave arrives back to the aorta after the late systole, resulting in a negative AIx and a smaller RI. In contrast, for the old man with stiffer arteries, the blood pulse velocity is higher and the reflected wave arrives back to the aorta in the early systole, leading to a positive AIx and a larger RI.

On the other hand, regarding the acquired current signal, the waveform comprises of five main parts, labeled as a-wave to e-wave: initially positive (A-wave), early negative (B-wave), re-increasing (C-wave), late re-decreasing (D-wave) and diastolic positive (E-wave). The height of each wave was measured from the baseline, with the values above the baseline being positive and those under it negative. A comparison of the current signal waveform of the young man with that of the old man indicates obvious differences between the two, especially that a positive C-wave is observed for the young man, while a negative C-wave for the old one. In order to quantify the difference, a parameter (C/A) was employed, which was defined by the ratio of the height of the C-wave to that of the A-wave. The C/A and AIx values were measured from 25 people with ages spanning from 26 to 70 years' old, as shown in Fig. 15.7d. It is interesting to find that the C/A index decreases with age, while AIx follows a reverse trend. These two parameters are reported to independently but complementarily indicate the vascular stiffness, that is, the AIx increases with the increasing of arterial stiffness, while the C/A index decreases. These statistical results highlight that the measured output voltage and current by BMS can reflect the condition of the cardiovascular system as well as the risk of cardiovascular disease. Thus, it can be utilized for non-invasive medical diagnosis.

Meanwhile, in order to obtain more information for cardiovascular risk estimation, a further step was made to measure the pulse wave velocity (PWV) by using multiple BMSs to monitor human pulse waves over different artery sites. The PWV is highly related to the vascular compliance and can be determined by measuring the pulse propagation time from the pressure waveforms at different sites along a vascular segment. A stiffer vessel will conduct the pulse wave faster than a more compliant one. Figure 15.7e presents the simultaneously acquired voltage signals when three BMSs were respectively attached to the carotid artery, left wrist and the chest. The inset is enlarged views of a cycle of output signals, and the characteristic dots indicate their cycle start points. It is worth noting that the start point of the signal acquired from chest comes first, which is 53 and 163 ms earlier than the start points of the signals from the carotid artery and left wrist, respectively. Here, the PWV is derived by dividing the pulse wave propagation distance between the two testing sites by the time delay over that distance. For the test subject, PWV1 = 5.66 m/s (along chest to carotid artery) and PWV2 = 5.9 m/s (along chest to left wrist) were experimentally obtained, which are characteristic values expected for a healthy adult.

To prove the capability of the BMS for self-powered noninvasive arterial pulse measurement, a set of practical applications was demonstrated, as shown in Fig. 15.7f. Three BMSs were directly attached to simultaneously monitor the pulse waves of the participant from his carotid artery, chest and wrist, and all of the essential information about the arterial physical situation, including pulse rate, PWV, AIx, and RI, were systematically obtained for a further medical diagnosis.

#### 15.3.1.2 Self-powered Throat-Attached Microphone

Throat microphone is becoming more and more popular since it can effectively pick up speech even in extremely noisy or windy environments, such as for a parachute jumper from sky, on a motorcycle or in a pandemoniac street, while it is beyond the capability of other types of microphones due to high levels of background noise and the limitation of current technology. Here, the BMS is also capable of acting as superior throat microphone, which is lightweight, small volume, portable, extremely low cost, easy to fabricate, more importantly, it can work without external power [14]. When the BMS was worn against the neck without speaking, a series of pulse waves with low frequency of 1.1 Hz were observed, corresponding to the silent state in Fig. 15.8a. While plentiful high frequency components were emerged and superimposed when the participant spoke the sentence "This is a self-powered throat microphone", which were reflected by the acquired voltage signals in the speaking state in Fig. 15.8a. From the Short-Time Fourier spectrum of the recorded data, the acquired voltage signal was analyzed, obviously indicating a majority of the frequency components spanning from 45 to 1500 Hz when peaking, as shown in Fig. 15.8b. Then, according to the frequency distribution, a 10 Hz cutoff-frequency low-pass finite impulse response digital filter was further used to get the low frequency components from the acquired voltage signal (Fig. 15.8c). While a ban-pass finite impulse response digital filter with a pass-band from 45 to 1500 Hz was applied to resolve the throat sound from the acquired voltage signal (Fig. 15.8d). Understandably, the low frequency components are mainly caused by the throat movement and the epidermis deformation during speech. While the high frequency components corresponds to vibrations transmitted through the skin during speech, which exhibits plenty high frequency detailed information and thus can be harnessed to rebuild/recover the speaking content, which is beyond the capability of previous report.

To prove the capability of the BMS as a superior self-powered throat microphone, a practical application was demonstrated, as shown in Fig. 15.8e. When the tester worn the BMS against her neck and spoke in a noisy environment, the speaking content can be successfully recovered. This is the first invention that a single self-powered sensor can exhibit a multi-functionality of simultaneously acting as a throat microphone and a non-invasive arterial pulse wave monitor,

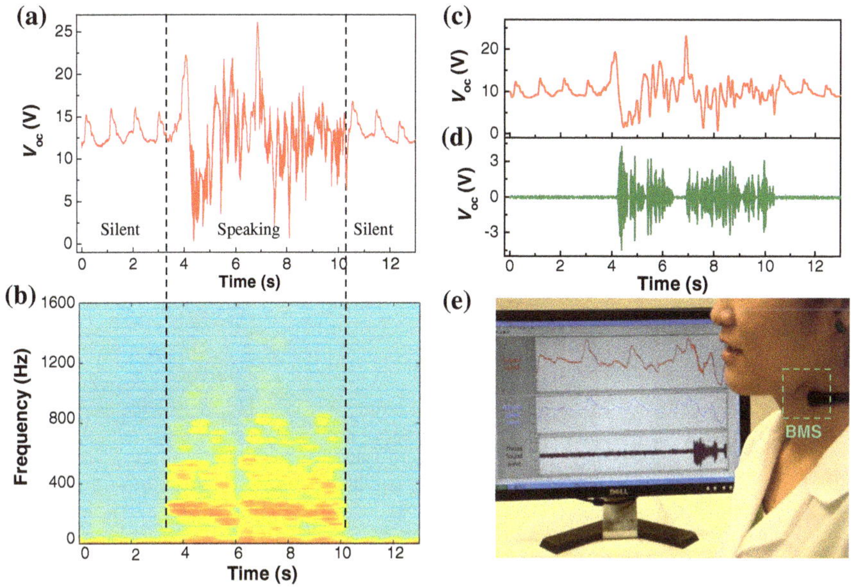


**Fig. 15.8** Bionic membrane sensor as a superior anti-interference self-powered throat microphone. **a** The real-time voltage output in response to the throat vibration during speaking when the sensor is worn against the neck. **b** Short-Time Fourier Transform of the acquired output voltage, which indicates a majority of the frequency components spanning from 45 to 1500 Hz. The low frequency components (**c**) and high frequency components (**d**) of the voltage signal, which are obtained respectively via a 10 Hz-cutoff-frequency low pass filter and a ban-pass filter with pass-band from 45 to 1500 Hz. The high frequency components elaborately illustrates the plenty output signals responding to the concrete dynamic change in the speaking, which assures the superior performance of the bionic membrane sensor as a throat microphone. **e** A photograph showing the bionic membrane sensor was worn against the participant's neck acting as a self-powered throat microphone. Reproduced with permission from Wiley [14]

which will attract extensive attentions, especially for astronaut, soldier, motorcycle rider, and those who has to keep quiet while communicating with others at a distance, such as during covert military operations.

#### 15.3.1.3 Arterial Pulse as a Biometric Recognition

Biometric recognition, an automatic identification and verification of a person based on the physiological or behavioral characteristics, is getting increasing research attention for that the current authentication system cannot be easily separated from the genuine owner. Alternative to the traditional approach is medical biometric. The advantage of the bio signal is that it gives some sort of intrinsic characteristics to avoid spoof attack, which can be easily combined with the traditional biometric, such as speech, without any additional cost. Here, we innovatively proposed a new medical biometric authentication technique by using the arterial pulse wave.

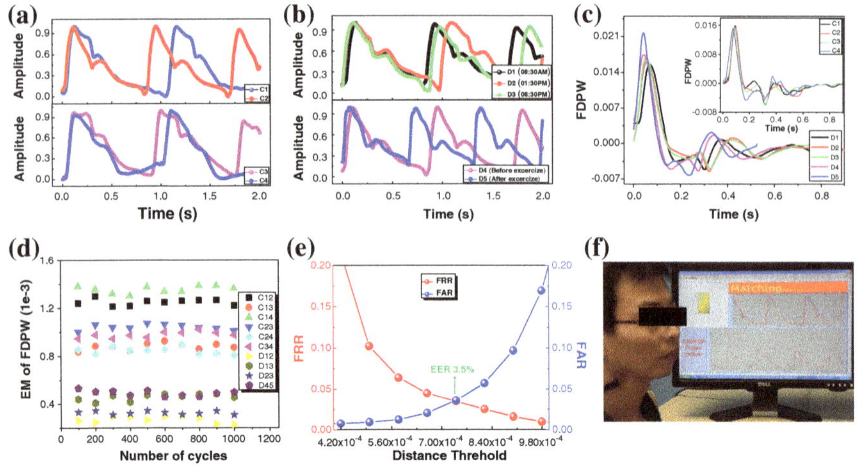


**Fig. 15.9** Bionic membrane sensor measured the arterial pulse wave for biometric recognition. **a** The normalized output voltage measured from the carotid artery of four participants as their heartbeat profiles. **b** Heartbeat profiles comparison of a person at different time of a day (up) as well as before and after exercise (down). **c** First-derivatives-of-the-pulse-waves (FDPWs) of a person under different activity states. Inset is the FDPWs of the four different participants. **d** Calculated Euclidean metrics (EMs) from the nine FDPWs for a comparison. **e** Performance evaluation of the measured heartbeat profiles for biometric authentication. The variation of FAR and FRR in relation to the threshold, which indicates a low EER value of 3.5 % at the threshold of 7.3e-4. **f** A photograph showing the bionic membrane sensor was worn against the participant's neck to measure the pulse wave for biometric recognition. Reproduced with permission from Wiley [14]

Due to the ultrasensitiveness of the BMS to external pressure, it can capture plenty of information to fully characterize each one's arterial pulse wave, which thus could be used for further verification and recognition purposes [14]. Figure 15.9a shows the normalized output voltages measured from the carotid arteries of four participants. A comparison of the obtained four set of pulse wave patterns indicates explicit differences not only in heart rates, but also in the arterial pulse wave shapes. This is mainly attributed to differences in age, gender, chest configuration and the position, size, as well as anatomy of the hearts of these individuals.

For quantitative differentiation, Euclidean Metric (EM) was calculated to measure the difference among pulse waves. Two pulse wave signals **X** and **Y** represent two different individuals. Let $\mathbf{X}_i = [x_{i1}, x_{i2}, \ldots, x_{iN}]$ and $\mathbf{Y}_i = [y_{i1}, y_{i2}, \ldots, y_{iM}]$ ($N < M$) respectively denote the $i$-th cycle of pulse waves **X** and **Y**. Then the average value of EM between **X** and **Y** over $K$ cycles can be defined as follows:

$$d_{\mathbf{XY}} = \frac{1}{K}\sum_{i=1}^{K}\frac{1}{N}\sum_{j=1}^{N}\left|x_{ij} - y_{ij}\right| \tag{15.1}$$

In the meanwhile, the average value of EM of an individual over different time periods can be expressed as:

$$d_{\mathbf{XX}} = \frac{1}{K}\sum_{i=1}^{K}\frac{1}{N}\sum_{j=1}^{N}\left|x_{ij} - x_{(i+1)j}\right| \tag{15.2}$$

Furthermore, the pulse waves also relate to the human activity states, such as emotional fluctuation, exercise, and so on. In this regard, as shown in Fig. 15.9b, the pulse waves of a same person at different times in a day (up) as well as before and after exercise (down) are systematically monitored, and the expected different time profiles are obtained. In this regards, a predicament emerges that a few EM values of individual pulse waves under different states are comparable with the EM values of pulse waves among different people. Consequently, it is difficult to distinguish individuals by simply using EMs of the pulse waves. The predicament cannot be walked out by resorting to the First-derivatives-of-the-pulse-waves (FDPWs), which is capable of representing the shape pattern of the pulse wave. Figure 15.9c shows the FDPWs of the same person under different activity states. Inset is the FDPWs of the four different individuals. Subsequently, a further step was taken to measure and compare the corresponding EMs of the FDPWs, as the results presented in Fig. 15.9d. It is inspiring to note that the EM values of the FDPWs among different individuals are obviously larger than that for the same person under different activity states. As a result, the distinguishable EMs of FDPWs can be used as the feature representation of pulse wave, acting as a threshold to separate the genuine pulse waves from the impostor ones, if an authentication system is established using pulse wave.

To evaluate the identification performance, 120 randomly chosen individuals were invited to record their arterial pulse waves. The collected 120 individual pulse waves were divided into 60 client patterns and 60 impostor patterns. And the performance of the pulse wave based authentication biometrics is characterized through two error rates: False Rejection Rate (FRR) and False Acceptance Rate (FAR). The variation of FAR and FRR in relation to the threshold is shown in Fig. 15.9e. The FAR is decreasing with the elevated EM of the FDPWs, while FRR follows a reverse trend. However, the FRR and FAR intersect at a certain threshold value of 7.3e-4, which indicates the EER point of the biometric authentication system with a low EER value of 3.5 %.

In addition, to prove the capability of the acquired pulse waves for biometric authentication, a practical application was demonstrated, as shown in Fig. 15.9f. When four individuals wearing the BMS against the necks, each one can be accurately identified and recognized from the stored templates in computer data base. Acting as a new biometric authentication technique, the measured pulse wave by BMS is an effective identifier, which is reliable, practical and thus a work towards developing a secured authentication system.

#### 15.3.1.4 Multi-modal Biometric Authentication

A biometric system that relies only on a single physiological or behavioral characteristic in making a personal identification is often not able to meet the desired performance requirements. Identification based on multiple biometrics represents an emerging trend. Here, relying on the superior sensing performance of BMS, a multimodal biometric authentication system was established using a single BMS, which, for the first time, integrates the pulse wave recognition and voice recognition in making a personal identification [14].

To demonstrate the capability of BMS for multi-modal biometric authentication, a first step was taken here to show its functionality for voice recognition, since the capability of BMS for pulse wave recognition was justified in the previous section. Figure 15.10a, b show the originally recorded voltage signals when two participants were continuously saying "one world" more than three times in each accustomed manner. Figure 15.10c, d are their corresponding high frequency components of the recorded voltage signals via a 45–1500-Hz band-pass filter, which correspond to the throat sound signals according to the frequency distribution. Enlarged views of one cycle of the voice signal for saying"one world" of two participants are respectively shown in Fig. 15.10e, f. It is clearly noted that these throat sound waves look apparently different from each other, which is mainly attributed to the differences of anatomy (e.g., size and shape of the throat) and behavioral patterns (e.g., voice pitch, speaking style).

Pearson correlation coefficient was utilized to quantitatively measure the difference between two throat sounds signals. Let $\mathbf{X} = [x_1, x_2, \ldots, x_N]$ and $\mathbf{Y} = [y_1, y_2, \ldots, y_N]$ denote two throat sound signals from two different individuals, the Pearson correlation coefficient between signals $\mathbf{X}$ and $\mathbf{Y}$ can be defined as:

$$r = \frac{\sum_{i=1}^{N} (x_i - \bar{x})(y_i - \bar{y})}{\sqrt{\sum_{i=1}^{N} (x_i - \bar{x})^2}\sqrt{\sum_{i=1}^{N} (y_i - \bar{y})^2}} \quad (15.3)$$

To evaluate the performance of the BMS for voice recognition, 120 participants were invited to independently say the phrase "one world" for more than four times in each accustomed manner with the BMS wearing against the necks. A calculation of the Pearson correlation coefficient between any two of the throat sound signals of 120 participants was performed. It is worth noting that most of the Pearson correlation coefficients from individual self-comparing are larger than 0.7, while the Pearson correlation coefficients from comparison among individuals are less than 0.7, which is indicative of that Pearson correlation coefficient can be harnessed to act as a threshold to separate the genuine voice from the impostor voice, if a verification system is thus established. The presented BMS based voice authentication system achieves an EER value of 2.7 % at the threshold of 0.55.

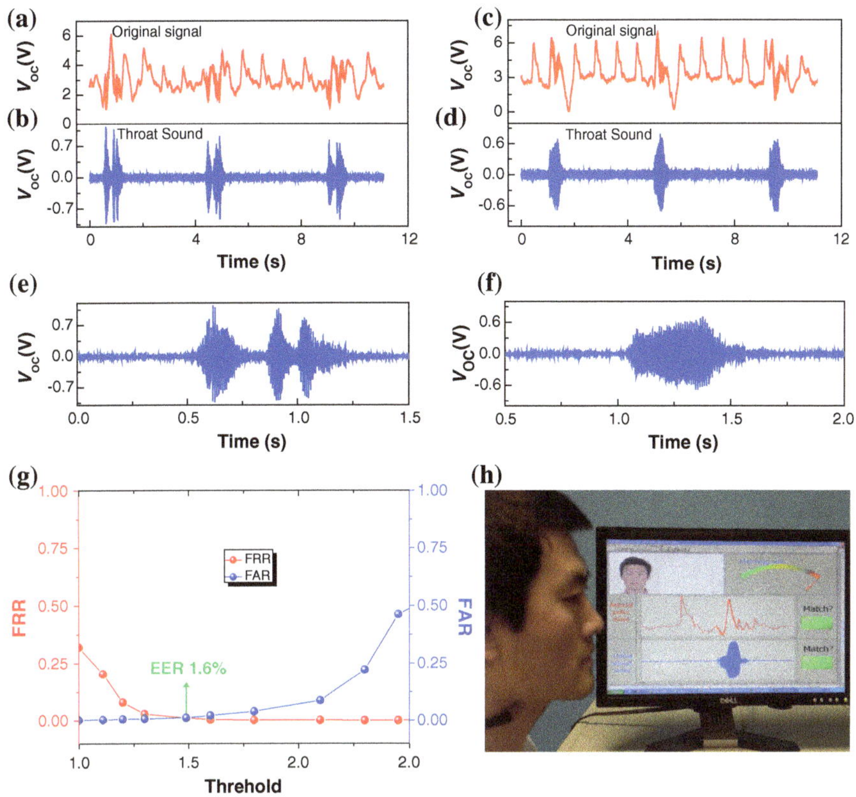


**Fig. 15.10** Bionic membrane sensor simultaneously recorded the arterial pulse and throat sound for single-sensor multi-modal biometric authentication. **a**, **b** The recorded voltage signals when Alex and Tom were continuously saying "one world" more than three times, respectively. **c**, **d** The high frequency components of the recorded voltage signals from Alex and Tom via a ban-pass filter, respectively. **e**, **f** The enlarged views of a cycle of the high frequency components from Alex and Tom, respectively. **g** Performance evaluation of the presented multi-modal biometric authentication system. The variation of FAR and FRR in relation to the threshold, which indicates a remarkably low EER value of 1.6 % at the threshold of 1.5. **h** A photograph showing the bionic membrane sensor was worn against the participant's neck to simultaneously record the arterial pulse and throat sound for a single-sensor multi-modal biometric authentication. Reproduced with permission from Wiley [14]

For an enhanced authentication performance, identification based on the integration of the pulse wave and voice recognition was evaluated as a multi-modal biometric. The new feature representation for the multi-modal biometric authentication system is a combination of EM of the FDPWs ($d_{\mathbf{XY}}$) from the pulse wave recognition and Pearson correlation coefficient ($r$) from the voice recognition. And the new feature representation is mathematically expressed by a feature vector $\mathbf{T} = [d_{\mathbf{XY}i}, r_i]$, where $i = 1, 2, 3, \ldots, u$, denoting the index of the user ($u$ is total number of users in the system database). A new threshold relying on the feature vector can be defined as:

$$TSM_i = w_1 \times 0.0018/d_{\mathbf{XY}i} + w_2 r_i \qquad (15.4)$$

The weight coefficients $w_1$ and $w_2$ are proportional to the authentication performance of the two uni-modal systems with a constraint of $w_1 + w_2 = 1$. A larger value of $TSM_i$ indicates a larger similarity between the user templates and client templates. As demonstrated in Fig. 15.10g, a remarkably low EER value of 1.6 % was achieved for the presented multi-modal biometric authentication system at a threshold value of 1.5. As demonstrated in Fig. 15.10h, the BMS was utilized for a multi-modal biometric authentication, by simultaneously recording the arterial pulse and throat sound, and only the genuine owner with matched both arterial pulse wave and throat sound can access into the computer. Given its authentication ability, a BMS is capable of simultaneously identifying personal characters physiologically as well as behaviorally. The results demonstrate that the identity established by such an integrated system is much more reliable than the identity established by a single biometrics, making it unique and practical for developing a highly secured biometrics authentication system.

## *15.3.2 Membrane-Based Triboelectric Sensor*

In this section, a membrane-based triboelectric sensor (M-TES) is introduced as a self-powered pressure change sensor. It generates a voltage induced by surface triboelectric charges in response to an air pressure change [15]. By integrating the M-TES with a signal processing unit, the M-TES can effectively respond to the pressure change caused by even heartbeats without an external power supply, which can be harnessed for bio-medical monitoring.

The M-TES has a multi-layered structure which is schematically illustrated in Fig. 15.11a. A piece of acrylic sheet with a thickness of 1.6 mm was prepared by laser cutting as a substrate. One layer of copper with a thickness of 100 nm was deposited on the substrate as a circular top electrode (a diameter of 2.0 cm) while another layer of copper was deposited on the back of the substrate as a back electrode. Then, a 125-μm-thick fluorinated ethylene propylene (FEP) film was securely attached onto the top electrode. A through air-conducting channel with a diameter of 2.0 mm was created at the center of the device. A latex membrane with a thickness of 50 μm was brought to a plane-plane intimate contact with the FEP film. Epoxy resin was used to seal the edges between the membrane and the FEP surface to prevent any air leakage. To promote the sensitivity of the M-TES, vertically aligned FEP nanorod (NR) arrays were created on the FEP surface by a top-down dry etching technique, as shown in the inset of Fig. 15.11a. As revealed

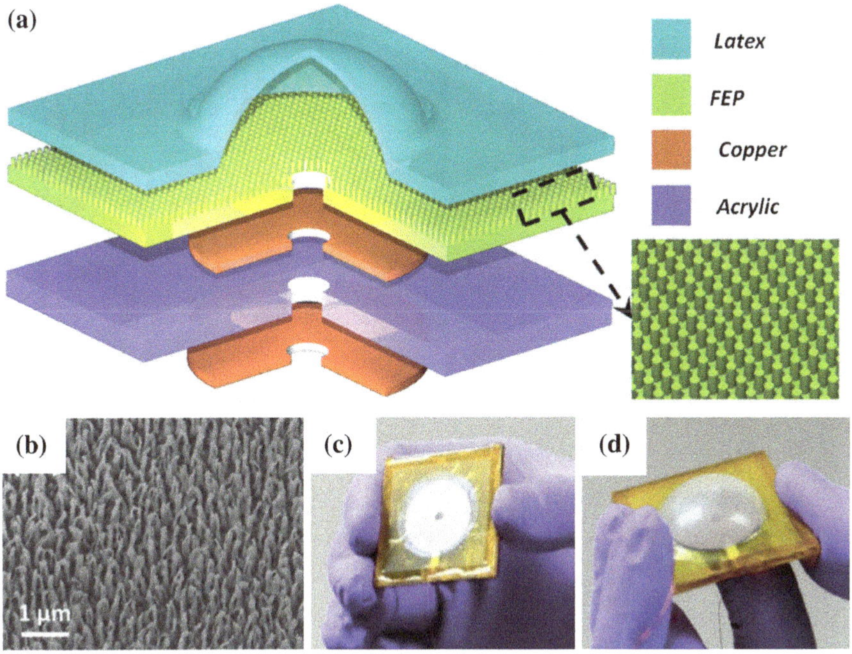


**Fig. 15.11** **a** Schematic of the M-TES. **b** SEM image of FEP nanorod arrays. **c** Photograph of an M-TES and **d** an M-TES with the latex membrane swelling. Reproduced with permission from Wiley [15]

by a scanning electron microscopy (SEM) image in Fig. 15.11b, the FEP NRs are uniformly distributed on the surface of the FEP film after being etched for five minutes. They have an average diameter of 130 nm and a length of 500 nm. Figure 15.11c exhibits a photograph of the M-TES with a circular apparent contact area of 3.2 $cm^2$. In response to a pressure change from the air-conducting channel, the latex membrane will swell, resulting in a separation away from the FEP film (Fig. 15.11d).

To demonstrate the ability of M-TESs as active sensors for self-powered bio-monitoring, two sets of practical applications were demonstrated. Firstly, a simple and low-cost system for respiration detection was developed by using the M-TESs. As sketched inserted in Fig. 15.12a, an air bag (18 × 22 × 1 cm) with air tubes was connected with an M-TES. An adjustable back brace is used to attach the air bag to the abdomen of a subject. The abdominal expansion and contraction of respiratory activity will result in a periodical air pressure change in the air bag which can be detected by the M-TES. Once the respiratory movement occurs, output voltage signals generated by the M-TES will be collected by a program developed by LabView. And the average breathing rate in real time displays on the

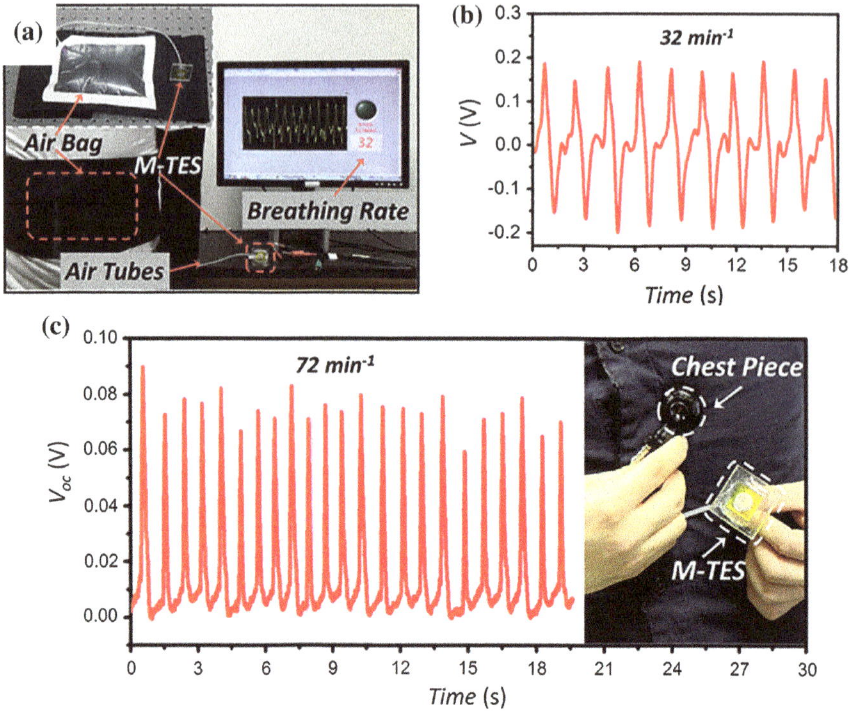


**Fig. 15.12** **a** Demonstration of respiration detection by utilizing the M-TES. **b** Output voltage signals of the M-TES at a breathing rate of 32 times per minute. **c** Output voltage signals of the M-TES as a heartbeat monitor, insert: setup of the M-TES as a heartbeat monitor. Reproduced with permission from Wiley [15]

monitor (Fig. 15.12a). Electric signals indicate a breathing rate of 32 times per minute when the subject carries on a normal breathing rate (Fig. 15.12b). Inspirations caused by pulmonary expansion are shown as maxima peaks, while expirations caused by pulmonary contraction are shown as minima ones. Heartbeat monitors has been largely used by performers of various types of physical exercise and medical purposes. Secondly, due to the high resolution, a heartbeat monitor was also demonstrated based on the M-TES. Air pressure change in a chest piece is caused by the heartbeat transmit via air-filled hollow tubes to the M-TES (inserted in Fig. 15.12c). Such a periodical air pressure change will result in output voltage signals for the M-TES. In such a case, each heartbeat can be recorded by a pulse signal. The output voltage signals around 0.06 V in Fig. 15.12c shows that the test subject has a stable heartbeat rate of 72 times per minute which is consistent with the counting result. Therefore, the obtained breathing and heartbeat rate can be successfully obtained by output voltage signals, which can be applied for easy handling detection systems required for daily home use.

## 15.4 Summary

In responding to mechanical triggering, TENG produces an electric response signal even without applying a power source. Such signals can not only be used for self-powered mechanical vibration sensing, but also for biomedical monitoring. This chapter reviews some of the existing studies in the self-powered or active vibration sensors.

Firstly, loaded with a seismic mass, the suspended 3D spiral structure based TENG can oscillate as a response to external disturbance at a high sensitivity for vibration detection and monitoring. And a contact-separation-enabled freestanding-TENG was developed to quantitatively measure the amplitude of the vibration. Furthermore, the sensing of acoustic waves is also another advantageous application of TENG. And a cantilever based triboelectric nanogenerator was reported for self-powered sound recording and sound source localization. In the meanwhile, a low-cost ultra-thin paper based self-powered microphone was also developed for self-powered recording purpose. And a rolled device holds a highly symmetric circle of the directional pattern, which indicates the output is independent of the sound wave incident direction. In addition, an eardrum inspired bionic membrane sensor was also developed for self-powered physiological and behavioral measurements, including arterial pulse wave measurement, acting as a self-powered throat-attached microphone, as well as using for biometric recognition. Lastly, a membrane-based triboelectric sensor was also reported for self-powered bio-medical monitoring.

To summary, in this chapter, the design, performance and applications of various triboelectric sensors have been presented. Their output voltage signal reflects the magnitude of the mechanical triggering, while the current signal indicate the velocity and acceleration at which the mechanical triggering occurs. Such technologies could be potentially important for environmental monitoring, infrastructure monitoring and biomedical applications.

## References

1. J. Chen, G. Zhu, W. Yang, Q. Jing, P. Bai, Y. Yang, T.-C. Hou, Z.L. Wang, Harmonic-resonator-based triboelectric nanogenerator as a sustainable power source and a self-powered active vibration sensor. Adv. Mater. **25**(42), 6094–6099 (2013)
2. Z.L. Wang, J. Chen, L. Lin, Progress in triboelectric nanogenerators as new energy technology and self-powered sensors. Energy Environ. Sci. **8**, 2250–2282 (2015)
3. J. Yang, J. Chen, Y. Yang, H. Zhang, W. Yang, P. Bai, Y. Su, Z.L. Wang, Broadband vibrational energy harvesting based on a triboelectric nanogenerator. Adv. Energy Mater. **4**(6), 1301322 (2014)
4. W. Yang, J. Chen, Q. Jing, J. Yang, X. Wen, Y. Su, G. Zhu, P. Bai, Z.L. Wang, 3D stack integrated triboelectric nanogenerator for harvesting vibration energy. Adv. Funct. Mater. **24** (26), 4090–4096 (2014)

5. W. Yang, J. Chen, G. Zhu, X. Wen, P. Bai, Y. Su, Y. Lin, Z.L. Wang, Harvesting vibration energy by a triple-cantilever based triboelectric nanogenerator. Nano Res. **6**(12), 880–886 (2013)
6. W. Yang, J. Chen, G. Zhu, J. Yang, P. Bai, Y. Su, Q. Jing, X. Cao, Z.L. Wang, Harvesting energy from the natural vibration of human walking. ACS Nano **7**(12), 11317–11324 (2013)
7. G. Zhu, B. Peng, J. Chen, Q. Jing, Z.L. Wang, Triboelectric nanogenerators as a new energy technology: from fundamentals, devices, to applications. Nano Energy **14**, 126–138 (2015)
8. H. Zhang, Y. Yang, Y. Su, J. Chen, K. Adams, S. Lee, C. Hu, Z.L. Wang, Triboelectric nanogenerator for harvesting vibration energy in full space and as self-powered acceleration sensor. Adv. Funct. Mater. **24**(10), 1401–1407 (2014)
9. S. Wang, S. Niu, J. Yang, L. Lin, Z.L. Wang, Quantitative measurements of vibration amplitude using a contact-mode freestanding triboelectric nanogenerator. ACS Nano **8**(12), 12004–12013 (2014)
10. Y. Hu, J. Yang, Q. Jing, S. Niu, W. Wu, Z.L. Wang, Triboelectric nanogenerator built on suspended 3d spiral structure as vibration and positioning sensor and wave energy harvester. ACS Nano **7**(11), 10424–10432 (2013)
11. X. Wang, J. Song, J. Liu, Z.L. Wang, Direct-current nanogenerator driven by ultrasonic waves. Science **316**(5821), 102–105 (2007)
12. J. Yang, J. Chen, Y. Liu, W. Yang, Y. Su, Z.L. Wang, Triboelectrification-based organic film nanogenerator for acoustic energy harvesting and self-powered active acoustic sensing. ACS Nano **8**(3), 2649–2657 (2014)
13. X. Fan, J. Chen, J. Yang, P. Bai, Z. Li, Z.L. Wang, Ultrathin, rollable, paper-based triboelectric nanogenerator for acoustic energy harvesting and self-powered sound recording. ACS Nano **9**(4), 4236–4243 (2015)
14. J. Yang, J. Chen, Y. Su, Q. Jing, Z. Li, F. Yi, X. Wen, Z. Wang, Z.L. Wang, Eardrum-inspired active sensors for self-powered cardiovascular system characterization and throat-attached anti-interference voice recognition. Adv. Mater. **27**(8), 1316–1326 (2015)
15. P. Bai, G. Zhu, Q. Jing, J. Yang, J. Chen, Y. Su, J. Ma, G. Zhang, Z.L. Wang, Membrane-based self-powered triboelectric sensors for pressure change detection and its uses in security surveillance and healthcare monitoring. Adv. Funct. Mater. **24**(37), 5807–5813 (2014)

# Chapter 16
# Self-powered Sensing for Tracking Moving Objects

Mechanical motions are generally characterized by a series of parameters, such as movement distance, velocity, accelerations, and angles. Sensors for the monitoring and detection of these parameters for the mechanical motions are vitally important for a lot of mechanical systems. With different structural designs, triboelectric nanogenerators have been developed to generate electricity from different types of mechanical motions, such as linear sliding and rotation [1]. Since the frequency, the amplitude and the total periods of the generated electrical signals are all directly related to the parameters of the input mechanical motions, the TENGs can serve as self-powered active motion sensors. Even if such motions are in concealed locations, TENGs can be utilized to track the trajectory and detect the instantaneously parameters of the motions. It output of voltage and current provides static and dynamic information of the mechanical triggering.

## 16.1 TENGs as Self-powered Linear Displacement Sensors

Measurement of displacement and speed at the micro- and nano-scale has ubiquitous applications in the scientific and industrial fields such as manufacturing, automation, robotics, and nano-manipulation. Most of the existing sensing technologies need pre-provided electrical or optical signals in order to detect the mechanical displacement, which inevitably requires external power sources. However, the increased number and density of portable electronic devices and sensing networks today desperately desires low power consumption and/or self-powered sensors. Based on the triboelectric nanogenerators, a self-powered, one-dimensional displacement and speed sensing technology has been demonstrated, which achieves high resolution, large dynamic range and long detecting distance [1].



Z.L. Wang et al., *Triboelectric Nanogenerators*,
Green Energy and Technology, DOI 10.1007/978-3-319-40039-6_16

The motion sensor consists of two micro gratings with identical patterns, as sketched in Fig. 16.1a. The grating at the bottom is made of an etched silicon wafer coated with aluminum as the bottom electrode and silicon dioxide as triboelectrically positive material; the grating on the top is made of patterned SU-8 film on a

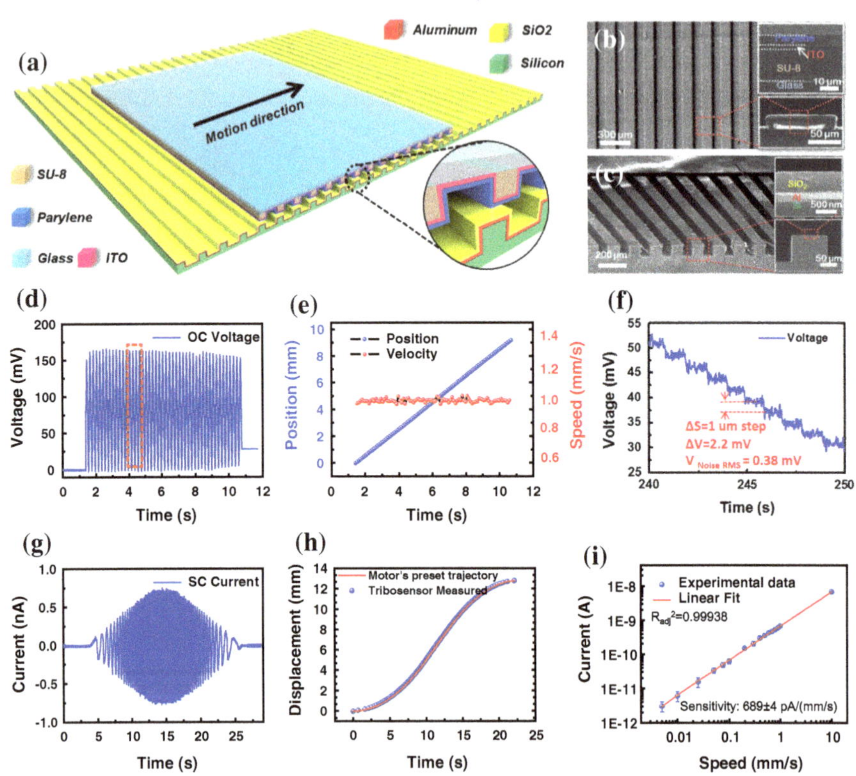


**Fig. 16.1** Micro-grated triboelectric nanogenerator as nanometer resolution self-powered linear displacement sensor. **a** Schematic showing the structure of a TENG-base self-powered displacement sensor, with a pair of micrograting*s*, the inset illustrates the detailed information of layers. **b** SEM images of the top micrograting with glass slide substrate. The *insets* show the cross section: the ITO layer on top of the patterned SU-8 photoresist serves as the top electrode, and the outmost layer is Parylene film serving as electronegative triboelectric layer. **c** SEM images of the bottom micrograting. The *insets* show the cross section profile: the etched Silicon is coated with Al as bottom electrode and $SiO_2$ as electropositive triboelectric layer. **d** The OC voltage signals acquired from a displacement of 9.2 mm at a preset speed of 1 mm $s^{-1}$. **e** The real time displacement and speed derived from the measured voltage signal. **f** Step motion with 1 μm per step in the sensitive region (10–190 μm) can be clearly resolved through the OC voltage signal. Given the RMS value of noise (0.38 mV), and the voltage change corresponding to 1 μm step motion (2.2 mV), the resolution can be calculated to be 173 nm. **g** The SC current signals acquired from a non-uniform motion (decelerate–uniform speed–deceleration). **h** The preset motor's trajectory and the real time displacement detected by the motion sensor. **i** A plot of SC current as a function of motion speed from 5 μm $s^{-1}$ to 10 mm $s^{-1}$ with a fitted sensitivity of 689 ± 4 pA (mm $s^{-1}$). Reproduced with permission from Wiley [1]

glass slide as supporting substrate and subsequently coated with ITO and Parylene film as the top electrode and triboelectrically negative material, respectively (Fig. 16.1a inset). The detailed microscopic structure of the as-fabricated devices is shown in Fig. 16.1b, c, respectively. The relative motion between two gratings leads to periodic separation of two micro-grated dielectric materials that are oppositely charged through triboelectrification. As a result, an alternating electric signal between the metal electrodes placed beneath the dielectric materials can be detected due to electrostatic induction.

During the movement, the open-circuit (OC) voltage measured between the two electrodes (Fig. 16.1d) alternates between 0 and $\sim$160 mV periodically. As aforementioned, since each cycle corresponds to one period of the gratings, which is 200 μm in this device, the real time displacement can be calculated by accumulating extra 200 μm once one more peak is detected. As shown in the blue curve in Fig. 16.1e, the detected displacement increases linearly with time. With time frame simultaneously recorded, the real time motion speed can also be derived by dividing the width of one pitch (200 μm in this device) by the time interval between two adjacent voltage peaks. Investigation on the voltage within a period can substantially enhance the resolution in displacement detection. Starting from a status with complete overlap, a step motion test with each step of 5 μm was performed. The sensitive region is found to be from 10 to 190 μm, where each step motion of 1 μm can be clearly differentiated from the voltage response, as shown in Fig. 16.1f. The change of the voltage for each step is about 2.2 mV, and the Root Mean Square (RMS) of the noise $V_{noise}$ at 1 Hz is 0.38 mV. Consequently, the displacement resolution at bandwidth of 1 Hz can be calculated to be 173 nm.

The measured short-circuit current can also serve as the sensing signal for the displacement distance and the speed. Since $I_{SC}$ is alternating between negative and positive periodically during the motion, as displayed in Fig. 16.1g, the real time displacement can be derived (shown by the blue dots in Fig. 16.1h) by counting the number of zero-crossings, each of which counted as 100 μm step of displacement. Compared with the preset trajectory as indicated in red curve, the measurement matched very well. On the other hand, the amplitude of the $I_{SC}$ can also help to quantitative determine the real time speed, given small aspect ratio of the insulating layer. Figure 16.1i shows the plot of the magnitude of the detected current against motion speed from 5 μm $s^{-1}$ to 10 mm $s^{-1}$ and a linear fit that gives its sensitivity of (679 $\pm$ 4) pA/(mm $s^{-1}$) with adjusted $R_2$ = 0.99938. The resolution in dynamic speed detection is calculated to be 1.2 μm/s, given the RMS value (0.8 pA) of the noise.

The resolution for both displacement and speed sensing can be further improved by reducing the grating period. This new approach for displacement/speed sensing distinguishes itself from the existing technologies by a combination of self-powered, nanometer resolution, long detecting range and non-optical compacted structure, showing extensive potential applications in automation, manufacturing, process control, portable device, etc.

Besides, this kind of idea can also be easily extended into 2-dimension displacement sensing. Jing et al. extends this idea and utilizes 2 grating-structured to do displacement sensing [2]. They report a robust and self-powered kinematic vector sensor fabricated using highly pliable organic films that can be bent to spread over curved and uneven surfaces. The device derives its operational energy from a close-proximity triboelectrification of two surfaces: a polytetrafluoroethylene film coated with a two-column array of copper electrodes that constitutes the mover and a polyimide film with the top and bottom surfaces coated with a two-column aligned array of copper electrodes that comprises the stator. During relative reciprocations, the electrodes in the mover generate electric signals of ±5 V to attain a peak power density of ≥65 mW $m^{-2}$ at a speed of 0.3 $ms^{-1}$ [2].

## 16.2 TENGs as Self-powered Active Rotation Sensors

On the basis of the attached-electrode sliding-mode of TENGs, a disk TENG with segmental structures for harvesting rotational mechanical energy has been demonstrated [3]. The electricity can be generated from rotations through cyclic in-plane charge separation between the segments that have distinct triboelectric polarities. Enabled by the distinct relationship between the electrical output (its frequency and amplitude) and the rotational speed, the disk TENG can also serve as a self-powered angular speed sensor for rotation motions.

Later on, a free-rotating disk TENG (FRD-TENG) has been developed based on the freestanding-triboelectric-layer mode, which could also be employed as a self-powered rotation sensor for simultaneously detecting the rotation speed and vertical displacement [4], as shown in Fig. 16.2a–c. The structure, material and basic operation of the FRD-TENG has been described in Sect. 5.3.2. Both the rotation speed (r) and the vertical distance ($d$) have impact on the output performance of the FRD-TENG, so that it could then be employed as self-powered sensor for the active detection of the two parameters.

To elucidate its impact of the vertical separation distance ($d$), the output performances of the FRD-TENG under a series of distances increasing from 0 to 5 mm were measured at a constant rotation speed of 500 rpm. The measured $J_{SC}$ shows a decreasing trend from 2.0 to 0.5 mA/$m^2$, due to the reduction of electrostatic induction from the negatively-charged FEP layer. This relationship implied that the FRD-TENG could be employed as a self-powered displacement sensor, which can also be visualized through direct powering of LEDs. The self-powered displacement sensor could have unique applications like monitoring the thickness of the braking pad in automobiles. On the other hand, the $J_{SC}$ under a series of rotation speeds ranging from 100 rpm to 1000 rpm was measured with a constant vertical distance at $d = 0.5$ mm, and it could be clearly found that the $J_{SC}$ increased linearly with the rotation speed, since the output current was the time differentiation of the transferred charges across the external load. In the meanwhile, it is also shown in the figure that the frequency of the $J_{SC}$ also exhibits perfect linear relationship to the rotation speed. Therefore, the FRD-TENG could be employed as a self-powered rotation speed sensor by analyzing both the magnitude and frequency of the

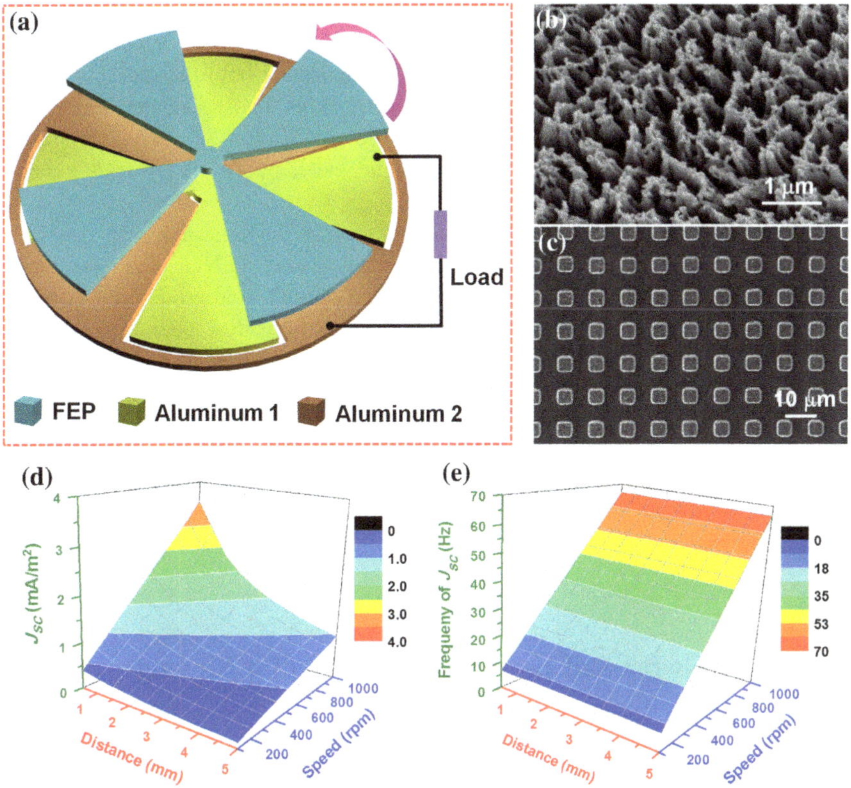


**Fig. 16.2** Self-powered vertical displacement and rotation speed sensor based on the free-rotating-disk triboelectric nanogenerator. **a** The schematic of the basic structure of the FRD-TENG composed of the freestanding FEP layer and the stationary aluminum layer. The *bottom inset* is the figure legend. **b** A 30°-tilted-view SEM image of the nanorod structure created on the surface of the FEP thin film. **c** A top view SEM image of the cubic micro-patterned structures on the aluminum foil. **d** A three-dimensional plot of the magnitude of the $J_{SC}$ depending on variable vertical displacements and rotation speeds. **e** A three-dimensional plot of the frequency of the $J_{SC}$ depending on variable vertical displacements and rotation speeds. Reproduced with permission from American Chemical Society [4]

measured $J_{SC}$. Thus, the rotation speed and the vertical separation distance were two input parameters influencing the output current signal, which could be characterized by the magnitude and the frequency. In order to illustrate this two-fold relationship and show a comprehensive working characteristic of the FRD-TENG as a self-powered sensor, two sets of experiments were carried out to give three-dimensional plots of the two-fold dependence of the magnitude (Fig. 16.2d) and frequency (Fig. 16.2e) on the vertical displacement and the rotation speed, respectively. As expected, the magnitude was affected by both the vertical displacement and the rotation speed, while the frequency was only affected by the rotation speed. From these results, the information about two important parameters of a mechanical motion could be solely determined by the measured profiles of the $J_{SC}$. Based on this principle, the FRD-TENG could be employed as a self-powered mechanical sensor in automobiles to detect both the radial and axial movement of a braking pad at the same time. This two-parameter monitoring approach provided us deeper insights on the application of nanogenerator in self-powered system and exhibited unique advantages than the existing single-parameter active sensors.

Through combing the structures of cylindrical TENGs that can serve as self-power rotation sensor [5] and case-encapsulated TENGs that can serve self-powered linear sliding sensors [6], a TENG-based dual-mode, self-powered velocity sensor for rectified linear and rotary motions has been reported [7]. Employing alternating Kapton-copper strips arranged in a spiral configuration on the inner and outer surfaces of two concentric cylinders, voltage assays for linear and rotary motions can be measured without applying an external power source. The triboelectric generated output signals when integrated with a digital circuit and a microcontroller unit (MCU) can be directly processed into velocity information. Linear speeds of 0.1–0.6 m/s and rotary speeds of 125–700 rpm have been measured with high stability.

## 16.3 TENGs for Self-powered Tracking of a Moving Object

By simply integrating multiple single-electrode triboelectric sensors into an array device, the self-powered motion tracking of an arbitrary moving object could be obtained with the information like the location, speed, and acceleration of the object. In this regard, the prototype of the first single-electrode TENG matrix has been developed for self-powered tracking of the moving object.

A self-powered triboelectric sensor (TES) structured with grounded electrodes has been reported by Yi et al. to monitor two-dimensional movements [8]. The trajectory and velocity of the moving object were detected through the measured open-circuit voltage and short-circuit current in real time. The velocity of the object in uniform motion can also be detected by utilizing the amplitude of the short-circuit current, and a sensitivity of $\sim$887 pA/(cm $s^{-1}$) was demonstrated.

As shown in Fig. 16.3a, a TES is composed of two components, with a 50 μm-thick layer of FEP at the top and patterns of 200 nm-thick Al films at the bottom. The upper surface of the FEP was dry-etched to produce aligned nanorod structures (Fig. 16.3b, c), which could increase the effective surface area and thus improve the triboelectric charge density on the surface. A photograph of an as-fabricated TES is presented in Fig. 16.3d, which has a device area of 10 cm × 10 cm. Each square Al electrode in the 4 × 4 matrix pattern has a side length of 1.5 cm, and the gap between two adjacent Al electrodes is 1 cm.

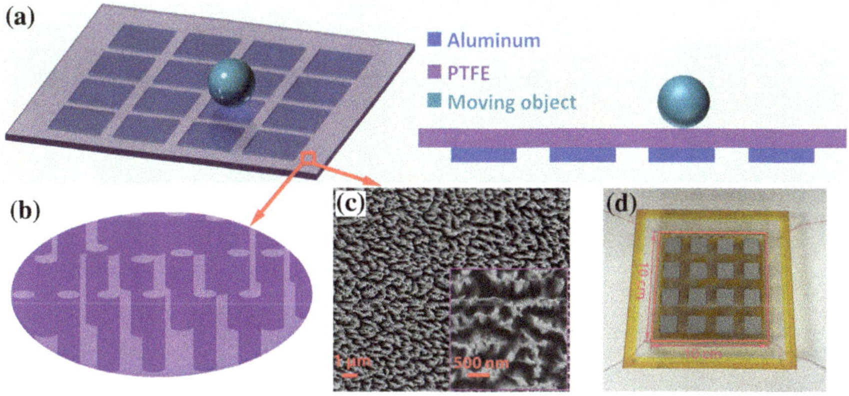


**Fig. 16.3** Structure of a triboelectric sensor. **a** Schematic diagram of the triboelectric sensor and its cross-section view. **b** Magnified schematic of the etched FEP surface. **c** FESEM image of the nanorod-structured FEP surface. The *inset* shows a view at a higher magnification. **d** Optical photograph of the triboelectric sensor, with each electrode having a size of 1.5 cm × 1.5 cm. Reproduced with permission from Wiley [8]

In order to quantitatively characterize the TES's response to a motion, the Al ball was attached to a linear motor, and the sensor was fixed under the ball to detect its movement. A multi-channel measurement system was utilized to record the output signals from each electrode separately. As exhibited in Fig. 16.4, when the Al ball moves on the TES along the path: electrode 11 (E11) → electrode 12 (E12) → electrode 13 (E13) → electrode 14 (E14) (Fig. 16.4a), the open-circuit voltage of the electrode that the ball passes through rises to a peak and then decreases (Fig. 16.4b). As discussed earlier, the peak value of the output voltage from a certain electrode occurs when the ball arrives at the point over the center of the electrode. In this regard, a two-dimensional contour plot of the values of output voltage from all electrodes is obtained when the ball gets to the center of an electrode, which explicitly indicates the moving path of the ball (Fig. 16.4c). In a similar way, when the ball moves along the path: E42 → E32 → E22 → E12 (Fig. 16.4d) and E11 → E22 → E33 → E44 (Fig. 16.4g), the recorded output voltage of the sensor (Fig. 16.4e, h) can clearly demonstrate the movement of the ball, and the mapping figures plotted at the moment when the ball arrives at the center of a particular electrode intuitively display the path that the ball passes through (Fig. 16.4f, i).

Since the interval between the centers of two neighboring electrodes is a constant that was determined when constructing the sensor, we can not only recognize when a moving object reaches which electrode but also acquire other information, such as the displacement, velocity, and acceleration of the object. Figure 16.5a–e show the measured results of the open-circuit voltage when the Al ball moves on the sensor with different velocities. Since there are hardly any responses from the electrodes out of the corresponding moving path, signals of those electrodes were omitted

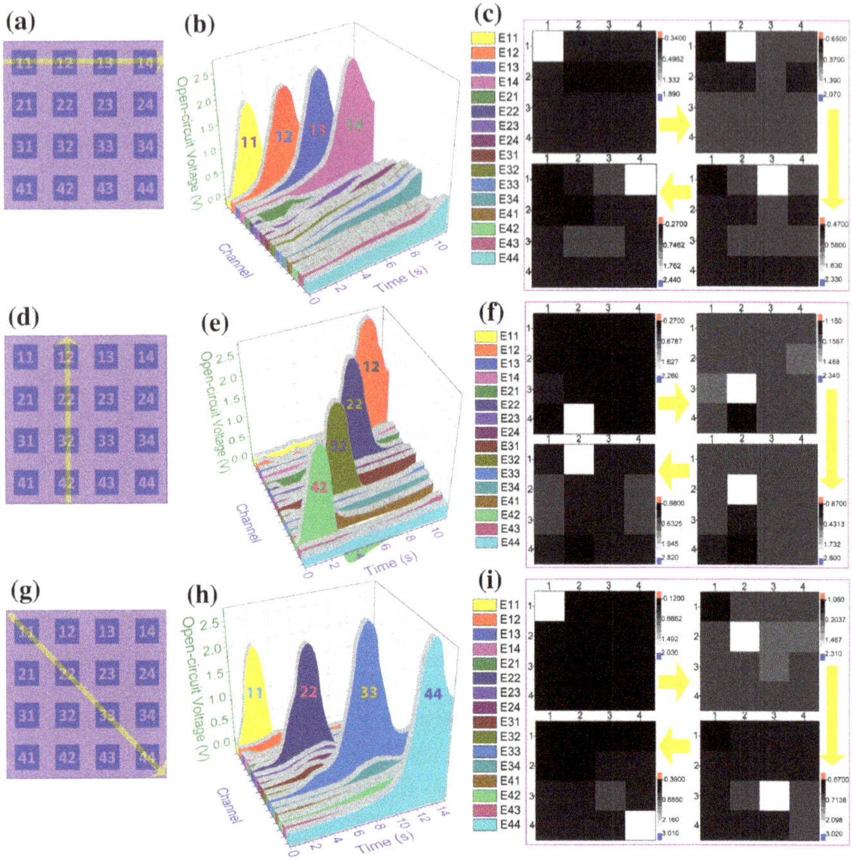

**Fig. 16.4** Motion imaging of an Al ball moving on the triboelectric sensor. **a**, **d**, **g** Illustration of the ball moving along a path (**a**) from E11 to E14, **d** from E42 to E12, **g** from E11 to E44. **b**, **e**, **h** Measured open-circuit voltage of the TES when the ball moves from (**b**) E11 to E14, (**e**) E42 to E12, (**h**) E11 to E44. **c**, **f**, **i** Mapping figures obtained from the values of open-circuit voltage when the ball gets to the point over the center of an electrode: (**c**) E11→E12→E13→E14, (**f**) E42→E32→E22→E12, (**i**) E11→E12→E13→E14. *Note* The Al ball moves at a preset speed of 20 cm $s^{-1}$. Reproduced with permission from Wiley [8]

from the graphs for simplicity. It should be noted that the same treatments were made to graphs discussed in the following paragraphs. By simply dividing the distance between two electrodes by the time taken to cover the distance, the velocity from one electrode to another can be derived. As presented in Fig. 16.5f, the detected velocity and the preset velocity of the ball between two electrodes have a very good match, with an average error of ~0.265 %. With the known location of each electrode on the sensor, the displacement of the moving ball over time can be easily derived, which is presented in Fig. 16.5g. Note that the origin of the displacement is at the edge of the device area. As the zero-crossing of the short-circuit

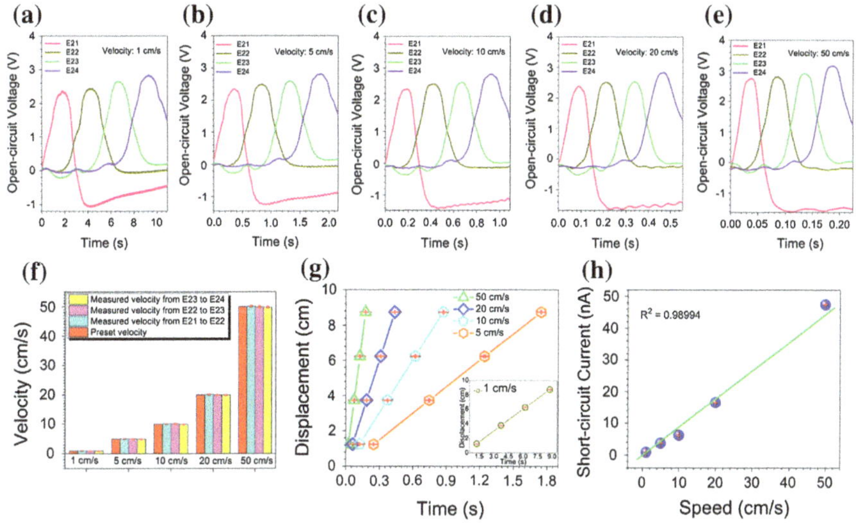


**Fig. 16.5** Detection of velocity and displacement. **a–e** Measured open-circuit voltage as an Al ball moves across a path (E21→E22→E23→E24) at a preset speed of (**a**) 1 cm $s^{-1}$, (**b**) 5 cm $s^{-1}$, (**c**) 10 cm $s^{-1}$, (**d**) 20 cm $s^{-1}$, (**e**) 50 cm $s^{-1}$. **f** Detected velocity of the ball by the triboelectric sensor and the preset velocity of the ball. **h** Plot of short-circuit current as a function of speed. Reproduced with permission from Wiley [8]

current is the instantaneous time at which the ball reaches the point over the center of the electrode, the velocity and displacement of the moving ball can also be detected by measuring the short-circuit current in a way similar to using the open-circuit voltage. Moreover, the velocity of the ball in uniform motion can be detected by analyzing the amplitude of the short-circuit current. The short-circuit current of the TES can be given as:

$$I_{sc} = dQ_{sc}/dt = (dQ_{sc}/dx) \times (dx/dt) = (dQ_{sc}/dx) \times v \qquad (16.1)$$

where $Q_{sc}$ is the short-circuit transferred charges, $x$ is the distance between the object and the electrode, and $v$ is the velocity of the object. For uniform motion, the amplitude of the short-circuit current can be given as:

$$I_{sc} - peak = (dQ_{sc}/dx)_{peak} \times v \qquad (16.2)$$

which means that the amplitude of the short-circuit current is linearly dependent on the velocity. The measured relationship between the amplitude of the short-circuit current and the velocity of the ball is exhibited in Fig. 16.5h, from which a sensitivity of ~887 pA/(cm $s^{-1}$) can be achieved with good linearity ($R^2 \approx 0.99$). This can be applied to detect the velocity of an object when recording time is not available, provided that the space charge distribution is the same. The value of the short-circuit current at a certain speed in Fig. 16.5h is the average peak value of short-circuit current from all the electrodes that the ball passes through during the travel.

Besides this work, a series of single-electrode triboelectric sensors [9] was integrated into a one-dimensional array of tube structures consisting of tube-shaped FEP thin film and grating copper electrodes for tracking the motion status (location, speed) of a steel ball moving inside the tubing, as shown in Fig. 16.6a. As the steel ball rolled across a Cu electrode, pulsed outputs of the open-circuit voltage and short-circuit current measured between the electrode and the ground reach 60 V and 0.4 μA, respectively. In order to track the traveling velocity of the steel ball inside the FEP tube, five Cu electrodes were evenly distributed along the FEP tube with a uniform interval and were connected in parallel to a positive probe of an electrometer. With the FEP tube having a tilt angle of 15° to the horizontal plane, the measured currents from devices having five grating electrodes were shown in Fig. 16.6b. The analysis in Fig. 16.6c provided two-fold information of the ball motion. First, the position of the ball could be identified based on the correspondence with the pulsed current signals; second, the average moving speed could be determined by the time interval between two consecutive current peaks. It could be found that the results from both analytic values and experimental data reached good agreement, indicating the validity of this measurement approach. The sensitivity and reliability of the TES were further evaluated with variable tilting angles. Moreover, to demonstrate the practical application of the TES in motion tracking, the device was developed into a seven-electrode structure for real time mapping of the ball location inside tubing, along with LED indicators from Labview software, as shown in Fig. 16.6d. From the real time current measurement result (Fig. 16.6e) and the turning on of LED indicators, it could be confirmed that this self-powered motion tracking system could clearly identify the motion of the steel balls inside tubing. This work had demonstrated the potential application of the TES in pipeline blockage inspection.

## 16.4 TENG as Self-powered Acceleration Sensors

Acceleration is one of the most important parameters in the characterization of an object's movement. Usually, the acceleration sensors are based on force sensing mechanisms, including piezoresistive or differential capacitance, which has a wide variety of applications for industrial engineering, biology and navigation. An external power source is indispensable to drive these sensors. For realizing the self-powered operation, a newly designed 3D-TENG has been developed as an active acceleration sensor [10]. The spherical TENG consists of two spheres, as shown in Fig. 16.7a. An ordinary rubber ball was coated with etched Perfluoroalkoxy-alkane (PFA) film to form a PFA ball, where the PFA film acts as a triboelectric polymer. The inner PFA ball can slide and bounce freely inside the transparent shell. A thin Al foil was adhered to the inner surface of the outer sphere, acting as both the triboelectric layer and the electrode. Figure 16.7b displays a

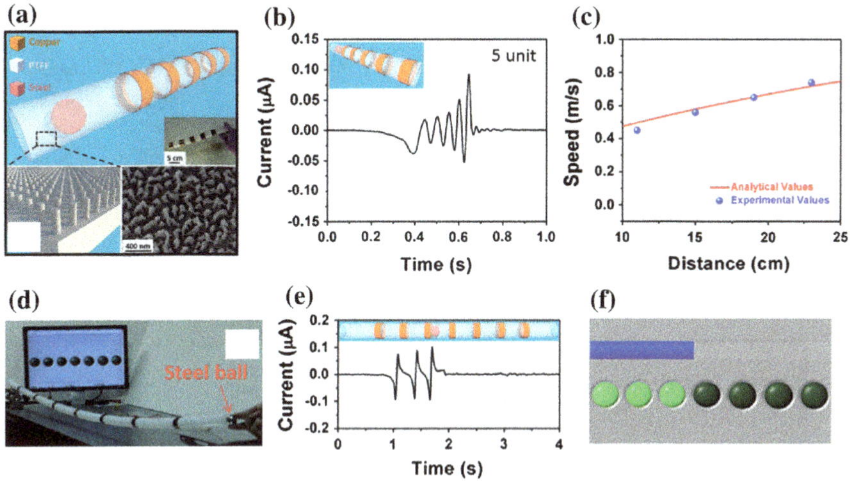


**Fig. 16.6** Triboelectric sensor for self-powered tracking of object motion inside a tube. **a** The schematic structure of the tube-shaped triboelectric sensor that could track the motion of an object inside the tube. **b** The measurement of the output current of the single-electrode triboelectric sensor with five grating electrodes. **c** The calculated motion speed of the steel ball with variable distances from both experimental and analytical results. **d** A photograph showing the real time measurement of the motion tracking of the steel ball inside tubing. **e** The measurement result of the output current when the steel ball passed the first three grating electrodes. **f** The real time location visualization of the ball motion inside tubing. Reproduced with permission from American Chemical Society [9]

corresponding photograph of TENG. In order to enhance the performance of the TENG, the PFA film was dry-etched using inductively coupled plasma (ICP) to create the nanoparticle-like structures on the surface. The fabricated TENG can scavenge the vibration energy in full space by working at a hybridization of both the contact-separation mode and the sliding mode.

When the spherical TENG serves as a self-powered acceleration sensor, the output voltages of the sensor at specific acceleration (1.3–2 g) are enumerated in Fig. 16.7c–g. It can be seen clearly that all of the output voltage signals are uniform and stable. The relationship between output voltage and acceleration is plotted in Fig. 16.7h, revealing a clear linear relationship by fitting the data, which is beneficial for practical applications of the sensors. The detection sensitivity of the acceleration was calculated to be about 15.56 V/g.

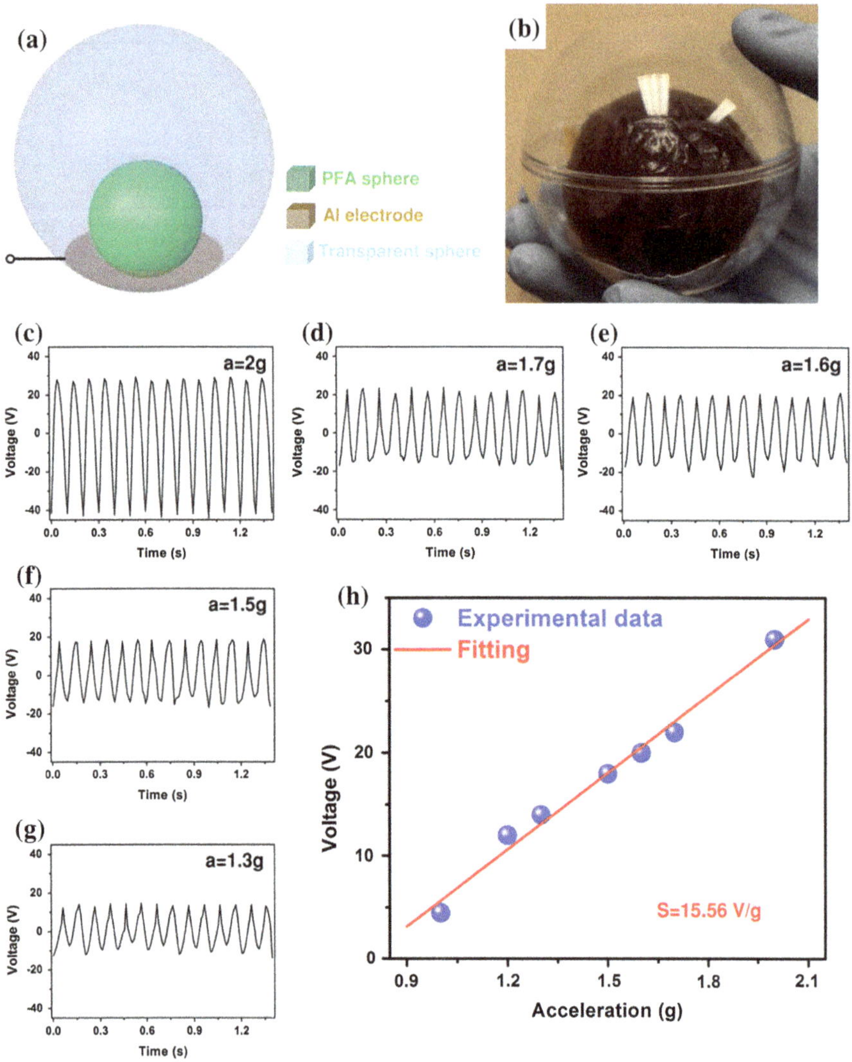


**Fig. 16.7** Triboelectric nanogenerator as a self-powered acceleration sensor. **a** The schematic diagram showing the device structure of a spherical TENG. **b** Photograph of a fabricated TENG. **c–g** Output voltage of the sensor subjected with certain acceleration. **h** Dependence of the output voltage on the acceleration. Reproduced with permission from Wiley [10]

## 16.5 Summary

Under mechanical triggering, TENGs can produce electrical signal even without applying a power source, which can be used for tracking a moving object. This chapter reviews some of the existing studies in the TENG sensors to sense displacement, rotation, velocity, and acceleration. Their working principle, design, performance and applications have been presented in this chapter. The tracking can be performed both qualitatively and quantitatively. The peak and/or zero-crossing position of $V_{OC}$ and $I_{SC}$ signal can provide qualitatively information of position and/or velocity. The magnitude of $V_{OC}$ and $I_{SC}$ signal has been shown to have a linear relationship with mechanical trigger intensity. Such technologies could be potentially important for environmental monitoring, infrastructure monitoring and biomedical applications.

## References


1. Y.S. Zhou, G. Zhu, S.M. Niu, Y. Liu, P. Bai, Q.S. Jing, Z.L. Wang, Nanometer resolution self-powered static and dynamic motion sensor based on micro-grated triboelectrification. Adv. Mater. **26**(11), 1719–1724 (2014)
2. Q.S. Jing, Y.N. Xie, G. Zhu, R.P.S. Han, Z.L. Wang, Self-powered thin-film motion vector sensor. Nat. Commun. **6**, 8031 (2015)
3. L. Lin, S.H. Wang, Y.N. Xie, Q.S. Jing, S.M. Niu, Y.F. Hu, Z.L. Wang, Segmentally structured disk triboelectric nanogenerator for harvesting rotational mechanical energy. Nano Lett. **13**(6), 2916–2923 (2013)
4. L. Lin, S.H. Wang, S.M. Niu, C. Liu, Y.N. Xie, Z.L. Wang, Noncontact free-rotating disk triboelectric nanogenerator as a sustainable energy harvester and self-powered mechanical sensor. ACS Appl. Mater. Inter. **6**(4), 3031–3038 (2014)
5. P. Bai, G. Zhu, Y. Liu, J. Chen, Q.S. Jing, W.Q. Yang, J.S. Ma, G. Zhang, Z.L. Wang, Cylindrical rotating triboelectric nanogenerator. ACS Nano **7**(7), 6361–6366 (2013)
6. Q.S. Jing, G. Zhu, P. Bai, Y.N. Xie, J. Chen, R.P.S. Han, Z.L. Wang, Case-encapsulated triboelectric nanogenerator for harvesting energy from reciprocating sliding motion. ACS Nano **8**(4), 3836–3842 (2014)
7. Q.S. Jing, G. Zhu, W.Z. Wu, P. Bai, Y.N. Xie, R.P.S. Han, Z.L. Wang, Self-powered triboelectric velocity sensor for dual-mode sensing of rectified linear and rotary motions. Nano Energy **10**, 305–312 (2014)
8. F. Yi, L. Lin, S.M. Niu, J. Yang, W.Z. Wu, S.H. Wang, Q.L. Liao, Y. Zhang, Z.L. Wang, Self-powered trajectory, velocity, and acceleration tracking of a moving object/body using a triboelectric sensor. Adv. Funct. Mater. **24**(47), 7488–7494 (2014)
9. Y.J. Su, G. Zhu, W.Q. Yang, J. Yang, J. Chen, Q.S. Jing, Z.M. Wu, Y.D. Jiang, Z.L. Wang, Triboelectric sensor for self-powered tracking of object motion inside tubing. ACS Nano **8**(4), 3843–3850 (2014)
10. H.L. Zhang, Y. Yang, Y.J. Su, J. Chen, K. Adams, S. Lee, C.G. Hu, Z.L. Wang, Triboelectric nanogenerator for harvesting vibration energy in full space and as self-powered acceleration sensor. Adv. Funct. Mater. **24**(10), 1401–1407 (2014)

# Chapter 17
# Self-powered Sensing for Chemical and Environmental Detection

In the last three chapters (Chaps. 14–16), we have discussed the applications of triboelectric nanogenerators (TENGs) as self-powered active sensors for detecting parameters of the stimulating mechanical input, such as pressure, displacement, speed, acceleration, and vibration, etc. The working principle of all those active sensors mainly relies on distance change of two triboelectric surfaces with respect to the external stimuli, which leads to variation of the open-circuit voltage and short-circuit current. The triboelectric charge density is another critical factor affecting the output performance of TENGs. It is an intrinsic parameter and usually taken as constant in prior active sensor devices. However, the surface charge density of the triboelectric surfaces can be tuned by changing the chemical or physical conditions of the surface materials. In this regard, the TENG could also be utilized as self-powered sensors for detection of certain chemical species in a solution, when the triboelectric surface has been modified in the solution. On the basis of this concept, this chapter summarizes a few chemical and environmental sensors that have been developed for detection of mercury ions, catechin, phenol, and UV intensity.

## 17.1 Self-powered Chemical Sensors

Mercury has been known to be highly toxic metal and can pose serious dangers to the environment and human health. Therefore, monitoring the mercury concentration is an extremely important issue to prevent such toxic metal from endangering human life. $Hg^{2+}$ ions are by far the most stable inorganic form of mercury, which are non-biodegradable and easy to be accumulated in organs. Even at a very low concentration, they can still be fatal to human brain, heart and kidney. Multiple approaches such as atomic absorption/emission spectrometry [1] and inductively



Z.L. Wang et al., *Triboelectric Nanogenerators*,
Green Energy and Technology, DOI 10.1007/978-3-319-40039-6_17

coupled plasma mass spectrometry (ICP-MS) [2] have been applied to detect the $Hg^{2+}$ ions in environmental and biological samples. Among these approaches, ICP-MS is the one that has the highest sensitivity and the widest linear range. However, the sample preparation before measurement is rather complicated and time-consuming, and requires expensive instrumentation and noble gas. Another drawback of this method is the difficulty in performing in-field analysis. In consideration of the above disadvantages, researchers have instead developed a number of optical methods (colorimetric [3, 4], fluorometric assays [5–7] and systems based on surface plasmon resonance [8], surface-enhanced Raman scattering [9] and electrochemical sensors [10, 11]) for the detection of $Hg^{2+}$ ions. Optical approaches are advantageous in high sensitivity and selectivity, and practicable for in-field analysis, but they are involved with sophisticated chemistry in incorporating organic probes, such as crown ethers, porphyrins, specific proteins, deoxyribonucleic acid (DNA) and polymers, which substantially limits their application range.

In order to overcome the limitations of these methods, the concept of self-powered nanosensors has recently been proposed and proved of potential toward toxic pollutant detection [12]. The working principle of self-powered nanosensors is based on combining/integrating nanogenerators with sensors. Nanogenerators here are harvesting energy from the environment [13–16] to power the sensors. Due to its convenient monitoring mechanism, the self-powered nanosensors could be the most desirable and promising prototype for environmental protection/detection in the near future because no battery is needed to power the device. In this work, it is proved that the principle of the TENG can be used as a sensor for the detection of $Hg^{2+}$ ions [17]. The first step is to improve the performance of the TENG through the assembly of Au nanoparticles (NPs) onto the metal plate. These assembled Au NPs not only act as steady gaps between the two plates at strain free condition, but also enable the function of enlarging the contact area of the two plates, which will increase the electrical output of the TENG. Through further modification of 3-mercaptopropionic acid (3-MPA) molecules on the assembled Au NPs, the high-output nanogenerator can become a highly sensitive and selective nanosensor for $Hg^{2+}$ ions detection because of the different triboelectric polarities of Au NPs and $Hg^{2+}$ ions. On the basis of this unique structure, the output voltage and current of the triboelectric nanosensor (TENG) reached 105 V and 63 μA with an effective dimension of 1 cm × 1 cm. Under optimum conditions, this TENG is selective for the detection of $Hg^{2+}$ ions, with a detection limit of 30 nM and linear range from 100 to 5 μM. The commercial LED lamp was tested as the indicator to replace the expensive electrometer and showed the potential to simplify the detection system. This study demonstrates an innovative and unique approach toward self-powered detection of $Hg^{2+}$.

The TENG has a layered structure based on two plates as schematically shown in Fig. 17.1a. Glass was selected as the material for substrate due to its strength, light weight and low cost. On the lower side, the metal plate is prepared. The metal plate consists of an Au thin film and assembled Au NPs, playing dual roles of electrode and contact material. By utilizing 1,3-propanedithiol as the linker molecules, 13, 32, and 56 nm Au NPs were uniformly assembled onto the surface of Au film

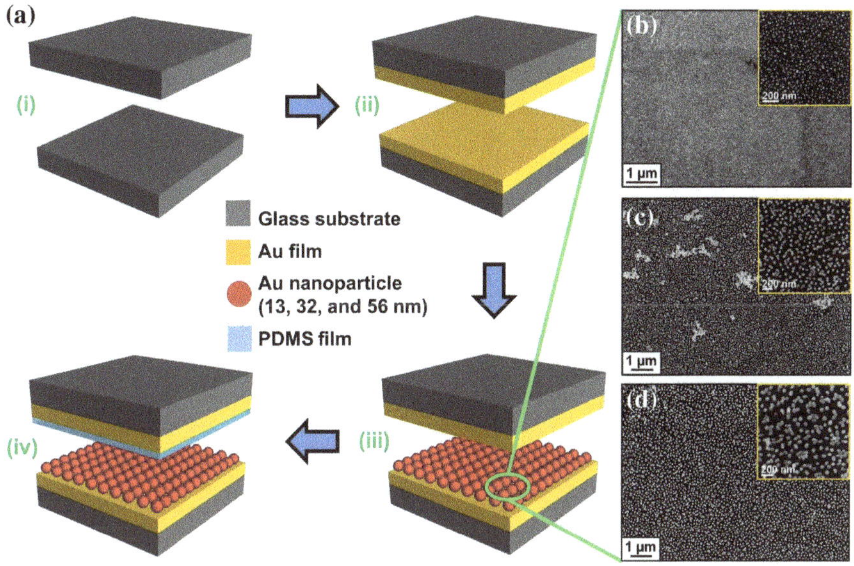


**Fig. 17.1** **a** Fabrication process of the triboelectric nanogenerator (TENG). **b–d** SEM images of the Au film modified with (**b**) 13 nm, (**c**) 32 nm, and (**d**) 56 nm AuNPs. The TENG has an effective dimension of 1 cm × 1 cm. Scale bars: (**b–d**) 1 mm; *insets* 200 nm. Reproduced with permission from Wiley [17]

(Fig. 17.1b–d). Au NPs were prepared by the reduction of $Au^{3+}$ ions with sodium citrate, which also acted as a capping agent to stabilize the as-prepared Au NPs. By carefully controlling the amount of sodium citrate in the synthesis process, different sized Au NPs were synthesized. Compared with other metal NPs, Au NPs are featured with greater long-term stability and modifiable surface chemistry, and can be synthesized with narrower size distributions. For the purpose of selective detection of $Hg^{2+}$ ions, 3-MPA is self-assembled onto the surface of Au NPs through strong Au-S interactions, and its "recognition" toward $Hg^{2+}$ ions induces $Hg^{2+}$ ions will selectively bind to the Au NP surface. On the other plate, another Au thin film is laminated between the glass substrate and a layer of polydimethylsiloxane (PDMS). PDMS is purposely chosen here for its advantages of flat surface and easy processing. In this designed device, PDMS and Au are of extreme differences in attracting and retaining electrons in the triboelectric series and can achieve the best results.

The TENG works in the vertical contact-separation mode, and its working principle has been extensively discussed in Chap. 2. The typical electrical output performance of the TENG was measured when it was triggered by a mechanical shaker. An open-circuit voltage and short-circuit current of up to 105 V and 63 $\mu A/cm^2$ could be achieved in the measurement, which also depend on the size of Au NPs. Both the $V_{OC}$ and $I_{SC}$ will be elevated with the size of Au NPs increasing from 13 to 56 nm.

The TENG was then employed as a self-powered device for detecting molecular species. Figure 17.2a shows the output of the TENG by using a full-wave rectifying bridge. For the selective detection of $Hg^{2+}$ ions, tris–borate (50 mM, pH 9) was used as the buffer solution. The Au plate was soaked in the buffer solution containing various concentrations of $Hg^{2+}$ ions and other metal ions at ambient temperature. The reaction time was optimized as 60 min. Then the Au plate was washed by water and dried at ambient temperature prior to the electrical measurement. After the interaction with $Hg^{2+}$ ions of 5 μM, the generated $I_{sc}$ density of TENG decreased from 63 to 8 μA/cm$^2$ (Fig. 17.2b). This is because the adsorbed molecular species modify the surface triboelectric polarity, using which we can detect the concentration of the $Hg^{2+}$. By integrating TENG with a light-emitting diode (LED) lamp (indicator), we could build a fully stand-alone and self-powered environmental sensing device. Inset in Fig. 17.2b shows the LED lamp totally went out when detecting the $Hg^{2+}$ ions of 5 μM. As a proof of concept, we demonstrated the TENG with the capability to be an environmental sensor without any supporting equipment (power source, capacitor, and electrometer). Figure 17.2c shows that the

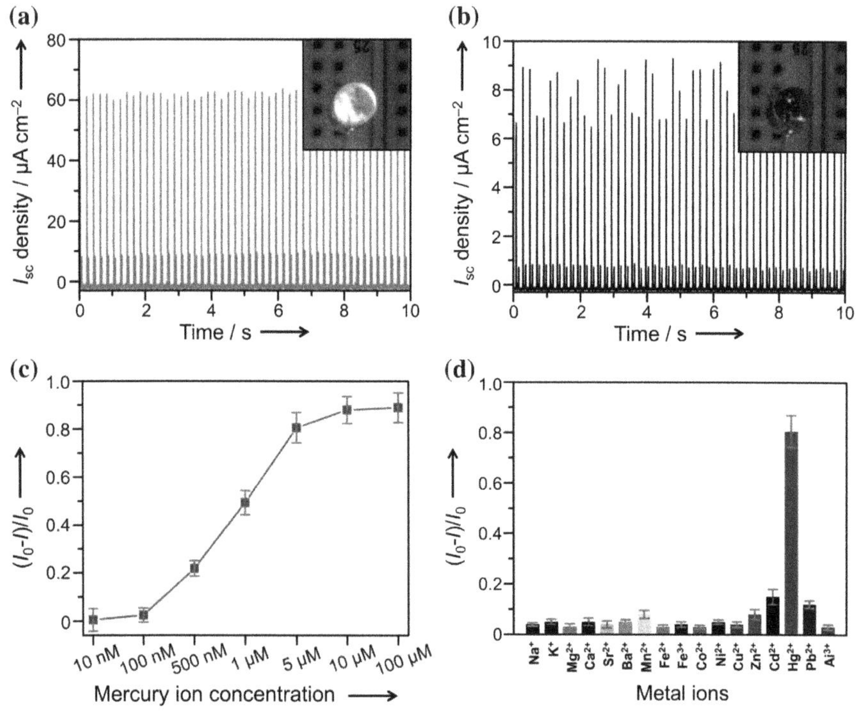


**Fig. 17.2** **a**, **b** Rectified Jsc of the as-developed TENG before (**a**) and after (**b**) the interaction with 5 mm $Hg^{2+}$ ions. *Insets* Photograph of the indicated LED lamp before (**a**) and after (**b**) interaction with 5 mm $Hg^{2+}$ ions, as an indication of detected concentration. **c** Sensitivity and **d** selectivity of the as-developed TENG for the detection of $Hg^{2+}$ ions. The concentration of all metal ions tested in the selectivity experiment was 5 mm. Reproduced with permission from Wiley [17]

electrical signal of the TENG decreased upon increasing the concentration of $Hg^{2+}$ ions, with a linear relationship between the short-circuit current ratio $((I_0 - I)/I_0)$ and the concentration of $Hg^{2+}$ ions ranging from 100 nM to 5 μM ($R^2 = 0.98$). This approach provides a detection limit at an S/N of 3 at 30 nM for $Hg^{2+}$ ions. Control experiments were carried out to test the selectivity of the developed system toward $Hg^{2+}$ ions detection as compared to other metal ions (each at a concentration of 5 μM). The results displayed in Fig. 17.2d reveals that the sensing system is specific to $Hg^{2+}$ ions, mainly because comparing to Au NPs, the $Hg^{2+}$ ions have lower tendency to transfer the electrons to PDMS. The high selectivity of 3-MPA toward $Hg^{2+}$ ions is resulted from the carboxylic acid which has much higher binding affinity towards $Hg^{2+}$ ions among all metal ions ($\log\beta_4 = 17.6$). The potential interference metal ions could not bind to 3-MPA modified Au NPs, resulting in negligible changes in the electrical signal of the TENG.

Above we have shown the first triboelectric effect-based sensor for the detection of $Hg^{2+}$ ions by using 3-MPA modified Au NPs as electrical performance enhancer and recognition element. Based on this concept, a few more works have been completed to demonstrate the application of triboelectric active sensors for detecting other chemical species, such as catechin [18]. The TENG for catechin detection has a layered structure based on two plates as schematically shown in Fig. 17.3a. On the lower side, a layer of $TiO_2$ nanomaterial array is prepared as one of the contact materials. $TiO_2$ nanowire (TNW) array and $TiO_2$ nanosheet (TNS) array grown on Ti foil were prepared via a two-step synthesis process. Firstly, $H_2Ti_2O_4(OH)_2$ nanowire array and $H_2Ti_2O_4(OH)_2$ nanosheet array were synthesized on Ti foil through a hydrothermal route in alkali solution combined with an ion-exchange process (acid treatment). Higher temperature causes a faster reaction between Ti foil and NaOH, hence formed a denser $Na_2Ti_2O_4(OH)_2$ nanowire array. Then TNW array and TNS array can be obtained via heat treatment of the as-prepared $H_2Ti_2O_4(OH)_2$ nanowire array and $H_2Ti_2O_4(OH)_2$ nanosheet array. The mean diameters of TNW and TNS determined from scanning electron microscopy (SEM) images are 67.2 nm (Fig. 17.3b) and 9.4 nm (Fig. 17.3c), respectively. The oriented nanowire array and nanosheet array are perpendicularly grown on the Ti foil, and the respective thicknesses of these films are 4.2 and 0.8 μm. On the other plate, polyethylene terephthalate (PET) was selected as the material for substrate due to its flexibility, light weight, and low cost. A Cu thin film is laminated between the PET substrate and a layer of polytetrafluoroethylene (PTFE). PTFE has the advantages of flat surface and easy processing. According to the triboelectric series, PTFE is triboelectric negative and has a large difference in ability to attract and retain electrons as compared to $TiO_2$, which will contribute a large electrical output of TENG.

It was then demonstrated that the as-developed TENG can be utilized as a self-powered nanosensor for catechin detection. Phosphate solution (20 mM, pH 7.0) was used as the buffer and the reaction time was optimized to be 20 min. After the reaction, the $TiO_2$ plate was washed by water and dried at ambient temperature prior to the electrical measurement. After reacting in 0.5 mM catechin solution for 20 min and then dried, the generated $V_{OC}$ and $J_{SC}$ were increased by 5 and 2.9

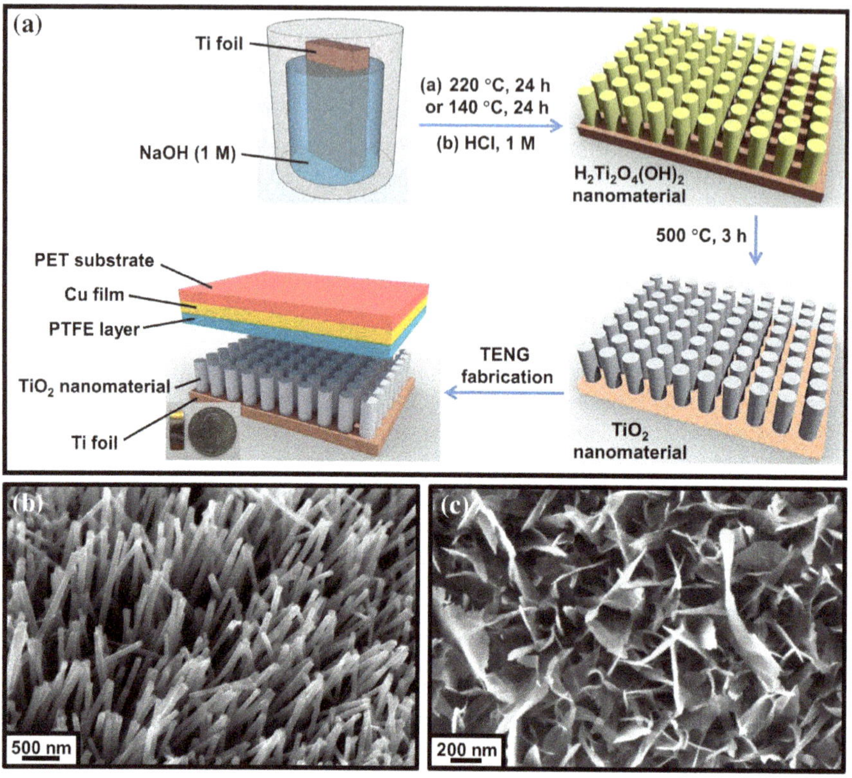


**Fig. 17.3** **a** Fabrication process of the TENG. **b** and **c** SEM images of the Ti foil covered with TNW array (**b**) and TNS array (**c**). Reproduced with permission from American Chemical Society [18]

times, respectively. Figure 17.4a, b manifest that the open-circuit voltage ratio $((V_{OC} - V^o_{OC})/V^o_{OC})$ of TNW array-based TENG and TNS array-based TENG both increased upon adding the catechin concentration. The linear relationships between the open-circuit voltage ratio and the catechin concentrations ranging from 10 μM to 0.5 mM ($R^2 = 0.98$) for TNW array-based TENG and 10 μM to 0.25 mM ($R^2 = 0.97$) for TNS array-based TENG. The relationships between the short-circuit current density ratio $((J_{SC} - J^o_{SC})/J^o_{SC})$ and catechin concentration are similar to that in Fig. 17.4a, b. The different linear ranges of TNW array-based TENG and TNS array-based TENG toward catechin detection is probably attributed to the different surface areas of TNW array and TNS array. $TiO_2$ nanomaterial array with larger surface area can adsorb more catechin molecules, which consequently contribute to the detection of catechin solution at higher concentration. The typical isotherms for nitrogen adsorption and desorption of the TNW array and TNS array are displayed in Fig. 17.4c–d, which categorize them as type IV according to IUPAC classification. The determined BET specific surface areas of TNW array and TNS array are 3.3 and 2.5 $m^2/g$, respectively, showing that the TNW array has

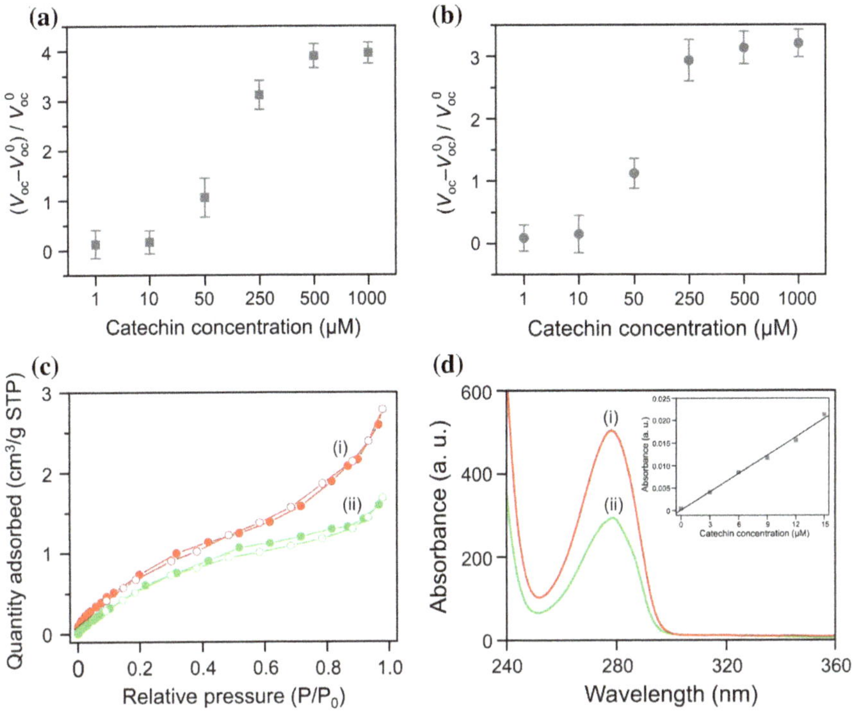


**Fig. 17.4** **a** and **b** The generated open-circuit voltage ratio $((V_{OC} - V_{OC,0})/V_{OC,0})$ of TNW array-based TENG (**a**) and TNS array based TENG (**b**) when sensing catechin solutions with different concentrations. **c** N2 adsorption and desorption isotherms of (*i*) TNW array and (*ii*) TNS array. *Solid circles* Adsorption; *hollow circles* Desorption. **d** Absorption spectra of the supernatants of catechin solutions containing (*i*) TNW array and (*ii*) TNS array. Inset of (**d**): the calibration curve of catechin concentration from absorption spectra. Reproduced with permission from American Chemical Society [18]

a larger surface area than that of TNS array. In the measurement of the saturated adsorption concentration of catechin, TNW array and TNS array were immersed in 10 μM solutions of catechin for 1 h. The absorption spectra of the related supernatants show the saturated adsorption concentrations of catechin on TNW array and TNS array are 6 and 3 μM, respectively. From these results, we can come to the conclusion that the TNW array has a larger surface area than that of TNS array, and hence contributes to the broader linear range toward catechin detection.

## 17.2 Self-powered UV Sensors

Ultraviolet (UV) photodetectors have been widely used in communications, biological and chemical analysis, environmental monitoring, remote control, memory storage, and optoelectronic circuits [19–21]. Previously, these self-powered UV photodetectors are mainly based on photoelectrochemical cell [22] and p-n junction [23]. The UV photodetectors based on photochemical cell have showed impressive performances in the literature, but they normally need to use $I^-/I_3^-$ redox couple as the electrolyte, which is not ideal for long-term operation. The $I^-/I_3^-$ redox couple is highly corrosive, photoreactive, volatile, and will interact with common metallic components and sealing materials. Moreover, the photodetectors based on p-n junction have the advantages of low applied field, fast response time, and no oxygen dependency. However, the sensitivity (6 mW $cm^{-2}$) could become a concern when detecting UV light with lower intensity, which is critical for environmental monitoring and medical therapy treatment.

To overcome the above disadvantages, a fully integrated photodetector for UV light sensing based on TENG conformation has been demonstrated [24]. 3D dendritic $TiO_2$ nanostructures synthesized using a chemical bath deposition method were served as not only the built-in photodetector and but also as one of the contact materials of the TENG. By utilizing polymethyl methacrylate (PMMA) and spring to assemble the TENG-based photodetector, a stand-alone and self-powered environmental sensing device is developed. Through a periodic impact on the TENG by a linear motor or biomechanical movement (finger tapping), a fast 18 ms rise and a 31 ms decay time, a high responsivity reaching 280 A $W^{-1}$, and a wide linear detection range starting from 20 μW $cm^{-2}$ to 7 mW $cm^{-2}$ were obtained.

The fabrication process of the active UV photodetector was detailed in Fig. 17.5a. The demonstrated device is constituted by a layered structure of two plates. On the lower side, the $TiO_2$ plate is prepared. Firstly, 3D dendritic $TiO_2$ nanostructures were designed to grow directly onto a glass substrate through a facile, scalable, and low cost chemical bath deposition method. Glass was chosen as the substrate due to its high transparency, strength, low cost and light weight. During the growth of $TiO_2$ crystal, Ti(III) species were gradually oxidized to Ti(IV) species by dissolved $O_2$ and then hydrolyzed to form $TiO_2$. Controlling the oxidation and hydrolysis of the Ti(III) precursor ($TiCl_3$) is the key factor to obtain $TiO_2$ nanomaterials with different morphologies. Unlike the previous approach, we prepared the 3D dendritic $TiO_2$ nanostructures instead of single layer of vertically aligned $TiO_2$ nanoplatelet arrays by increasing the reaction time and growth cycles. Figure 17.5b displays the SEM image and photograph of 3D dendritic $TiO_2$ nanostructures coated glass, which can be easily observed with naked eyes that the nanostructures are uniformly distributed on the substrate. The enlarged SEM image (Fig. 17.5c) further reveals the nanostructures are dendrite-like particles with an average size of 2 μm and composed by nanoplatelets with an average width of 28 nm and length of 160 nm, respectively. The crystal phase of the as-prepared 3D dendritic $TiO_2$ nanostructures is identified by Raman spectroscopy (Fig. 17.5d). The strong Raman bands appearing at 131, 243, 440, and 606 $cm^{-1}$ can be ascribed to the $B_{1g}$, two-phonon scattering, $E_g$, and $A_{1g}$ modes of rutile phase, indicating the structure is mainly constructed by rutile $TiO_2$ nanoplatelets. The 3D nanostructures fabricated in this study are of great interest not only unique in the structural characteristics, but also remarkable in the electrical, chemical and optical properties. Then a colorless thin film of indium tin oxide (ITO) was deposited on the back of

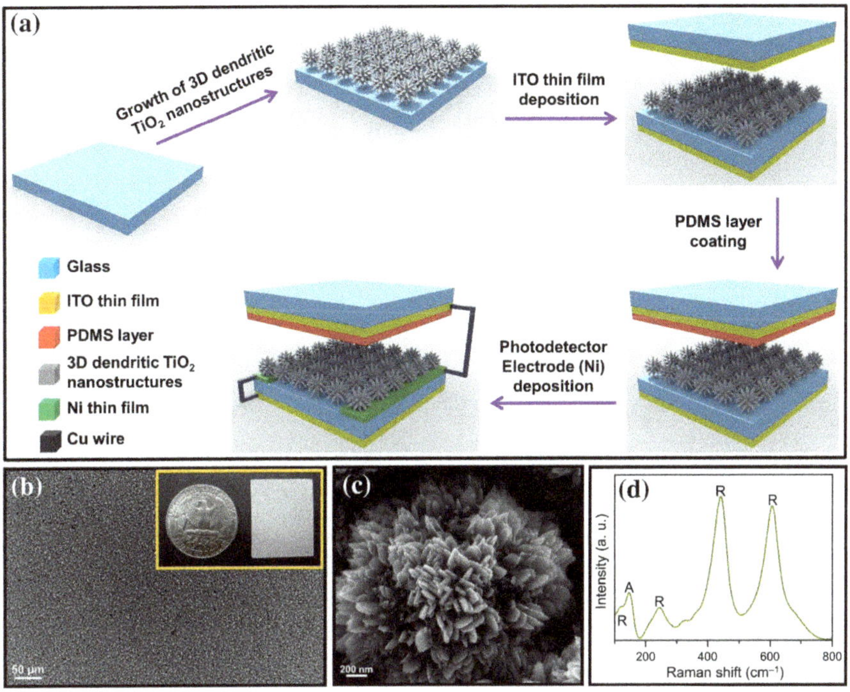


**Fig. 17.5** **a** Fabrication process of the self-powered UV photodetector. **b**, **c** SEM images and (**d**) Raman spectrum of the 3D dendritic $TiO_2$ nanostructures coated glass substrate. Reproduced with permission from Wiley [24]

the 3D dendritic $TiO_2$ nanostructures coated glass substrate as the conducting electrode. On the other plate, another ITO thin film is deposited between a glass substrate and a layer of polydimethylsiloxane (PDMS). PDMS is purposely chosen here for its advantages of high transparency, easy processing, and flat surface. Finally, the Ni thin film was selected to deposit as the electrodes of the built-in photodetector to ensure the Schottky contact can be created. The photodetector with Schottky contact has been proved to have the advantages of low dark current, high responsivity, and direct compatibility with modern high-speed integrated circuitry. The built-in photodetector is then connected with the TENG to construct the active UV photodetector. The built-in photodetector in the device can function as a photoresistor. The resistance of the built-in photodetector varies upon UV light irradiation with different intensities, which will consequently affect the capability of the surface charge transfer during triboelectrification process and the output of the TENG. Therefore, the light intensity can be determined by monitoring the output of the TENG.

The stability of the active UV photodetector is first clarified by measuring the photoresponse switching behavior of the built-in photodetector. As serially connecting the build-in photodetector to TENG, a new type of self-powered UV

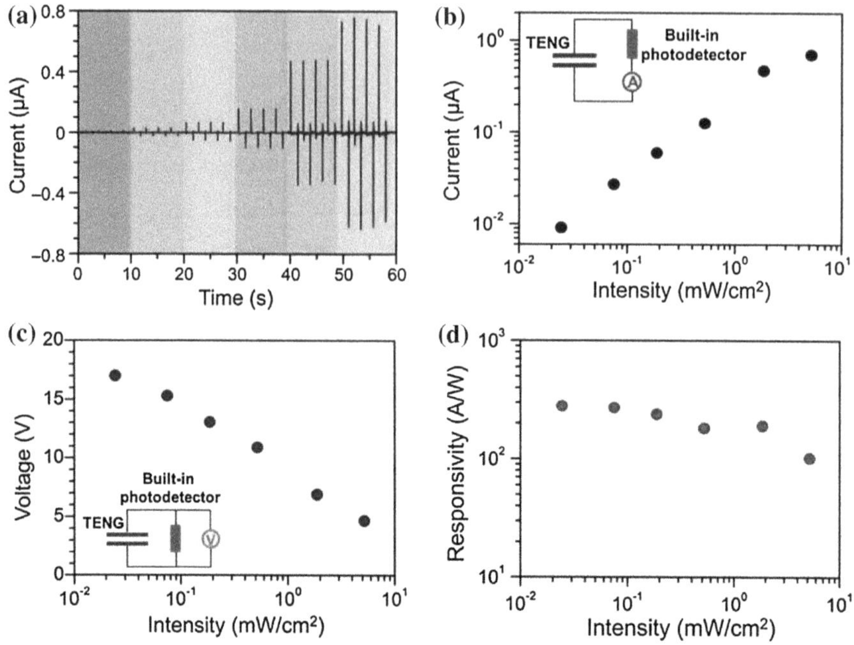


**Fig. 17.6** **a** Output current of the self-powered photodetector under UV light illumination with various power intensities. **b–d** The dependence of output current (**b**), output voltage (**c**), and responsivity (**d**) of the self-powered photodetector on UV light illumination with various power intensities. Reproduced with permission from Wiley [24]

photodetector was demonstrated. In the dark, the resistance of built-in photodetector is larger, hence the output current will be lower and the output voltage will be higher and closed to the open circuit voltage of $V_{OC}$. While under UV light irradiation, the resistance of built-in photodetector drops, and the output current of the self-powered photodetector when sensing UV light at different intensity is displayed in Fig. 17.6a. The output current increases from 2 nA (in the dark) to 0.7 μA when the power intensity of incident UV light is varied to 7 mW $cm^{-2}$. The dependence of the output current on the incident light power density can be seen more clearly from Fig. 17.6b. It is found that the output current has a linear relationship with the incident light power density from 20 μW $cm^{-2}$ to 7 mW $cm^{-2}$.

Comparatively, the output voltage diminishes to 4.8 V when the power intensity of incident UV light is increased to 7 mW $cm^{-2}$. Although a linear relationship between the output voltage and incident light power intensity is also observed (Fig. 17.6c), the change of the output current toward incident UV light with different intensities can reach up to 2 orders and is more convenient to be used to determine the UV light intensity. The responsivity of the active UV photodetector is displayed in Fig. 17.6d. This active UV photodetector exhibits an excellent capacity to detect the incident light intensity from 20 μW $cm^{-2}$ to 7 mW $cm^{-2}$, which is more sensitive than the self-powered UV photodetector based on p-n junction. Finally PMMA and spring were used to assemble the self-powered UV photodetector and demonstrate as a portable device for UV light detection. Under periodic external mechanical deformation by biomechanical movements (finger tapping, 23 N), a linear relationship between the output current and power intensity of incident light similar to Fig. 17.6b is observed. All of these reported results indicate that the as-developed active UV photodetector based on TENG shows great potential as a self-powered UV photodetector in the near future.

In summary, an integrated and active UV photodetector based on TENG configuration has been discussed. Dendritic $TiO_2$ nanostructures were applied as the built-in photodetector and contact material of the TENG. The easy-fabricated, cost-effective, and robust TENG-based photodetector provides a rapid response time (rise time of 18 ms and decay time of 31 ms), an excellent responsivity reaching 280 $AW^{-1}$, and good linearity in a wide light intensity range from 20 $\mu W\ cm^{-2}$ to 7 $mW\ cm^{-2}$.

## 17.3 Self-powered Environmental Monitoring

The working principle of TENG for self-powered environmental monitoring is almost the same as that of self-powered chemical sensor. Here we will introduce two examples to show its applications for detecting ingredients that have potential hazard to the ambient environment. As the first example, phenol is corrosive to human eyes, skins, and the respiratory tract due to its high toxicity. Repeated or prolonged human contact with phenol may cause harmful effects on the liver, kidneys, heart and central nervous system, resulting in dysrhythmia, seizures and coma. In almost all cases, phenol is damaging not only to individual species and populations, but also to the natural biological communities [25]. Disturbingly, with plenty of discharging sources, such as chemical plants, pharmaceutical plants and petroleum refineries, phenol is one of the most serious and persistent organic pollutants widely existed in ambient environment, especially in the surface waters [26]. Considerable effort has been committed to develop various techniques for phenol detection and degradation, including chromatographic [27], spectrophotometric, photocatalytic [28], adsorptive [29], electrochemical analyses [30] and so on. However, widely adoption of these techniques may be shadowed by the limitations such as sophisticated and expensive instruments, complex and time-consuming procedures, high energy consumption as well as high operating cost [31].

In this example, a unique route has been reported that uses the energy harvested via the β-cyclodextrin enhanced triboelectrification for a self-powered phenol treatment, which is an efficient integration of phenol detection and degradation [32]. Relying on the β-cyclodextrin (β-CD) as the recognition element, the as-fabricated nanosensors can selectively capture and detect phenol molecules with a sensitivity of 0.01/μM in a sensing range of 10–100 μM. The presented nanosenors are proved reusable after being refurbished with ethyl alcohol. Meanwhile, the β CD enhanced triboelectrification was creatively utilized to harvest the kinetic impact energy from wastewater waves, acting as a direct power source to electrochemically degrade the phenol. The β-CD enhanced triboelectrification offers a unique route to treat the environmental phenol with a systematic integration of detection and degradation in a self-powered manner. With a collection of compelling features, such as high detection sensitivity and degradation efficiency, extremely low cost, simplicity and reusability, the presented work not only provides a new and efficient pathway for environmental chemical treatment, but also a solid "green" advancement in the fields of wastewater treatment, ecological sanitation, environmental degradation, monitoring, assessment and sustainability.

The systematic phenol treatment by using the β-CD enhanced triboelectrification consisted of two steps, detection and degradation. Regarding phenol detection, the as-fabricated nanosensor holds a multilayered structure with acrylic as supporting substrates, as schematically shown in Fig. 17.7a. On the upper substrate, a layer of polytetrafluoroethylene (PTFE) film was adhered acting as one contact surface with coated copper as the back electrode. On the lower substrate, a layer of Ti foil with surface grown $TiO_2$ nanowires is laminated as another contact surface. $TiO_2$

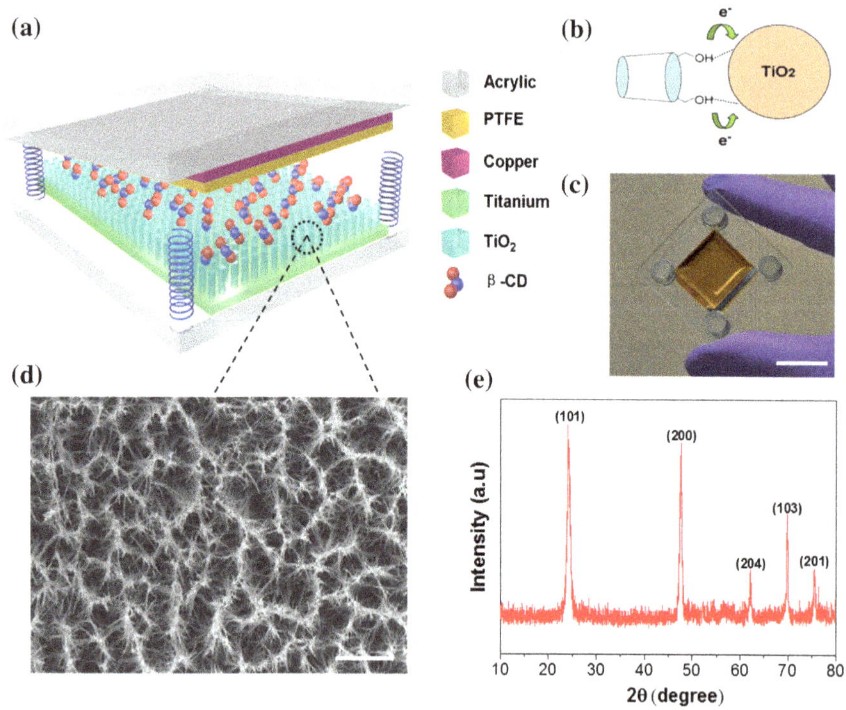


**Fig. 17.7** **a** A sketch of the triboelectrification with b-cyclodextrin surface modification for phenol detection. **b** A SEM image of the $TiO_2$ nanowires. The scale bar is 5 mm. **c** The XRD spectra of the grown $TiO_2$ nanowires. **d** Schematic diagram for illustrating the charge transfer from the hydroxyl groups of b-cyclodextrin to $TiO_2$ nanowires. **e** An image of the as-fabricated device for phenol detection. The scale bar is 2 cm. Reproduced with permission from Royal Society of Chemistry [32]

nanowires on the Ti foil were prepared via a coupling of hydrothermal route in alkali solution with ion-exchange process. A scanning electron microscopy (SEM) image of grown $TiO_2$ nanowires on the Titanium foil is presented in Fig. 17.7b with measured average nanowire diameter and length of 73.4 nm and 1.92 μm, respectively. And the corresponding X-ray diffraction (XRD) pattern of $TiO_2$ nanowires is shown in Fig. 17.7c. β-CD molecules were then assembled onto the $TiO_2$ nanowires as a surface chemical modification. Figure 17.7d illustrates the charges transfer from the hydroxyl groups of β-CD to $TiO_2$ nanowires. Here, β-CD plays dual roles of a phenol recognition element and an electrical performance enhancer. A detailed description of the function of β-CD is presented in the following working principle section. A photograph of the as-fabricated phenol nanosensor is shown in Fig. 17.7e. The device has a dimension of 4 cm × 4 cm.

Designed to use β-CD as the phenol recognition element as well as the electrical performance enhancer, the working principle of the devices based on β-CD enhanced triboelectrification can be elucidated in two aspects, namely, β-CD as surface chemical modification, and triboelectrification for phenol detection and degradation. On one hand, β-CD is cyclic oligosaccharides of seven α-D-glucose units connected through glycosidic α-1,4 bonds, which are composed of hydrophobic internal cavity and hydrophilic external surface. This special molecular structure allows it to form host/guest inclusion complex via various guest molecules with suitable polarity and dimension. The cavity diameter of β-CD has been found to be the most appropriate size for selective adsorption of phenol, which explains the β-CD as surface chemical modification for phenol recognition. On the other hand, the triboelectrification for phenol detection is a two-way coupling of triboelectric effect and electrostatic induction, as we have discussed in Chap. 2. Given a consistent and cyclical operation of the two plates of the as-fabricated nanosensor at a fixed surface concentration of β-CD, the acquired output electrical signals in the external circuit is determined by the phenol concentration absorbed by the β-CD on the surface of $TiO_2$ nanowires, which is the cornerstone of the β-CD enhanced triboelectrification for phenol detection. Furthermore, the electrical signals generated by the β-CD enhanced triboelectrification can also act as a direct power source to electrochemically degrade the phenol in the wastewater. Since the electricity is converted from the motion of wastewater wave, which renders a self-powered manner for phenol degradation.

After looking into the impact of β-CD concentrations on the electrical output, a further step was taken to evaluate the performance of β-CD enhanced triboelectrification for self-powered phenol detection. Under a fixed β-CD concentration of 80 μM, phenol solutions with various concentrations but constant volumes (20 μL) were dropped onto the β-CD modified $TiO_2$ nanowires surface. Prior to the further electrical measurement, the phenol treated nanosensors were dried at ambient temperature. The dependence of positive peak values of the current and voltage outputs on the phenol concentrations are presented in Fig. 17.8a, b, respectively. In a certain phenol concentration region of 10–100 μM, both the current and voltage outputs are a monotonically decreasing function of phenol concentrations throughout the experimental time windows. The decrease is mainly attributed to a

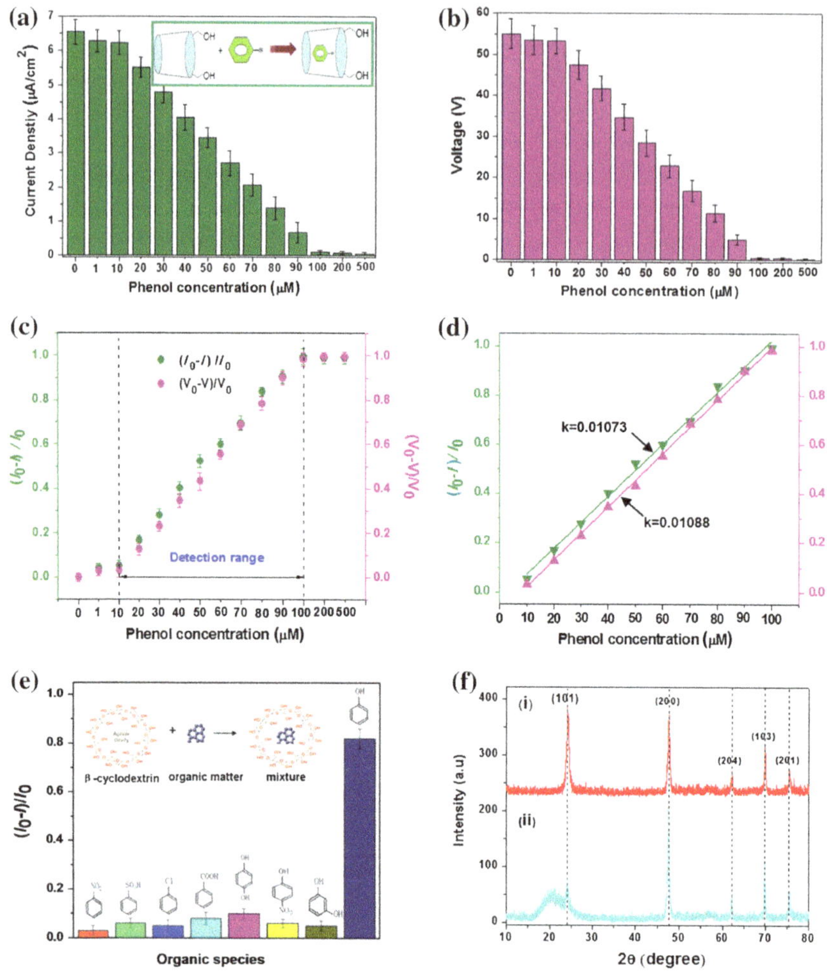


**Fig. 17.8** Under a fixed 80 mM b-cyclodextrin surface modification, dependence of the current (**a**) and voltage (**b**) output on the phenol concentrations. *Inset* shows the reaction mechanism between b-cyclodextrin and a phenol molecule. **c** The sensitivity and detection range of the as-developed b-CD enhanced triboelectrification for phenol detection in terms of both current and voltage output. **d** A sensitivity of 0.01/mM was simultaneously achieved from both current and voltage signals with a detection range of 10–100 mM. **e** Selectivity of the as-developed b-CD enhanced triboelectrification for phenol detection. *Inset* shows an illustration of the reaction mechanism between b-cyclodextrin and different types of organic species. **f** A comparison of the XRD pattern of the $TiO_2$ nanowires: (*i*) as-grown and (*ii*) refurbished with 20 mL ethyl alcohol after phenol detection. Reproduced with permission from Royal Society of Chemistry [32]

modified surface triboelectric behavior due to the adsorbed phenol molecules. The phenol will replace the position of $TiO_2$ to contact with PTFE. In comparison to $TiO_2$, the phenol molecules have lower tendency to transfer the electrons to PTFE, resulting in a phenol concentrations depended electrical output. In order to render a direct view of the sensing performance of the β-CD enhanced triboelectrification for phenol detection, both the current ratio ($(I_0 - I)/I_0$) and voltage ratio ($(V_0 - V)/V_0$) were plotted versus the phenol concentrations, as shown in Fig. 17.8c, d. These results reveal the designed nanosensor is sensitive to the phenol molecules with a sensitivity of 0.01/μM in the sensing range of 10–100 μM.

It is worth noting that, triboelectric charges are also very dependent on the surface properties (e.g. roughness) of the two contacting surfaces. A control experiment was conducted to validate the changes of triboelectric signals were mainly caused by the binding between β-CD and phenol. Different phenol solutions with various concentrations with a constant volume (20 μL) were directly dropped onto the $TiO_2$ surface without β-CD modification. As can be seen, increasing the concentration of phenol would increase the output of the as-fabricated device, which was attributed to the surface roughness alteration due to the solid materials remaining on the $TiO_2$ surface after evaporation. But such effect seemed to be small and insignificant in comparison to the effect from β-CD surface modification, especially when the phenol concentration was in a lower concentration range from 0 to 100 μM. This verified the dominant role of β-CD to alter triboelectric signals in the phenol detection processing.

Furthermore, control experiments were carried out to test the selectivity of the as-developed β-CD enhanced triboelectrification toward phenol detection as compared to other organic species. With a constant concentration of 50 μM for all the testing organic species, the obtained current ratio from the phenol absorbed nanosensor was far larger than other organic species (Fig. 17.8e), which indicated that the β-CD enhanced triboelectrification is an effective means for phenol detection with excellent selectivity. Additionally, besides the compelling features, such as high sensitivity, selectivity, extremely low cost, environmentally friendly and working in a self-powered manner, the β-CD enhanced triboelectrification based nanosensor is also reusable and can be refurbished by ethyl alcohol rinsing after phenol detection. The measured current and voltage output of a refurbished nanosensors shows no observable output degradation after refurbishment. And also, the XRD pattern (Fig. 17.8f) shows no chemical composition or surface morphology change after refurbishment, which proves a good reusability of the device for phenol detection.

Admittedly, the presence of water or environmental humidity is still a challenge to all the reported triboelectrification based energy harvesters or sensors because it largely affect the preservation of triboelectricity on the surface. Experimentally, from the device level, the reported results indicated a negative impact on the output performance from the water. However, theoretically, from the fundamental level of the study of the triboelectrification, the mechanism of contact electrification as well as the influence of humidity on it remains poorly understood. To date, the results of some studies have proven inconclusive suggesting that depending on humidity and the properties of contacting materials, water contained in air can either promote or inhibit contact electrification.

In another work, a fundamentally new working principle is introduced into the field of heavy metal ions treatment by reporting a unique route that worked in a self-powered manner by harnessing the ambient energy using the triboelectrification effect [33]. Relying on the modified Anodic Aluminum Oxide (AAO) nanopores surface with a layer of appropriate ligand molecules as recognition element, the as-developed tribo-nanosensors can selectively capture and detect $Cu^{2+}$, $Pb^{2+}$ and $Cr^{3+}$, which are commonly existing toxic heavy metal ions in industrial wastewater, in a sensing range of 0–200 μM with a sensitivity of 0.005, 0.003 and 0.004 $\mu M^{-1}$ respectively. The presented tribo-nanosensors are also proved to possess good stability after continuous working for up to 50,000 cycles. Moreover, the ambient triboelectrification effect was further utilized to develop a water-driven triboelectric nanogenerator (WD-TENG) as a sustainable power source for heavy metal ions removal by recycling the kinetic energy from flowing wastewater. The self-provided electric field can boost the migration and combination of ions as well as the electrolysis effect. The later induced a generation of large amount of $OH^-$ at the cathode in the wastewater, which promoted the precipitation of heavy metal ions. By controlling the wastewater pH values, $Cu^{2+}$, $Pb^{2+}$ and $Cr^{3+}$ were demonstrated to be fractionally precipitated from the wastewater. Under a fixed water flow rate of 3 L $min^{-1}$ and initial heavy metal ions concentration of 100 μM, the self-powered cleaning system was capable of removing 97.4 % of the heavy metal ions in the wastewater in 100 min. In addition, a further step was taken to recycle and collect the precipitated metals. Through a filtration, acidification and chemical reduction process, pure metals are respectively obtained, which realizes the clean production and recycling economy.

The triboelectrification enabled self-powered heavy metal ions treatment systematically consists of two steps, a tribo-nanosensor for metal ions detection and a water-driven triboelectric nanogenerator for metal ions removal. As demonstrated in Fig. 17.9a, the as-developed tribo-nanosensor holds a multilayered structure with acrylic as supporting substrates. On the upper substrate, a layer of PTFE film was adhered as one contact surface with back-coated copper as the electrode. PTFE nanowires arrays were created on the exposed PTFE surface by a top-down method through reactive ion etching. A SEM image of the PTFE nanowires is displayed in Fig. 17.9b, which indicates an average diameter of $34 \pm 3$ nm and an average length of $1.1 \pm 0.4$ μm. On the lower substrate, a layer of Al foil with surface grown Anodic Aluminum Oxide (AAO) nanopores is laminated as another contact surface. An SEM image of the nanopores is presented in Fig. 17.9c. The nanopores were uniformly distributed on the aluminum surface with an average diameter of $80 \pm 5$ nm and a pore depth of $40 \pm 10$ nm with a distribution density of 210 $\mu m^{-2}$.

The working principle of the tribo-nanosensor for heavy metal ions detection is schematically illustrated in Fig. 17.9d, which can be elucidated in two aspects, namely, ligand molecules as surface chemical modifications and triboelectrification for heavy metal ions detection. To operate, AAO nanopores were firstly chemically modified with appropriate ligand molecules via physical adsorption, which enables the AAO nanopores to possess good chelating properties towards different metal ions. These modifying ligand molecules are capable of selectively adsorbing the

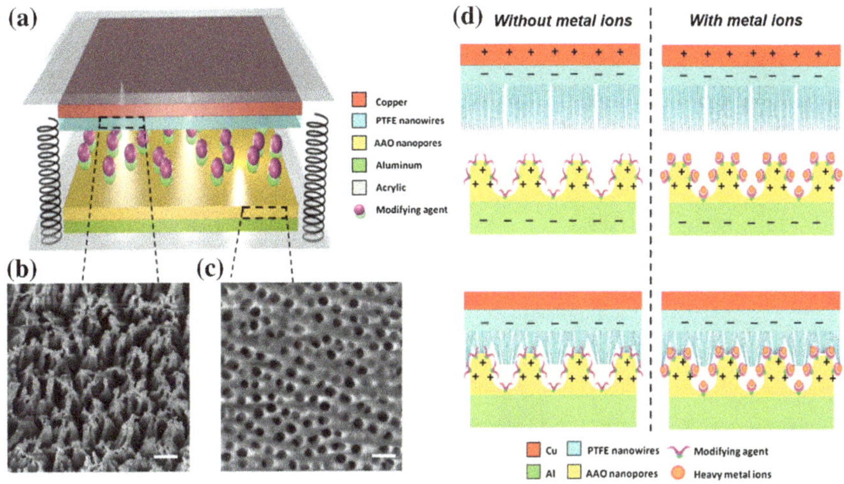


**Fig. 17.9** Structure design and working principle of the as-developed tribo-nanosensor for heavy metal ions detection. **a** A sketch to show the structural design of the tribo-nanosensor. **b** A SEM image of the PTFE nanowires. The scale bar is 200 nm. **c** A SEM image of the Anodic Aluminum Oxide nanopores. The scale bar is 500 nm. **d** Schematic diagram of the working principle of the tribo-nanosensor for heavy metal ions detection, which consists of a triboelectrification and a complexation reaction process. Reproduced with permission from Wiley [33]

heavy metal ions as their recognizing elements. A difference of the complexation constant between heavy metal ions and ligand molecules contributes to a different affinity to the metal ions. Chemically, a higher complexation will lead to a stronger ability of metal ions adsorption, while very limited amount of metal ions can be adsorbed in the presence of a weak complexation. For the sensing, the adsorbed metal ions on the AAO surface will highly influence the electric output of the as-developed tribo-nanosensor. Since an increasing amount of metal ions will gradually reduce the triboelectrification between the PTFE nanowires and the AAO nanopores, and thus the acquired electric output of the tribo-nanosensor. Subsequently, by analyzing the electric output signals of the as-fabricated tribo-nanosensor, the detection of the heavy metal ions can be realized. Here, AAO is selected owing to its relatively high adsorption capacity, strong mechanical structure, and more importantly, low cost and easy fabrication.

To demonstrate the capability of the as-developed tribo-nanosensor for heavy metal ions detection, $Pb^{2+}$, $Cr^{3+}$, $Cu^{2+}$ were selected for the test due to their high toxicity and common existence in the industrial wastewater. And dithizone, diphenylcarbazide, sodium diethyldithiocarbamate were respectively assembled onto the AAO nanopores as recognition marks for $Pb^{2+}$, $Cr^{3+}$, and $Cu^{2+}$, respectively. Experimentally, heavy metal ion solutions with various concentrations but constant volumes of 20 μL were dropped onto the modified AAO nanopores surface after being modified by its corresponding agent with a fixed concentration of 100 μM. A drying process at ambient temperature was performed before a further

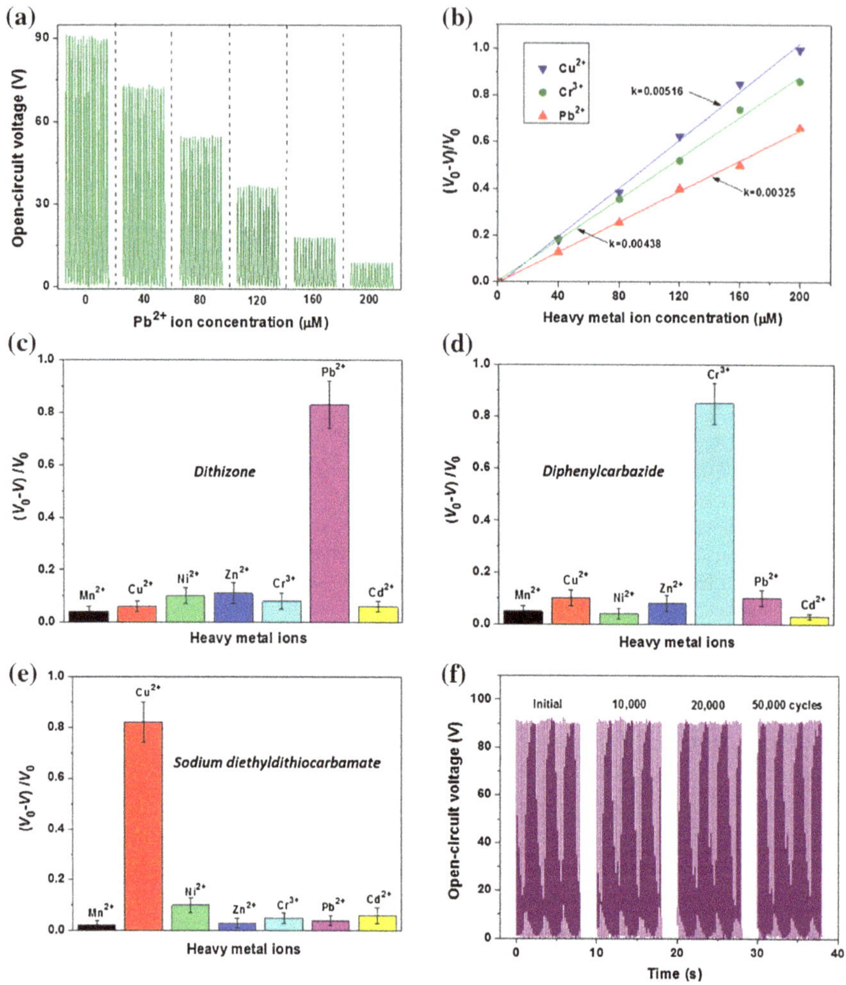


**Fig. 17.10** Performance characterization of the tribo-nanosensor for heavy metal ions detection. **a** Dependence of the open-circuit voltage of the tribo-nanosensor on the $Pb^{2+}$ concentration, under a fixed amount of the dithizone molecules as the surface modifying agent. **b** Tests of the sensitivity of the as-developed tribo-nanosensor for $Pb^{2+}$, $Cu^{2+}$, $Cr^{3+}$ heavy metal detection. **c** Test of the selectivity of the as-developed tribo-nanosensor for $Pb^{2+}$ detection by using the dithizone as the surface modifying agent. **d** Test of the selectivity of the as-developed tribo-nanosensor for $Cr^{3+}$ detection by using the diphenylcarbazide as the surface modifying agent. **c** Test of the selectivity of the as-developed tribo-nanosensor for $Cu^{2+}$ detection by using the sodium diethyldithiocarbamate as the surface modifying agent. **f** A test of the stability of the as-developed tribo-nanosensor for heavy metal ions detection. Reproduced with permission from Wiley [33]

electrical measurement. The dependence of the open-circuit voltage output on the $Pb^{2+}$ concentration is presented in Fig. 17.10a. In a certain $Pb^{2+}$ concentration region of 0–200 μM, the open-circuit voltage is a monotonically decreasing function of the $Pb^{2+}$ concentrations throughout the experimental time windows. The dependence of short-circuit current output on the $Pb^{2+}$ concentrations was also investigated, showing the same trend. A decrease of the electric output is mainly attributed to the surface adsorbed $Pb^{2+}$ molecules, which will partially replace the position of AAO to contact with PTFE. Given a lower tendency of the $Pb^{2+}$ molecules to lose electrons, a decreasing of the electric output is thus observed with the increasing of the $Pb^{2+}$ concentrations. Likewise, $Cu^{2+}$ and $Cr^{3+}$ exhibited similar changing trends between the electric output and the applied concentration. The sensing performance of the tribo-nanosensor for heavy metal ions detection was evaluated in term of open-circuit voltage ratio $((V_0 - V)/V_0)$, as shown in Fig. 17.10b. The experimental observations reveal that the tribo-nanosensor is capable of effectively detecting the applied $Cu^{2+}$, $Pb^{2+}$ and $Cr^{3+}$ in a sensing range of 0–200 μM. And more notably, in a self-powered working manner, a superior sensitivity of 0.005, 0.003, 0.004 $\mu M^{-1}$ was respectively achieved towards the ambient $Cu^{2+}$, $Pb^{2+}$ and $Cr^{3+}$ detection.

As a critical feature of a sensor, the selectivity of the as-developed tribo-nanosensor was further explored. In this regard, firstly, a set of control experiments were carried out toward $Pb^{2+}$ detection. To start, the tribo-nanosensors were treated with dithizone as the modifying agent and $Pb^{2+}$ recognition element. As shown in Fig. 17.10c, under the same testing condition, the open-circuit voltage ratio of the tribo-nanosensor with $Pb^{2+}$ absorption was far larger than that for other heavy metal ions, which indicated an excellent selectivity of the tribo-nanosensor for $Pb^{2+}$ detection. Likewise, being respectively treated with diphenylcarbazide and sodium diethyldithiocarbamate, the tribo-nanosensors were also showing remarkable selectivity towards $Cr^{3+}$ and $Cu^{2+}$ detection, as demonstrated in Fig. 17.10d, e.

For a systematic investigation, a further step was taken to evaluate the stability of tribo-nanosensor for heavy metal ions detection. To perform it, an electrodynamic shaker that provides a sinusoidal wave was used as an impact source to operate the tribo-nanosensor. After a 50,000-cycle continuous operation, no degradation of the electric output was observed, as shown in Fig. 17.10f, indicating a good stability of the device. Featured as high sensitivity, selectivity and stability, the reported tribo-nanosensor represented a self-powered technique, which could also be extensively applied to other heavy metal ions detection.

## 17.4 Summary

In this chapter, we have introduced one unique type of self-powered active sensor that has distinct working mechanism compared with the self-powered sensors we discussed in prior chapters. Metal ions or organic species in solutions could be assembled onto the triboelectric surfaces to tune triboelectric charge density when they are supplied with a contact event. Hence, the output performance of the TENG will be dependent on the concentration of the chemical species to study. With this indirect approach, many different types of chemical ingredients could be precisely detected with its matched ligands assembled on the triboelectric surfaces. The chemical sensing and environmental monitoring has delivered high sensitivity, high selectivity, and extremely low power consumption. Moreover, it has also been demonstrated that the TENG could be used to collect the water wave energy of waste water for degradation of the hazardous components, as has been covered in Chap. 13. By integrating built-in photodetector into the external load of a TENG, it can be further applied as a self-powered UV sensor for detecting intensity of UV illumination.

## References


1. C.P. Hanna, J.F. Tyson, S. Mcintosh, Determination of total mercury in waters and urine by flow-injection atomic-absorption spectrometry procedures involving online and off-line oxidation of organomercury species. Anal. Chem. **65**(5), 653–656 (1993)
2. M. Leermakers, W. Baeyens, P. Quevauviller, M. Horvat, Mercury in environmental samples: speciation, artifacts and validation. Trac-Trend Anal. Chem. **24**(5), 383–393 (2005)
3. J.S. Lee, M.S. Han, C.A. Mirkin, Colorimetric detection of mercuric ion (Hg2+) in aqueous media using DNA-functionalized gold nanoparticles. Angew. Chem. Int. Ed. **46**(22), 4093–4096 (2007)
4. B. Leng, L. Zou, J.B. Jiang, H. Tian, Colorimetric detection of mercuric ion (Hg(2+)) in aqueous media using chemodosimeter-functionalized gold nanoparticles. Sensor Actuat. B-Chem. **140**(1), 162–169 (2009)
5. S. Yoon, E.W. Miller, Q. He, P.H. Do, C.J. Chang, A bright and specific fluorescent sensor for mercury in water, cells, and tissue. Angew. Chem. Int. Ed. **46**(35), 6658–6661 (2007)
6. C.C. Huang, Z. Yang, K.H. Lee, H.T. Chang, Synthesis of highly fluorescent gold nanoparticles for sensing Mercury(II). Angew. Chem. Int. Ed. **46**(36), 6824–6828 (2007)
7. Z. Gu, M.X. Zhao, Y.W. Sheng, L.A. Bentolila, Y. Tang, Detection of mercury ion by infrared fluorescent protein and its hydrogel-based paper assay. Anal. Chem. **83**(6), 2324–2329 (2011)
8. D. Li, A. Wieckowska, I. Willner, Optical analysis of Hg(2+) ions by oligonucleotide-gold-nanoparticle hybrids and DNA-based machines. Angew. Chem. Int. Ed. **47**(21), 3927–3931 (2008)
9. T. Senapati, D. Senapati, A.K. Singh, Z. Fan, R. Kanchanapally, P.C. Ray, Highly selective SERS probe for Hg(II) detection using tryptophan-protected popcorn shaped gold nanoparticles. Chem. Commun. **47**(37), 10326–10328 (2011)
10. Z.Q. Zhu, Y.Y. Su, J. Li, D. Li, J. Zhang, S.P. Song, Y. Zhao, G.X. Li, C.H. Fan, Highly sensitive electrochemical sensor for mercury(II) ions by using a mercury-specific oligonucleotide probe and gold nanoparticle-based amplification. Anal. Chem. **81**(18), 7660–7666 (2009)
11. D. Wen, L. Deng, S.J. Guo, S.J. Dong, Self-powered sensor for trace Hg2+ detection. Anal. Chem. **83**(10), 3968–3972 (2011)
12. M. Lee, J. Bae, J. Lee, C.S. Lee, S. Hong, Z.L. Wang, Self-powered environmental sensor system driven by nanogenerators. Energy Environ. Sci. **4**(9), 3359–3363 (2011)
13. Z.L. Wang, Triboelectric nanogenerators as new energy technology for self-powered systems and as active mechanical and chemical sensors. ACS Nano **7**(11), 9533–9557 (2013)
14. Z.L. Wang, Triboelectric nanogenerators as new energy technology and self-powered sensors —principles, problems and perspectives. Faraday Discuss. **176**, 447–458 (2014)
15. Z.L. Wang, J. Chen, L. Lin, Progress in triboelectric nanogenerators as a new energy technology and self-powered sensors. Energy Environ. Sci. **8**(8), 2250–2282 (2015)

16. F.R. Fan, Z.Q. Tian, Z.L. Wang, Flexible triboelectric generator! Nano Energy **1**(2), 328–334 (2012)
17. Z.H. Lin, G. Zhu, Y.S. Zhou, Y. Yang, P. Bai, J. Chen, Z.L. Wang, A self-powered triboelectric nanosensor for mercury ion detection. Angew. Chem. Int. Ed. **52**(19), 5065–5069 (2013)
18. Z.H. Lin, Y.N. Xie, Y. Yang, S.H. Wang, G. Zhu, Z.L. Wang, Enhanced triboelectric nanogenerators and triboelectric nanosensor using chemically modified $TiO_2$ nanomaterials. ACS Nano **7**(5), 4554–4560 (2013)
19. D. Walker, M. Razeghi, The development of nitride-based UV photodetectors. Opto-Electron. Rev. **8**(1), 25–42 (2000)
20. E. Monroy, F. Omnes, F. Calle, Wide-bandgap semiconductor ultraviolet photodetectors. Semicond. Sci. Tech. **18**(4), R33–R51 (2003)
21. B.Z. Tian, C.M. Lieber, Design, synthesis, and characterization of novel nanowire structures for photovoltaics and intracellular probes. Pure Appl. Chem. **83**(12), 2153–2169 (2011)
22. X.D. Li, C.T. Gao, H.G. Duan, B.G. Lu, X.J. Pan, E.Q. Xie, Nanocrystalline $TiO_2$ film based photoelectrochemical cell as self-powered UV-photodetector. Nano Energy **1**(4), 640–645 (2012)
23. S.M. Hatch, J. Briscoe, S. Dunn, A self-powered ZnO-Nanorod/CuSCN UV photodetector exhibiting rapid response. Adv. Mater. **25**(6), 867–871 (2013)
24. Z.H. Lin, G. Cheng, Y. Yang, Y.S. Zhou, S. Lee, Z.L. Wang, Triboelectric nanogenerator as an active UV photodetector. Adv. Funct. Mater. **24**(19), 2810–2816 (2014)
25. J. Wang, J.M. Lu, S.Y. Ly, M. Vuki, B.M. Tian, W.K. Adeniyi, R.A. Armendariz, Lab-on-a-Cable for electrochemical monitoring of phenolic contaminants. Anal. Chem. **72** (11), 2659–2663 (2000)
26. T. Zhao, J. He, X.X. Wang, B.Z. Ma, X.X. Wang, L. Zhang, P. Li, N. Liu, J. Lu, X.L. Zhang, Rapid detection and characterization of major phenolic compounds in Radix Actinidia chinensis Planch by ultra-performance liquid chromatography tandem mass spectrometry. J. Pharmaceut. Biomed. **98**, 311–320 (2014)
27. X.X. Han, L. Chen, U. Kuhlmann, C. Schulz, I.M. Weidinger, P. Hildebrandt, Magnetic titanium dioxide nanocomposites for surface-enhanced resonance raman spectroscopic determination and degradation of toxic anilines and phenols. Angew. Chem. Int. Ed. **53**(9), 2481–2484 (2014)
28. M.M. Ali, K.Y. Sandhya, Visible light responsive titanium dioxide-cyclodextrin-fullerene composite with reduced charge recombination and enhanced photocatalytic activity. Carbon **70**, 249–257 (2014)
29. W.H. Ma, J. Li, X. Tao, J. He, Y.M. Xu, J.C. Yu, J.C. Zhao, Efficient degradation of organic pollutants by using dioxygen activated by resin-exchanged Iron(II) bipyridine under visible irradiation. Angew. Chem. Int. Ed. **42**(9), 1029 (2003)
30. H.Y. Li, Y. Chen, Y.H. Zhang, W.Q. Han, X.Y. Sun, J.S. Li, L.J. Wang, Preparation of Ti/$PbO_2$-Sn anodes for electrochemical degradation of phenol. J. Electroanal. Chem. **689**, 193–200 (2013)
31. G.H. Chen, Electrochemical technologies in wastewater treatment. Sep. Purif. Technol. **38**(1), 11–41 (2004)
32. Z.L. Li, J. Chen, J. Yang, Y.J. Su, X. Fan, Y. Wu, C.W. Yu, Z.L. Wang, Beta-cyclodextrin enhanced triboelectrification for self-powered phenol detection and electrochemical degradation. Energy Environ. Sci. **8**(3), 887–896 (2015)
33. Z.L. Li, J. Chen, H.Y. Guo, X. Fan, Z. Wen, M.-H. Yeh, C.W. Yu, X. Cao, Z.L. Wang, Triboelectrification enabled self-powered detection and removal of heavy metal ions in wastewater. Adv. Mater. (2015). doi:10.1002/adma.201504356

# Journal Articles by Wang's Group on Triboelectric Nanogenerators (2012–2016)

## *2016*

[1] Yunlong Zi, Hengyu Guo, Zhen Wen, Min-Hsin Yeh, Chenguo Hu, Zhong Lin Wang, "Harvesting Low-Frequency (<5 Hz) Irregular Mechanical Energy: A Possible Killer-Application of Triboelectric Nanogenerator", ACS Nano, 10 (2016) 4797–4805.

[2] Keun Young Lee, Hong-Joon Yoon, Tao Jiang, Xiaonan Wen, Wanchul Seung, Sang-Woo Kim, Zhong Lin Wang, "Fully packaged self-powered triboelectric pressure sensor using hemispheres-array", Adv. Energy Mater., 6 (2016) 1502566.

[3] Xiangyu Chen, Tao Jiang, Yanyan Yao, Liang Xu, Zhenfu Zhao, Zhong Lin Wang, "Stimulating acrylic elastomer by triboelectric nanogenerator-toward self-powered electronic skin and artificial muscle", Adv. Func. Mater., 26 (2016) 4906–4913.

[4] Xin Zheng, Jingzhen Su, Xianjun Wei, Tao Jiang, Shuyan Gao, Zhong Lin Wang, "Self-powered Electrochemistry on Oxidation of Organic Molecules by a Cross-Linked Triboelectric Nanogenerator", Adv. Mater., 28 (2016) 5188–5194.

[5] Tao Jiang, Wei Tang, Xiangyu Chen, Chang Bao Han, Long Lin, Yunlong Zi, and Zhong Lin Wang, "Figure-of-Merits for Rolling-Friction-Based Triboelectric Nanogenerators", Adv. Mater. Technologies, 1 (2016) 1600017.

[6] Weibo Guo, Xiaodi Zhang, Xin Yu, Shu Wang, Jichuan Qiu,Wei Tang, Linlin Li, Hong Liu, Zhong Lin Wang, "Self-powered electrical stimulation for enhancing neural differentiation of MSCs on graphene-PEDOT hybrid microfibers", ACS Nano, 10 (2016) 5086–5095

[7] Tao Zhou, Li Min Zhang, Fei Xue, Wei Tang, Chi Zhang, Zhong Lin Wang "Multilayered Electret Films based Triboelectric Nanogenerator", Nano Res., 9 (2016) 1442–1451.

 
Z.L. Wang et al., *Triboelectric Nanogenerators*,
Green Energy and Technology, DOI 10.1007/978-3-319-40039-6

[8] Kesong Hu, Rui Xiong, Hengyu Guo, Ruilong Ma, Shuaidi Zhang, Zhong Lin Wang, Vladimir V. Tsukruk, "Self-Powered Electronic Skin with Biotactile Selectivity", Adv. Mater., 28 (2016) 3549–3556.

[9] Fang Yi, Xiaofeng Wang, Simiao Niu, Shengming Li, Guangjie Zhang, Long Lin, Zhen Wen, Hengyu Guo, Jie Wang, Min-Hsin Yeh, Yunlong Zi, Qingliang Liao, Zheng You, Yue Zhang, and Zhong Lin Wang, "A highly shape-adaptive, stretchable and conductive liquid-electrode based triboelectric nanogenerator for energy harvesting and self-powered biomechanical monitoring" Sci. Adv., 2 (2016): e1501624.

[10] Feng Ru Fan, Wei Tang, Zhong Lin Wang, "Flexible Nanogenerators for Energy Harvesting and Self-Powered Electronics", Adv. Mater., 28(2016) 4283–4305.

[11] Shuwen Chen, Long Ma, Magnus Willander, Huarui Zhu, Xia Cao, Zhong Lin Wang, "Triboelectric Nanogenerator for Sustainable Wastewater Treatment via a Self-powered Electrochemical Process", Adv. Energy Mater., 6 (2016) 1501778.

[12] Tao Jiang, Xiangyu Chen, Keda Yang, Chang Bao Han, Wei Tang, Zhong Lin Wang, "Theoretical Study on Rotary-Sliding Disk Triboelectric Nanogenerators in Contact and Non-contact modes", Nano Res., 9 (2016) 1057–1070.

[13] Zhaoling Li, Jun Chen, Hengyu Guo, Xing Fan, Zhen Wen, Min-Hsin Yeh, Chongwen Yu, Xia Cao, Zhong Lin Wang, "Triboelectrification Enabled Self-powered Detection and Removal of Heavy Metal Ions in Wastewater", Adv. Mater., 28 (2016) 2983–2991.

[14] Aifang Yu, Xiangyu Chen, Rui Wang, Jingyu Liu, Jianjun Luo, Libo Chen, Yang Zhang, Wei Wu, Caihong Liu, Mingzeng Peng, Weiguo Hu, Junyi Zhai, Zhong Lin Wang, "Triboelectric nanogenerator as a self-powered communication unit for receiving and transmitting information", ACS Nano, 10 (2016) 3944–3950.

[15] Jie Wang, Zhen Wen, Yunlong Zi, Long Lin, Changsheng Wu, Hengyu Guo, Yi Xi, Youlong Xu, Zhong Lin Wang, "Self-powered electrochemical synthesis of polypyrrole from pulsed output of triboelectric nanogenerator as a sustainable energy system", Adv. Funct. Mater., 26 (2016) 3542–3548.

[16] Sihong Wang, Yunlong Zi, Yu Sheng Zhou, Shengming Li, Fengru Fan, Long Lin, Zhong Lin Wang "Molecular Surface Functionalization to Enhance Power Output of Triboelectric Nanogenerators", J. Mater. Chem. A, 4 (2016) 3728–3734.

[17] Yunlong Zi, Jie Wang, Sihong Wang, Shengming Li, Zhen Wen, Hengyu Guo, Zhong Lin Wang, "Effective Energy Storage from Triboelectric Nanogenerator", Nat. Commun., 7 (2016) 10987.

[18] Xian Song Meng, Zhong Lin Wang, Guang Zhu, "Triboelectric-potential-regulated charge transport through p-n junctions for area-scalable energy harvesting", Adv. Mater., 28 (2016) 668–676.

[19] Xiao Hui Li, Chang Bao Han, Tao Jiang, Chi Zhang, Zhong Lin Wang, "A ball-bearing structured triboelectric nanogenerator for nondestructive damage and rotating speed measurement", Nanotechnology 27 (2016) 085401.

[20] Qiang Zheng, Yang Zou, Yalan Zhang, Zhuo Liu, Bojing Shi, Xinxin Wang, Yiming Jin, Zhou Li, Zhong Lin Wang, "Biodegradable triboelectric nanogenerator as a life-time designed power source for implantable medical devices", Sci. Adv., 2 (2016) e1501478.

[21] Zhaoling Li, Jun Chen, Jiajia Zhou, Li Zheng, Ken C. Pradel, Xing Fan, Hengyu Guo, Zhen Wen, Min-Hsin Yeh, Chongwen Yu, Zhong Lin Wang, "High-efficiency Ramie Fiber Degumming and Self-powered Degumming Wastewater Treatment Using Triboelectric Nanogenerator", Nano Energy, 22 (2016) 548–557.

[22] Zhenfu Zhao, Xiong Pu, Chunhua Du, Linxuan Li, Chunyan Jiang, Weiguo Hu, Zhong Lin Wang, "Freestanding Flag-type Triboelectric Nanogenerator for Harvesting High-Altitude Wind Energy from Arbitrary Directions", ACS Nano, 10 (2016) 1780–1787.

[23] Li Min Zhang, Chang Bao Han, Tao Jiang, Tao Zhou, Xiao Hui Li, Chi Zhang, and Zhong Lin Wang, "Multilayer Wavy-structured Robust Triboelectric Nanogenerator for Harvesting Water Wave Energy", Nano Energy, 22 (2016) 87–94.

[24] Shengming Li, Yu Sheng Zhou, Yunlong Zi, Gong Zhang, Zhong Lin Wang, "Excluding Contact Electrification in Surface Potential Measurement Using Scanning Kelvin Probe Microscopy", ACS Nano, 10 (2016) 2528–2535.

[25] Min-Hsin Yeh, Hengyu Guo, Long Lin, Zhen Wen, Zhaoling Li, Chenguo Hu and Zhong Lin Wang, "Rolling friction enhanced free-standing triboelectric nanogenerators and its applications in self-powered electrochemical recovery system", Adv. Func. Mater. 26 (2016) 1054–1062.

[26] Jie Wang, Zhen Wen, Yunlong Zi, Pengfei Zhou, Jun Lin, Hengyu Guo, Youlong Xu, Zhong Lin Wang, "All-Plastic-Materials Based Self-charging Power System Composed of Triboelectric Nanogenerators and Supercapacitors", Adv. Func. Mater. 26 (2016) 1070–1076.

[27] Lei Zhang, Binbin Zhang, Jun Chen, Long Jin, Weili Deng, Junfeng Tang, Haitao Zhang, Hong Pan, Minhao Zhu, Weiqing Yang, Zhong Lin Wang, "Lawn Structured Triboelectric Nanogenerators for Scavenging Sweeping Wind Energy on Rooftop", Adv. Mater. 28 (2016) 1650–1656.

[28] Bojing Shi, Qiang Zheng, Wen Jiang, Ling Yan, Xinxin Wang, Hong Liu, Yan Yao, Zhou Li, Zhong Lin Wang, "Packaged self-powered system with universal connectors based on hybridized nanogenerators", Adv. Mater. 28 (2016) 846–852.

[29] Xiong Pu, Linxuan Li, Mengmeng Liu, Chunyan Jiang, Chunhua Du, Zhenfu Zhao, Weiguo Hu, Zhong lin Wang, "Wearable self-charging power-textile based on flexible yarn supercapacitors and fabric nanogenerators", Adv. Mater., 28 (2016) 98-105.

[30] Zuci Quan, Chang Bao Han, Tao Jiang, Zhong Lin Wang, "Robust Thin Films Based Triboelectric Nanogenerator Arrays for Harvesting Bidirectional Wind Energy", Adv. Energy Mater., 6 (2016) 1501799.

[31] Hengyu Guo, Zhen Wen, Yunlong Zi, Min-hsin Yeh, Jie Wang, Liping Zhu, Chenguo Hu, Zhong Lin Wang, "A Water-Proof Triboelectric-Electromagnetic Hybrid Generator for Energy Harvesting in Harsh Environments", Adv. Energy Mater. 6 (2016) 1501593.

[32] Xiong Pu, Mengmeng Liu, Linxuan Li, Chi Zhang, Yaokun Pang, Chunyan Jiang, Lihua Shao, Weiguo Hu, Zhong lin Wang, "Efficient Charging of Li-ion Batteries with Pulsed Output Current of Triboelectric Nanogenerators", Adv. Sci., 3 (2016) 1500255.

## *2015*

[33] Jianjun Luo, Feng Ru Fan, Tao Jiang, Zhiwei Wang, Wei Tang, Cuiping Zhang, Mengmeng Liu, Zhong Lin Wang, "Integration of micro-supercapacitors with triboelectric nanogenerators for a flexible self-charging power unit", Nano Res., 8 (2015) 3934-3943.

[34] Chang Bao Han, Tao Jiang, Chi Zhang, Xiaohui Li, Chaoying Zhang, Xia Cao' and Zhong Lin Wang, "Self-powered particulate filtering from automobile exhaustion for clean air by triboelectrification effect", ACS Nano, 9 (2015) 12552–12561.

[35] Li Su, Zhen Xuan Zhao, Hua Yang Li, Jian Yuan, Zhong Lin Wang, Guo Zhong Cao, Guang Zhu, "High-Performance Organolead Halide Perovskite-Based Self-Powered Triboelectric Photodetector", ACS Nano, 9 (2015) 11310–11316.

[36] Ting Quan, Xue Wang, Zhong Lin Wang, Ya Yang, "Hybridized Electromagnetic-Triboelectric Nanogenerator for Self-Powered Electronic Watch", ACS Nano, 9 (2015) 12301–12310.

[37] Tao Jiang, Li Min Zhang, Xiangyu Chen, Chang Bao Han, Wei Tang, Chi Zhang, Liang Xu, Zhong Lin Wang, "Structural Optimization of Triboelectric Nanogenerator for Harvesting Water Wave Energy", ACS Nano, 9 (2015) 12562–12572.

[38] Jun Chen, Jin Yang, Xing Fan, Hengyu Guo, Zhaoling Li, Li Zheng, Yuanjie Su, Zhen Wen, Changsheng Wu, Zhong Lin Wang, "Automatic mode transition enabled robust triboelectric nanogenerators", ACS Nano, 9 (2015) 12334–12343.

[39] Xiaofeng Wang, Simiao Niu, Yajiang Yin, Fang Yi, Zheng You, Zhong Lin Wang, "Triboelectric Nanogenerator based on Fully-enclosed Rolling Spherical Structure for Harvesting Low-frequency Water Wave Energy", Adv. Energy Mater., 5 (2015) 1501467.

[40] Xiuhan Li, Min-Hsin Yeh, Zong-Hong Lin, Hengyu Guo, Po-Kang Yang, Jie Wang, Sihong Wang, Ruomeng Yu, Tiejun Zhang, Zhong Lin Wang, "Self-powered Triboelectric Nanosensor for Microfluidics and Cavity-Confined Solution Chemistry", ACS Nano, 9 (2015) 11056–11063.
[41] Huarui Zhu, Ying Xu, Yu Han, Shuwen Chen, Tao Zhou, Magnus Willander, Xia Cao, Zhong Lin Wang, "Self-powered electrochemical anodic oxidation: a new method for preparation of mesoporous $Al_2O_3$ without applying electricity" Nano Res. 8 (2015) 3604–3611.
[42] Li Zheng, Gang Cheng, Jun Chen, Long Lin, Jie Wang, Yongsheng Liu, Zhong Lin Wang, "Hybridized Power Panel to Simultaneously Generate Electricity from Sunlight, Rain Drops and Wind Around the Clock", Adv. Energy Mater., 5 (2015) 1501152.
[43] Simiao Niu, Xiaofeng Wang, Fang Yi, Yu Sheng Zhou, Zhong Lin Wang, "A Universal Self-charging Power System Driven by Random-body-motion for Regulated and Sustainable Operation of Mobile and Wearable Electronics", Nat. Commun., 6 (2015) 8975.
[44] Shuhua Wang, Xiaojing Mu, Xue Wang, Alex Yuandong Gu, Zhong Lin Wang, Ya Yang, "Elasto-Aerodynamics-Driven Triboelectric Nanogenerator for Scavenging Air-Flow Energy", ACS Nano, 9 (2015) 9554–9563.
[45] Lu Zhang, Suo Bai, Chen Su, Youbin Zheng, Yong Qin. Chen Xu, Zhong Lin Wang, "High-reliability Kevlar fiber-ZnO nanowires hybrid nanogenerator and its application on self-powered UV detection", Adv. Func. Mater. 25 (2015) 5794–5798.
[46] Jinsung Chun, Jin Woong Kim, Woo-suk Jung, Chong-Yun Kang, Sang-Woo Kim, Zhong Lin Wang, Jeong Min Baik, "Mesoporous Pores Impregnated with Au Nanoparticles As Effective Dielectrics for Enahncing Triboelectric Nanogenerator Performance under Hash Environment", Energy Environ. Sci., 8 (2015) 3006–3012.
[47] Xiao Yan Wei, Shuang Yang Kuang, Hua Yang Li, Caofeng Pan, Guang Zhu, Zhong Lin Wang, "Interface-Free Area-Scalable Self-Powered Electroluminescent System Driven by Triboelectric Nanogenerator", Sci. Rep. 5 (2015) 13658.
[48] Xue Wang, Shuhua Wang, Ya Yang, Zhong Lin Wang, "Hybridized Electromagnetic-Triboelectric Nanogenerator for Scavenging Air-Flow Energy to Sustainably Power Temperature Sensors", ACS Nano, 9 (2015) 4553–4562.
[49] Yao Kun Pang, Xiao Hui Li, Meng Xiao Chen, Chang Bao Han, Chi Zhang, Zhong Lin Wang, "Triboelectric Nanogenerators as a Self-Powered 3D Acceleration Sensor", ACS Appl. Mater. Interfaces, 7 (2015), 19076–19082.
[50] Wei Tang, Jingjing Tian, Qiang Zheng, Lin Yan, Jiangxue Wang, Zhou Li, Zhong Lin Wang, "Implantable Self-Powered Low-Level Laser Cure System for Mouse Embryonic Osteoblasts' Proliferation and Differentiation", ACS Nano, 9 (2015) 7867–7873.

[51] Yunlong Zi, Simiao Niu, Jie Wang, Zhen Wen, Wei Tang, Zhong Lin Wang, "Standards and Figure of Merits for Quantifying the Performance of Triboelectric Nanogenerators", Nat. Commun., 6 (2015) 8376.

[52] Jie Wang, Xiuhan Li, Yunlong Zi, Sihong Wang, Zhaoling Li, Li Zheng, Fang Yi, Shengming Li, Zhong Lin Wang, "Flexible fiber-based supercapacitor-triboelectric nanogenerator power system for wearable electronics", Adv. Mater. 25 (2015) 4830–4836.

[53] Hua Yang Li, Li Su, Shuang Yang Kuang, Cao Feng Pan, Guang Zhu, Zhong Lin Wang, "Significant Enhancement of Triboelectric Charge Density by Fluorinated Surface Modification in Nanoscale for Converting Mechanical Energy", Adv. Funct. Mater., 25 (2015) 5691–5697.

[54] Yang Jie, Ning Wang, Xia Cao, Ying Xu, Tao Li, Xueji Zhang, Zhong Lin Wang, "Self-Powered Triboelectric Nanosensor with Poly(tetrafluoroethylene) Nanoparticle Arrays for Dopamine Detection", ACS Nano, 9 (2015) 8376–8383.

[55] Xiao Hui Li, Chang Bao Han, Li Min Zhang, Zhong Lin Wang, "Cylindrical Spiral Triboelectric Nanogenerator", Nano Res., 8 (2015) 3197–3204.

[56] Xiangyu Chen, Dai Taguchi, Takaaki Manaka, Mitsumasa Iwamoto, Zhong Lin Wang, "Direct probing of contact electrification by using optical second harmonics generation technique", Sci. Rep., 5 (2015) 13019.

[57] Yuanxing Fang, Jinhui Tong, Qize Zhong, Qiao Chen, Jun Zhou, Yinhua Zhou, Zhong Lin Wang, Bin Hu, "Solution Processed Flexible Hybrid Cell for Scavenging Solar and Mechanical Energies", Nano Energy, 16 (2015) 301–309.

[58] Qingshen Jing, Yannan Xie, Guang Zhu, Ray P.S. Han, Zhong Lin Wang, "Self-Powered Thin-Film Motion Vector Sensor", Nat. Commun., 6 (2015) 8031.

[59] Shengming Li, Sihong Wang, Yunlong Zi, Zhen Wen, Long Lin, Gong Zhang, Zhong Lin Wang, "Largely Improving the Robustness and Lifetime of Triboelectric Nanogenerators through Automatic Transition between Contact and Non-Contact Working States", ACS Nano, 9 (2015) 7479–7487.

[60] Kyeong Nam Kim, Jinsung Chun, Jin Woong Kim, Keun Young Lee, Jang-Ung Park, Sang-Woo Kim, Zhong Lin Wang, Jeong Min Baik, "Highly Stretchable 2D Fabrics for Wearable Triboelectric Nanogenerator under Harsh Environments", ACS Nano, 9 (2015) 6394–6400.

[61] Xiaodai Zhong, Ya Yang, Xue Wang, Zhong Lin Wang, "Rotating-Disk-Based Hybridized Electromagnetic-Triboelectric Nanogenerator for Scavenging Biomechanical Energy as a Mobile Power Source", Nano Energy, 13 (2015) 771–780.

[62] Zhong Lin Wang, Jun Chen, Long Lin, "Progress in triboelectric nanogenertors as new energy technology and self-powered sensors", Energy Environ. Sci., 8 (2015) 2250–2282.

[63] Woo-Suk Jung, Min Gyu Kang, Hi Gyu Moon, Seok-Jing Yoon, Zhong-Lin Wang, Sang-Woo Kim, Chong-Yun Kang, "High-output hybrided piezo/triboelectric generator", Sci. Rep., 5 (2015) 9309.
[64] Fang Yi, Long Lin, Simiao Niu, Po Kang Yang, Zhaona Wang, Jun Chen, Yusheng Zhou, Yunlong Zi, Jie Wang, Qingliang Liao, Yue Zhang, Zhong Lin Wang, "Stretchable-rubber-based triboelectric nanogenerator and its application as self-powered body motion sensors", Adv. Fun. Mater., 25 (2015) 3688–3696.
[65] Xue Jiao Zhao, Guang Zhu, Zhong Lin Wang, "Coplanar Induction Enabled by Asymmetric Permittivity of Dielectric Materials for Mechanical Energy Conversion", ACS Appl. Mater. Interfaces. 7 (2015) 6025–6029.
[66] Wei Tang, Tao Jiang, Feng Ru Fan, Ai Fang Yu, Chi Zhang, Cao Xia, Zhong Lin Wang, "Liquid metal as an effective contact electrode for high performance triboelectric nanogenerator at an instantaneous energy conversion efficiency of 70.6% and its application in harvesting vibration energy", Adv. Func. Mater., 25 (2015) 3718–3725.
[67] Po-Kang Yang, Long Lin, Fang Yi, Xiuhan Li, Ken C. Pradel, Yunlong Zi, Chih-I Wu, Jr-Hau He, Yue Zhang, Zhong Lin Wang, "A flexible, stretchable and shape-adaptive approach for versatile energy conversion and self-powered biomedical monitoring", Adv. Mater. 27 (2015) 3817–3824.
[68] Zhen Wen, Jun Chen, Min-Hsin Yeh, Hengyu Guo, Zhaoling Li, Xing Fan, Tiejun Zhang, Liping Zhu, Zhong Lin Wang, "Blow-Driven Triboelectric Nanogenerator as an Active Alcohol Breath Analyzer", Nano Energy, 16 (2015) 38–46.
[69] Hengyu Guo, Jun Chen, Min-Hsin Yeh, Xing Fan, Zhen Wen, Zhaoling Li, Chenguo Hu, Zhong Lin Wang, "An Ultra-Robust High-Performance Triboelectric Nanogenerator Based on Charge Replenishment", ACS Nano, 9 (2015) 5577–5584.
[70] Min-Hsin Yeh, Long Lin, Po-Kang Yang, Zhong Lin Wang, "Motion-driven Electrochromic Reactions for Self-powered Smart Window System", ACS Nano, 9 (2015) 4757–4765.
[71] Ananthakumar Ramadoss, Balasubramaniam Saravanakumar, Sang Jae Kim, Seung Woo Lee, Young-Soo Kim, Zhong Lin Wang, "Self-Charging Supercapacitor Power Cell", ACS Nano, 9 (2015) 4337–4345
[72] Tao Jiang, Xiang Yu Chen, Chang Bao Han, Wei Tang, Zhong Lin Wang, "Theoretical Study of Rotary Freestanding Triboelectric Nanogenerators", Adv. Func. Mater., 25 (2015) 2928–2938.
[73] Xiong Pu, Linxuan Li, Huanqiao Song, Chunhua Du, Zhengfu Zhao, Chunyan Jiang, Weiguo Hu, Guozhong Cao, Zhong Lin Wang, "Self-charging power unit by integrating textile triboelectric nanogenerator and flexible lithium-ion battery for wearable electronics", Adv. Mater., 27 (2015) 2472–2478.

[74] Xing Fan, Jun Chen, Jin Yang, Peng Bai, Zhaoling Li, Zhong Lin Wang, "Ultrathin, Rollable, Paper-Based Triboelectric Nanogenerator for Acoustic Energy Harvesting and Self-Powered Sound Recording", ACS Nano, 9 (2015) 4236–4243.

[75] Yunlong Zi, Long Lin, Jie Wang, Sihong Wang, Jun Chen, Xing Fan, Po-Kang Yang, Fang Yi, Zhong Lin Wang, "Triboelectric-Pyroelectric-Piezoelectric Hybrid Cell for High-Efficient Energy-Harvesting and Self-Powered Sensing", Adv. Mater., 27 (2015) 2340–2347.

[76 Aifang Yu, Libo Chen, Xiangyu chen, Aihua Zhang, Fengru Fan, Yan Zhang, Zhong Lin Wang, "Triboelectric sensor as self-powered signal reader for scanning probe surface topography imaging", Nanotechnology, 26 (2015) 165501.

[77] Kewei Zhang, Xue Wang, Ya Yang, Zhong Lin Wang, "Hybridized Electromagnetic-Triboelectric Nanogenerator for Scavenging Biomechanical Energy for Sustainably Powering Wearable Electronics", ACS Nano, 9 (2015) 3521–3529.

[78] Gang Cheng, Li Zheng, Zong-Hong Lin, Jin Yang, Zuliang Du, Zhong Lin Wang, "Multilayered-Electrodes Based Triboelectric Nanogenerator with Managed Output Voltage and Multi-fold Enhanced Charge Transport", Adv. Energy Mater., 5 (2015) 1401452.

[79] Ying Wu, Qingshen Jing, Jun Chen, Peng Bai, Junjie Bai, Guang Zhu, Yuanjie Su, Zhong Lin Wang, "A Self-Powered Angle Measurement Sensor Based on Triboelectric Nanogenerator", Adv. Funct. Mater., 25 (2015) 2166–2174.

[80] Nuanyang Cui, Long Gu, Jinmei Liu, Suo Bai, Jiawen Qiu, Jiecai Fu, Xinli Kou, Hong Liu, Yong Qin, Zhong Lin Wang, "High performance sound driven triboelectric nanogenerator for harvesting noise energy", Nano Energy, 15 (2015) 321–328.

[81] Qianwen Jiang, Yu Han, Wei Tang, Huarui Zhu, Caizhen Gao, Shuwen Chen, Magnus Willander, Xia Cao, Zhong Lin Wang, "Self-powered seawater desalination and electrolysis using flowing kinetic energy", Nano Energy, 15 (2015) 266–274.

[82] Zong-Hong Lin, Gang Cheng, Xiuhan Li, Po-Kang Yang, Xiaonan Wen, Zhong Lin Wang, "A Multi-layered Interdigitative-Electrodes-Based Triboelectric Nanogenerator for Harvesting Hydropower", Nano Energy, 15 (2015) 256–265.

[83] Hua Rui Zhu, Wei Tang, Cai Zhen Gao, Yu Han, Tao Li, Xia Cao, Zhong Lin Wang, "Self-powered Metal Surface Anti-corrosion Protection Using Energy Harvested from Rain Drops and Wind", Nano Energy, 14 (2015) 193–200.

[84] Simiao Niu, Zhong Lin Wang, "Theoretical systems of triboelectric nanogenerators", Nano Energy, 14 (2015) 161–192.

[85] Zhong Lin Wang, Xudong Wang, "Nanogenerators and Piezotronics-Preface", Nano Energy, 14 (2015) 1–2.

[86] Shu Wen Chen, Cai Zhen Gao, Wei Tang, Hua Rui Zhu, Yu Han, Qian Wen Jiang, Tao Li, Xia Cao, Zhong Lin Wang, "Self-powered cleaning of air pollution by wind driven triboelectric nanogenerator", Nano Energy, 14 (2015) 217–225.

[87] Xin Yu, Xun Han, Zhenhuan Zhao, Jian Zhang, Weibo Guo, Caofeng Pan, Aixue Li, Hong Liu, Zhong Lin Wang, "Hierarchical $TiO_2$ nanowire/graphite fiber photoelectrocatalysis setup powered by a wind-driven nanogenerator: A highly efficient photoelectrocatalytic device entirely based on renewable energy", Nano Energy, 11 (2015) 19–27.

[88] Ya Yang, Zhong Lin Wang, "Hybrid energy cells for simultaneously harvesting multi-types of energies", Nano Energy, 14 (2015) 245–256.

[89] Guang Zhu, Peng Bai, Jun Chen, Qingshen Jing, Zhong Lin Wang, "Triboelectric nanogenerators as a new energy technology: From fundamentals, devices, to applications", Nano Energy, 14 (2015) 126–138.

[90] Paul S. Weiss, Zhong Lin Wang, "A conversation with Prof. Zhong Lin Wang, Energy Harvester", ACS Nano, 9 (2015) 2221-2226.

[91] Jianjun Luo, Feng Ru Fan, Tao Zhou, Wei Tang, Fei Xue, Zhong Lin Wang, "Ultrasensitive Self-powered Pressure Sensing System", Extreme Mech. Lett., 2 (2015) 28–36.

[92] Chang Bao Han, Chi Zhang, Jingjing Tian, Xiao Hui Li, Limin Zhang, Zhou Li, Zhong Lin Wang, "Triboelectrification Induced UV Emssion from Gas Plasmon", Nano Res., 8 (2015) 219–226.

[93] Jun Chen, Jin Yang, Zhaoling Li, Xing Fan, Yunlong Zi, Qingshen Jing, Hengyu Guo, Zhen Wen, Ken C. Pradel, Simiao Niu, Zhong Lin Wang, "Networks of Triboelectric Nanogenerators for Harvesting Water Wave Energy—a potential approch towad blue energy", ACS Nano, 9 (2015) 3324–3331.

[94] Jin Yang, Jun Chen, Yuanjie Su, Qingshen Jing, Zhaoling Li, Fang Yi, Xiaonan Wen, Zhaona Wang and Zhong Lin Wang "Eardrum-inspired active sensors for self-powered cardiovascular system characterization and throat-attached anti-interference voice recognition", Adv. Mater., 27 (2015) 1316.

[95] Zhaoling Li, Jun Chen, Jin Yang, Yuanjie Su, Xing Fan, Ying Wu, Chongwen Yu, Zhong Lin Wang, "β-cyclodextrin Enhanced Triboelectricfication for Self-Powered Phenol Detection and Electrochemical Degradation", Energy Environ. Sci., 8 (2015) 887–896.

[96] Simiao Niu, Ying Liu, Yu Sheng Zhou, Sihong Wang, Long Lin, Zhong Lin Wang, "Optimization of charging performance in triboelectric nanogenerators for efficient energy harvesting", IEEE Trans. Electron Dev., 64 (2015) 641–647.

[97] Aifang Yu, Ming Song, Yan Zhang, Yang Zhang, Libo Chen, Junyi Zhai, Zhong Lin Wang, "Self-Powered Acoustic Source Locator in Underwater Environment Based on Organic Film Triboelectrific Nanogenerator", Nano Res., 8 (2015) 765–773.

[98] Xian Song Meng, Hua Yang Li, Guang Zhu, Zhong Lin Wang, "Fully enclosed bearing-structured self-powered rotation sensor based on electrification at rolling interfaces for multi-tasking motion measurement", Nano Energy, 12 (2015) 606–611.

[99] Simiao Niu, Ying Liu, Xiangyu Chen, Sihong Wang, Yu Sheng Zhou, Long Lin, Yannan Xie, Zhong Lin Wang, "Theory of freestanding triboelectric-layer-based nanogenerators", Nano Energy, 12 (2015) 760–774.

[100] Chang Bao Han, Chi Zhang, Wei Tang, Xiao Hui Li, Zhong Lin Wang, "High Power Triboelectric Nanogenerator Based on Printed Circuit Board (PCB) Technology", Nano Res., 8 (2015) 722–730.

[101] Yingchun Wu, Xue Wang, Ya Yang, Zhong Lin Wang, "Hybrid Energy Cell for Harvesting Mechanical Energy from One Motion Using Two Approaches", Nano Energy, 11 (2015) 162–170.

[102] Jun Chen, Guang Zhu, Jin Yang, Qingshen Jing, Peng Bai, Weiqing Yang, Xuewei Qi, Yuanjie Su, Zhong Lin Wang, "Personalized Keystroke Dynamics for Self-powered Human-Machine Interfacing", ACS Nano, 9 (2015) 105–116.

[103] Long Lin, Yannan Xie, Simiao Niu, Sihong Wang, Po-Kang Yang, Zhong Lin Wang, "Robust triboelectric nanogenerator based on rolling electrification and electrostatic induction at an instantaneous energy conversion efficiency of ~55 %", ACS Nano, 9 (2015) 922–930.

[104] Po-Kang Yang, Zong-Hong Lin, Ken C. Pradel, Long Lin, Xiuhan Li, Xiaonan Wen, Jr-Hau He, Zhong Lin Wang, "Paper-based Origami Triboelectric Nanogenerators and Self-Powered Pressure Sensors", ACS Nano, 9 (2015) 901–907.

[105] Xiangyu Chen, Mitsumasa Iwamoto, Zhemin Shi, Limin Zhang, Zhong Lin Wang, "Self-powered Trace Memorization by Conjunction of Contact-electrification and Ferroelectricity", Adv. Funct. Mater., 25 (2015) 739–747.

[106] Chi Zhang, Wei Tang, Yaokun Pang, Changbao Han, Zhong Lin Wang, "Active Micro-Actuators for Optical Modulation Based on Planar Sliding Triboelectric Nanogenerator", Adv. Mater., 27 (2015) 719–726.

[107] Wei Tang, Yu Han, Chang Bao Han, Cai Zhen Gao, Xia Cao, Zhong Lin Wang, "Self-Powered Water Splitting Using Flowing Kinetic Energy", Adv. Mater., 27 (2015) 272–276.

[108] Shuhua Wang, Xiaojing Mu, Ya Yang, Alex Yuandong Gu, Zhong Lin Wang, "Flow-Driven Triboelectric Nanogenerator For Directly Powering a Wireless Sensor Node", Adv. Mater., 27 (2015) 240–248.

[109] Peng Bai, Guang Zhu, Qingshen Jing, Ying Wu, Jin Yang, Jun Chen, Jusheng Ma, Gong Zhang, Zhong Lin Wang, "Transparent and Flexible Barcode Based on Sliding Electrification for Self-Powered Identification Systems", Nano Energy, 12 (2015) 278–286.

[110] Sihong Wang, Long Lin, Zhong Lin Wang, "Triboelectric Nanogenerators as Self-Powered Active Sensors", Nano Energy, 11 (2015) 436–462.

[111] Yufang Li, Gang Cheng, Zong-Hong Lin, Jin Yang, Long Lin, Zhong Lin Wang, "Single-Electrode-Based Rotatinary Triboelectric Nanogenerator and its Applications as Self-Powered Contact Area and Eccentric Angle Sensors", Nano Energy, 11 (2015) 323–332.

## 2014

[112] Bo Zhang, Zemin Xiang, Siwei Zhu, Qiyi Hu, Yuanzhi Cao, Junwen Zhong, Qize Zhong, Bo Wang, Yunsheng Fang, Jun Zhou, Bin Hu, Zhong Lin Wang, "Dual functional transparent film for proximity and pressure sending", Nano Res., 7 (2014) 1488–1496.
[113] Wei Tang, Tao Zhou, Chi Zhang, Feng Ru Fan, Chang Bao Han, Zhong Lin Wang, "A power-transformed-and-managed triboelectric nanogenerator and its applications in a self-powered wireless sensing node", Nanotechnology, 25 (2014) 225402.
[114] Zhong Lin Wang, "Triboelectric nanogenerators as new energy technology and self-powered sensors—principles, problems and perspectives", Faraday Discuss., 176 (2014) 447–451.
[115] Sihong Wang, Simiao Niu, Jin Yang, Long Lin, Zhong Lin Wang, "Quantitative Measurements of Vibration Amplitude Using a Contact-Mode Freestanding Triboelectric Nanogenerator", ACS Nano, 8 (2014) 12004–12013.
[116] Zhe Wang, Li Cheng, Youbin Zheng, Yong Qin, Zhong Lin Wang, "Enhancing the performance of triboelectric nanogenerator through prior-charge injection and its application on self-powered anticorrosion", Nano Energy, 10 (2014) 37–43.
[117] Wenxi Guo, Xiaoyi Li, Mengxiao Chen, Lu Xu, Xia Cao, Wei Tang, Lin Dong, Changjian Lin, Caofeng Pan, Zhong Lin Wang, "Electrochemical Cathodic Protection Powered by Triboelectric Nanogenerator", Adv. Func. Mater., 24 (2014) 6691–6699.
[118] Tae Il Lee, Woo Soon Jang, Eungkyu Lee, Youn Sang Kim, Zhong Lin Wang, Hong Koo Baik, Jae Min Myoung, "Ultrathin Self-Powered Artificial Skin", Energy Environ. Sci., 7 (2014) 3994–3999.
[119] Chang Kyu Jeong, Kwang Min Baek, Simiao Niu, Tae Won Nam, Yoon Hyung Hur, Dae Yong Park, Geon-Tae Hwang, Myeonghwan Byun, Zhong Lin Wang, Yeon Sik Jung, Keon Jae Lee, "Topographically-Designed Triboelectric Nanogenerator via Block Copolymer Self-Assembly", Nano Lett., 14 (2014) 7031–7038.
[120] Fang Yi, Long Lin, Simiao Niu, Jin Yang, Wenzhuo Wu, Sihong Wang, Qingliang Liao, Yue Zhang, Zhong Lin Wang, "Self-powered, trajectory, velocity and acceleration tracking of a moving object/body using a triboelectric sensor", Adv. Funct. Mater., 24 (2014) 7488–7494.

[121] Yannan Xie, Sihong Wang, Simiao Niu, Long Lin, Qingshen Jing, Jin Yang, Zhengyun Wu, Zhong Lin Wang, "Grating-Structured Freestanding Triboelectric-layer Nanogenerator for Harvesting Mechanical Energy at 85% Total Conversion Efficiency", Adv. Mater., 26 (2014) 6599–6607.

[122] Wei Tang, Chi Zhang, Chang Bao Han, Zhong Lin Wang, "Enhancing output power of cylindrical triboelectric nanogenerator by segmentation design and multi-layer integration", Adv. Funct. Mater., 24 (2014) 6684–6690.

[123] Yingchun Wu, Xiandai Zhong, Xue Wang, Ya Yang, Zhong Lin Wang, "Hybrid Energy Cell for Simultaneously Harvesting Wind, Solar, and Chemical Energies", Nano Res., 7 (2014) 1631–1639.

[124] Ya Yang, Hulin Zhang, Zhong Lin Wang, "Direct-Current Triboelectric Nanogenerator", Adv. Funct. Mater., 24 (2014) 3745–3750.

[125] Chi Zhang, Tao Zhou, Wei Tang, Changbao Han, Limin Zhang, Zhong Lin Wang, "Rotating Disk Based Direct-Current Triboelectric Nanogenerator", Adv. Energy Mater., 4 (2014) 1301798.

[126] Xiuhan Li, Zong-Hong Lin, Gang Cheng, Xiaonan Wen, Ying Liu, Simiao Niu, Zhong Lin Wang, "3D Fiber-Based Hybrid Nanogenerator for Energy Harvesting and as a Self-Powered Pressure Sensor", ACS Nano, 8 (2014) 10674–10681.

[127] Sihong Wang, Yannan Xie, Simiao Niu, Long Lin, Chang Liu, Yu Sheng Zhou, Zhong Lin Wang, "Maximum Surface Charge Density for Triboelectric Nanogenerators Achieved by Ionized-Air Injection: Methodology and Theoretical Understanding", Adv. Mater., 26 (2014) 6720–6728.

[128] Peng Bai, Guang Zhu, Qingshen Jing, Jin Yang, Jun Chen, Yuanjie Su, Jusheng Ma, Gong Zhang, and Zhong Lin Wang, "Membrane-Based Self-Powered Triboelectric Sensors for Air Pressure Detection and Its Uses in Security Surveillance and Healthcare Monitoring", Adv. Funct. Mater., 24 (2014) 5807–5813.

[129] Qingshen Jing, Guang Zhu, Wenzhuo Wu, Peng Bai, Yannan Xie, Ray P. S. Han, Zhong Lin Wang, "Self-powered triboelectric velocity sensor for dual-mode sensing of rectified linear and rotary motions", Nano Energy, 10 (2014) 305–312.

[130] Qiang Zheng, Bojing Shi, Fengru Fan, Xinxin Wang, LingYan, Weiwei Yuan, Sihong Wang, Zhou Li, Zhong Lin Wang, "In-vivo powering of pacemaker by breathing-driven implanted-triboelectric-nanogenerator", Adv. Mater., 26 (2014) 5851–5856.

[131] Mengxiao Chen, Xiaoyi Li, Long Lin, Weiming Du, Xun Han, Jing Zhu, Caofeng Pan, Zhong Lin Wang, "Triboelectric nanogeneraters as a self-powered motion tracking system", Adv. Funct. Mater., 24 (2014) 5059–5066.

[132] Jihyun Bae, Jeongsu Lee, SeongMin Kim, Jeawook Ha, Byoung-Sun Lee, YoungJun Park, Chweelin Choong, Jinbeak Kim, Zhong Lin Wang, Ho-Young Kim, Jong-Jin Park, U-In Chung, "Flutter-driven triboelectrification for harvesting wind energy", Nat. Commun., 5 (2014) 4929.

[133] Xiao Yan Wei, Guang Zhu, Zhong Lin Wang, "Contact electrification between identical materials enabled by surface-charge engineering: application in energy conversion and fundamental mechanism", Nano Energy, 10 (2014) 83–89.

[134] Tao Zhou, Chi Zhang, Chang Bao Han, Feng Ru Fan, Wei Tang, and Zhong Lin Wang "Woven structured triboelectric nanogenerator for wearable devices", ACS Appl. Mater. Interfaces, 6 (2014) 14695–14701.

[135] Yuanjie Su, Xiaonan Wen, Guang Zhu, Jin Yang, Jun Chen, Peng Bai, Zhiming Wu, Yadong Jiang, Zhong Lin Wang, "Hybrid triboelectric nanogenerator for harvesting water wave energy and as a self-powered distress signal emitter", Nano Energy, 9 (2014) 186–195.

[136] Feng Ru Fan, Jianjun Luo, Wei Tang, Chaoyu Li, Cuiping Zhang, Zhongqun Tian, Zhong Lin Wang, "Highly transparent and flexible triboelectric nanogenerators: performance improvements and fundamental mechanisms", J. Mater. Chem. A, 2 (2014) 13219–13225.

[137] Wei Tang, Chang Bao Han, Chi Zhang, and Zhong Lin Wang "Cover-sheet-based nanogenerator for charging mobile electronics using low-frequency body motion/vibration", Nano Energy, 9 (2014) 121–127.

[138] Simiao Niu, Yu Sheng Zhou, Sihong Wang, Ying Liu, Long Lin, Yoshio Bando, Zhong Lin Wang "Simulation method for optimizing the integrated performance of a triboelectric nanogenerator power source", Nano Energy, 8 (2014) 150–156.

[139] Chang Bao Han, Chi Zhang, Xiao Hui Li, Limin Zhang, Tao Zhou, Weiguo Hu, Zhong Lin Wang, "Self-powered velocity and trajectory tracking sensor array made of planar triboelectric nanogenerator pixels", Nano Energy, 9 (2014) 325–333.

[140] Li Zheng, Zong-Hong Lin, Gang Cheng, Wenzhuo Wu, Xiaonan Wen, Sangmin Lee, Zhong Lin Wang, "Silicon-based hybrid cell for harvesting both solar energy and raindrop electrostatic energy", Nano Energy, 9 (2014) 291–300.

[141] Li Min Zhang, Fei Xue, Weiming Du, Chang Bao Han, Chi Zhang, Zhong Lin Wang, "Transparent paper-based triboelectric nanogenerator as page mark and anti-theft sensor", Nano Res., 7 (2014) 1215–1223.

[142] Xiaonan Wen, Weiqing Yang, Qingshen Jing, Zhong Lin Wang, "Harvesting Broadband Kinetic Impact Energy from Mechanical Triggering/Vibration and Water Waves", ACS Nano, 8 (2014) 7405–7412.

[143] Youfan Hu, Jin Yang, Simiao Niu, Wenzhuo Wu, Zhong Lin Wang, "Hybridizing triboelectrification and electromagnetic-induction effects for high efficient mechanical energy harvesting", ACS Nano, 8 (2014) 7442–7450.

[144] Zong-Hong Lin, Gang Cheng, Wenzhuo Wu, Ken C. Pradel, Zhong Lin Wang, "Dual-mode triboelectric nanogenerator for harvesting water energy and as a self-powered ethanol nanosensor", ACS Nano, 8 (2014) 6440–6448.
[145] Weiqing Yang, Jun Chen, Xiaonan Wen, Qingshen Jing, Jin Yang, Yuanjie Su, Guang Zhu, Wenzuo Wu, Zhong Lin Wang, "Triboelectrification based motion sensor for human-machine interfacing", ACS Appl. Mater. Interfaces, 6 (2014) 7479–7484.
[146] Xian Song Meng, Guang Zhu, Zhong Lin Wang, "A robust thin-film generator based on segmented contact-electrification for harvesting wind energy", ACS Appl. Mater. Interfaces, 6 (2014) 8011–8016.
[147] Junwen Zhong, Yan Zhang, Qize Zhong, Qiyi Hu, Bin Hu, Zhong Lin Wang, Jun Zhou, "Fiber-based generator for wearable electronics and mobile medication", ACS Nano, 8 (2014) 6273–6280.
[148] Wei Tang, Tao Zhou, Chi Zhang, Feng Ru Fan, Chang Bao Han, Zhong Lin Wang, "Power transform and management for triboelectric nanogenerators by a self-triggering process", Nanotechnology, 25 (2014) 225402.
[149] Guang Zhu, Yuanjie Su, Peng Bai, Jun Chen, Qingshen Jing, Weiqing Yang, Zhong Lin Wang, "Harvesting water wave energy by asymmetric screening of electrostatic charges on nanostructured hydrophobic thin-film surfaces", ACS Nano, 8 (2014) 6031–6037.
[150] Guang Zhu, Wei Qing Yang, Tiejun Zhang, Qingshen Jing, Jun Chen, Yu Sheng Zhou, Peng Bai, Zhong Lin Wang, "Self-powered, ultra-sensitive, flexible tactile sensors based on contact electrification", Nano Lett. 14 (2014) 3208–3213.
[151] Zong-Hong Lin, Gang Cheng, Sangmin Lee, Zhong Lin Wang, "Harvesting water drop energy by a sequential contact-electrification and electrostatic-induction process", Adv. Mater., 26 (2014) 4690–4696.
[152] Peng Bai, Guang Zhu, Yusheng Zhou, Sihong Wang, Jusheng Ma, Gong Zhang, Zhong Lin Wang, "Dipole-moment-induced effect on contact electrification for triboelectric nanogenerator", Nano Res., 7 (2014) 990–997.
[153] Chi Zhang, Wei Tang, Changbao Han, Fengru Fan, and Zhong Lin Wang, "Theoretical comparison, equivalent transformation and conjunction operations of electromagnetic induction generator and triboelectric nanogenerator for harvesting mechanical energ", Adv. Mater., 26 (2014) 3580–3591.
[154] Weiming Du, Xun Han, Long Lin, Mengxiao Chen, Xiaoyi Li, Caofeng Pan and Zhong Lin Wang "Three dimensional multi-layered sliding triboelectric nanogenerator", Adv. Energy Mater., 4 (2014) 1301592.
[155] Guang Zhu, Yu Sheng Zhou, Peng Bai, Xiansong Meng, Qingshen Jing, Jun Chen, Zhong Lin Wang, "A shape-adaptive thin-film-based approach for a 50% high-efficiency energy generation through micro-grating sliding electrification", Adv. Mater., 26 (2014) 3788–3796.
[156] Weiqing Yang, Jun Chen, Qingshen Jing, Jin Yang, Xiaonan Wen, Yuanjie Su, Guang Zhu, Peng Bai, Zhong Lin Wang, "3D stack integrated triboelectric nanogenerator for harvesting vibration energy", Adv. Funct. Mater., 24 (2014) 4090–4096.

[157] Simiao Niu, Ying Liu, Sihong Wang, Long Lin, Yu Sheng Zhou, Youfan Hu, Zhong Lin Wang, "Theoretical investigation and structure optimization of single-electrode triboelectric nanogenerators", Adv. Funct. Mater., 24 (2014) 3332–3340.

[158] Simiao Niu, Sihong Wang, Ying Liu, Yu Sheng Zhou, Long Lin, Youfan Hu, Zhong Lin Wang, "Theoretical study of grating structured triboelectric nanogenerators", Energy Environ. Sci., 7 (2014) 2339–2349.

[159] Chang Bao Han, Weiming Du, Chi Zhang, Wei Tang, Fengru Fan, Aifang Yu, Limin Zhang, Zhong Lin Wang, "Harvesting energy from automobile brake in contact and non-contact mode by conjunction of triboelectrication and electrostatic-induction processes", Nano Energy, 6 (2014) 59–65.

[160] Zong-Hong Lin, Gang Cheng, Ya Yang, Yu Sheng Zhou, Sangmin Lee, Zhong Lin Wang, "Triboelectric nanogenerator as a self-powered active UV photodetector", Adv. Funct. Mater., 24 (2014) 2810–2816.

[161] Gang Cheng, Zong-Hong Lin, Zuliang Du, Zhong Lin Wang, "Increase output energy and operation frequency of a triboelectric nanogenerator by two grounded electrodes approach", Adv. Funct. Mater., 24 (2014) 2892–2898.

[162] Sihong Wang, Yannan Xie, Simiao Niu, Long Lin, Zhong Lin Wang, "Freestanding-triboelectric-layer based nanogenerators for harvesting energy from a moving object or human motion in contact and non-contact modes", Adv. Mater., 26 (2014) 2818–2824.

[163] Jin Yang, Ya Yang, Jun Chen, Hulin Zhang, Weiqing Yang, Peng Bai, Yuanjie Su, Zhong Lin Wang, "Broadband vibration energy harvesting based on triboelectric nanogenerator", Adv. Energy Mater., 4 (2014) 1301322.

[164] Xinyu Xue, Bin He, Ping Deng, Yuxin Nie, Lili Xing, Yan Zhang, Zhong Lin Wang, "Flexible self-charging power cell for one-step energy conversion and storage", Adv. Energy Mater., 4 (2014) 1301329.

[165] Yuanjie Su, Guang Zhu, Weiqing Yang, Jin Yang, Jun Chen, Qingshen Jing, Zhiming wu, Yadong Jiang, Zhong Lin Wang, "triboelectric sensor for self-powered tracking of object motion inside tubing", ACS Nano, 8 (2014) 3843–3850.

[166] Yannan Xie, Sihong Wang, Simiao Niu, Long Lin, Qingshen Jing, Yuanjie Su, Zhengyun Wu, Zhong Lin Wang, "Multi-layered disk triboelectric nanogenerator for harvesting hydropower", Nano Energy, 6 (2014) 129–136.

[167] Qingshen Jing, Guang Zhu, Peng Bai, Yannan Xie, Jun Chen, Ray P. S. Han, Zhong Lin Wang, "Case-encapsulated triboelectric nanogenerator for harvesting energy from reciprocating sliding motion", ACS Nano, 7 (2013) 10424–10432.

[168] Yu Sheng Zhou, Guang Zhu, Simiao Niu, Ying Liu, Peng Bai, Qingsheng Jing, Zhong Lin Wang, "Nanometer resolution self-powered static and dynamic motion sensor based on micro-grated triboelectrification", Adv. Mater., 26 (2014) 1719–1724.

[169] Hulin Zhang, Ya Yang, Yuanjie Su, Jun Chen, Katherine Adams, Sangmin Lee, Chenguo Hu, Zhong Lin Wang, "Triboelectric nanogenerator for harvesting vibration energy in full space and as self-powered acceleration sensor", Adv. Funct. Mater., 24 (2014) 1401–1407.

[170] Ya Yang, Hulin Zhang, Xiandai Zhong, Fang Yi, Ruomeng Yu, Yue Zhang, Zhong Lin Wang, "Electret film enhanced triboelectric nanogenerator matrix for self-powered instantaneous tactile imaging", ACS Appl. Mater. Interfaces, 6 (2014) 3680−3688.

[171] Yu Sheng Zhou, Sihong Wang, Ya Yang, Guang Zhu, Simiao Niu, Zong-Hong Lin, Ying Liu, Zhong Lin Wang, "Manipulating nano-scale contact electrification by applied electric field", Nano Lett., 14 (2014) 1567–1572.

[172] Feng-Ru Fan, Wei Tang, Yan Yao, Jianjun Luo, Chi Zhang, Zhong Lin Wang, "Complementary power output characteristics of electromagnetic generator and triboelectric generator", Nanotechnology, 25 (2014) 135402.

[173] Jin Yang, Jun Chen, Ying Liu, Weiqing Yang, Yuanjie Su, Zhong Lin Wang, "Triboelectrification based organic film nanogenerator for acoustic energy harvesting and self-powered active acoustic sensing", ACS Nano, 8 (2014) 2649–2657.

[174] Guang Zhu, Jun Chen, Tiejun Zhang, Qingshen Jing, Zhong Lin Wang, "Radial-arrayed rotary electrification for high-performance triboelectric generator", Nat. Commun., 5 (2014) 3456.

[175] Gang Cheng, Zong-Hong Lin, Zu-liang Du, Zhong Lin Wang, "Simultaneously harvesting electrostatic and mechanical energies from flowing water by a hybridized triboelectric nanogenerator" ACS Nano, 8 (2014) 1932–1939.

[176] Long Lin, Sihong Wang, Simiao Niu, Chang Liu, Yannan Xie, Zhong Lin Wang, "Non-contact free-rotating disk triboelectric nanogenerator as a sustainable energy harvester and self-powered mechanical sensor", ACS Appl. Mater. Interfaces, 6 (2014) 3031–3038.

[177] Xiaonan Wen, Yuanjie Su, Ya Yang, Hulin Zhang, Zhong Lin Wang, "Applicability of triboelectric generator over a wide range of temperature", Nano Energy, 4 (2014) 150–156.

[178] Hulin Zhang, Ya Yang, Xiandai Zhong, Yuanjie Su, Yusheng Zhou, Chenguo Hu, Zhong Lin Wang, "Single-electrode-based rotating triboelectric nanogenerator for harvesting energy from tires", ACS Nano, *8* (2014) 680–689.

[179] Yuanjie Su, Ya Yang, Xiandai Zhong, Hulin Zhang, Zhiming Wu, Yadong Jiang, Zhong Lin Wang, "Fully Enclosed Cylindrical Single-Electrode-Based Triboelectric Nanogenerator", ACS Appl. Mater. Interfaces, 6 (2014) 553–559.

## 2013

[180] Sihong Wang, Zong-Hong Lin, Simiao Niu, Long Lin, Yannan Xie, Zhong Lin Wang, "Motion Charged Battery as Sustainable Flexible-Power-Unit", ACS Nano, 7 (2013) 11263–11271.

[181] Weiqing Yang, Jun Chen, Guang Zhu, Xiaonan Wen, Peng Bai, Yuanjie Su, Yuan Lin, Zhong Lin Wang, "Harvesting vibration energy by a triple-cantilever based triboelectric nanogenerator", Nano Res., 6 (2013) 880–886.

[182] Weiqing Yang, Jun Chen, Guang Zhu, Jin Yang, Peng Bai, Yuanjie Su, Qingshen Jing, Zhong Lin Wang, "Harvesting energy from natural vibration of a human walking", ACS Nano, 7 (2013) 11317–11374.

[183] Youfan Hu, Jin Yang,Qingshen Jing, Simiao Niu, Wenzhuo Wu, Zhong Lin Wang, "Triboelectric Nanogenerator Built on Suspended 3D Spiral Structure as Vibration and Positioning Sensor and Wave Energy Harvester", ACS Nano, 7 (2013) 10424–10432.

[184] Ya Yang, Yu Sheng Zhou, Hulin Zhang, Ying Liu, Sangmin Lee, Zhong Lin Wang, "Single-Electrode Based Triboelectric Nanogenerator as Self-Powered Tracking System", Adv. Mater., 25 (2013) 6594–6601.

[185] Ya Yang, Hulin Zhang, Ruoyu Liu, Xiaonan Wen, Te-Chien Hou, Zhong Lin Wang, "Fully Enclosed Triboelectric Nanogenerators for Applications in Water and Hash Environment", Adv. Energy Mater., 3 (2013) 1563–1568.

[186] Zong-Hong Lin, Gang Cheng, Long Lin, Sangmin Lee, Zhong Lin Wang, "Water-Solid Surface Contact Electrification and its Use for Harvesting Liquid Wavy Energy", Angew. Chem. Int. Ed., 52 (2013) 12545–12549.

[187] Ya Yang, Guang Zhu, Hulin Zhang, Jun Chen, Xiandai Zhong, Zong-Hong Lin, Yuanjie Su, Peng Bai, Xiaonan Wen, Zhong Lin Wang, "Triboelectric Nanogenerator for Harvesting Wind Energy and as Self-Powered Wind Vector Sensor System", ACS Nano, 7 (2013) 9461–9468.

[188] Zhong Lin Wang, "Triboelectric Nanogenerators as New Energy Technology for Self-Powered Systems and as Active Mechanical and Chemical Sensors", ACS Nano, 7 (2013) 9533–9557.

[189] Long Lin, Yannan Xie, Sihong Wang, Wenzhuo Wu, Simiao Niu, Xiaonan Wen, Zhong Lin Wang, "Triboelectric active sensor array for self-powered static and dynamic pressure detection and tactile imaging", ACS Nano, 7 (2013) 8266–8274.

[190] Ya Yang, Hulin Zhang, Zong-Hong Lin, Yusheng Zhou, Qingshen Jing, Yuanjie Su, Jin Yang, Jun Chen, Chenguo Hu, Zhong Lin Wang, "Human Skin Based Triboelectric Nanogenerators for Harvesting Biomechanical Energy and as Self-Powered Active Tactile Sensor System", ACS Nano, 7 (2013) 9213–9222.

[191] Sangmin Lee, Yean Lee, Dongseob Kim, Ya Yang, Long Lin, Woonbong Hwang, Zhong Lin Wang, "Triboelectric Nanogenerator for Harvesting Pendulum Oscillation Energy", Nano Energy, 2 (2013) 1113–1120.

[192] Te-Chien Hou, Ya Yang, Hulin Zhang, Jun Chen, Lih-Juann Chen, Zhong Lin Wang "Triboelectric nanogenerator built inside shoe insole for harvesting walking energy", Nano Energy, 2 (2013) 856–862.

[193] Jun Chen, Guang Zhu, Weiqing Yang, Qingshen Jin, Peng Bai, Ya Yang, Te-Chien Hou, Zhong Lin Wang, "Harmonic Resonator Based Triboelectric Nanogenerator as Sustainable Power Source and Self-powered Active Vibration Sensor", Adv. Mater., 25 (2013) 6094–6099.

[194] Simiao Niu, Ying Liu, Sihong Wang, Long Lin, Yu Sheng Zhou, Youfan Hu, Zhong Lin Wang, "Theory of Sliding-mode Triboelectric Nanogenerators", Adv. Mater., 25 (2013) 6184–6193.

[195] Simiao Niu, Sihong Wang, Long Lin, Ying Liu, Yu Sheng Zhou, Youfan Hu, Zhong Lin Wang, "Theoretical Study of the Contact-Mode Triboelectric Nanogenerators as Effective Power Source", Energy Environ. Sci., 6 (2013) 3576–3583.

[196] Hulin Zhang, Ya Yang, Yuanjie Su, Jun Chen, Chenguo Hu, Zhenkun Wu, Yan Liu, Ching Ping Wong, Yoshio Bando, Zhong Lin Wang, "Triboelectric nanogenerator as self-powered active sensors for detecting liquid/gaseous water/ethanol", Nano Energy, 2 (2013) 693–701.

[197] Guang Zhu, Peng Bai, Jun Chen, Zhong Lin Wang, "Power-generating shoe insole based on triboelectric nanogenerators for self-powered consumer electronics", Nano Energy, 2 (2013) 688–692.

[198] Hulin Zhang, Ya Yang, Te-Chien Hou, Yuanjie Su, Chenguo Hu, Zhong Lin Wang, "Triboelectric Nanogenerator Built Inside Clothes for Self-powered Glucose Biosensors", Nano Energy, 2 (2013) 1019–1024.

[199] Gang Cheng, Zong-Hong Lin, Long Lin, Zuliang Du, Zhong Lin Wang, "Pulsed Nanogenerator with Huge Instantaneous Output Power Density", ACS Nano, 7 (2013) 7383–7391.

[200] Ya Yang, Hulin Zhang, Jun Chen, Qingshen Jing, Yu Sheng Zhou, Xiaonan Wen, Zhong Lin Wang, "Single-Electrode Based Sliding Triboelectric Nanogenerator for Self-Powered Displacement Vector Sensor System", ACS Nano, 7 (2013) 7342–7351.

[201] Yuanjie Su, Ya Yang, Hulin Zhang, Yannan Xie, Zhiming Wu, Yadong Jiang, Naoki Fukata,Yoshio Bando, Zhong Lin Wang, "Enhanced Photodegradation of Methyl Orange using $TiO_2$ Nanoparticles by Triboelectric Nanogenerator", Nanotechnology, 24 (2013) 295401.

[202] Yannan Xie, Sihong Wang, Long Lin, Qingshen Jing, Zong-Hong Lin, Simiao Niu, Zhengyun Wu, Zhong Lin Wang, "Rotary Triboelectric Nanogenerator Based on a Hybridized Mechanism for Harvesting Wind Energy", ACS Nano, 7 (2013) 7119–7125.

[203] Ya Yang, Hulin Zhang, Sangmin Lee, Te-Chien Hou, Zhong Lin Wang, "Simultaneously Harvesting Mechanical and Chemical Energies by a Hybrid Cell for Self-Powered Biosensors and Personal Electronics", Energy Environ. Sci., 6 (2013) 1744–1749.

[204] Peng Bai Guang Zhu, Ying Liu, Jun Chen, Qingshen Jing, Weiqing Yang, Jusheng Ma, Gong Zhang, Zhong Lin Wang, "Cylindrical Rotating Triboelectric Nanogenerator", ACS Nano, 7 (2013) 6361–6366.

[205] Ya Yang, Hulin Zhang, Zong-Hong Lin, Yan Liu, Jun Chen, Ziyin Lin, Yusheng Zhou, Ching Ping Wong, Zhong Lin Wang, "Hybrid Energy Cell for Self-Powered Water Splitting", Energy Environ. Sci., 6 (2013) 2429–2434.

[206] Junwen Zhong, Qize Zhong, Fengru Fan, Yan Zhang, Shihong Wang, Bin Hu, Jun Zhou, Zhong Lin Wang, "Finger typing driven triboelectric nanoenerator and its use for instantaneously lighting up LEDs", Nano Energy, 2 (2013) 491–497.

[207] Yu Sheng Zhou, Ying Liu, Guang Zhu, Zong-Hong Lin, Caofeng Pan, Qingshen Jing, Zhong Lin Wang, "In-situ Quantitative Study of Nanoscale Triboelectrification and Patterning", Nano Lett., 13 (2013) 2771–2776.

[208] Long Lin, Sihong Wang, Yannan Xie, Qingshen Jing, Simiao Niu, Youfan Hu, Zhong Lin Wang, "Segmentally Structured Disk Triboelectric Nanogenerator for Harvesting Rotational Mechanical Energy", Nano Lett., 13 (2013) 2916–2923.

[209] Zong-Hong Lin, Yannan Xie, Ya Yang, Sihong Wang, Guang Zhu, Zhong Lin Wang, "Enhanced Triboelectric Nanogenerators and Triboelectric Nanosensor using Chemically Modified $TiO_2$ Nanomaterials", ACS Nano, 7 (2013) 4554–4560.

[210] Qize Zhong, Junwen Zhong, Bin Hu, Qiyi Hu, Jun Zhou, Zhong Lin Wang, "A Paper-Based Nanogenerator as Power Source and Active Sensor", Energy Environ. Sci., 6 (2013) 1779–1784.

[211] Guang Zhu, Jun Chen, Ying Liu, Peng Bai, Yu Sheng Zhou, Qingshen Jing, Caofeng Pan, Zhong Lin Wang "Linear-grating triboelectric generator based on sliding electrification", Nano Lett., 13 (2013) 2282–2289.

[212] Sihong Wang, Long Lin, Yannan Xie, Qingshen Jing, Simiao Niu, Zhong Lin Wang, "Sliding-Triboelectric Nanogenerators Based on In-Plane Charge-Separation Mechanism", Nano Lett., 13 (2013) 2226–2233.

[213] Zong-Hong Lin, Guang Zhu, Yu Sheng Zhou, Ya Yang, Peng Bai, Jun Chen, Zhong Lin Wang, "Self-Powered Triboelectric Nanosensor for Mercury Ion Detection", Angew. Chem. Int. Ed., 52 (2013) 5065–5069.

[214] Peng Bai, Guang Zhu, Zong-Hong Lin, Qingshen Jing, Jun Chen, Gong Zhang, Jusheng Ma, Zhong Lin Wang, "Integrated Multilayered Triboelectric Nanogenerator for Harvesting Biomechanical Energy from Human Motions", ACS Nano, 7 (2013) 3713–3719.

[215] Ya Yang, Hulin Zhang, Yan Liu, Zong-Hong Lin, Sangmin Lee, Ziyin Lin, Ching Ping Wong, Zhong Lin Wang, "Silicon-Based Hybrid Energy Cell for Self-Powered Electrodegradation and Personal Electronics", ACS Nano, 7 (2013) 785–790.

[216] Ya Yang, Hulin Zhang, Sangmin Lee, Dongseob Kim, Woonbong Hwang, Zhong Lin Wang, "Hybrid Energy Cell for Degradation of Methyl Orange by Self-Powered Electrocatalytic Oxidation", Nano Lett., 13 (2013) 803–808.
[217] Guang Zhu, Zong-Hong Lin, Qingshen Jing, Peng Bai, Caofeng Pan, Ya Yang, Yusheng Zhou, Zhong Lin Wang, "Toward Large-Scale Energy Harvesting by Nanoparticle-Enhanced Triboelectric Nanogenerator", Nano Lett., 13 (2013) 847–853.

## *2012*

[218] Zhong Lin Wang, Guang Zhu, Ya Yang, Sihong Wang, Caofeng Pan, "Progress in Nanogenerators for Self-Powered Mobile/Portable Electronics", Mater. Today, 15 (2012) 532–543.
[219] Ya Yang, Long Lin, Yue Zhang, Qingshen Jing, Te-Chien Hou, Zhong Lin Wang, "Self-Powered Magnetic Sensor Based on Triboelectric Nanogenerator", ACS Nano, 6 (2012) 10378–10383.
[220] Sihong Wang, Long Lin, Zhong Lin Wang, "Nanoscale-triboelectric-effect enabled energy conversion for sustainably powering of portable electronics", Nano Lett., 12 (2012) 6339–6346.
[221] Xiaohong Yang, Guang Zhu, Sihong Wang, Rui Zhang, Long Lin, Wenzhou Wu, Zhong Lin Wang, "Self-powered Electrochromic Device Driven by Nanogenerator", Energy Environ. Sci., 5 (2012) 9462–9466.
[222] Guang Zhu, Caofeng Pan, Wenxi Guo, Chih-Yen Chen, Yusheng Zhou, Ruomeng Yu, Zhong Lin Wang, "Triboelectric-generator-driven pulse electrodeposition for micro-patterning", Nano Lett., 12 (2012) 4960–4965.
[223] Feng-Ru Fan, Long Lin, Guang Zhu, Wenzhuo Wu, Rui Zhang, Zhong Lin Wang, "Transparent Triboelectric Nanogenerators and Self-powered Pressure Sensors Based on Micro-patterned Plastic Films", Nano Lett., 12 (2012) 3109–3114.
[224] Fengru Fan, Zhong-Qun Tian, Zhong Lin Wang, "Flexible triboelectric generator", Nano Energy, 1 (2012) 328–334.

# Index





Z.L. Wang et al., *Triboelectric Nanogenerators*,
Green Energy and Technology, DOI 10.1007/978-3-319-40039-6

GPSR Compliance
The European Union's (EU) General Product Safety Regulation (GPSR) is a set of rules that requires consumer products to be safe and our obligations to ensure this.

If you have any concerns about our products, you can contact us on

ProductSafety@springernature.com

In case Publisher is established outside the EU, the EU authorized representative is:

Springer Nature Customer Service Center GmbH
Europaplatz 3
69115 Heidelberg, Germany

www.ingramcontent.com/pod-product-compliance
Ingram Content Group UK Ltd.
Pitfield, Milton Keynes, MK11 3LW, UK
UKHW022217051026
15624UKWH00001B/208
*9783319400389*